Benjamin Koetz, Zoltán Vekerdy, Massimo Menenti and
Diego Fernández-Prieto (Eds.)

Earth Observation for Water Resource Management in Africa

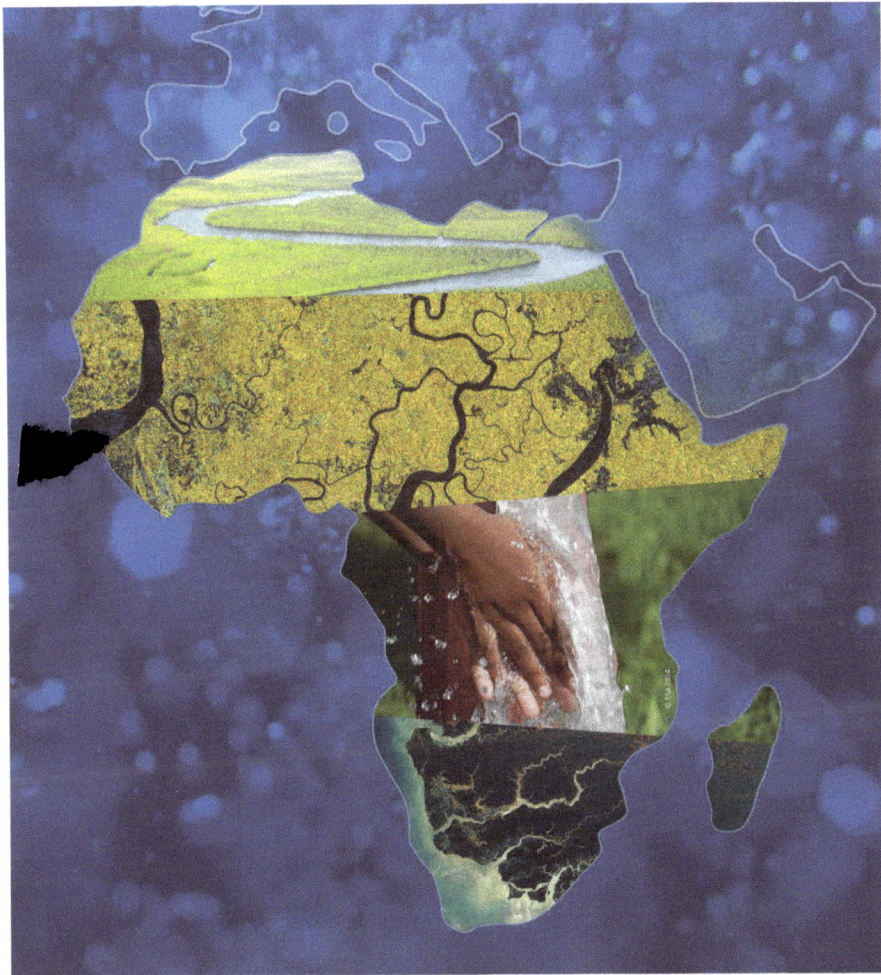

MDPI

This book is a reprint of the special issue that appeared in the online open access journal *Remote Sensing* (ISSN 2072-4292) in 2014 (available at: http://www.mdpi.com/journal/remotesensing/special_issues/water_management_in_africa).

Guest Editors
Benjamin Koetz
European Space Agency
Italy

Zoltán Vekerdy
University of Twente
The Netherlands

Massimo Menenti
Delft University of Technology
The Netherlands

Diego Fernández-Prieto
European Space Agency
Italy

Editorial Office	*Publisher*	*Managing Editor*
MDPI AG	Shu-Kun Lin	Elvis Wang
Klybeckstrasse 64		
Basel, Switzerland		

1. Edition 2015

MDPI • Basel • Beijing • Wuhan

ISBN 978-3-03842-153-5 (Hbk)
ISBN 978-3-03842-154-2 (PDF)

Table of Contents

Chapter 3: Evapotranspiration

Chapter 4: Surface Water Hydrology

List of Contributors

Nadia Akdim: Faculty of Sciences, Chouaib Doukkali University, BD Jabran Khalil Jabran B.P 299, 24000 EL Jadida, Morocco; Geosciences and Remote Sensing Department, Delft University of Technology, Stevinweg 12628 CN Delft, The Netherlands.

Henok Alemu: Geospatial Sciences Center of Excellence (GSCE), South Dakota State University, Brookings, 57007 SD, USA.

Silvia Maria Alfieri: Geosciences and Remote Sensing Department, Delft University of Technology, Stevinweg 12628 CN Delft, The Netherlands; Institute for Mediterranean Agricultural and Forest Systems, Italy (ISAFOM), Ercolano 80056, Italy.

Donato Amitrano: Department of Electrical Engineering and Information Technology, University of Napoli Federico II, Via Claudio 21, 80125 Napoli, Italy.

Rim Amri: CESBIO/UMR 5126, 18 av. Edouard Belin, bpi 2801, 31401 Toulouse Cedex 9, France; GREF, Université de Carthage/INAT, 43, Avenue Charles Nicolle 1082-Tunis-Mahrajène, Tunisia.

Frank O. Annor: Faculty of Civil Engineering and Geosciences, Delft University of Technology, Stevinweg 1, 2628 CN Delft, The Netherlands; Department of Civil Engineering, Kwame Nkrumah University of Science and Technology, Kumasi, Ghana.

Wim G.M. Bastiaanssen: International Water Management Institute (IWMI), Colombo, Sri Lanka and Vientiane, Laos; UNESCO-IHE, 2611 AX Delft, The Netherlands; Delft University of Technology, 2628 CN Delft, The Netherlands.

Peter Bauer-Gottwein: Department of Environmental Engineering, Technical University of Denmark, Bygningstorvet, B115, 2800 Kgs. Lyngby, Denmark.

Bernhard Bauer-Marschallinger: Department of Geodesy and Geoinformation, Vienna University of Technology, Gußhausstraße 27-29, Vienna 1040, Austria.

Mélanie Becke: UAG/ESPACE-DEV, Route de Montabo, Cayenne 97300, French Guiana.

Muriel Bergé-Nguyen: CNES/Legos, 14 Av Edouard Belin, 31400 Toulouse, France.

Stefano Bocchi: Department of Agricultural and Environmental Science, Università degli Studi di Milano, Milano 20133, Italy.

Mirco Boschetti: Institute of Electromagnetic Sensing of Environment, National Research Council of Italy (CNR-IREA), Via Bassini 15, Milan 20133, Italy.

Gilles Boulet: CESBIO/UMR 5126, 18 av. Edouard Belin, bpi 2801, 31401 Toulouse Cedex 9, France.

Alessandro Brivio: Institute of Electromagnetic Sensing of Environment, National Research Council of Italy (CNR-IREA), Via Bassini 15, Milan 20133, Italy.

Stéphane Calmant: IRD/LEGOS, 14 Av. Edouard Belin, Toulouse 31400, France.

Jean Christophe Calvet: CNRM-GAME, Météo-France, CNRS, URA 1357, 42 avenue Gaspard Coriolis, 31057 Toulouse Cedex 1, France.

Gabriele Candiani: Institute of Electromagnetic Sensing of Environment, National Research Council of Italy (CNR-IREA), Via Bassini 15, Milan 20133, Italy.

Elijah Cheruiyot: Geosciences and Remote Sensing Department, Delft University of Technology, Stevinweg 12628 CN Delft, The Netherlands; Department of Physics, University of Nairobi, P.O. Box 30197, 00100 Nairobi, Kenya.

Abdeloihab Choukri: Faculty of Sciences, Chouaib Doukkali University, BD Jabran Khalil Jabran B.P 299, 24000 EL Jadida, Morocco.

Jean-François Crétaux: CNES/Legos, 14 Av Edouard Belin, 31400 Toulouse, France.

Joecila Santos da Silva: UEA/CESTU, Av. Djalma Batista 3578, Manaus 69058-807, Brazil.

Alena Dostálová: Department of Geodesy and Geoinformation, Vienna University of Technology, Gußhausstraße 27-29, Vienna 1040, Austria.

Marcela Doubková: Department of Geodesy and Geoinformation, Vienna University of Technology, Gusshausstrasse 27-29/E122, 1040 Vienna, Austria.

Zheng Duan: Delft University of Technology, Stevinweg 1, 2628 CN Delft, The Netherlands.

Sebinasi Dzikiti: Natural Resource and Environment, Council for Scientific and Industrial Research, P.O. Box 395, Pretoria 0001, South Africa.

Dirk Eilander: Faculty of Civil Engineering and Geosciences, Delft University of Technology, Stevinweg 1, 2628 CN Delft, The Netherlands.

Salah Er-Raki: LP2M2E, Faculté des Sciences et Techniques, Université Cadi Ayyad de Marrakech, Marrakech 40000, Morocco; LMI TREMA laboratory.

Frédéric Frappart: Groupe de Recherche en Géodésie Spatiale (GRGS), Observatoire Midi-Pyrénées (OMP), GET-UMR5563 CNRS/IRD/UPS, 14, Avenue Edouard Belin, 31400 Toulouse, France.

Ben Gorte: Department of Geoscience and Remote Sensing, Delft University of Technology, P.O. Box 5048, 2600 GA Delft, The Netherlands.

Webster Gumindoga: Department of Civil Engineering, University of Zimbabwe, Box MP 167, Harare, Zimbabwe; Department of Water Resources, Faculty of Geo-information Science and Earth Observation (ITC), University of Twente, P.O. Box 6, AA Enschede 7500, The Netherlands.

Radoslaw Guzinski: DHI GRAS, DK-2970 Hørsholm, Denmark.

Emad Habib: Faculty of Sciences, Chouaib Doukkali University, BD Jabran Khalil Jabran B.P 299, 24000 EL Jadida, Morocco; Department of Civil Engineering, University of Louisiana at Lafayette, Lafayette, LA 70504, USA.

Olivier Hagolle: CESBIO, Unité Mixte de Recherche (CNRS, UPS, CNES, IRD), Toulouse 31000, France.

Alemseged Tamiru Haile: International Water Management Institute, Nile Basin and East Africa Sub-Regional Office, P.O. Box 5689 Addis Ababa, Ethiopia.

Arjen Y. Hoekstra: Twente Water Centre, Water Management Group, University of Twente, Enschede NL-7500 AE, The Netherlands.

Silvia Huber: DHI GRAS, DK-2970 Hørsholm, Denmark.

Mireille Huc: CESBIO, Unité Mixte de Recherche (CNRS, UPS, CNES, IRD), Toulouse 31000, France.

Lorenzo Iannini: Faculty of Civil Engineering and Geosciences, Delft University of Technology, Stevinweg 1, 2628 CN Delft, The Netherlands.

Antonio Iodice: Department of Electrical Engineering and Information Technology, University of Napoli Federico II, Via Claudio 21, 80125 Napoli, Italy.

Lionel Jarlan: CESBIO, Unité Mixte de Recherche (CNRS, UPS, CNES, IRD), Toulouse 31000, France; LMI TREMA laboratory.

Iris Hedegaard Jensen: Department of Environmental Engineering, Technical University of Denmark, Bygningstorvet, B115, 2800 Kgs. Lyngby, Denmark.

Nebo Jovanovic: Natural Resource and Environment, Council for Scientific and Industrial Research, P.O. Box 395, Pretoria 0001, South Africa.

Armel T. Kaptue: Geospatial Sciences Center of Excellence (GSCE), South Dakota State University, Brookings, 57007 SD, USA.

Poolad Karimi: International Water Management Institute (IWMI), Colombo, Sri Lanka and Vientiane, Laos; UNESCO-IHE, 2611 AX Delft, The Netherlands; Delft University of Technology, 2628 CN Delft, The Netherlands.

Mohamed Kasbani: CESBIO, Unité Mixte de Recherche (CNRS, UPS, CNES, IRD), Toulouse 31000, France; LMI TREMA laboratory.

Steve Kass: GeoVille, Sparkassenplatz 2, 6020 Innsbruck, Austria.

Saïd Khabba: LMME, Faculté des Sciences Semlalia, Université Cadi Ayyad de Marrakech, Marrakech 40000, Morocco; LMI TREMA laboratory.

M. Hakim Kharrou: ORMVAH, Office Régional de Mise en Valeur Agricole du Haouz, Marrakech 40000, Morocco; LMI TREMA laboratory.

Roderik Koenders: Department of Geoscience and Remote Sensing, Delft University of Technology, P.O. Box 5048, 2600 GA Delft, The Netherlands.

Toshio Koike: Department of Civil Engineering, The University of Tokyo, Tokyo 113-8656, Japan.

Valeriy Kovalskyy: Geospatial Sciences Center of Excellence (GSCE), South Dakota State University, Brookings, 57007 SD, USA.

Maarten S. Krol: Twente Water Centre, Water Management Group, University of Twente, Enschede NL-7500 AE, The Netherlands.

Kamal Labbassi: Faculty of Sciences, Chouaib Doukkali University, BD Jabran Khalil Jabran B.P 299, 24000 EL Jadida, Morocco.

Michel Le Page: CESBIO, Unité Mixte de Recherche (CNRS, UPS, CNES, IRD), Toulouse 31000, France; LMI TREMA laboratory.

Xin Li: Cold and Arid Regions Environmental and Engineering Research Institute, Chinese Academy of Sciences, Lanzhou 730000, China.

Zohra Lili-Chabaane: GREF, Université de Carthage/INAT, 43, Avenue Charles Nicolle 1082-Tunis-Mahrajène, Tunisia.

Laurent Linguet: UAG/ESPACE-DEV, Route de Montabo, Cayenne 97300, French Guiana.

Nobuhle Majozi: Natural Resource and Environment, Council for Scientific and Industrial Research, P.O. Box 395, Pretoria 0001, South Africa.

Gerardo Di Martino: Department of Electrical Engineering and Information Technology, University of Napoli Federico II, Via Claudio 21, 80125 Napoli, Italy.

Renaud Mathieu: Natural Resource and Environment, Council for Scientific and Industrial Research, P.O. Box 395, Pretoria 0001, South Africa; Department of Geography, Geoinformatics and Meteorology, University of Pretoria, Private Bag X20 Hatfield, Pretoria 0028, South Africa.

Massimo Menenti: Geosciences and Remote Sensing Department, Delft University of Technology, Stevinweg 12628 CN Delft, The Netherlands.

Francesco Mitidieri: Department of Civil Engineering, University of Salerno, Via Giovanni Paolo II 132, 84084 Fisciano (SA), Italy.

Collins Mito: Department of Physics, University of Nairobi, P.O. Box 30197, 00100 Nairobi, Kenya.

Yasir A. Mohamed: UNESCO-IHE Institute for Water Education, P.O. Box 3015, 2601 DA Delft, The Netherlands; Hydraulic Research Station, P.O. Box 318, Nile Street, Wad Medani, Sudan; Delft University of Technology, Stevinweg 1, 2628 CN Delft, The Netherlands.

Abdelilah El Moutamanni: CESBIO, Unité Mixte de Recherche (CNRS, UPS, CNES, IRD), Toulouse 31000, France; LMI TREMA laboratory.

Felix Mtalo: Department of Water Resources Engineering, University of Dar es Salaam, P.O. Box 35131, 14115 Dar es Salaam, Tanzania.

Joseph Mtamba: Department of Water Resources Engineering, University of Dar es Salaam, P.O. Box 35131, 14115 Dar es Salaam, Tanzania; Department of Water Resources, Faculty of Geo-Information and Earth Observation (ITC), University of Twente, P.O. Box 6, AA Enschede7500, The Netherlands.

Eric Muala: Water Resources Commission, P.O. Box CT 5630, Cantonment, Accra, Ghana.

Lal Muthuwatte: International Water Management Institute (IWMI), Colombo, Sri Lanka and Vientiane, Laos.

Vahid Naeimi: Department of Geodesy and Geoinformation, Vienna University of Technology, Gusshausstrasse 27-29/E122, 1040 Vienna, Austria.

Preksedis Ndomba: Department of Water Resources Engineering, University of Dar es Salaam, P.O. Box 35131, 14115 Dar es Salaam, Tanzania.

Innocent Nhapi: Department of Environmental Engineering, Chinhoyi University of Technology, P. Bag 772, Chinhoyi, Zimbabwe.

Alecia Nickless: Natural Resource and Environment, Council for Scientific and Industrial Research, P.O. Box 395, Pretoria 0001, South Africa.

Francesco Nutini: Institute of Electromagnetic Sensing of Environment, National Research Council of Italy (CNR-IREA), Via Bassini 15, Milan 20133, Italy.

Nicolina Papa: Department of Civil Engineering, University of Salerno, Via Giovanni Paolo II 132, 84084 Fisciano (SA), Italy.

Guillaume Ramillien: Centre National de la Recherche Scientifique (CNRS), GET-UMR5563 CNRS/IRD/UPS, 14, Avenue Edouard Belin, 31400 Toulouse, France; Groupe de Recherche en Géodésie Spatiale (GRGS), 14, Avenue Edouard Belin, 31400 Toulouse, France.

Abel Ramoelo: Natural Resource and Environment, Council for Scientific and Industrial Research, P.O. Box 395, Pretoria 0001, South Africa; Risk and Vulnerability Assessment Centre, University of Limpopo, Sovenga 0727, South Africa.

Mohamed Rasmy: Department of Civil Engineering, The University of Tokyo, Tokyo 113-8656, Japan.

Lisa-Maria Rebelo: International Water Management Institute (IWMI), Colombo, Sri Lanka and Vientiane, Laos.

Daniele Riccio: Department of Electrical Engineering and Information Technology, University of Napoli Federico II, Via Claudio 21, 80125 Napoli, Italy.

Tom Rientjes: Department of Water Resources, Faculty of Geo-information Science and Earth Observation (ITC), University of Twente, P.O. Box 6, AA Enschede 7500, The Netherlands.

Vivien Robinet: UAG/ESPACE-DEV, Route de Montabo, Cayenne 97300, French Guiana.

Mireia Romaguera: Faculty of Geo-Information Science and Earth Observation, University of Twente, 7500 AE Enschede, The Netherlands; Twente Water Centre, Water Management Group, University of Twente, Enschede NL-7500 AE, The Netherlands.

Giuseppe Ruello: Department of Electrical Engineering and Information Technology, University of Napoli Federico II, Via Claudio 21, 80125 Napoli, Italy.

Donald Tendayi Rwasoka: Upper Manyame Subcatchment Council, Box 1892, Harare, Zimbabwe.

Daniel Sabel: Department of Geodesy and Geoinformation, Vienna University of Technology, Gußhausstraße 27-29, Vienna 1040, Austria.

Mhd. Suhyb Salama: Faculty of Geo-Information Science and Earth Observation, Department of Water Resources, University of Twente, Enschede NL-7500 AE, The Netherlands.

Nazmus Sazib: Department of Civil Engineering, University of Louisiana at Lafayette, Lafayette, LA 70504, USA; Current Address: Department of Civil and Environmental Engineering, Utah State University, Logan, UT 84322, USA.

Gabriel Senay: Geospatial Sciences Center of Excellence (GSCE), South Dakota State University, Brookings, 57007 SD, USA; Earth Resources Observation and Science (EROS) Center, U.S. Geological Survey, Sioux Falls, SD 57198, USA.

Lucia Seoane: Groupe de Recherche en Géodésie Spatiale (GRGS), 14, Avenue Edouard Belin, 31400 Toulouse, France; Observatoire Midi-Pyrénées (OMP), GET-UMR5563 CNRS/IRD/UPS, 14, Avenue Edouard Belin, 31400 Toulouse, France.

Frédérique Seyler: IRD/ESPACE-DEV, Route de Montabo, Cayenne 97300, French Guiana.

Munyaradzi Davis Shekede: Department of Geography and Environmental Science, University of Zimbabwe, Box MP 167, Harare, Zimbabwe.

Vladimir Smakhtin: International Water Management Institute (IWMI), Colombo, Sri Lanka and Vientiane, Laos.

Zhongbo Su: Faculty of Geo-Information Science and Earth Observation, Department of Water Resources, University of Twente, Enschede NL-7500 AE, The Netherlands.

Camille Szczypta: CESBIO/UMR 5126, 18 av. Edouard Belin, bpi 2801, 31401 Toulouse Cedex 9, France.

Adrien Tavernier: CESBIO, Unité Mixte de Recherche (CNRS, UPS, CNES, IRD), Toulouse 31000, France; LMI TREMA laboratory.

Christian Tottrup: DHI GRAS, DK-2970 Hørsholm, Denmark.

Jihad Toumi: LMME, Faculté des Sciences Semlalia, Université Cadi Ayyad de Marrakech, Marrakech 40000, Morocco; LMI TREMA laboratory.

Nick van de Giesen: Faculty of Civil Engineering and Geosciences, Delft University of Technology, Stevinweg 1, 2628 CN Delft, The Netherlands.

Rogier van der Velde: Department of Water Resources, Faculty of Geo-Information and Earth Observation (ITC), University of Twente, P.O. Box 6, AA Enschede7500, The Netherlands.

Pieter van der Zaag: UNESCO-IHE Institute for Water Education, P.O. Box 3015, 2601 DA Delft, The Netherlands; Delft University of Technology, Stevinweg 1, 2628 CN Delft, The Netherlands.

Zoltán Vekerdy: Department of Water Resources, Faculty of Geo-Information and Earth Observation (ITC), University of Twente, P.O. Box 6, AA Enschede7500, The Netherlands; Department of Water and Waste Management, Szent István University, Gödöllő, 2100 Páter Károly utca 1., Hungary.

Wolfgang Wagner: Department of Geodesy and Geoinformation, Vienna University of Technology, Gußhausstraße 27-29, Vienna 1040, Austria.

Andreas Walli: GeoVille, Sparkassenplatz 2, 6020 Innsbruck, Austria.

Mohamed Yousfi: ORMVAH, Office Régional de Mise en Valeur Agricole du Haouz, Marrakech 40000, Morocco.

Yu Zhang: NOAA/NWS/OHD, Silver Spring, MD 20910, USA.

Mehrez Zribi: CESBIO/UMR 5126, 18 av. Edouard Belin, bpi 2801, 31401 Toulouse Cedex 9, France.

About the Guest Editors

Benjamin Koetz works as Application Scientist in the Earth Observation Directorate of the European Space Agency. His tasks focus on the development of Earth Observation (EO) applications in close collaboration with relevant user communities, scientists, and EO service providers. In particular, he is responsible for the TIGER initiative dealing with EO for water resource management in Africa and is involved as co-lead in the GEO-Global Agricultural Monitoring Initiative (GEOGLAM). Benjamin Koetz received his M.S. degree in Environmental Sciences with a major in Remote Sensing from the University of Trier, Germany. He also holds a Ph.D. with a specialization in Earth Observation from the University of Zürich, Switzerland. His scientific expertise focuses on the development of physically-based methodologies to derive geo-biophysical EO products.

Zoltán Vekerdy is a hydrologist and remote sensing specialist, who holds the position of Assistant Professor at the ITC Faculty of the University of Twente, Netherlands. He also works as Scientific Advisor at the Szent István University, Hungary. He started his research carrier in the 1980s at the Water Resources Research Centre (VITUKI) in Hungary. Since the beginning, his field of interest has been the application of Earth Observation for water management, with focus on environmental and agricultural aspects. He did research on a number of wetlands around the world, including, among others, in Iran, China, Mexico, and several countries of Africa. Throughout his carrier, he has been supervising several young researchers at PhD and MSc levels at universities of the US, Netherlands, Hungary, and Zambia. He (co-)authored more than hundred scientific publications, including peer-reviewed articles, book chapters and scientific reports. Since 2008, as the Principal Investigator of the TIGER Capacity Building Facility funded by the European Space Agency, he has been coordinating the network of several hundreds of researchers working on more than fifty Earth observation research projects in the water sector of Africa.

Massimo Menenti is professor of Optical and Laser Remote Sensing at the Delft University of Technology in The Netherlands and an internationally renowned scientist in the fields of Earth Observation and Global Terrestrial Water Cycle. His achievements have been attained in the retrieval of land surface properties from remote sensing, including the estimation of evapotranspiration (ET), time series analysis of remote sensing products, and the application of remote sensing technology in hydrology and climate models. He is one of the earliest researchers in using laser technology to measure surface aerodynamic roughness. He initiated the use of time series analysis techniques to extract information from satellite data. He developed the surface energy balance index (SEBI) theory for ET estimation, which is the prototype of the following S-SEBI, SEBS, and SEBAL models. He held senior research positions in the Netherlands, France, the USA, and Italy, and has coordinated many large European projects with participants from Europe, Asia, America, and Africa. He has recently been granted a National Distinguished Foreign Expert award by the People Republic of China.

Diego Fernández-Prieto received his B.S Degree in Physics from the University of Santiago de Compostela, Spain, in 1994. In 1994 and 1996, he was with the "Istituto per la Matematica Applicata" (I.M.A) of the National Research Council (C.N.R), Genoa, Italy. In 1997, he received his Master Degree in Business Administration (MBA) from the University of Deusto, Spain, and the University of Kent, United Kingdom. In 2001, he received his Ph.D degree in Electronic Engineering and Computer Science from the Department of Biophysical and Electronic Engineering at the University of Genoa, Italy. Since 2001, he has been with the Earth Observation (EO) Science, Applications and Future Technologies Department of the European Space Agency (ESA). At present, he is program manager of the Support To Science Element (STSE), aimed at addressing the scientific needs and requirements of the Earth system science community in terms of novel mission concepts, new algorithms, and products and innovative Earth science results.

Preface

Reliable access to water, managing the spatial and temporal variability of water availability, ensuring the quality of freshwater, and responding to climatological changes in the hydrological cycle are prerequisites for the development of countries in Africa. Water, being an essential input for biomass growth and for renewable energy production (e.g. biofuels and hydropower schemes), plays an integral part in ensuring food and energy security for any nation. Water, as a source of safe drinking water, is furthermore the basis for ensuring the health of citizens and plays an important role in urban sanitation. In view of the transversal importance of water in our society, the United Nations has recently announced a dedicated goal on Water and Sanitation as part of the new Sustainable Development Goals (SDG).

The concept of Integrated Water Resource Management (IWRM) is seen as an opportunity to deal with water variability and the wide spread water scarcity in Africa. One key component missing from IWRM in Africa is knowledge of the available extent and quality of water resources at a basin level. Earth Observation (EO) technology can help fill this information gap by assessing and monitoring water resources at adequate temporal and spatial scales.

The goal of this Special Issue is to understand and demonstrate the contribution that satellite observations, consistent over space and time, can bring to improve water resource management in Africa. Possible EO products and applications range from catchment characterization, water quality monitoring, soil moisture assessment, water extent and level monitoring, irrigation services, urban and agricultural water demand modeling, evapotranspiration estimation, ground water management, to hydrological modeling and flood mapping/forecasting. Some of these EO applications have already been developed by African scientists within the ten-year lifetime of the TIGER initiative: Looking after Water in Africa (http://www.tiger.esa.int), whose contributions was the starting point of this Special Issue but is only one example of the wide range of activities in the field. The total of 22 papers in this Special Issue gives access to wide range of expertise from the entire African and international scientific community, dealing with the challenges of water resource management in Africa. Several papers also addressed the latest developments in terms of new missions (such as the Sentinel missions), as well as related EO products and techniques that are now available to improve IWRM in Africa.

Benjamin Koetz, Zoltán Vekerdy,
Massimo Menenti and Diego Fernández-Prieto
Guest Editors

Chapter 1:
Water Resource Management

Enabling the Use of Earth Observation Data for Integrated Water Resource Management in Africa with the Water Observation and Information System

Radoslaw Guzinski, Steve Kass, Silvia Huber, Peter Bauer-Gottwein, Iris Hedegaard Jensen, Vahid Naeimi, Marcela Doubková, Andreas Walli and Christian Tottrup

Abstract: The Water Observation and Information System (WOIS) is an open source software tool for monitoring, assessing and inventorying water resources in a cost-effective manner using Earth Observation (EO) data. The WOIS has been developed by, among others, the authors of this paper under the TIGER-NET project, which is a major component of the TIGER initiative of the European Space Agency (ESA) and whose main goal is to support the African Earth Observation Capacity for Water Resource Monitoring. TIGER-NET aims to support the satellite-based assessment and monitoring of water resources from watershed to cross-border basin levels through the provision of a free and powerful software package, with associated capacity building, to African authorities. More than 28 EO data processing solutions for water resource management tasks have been developed, in correspondence with the requirements of the participating key African water authorities, and demonstrated with dedicated case studies utilizing the software in operational scenarios. They cover a wide range of themes and information products, including basin-wide characterization of land and water resources, lake water quality monitoring, hydrological modeling and flood forecasting and mapping. For each monitoring task, step-by-step workflows were developed, which can either be adjusted by the user or largely automatized to feed into existing data streams and reporting schemes. The WOIS enables African water authorities to fully exploit the increasing EO capacity offered by current and upcoming generations of satellites, including the Sentinel missions.

Reprinted from *Remote Sens.* Cite as: Guzinski, R.; Kass, S.; Huber, S.; Bauer-Gottwein, P.; Jensen, I.H.; Naeimi, V.; Doubková, M.; Walli, A.; Tottrup, C. Enabling the Use of Earth Observation Data for Integrated Water Resource Management in Africa with the Water Observation and Information System. *Remote Sens.* **2014**, *6*, 7819-7839.

1. Introduction

Despite having experienced more than 10 years of continuous economic growth, Africa today faces great water resource management challenges. With 10% of the world's renewable water resources, more than 60 trans-boundary basins, a low level of water development and utilization and increasing population, Africa's future economic growth will continue to be constrained by the development of its water resources. Today, in many African countries, water policies and management decisions are based on sparse and unreliable information. In this challenging context, water information systems are fundamental for improving water governance and implementing integrated water resource management (IWRM) successfully. This water information gap is a major limitation for putting in practice IWRM plans to face the current and coming challenges of the African water sector. Recognizing the utility of satellite data for IWRM, the European Space Agency (ESA), through its participation in the Committee on Earth Observation Satellites (CEOS), launched the TIGER initiative in 2002 [1]. The TIGER initiative supports water authorities, technical centers and other stakeholders in the African water sector to enhance their capacity to collect and use water-relevant geo-information to better monitor, assess and inventorize their water resources by exploiting Earth Observation (EO) products and services [2]. Currently, the TIGER initiative consists mainly of the TIGER Capacity Building Facility (including support for selected research projects) and the TIGER-NET project.

The aim of TIGER-NET is to build a pre-operational capacity for water resources monitoring based on EO technologies at mandated African water authorities. The initial key host institutions already actively involved in TIGER-NET encompass major river basin authorities (Nile Basin Initiative, Lake Chad Basin Commission, Zambezi Watercourse Commission and Volta Basin Authority), national ministries and agencies (Department of Water Affairs South Africa; the Hydrologic Division of the Namibian Ministry of Agriculture, Water and Forestry; the Department of Water Affairs of the Zambian Ministry of Mines, Energy and Water Development; DR Congo National Agency of Meteorology and Teledetection by Satellite; Instituto Nacional de Meteorologia of Mozambique), as well as international research and humanitarian organizations (International Water Management Institute, United Nations World Food program and Action Against Hunger).

The TIGER-NET project builds on the 10 years of experience gained within TIGER demonstration and capacity building activities in order to develop practices and tools required for an eventual transfer of EO information into the day-to-day work of water authorities. A steering committee consisting of experts from the African Water Facility, African Ministers' Council on Water-Technical Advisory Committee (AMCOW-TAC), the Water Research Commission of South Africa, United Nations' Economic Commission for Africa (UN-ECA) and United Nations Educational, Scientific and Cultural Organization's International Hydrological Programme (UNESCO IHP), provides guidance with regard to the African water sector priorities. The major focus of the project is on developing, demonstrating and training a user-driven, open-source Water Observation and Information System (WOIS), which enables the production and application of a range of satellite EO-based information products needed for IWRM in Africa. Importantly, one of the aims is to develop the necessary local capacity for accessing and exploiting historic satellite data, as well as future Sentinel observations [3]. Free data access, free licensing and the ability to integrate with existing systems are key advantages of the WOIS, which should enable its extension to other countries and regions in Africa and beyond, as well as encourage user-driven sustainability in terms of funding and operation.

Against this background, this paper outlines the development framework of the WOIS software to illustrate current features of the system and to review selected application cases demonstrating the real impact of the system on enhancing water management and integrated water resource management plans in Africa.

2. Technical Development and WOIS Design

2.1. User Driven Design and Development

The WOIS has been designed in direct response to user requirements, *i.e.*, based on extensive consultation, review and analysis of the user needs in terms of their current technological and personnel capacity, application-specific monitoring demands, as well as geo-information and system needs. In general, the common requirement was for an easy-to-operate, open-source end-to-end system enabling a full capacity to establish water-related information for monitoring, analysis and reporting (maps, tables and graphs) per sub-watershed for IWRM. While the system requirements were found to be very common among the host institutions, the specific application requirements and information demands varied according to the variety of IWRM challenges faced in the different river basins of Africa. Those applications included mapping and monitoring of lake water quality, flood monitoring, land degradation and land cover characterization, water bodies and wetlands mapping, hydrological modeling, hydrological characterization (soil moisture, precipitation and evapotranspiration), soil erosion potential indicators, as well as urban water supply and sanitation planning support.

The users have also been part of the actual WOIS development, which has followed the agile principles for software development in which the developers stay flexible and responsive to the latest issues reported by the users [4]. The work has progressed via feedback loops where the developers have tackled any outstanding issues, prioritized based on their importance to the users, before testing the solutions and integrating them into the software system. At the end of each loop, a working product was delivered to the users, who would then provide more feedback to the developers. In the case of WOIS software, the initial users were the EO specialists involved in the system design and application creation, and later, during the system installation and demonstration, the development was driven directly by feedback from the African water management authorities.

2.2. System Architecture and Functionality

As no single software package could provide all of the requested functionality, the underlying design principle was to develop a system that uses dedicated software for specific tasks and where the various software components are integrated into a single graphical user interface (GUI). All of the WOIS software components (Figure 1) are based on proven and stable open-source (free) software and include:

- QGIS 2.2 [5]: extensive and user friendly GIS (software website: qgis.org (accessed on 9 March 2014));
- GRASS GIS 6.4.3 [6]: modular GIS consisting of raster and vector analysis algorithms (software website: grass.osgeo.org (accessed on 9 March 2014));
- BEAM 5.0 [7]: processing of optical and thermal ESA data products (software website: brockmann-consult.de/cms/web/beam (accessed on 9 March 2014));
- NEST 5.1 [8]: processing of radar ESA data products (software website: earth.esa.int/web/nest/home (accessed on 9 March 2014));

- Orfeo Toolbox 4.0.0 [9]: high resolution image processing (software website: orfeo-toolbox.org (accessed on 9 March 2014));
- Soil Water Assessment Tool (SWAT) 2.9 [10]: hydrological modeling (software website: swat.tamu.edu (accessed on 9 March 2014));
- R 3.1.0 language scripts [11]: statistical and graphical tools (software website: www.r-project.org (accessed on 9 March 2014));
- PostGIS 2.1.3 [12]: geospatial database (software website: postgis.net (accessed on 9 March 2014)).

Figure 1. Open-source software packages integrated as part of the Water Observation and Information System (WOIS).

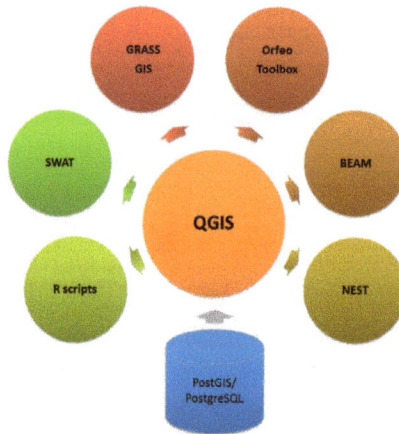

In addition, Python scripts [13] were used for automating certain tasks and integrating the different software. WOIS combines full versions of the component software into a multipurpose system consisting of a storage container for the geodata, extraction and processing of the EO data through customized processing facilities and integrative tools and models aimed at decision support, e.g., hydrological modeling and GIS-embedded visualization and analysis tools.

Selected examples of generic WOIS capabilities are georeferencing, reprojection and radiometric calibration of optical and SAR data obtained by (among others) MERIS and ASAR sensors onboard the Envisat satellite and the SAR sensor onboard the RADARSAT-2 satellite, terrain analysis, image classification and change detection, time-series analysis, interactive data exploration and export (tables and graphs), map composing and 3D visualization. WOIS also provides a hydrological modeling framework for scenario-based model development and operational simulation and forecasting. Furthermore, a PostGIS database enables centralized or distributed storage of vector data, while a library of import/export functions ensures the ability to integrate and/or connect to external IT infrastructures and databases.

There are no minimum system requirements for using WOIS, and the system performance depends on the size of the raster and vector data sets that are to be analyzed and the computational complexity of the analysis tasks to be performed. Therefore, for optimal performance it is recommended for the host computer to have at least an Intel Core i5-3570 processor, 8 GB of

RAM, 1 TB of hard disk space and to be running Windows 7 (64 bit) or higher. However, WOIS has been successfully installed and operated on 32- and 64-bit computers falling far below the above specifications, with Windows versions ranging from XP to 8.

2.3. Component Integration

QGIS was chosen as the central integrating platform, due to its clear and accessible GUI, strong development community, ease of implementing additional functionalities through Python plugins and its high level of interoperability with major GIS data formats through the use of the Geospatial Data Abstraction Library (GDAL/OGR) library [14]. Moreover, the integrated Processing toolbox, formerly known as SEXTANTE [15], brings the ability to easily incorporate geoprocessing algorithms from various applications into QGIS. It acts as a joint repository for a wide range of algorithms, some native to QGIS and others imported from external applications, such as GRASS or the Orfeo Toolbox. It also allows for easy incorporation of R and Python scripts. The algorithms included in the Processing toolbox integrate seamlessly with the QGIS capabilities of data I/O, rendering or map creation.

The Processing toolbox is based on modular architecture with limited core functionality and the ability to easily add geoalgorithms from different applications through provider modules. The core functionality is responsible for, for example, data passing to and from QGIS or automatic GUI creation for each algorithm. The provider modules take care of exposing the algorithms to the toolbox, communicating with the external applications and setting up the correct environment for algorithm execution. The external communication is mostly performed through command line-based instructions, although it is also possible to engage the external applications through their Python bindings.

The Processing toolbox already included modules linking with many of the WOIS software components. However, an algorithm provider for BEAM and the Next ESA SAR Toolbox (NEST) had to be developed as an additional QGIS plugin. Since NEST is built on top of BEAM's core libraries, it was possible to create a common provider for the two applications. The communication with BEAM and NEST is performed through the Graph Processing Framework (GPF), which takes care of low level issues, such as efficient data input and output or multi-threading. The GPF can be called on a command line, and through passing of an XML file a chosen operator can be executed with the given settings. Since the toolboxes for the upcoming Sentinel missions will be based on BEAM and NEST [16,17], the use of GPF ensures an easy implementation path for Sentinel algorithms into WOIS.

Similarly, a QGIS plugin was developed for incorporating SWAT modeling inside QGIS processing. The plugin has functionality for setting up and calibrating SWAT models, acquiring climate data from outside sources, running the models, assimilating observations and plotting the results.

2.4. Processing Workflows

One of the features of the QGIS Processing toolbox is the modeler functionality, which enables the creation of models combining any of the algorithms present in the toolbox. The modeler comes with an easy to use drag and drop GUI, making it possible to quickly create advanced processing models. A similar functionality was developed as part of WOIS inside a new QGIS plugin, to enable the creation of processing workflows through an easy to use GUI.

Figure 2. The WOIS graphical user interface, including the embedded workflow library (**center**) and wizard-based processing workflow (**right**).

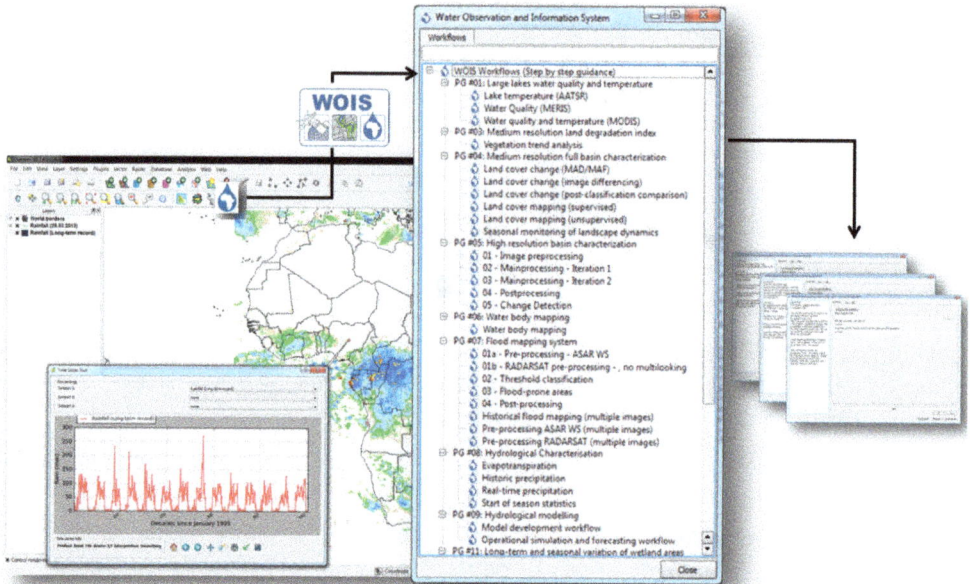

The workflows transparently combine algorithms from the different providers and guide the users with wizard-like, step-by-step instructions through the available processing chains. They are intended for novice and intermediate users, as an introduction to the theory and practice of using EO data for various tasks related to their field of interest. Therefore, they were designed to be used with minimal technical skills, although in some cases, expert local knowledge or GIS/modelling experience is still required. The workflows are accessible from the WOIS toolbox, which is available through the QGIS GUI (Figure 2) and functions as a workflow library. More advanced users may choose to explore the full suite of algorithms and tools available from the Processing toolbox in order to create their own workflows or models.

3. Water Resource Applications

The operational and practical use of the WOIS to support IWRM in Africa has been demonstrated via a series of user-specific demonstration cases, some of which are described in this section and summarized in Table 1 [18–20]. They show the depth and versatility of WOIS for performing numerous tasks related to water resource management and the advantages of combining the capabilities of the different WOIS component software.

The demonstration cases had several stages. First, customized end-to-end processing workflows were developed for the requested use cases. The developed workflows were subsequently used for product derivation over significant areas and time periods, as requested by the users. Continental-scale products at 1–25 km resolution are already provided on an operational basis. In addition, trans-boundary products at 150–500 m covering in total over 17,000,000 km^2, basin-scale products at 2.5–30 m covering in total around 120,000 km^2 and local-scale products at 0.5–2.5 m covering in total some 300 km^2 were demonstrated with the WOIS to date on a number of African subsets chosen by the participating host institutions. The final step involved the testing of the workflows' stability/performance and ease of use, as well as evaluating the validity and usefulness of the outcome products in close dialogue with the users.

Table 1. Summary of the WOIS demonstration cases described in this paper.

Name	Key Output Variables	Region of Application	Accuracy/Performance	Limitations	Required User Skills
Large lakes water quality and temperature monitoring	Water surface temperature, chlorophyll concentration, suspended sediments concentration.	Lake Victoria, Lake Chad	Spatiotemporal variation in accordance with expected patterns. MODIS-derived water quality is of lesser accuracy.	Works on medium to coarse resolution data, so not applicable to small lakes. Operational use dependent on Sentinel 3.	Minimal.
Medium resolution full-basin characterization	Land cover/use maps and change statistics.	Volta Basin, Lake Chad area	Overall accuracy of 80%. Kappa coefficient exceeding 0.7.	Designed for medium and coarse resolution data, so cannot resolve small-scale changes.	Minimal technical skills. Expert local knowledge needed for selection/labelling of classes.
Medium resolution land degradation index	Maps of areas with rainfall-independent, statistically-significant vegetation change.	Volta Basin, Lake Chad area	Vegetation trends were confirmed by local experts and other studies [18].	Applicable in rainfall limited ecosystems only.	Minimal.
Hydrological characterization	Historic and real-time precipitation, evapotranspiration.	Whole of Africa	Uses well-established datasets with documented accuracy [19,20].	Coarse spatial resolution.	Minimal.
High resolution basin characterization	Land cover/use maps.	Lake Chad area, South Africa, Namibia, Zambia.	Overall accuracy above 80%. Kappa coefficient exceeding 0.8.	Requires expert local knowledge or reference data.	Intermediate technical skills. Expert local knowledge needed for selection/labelling of classes.

10

Table 1. *Cont.*

Name	Key Output Variables	Region of Application	Accuracy/Performance	Limitations	Required User Skills
Water body mapping	Water extent mask.	Volta Basin, Lake Chad area, Zambia	Overall accuracy above 90%. Kappa coefficient exceeding 0.8.	Requires NIR and SWIR spectral information.	Intermediate technical skills.
Hydrological modelling	River discharge forecasts.	Kavango, Mokolo, Volta and Zambezi basins	Nash-Sutcliffe efficiency of 0.96 for 1-day forecast, 0.77 for 7-day forecast.	Requires field measurements of discharge for model calibration.	Advanced technical skills for model setup. Minimal technical skills for operational forecasting.
Flood mapping	Historical and real-time flood maps.	Nile basin in Sudan and Lake Chad basin	Overall accuracy of 0.95 to 0.99. Kappa coefficient between 0.64 and 0.75.	Lower accuracy in rough water surfaces, areas with partially submerged vegetation or desert regions.	Minimal

The following sections review five application cases in order to illustrate the use of WOIS for various tasks related to water resource management: monitoring of lake water quality, basin-wide land and hydrological characterization, high-resolution land and water characterization, hydrological modeling and flood monitoring.

3.1. Large Lakes Water Quality Monitoring

The provision of clean fresh water is a serious environmental challenge, and optical remote sensing has become an increasingly important tool for monitoring water quality on a regular basis. Therefore, WOIS provides workflows for estimating operational and historical, satellite-derived, water quality monitoring products for major lakes in Africa (Figure 3). The products can be used for, e.g., potential identification of point sources of pollution, the establishment of possible correlations with regular cholera outbreaks, better understanding of eutrophication processes and regular reporting obligations.

Under TIGER-NET, monitoring information about water quality and temperature is provided for Lake Chad and Lake Victoria using Envisat MERIS and AATSR (historic information) and MODIS AQUA (current information). Envisat data are processed using WOIS-embedded BEAM functionalities, including the eutrophic lakes processor, to derive water quality parameters (e.g., concentrations of chlorophyll and total suspended matter) from MERIS [21], and the Sea Surface Temperature (SST) processor, to obtain surface water temperature from AATSR data. Due to the failure of Envisat satellite in April 2012, the MODIS sensor on the AQUA satellite is being temporarily used for operational lake water quality and temperature observations. The MODIS data are processed by the TIGER-NET consortium using the L2 data processors available in SeaWiFS Data Analysis System (SeaDAS) [22] and then delivered to the WOIS database.

Figure 3. An example product from the WOIS workflow for monitoring lake water quality.

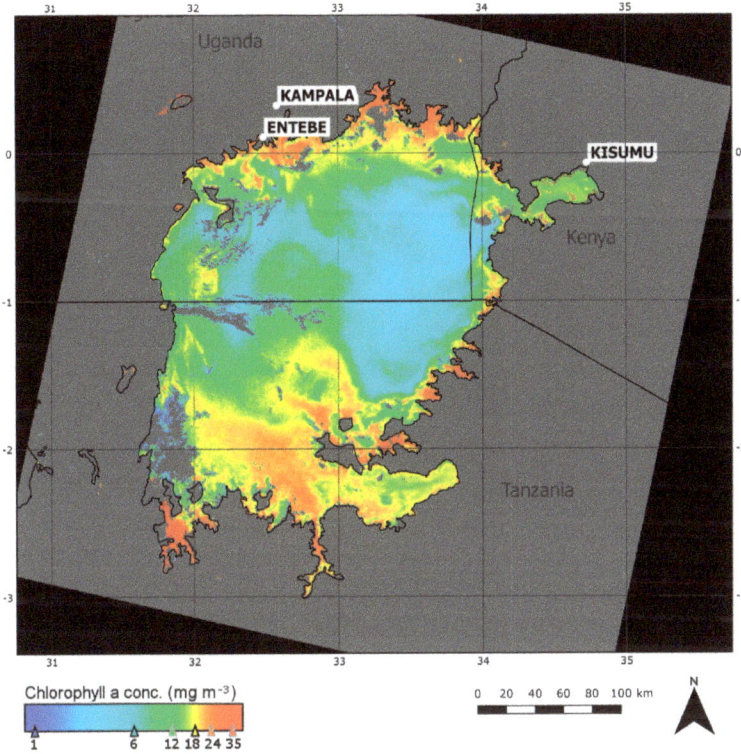

The validation of the water quality and temperature products has shown spatiotemporal variation in accordance with expected patterns. Especially the seasonal variation in lake surface temperature over both lakes is well captured in both historical and operational mode, hence underpinning the strong similarity of AATSR and MODIS AQUA temperature products. For the water quality products, the outcome is more ambiguous, as it depends on the performance of the processor for the specific lake. Looking past the extreme cases, the MERIS-derived concentrations of chlorophyll and total suspended matter exhibit spatial and temporal consistency with absolute values residing within the range of published numbers for both Lake Chad and Lake Victoria. The operational MODIS outputs show spatiotemporal patterns similar to the MERIS outputs over Lake Victoria, yet the output values are an order of magnitude lower, while the operational delivery of water quality products over Lake Chad is either impossible or inconsistent at best. The divergence between the two data sources is explained by the calibration range of the input algorithm for MODIS, which is designed for ocean color mapping and, thus, not ideal for inland lakes. The situation is expected to be rectified in the future, where data from the Sentinel 3 mission (expected to launch in 2015) will be used for the provision of water quality monitoring information through dedicated WOIS workflows.

12

3.2. Basin-Wide Characterization of Land and Water Resources

The basin-wide assessment and monitoring of hydrological system components and their interactions is very important for water resource management. Such components include large-scale land use changes, as well as regional precipitation, evapotranspiration and soil moisture estimates (including soil water index products), all of which are important for basin hydrology (e.g., by impacting runoff, streamflow or water availability) and for the current and future utilization potential of the land.

The WOIS includes six workflows, based mostly on the Orfeo Toolbox functionality, for basin-wide land use characterization and change detection analysis. For example, basin-wide land cover and land use maps can be derived from medium resolution imagery using either the supervised support vector machine [23] or the unsupervised *k*-means classifiers (Figure 4). Spectral changes between multi-temporal imagery can be analyzed using simple change detection algorithms, such as image differencing, as well as more advanced techniques, such as multivariate alteration detection and the maximum autocorrelation factor [24]. Thematic changes can be reported using a post-classification workflow, which returns the cross-tabulation of two input classification maps.

Figure 4. Recent land cover map of the Volta Basin derived using WOIS workflows for land cover mapping.

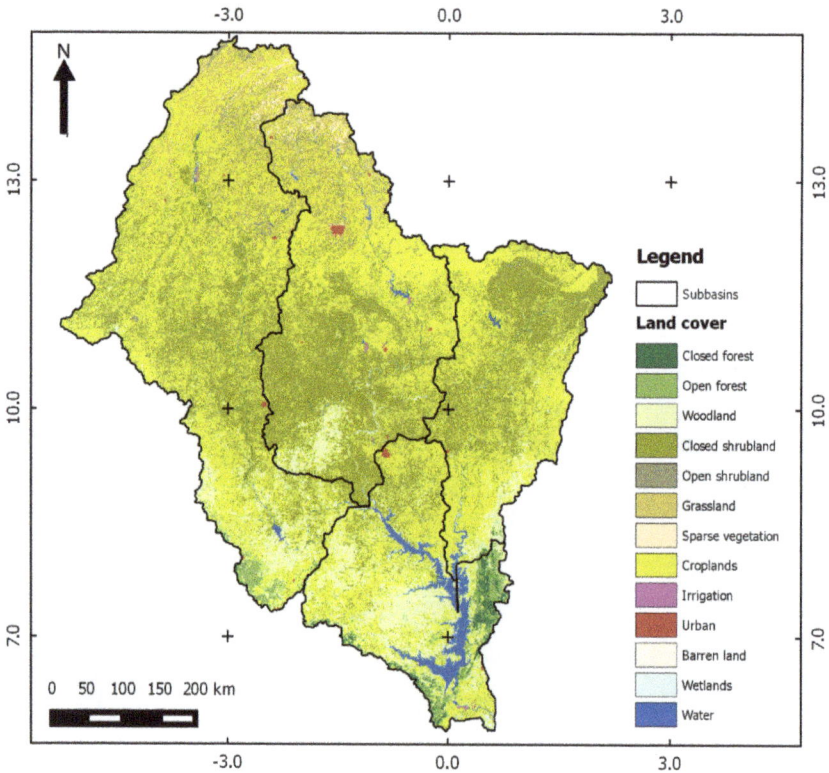

Basin-wide land degradation mapping can also be performed using a WOIS-embedded workflow. The WOIS implementation of the mapping method for land degradation uses mostly GRASS modules, with additional Python scripts to facilitate the processing of time-series data according to principles put forward in Huber *et al.* (2011) [25] and Hellden and Tottrup (2008) [26]. The workflow ingests gap-filled time series of NDVI (as a proxy for vegetation biomass [27]) and rainfall estimates in order to analyze vegetation/rainfall correlations and to control NDVI trends for variability in rainfall. NDVI residual time series, originating from regressing NDVI on rainfall, is subsequently searched for significant long-term trends in vegetation productivity, which is not related to rainfall, but possibly contributable to humans (e.g., population growth, changing land use practices, deforestation, infrastructure developments, as well as rural exodus and urbanization). Full basin assessments of land cover and land use changes, as well as land degradation processes have been successfully demonstrated for Lake Chad and Volta Basin using medium resolution imagery from MODIS and SPOT VGT. When evaluated against higher resolution imagery (e.g., Landsat and Google Earth), the overall accuracy of the land cover/land use products was assessed to be around 80%, with a kappa coefficient of agreement exceeding 0.7. High resolution imagery also supported the validation of the land degradation analysis, yet the causes behind the observed vegetation trends are often manifold, and hence, local experts were consulted to verify and give reasons for distinctive negative or positive vegetation trends. The local experts were able to explain most of the negative vegetation trends with urbanization, dam constructions and deforestation, while positive vegetation trends were mostly associated with protected areas and irrigated lands. A particular interesting trend pattern was observed along the border area of Chad and Sudan. Here, large areas with strong positive vegetation trends appear on the Sudanese side, while pockets of negative vegetation trends are spotted on the Chad side. The reasoning behind this pattern is explained by population displacement as a consequence of the conflict in Darfur (Figure 5) and as corroborated by other studies [18].

The WOIS workflows for basin-wide land characterizations have proved useful for the provision of ground cover information needed for water resource management and planning, as well as establishing the baseline information from which monitoring activities can be performed. Still, the workflows are designed for being used with medium to coarse resolution data, and hence, both land cover transitions and land cover changes may be obscured by the resolving power of the data. Results should therefore not be interpreted as undeniable facts and the area measurements provided certainly not perceived as accurate, but they do indicate a trend that is likely to be real and most likely in the right order of magnitude.

Contrary to the land characterization products, which are the result of dedicated image processing workflows, the integration of the hydrological characterization products into the WOIS database is mainly based on facilitating linkages to external data providers. For example, the near-real-time rainfall data product is downloaded directly from the NOAA Climate Prediction Center [19] (http://nomads.ncep.noaa.gov/ (accessed on 10 June 2014)) through a WOIS workflow, which also allows the user to calculate accumulated rainfall or subset the downloaded images, while the Land Surface Analysis Satellite Applications Facility (LSA-SAF) evapotranspiration product [20] (http://landsaf.meteo.pt/algorithms.jsp?seltab=7&starttab=7 (accessed on 10 June

2014)) is first preprocessed (subsetted and reprojected) by the TIGER-NET consortium before being made available on the TIGER-NET FTP server. All hydrological characterization products have Pan-African coverage and, hence, are available to all users who can download the products using a WOIS-embedded workflow.

Figure 5. Land degradation in Eastern Chad caused by the war in Sudan's Darfur region. Since 2003, over 3000 villages have been destroyed and hundreds of thousands of people have been displaced into refugee camps in neighboring Chad. These areas are clearly visible in satellite data, as growing camp sites and use of natural resources have caused a vegetation decline. On the other hand, the Sudan side shows signs of vegetation greening caused by agricultural land abandonment as forced by the population displacement.

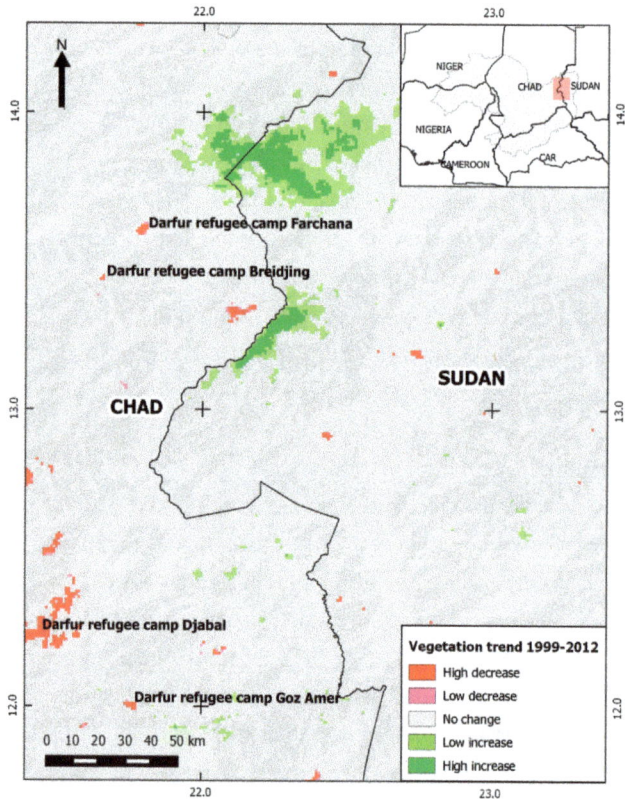

3.3. High Resolution Land and Water Characterization

Mapping land cover at the sub-basin level with high resolution (5–20 m pixel size) EO observations has many practical applications in water management and water resource accounting. Those applications include tracking seasonal and long-term land cover changes (disappearance of vegetation, change of mining or cropland areas), observing the capacity and location of small water bodies and delineating lake shorelines and wetlands. From the regional water demand perspective, accurate mapping of "cultivated areas" (irrigated and non-irrigated) and "urban areas" (residential and commercial/industrial) was deemed of high importance by the participating water authorities.

The methodology implemented in the relevant WOIS workflows follows an automated, hybrid pixel- and object-based EO image classification approach, based on the multi-spectral and spatial properties of the satellite imagery, followed by stringent post-processing rules for refinement of the results. As the main pixel-based approach, an unsupervised k-means classification method was selected [28]. The segment-based classification process consists of two steps: image segmentation and classification, both controlled by a dedicated rule set aimed at being as simple as possible to ease method transfer to other regions, but as complex as necessary for the desired results [29]. The outputs of the two approaches were fused, based on spatial statistics per land cover type, thus combining the advantages of both classification methods. The workflows allow the possibility of including point sampling data in the processing chain, thus ensuring the participation of local experts during the production and validation phases.

The WOIS high resolution land and water mapping tools were so far successfully implemented for seasonal small water body mapping in the Volta Basin and for mapping water demand-related land cover changes in sub-basins of South Africa. They are currently being implemented for, among others, flood vulnerability mapping in Namibia, as well as for dam monitoring in sub-basins of Zambia. The system components have further been employed by the Lake Chad Basin Commission for assessing in detail the historic changes of the Lake Chad area extent (Figure 6) and its surrounding basin land cover changes, documented in the first Lake Chad Biannual Environmental Report. The historic water area extent has been estimated for a number of selected years (Figure 6a) from the maximum water extent derived from a composite of high resolution images for each year, taken predominantly during the dry season (Figure 6b). It has been shown that despite the significant decrease of Lake Chad in the 1980s, the area of water bodies has nearly doubled from 1986 to 2011, resulting in a significant change in vegetation cover and land use in the basin originally occupied by the lake. The results are directly employed to control and evaluate water management regulations in the basin.

High resolution land cover characterization remains challenging, and the provided tools do not compensate for good user skills regarding image interpretation and classification. The tools provide instruments to derive and characterize, leaving it up to the user to choose the best fitting method and combination in order to achieve adequate results.

3.4. Hydrological Modeling Framework for Real-Time Water Discharge and Flood Forecasting

Hydrological models (HMs) are key decision support tools for integrated water resources management. HMs are quantitative computer simulation engines used to reproduce and analyze the interactions of all relevant hydrological processes and water users in a river basin. They provide answers to "what-if" questions, both in the context of long-term planning and real-time operational management decisions. Long-term planning problems arise because land-use practices, water demands and water-related risks are constantly changing over time. Moreover, as a consequence of global climate and land use changes, the probability distributions for many hydrological variables are starting to change (e.g., [30]). Real-time management problems arise because of the occurrence of extreme hydrological events. The optimal response to such extreme events depends on the actual state of the hydrological system, and real-time information on the system state is thus essential.

Figure 6. (a) Lake Chad historic water extent (indicated in blue) as determined using WOIS. The numbers in brackets on top of each image indicate the months of acquisition of high resolution images used for deriving the water extent for a given year. For the extent in year 2011, images from 2011 and 2012 were used. **(b)** Area statistics of Lake Chad historic water extent shown in (a). The grey bar indicates water area in km² (left axis) with the percentage above each bar showing the size of the area relative to year 1973. Red diamonds and blue dots indicate the number of images in dry and wet seasons, respectively, used to estimate the water extent in a given year (right axis). Note that images from 2012 were used for estimating water extent in 2011.

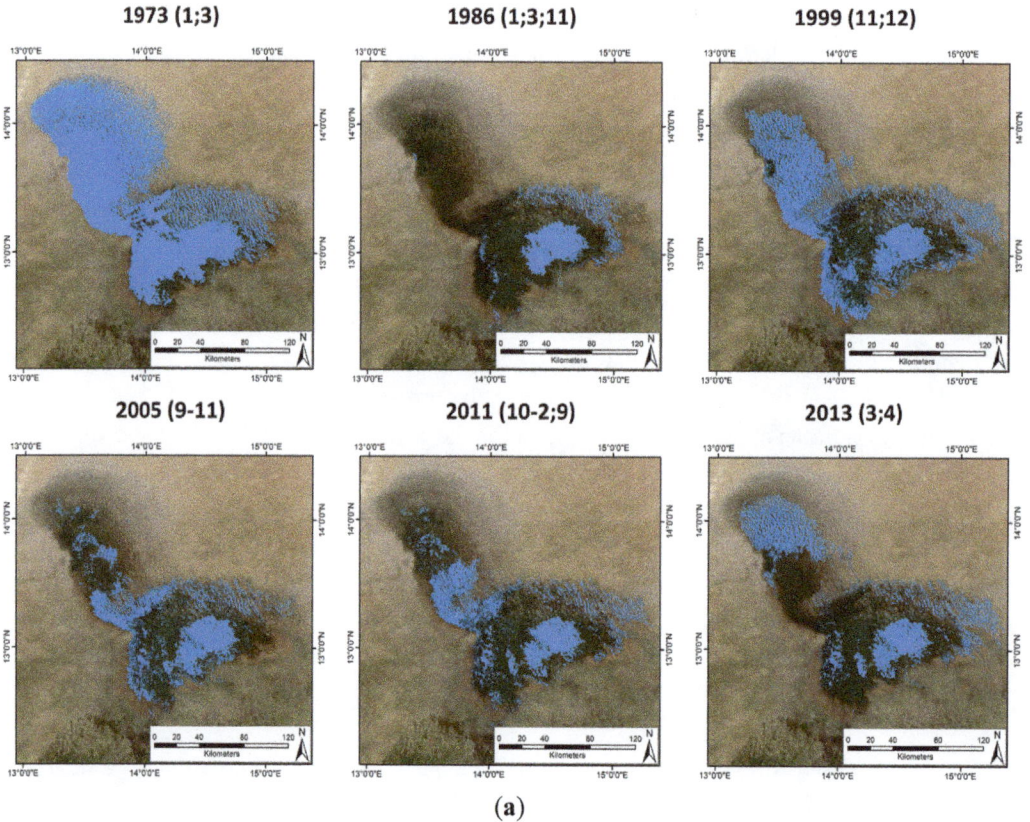

(a)

Figure 6. *Cont.*

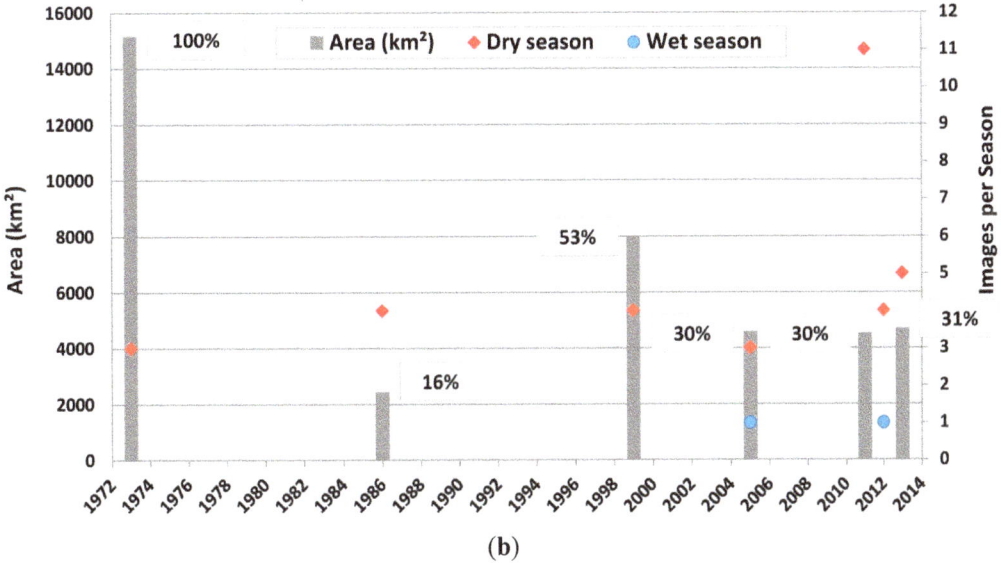

(b)

In the context of real-time operational water resources management, data assimilation (DA) has become the state-of-the-art technique to merge model predictions with the latest available data from a variety of sensors, including *in situ* and satellite-borne instruments. Assimilation of *in situ* data has become standard practice in most operational flood forecasting models (e.g., [31]). Many operational hydrological forecasting systems use variants of the Kalman filter [32] for data assimilation. In particular, the extended Kalman filter (EKF, [33]) and the ensemble Kalman filter (EnKF, [34]) are widely used in hydrological applications, since they are suitable for non-linear problems.

The HM implemented in the WOIS is the SWAT model, which is an open-source, physically-based, semi-distributed hydrological model developed and maintained by the U.S. Department of Agriculture [35]. SWAT hydrological response is not computed on grid cells, but instead on variably sized hydrological response units (HRU), which are portions of the sub-basins having unique combinations of slope, land cover and soil type. WOIS SWAT models are parameterized with global elevation, land-cover and soil type datasets and are forced with climate data from European Centre for Medium-Range Weather Forecasts (ECMWF) [36], Famine Early Warning Systems Network-Rainfall Estimate (FEWS-RFE) [37] or National Oceanic and Atmospheric Administration-Global Forecast System (NOAA-GFS) [38]. Automatic SWAT model calibration is performed with the public-domain software, PEST [39,40]. PEST provides a local gradient search algorithm, as well as a shuffled complex evolution algorithm for global search.

Figure 7. Example river discharge seven-day forecast for low flow conditions (**top right**) and high flow condition (**bottom right**) for the station Rundu on the Kavango River in Namibia, issued for October and March 2009 respectively. The solid green line is the central model forecast, and the green shaded area is the confidence interval of the forecast. Red dots are assimilated observations, and blue dots are daily observations after the issue date of the forecast.

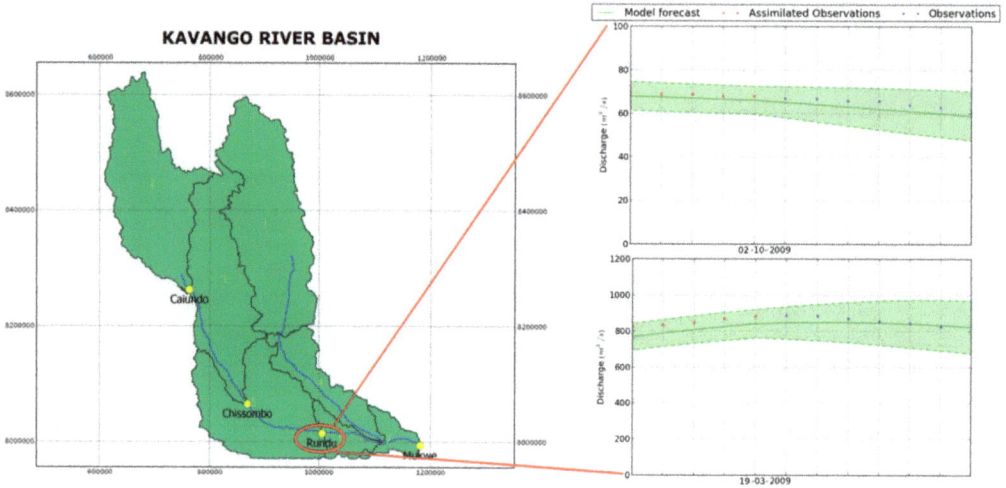

The WOIS operational forecasting approach (Figure 7) uses the EKF to assimilate water discharge measurements from any available monitoring stations into the SWAT hydrological model and is driven by NOAA-GFS eight-day ahead atmospheric forecasts. The approach is presented in detail in [41]. The set-up and calibration of WOIS SWAT models for a number of case study basins are documented in [42–44]. The WOIS operational forecasting approach has been implemented for the Kavango and Mokolo basins and is presently being implemented for the Volta and Zambezi basins. Daily Kavango forecasts are used operationally by the Namibian Ministry of Agriculture, Water and Forestry. In Kavango, forecast skill ranges from a Nash-Sutcliffe efficiency (NSE) of 0.96 for the one-day horizon to 0.77 for the seven-day horizon. The quality of the precipitation forcing product has the most significant impact on forecast skill. Key assumptions in the forecasting system are related to the representation of modeling and observation errors.

3.5. Historic and Real-Time Flood Mapping and Monitoring

With a constantly increasing density of population, flood-related economic and social risks increase. The monitoring of floods using data from synthetic aperture radar (SAR) has been exploited during the last thirty years and has proven to be well suited for understanding the spatio-temporal flood characteristics. The major advantage of using SAR compared to optical and infrared imagery lies in its ability to penetrate clouds and vegetation cover. In addition, it presents a significant improvement in spatial resolution when compared to coarse resolution microwave products (*i.e.*, ASCAT, AMSR-E, SMOS).

The flood mapping methodology used in the WOIS uses primarily ASAR Wide Swath (WS) mode data at 150-m resolution for historic flood mapping and RADARSAT imagery for near-real-time flood mapping. The methodology workflow consists of pre-processing, classification and post-processing steps. In the pre-processing step, precise orbit vectors and range-Doppler terrain correction are applied to obtain a georeferenced SAR image. The classification module of the WOIS workflow relies on the specular reflection properties of calm water surfaces, which appear dark in the resulting SAR imagery. Within the WOIS, both an automatic and a manual thresholding approach are implemented. In the case of manual thresholding, the user can plot the histogram of the SAR image reflectance, which helps to determine a suitable threshold value. The automatic thresholding comprises a combination of image tiling inspired by Martinis *et al.* [45] and Otsu's histogram thresholding method [46]. Finally, to mask out areas that are not prone to flooding and to remove pixels falsely classified as water due to topography-induced radar shadows, the Height Above Nearest Drainage (HAND) [47] index is used, which consists of the relative height of a cell in the digital elevation model (DEM) w.r.t. the closest DEM cell pertaining to the drainage network. The distance to the drainage network is measured along the flow lines of water in the DEM. The HAND index was based on the HYDROSHEDS database [48].

The demonstration cases for the historical flood mapping in the TIGER-NET project were the southern Nile Basin (NB) in Sudan and the Lake Chad Basin (LCB). The total accuracy of the derived product when compared with water maps derived from the NDVI-NDWI indices retrieved from LANDSAT-7 imagery were 0.95 and 0.99 over Sudan and Chad, respectively. The kappa coefficients for the validated scenes were on average 0.75 in the NB and 0.64 in the LCB demonstration case. As an example, Figure 8 shows a significant flood event near Khartoum city along the Nile River and surroundings captured by ASAR on 20 August 2006. Figure 8, right, illustrates the flood scene extracted from the Landsat-7 acquisition from the day before. According to the reports, the flood started at the beginning of August, due to heavy rain, and increased to a large-scale emergency by August 25. Twenty seven people were killed, and about 10,000 houses were damaged [49,50].

It was found that the accuracy of the final products deteriorates with roughening of the water surface or with partially submerged vegetation. Furthermore, in the desert regions, the low differences in backscatter levels between bare ground and water surfaces may exert risks on the quality of the final flood product.

Figure 8. Northeast of Khartoum city, Sudan. Comparison of the ASAR flood map from 20 August 2006, with the Landsat-7 ETM+ water map produced by thresholding of the NDWI-NDVI index from 19 August 2006.

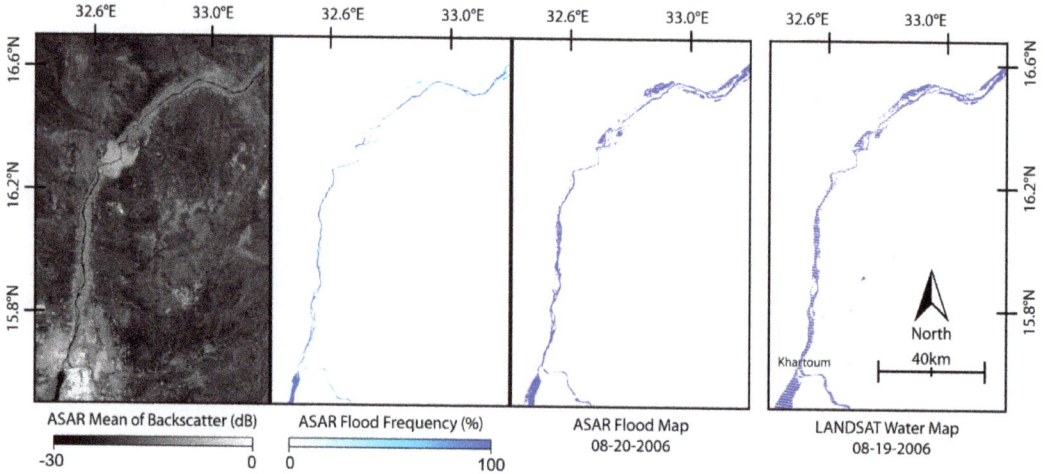

4. Outlook and Conclusions

Current water management practices in Africa are hampered by sparse and unreliable information on water resource availability. The Water Observation and Information System (WOIS) was created to support African institutions in improving their Integrated Water Resource Management (IWRM) by exploiting the advantages of Earth observation (EO) technology. The WOIS has been designed and developed as a user-friendly, yet powerful multipurpose system supporting the full range of EO products and models needed for assessing, monitoring and inventorying water resources from sub-catchment to river basin levels. It contains over 40 workflows to guide the less experienced users through EO data processing and GIS analysis in order to derive products required for IWRM. The validity and accuracy of those products has been assessed through numerous demonstration cases. For example, medium resolution land cover maps derived with WOIS have been shown to have a kappa coefficient above 0.7; high resolution water body mapping achieved kappa exceeding 0.8; and SAR-derived flood maps reached an overall accuracy of 0.95 to 0.99, while hydrological modeling resulted in forecast skill with a Nash–Sutcliffe efficiency of 0.77 for a seven-day forecast.

The development of the WOIS represents a successful example of a user-driven and collaborative development model, where functionalities have been designed, developed and evaluated through user-designated cases in order to demonstrate the real impact of the system on enhancing water management and integrated water resource management plans. The WOIS is already implemented in major African river basin authorities, several African ministries and agencies, as well as in research and humanitarian organizations, and new users are expected once the source code is released. It will therefore continue to develop in response to continued user requirements for new functionalities and functional improvements and due to general software, algorithm and method enhancements. A particular focus will be to ensure the support and implementation of processing capacity for the upcoming Sentinel satellite systems by integrating the ESA Sentinel toolboxes and developing dedicated production workflows, which will turn WOIS into a fully-operational monitoring system.

Through provision of this free, powerful and extendable system in combination with continued capacity building and training efforts, the TIGER-NET project strives to build the basis for an extension, *i.e.*, to roll-out to other countries and regions in Africa and beyond. Another major aim is the continued support of the users and stakeholders in order to reach sustainability by attracting external funding opportunities to enable operational utilization of satellite data for Integrated Water Resource Management in Africa. More information about the WOIS software and the TIGER-NET project can be found on the project's website: tiger-net.org.

Acknowledgments

The TIGER-NET project is part of ESA's TIGER initiative, funded by the Strategic Initiative and run within the Data User Element program.

Author Contributions

Radoslaw Guzinski has been responsible for the WOIS system design, development and implementation. Steve Kass has been responsible for the high resolution land and water characterization component. Silvia Huber has been responsible for large lake water quality monitoring and land degradation mapping. Peter Bauer-Gottwein and Iris Hedegaard Jensen have been responsible for the implementation of the WOIS hydrological modeling and forecasting component. Vahid Naeimi and Marcela Doubková have been responsible for the implementation of historic and real-time flood mapping and monitoring component. Andreas Walli has been the TIGER-NET project manager and responsible for eliciting and documenting user requirements. Christian Tottrup has been responsible for the overall design and system engineering, as well as the implementation of the basin-wide characterization of land and water resources.

Conflicts of Interest

The authors declare no conflict of interest.

References

1. TIGER 2012 Report. Available online: http://www.tiger.esa.int/files/pdf/tiger_report_single_ pages_lowres.pdf (accessed on 18 August 2014).
2. Fernando-Prieto, D.; Palazzo, F. The role of Earth observation in improving water governance in Africa: ESA's TIGER initiative. *Hydrogeol. J.* **2007**, *15*, 101–104.
3. Berger, M.; Moreno, J.; Johannessen, J.A.; Levelt, P.F.; Hansen, R.F. ESA's sentinel missions in support of Earth system science. *Remote Sens. Environ.* **2012**, *120*, 84–90.
4. Martin, R.C. *Agile Software Development: Principles, Patterns, and Practices*; Prentice Hall PTR: Upper Saddle River, NJ, USA, 2003.

5. Graser, A. *Learning QGIS 2.0*; Packt Publishing Ltd.: Birmingham, UK, 2013.

6. Neteler, M.; Bowman, M.H.; Landa, M.; Metz, M. GRASS GIS: A multi-purpose open source GIS. *Environ. Model. Softw.* **2012**, *31*, 124–130.

7. Fomferra, N.; Brockmann, C. BEAM—The ENVISAT MERIS and AATSR toolbox. In Proceedings of the MERIS (A) ATSR Workshop, Frascati, Italy, 26–30 September 2005; Volume 597, p. 13.

8. Engdahl, M.; Minchella, A.; Marinkovic, P.; Veci, L.; Lu, J. NEST: An ESA open source toolbox for scientific exploitation of SAR data. In Proceedings of the 2012 IEEE International Geoscience and Remote Sensing Symposium (IGARSS), Munich, Germany, 22–27 July 2012; pp. 5322–5324.

9. Inglada, J.; Christophe, E. The Orfeo toolbox remote sensing image processing software. In Proceedings of the 2009 IEEE International Geoscience and Remote Sensing Symposium (IGARSS), Cape Town, South Africa, 12–17 July 2009; Volume 4, pp. 733–736.

10. Neitsch, S.L.; Arnold, J.G.; Kiniry, J.R.; Williams, J.R. *Soil and Water Assessment Tool Theoretical Documentation Version 2009*; Texas Water Resources Institute: Pleasanton, TX, USA, 2011. Available online: http://hdl.handle.net/1969.1/128050 (accessed on 18 August 2014).

11. Ihaka, R.; Gentleman, R. R: A language for data analysis and graphics. *J. Comput. Graph. Stat.* **1996**, *5*, 299–314.

12. Obe, R.; Hsu, L. *PostGIS in Action*; Manning Publications Co.: Greenwich, CT, USA, 2011.

13. Sanner, M.F. Python: A programming language for software integration and development. *J. Mol. Graph. Model.* **1999**, *17*, 57–61.

14. GDAL Development Team. GDAL—Geospatial Data Abstraction Library. 2014. Available online: http://gdal.org (accessed on 18 August 2014).

15. Olaya, V. SEXTANTE, a free platform for geospatial analysis. *OSGeo J.* **2008**, *6*, 32–39.

16. Sentinel-2 MSI Software Tools. Available online: https://sentinel.esa.int/web/sentinel/user-guides/sentinel-2-msi/software-tools (accessed on 10 March 2014).

17. Sentinel-1 SAR Software Tools. Available online: https://sentinel.esa.int/web/sentinel/user-guides/sentinel-1-sar/software-tools (accessed on 10 March 2014).

18. Boschetti, M.; Nutini, F.; Brivio, P.A.; Bartholomé, E.; Stroppiana, D.; Hoscilo, A. Identification of environmental anomaly hot spots in West Africa from time series of NDVI and rainfall. *ISPRS J. Photogramm. Remote Sens.* **2013**, *78*, 26–40.

19. Xie, P.; Arkin, P.A. Global precipitation: A 17-year monthly analysis based on gauge observations, satellite estimates, and numerical model outputs. *Bull. Am. Meteorol. Soc.* **1997**, *78*, 2539–2558.

20. Gellens-Meulenberghs, F.; Arboleda, A.; Ghilain, N. Towards a continuous monitoring of evapotranspiration based on MSG data. In Proceedings of Symposium HS3007 at IUGG2007, Perugia, Italy, 7–13 July 2007; pp. 228–234.

21. Doerffer, R.; Schiller, H. The MERIS Case 2 water algorithm. *Int. J. Remote Sens.* **2007**, *28*, 517–535.

22. O'Reilly, J.E.; Maritorena, S.; Mitchell, B.G.; Siegel, D.A.; Carder, K.L.; Garver, S.A.; Kahru, M.; McClain, C. Ocean color chlorophyll algorithms for SeaWiFS. *J. Geophys. Res. Oceans (1978–2012)* **1998**, *103*, 24937–24953.

23. Huang, C.; Davis, L.S.; Townshend, J.R.G. An assessment of support vector machines for land cover classification. *Int. J. Remote Sens.* **2002**, *23*, 725–749.

24. Nielsen, A.A.; Conradsen, K.; Simpson, J.J. Multivariate alteration detection (MAD) and MAF postprocessing in multispectral, bitemporal image data: New approaches to change detection studies. *Remote Sens. Environ.* **1998**, *64*, 1–19.

25. Huber, S.; Fensholt, R.; Rasmussen, K. Water availability as the driver of vegetation dynamics in the African Sahel from 1982 to 2007. *Glob. Planet. Chang.* **2011**, *76*, 186–195.

26. Helldén, U.; Tottrup, C. Regional desertification: A global synthesis. *Glob. Planet. Chang.* **2008**, *64*, 169–176.

27. Tucker, C.J. Red and photographic infrared linear combinations for monitoring vegetation. *Remote Sens. Environ.* **1979**, *8*, 127–150.

28. Jain, A.K.; Duin, R.P.W.; Mao, J. Statistical pattern recognition: A review. *IEEE Trans. Pattern Anal. Mach. Intell.* **2000**, *22*, 4–37.

29. Blaschke, T. Object based image analysis for remote sensing. *ISPRS J. Photogramm. Remote Sens.* **2010**, *65*, 2–16.

30. Alexander, L.V.; Zhang, X.; Peterson, T.C.; Caesar, J.; Gleason, B.; Klein Tank, A.M.G.; Vazquez-Aguirre, J.L. Global observed changes in daily climate extremes of temperature and precipitation. *J. Geophys. Res. Atmos. (1984–2012)* **2006**, *111*, D05109.

31. Madsen, H.; Skotner, C. Adaptive state updating in real-time river flow forecasting—A combined filtering and error forecasting procedure. *J. Hydrol.* **2005**, *308*, 302–312.

32. Kalman, R.E. A new approach to linear filtering and prediction problems. *J. Basic Eng.* **1960**, *82*, 35–45.

33. Chui, C.K.; Chen, G. *Kalman Filtering with Real-Time Applications*, 4th ed.; Springer: Berlin, Germany, 2009.

34. Evensen, G. Sequential data assimilation with a nonlinear quasi-geostrophic model using Monte Carlo methods to forecast error statistics. *J. Geophys. Res.: Oceans (1978–2012)* **1994**, *99*, 10143–10162.

35. Neitsch, S.L.; Arnold, J.R.; Kiniry, J.R.; Williams, J.R. *Soil and Water Assessment Tool, Theoretical Documentation*, Version 2009; Texas Water Resources Institute Technical Report No. 406; Texas A&M University System: College Station, TX, USA, 2011.

36. ECMWF. ECMWF Data Server. 2013. Available online: http://data-portal.ecmwf.int/ (accessed on 15 May 2013).

37. NOAA (National Oceanic and Atmospheric Administration)-CPC (Climate Prediction Center). *African Rainfall Estimation Algorithm Version 2*; NOAA-CPC: College Park, MD, USA, 2013. Available online: http://www.cpc.ncep.noaa.gov/products/fews/RFE2.0_tech.pdf (accessed on 15 May 2013).

38. NOMADS—NOAA Operational Model Archive and Distribution System. Available online: http://nomads.ncep.noaa.gov/ (accessed on 15 May 2013).

39. Doherty, J. *Addendum to the PEST Manual*; Watermark Numerical Computing: Brisbane, QL, Australia, 2012.

40. Doherty, J. *PEST: Model Independent Parameter Estimation—Fifth Edition of User Manual*; Watermark Numerical Computing: Brisbane, QL, Australia, 2004.

41. Michailovsky, C.I.; Milzow, C.; Bauer-Gottwein, P. Assimilation of radar altimetry to a routing model of the Brahmaputra River. *Water Resour. Res.* **2013**, *49*, 4807–4816.

42. Jensen, I.H. Operational Modelling of Water Availability for the Mokolo Catchment, South Africa. Master's Thesis, Technical University of Denmark, Lyngby, Denmark, 2013; p. 54.

43. Denager, T. Using Remote Sensing to Inform a Large-Scale River Basin Model of the Chari-Logone. Master's Thesis, Technical University of Denmark, Lyngby, Denmark, 2013; p. 80.

44. Hansen, S. Operational Hydrological Modelling of the Kavango River Basin. Master's Thesis, Technical University of Denmark, Lyngby, Denmark, 2013; p. 59.

45. Martinis, S.; Twele, A.; Voigt, S. Towards operational near real-time flood detection using a split-based automatic thresholding procedure on high resolution TerraSAR-X data. *Nat. Hazards Earth Syst. Sci.* **2009**, *9*, 303–314.

46. Otsu, N. A Threshold selection method from gray-level histograms. *IEEE Trans. Syst. Man Cybern.* **1979**, *9*, 62–66.

47. Rennó, C.D.; Nobre, A.D.; Cuartas, L.A.; Soares, J.V.; Hodnett, M.G.; Tomasella, J.; Waterloo, M.J. HAND, a new terrain descriptor using SRTM-DEM: Mapping terra-firme rainforest environments in Amazonia. *Remote Sens. Environ.* **2008**, *112*, 3469–3481.

48. Lehner, B.; Verdin, K.; Jarvis, A. New global hydrography derived from spaceborne elevation data. *Eos Trans. AGU* **2008**, *89*, 93–94.

49. UNOSAT. Flooding in Sudan, International Charter Space & Major Disasters. 2006. Available online: http://www.disasterscharter.org/web/charter/activation_details?p_r_p_1415474252_assetId=ACT-130 (accessed on 18 August 2014).

50. Brakenridge, G.R. *Global Active Archive of Large Flood Events*; Dartmouth Flood Observatory, University of Colorado, 2010. Available online: http://floodobservatory.colorado.edu/Archives/index.html (accessed on 18 August 2014).

Application of the Regional Water Mass Variations from GRACE Satellite Gravimetry to Large-Scale Water Management in Africa

Guillaume Ramillien, Frédéric Frappart and Lucia Seoane

Abstract: Time series of regional 2° × 2° Gravity Recovery and Climate Experiment (GRACE) solutions of surface water mass change have been computed over Africa from 2003 to 2012 with a 10-day resolution by using a new regional approach. These regional maps are used to describe and quantify water mass change. The contribution of African hydrology to actual sea level rise is negative and small in magnitude (*i.e.*, −0.1 mm/y of equivalent sea level (ESL)) mainly explained by the water retained in the Zambezi River basin. Analysis of the regional water mass maps is used to distinguish different zones of important water mass variations, with the exception of the dominant seasonal cycle of the African monsoon in the Sahel and Central Africa. The analysis of the regional solutions reveals the accumulation in the Okavango swamp and South Niger. It confirms the continuous depletion of water in the North Sahara aquifer at the rate of −2.3 km^3/y, with a decrease in early 2008. Synergistic use of altimetry-based lake water volume with total water storage (TWS) from GRACE permits a continuous monitoring of sub-surface water storage for large lake drainage areas. These different applications demonstrate the potential of the GRACE mission for the management of water resources at the regional scale.

Reprinted from *Remote Sens.* Cite as: Ramillien, G.; Frappart, F.; Seoane, L. Application of the Regional Water Mass Variations from GRACE Satellite Gravimetry to Large-Scale Water Management in Africa. *Remote Sens.* **2014**, *6*, 7379-7405.

1. Introduction

Satellite gravimetry remains the only technique that provides information on the total water storage change at continental scales and gives access to groundwater variations when *a priori* information on surface and sub-surface reservoirs is available [1–3]. Data of the Gravity Recovery and Climate Experiment (GRACE) mission were widely used to estimate changes in land water storage and fluxes over Africa at basin to regional scales. By using the first two years (April 2002 to May 2003) of GRACE data, the study of [4] show that seasonal total water storage (TWS) variations vary between ±50 mm of TWS in the Congo and Niger basins. Time series of GRACE data allow us to estimate inter-annual variations and trends in TWS, as well as the contributions of TWS to sea level change for the largest drainage basins and lakes of Africa [5–12]. They were also used for comparisons, validation and calibration of hydrological model outputs at the basin scale and over large bio-climatic regions, such as the Sahel or West Africa [13–16]. Combined with external datasets *in situ*, GRACE data offer the unique opportunity to estimate groundwater storage variations for lake drainage areas [17,18] and large river basins [19] or at a regional scale [20], river discharges [21], evapotranspiration at the basin scale [22–24] and the water budget [25]. They were also used to determine the specific yield [26,27] and loading effects in the Sahel [28].

In the following section, the principle of classical satellite gravimetry and the particularities of the regional method for recovering water mass variations from GRACE satellite data are presented. Time series of 10-day 2° by 2° maps of water mass variations over Africa (30°W–60°E; 40°S–40°N) are computed from accurate K-band range rate (KBRR) residuals along daily GRACE orbits for the whole period of GRACE (2003–2012). Then, global solutions used for comparing our regional solutions computed over Africa are listed. The spatial averages of our African solutions over hydrological units (see the drainage basins and desert aquifer area in Figure 1) *versus* time are computed to establish water mass balances, and, thus, sea level contributions, for the recent period covered by the GRACE mission. For isolating the African regions that produce the largest contributions to the sea level mass balance (and the ones in water deficit), the first space and time modes of the variability of GRACE data are extracted using a principal component analysis (PCA). Then, the PCA modes are compared to pure seasonal, semi-seasonal and multi-year linear trend variations (e.g., African monsoon), so that multi-year water mass gains (or losses) can be located in Africa and quantified. Finally, combining water volumes derived from regional TWS solutions and radar altimetry measurements of the level of the lakes enables us to estimate the soil and groundwater variations over the East African Great Lakes.

Figure 1. Geographical locations of the main drainage basins of Africa used in this study: (1) Congo (~4 million square kilometers); (2) Nile (~3.4 million km²); (3) Niger (~2.1 million km²); (4) Zambezi (~1.4 million km²); (5) Orange (~0.97 million km²); (6) Volta (~408,000 km²); (7) Senegal (~270,000 km²); as well as the areas contributing to the Atlantic Ocean (blue dots), India Ocean (red dots), Mediterranean Sea (green dots) and the endorheic ensemble (purple dots). The driest part of the Sahara Desert area in South Algeria (8) (~1.1 million km²) and (9) the North Western Sahara Aquifer System (NWSAS) (~1.6 million km²) are also displayed.

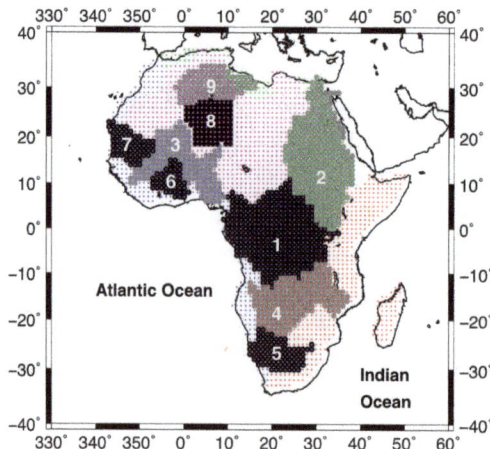

2. Recovery of Surface Water Mass Changes by Space Gravimetry

Since its launch in 2002, the Gravity Recovery and Climate Experiment (GRACE) space mission has measured, for the first time, changes of total water storage (TWS), including surface water, soil, moisture and groundwater, with unprecedented centimeter accuracy in terms of geoid height. GRACE data have already demonstrated a strong potential for estimating hydrological system information, such as river discharges [29], evapotranspiration rate [22,30], groundwater variations [31–33] and the detection of extreme climate events, such as floods and droughts [34–40]. Several analysis centers, such as the Center for Space Research, University of Texas (CSR) in Austin (TX), the Jet Propulsion Laboratory (JPL) in Pasadena (CA), the GeoForschungsZentrum (GFZ) in Potsdam (Germany) and the Groupe de Recherche en Géodésie Spatiale (GRGS) in Toulouse (France), use Level 1-B GRACE observations to produce lists of monthly and 10-day global Stokes coefficients (*i.e.*, spherical harmonics of the geopotential) up to harmonic degree 60 for CSR Release 5, 80 for GRGS RL03 and 90 for JPL and GFZ RL05; in other words, at the maximum surface resolution of 250–300 km. During the estimation process of the dimensionless Stokes coefficients from orbit data, the static gravity field and its time variations (*i.e.*, atmospheric and ocean mass changes, including the effects of the periodic tides) are removed through *a priori* models describing these known gravitational accelerations. Therefore, the residuals correspond to the unmodeled contributions of mass to the observed gravity field and, mainly, the continental hydrology component.

Unfortunately, the correction models remain imperfect due to their lack of completeness in the description of water mass movements by omission and/or the lack of resolution, which represent important sources of error in the recovery of continental hydrology variations. As GRACE-based residual Stokes coefficients are averages over constant time intervals of 10 days or a month, errors in the correction models with periods from hours to days contaminate these GRACE solutions by aliasing and, thus, degrade their accuracy [41]. These effects of signal distortion deteriorate the quality of true water mass signals into other time frequencies and make these signals indistinguishable by sampling.

In the case of the GRACE orbit, hydrology-related signals are measured mainly along satellite tracks in the nearly latitudinal direction, but they are projected onto global spherical harmonics (SH) functions, which ensure the best spatial frequency representation. Because of this polar plane geometry of the GRACE orbit, this particular distribution of measurements creates north-south "stripes" in the 10-day and monthly GRACE solutions. Moreover, the determination of the SH coefficients leads to underdetermined systems of normal equations to be solved by creating correlations between SH coefficients of high degrees (*i.e.*, >10–15) [42] and amplification of this orbit error and data noise [43].

Another problem while using SH is the "leakage" of energetic signals propagating over the entire sphere, as these global undulations come to pollute the water mass estimated in the region of interest. This is particularly the case of small regions that are not fully represented by the degree 60 truncated SH spectra of the GRACE solution (*i.e.*, error by omission). Besides, different low-pass filtering techniques have been proposed, but they can partly cancel some of these effects [42,44,45]. The simplest way to increase the signal-to-noise ratio in estimation remains to average the signals over large surfaces of more than one million square kilometers, such as tropical drainage basins, to cancel the effects of the short wavelength SH undulations.

An alternative approach for estimating surface water mass densities in a region from GRACE data has been recently proposed by [46,47]. This new strategy is based on the optimal localization in space, instead of the best localization in spatial frequency, and leads theoretically to better spatial localization and resolution [48]. The authors of [47] have shown that this regional method offers a reduction of both north-south striping due to the distribution of GRACE satellite tracks and the temporal aliasing of correction models over South America [36] and Australia [49]. According to these two latter studies, regional maps present more realistic spatial and temporal patterns than the global solutions when compared to independent datasets of rainfall. The main modes of variability in South America coincide with the geographical limits of known hydrological units, such as individual groundwater layers [36]. In the present article, 10-day regional solutions over Africa are analyzed and compared to other datasets.

3. Methodology of the Regional Approach

The two main steps of this regional method are: (1) using the principle of mechanical energy conservation to deduce the variations of difference potential anomalies (DPA) between the twin GRACE satellites, representing mainly the continental hydrology contribution, from the accurate along-track KBRR measurements; and (2) adjusting the equivalent water heights (EWH) of a network of juxtaposed 2° by 2° surface tiles by the linear inversion of the DPA passing over the considered region every 10 days [46,47]. This regional approach differs from the NASA "mascons" [50–52] as, instead of classical band-limited SH, the regional method imposes the geometry and the best spatial localization of surface hydrology structures by construction.

In the first step, KBRR observations are reduced by removing the contributions of known gravitational accelerations related to large-scale mass variations (*i.e.*, atmosphere and ocean mass variations, polar movements, solid and oceanic tides, as well as the static gravity field of the Earth that represents 99% of the observed signals). This operation is made by iterative least squares adjustment of daily dynamical orbits using the Géodésie par Intégrations Numériques Simultanées (GINS) software [53,54]. Thanks to the measurements of on-board GRACE accelerometers, the effects of non-conservative forces are also removed from the KBRR observations in the orbit adjustment. KBRR residuals represent the cumulated contributions of unmodeled phenomena and, mainly, water storage change over continents. These residuals are related to the different accelerations of the two GRACE vehicles resulting from the gravity signals of continental hydrology, and they are easily converted into variations of kinetic energy differences. According to the principle of energy conservation, these kinetic energy difference variations directly correspond to potential energy differences, or in other words, DPA. To reduce unrealistic orbit errors at fractions of the satellite revolution periods and, thus, to avoid numerical instabilities in the following linear inversion, DPA arcs passing over Africa are linearly de-trended. It locally absorbs orbit error and keeps a subset of DPA short and medium wavelengths that are less than the latitudinal dimension of the considered region. The missing long-wavelength information of water mass change is from the first degrees of the GRGS solutions, and these large undulations are added to the DPA-derived regional solutions to complete the water mass signals after the inversion of residual DPA [36,47,49].

Figure 2. Schematic view of the processing for estimating global and regional solutions from a given 10-day or monthly period of Gravity Recovery and Climate Experiment (GRACE) observations. GINS, Géodésie par Intégrations Numériques Simultanées; GRGS, Groupe de Recherche en Géodésie Spatiale; SVD, singular value decomposition; KBRR, K-band range rate.

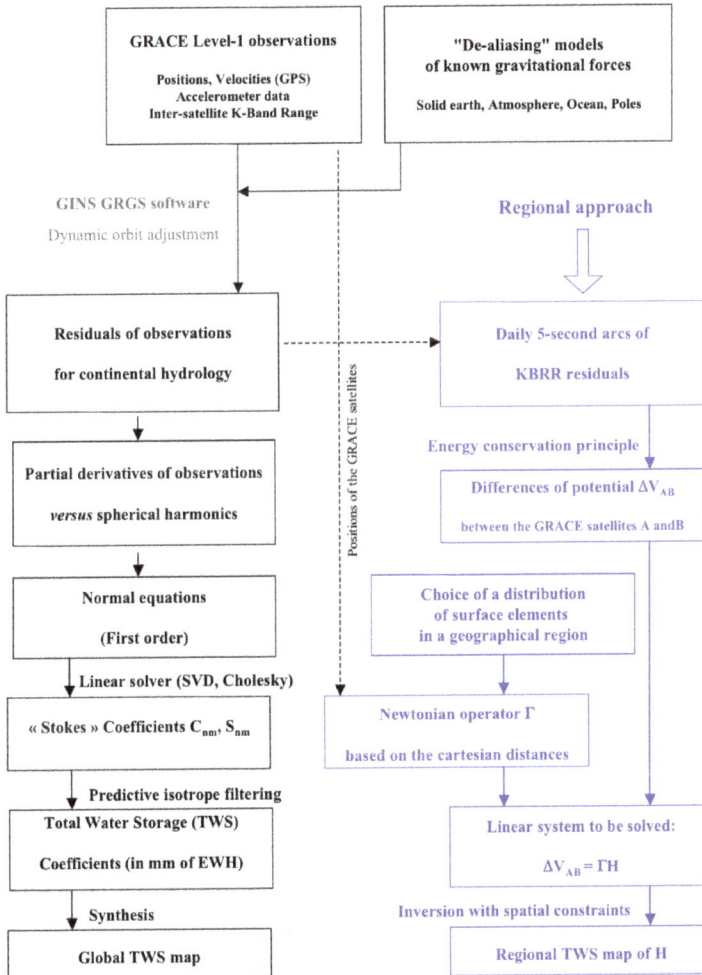

In the second step of the method, the Newtonian matrix *A* is defined from the positions of the two GRACE satellites and of the surface tiles, in a geocentric reference frame, according to Newton's first law of attraction. This matrix relates each unknown EWH to the DPA observations inside the region during 10 successive days. As gravimetry inversion does not usually provide a unique solution, regularization strategies should be applied to find numerically-stable solutions, either based on the truncation of singular values [46] or by introducing an averaging radius [47].

This latter type of regularization consists of adding a spatial constraint matrix block C to the Newtonian matrix A. The coefficients of this extra matrix C are obtained by imposing each equivalent water height to be a linear combination of its neighbors weighted by the inverse of their angular distances and inside a maximum geographical radius r. In the case of spatial "averaging", the coefficients for a given radius r should equal $1/P$, as P is the number of surface tiles located at a distance lesser than r, and 0 elsewhere. Introduction of these linear constraints enables the ill-conditioned matrix A to be inverted. A good compromise for keeping enough hydrological details with no smoothing and limiting the increase of numerical noise was earlier found by considering a radius of $r = 600$ km over continental areas [47]. A simplified flowchart summarizing the estimation process of regional solutions is presented in the following Figure 2.

4. Datasets Used in This Study

4.1. 10-Day Regional GRACE Solutions for Africa

Daily arcs of five-second sampled K-band range (KBR) measurements of the inter-satellite velocity have been used in the GINS software [53,54] to adjust dynamical reference orbits. GRACE data are corrected from the known gravitational accelerations related to atmosphere and ocean mass redistributions, including tides and polar movements, using a priori global models. KBR rate residuals that represent mainly the continental hydrology have been converted into residual differences of potential (RDP), according to the conservation of the mechanical energy of the two GRACE vehicles versus time. Satellite tracks flying over Africa are selected, and each one is corrected from a least squares-adjusted linear trend. Following the two-step regional method explained in the previous section, time series of successive 10-day and 2° by 2° solutions of water mass change have been inverted over the whole African continent (30°W–60°E; 40°S–40°N) from RDP and, then, completed with the long wavelengths (>6700 km) of the GRGS GRACE solutions, or equivalently, the SH of degrees less than six, for the period 2003–2012. One complete year of regional solutions is displayed in Figure 3.

4.2. Global GRACE Solutions from Official Centers CSR, GFZ and JPL

Three processing centers, including the Center for Space Research (CSR), Austin, TX, USA, the GeoForschungsZentrum (GFZ), Potsdam, Germany, the Jet Propulsion Laboratory (JPL), Pasadena, CA, USA, and the Science Data Center (SDC) are in charge of the processing of the GRACE data and the production of Level-1 and Level-2 products. These products are distributed by the GFZ's Integrated System Data Center (ISDC) [55] and the JPL's Physical Oceanography Distributive Active Data Center (PODAAC) [56]. Preprocessing of Level-1 GRACE data (i.e., positions and velocities measured by GPS, accelerometer data and KBR inter-satellite measurements) is routinely made by the SDC, as well as monthly global GRACE gravity solutions (Level 2). These latter solutions consist of time series of monthly averages of Stokes coefficients (i.e., dimensionless spherical harmonics coefficients of geopotential) developed up to a degree between 50 and 120 that are adjusted from along-track GRACE measurements. A dynamical approach, based on the Newtonian formulation of the satellite's equation of motion in an inertial reference frame, centered at the Earth's center, combined with dedicated modeling of the gravitational and non-conservative forces acting on the spacecraft, is used to compute the monthly GRACE solutions [57]. During the estimation process, atmospheric and ocean barometric redistribution of mass variations are removed from the GRACE coefficients using European Centre for Medium-range Weather Forecasts (ECMWF) and National Centers for Environmental Prediction (NCEP) reanalysis for atmospheric mass variations and ocean tides, as well as global ocean circulation models. The GRACE coefficients are hence residuals that should represent mainly continental water storage, but also errors from the correction models and noise. The monthly GRACE solutions differ from one official provider from another due to the differences in the data processing, the choice of the correction models and the data selection for computing the monthly averages.

Figure 3. Example of one-year series of regional maps of water mass changes presented at 10-day intervals from June 2007 to June 2008 (see also the next section), for the two semesters.

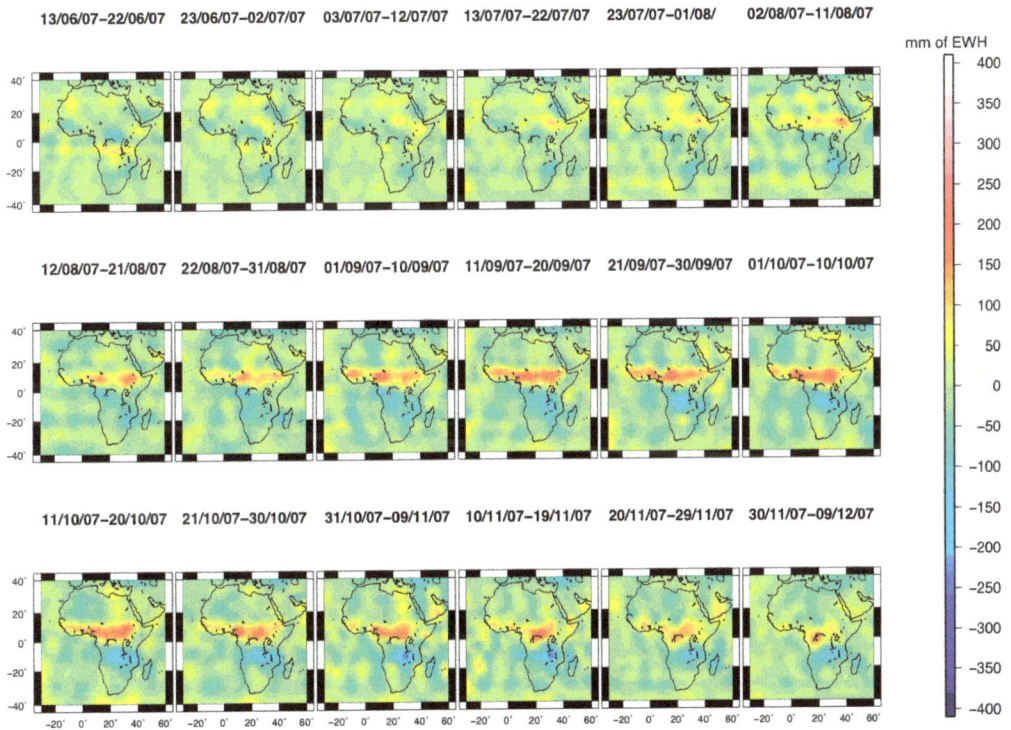

Figure 3. *Cont.*

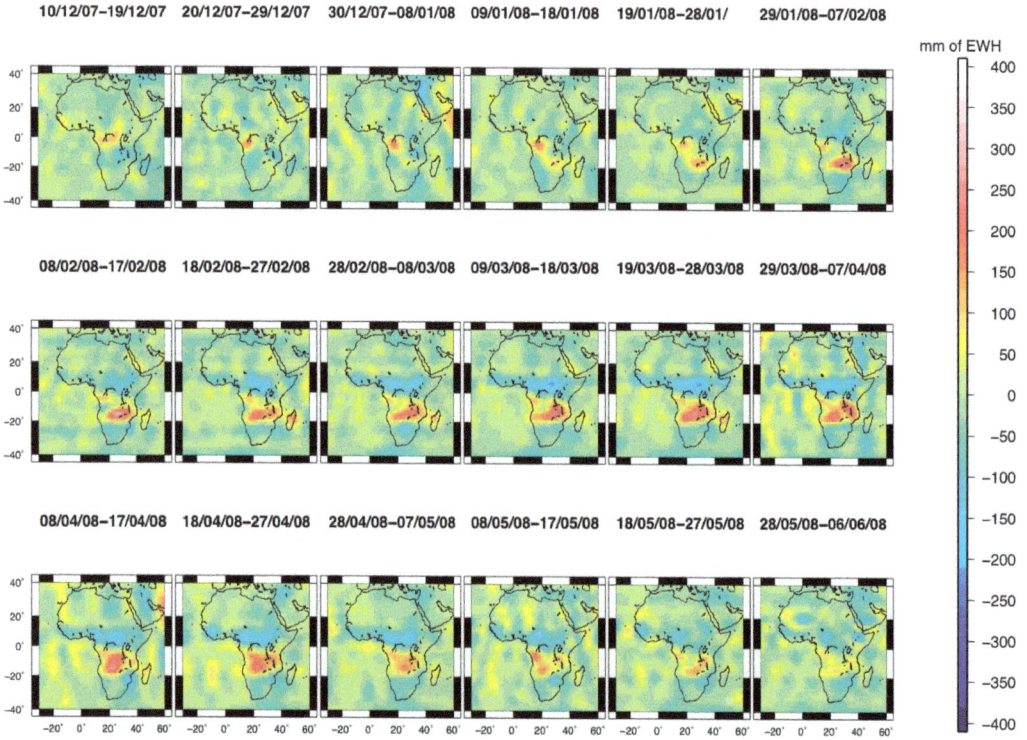

4.3. Global GRACE Solutions Provided by GRGS

These Level-2 European Improved Gravity model of Earth by New techniques solely from GRACE Satellite data (EIGEN) Release 04 10-day gravity models are derived from Level-1 GRACE measurements, including KBRR, from LAser GEOdynamics Satellites (LAGEOS) 1 and 2 Satellite Laser Ranging (SLR) data for the enhancement of lower harmonic degrees [47] and using an empirical stabilization approach without any post-processing smoothing or filtering. The 10-day Stokes coefficients are converted into terms of water mass coefficients from degree 2 up to degree 50–60 (*i.e.*, spatial resolution of 400 km), expressed in EWH. Regular one-degree 10-day maps of surface water mass for the period 2002–2012 are derived from these latter SH water mass coefficients and made available for the last release (RL03) [58].

4.4. Independent Component Analysis of CSR, JPL and GFZ GRACE Solutions

Since the global GRACE solutions are unfortunately dominated by striping, our idea is to combine monthly solutions from different centers of analysis to extract the continental hydrology component from noisy sources by redundancy. A post-processing method based on independent component analysis (ICA) was applied to the Level-2 GRACE solutions from official providers (*i.e.*, University of Texas–Center for Space Research (UTCSR), JPL and GFZ). Pre-filtered with 400-km radius Gaussian filters before applying an ICA, the Level-2 GRACE solutions need to be somehow low-passed filtered to not have a Gaussian distribution. When they are not filtered enough, they are still dominated by striping, and their distribution keeps being Gaussian. If they are too low-pass filtered, they correspond to the long wavelengths of the continental hydrology, and they also exhibit Gaussian properties. A compromise of ~400 km to ensure no Gaussianity, and, thus, an efficient separation, has been proposed by [59] after several tests to extract most of the parts of the continental hydrology. Time series of ICA-based global maps of continental water mass changes computed over the period of March 2003–December 2010, are used for the comparison in this study [45]. For a given month, the ICA 400-km filtered solutions only differ by a scaling factor, so that only the GFZ-derived ICA 400-km filtered ones are presented.

4.5. Time Series of Altimetry-Derived Water Level of Lakes in East Africa

Satellite altimetry was originally designed to provide accurate measurements of the dynamic topography of the ocean [60]. Radar altimeters demonstrated strong capabilities to accurately estimate water levels over land and are now used for systematic monitoring of lakes [61,62], large rivers [63,64], wetlands and floodplains [65].

In this study, we used the time series of water levels derived from satellite altimetry measurements made available by the Hydroweb database at Laboratoire d'Etudes en Géophysique et Océanographie Spatiales (LEGOS)–Observatoire Midi-Pyrénées (OMP) [66] for four large lakes of East Africa (*i.e.*, lakes Turkana, Victoria, Tanganyika and Malawi). All details about the processing of altimetry data and the computation of time series of water levels can be found in [61]. As these lakes do not have significant changes in area, the time series of water levels were simply converted into time series of water volumes using the mean surface of the lakes, as in [18]. The mean surfaces of the lakes are 8860, 68,800, 32,600 and 22,490 km^2 for Turkana, Victoria, Tanganyika and Malawi lakes, respectively [67].

4.6. TRMM 3B43 Monthly Rainfall

In this study, we used the Tropical Rainfall Measuring Mission (TRMM) 3B43 product which is a combination of monthly rainfall at a spatial resolution of 0.25° from January 1998 to December 2012, and other data sources. This dataset is obtained by combining satellite information from the passive microwave imager (TMI) and precipitation radar (PR) onboard the Tropical Rainfall Measuring Mission (TRMM), a Japan-U.S. satellite launched in November 1997, the Visible and Infrared Scanner (VIRS) onboard the Special Sensor Microwave Imager (SSM/I) and rain gauge observations. The dataset results from the merging of the TRMM 3B42-adjusted merged infrared precipitation with the monthly accumulated Climate Assessment Monitoring System or Global Precipitation Climatology Center Rain Gauge analyses [68,69].

5. Results and Discussion

5.1. Residual Errors Estimated in the Arid Sahara Region

Figure 4 presents a time series of TWS in a desert area in the south of Algeria. As no hydrological variations are expected in such a very dry region, the residual water mass signals can be interpreted as a good indicator of the error in estimating water mass variations from GRACE orbit data. These recovery errors do not exceed 15–19 mm EWH RMS when GRACE solutions are averaged over a surface of ~1 million square kilometers, consistently found by [70] from one year of global solutions. They also exhibit a seasonal cycle, in particular for the global and regional solutions (see Figure 4), which may result from leakage effects from surrounding areas polluting stronger water mass signals from the neighboring drainage basins, as well as errors from the *a priori* models used for correcting GRACE data and isolating the continental hydrology variations. For the global solutions, this latter effect is due to the spectral SH truncation of the global solutions at degree n = 60–90 (and of the underlying correcting models developed in SH for our regional solutions); in other words, the lack of spatial resolution to describe small objects, which creates unrealistic undulations propagating over the entire terrestrial sphere (*i.e.*, leakage), as explained in Section 2.

Figure 4. Total water storage (TWS) time series for the dry region of South Algeria (8 in Figure 1) considering the regional GRACE solutions (**top**) and global solutions from different pre-processing centers (**bottom**). Amplitudes over this region are typically less than 20 mm of equivalent water height (EWH) RMS. RL, release.

Times series of TWS GRACE means over Sahara desert

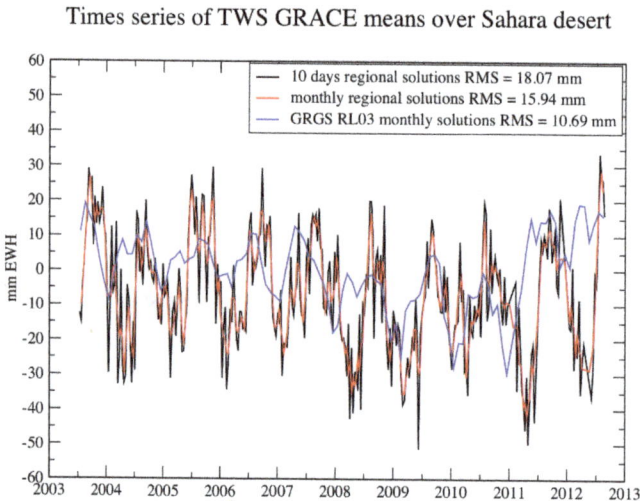

Figure 4. *Cont.*

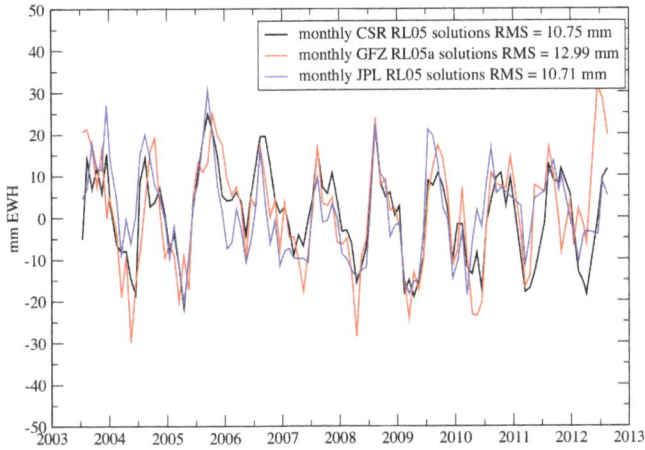

5.2. Pre-Analysis of the 10-Day Maps of Water Mass over Africa

Annual, semi-annual amplitudes and linear trends for each surface element have been least squares adjusted from the complete series of 10-day regional solutions over the period 2003–2012. Dominant seasonal amplitudes of ±200 mm EWH are well located in the Sahel latitudinal band (*i.e.*, 5°N–15°N), as expected, and in the Congo Basin (Figure 5).

Figure 5. Maps of the seasonal amplitudes of the water mass changes adjusted by least squares adjustment of a pure annual sinusoid at each grid cell of the 10-day regional solutions over Africa (2003–2012): (**a**) Regional solutions; (**b**) GRGS RL03 global solutions; (**c**) ICA 400-km filtered solutions. Units are mm of EWH. Note the strong signals due to the African monsoon in the Sahel latitudinal band.

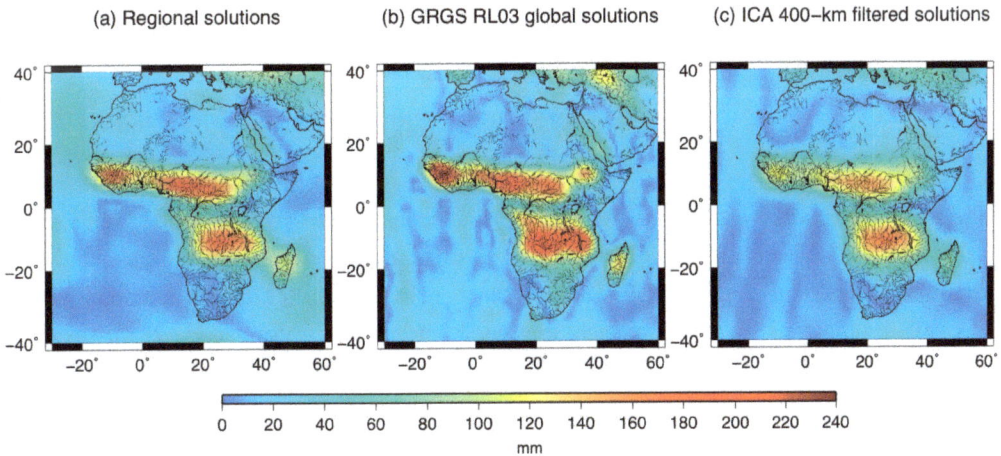

36

Figure 6. Maps of the linear trends of the water mass changes by fitting a linear trend for each grid cell from all of the 10-day regional solutions over Africa for 2003–2012: **(a)** Regional solutions; **(b)** GRGS RL03 global solutions; **(c)** ICA 400-km filtered solutions; **(d)** TRMM Precipitation 2003–2012. Units are mm of EWH per year.

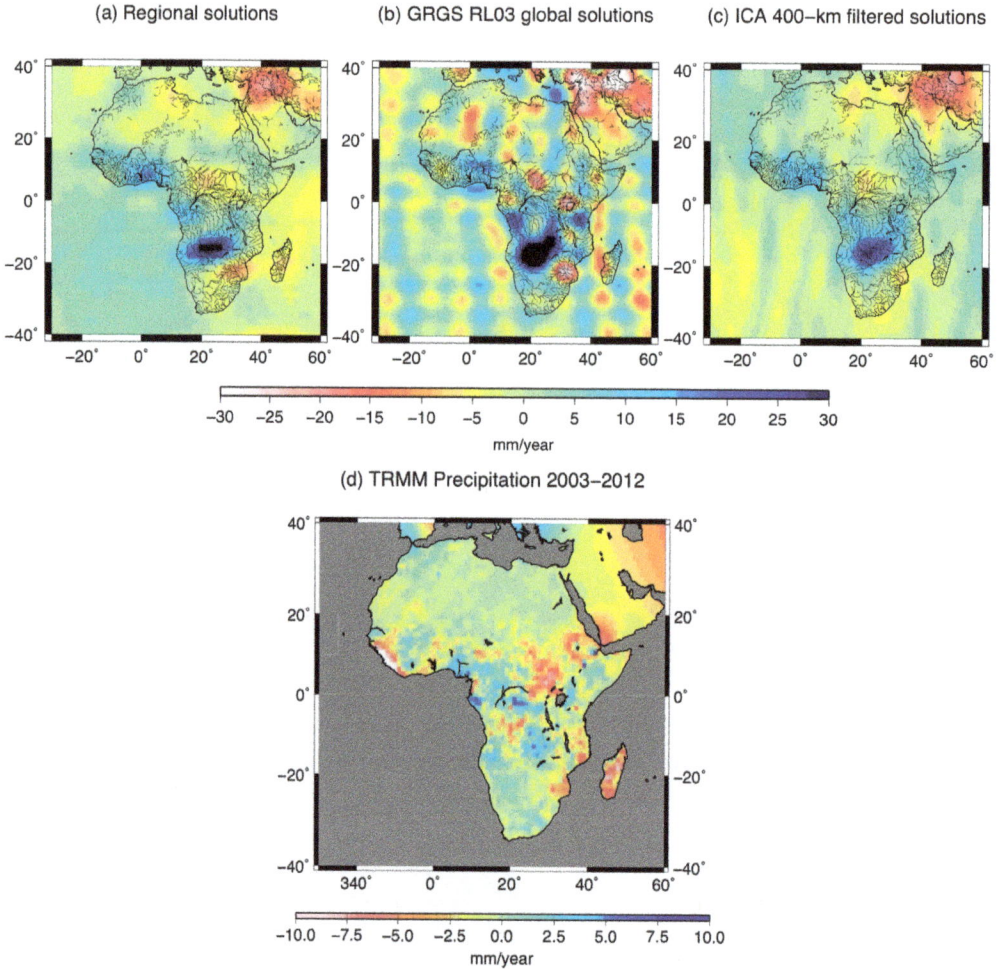

The seasonal signature of the West African monsoon can also be seen [14]. It is slightly greater for the global GRGS and 400-km ICA solutions. Besides, negative linear trends are found in the Tigris-Euphrates region of −25 mm EWH per year, consistent with the recent depletion of water mass shown by previous studies [71]; there are important gains of water mass over the Okavango swamps, reaching more than +30 mm EWH per year (Figure 6). In the Niger basin, the gain remains close to +15 mm EWH per year. There are also slight depletions in the south of Lake Chad and in the south of Mozambique. Unrealistic north-south striping in the trend estimates is more important for the global GRGS solutions than for 400-km ICA solutions, as the combination of different global solutions by ICA reinforces the hydrological signals *versus* the noise. Striping does not appear in the linear trend map related to the regional solutions at all (e.g., see the residuals over the oceans).

5.3. Principal Component Analysis of the GRACE Datasets

Principal component analysis (PCA), or discrete Karhunen–Loève transform, consists of decomposing the time series of water mass maps (*i.e.*, maps of regional water mass over Africa) into main space and time "modes" that are projections onto orthogonal directions (or the principal axis) of variability [72] (e.g., see [73] for the computational aspects). Before PCA decomposition, the time series of the regional solutions has been corrected from the dominant seasonal oscillation related to each surface element (see Figure 5) using 13-month window averaging.

5.3.1. First Mode of Variability: The Long-Term Behavior

This mode corresponds to multi-year variations of water mass, since its temporal mode is characterized by a regular increase from 2006 to 2010 (Figure 7). All of the spatial modes range in ±100 mm EWH. This represents 50%–70% of the explained variance of the regional and smooth ICA solutions and 71% of the GRGS solutions.

Figure 7. The first spatial and temporal modes of principal component analysis (PCA) of the regional, global GRGS and 400-km ICA GRACE solutions (**top**). The corresponding linear trends of the precipitation from TRMM for comparison (**bottom**).

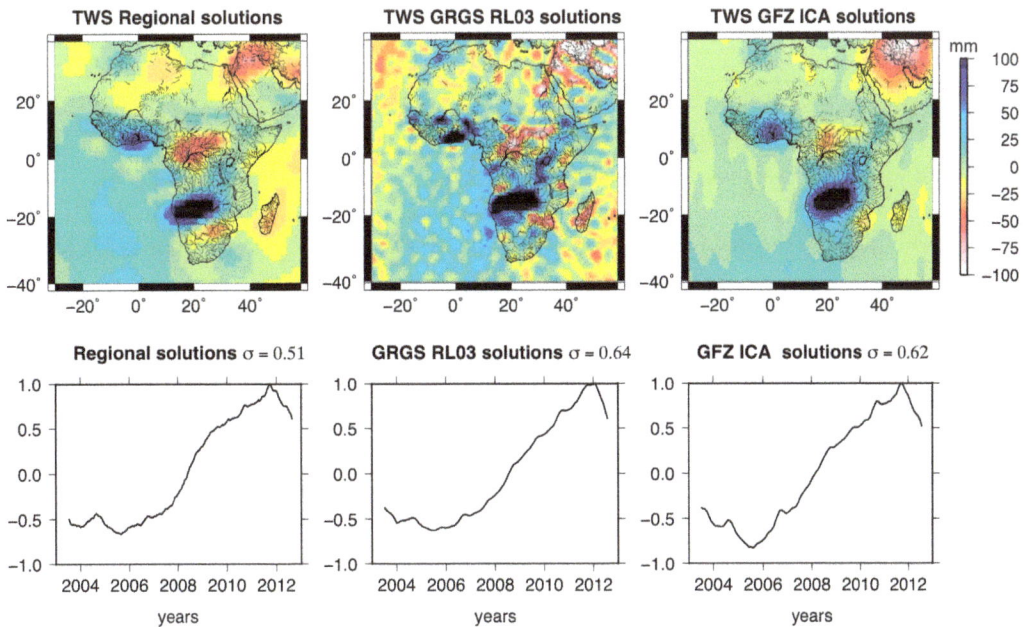

38

Figure 7. *Cont.*

TRMM Precipitation --- Mode 1

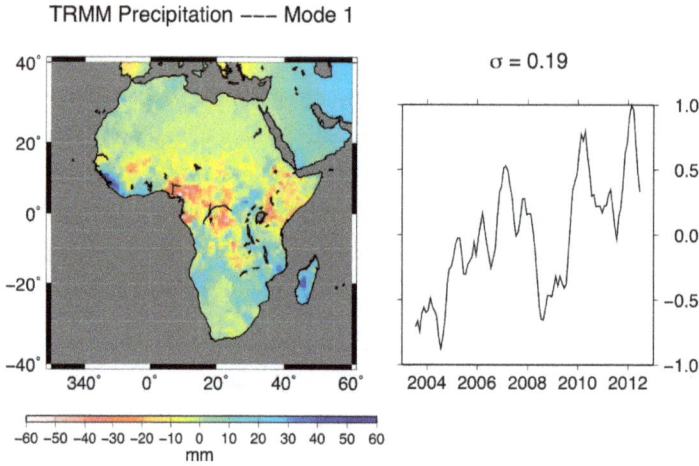

Thus, the corresponding spatial modes and the linear trend map presented in Figure 6 show similar patterns. The spatial mode associated with the monthly GRGS solutions contains unrealistic, short-scale undulations on the oceanic areas, as they are now derived from an empirical singular value decomposition (SVD) process of stabilization, and not a low-pass filtering, as for the GFZ, CSR and JPL solutions, which eliminates short wavelength more efficiently. The spatial components of the global solutions are affected by striping, especially the GRGS solutions, whereas no striping is visible in the regional water mass maps. There has been a dominant increase of mass in the Okavango swamp in the Zambezi River basin since 2006. The Okavango water system is an endorheic basin with no outlet to the sea, and it empties into the Kalahari Desert (~18,000 square kilometers), known as the Okavango Delta. Water storage also increases regularly over Volta Lake in Ghana, which is the largest artificial lake in the world. An important mass loss is observed in the Congo basin, centered on the extensive floodplains known as the "Cuvette Centrale" (see [74] for their spatial extent), especially in the regional solutions. The signature of water mass depletion in the NWSAS region is also clearly visible in this first PCA mode associated with the regional solutions. The temporal component of the first mode of the TRMM rainfall also exhibits multi-annual variations (Figure 7). The corresponding spatial mode of the precipitation shows patterns that coincide with the modes of the PCA of regional/global GRACE solutions, in particular the long-term deficit of water for the Niger Delta and further over the land of Cameroon, as well as the increase along the coast of Guinea, Sierra Leone, Liberia, Ivory Coast and Ghana.

Figure 8. Second spatial (**top**) and temporal (**bottom**) modes of the PCA of the regional, global GRGS and 400-km ICA GRACE solutions. Units are mm EWH.

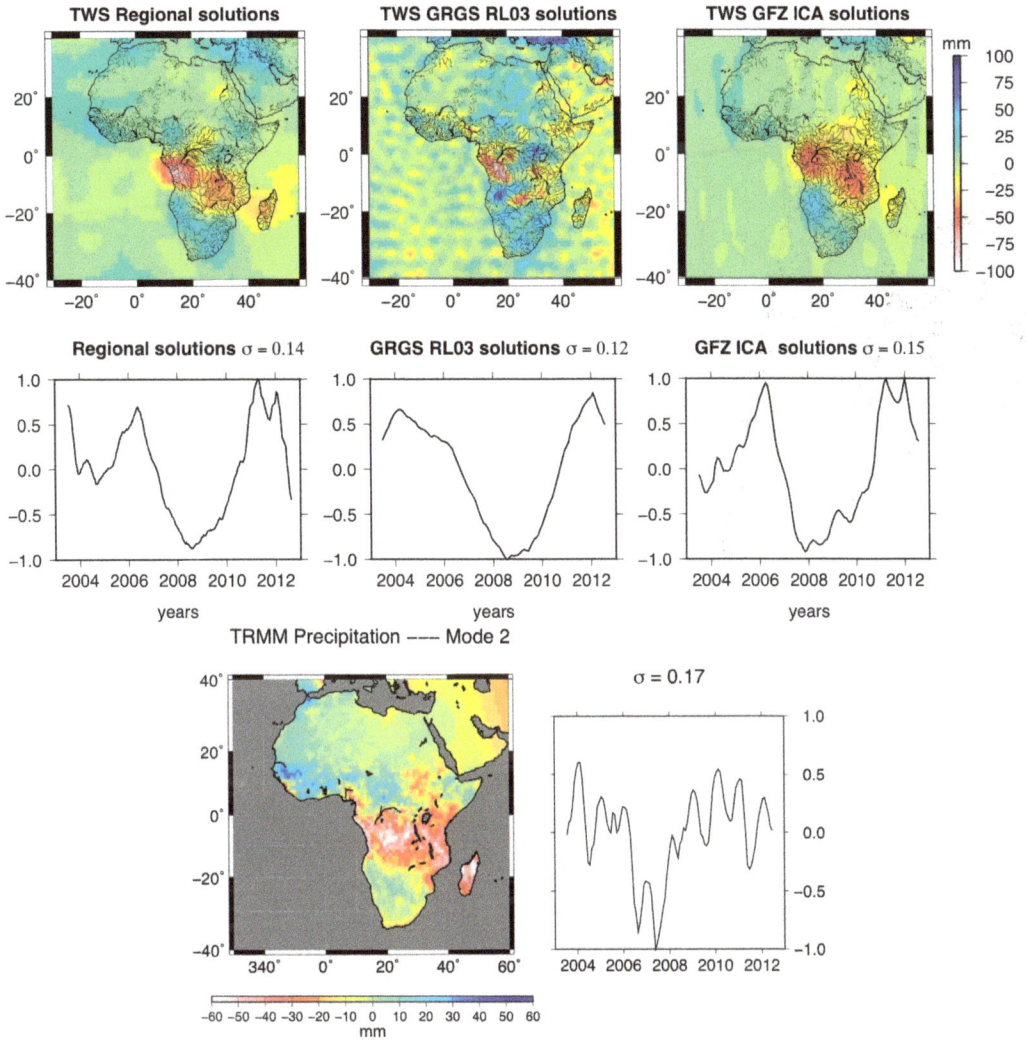

5.3.2. Second and Third Modes of Variability

Second and third modes represent only 10%–15% and 6%–10% of the explained variances of the water mass signals, respectively (Figures 8 and 9). As the amplitudes of the third modes are smaller, it is more complicated to interpret its spatial and temporal patterns. The modes of global solutions still suffer from unrealistic striping that is visible over oceanic areas. In particular, the modes related to the GRGS solutions are affected by short wavelength noise. Except the JPL solutions, the temporal characteristics of the second PCA modes are two-year oscillations, as well as the maximum amplitudes of the spatial patterns on Central and Western Africa (*i.e.*, northern part of the Congo River basin), including the Sahel, and along the Ivory Coast. As seasonal and semi-annual components have been removed before applying PCA, these patterns correspond to the bi-annual or quadrennial water mass variations related to the West African monsoon. Residual six-month and even multi-year oscillations appear for temporal modes of JPL.

Figure 9. Third spatial (**top**) and temporal (**bottom**) modes of the PCA of the regional, global GRGS and 400-km ICA GRACE solutions. Units are mm EWH.

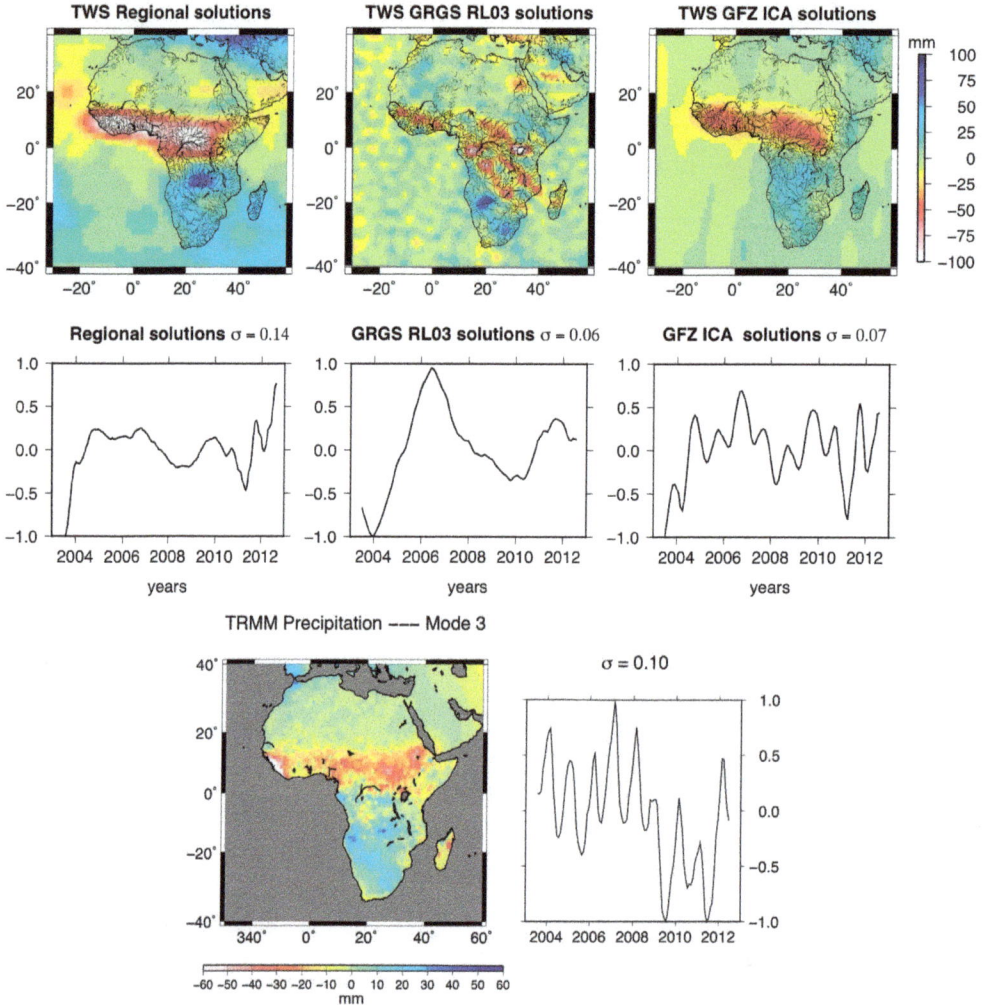

As for the first mode of variability, similarities between PCA modes of GRACE and TRMM rainfall exist. The second PCA mode of the TRMM data is characterized by important amplitudes of precipitation in the southern tropics, in particular over the Congo Basin. This strong signal also appears in the second modes of GRACE datasets, but it is less visible in the case of GRGS solutions. The temporal mode of the rainfall data is shifted; it occurs ~6 months sooner than the GRACE solutions (Figure 8). The third mode of rainfall is related to the African monsoon, as its main signature is located in the Sub-Saharan band, as for the spatial PCA modes of GRACE solutions (Figure 9).

5.4. Time Series of TWS Averaged over African River Drainage Basins

The masks used in this study come from the five-minute, 1/2° and 1° datasets of the continental watersheds and river networks for use in regional and global hydrologic and climate system modeling studies [75]. The 1° dataset is used for the drainage regions and the river watersheds, while the 0.5° dataset is used to define the drainage regions around the lakes. The coordinates of the drainage basin limits for each lake are obtained using the lake boundaries and the drainage network at 1/2°.

Time series of water mass for the main drainage basins of Africa over 2004–2012 (see Figure 1) have been computed as masked averages *versus* time and then converted into mm of equivalent sea level (ESL) profiles. In this latter operation, the water volume variations (*i.e.*, equivalent water heights times the basin surface) are divided by the surface of the oceans (~360 million square kilometers) and multiplied by −1. The per-basin time series of ESL are presented in Figure 10. Positive multi-year contributions to the sea level concern the Congo, Nile and Orange river basins, and they represent +0.044 mm/y in total, whereas the total negative contribution from the other basins is more important in magnitude (*i.e.*, −0.123 mm/y), where the multi-year sea level contribution of the Zambezi Basin is the largest in magnitude (*i.e.*, −0.1 mm/y). This region centered on the Okavango swamps and floodplains has been storing more and more water during the last decade.

As displayed in Figure 11, the sea level contributions averaged over large drainage areas to the Atlantic and Indian oceans and the Mediterranean Sea exhibit clear seasonal oscillations of amplitudes ±2, ±1 and ±0.5 mm of ESL, respectively. The seasonal contribution of land waters to the Indian Ocean (*i.e.*, runoff along the southeastern coast of Africa) is in opposite phase with the two others. The sum of the contribution of these three large areas is negative (*i.e.*, −0.088 mm per year); thus, the water mass balance of Africa for the period 2004–2012 indicates a gain of mass on the continent. In particular, the strongest multi-year trend magnitude is the one related to the Indian Ocean (*i.e.*, −0.094 mm/y), as it shows a clear acceleration of the gain of water mass on land during recent years, passing from −0.034 mm/y before 2008 to −0.123 mm/y after 2008. A comparison with Figure 5 indicates that the latter contribution is driven by the long-term variations of the hydrology in the Zambezi River Basin. For the Atlantic and Mediterranean contributions, the positive ESL trends have been decreasing since 2008 from +0.138 mm/y and +0.030 mm/y down to +0.071 mm/y and +0.027 mm/y, respectively. Once the global isostatic adjustment (GIA) of the solid Earth from mainly post-glacial rebound and representing −0.3 mm/y is corrected [76,77], the net contribution of African rivers to sea level for 2004–2012 remains small, since it represents only ~3% of the global increase of the sea level of 3.3 ± 0.4 mm/y measured by altimetry. This latter comparison suggests that the multi-year contribution of Africa hydrology remains in the total sea level error bar.

Figure 10. Time series of equivalent sea level (ESL) for the chosen main African river basins (solid line) and the corresponding 13-month low pass filtered profiles (dashed line).

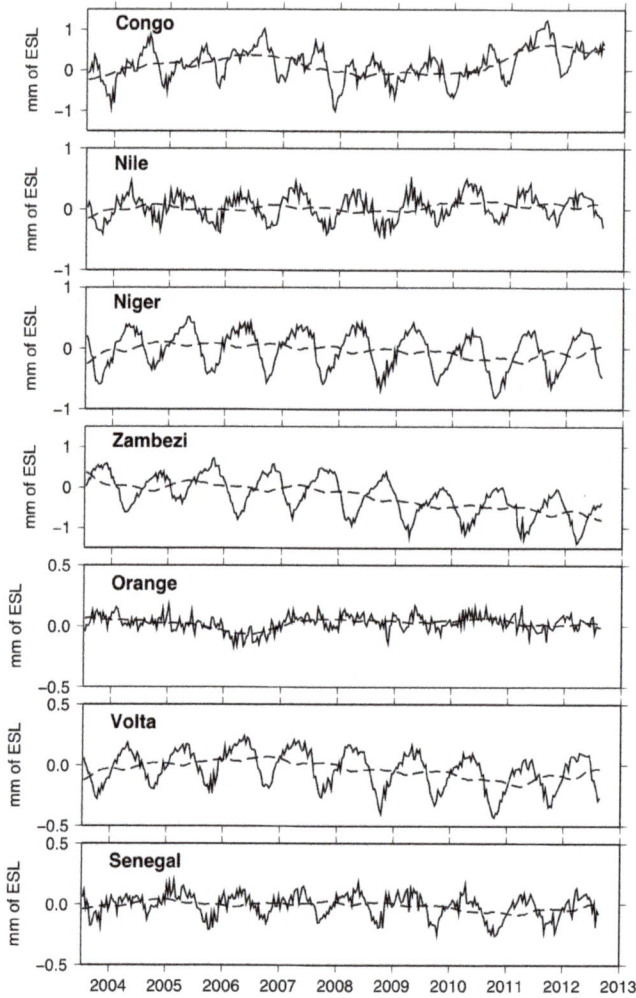

Figure 11. Time series of the main African river basins contributions to the Atlantic and Indian oceans, as well as to the Mediterranean Sea, expressed in mm of ESL.

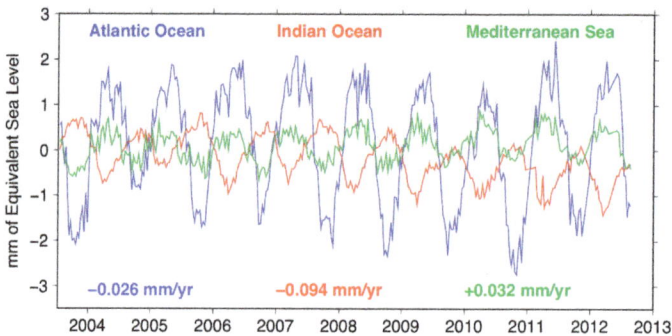

5.5. Detection of Recent Groundwater Withdraw of NWSAS Aquifer

The NWSAS presented in Figure 1 (Area 9) is characterized by a transboundary "fossil" aquifer (*i.e.*, with no meteoric water recharge) shared by three countries (*i.e.*, 60% Algerian, 30% Libyan and 10% Tunisian surfaces) and in a critical situation of depletion by intense water pumping. While the groundwater imbalance is −2.2 km^3 per year as estimated from historical records of piezometry before the year 2000 (see the report of [78]), our regional GRACE solutions averaged over the NWSAS indicate that the depletion of groundwater for the most recent decade is twice that, suggesting a worrying acceleration of the withdrawing of drinking water (Figure 12). In particular, a sudden loss of −25 km^3 lasting a few months appears in the year 2007 between the periods January 2004–December 2007, and January 2008–January 2012. A comparison with the latter trends estimated considering the total GRACE period suggests that the rapid drop of groundwater in 2007–2008 remains exceptional (according to the dashed line). These are consistent with the recent estimates of groundwater loss from −2.2 km^3/y in 2000 [78] to −2.75 km^3/y in 2010 [79]. Time variations of TWS from GRACE appear reliable to directly and efficiently estimate the region-wide groundwater changes in a large aquifer system in arid areas and to provide useful information on groundwater recharge.

Figure 12. TWS time series over the NWSAS region (Libya-Tunisia-Algeria) aquifer (9 in Figure 1) (solid line) and the corresponding 13-month averaged profile (dashed line).

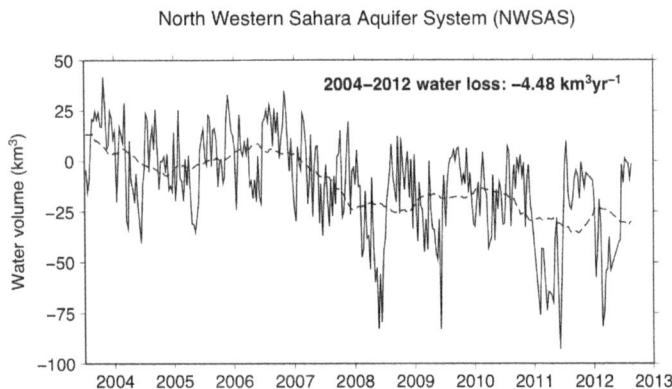

5.6. Subsurface Water Storage Changes by Combining Regional Solutions and Radar Altimeter Data

Time series of the subsurface waters (*i.e.*, soil plus ground waters) were estimated by removing altimetry-based water levels of the main African lake, or reservoir waters, from the TWS measured by GRACE. As altimetry-based water levels from [78] have a temporal resolution of one month, the ten-day regional solutions were consequently averaged on monthly time periods before computing the mere difference. Anomalies of these residual subsurface waters, including groundwater, were estimated at a monthly time-scale as the difference between TWS and surface storage over the common period of availability of two datasets (*i.e.*, mid-2003–2012). These anomalies are presented in Figure 13 for the largest East African lakes: Turkana, Victoria, Tanganyika and Malawi. They temporally and/or spatially extend the previous studies from [17,18,80], but using regional solutions instead of classical global ones.

Figure 13. Time-series of the anomaly of the water storage of TWS (black), of the surface reservoir (blue) and of the subsurface reservoir, including groundwater (red), in the left column, and their inter-annual variations, in the right column, for Lake Turkana (**a**), Lake Victoria (**b**), Lake Tanganyika (**c**) and Lake Malawi (**d**).

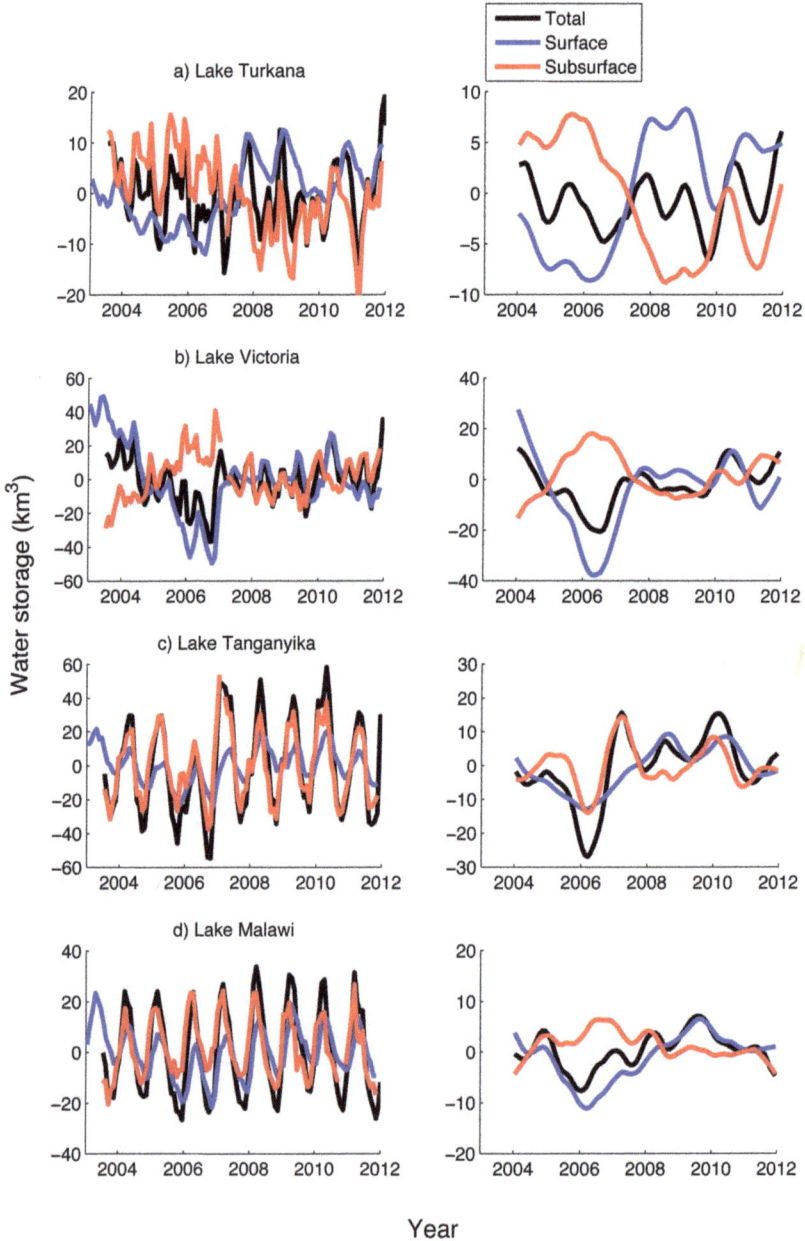

Time variations of TWS and subsurface waters are well correlated for the three largest lakes, as the correlation coefficient R equals 0.75, 0.76 and 0.76 for Lake Victoria, Lake Tanganyika and Lake Malawi, but not for the smaller Lake Turkana, with R equal to 0.28. If the surface water storage represents only a small part of TWS variations, as for Lake Turkana over 2003–2013, its inter-annual variations are larger than the inter-annual TWS signals, similarly to what was found by [18]. The inter-annual signals of subsurface water presents three successive phases: from mid-2005 to mid-2008: they decreased at an average rate of −4 km^3/y from mid-2008 to mid-2009 and remained stable; and since then, they present a bi-annual cycle of ±5 km^3 of amplitude (Figure 13a). For Lake Victoria, the variations of the anomaly of water storage are greater for the surface reservoir than for the TWS. A larger decrease is observed for the anomaly of surface storage (−60 km^3/y from 2004 to the beginning of 2006) than for TWS (−70 km^3/y from 2004 to mid-2006) and the opposite up to the end of 2007, with increases of 40 and 20 km^3/y for the surface water storage and TWS, respectively. Consequently, subsurface water is varying as the difference over the same time periods (Figure 13b). The results obtained with the regional solutions are very similar to those obtained by [18,81]. Nevertheless, [18] found larger variations of TWS than these from surface water, most likely because these authors included in the drainage areas of Lake Victoria the Kivu, Edouard, Albert and Kyoga lakes and their drainage areas, without adding their volume variations to the surface storage.

The TWS variations are dominated by the subsurface water component for Lake Tanganyika. A large variation of TWS, strongly impacting the subsurface water reservoir, is observed from 2005 to 2008, with a minimum reaching −40 km^3 in the beginning of 2006 and a maximum of 30 km^3 in the beginning of 2007. Smoother variations are present in the surface reservoir (Figure 13c).

As for Lake Victoria, TWS variations of Lake Malawi are dominated by the surface component, but the amplitudes of the surface reservoir are slightly greater than these from TWS at an inter-annual time scale. The time variations of the surface reservoir exhibit a steep decrease from 2004 to the beginning of 2006, followed by a significant increase up to mid-2009, whereas they are lower for TWS (Figure 13d).

These results demonstrate the strong capacities of multi-satellite observations to monitor quantitatively the changes in the storage of the surface and sub-surface reservoirs associated with climate variability and human activities at a regional scale. These remotely-sensed datasets are likely to have huge importance in regions where *in situ* data are sparse, while they have huge importance in terms of water supply for dense human populations. In the case of Lake Victoria, it directly supports 30 million people in terms of a freshwater supply [8] and indirectly another 340 million people along the Nile Basin [82], being the source of the White Nile.

6. Conclusions

In this paper, we present 10-day regional solutions of water mass change over Africa for the period 2003–2012, revealing the dominant seasonal and African monsoon signals (±250 mm EWH). Principal component analysis (PCA) of the GRACE datasets provided the main modes of variability of the surface water mass. Temporal and spatial patterns are consistent for the regional and global solutions. However, the regional solutions offer a better geographical localization of hydrological structures, while global solutions remain affected by aliasing errors (*i.e.*, north-south striping). This is probably due to the benefit of details brought by GRACE short tracks into the regional solution, instead of considering band-limited spherical harmonics defined on the entire Earth.

Monitoring water supply by using these regional solutions enables us to confirm the long-term drought of the NWSAS aquifer and even to reveal the sudden water loss occurring in early 2008. In terms of large-scale mass balance, the small contribution of the African hydrology changes to global sea level rise for the period 2004–2012 remains negative, especially due to the gain of water mass in the swamp regions of the Zambezi River basin that represents −0.123 mm/y of ESL as a continuous deficit of the level for the Indian Ocean. Principal component analysis of the complete time series of regional GRACE solutions has been made to identify the separate contributions of the different African regions. In particular, the first mode of PCA, representing 50%–70% of the explained variance, reveals the long-term drought in East Africa and in the region of Lake Chad and, alternatively, the increase of water mass in the Okavango and Niger regions. The second PCA mode of 10% corresponds to an oscillation of two years over the Sahel and the central African regions. This latter mode is clearly related to the residuals of the dominant African monsoon. GRACE-derived TWS variations demonstrated strong capabilities for land water management in terms of monitoring the sub-surface changes alone over arid and semi-arid areas, such the NWAS basins, or in combination with altimetry-based water levels for lake drainage areas.

Acknowledgements

This work was supported by the CNES TOSCA grant "Surcharges et Propagation". The authors would like to thank four anonymous reviewers for helping us in improving the quality of our manuscript.

Author Contributions

All three authors of the present work contributed to the writing of the manuscript, as well as the discussion of the results. The methodology for deriving regional solutions from accurate GRACE measurements had been previously developed by Guillaume Ramillien. Lucia Seoane computed the 10-day two-degree regional maps from daily GRACE orbits for the multi-year period, 2003–2012. Along-track residuals of gravity potential differences for continental hydrology were obtained using the GINS software developed by the GRGS group in Toulouse. In particular, Frédéric Frappart made the combination of the regional GRACE solutions with altimetry to derive the measured gravity contribution of sub-surface waters and to analyze their dynamics.

Conflicts of Interest

The authors declare no conflict of interest.

References

1. Famiglietti, J.S.; Lo, M.; Ho, S.L.; Bethune, J.; Anderson, K.J.; Syed, T.H.; Swenson, S.C.; de Linage, C.R.; Rodell, M. Satellites measure recent rates of groundwater depletion in California's Central Valley. *Geophys. Res. Lett.* **2011**, doi:10.1029/2010GL046442.
2. Scanlon, B.R.; Faunt, C.C.; Longuevergne, L.; Reedy, R.C.; Alley, W.M.; McGuire V.L.; MacMahon, P.B. Groundwater depletion and sustainability of irrigation in the US High Plains and Central Valley. *Proc. Natl. Acad. Sci. USA* **2012**, *109*, 9320–9325.
3. Scanlon, B.R.; Longuevergne, L.; Long, D. Ground referencing GRACE satellite estimates of groundwater storage changes in the California Central Valley, USA. *Water Resour. Res.* **2012**, doi:10.1029/2011WR011312.
4. Ramillien, G.; Frappart, F.; Cazenave, A.; Güntner, A. Time variations of land water storage from inversion of 2-years of GRACE geoids. *Earth Planet. Sci. Lett.* **2005**, *235*, 283–301.
5. Crowley, J.W.; Mitrovica, J.X.; Bailey, R.C.; Tamisiea, M.E.; Davis, J.L. Land water storage within the Congo basin inferred from GRACE satellite gravity data. *Geophys. Res. Lett.* **2006**, doi:10.1029/2006GL027070.
6. Papa, F.; Güntner, A.; Frappart, F.; Prigent, C.; Rossow, W.B. Variations of surface water extent and water storage in large river basins: A comparison of different global data sources. *Geophys. Res. Lett.* **2008**, doi:10.1029/2008GL033857.
7. Ramillien, G.; Bouhours, S.; Lombard, A.; Cazenave, A.; Flechtner, F.; Schmidt, R. Land water storage contribution to sea levelfrom GRACE geoid data over 2003–2006. *Glob. Planet. Chang.* **2008**, *60*, 381–392.
8. Lee, H.; Beighley, R.E.; Alsdorf, D.; Jung, H.; Shum, C.; Duan, J.; Guo, J.; Yamazaki, D.; Andreadis, K. Characterization of terrestrial water dynamics in the Congo Basin using GRACE and satellite radar altimetry. *Remote Sens. Environ.* **2011**, *115*, 3530–3538.
9. Awange, J.L.; Sharifi, M.A.; Ogonda, G.; Wickert, J.; Grafarend, E.W.; Omulo, M.A. The falling Lake Victoria water level: GRACE, TRIMM and CHAMP satellite analysis of the lake basin, *Water Resour. Manag.* **2008**, *22*, 775–796.
10. Awange, J.L.; Ong'ang'a, O. *Lake Victoria: Ecology Resource and Environment*; Springer: Berlin, Germany, 2006; p. 354.
11. Deus, D.; Gloaguen, R.; Krause, P. Water balance modeling in a semi-arid environment with limited *in situ* data using remote sensing in Lake Manyara, East African Rift, Tanzania. *Remote Sens.* **2013**, *5*, 1651–1680.
12. Jensen, L.; Rietbrock, R.; Kusche, J. Land water contribution to sea level from GRACE and Jason-1 measurements. *J. Geophys. Res.: Oceans* **2013**, *118*, 212–226.
13. Klees, R.E.; Zapreeva, A.; Winsemius, H.C.; Savenije, H.H.C. The bias in GRACE estimates of continental water storage variations. *Hydrol. Earth Syst. Sci.* **2007**, *11*, 1227–1241.
14. Grippa, M.; Kergoat, L.; Frappart, F.; Araud, Q.; Boone, A.; de Rosnay, P.; Lemoine, J.-M.; Gascoin, S.; Balsamo, G.; Ottlé, C.; *et al.* Land water storage over West Africa estimated by GRACE and land surface model. *Water Resour. Res.* **2011**, doi:10.1029/2009WR008856.

15. Milzow, C.; Krogh, P.E.; Bauer-Gottwein, P. Combining satellite radar altimetry, SAR surface soil moisture and GRACE total storage changes for hydrological calibration in a large poorly gauged catchment. *Hydrol. Earth Syst. Sci.* **2011**, *15*, 1729–1743.

16. Xie, H.; Longuevergne, L.; Ringler, C.; Scanlon, B.R. Calibration and evaluation of a semi-distributed watershed model of a sub-saharan Africa using GRACE data. *Hydrol. Earth Syst. Sci.* **2012**, *16*, 3083–3099.

17. Swenson, S.; Wahr, J. Monitoring the water balance of Lake Victoria, East Africa, from space. *J. Hydrol.* **2009**, doi:10.1016/j.jhydrol.2009.03.008.

18. Becker, M.; Llovel, W.; Cazenave, A.; Güntner, A.; Crétaux, J.-F. Recent hydrological behaviour of the East African Great Lakes region inferred from GRACE, satellite altimetry and rainfall observations. *Comptes Rendus Géosci.* **2010**, doi:10.1016/j.crte.2009.12.010.

19. Bonsor, H.C.; Mansour, M.M.; MacDonald, A.M.; Hughes, A.G.; Hipkin, R.G.; Bedada, T. Interpretation of GRACE data of the Nile Basin using a groundwater recharge model. *Hydrol. Earth Syst. Sci.* **2010**, *7*, 4501–4533.

20. Henry, C.M.; Allen, D.M.; Huang, J. Groundwater storage variability and annual recharge using well-hydrograph and GRACE satellite data. *Hydrogeol. J.* **2011**, *19*, 741–755.

21. Syed, T.H.; Famiglietti, J.S.; Chambers, D.P. GRACE-based estimates of terrestrial freshwater discharge from basin to continental scales. *Hydrogeol. J.* **2009**, *10*, 22–40.

22. Ramillien, G.; Frappart, F.; Güntner, A.; Ngo-Duc, T.; Cazenave, A.; Laval, K. Time variations of regional evapotranspiration rate from Gravity Recovery and Climate Experiment (GRACE) satellite gravimetry. *Water Resour. Res.* **2006**, doi:10.1029/2005WR004331.

23. Rodell, M.; McWilliams, E.B.; Famiglietti, J.S.; Beaudoing, H.K.; Nigro, J. Estimating evapotranspiration using an observation based terrestrial water budget. *Hydrol. Process.* **2011**, *25*, 4083–4092.

24. Long, D.; Longuevergne, L; Scanlon, B.R. Uncertainty in evapotranspiration from land surface modeling, remote sensing, and GRACE satellites. *Water Resour. Res.* **2014**, *50*, 1131–1151.

25. Trenberth, K.E.; Fasullo, J.T. Regional energy and water cycles: Transports from ocean to land. *J. Clim.* **2013**, *26*, 7837–7851.

26. Pfeffer, J.; Boucher, M.; Hinderer, J.; Favreau, G.; Boy, J.-P.; de Linage, C.; Cappelaere, B.; Luck, B.; Oi, M.; le Moigne, M. Local and global hydrological contribution to time variable gravity in southwest Niger. *Geophys. J. Int.* **2011**, *184*, 661–672.

27. Hector, B.; Seguis, L.; Hinderer, J.; Descloitres, M.; Vouillamoz, J.-M.; Wubda, M.; Boy, J.-P.; Luck, B.; le Moigne, N. Gravity effect of water storage changes in a weathered hard-rock aquifer in West Africa: Results from joint absolute gravity, hydrological monitoring and geophysical prospection. *Geophys. J. Int.* **2013**, *194*, 737–750.

28. Nahmani, S.; Bock, O.; Bouin, M.-N.; Santamaría-Gómez, A.; Boy, J.-P.; Collilieux, X.; Métivier, L.; Panet, I.; Genthon, G.; de Linage, C.; *et al.* Hydrological deformation induced by the West African Monsoon: Comparison of GPS, GRACE and loading models. *J. Geophys. Res.: Solid Earth* **2012**, doi:10.1029/2011JB009102.

29. Syed, T.H.; Famiglietti, J.S.; Rodell, M.; Chen, J.; Wilson, C.R. Analysis of terrestrial water storage changes from GRACE and GLDAS. *Water Resour. Res.* **2008**, doi:10.1029/2006WR005779.

30. Rodell, M.; Famiglietti, J.S.; Chen, J.; Seneviratne, S.I.; Viterbo, P.; Holl, S.; Wilson, C.R. Basin sacle estimate of evapotranspiration using GRACE and other observations. *Geophys. Res. Lett.* **2004**, doi:10.1029/2004GL020873.

31. Rodell, M.; Chen, J.; Kato, H.; Famiglietti, J.S.; Nigro, J.; Wilson, C.R. Estimating grounwater storage changes in the Mississippi River basin (USA) using GRACE. *Hydrogeol. J.* **2007**, *15*, 159–166.

32. Leblanc, M.; Tregoning, P.; Ramillien, G.; Tweed, S.; Fakes, A. Basin scale, integrated observations of the early 21st century multi-year drought in Southeast Australia. *Water Resour. Res.* **2009**, doi:10.1029/2008WR007333.

33. Taylor, R.G.; Scanlon, B.R.; Döll, P.; Rodell, M.; van Beek, R.; Wada, Y.; Longuevergne, L.; Leblanc, M.; Famiglietti, J.S.; Edmunds, M.; *et al.* Ground water and climate change. *Nat. Clim. Chang.* **2013**, doi:10/1038/nclimate1744.

34. Andersen, O.B.; Seneviratne, S.I.; Hinderer, J.; Viterbo, P. GRACE-derived terrestrial water storage depletion associated with the 2003 European heat wave. *Geophys. Res. Lett.* **2005**, doi:10.1029/2005GL023574.

35. Frappart, F.; Papa, F.; Santos da Silva, J.; Ramillien, G.; Prigent, C.; Seyler, F.; Calmant, S. Surface freshwater storage and dynamics in the Amazon basin during the 2005 exceptional drought. *Environ. Res. Lett.* **2012**, doi:10.1088/1748-9326/7/4/044010.

36. Frappart, F.; Seoane, L.; Ramillien, G. Validation of GRACE-derived water mass storage using a regional approach over South America. *Remote Sens. Environ.* **2013**, *137*, 69–83.

37. Houborg, R.; Rodell, M.; Li, B.; Reichle, R.; Zaitchik B.F. Drought indicators based on model-assimilated Gravity Recovery and Climate Experiment (GRACE) terrestrial water storage observations. *Water Resour. Res.* **2012**, doi:10.1029/2011WR011291.

38. Chen, J.L.; Wilson, C.R.; Tapley, B.D. The 2009 exceptional Amazon flood and interannual terrestrial water storage change observed by GRACE. *Water Resour. Res.* **2010**, doi:10.1029/2010WR009383.

39. Chen, J.L.; Wilson, C.R.; Tapley, B.D.; Longuevergne, L.; Yang, Z.L.; Scanlon, B.R. Recent La Plata basin drought conditions observed by satellite gravimetry. *J. Geophys. Res.: Atmos.* **2010**, doi:10.1029/2010JD014689.

40. Long, D.; Scanlon, B.R.; Longuevergne, L; Sun, A.; Fernando, N.; Save, H. GRACE satellite monitoring of large depletion in water storage in response to the 2011 drought in Texas. *Geophys. Res. Lett.* **2013**, *40*, 3395–3401.

41. Duan, J.; Shum, C.K.; Guo, J.; Huang, Z. Uncovered spurious jumps in the GRACE atmospheric de-aliasing data: Potential contamination of GRACE observed mass change. *Geophys. J. Int.* **2012**, *191*, 83–87.

42. Swenson, S.C.; Wahr, J. Post-processing removal of correlated errors in GRACE data. *Geophys. J. Int.* **2006**, doi:10.1029/2005GL025285.

43. Himanshu, S.; Bettadpur, S.; Tapley, B.D. Reducing errors in the GRACE gravity solutions using regularization. *J. Geodesy* **2012**, *86*, 695–711.

44. Klees, R.; Revtova, E.A.; Gunter, B.C.; Ditmar, P.; Oudman, E.; Winsemius, H.C.; Savenije, H.H.G. The design of an optimal filter for monthly GRACE gravity models. *Geophys. J. Int.* **2008**, *175*, 417–432.

45. Frappart, F.; Ramillien, G.; Leblanc, M.; Tweed, S.O.; Bonnet, M.-P.; Maisongrande, P. An independent component analysis approach for filtering continental hydrology in the GRACE gravity data. *Remote Sens. Environ.* **2011**, *115*, 187–204.

46. Ramillien, G.; Biancale, R.; Gratton, S.; Vasseur, X.; Bourgogne, S. GRACE-derived surface mass anomalies by energy integral approach. Application to continental hydrology. *J. Geodesy* **2011**, *85*, 313–328.

47. Ramillien, G.; Seoane, L.; Frappart, F.; Biancale, R.; Gratton, S.; Vasseur, X.; Bourgogne, S. Constrained regional recovery of continental water mass time-variations from GRACE based geopotential anomalies over South America. *Surv. Geophys.* **2012**, *33*, 887–905.

48. Freeden, W.; Schreiner, M. *Spherical Functions of Mathematical Geosciences: A Scalar, Vectorial and Tensorial Setup*; Springer: Berlin, Germany, 2008.

49. Seoane, L.; Ramillien, G.; Frappart, F.; Leblanc, M. Regional GRACE-based estimates of water mass variations over Australia: Validation and interpretation. *Hydrol. Earth Syst. Sci.* **2013**, *17*, 4925–4939.

50. Rowlands, D.D.; Ray, R.D.; Chinn, D.S.; Lemoine, F.G. Short-arc analysis of intersatellite tracking data in a gravity mapping mission. *J. Geodesy* **2002**, doi:10.1007/s00190-002-0255-8.

51. Rowlands, D.D.; Luthcke, S.B.; Klosko, S.M.; Lemoine, F.G.; Chinn, D.S.; McCarthy, J.J.; Cox, C.M.; Anderson, O.B. Resolving mass flux at high spatial and temporal resolution using GRACE intersatellite measurements. *Geophys. Res. Lett.* **2005**, doi:10.1029/2004GL021908.

52. Lemoine, F.G.; Luthcke, S.B.; Rowlands, D.D.; Chinn, D.S.; Klosko, S.M.; Cox, C.M. The use of mascons to resolve time-variable gravity from GRACE. In *Dynamic Planet*; Springer: Berlin, Germany, 2007; pp. 231–236.

53. Lemoine, J.-M.; Bruinsma, S.; Loyer, S.; Biancale, R.; Marty, J.-C.; Pérosanz, F.; Balmino, G. Temporal gravity field models inferred from GRACE data. *Adv. Space Res.* **2007**, *39*, 1620–1629.

54. Bruinsma, S.; Lemoine, J.-M.; Biancale, R.; Valès, N. CNES/GRGS 10-day gravity models (release 2) and their evaluation. *Adv. Space Res.* **2010**, *45*, 587–601.

55. GFZ Potsdam Information Systems & Data Center. Available online: http://isdc.gfz-potsdam.de (accessed on 22 July 2014).

56. Physical Oceanography Distributed Active Archive Data. Available online: http://podaac-www.jpl.nasa.gov (accessed on 31 July 2014).

57. Schmidt, R.; Flechtner, F.; Meyer, U.; Neumayer, K.-H.; Dahle, Ch.; Koenig, R.; Kusche, J. Hydrological signals observed by GRACE satellites. *Surv. Geophys.* **2008**, *29*, 319–334.

58. Groupe de Recherche en Géodésie Spatiale. Available online: http://grgs.obs-mip.fr (accessed on 22 July 2014).

59. Frappart, F.; Ramillien, G.; Maisongrande, P.; Bonnet, M.-P. Denoising satellite gravity signals by independent component analysis. *IEEE Geosci. Remote Sens. Lett.* **2010**, *7*, 421–425.

60. Fu, L.L.; Cazenave, A. *Satellite Altimetry and Earth Sciences, A Handbook of Techniques and Applications, International Geophysics Series*; Academic Press: San Diego, FL, USA, 2001; Volume 69.

61. Crétaux, J.-F.; Jelinski, W.; Calmant, S.; Kouraev, A.; Vuglinski, V.; Bergé-Nguyen, M.; Gennero, M.-C.; Niño, F.; Abarca del Rio, R.; Cazenave, A.; *et al.* SOLS: A lake database to monitor in near real time water level and storage variations from remote sensing data. *J. Adv. Space Res.* **2011**, doi:10.1016/j.asr.2011.01.004.

62. Ričko, M.; Birkett, C.M.; Carton, J.A.; Crétaux, J.-F. Intercomparison and validation of continental water level products derived from satellite radar altimetry. *J. Appl. Remote Sens.* **2012**, doi:10.1117/1.JRS.6.061710.

63. Frappart, F.; Calmant, S.; Cauhopé, M.; Seyler, F.; Cazenave, A. Preliminary results of ENVISAT RA-2 derived water levels validation over the Amazon basin. *Remote Sens. Environ.* **2006**, *100*, 252–264.

64. Santos da Silva, J.; Calmant, S.; Seyler, F.; Rottuno Filho, O.C.; Cochonneau, G.; Mansur, W.J. Water levels in the Amazon basin derived from the ERS-2 and ENVISAT radar altimetry missions. *Remote Sens. Environ.* **2010**, *114*, 2160–2181.

65. Frappart, F.; Seyler, F.; Martinez, J.-M.; Leon, J.G.; Cazenave, A. Floodplain water storage in the Negro River basin estimated from microwave remote sensing of inundation area and water levels. *Remote Sens. Environ.* **2005**, *99*, 387–399.

66. Hydrology from Space. Lakes, Rivers and Wetlands Water Levels from Satellite Altimetry. Available online: http://www.legos.obs-mip.fr/soa/hydrologie/hydroweb/ (accessed on 22 July 2014).

67. Spigel, R.H.; Coulter, G.W. Comparison of hydrology and physical limnology of the East African Great lakes: Tanganyika, Malawi, Victoria, Kivu and Turkana (with references to some North American Great lakes). In *The Limnology, Climatology and Paleoclimatology of the East African Lakes*; Johnson, T.C., Odada, E.O., Eds.; Gordon & Breach Publishers: Amsterdam, The Netherlands, 1996; pp. 103–135.

68. Huffmann, G.J.; Adler, R.F.; Rudolf, B.; Schneider, U; Keehn, P.R. Global precipitation estimates based on a technique for combining satellite-based estimates, rain gauge analysis, and NWP model precipitation information. *J. Clim.* **1995**, *8*, 1284–1295.

69. Huffmann, G.J.; Adler, R.F.; Bolvin, D.T.; Gu, G.; Nelkin, E.J.; Bowman, K.P.; Hong, Y.; Stocker, E.F.; Wolf, D.B. The TRMM Multi-satellite Precipitation Analysis (TMPA): Quasi-global, multi-year, combined-sensor precipitation estimates at fine scale. *J. Hydrometeorol.* **2007**, *8*, 38–55.

70. Seo, K.-W.; Wilson, C.R.; Famiglietti, J.S.; Chen, J.L.; Rodell, M. Terrestrial water mass load changes from Gravity Recovery and Climate Experiment (GRACE). *Water Resour. Res.* **2006**, doi:10.1029/2005WR004255.

71. Voss, K.A.; Famiglietti, J.S.; Lo, M.; de Linage, C.; Rodell, M.; Swenson, S.C. Groundwater depletion in the Middle East from GRACE with implications for transboundary water management in the Tigris-Euphrates-Western Iran region. *Water Resour. Res.* **2013**, *49*, 904–914.

72. Preisendorfer, R. *Principal Component Analysis in Meteorology and Oceanography*; Elsevier: Amsterdam, The Netherlands, 1988.
73. Toumazou, V.; Crétaux, J.F. Using a Lanczos eigensolver in the computation of empirical orthogonal functions. *Mon. Month Rev.* **2001**, *129*, 1243–1250.
74. Betbeder, J.; Gond, V.; Frappart, F.; Baghdadi, N.; Briant, G.; Bartholomé, E. Mapping of Central Africa forested wetlands using remote sensing. *IEEE J. Sel. Top. Earth Observ. Remote Sens.* **2014**, *7*, 531–542.
75. Graham, S.T.; Famiglietti, J.S.; Maidment, D.R. 5-minute, 1/2° and 1° data sets of continental watersheds and river networks for use in regional and global hydrologic and climate system modeling studies. *Water Resour. Res.* **1999**, *35*, 583–587.
76. Cazenave, A.; Dominh, K.; Guinehut, S.; Berthier, E.; Llovel, W.; Ramillien, G.; Ablain, M. Larnicol, G. Sea level budget over 2003–2008: A reevaluation from GRACE space gravimetry, satellite altimetry and Argo. *Glob. Planet. Chang.* **2009**, *65*, 83–88.
77. Cazenave, A.; Llovel, W. Contemporary sea level rise. *Annu. Rev. Mar. Sci.* **2010**, *2*, 145–173.
78. Sappa, G.; Rossi, M. The North West Sahara Aquifer System: The complex management of a strategic transboundary resource. In Proceedings of the International Conferences "Transboundary Aquifers: Challenges and New Directions", Paris, France, 6–8 December 2010.
79. Gonçalvès, J.; Petersen, J.; Deschamps, P.; Hamelin, B.; Baba-Sy, O. Quantifying the modern recharge of the "fossil" Sahara aquifers. *Geophys. Res. Lett.* **2013**, *40*, 2673–2678.
80. Awange, J.L.; Anyah, R.; Agola, N.; Forootan, E.; Omondi, P. Potential impacts of climate and environmental change on the stored water of Lake Victoria Basin and economic implications. *Water Resour. Res.* **2013**, *49*, 8160–8173.
81. Awange, J.L.; Forootan, E.; Kusche, J.; Kiema, J.B.K.; Omondi, P.; Heck, B.; Fleming, K.; Ohanya, S.O.; Gonçalves, R.M. Understanding the decline of water storage across the Ramser-Lake Naivasha using satellite-based methods. *Adv. Water Resourc.* **2013**, *60*, 7–23.
82. Sutcliffe, J.V.; Parks, Y.P. *The Hydrology of the Nile*; IAHS Special Publication Number 5; IAHS Press: Oxfordshire, UK, 1999; p. 180.

Chapter 2:
Hydrological Modeling

Use of Radarsat-2 and Landsat TM Images for Spatial Parameterization of Manning's Roughness Coefficient in Hydraulic Modeling

Joseph Mtamba, Rogier van der Velde, Preksedis Ndomba, Zoltán Vekerdy and Felix Mtalo

Abstract: Vegetation resistance influences water flow in floodplains. Characterization of vegetation for hydraulic modeling includes the description of the spatial variability of vegetation type, height and density. In this research, we explored the use of dual polarized Radarsat-2 wide swath mode backscatter coefficients ($\sigma°$) and Landsat 5 TM to derive spatial hydraulic roughness. The spatial roughness parameterization included four steps: (i) land use classification from Landsat 5 TM; (ii) establishing a relationship between $\sigma°$ statistics and vegetation parameters; (iii) relative surface roughness (K_s) determination from Synthetic Aperture Radar (SAR) backscatter temporal variability; (iv) derivation of the spatial distribution of the spatial hydraulic roughness both from Manning's roughness coefficient look up table (LUT) and relative surface roughness. Hydraulic simulations were performed using the FLO-2D hydrodynamic model to evaluate model performance under three different hydraulic modeling simulations results with different Manning's coefficient parameterizations, which includes SWL1, SWL2 and SWL3. SWL1 is simulated water levels with optimum floodplain roughness (n_p) with channel roughness $n_c = 0.03$ m$^{-1/3}$/s; SWL2 is simulated water levels with calibrated values for both floodplain roughness $n_p = 0.65$ m$^{-1/3}$/s and channel roughness $n_c = 0.021$ m$^{-1/3}$/s; and SWL3 is simulated water levels with calibrated channel roughness n_c and spatial Manning's coefficients as derived with aid of relative surface roughness. The model performance was evaluated using Nash-Sutcliffe model efficiency coefficient (E) and coefficient of determination (R^2), based on water levels measured at a gauging station in the wetland. The overall performance of scenario SWL1 was characterized with E = 0.75 and R^2 = 0.95, which was improved in SWL2 to E = 0.95 and R^2 = 0.99. When spatially distributed Manning values derived from SAR relative surface values were parameterized in the model, the model also performed well and yielding E = 0.97 and R^2 = 0.98. Improved model performance using spatial roughness shows that spatial roughness parameterization can support flood modeling and provide better flood wave simulation over the inundated riparian areas equally as calibrated models.

Reprinted from *Remote Sens.* Cite as: Mtamba, J.; van der Velde, R.; Ndomba, P.; Zoltán, V.; Mtalo, F. Use of Radarsat-2 and Landsat TM Images for Spatial Parameterization of Manning's Roughness Coefficient in Hydraulic Modeling. *Remote Sens.* **2015**, *7*, 836-864.

1. Introduction

Riparian wetland ecosystems are flooded on multiple occasions every year, as floods are generated from excess rainfall in the catchment. The alteration of hydraulic regime in the river affects several ecological processes in the wetland floodplains, e.g., ecosystem productivity, species distribution and occurrence, nutrients and sediment dynamics [1–4]. To predict the impacts of flood wave on the floodplain, the hydraulic processes of the river have to be assessed at optimal temporal and spatial scales using hydraulic models. Several efforts have been made to develop hydraulic models that can simulate flow patterns and predict extreme flood levels in rivers and floodplains [5–10]. The development of an accurate and reliable hydraulic model that well describes surface water flow across a large wetland floodplain depends on topographic data and hydraulic roughness, *viz.* Manning's roughness of the floodplain [11,12]. Manning's roughness is one of the key variables of a hydraulic model; vegetation component plays a crucial role in the total resulting roughness, especially in vegetated floodplains. High hydraulic roughness values reduce wave celerity and rise flood depth [13,14]. The uncertainty of roughness parameterization leads to errors in water level estimation and affects the hydrograph characteristics.

The hydraulic roughness of a floodplain depends on factors, such as the type and structure of vegetation, the cross section area, obstructions in the channel and floodplain as well as the degree of meandering [15,16]. Different approaches have been used to quantify the hydraulic roughness across floodplains, e.g., using tabulated reference roughness coefficients, visually inspecting and comparing photographs of floodplain reaches and assigning the appropriate roughness values to similar floodplains [16] and using a momentum balance to define the hydraulic impacts of vegetation [11,17,18]. On small floodplain areas conventional ground surveys have been used [19]. Systematic detection, identification and assessment of riparian vegetation using conventional field sampling are often unachievable as these techniques are time-consuming and expensive.

Remote sensing techniques, e.g., vegetation mapping, terrestrial laser scanning and digital parallel photography, have been proven useful in determination of vegetation types, density and height [19–23]. Different techniques of remote sensing of hydraulic roughness have been discussed in detail by Forzieri *et al.* [24]. These techniques include classification-derived hydraulic maps and estimation of vegetation hydrodynamic properties. The vegetation maps are obtained from classification of digital satellite images, and a hydraulic roughness value is assigned to each vegetation type then, based on a look-up table (LUT). This approach does not provide spatial variability of hydraulic roughness within a vegetation class. The second method of estimating hydraulic roughness is based on *in-situ* defined biomechanical properties of vegetation, e.g., height, density, but this requires extended complementary ground surveys that can be feasible only at local scale [19,23]. The benefits of calibrating of hydraulic model using this approach are to support determination of spatial hydraulic indices more accurately for other studies that requires calibrated hydraulic model, e.g., sedimentation deposition assessment.

Previous studies in riparian environments using spectral remote sensing data have focused on classifying vegetation [25–30]. The applied LUTs were based on literature (e.g., [8,15,16]). Despite some encouraging results, the problem of the lack of within-class variability of the roughness coefficients was obvious, since differences in vegetation height and density within the class should have yielded different Manning's roughness values [31]. In fact, different arboreal patterns, bushes and meadows, can have similar spectral signatures and cannot be always differentiated in optical satellite imagery. There is, therefore, a need to define new approaches that can support characterizing the spatial variability of the hydraulic roughness for large floodplains based on vegetation biophysical characteristics.

Synthetic Aperture Radar (SAR) is a remote sensing technique that enables the construction of high-resolution images from active microwave observations. In essence, these active microwave instruments quantify the apparent roughness of the target surface that is largely defined by the geometry and to lesser extent by the moisture content of the surface. As such, SAR observations have been used in the past for soil moisture monitoring [32,33], flood mapping [34–36] and vegetation mapping [37–43]. The SAR backscatter is controlled by vegetation structure, surface geometry and dielectric properties of the ground targets [44]. The dielectric properties are influenced by soil moisture and vegetation water content [32,44]. Microwave radiation penetrates vegetation canopies interacting with the canopy scatterers, *i.e.*, leaves, twigs, branches, and the trunk [45]. Cross-polarization ratio can be used to determine relative surface roughness (K_s) variations within the vegetation class, identifying different areas with varying vegetation densities and types. Estimation of vegetation structure depends on multi-polarization scattering of SAR signals. Mattia *et al.* [46] studied the effect of surface roughness on multi frequency polarimetric L and C band data and found that depolarization ratio can be used to discriminate different types of vegetation canopies especially forest and non-forest areas. The depolarization ratio is equivalent to cross-polarization ratio in other literature. The cross-polarization ratio is characterized by surface and volume scattering mechanism at the vegetation canopy surface, while co-polarization ratio is characterized by surface and volume scattering (including tree trunk and branches mechanism). A study on vegetation canopy cover by Mathiew *et al.* [22] on savannah woody vegetation using C-band Quad- Polarization Fine Beam RADARSAT-2 imagery found that dry period HV polarization can be effectively used to predict structural metrics. The added advantage of the use of SAR in this research is to derive relative surface roughness that is correlated to biophysical properties of vegetation to support spatial variability of the hydraulic roughness within vegetation types.

The main objective of this paper is to investigate the possible advantage of spatially distributed hydraulic roughness parameterization for hydraulic modeling of floodplains in comparison to a single hydraulic roughness value for the whole floodplain and a set of hydraulic roughness values defined per vegetation type.

2. Material

2.1. Study Area

The Mara wetland (Figure 1) is one of the largest tropical wetland systems in East Africa, which has drawn attention for conservation. To understand hydrology and hydrodynamic processes in the wetland floodplain reliable models need to be developed. Mara wetland receives most of its water from the Mara River, originating from the Mau escarpment forest, passing through vast low-lying plains including the transboundary Masai Mara/Serengeti national parks, and releasing water into Lake Victoria. Floods typically occur during the months of November/December and April/May with water depths varying from 0.5 to 2 m in the floodplain. The riverbeds of the main Mara River and its tributaries are fairly well defined at their upper reaches, but become increasingly meandering as they approach the confluence with the Tigithe River. The decreasing slope reduces the downstream river velocity and increases deposition, a characteristic for most fluvial systems.

58

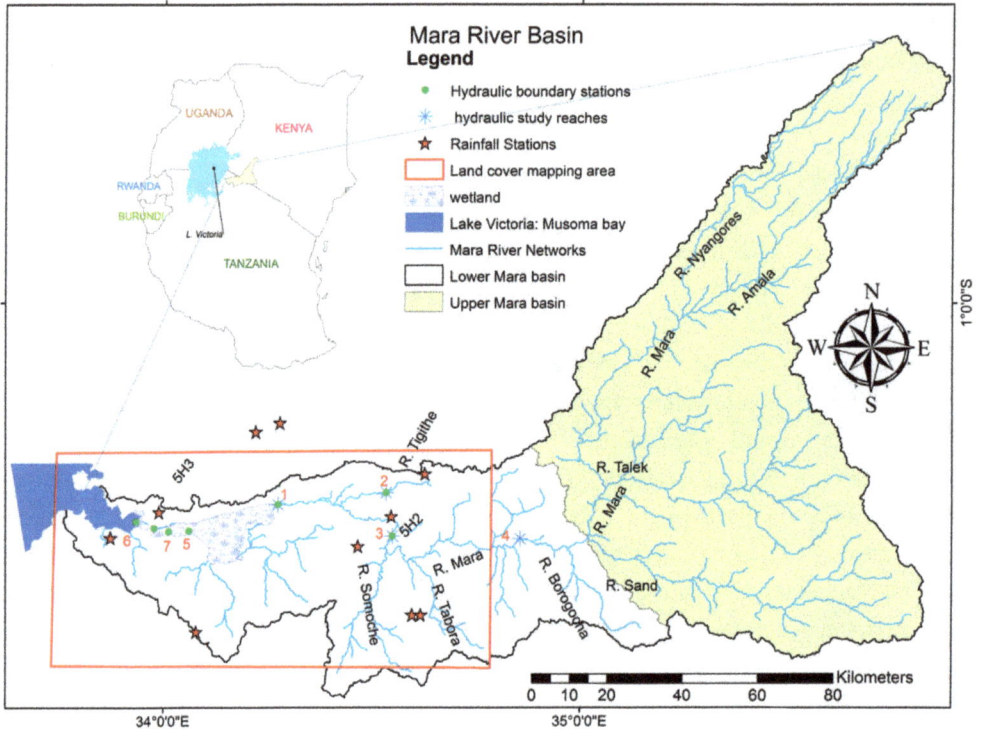

Figure 1. Location of the study area (rectangle) within the lower Mara River basin.

The Lower Mara Basin has ten rain-gauging stations and five river water level measuring stations at Bisarwi (1), Tigithe bridge (2), Mara mine (5H2) (3), Kogatende (4) and Kirumi ferry (5H3). The development of rating curves for stations (1)–(4) is described in Mtamba *et al.* [47]. The Mara mines and Kirumi ferry gauging stations have been in operation since 1969. The other stations were specifically installed to support the flood modeling and understanding reach-scale hydraulics in 2011–2013 [47,48]. The figure below shows historical gauging stations and temporary stations for hydraulic study (1–4). Mara mine (5H2) falls in both categories.

2.2. Field Measurements and Remote Sensing Data

2.2.1. Rainfall and River Discharge Measurements

Rainfall data was obtained from the Lake Victoria Basin Office at Musoma office. The rainfall gauging stations are equipped with standard rain gauges. The measurement accuracy is about 0.5 mm. The rainfall data from ten gauging stations in the lower Mara Basin were analyzed and averaged using Thiessen polygon approach to obtain daily rainfall in mm/day.

Discharge at 5H2 gauge station and aerial rainfall in Lower Mara basin

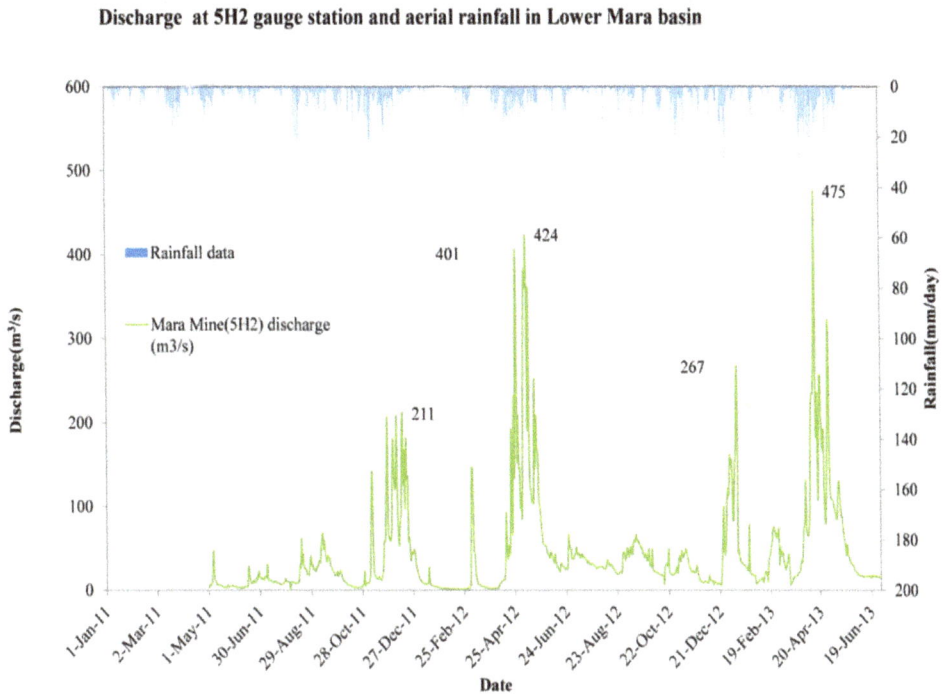

Figure 2. Discharge and areal rainfall at Mara mine gauging station. The values show four flood events during November/December and April/May.

The gauging stations in the study area were equipped with staff gauges. Mara Mine and Tigithe bridge gauging station is located upstream of the wetland floodplain. The stations are equipped with standard gauge staff plates of 1m for water level measurements in the river. The measurements are taken manually twice a day at 9:00 and 15:00 h. The daily levels records are averaged to obtain average daily values. These water levels are converted to river discharge using a calibrated rating curve [47]. The two-year flow hydrograph for Mara Mine (Figure 2) highlights two distinct flood periods, namely in November/December and April/May. During these events a maximum discharge of 475 m^3·s^{-1} was recorded. The floods are generated by surface runoff in response to convective rain events of several mm per day across the Mara River Basin.

2.2.2. Vegetation Characterization

A detailed vegetation survey was done in October 2010 and September 2013. During the 2010 field campaign, the vegetation data collected in each site included type of vegetation/land use (e.g., grassland, dense green papyrus and shrub/thicket). It was not possible to undertake the collection of the wetland ground reference data concurrently with the acquisition dates of the images. However, information from locals suggests that there were no significant changes on land cover between the image acquisition and the fieldwork dates. It was considered that the wetland floodplain grassland changes significantly between wet season and dry season due to rainfall and grazing by herds of cattle. The main objective of the 2010 field visit was to collect data to support wetland vegetation

60

classification for the study area. 300 ground truth points were recorded and classified in the field (Table 1). Figure 3 provides sample photos of vegetation in the area.

Table 1. Field vegetation classification in 2010.

S/No	Land Cover Class	Land Cover Characteristics
1	Water	Open water: Water without plant cover (in rivers, lagoons, oxbow lakes, ponds *etc.*)
2	Swamps	Swamps: Sediment deposited areas, burned papyrus vegetation, sediment laden waters
3	Partially submerged vegetation	Sparse green vegetation: vegetation which include papyrus, grassland with water background or partially submerged
4	Papyrus/thicket	Dense green vegetation: Dark green vegetation which includes green papyrus, trees, bushes
5	Regenerated papyrus/shrub	Very sparse green vegetation: Regenerated vegetation which include papyrus, grassland with water background or partially submerged
6	Eichorrhoea crassipes/grassland	Grassland: Eichoria crassipes vegetation, lush grassland,
7	Agriculture/bare land	Bare land: open Bare land, settlement, houses, open farms, dry grass, clouds
	Total	

During the 2013 field visit, the objective was to collect independent data sets for vegetation attributes to derive Plant Area Index (PAI) for the herbaceous pattern vegetation (papyrus and reeds) and Fractional Vegetation Cover (FVC) for the arboreal and shrub classes (shrub/thicket and floodplain forest).

2.2.3. Landsat 5 TM Imagery and Preprocessing

Landsat Thematic Mapper (TM) is a multispectral scanning radiometer that was carried onboard Landsat 4 and 5. The TM sensors have provided nearly continuous data in the period of 1 March 1984–05 June 2013. TM data was collected in seven bands, which cover the visible, near-infrared, shortwave, and thermal infrared spectral bands of the electromagnetic spectrum. All bands have a resolution of 30 m except the thermal band that has a resolution of 120 m, with a swath width of 185 km. The thermal band was excluded during the present analysis due to its coarser spatial resolution and the different physical processes involved in the observation. Landsat provides data acquisition over the Earth's landmass with a repeat cycle of 16 days.

Figure 3. Photos showing different land cover types in the area.

From 2009 to 2013, there are several Landsat images present in USGS Global Visualization Viewer (GLOVIS) database (http://glovis.usgs.gov). Preliminary observation was used to select cloud free images during non-flooded seasons. Normalized Difference Vegetation Index (NDVI) for each image was determined and used to select images that can be used to describe vegetation classes in the study area. The Landsat TM acquired on 11 June 2009 was selected for vegetation mapping in the Lower Mara Basin. The image was atmospherically corrected using the ATCOR module in the ERDAS Imagine software and geometrically transformed with tie points onto a topographic map of the area. A subset of the area was extracted for the further analysis.

2.2.4. Radarsat 2 Imagery and Preprocessing

The RADARSAT-2 satellite was launched on 14 December 2007 by the Canadian Space Agency (CSA). It is carrying a SAR sensor operating at a center frequency of 5.405 GHz (C-band). The system is capable of acquiring backscatter observations ($\sigma°$) in various imaging modes including single (HH or VV) polarization; dual co- and cross-polarization combinations (HH/HV, VV/VH) with several image swaths. The imagery used for the present study is comprised of a series of 12 VV/VH polarized RADARSAT-2 wide swath mode images from April 2011 to August 2012. The SAR images were received as preprocessed level-1 SAR Georefferenced Fine Product (1SGF) with 25 m × 28 m resolution for a 150 km × 150 km area covering a view angle range of 20–45 degrees.

The Next ESA SAR Toolbox (NEST) was utilized to process the Radarsat-2 data sets to calibrated backscatter (σ°) products through application of the Range Doppler terrain correction utility. A median filter with a 5 × 5 kernel size was applied to the σ° data sets to suppress the speckle noise inherent to the SAR data. After filtering, the images were reprojected to UTM zone 36 WGS84 coordinate system subsetted for the area of interest and automatically co registered.

3. Methods

3.1. Overall Framework

Developing large-scale hydraulic models requires appropriate data to describe flow resistance in the model domain. There is a need to investigate how friction component can be parameterized to improve model performance hence to minimize the efforts for calibration and validation of hydraulic models. The approach used in this study includes derivation of Manning's coefficients derived from the spectral and radar images by combining Manning's LUT and relative surface roughness. Landsat imagery was used to derive vegetation classes to which minimum and maximum Manning's roughness coefficients were assigned from literature. The SAR imagery was used to determine K_s from cross-polarization ratio. The spatial distribution of the hydraulic roughness values was then determined from combining K_s maps and Manning's LUT values. The roughness was averaged per vegetation class to obtain mean roughness values per vegetation class. A FLO-2D hydraulic model was set up to check the improvement of model performance when different scenarios of roughness parameterization were implemented. The overall framework is summarized in Figure 4 below.

3.2. PAI and FVC Retrieval

The PAI and FVC are ratios describing the fraction of area covered by vegetation per unit area of ground surface. The higher the PAI or the FVC the higher the vegetation density is, hence the higher is the hydraulic roughness. PAI was determined from 35 sites for herbaceous vegetation using the gap fraction approach. In papyrus and reed vegetation, four to five vertical photographs were taken below the canopy at each site; an approach of determining the vegetation gap fraction. This method is based on simplified models of light transmission into the canopy. A variety of approaches for estimating PAI using direct and indirect methods were presented by Bréda [49], Jonckheere et al. [50] and Weiss et al. [51]. Vegetation gap fraction (G) is defined as the fraction of sky seen from below the canopy [49–52]. Gap fraction is related to PAI $(PAI = 1 - G)$. Images were processed using a segmentation algorithm implemented in the ILWIS software. The segmentation process was developed to classify open sky pixels and cloudy sky pixels according to their chromatic values. The segmentation process is simply based on the pixels intensity assuming that vegetation elements appear darker than the sky. Figure 5a shows a sample image of vertical photographs was taken below the vegetation canopy. Figure 5b shows the results of image after segmentation process, dark areas are areas covered by vegetation and white areas are clasified as open sky. Areal average PAI was calculated for each sampled location. FVC was determined for 10 plots of 100 m × 100 m in the arboreal and shrub land cover classes. The crown base diameters of each vegetation type were measured with measuring tape at each plot. The FVC was calculated as the fraction of total surface area covered by arboreal and shrub vegetation. Results are summarized in Table 2.

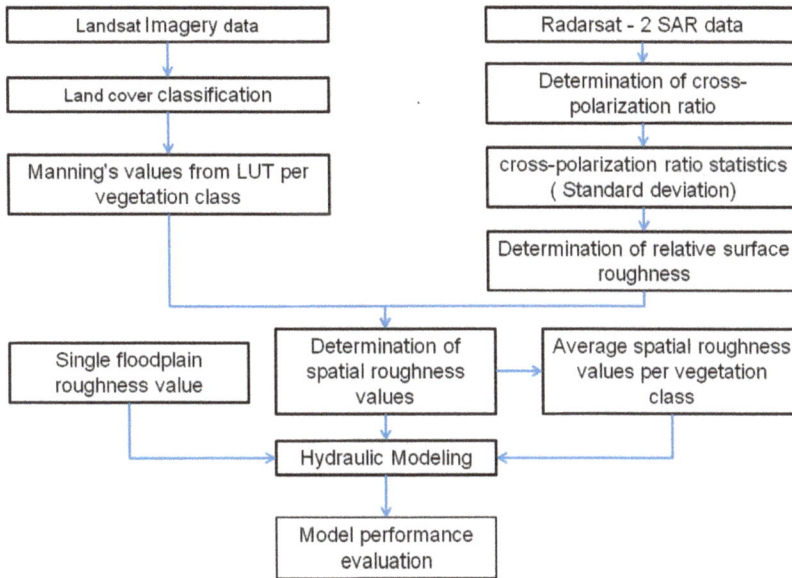

Figure 4. Overall framework of research approach.

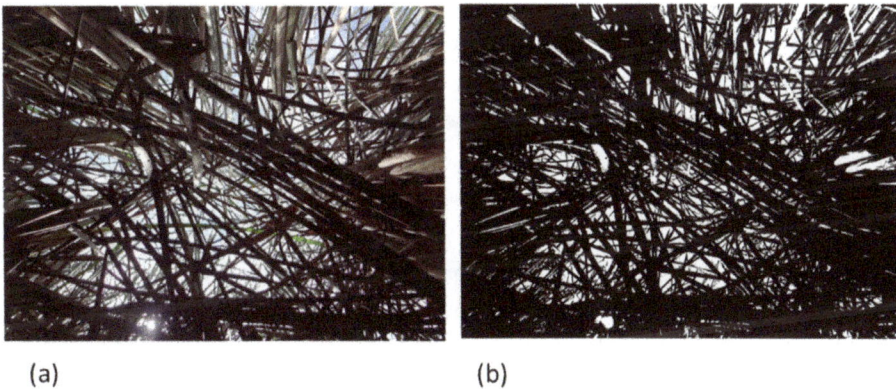

(a)   (b)

Figure 5. Canopy photos at a reed site **(a)** Vertical photo taken below the canopy **(b)** image after segmentation.

Table 2. Plant Area Index (PAI) and Fractional Vegetation Cover (FVC) measured *in situ* at different sites.

	Vegetation Type	No of Samples	PAI (1) and FVC (2)			
			Min	Max	Mean	Standard Deviation
1	Papyrus	15	0.335	0.867	0.710	0.121
	Reeds	20	0.223	0.871	0.725	0.145
2	Shrub/Thicket	10	0.401	0.860	0.674	0.140

3.3. Vegetation Classification from Landsat TM Imagery

Supervised classification was done to determine vegetation types in the study area. Visual interpretation of the Landsat TM true color composite bands (321) and combination of NDVI was used to assist in the collection of the training set for classification. Training sites for each vegetation class were identified and a classification scheme prepared for classification process. Maximum likelihood classification was done using the ERDAS Imagine 2013 software, resulting in 7 vegetation classes. The result was cleaned by applying a median filter of 5 × 5 pixels to remove isolated pixels. The accuracy assessment was done as recommended by Congalton and Green [53], using 300 data points from the field survey, which were not used in the training process.

3.4. Vegetation Class Characterization Based on Backscatter Statistics

3.4.1. SAR Images

It is well understood that the backscatter ($\sigma°$) is affected by the dielectric properties of the scattered surface or volume (in our case the wetland) and the scattering geometry. Due to the large difference between the dielectric constants of water and the other materials, the backscatter is determined by the water content in as well as on the vegetation and the soil, furthermore, the vegetation morphology and the land surface roughness. To investigate the vegetation dynamics in the study area and period visually, an RGB color composite for Vertically transmitted and Vertically received signals (VV polarization) and Vertically transmitted and Horizontally received signals (VH polarization) (Green—VH polarization; Red and Blue—VV polarization) was made each of the available twelve images in total (Figure 6a–l). It can be observed that within the wetland, the greenish colors represent small differences between VV and VH backscatter signals, which indicate the presence of vegetation [22]. The reddish color highlights large VV *versus* VH backscatter, associated with bare land. The black areas refer to low VV and VH backscatter, related to smooth surfaces, usually open water bodies and flooded areas. In order to avoid the bare lands and the effect of temporary flooding on the backscatter of vegetated areas, five SAR scenes with high level of reddish and black coloration on floodplain were excluded from the further analysis. Rainfall and river discharge data has been used also to confirm flooded situations in the image selection.

From the visual interpretation of these SAR images, backscatter variations in densely vegetated, deep-rooted areas are more stable than in less dense areas (e.g., grass) during different seasons. In densely vegetated areas the backscatter response is more or less constant during all seasons because in the floodplain, they can maintain their water content even during the dry season. Floods and rainfall events are also a major cause of high backscatter variation in flood plains as can be seen in the images acquired on the dates of 12 March 2011 and 18 May 2012.

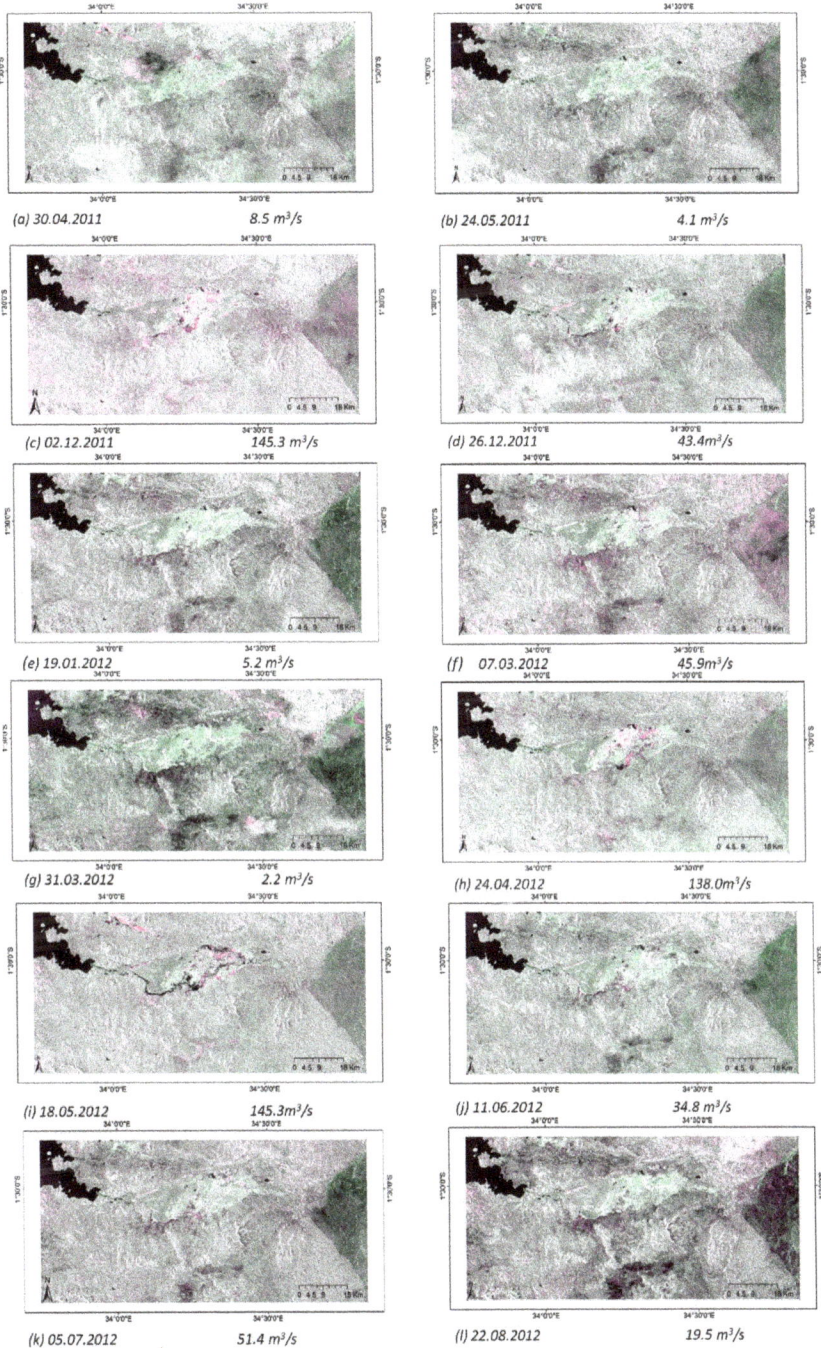

Figure 6. Sub figures **a–l** above shows RGB color composite images for 12 Radarsat-2 SAR images acquired over the Lower Mara Basin. Image aquisation dates and average daily river discharge values are shown in the bottom of each image. RGB color composite imageries are attributting to VV, VH and VV polarization for Red, Green and Blue channels respectively.

66

Figure 6a–l shows RGB color composite for 12 SAR imagery used in this study and river discharge data for the Mara Mine gauging station. These data were used to aid selection of images for the subsequent analysis of vegetation attributes. Initial temporal variability assessment was done on all images and it was observed that vegetated areas affected by flood cannot be separated clearly from grassland areas. Therefore uniformity of rainfall and flood spread within the images was used as the key criteria of selection of images. The minimum number of images for multitemporal analysis was also limited as recommended by Maghsoudi *et al.*, (2011) [54]. From Figure 6, SAR scenes of 2 December 2011, 7 March 2012, 24 April 2012, 18 May 2012 and 22 August 2012 were excluded from the vegetation analysis because of the influence of floods and rainfall that caused high temporal variability of the backscatter.

3.4.2. Backscatter Characterization and Statistics

The estimators of temporal variability of multitemporal SAR data sets include standard deviation, normalized standard deviation, saturation, maximum-minimum ratio of σ° in dB [55]. In this study, we used mean, maximum and standard deviation of the seven selected SAR images. The mean VV and VH polarization does not provide clear separation between vegetation classes. In Figure 7a,b, the standard deviation identifies the flooded vegetation formations and monitor non-flooded vegetation phenology. The Figure 7a,b for standard deviation maps of VV and VH polarized σ° respectively shows a separation between densely vegetated areas (papyrus, floodplain forest) and other less vegetated areas. Standard deviation of σ° measures the variability encountered in the polarimetric SAR images. In the agricultural fields, the temporal variability is high because of the influence of soil moisture changes, vegetation phenology and growth, cultivation practice (e.g. grazing).

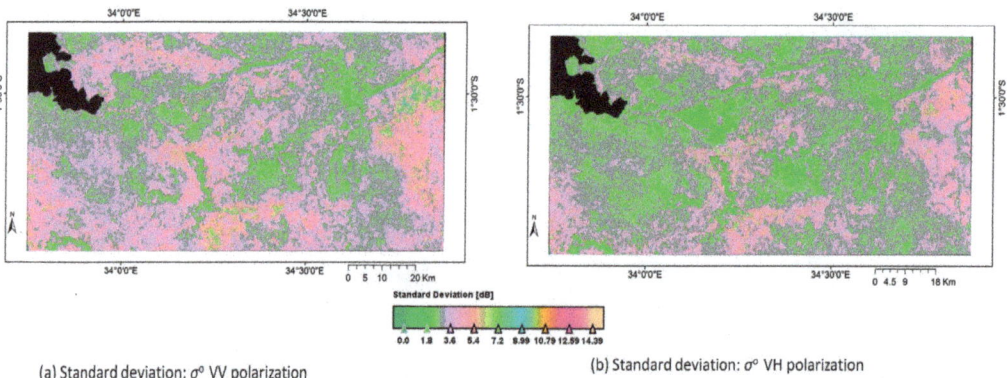

(a) Standard deviation: σ° VV polarization

(b) Standard deviation: σ° VH polarization

Figure 7. Standard deviation of the backscatter (in dB) for (**a**) Vertically transmitted and Vertically received signals (VV polarization) and (**b**) Vertically transmitted and Horizontally received signals (VH polarization) on the study area characterizing temporal variability of the backscatter. Areas with higher vegetation have low standard deviation than areas with shorter vegetation e.g., grasslands.

High mean σ° and low standard deviation was observed in aquatic vegetation and forest/thicket/papyrus vegetation areas. In floodplain grassland and less dense vegetation intermediate σ° and high standard deviation were observed. Agricultural/grassland areas outside the floodplain are characterized by low σ° and high standard deviation. Figure 7 shows the temporal variability statistic (standard deviation) for the analyzed SAR imagery.

Figure 8 shows temporal variation of backscatter statistic on vegetation types in some specific selected sites. It can be observed that standard deviation provides a clearer separation between vegetation types than the mean backscatter values.

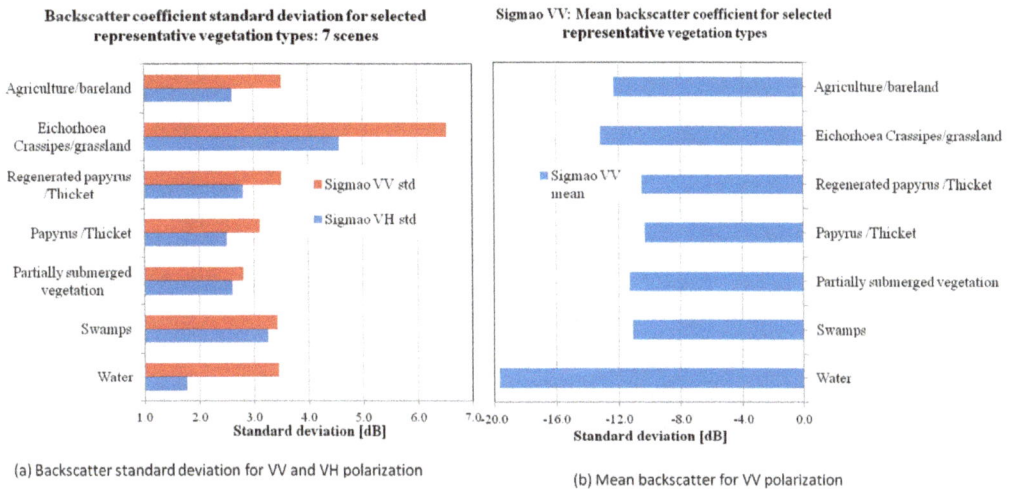

(a) Backscatter standard deviation for VV and VH polarization

(b) Mean backscatter for VV polarization

Figure 8. Backscatter statistics for different vegetation types for (**a**) Standard deviation (std) for VV and VH polarization (**b**) Mean for VV polarization. Sigmao represents σ°. Sigmao VV std is standard deviation of VV polarisation; Sigmao VV mean is average of VV polarisation and Sigmao VH std is standard deviation of VH polarization for the selected seven images.

3.5. Determination of the Spatial Distribution of the Hydraulic Roughness

3.5.1. Concept

Hydraulic vegetation roughness refers to the resistance force exerted by vegetation on water flowing over or through it [23]. Rough vegetation reduces water flow velocity and leads to higher water levels and thus increases flood risks. Height and density of submerged vegetation and density of emergent vegetation are the key characteristics from which roughness parameters in hydraulic models are derived. The parameter describing emergent vegetation, such as forest, is vegetation density [56]. Hydraulic vegetation density is the sum of the projected plant areas in the direction of the flow per unit volume. The parameters describing submerged vegetation, such as grassland, reed, or herbs are vegetation height and density [57]. The spatial distribution of these vegetation characteristics is an essential input for hydrodynamic models.

SAR data was investigated for quantitative assessment of floodplain relative hydraulic roughness, which is related to vegetation height and density. Grassland is hydraulically smooth, forests and dense shrubs are hydraulically rough [15]. This concept is analogous where grassland had high temporal variability of backscatter and forests have low temporal variability of VH polarized $\sigma°$ which describes volume scattering mechanism (dB·m^{-3}). In addition variability of VV polarized $\sigma°$ can describe surface scattering mechanism (dB·m^{-2}). The variability of the ratio of volume to surface backscattering can describe vegetation density differences (m^{-1}) within a vegetation type. This ratio is described by the cross-polarization ratio and its exponential inversion provides values at a scale of 0 to 1, describing smooth to rough surface within the vegetation class, further on referred to a relative surface roughness.

3.5.2. Implementation

Cross-polarization ratio of the seven SAR images and their descriptive statistical parameters were calculated. It was observed that the standard deviation of the cross-polarization ratio can be used to determine relative surface roughness variations within the vegetation class. The inversion was based on the exponential Markov Random Fields (MRF) model. The MRF is a global model uniquely determined by a local statistical description of a single image or multispectral images for image pattern analysis, texture modeling and image classification [58–64]. In this paper, we use MRF model to determine relative surface roughness from standard deviation of cross-polarized ratio of series of SAR imagery. The empirical inversion technique used for series of cross-polarization ratio images based on Equation (1). The equation assumes that the relative roughness parameters follow the Markovian probability density function (pdf). Equation (1) was used to determine the relative surface roughness map of the study area.

$$K_s = \frac{1}{Z} e^{-\rho std} \tag{1}$$

$$\rho_{std} = \sqrt{\frac{1}{N} \sum_{i=1}^{N} (\rho_i - \mu)^2} \tag{2}$$

$$\rho_i = \frac{\sigma^0_{VH_i}}{\sigma^0_{VV_i}} \tag{3}$$

where;

K_s = relative surface roughness [-]

z = correction factor [-], for this case considered to be 1 for the Markovian pdf

ρ_{std} = standard deviation of cross-polarized ratio of series of SAR imagery [dB]

N = number of SAR imagery used in analysis

ρ_i = cross-polarization ratio of SAR imagery i

μ = mean of cross polarization ratio of SAR imagery used in analysis

$\sigma^0_{VH_i}$ = VH polarization of SAR imagery i [dB]

$\sigma^0_{VV_i}$ = VV polarization of SAR imagery i [dB]

3.5.3. Hydraulic Roughness Map

The relative roughness was used to aid the spatial parameterization of the Manning's roughness within the vegetation classes. Numerous researchers give guidance on selection of Manning's values for various vegetation classes [6,8,12,15,65]. After a careful review and investigation of Manning's values by comparing the study site vegetation characteristics and those given in the literature, the value ranges shown in Table 3 were adopted for the study area. These values should not be considered as the actual, since their determination is based on vegetation classes and Manning's values found in the literature. Table 3 is LUT for Manning's coefficient ranges for each vegetation class.

Table 3. Manning's roughness ranges for different vegetation classes referred from Chow [15] and O'Brien [8].

		a	b	c
No	Vegetation Class	$n_{min,c}$ (m$^{-1/3}$/s)	$n_{ave,c}$ (m$^{-1/3}$/s)	$n_{max,c}$ (m$^{-1/3}$/s)
1	Water	0.02	0.03	0.085
2	Swamps	0.09	0.2	0.34
3	Partially submerged vegetation	0.17	0.3	0.48
4	Papyrus/ Thicket	0.17	0.3	0.8
5	Regenerated papyrus/Shrub	0.2	0.4	0.4
6	Eichorrhoea Crassipes/ Grassland	0.2	0.3	0.3
7	Agriculture/ bare soil	0.1	0.2	0.3

Spatial hydraulic roughness was calculated by Equation (4) below. This equation assumes a linear increase of Manning's coefficient with relative surface roughness within the vegetation class.

$$n_{i,c} = n_{min,c} + (n_{max,c} - n_{min,c}) \frac{(K_s)_{i,c} - (K_s)_{min,c}}{(K_s)_{max,c} - (K_s)_{min,c}} \tag{4}$$

where;

$n_{i,c}$ = Manning's roughness value for a cell within a vegetation class (m$^{-1/3}$/s)
$n_{min,c}$ = minimum Manning's roughness value for a cell within a vegetation class (m$^{-1/3}$/s)
$n_{max,c}$ = maximum Manning's roughness value for a cell within a vegetation class (m$^{-1/3}$/s)
$(K_s)_{i,c}$ = relative surface roughness value for a cell within a vegetation class (-)
$(K_s)_{min,c}$ = minimum relative surface roughness value for a cell within a vegetation class (-)
$(K_s)_{max,c}$ = maximum relative surface roughness value for a cell within a vegetation class (-)

3.6. Hydraulic Modeling

The derived hydraulic parameters were used in the FLO-2D River modeling software. FLO-2D model is a two-dimensional (2D) hydrodynamic model developed by FLO-2D Software *Inc*—Arizona, USA (http://www.flo-2d.com). Flood routing in two dimensions is handled through a numerical integration of the equations of motion and volume conservation within each cell in eight directions. The simple volume conservation governing equations are written as general constitutive fluid equations, which include the continuity equation and the dynamic wave momentum equation [8].

The 2D hydrodynamic model was used to develop a floodplain inundation model for a 120 km river reach at the outlet of Mara river basin (Figure 9). The main channel was represented by a one-dimensional linear channel based on fifty surveyed river cross-sections and the floodplain topography was represented by a calibrated digital terrain model based on Advanced Space borne Thermal Emission and Reflection Radiometer Global Digital Elevation Model version 2 (ASTERDEM). The ASTER DEM was pre-processed to remove noise by using 2D Kalman filtering techniques [66,67]. The floodplain model domain was discretized into 100 m by 100 m grid cells. The upstream model boundary was represented as a daily time-flow boundary condition at Mara Mine (5H2) station (3) and at Tigithe Bridge (2) while the downstream boundary condition as the levels of Lake Victoria (6). An internal river gauge at point 5 was used for model performance assessment (Figure 1).

The model was calibrated by determining optimum floodplain n_p and channel n_c roughness parameters. The optimum floodplain plain roughness parameter was found by setting river channel roughness at 0.03 $m^{-1/3}$/s. The optimum value of river channel roughness was obtained from instream sediment characteristics using median sediment particle diameter in meter and applying strickler formula developed in 1923 [68] and then adjusted accordingly by factors proposed by Arcrement and Schneider [16]. After which the calibrated river channel roughness parameter was determined through sensitivity analysis by varying its parameters while maintaining floodplain roughness at optimized value and evaluating the model performance. The model performance was evaluated at the internal gauging station using daily water level records. Model performance was evaluated based on the fit between simulated and observed water levels, using the Root Mean Square Error (RMSE), the Nash-Sutcliffe model efficiency coefficient (E) and the index of agreement (d) [69,70]. Different parameterizations of surface roughness were evaluated by comparing model performance. The three water surface simulations from the hydraulic model were analyzed to find out if use of spartial hydraulic roughness improves the modeling results hence a better model. The model results analyzed to show model improvement included; (1) Optimum floodplain roughness (n_p) with channel roughness n_c = 0.03 $m^{-1/3}$/s; (2) calibrated floodplain roughness (n_p) and channel roughness n_c; and (3) calibrated channel roughness n_c and spatial Manning's coefficients as derived with aid of relative surface roughness. A six-month period was simulated from 30 January 2004 to 30 June 2012 including a major flood event to capture model behavior during floods.

4. Results

4.1. Riparian Vegetation Mapping

The vegetation mapping accuracy assessment results are shown in Table 4. The overall accuracy for the seven classes was 69% with a Kappa Coefficient of 0.624. The resulting land cover map of the study area is presented in Figure 9. Only a relatively low overall accuracy could be achieved, due to the mixed character of some vegetation classes (Table 4). Especially the papyrus, the most dominant vegetation type can be mixed with other types. At some areas of the floodplain, dense and thick papyrus is the dominant cover (Papyrus/thicket), mapped with a relatively high producers accuracy. In the upper reach, dense thicket and shrub areas could not be separated well from the papyrus and regenerated papyrus, due to their spectral signature resemblance and patchy distribution. In Figure 9, it can be seen that the dense papyrus vegetation is dominantly present near the main river channel. The class of partially submerged vegetation and papyrus has lower mapping accuracies due to its patchy distribution, especially in the upstream part of the wetland. Swampy areas are areas dominated by frequently flooded grasslands, so during the dry season these areas characterized by healthy grass, which could be mapped with the lowest accuracies among the natural wetland vegetation classes. Vegetation class 7 (agriculture/bare land) was virtually having the highest classification accuracy, but this is represented with a very small proportion in the wetland. Due to the above-described limitations, we accepted this result, since the classification accuracy of all classes was above 60%. Some other methods, such as data fusion between optical and SAR imagery may have produced better results, since SAR can provide structural attributes of different vegetation patterns [71]. Understanding the limitations, these results were used in the roughness parameterization.

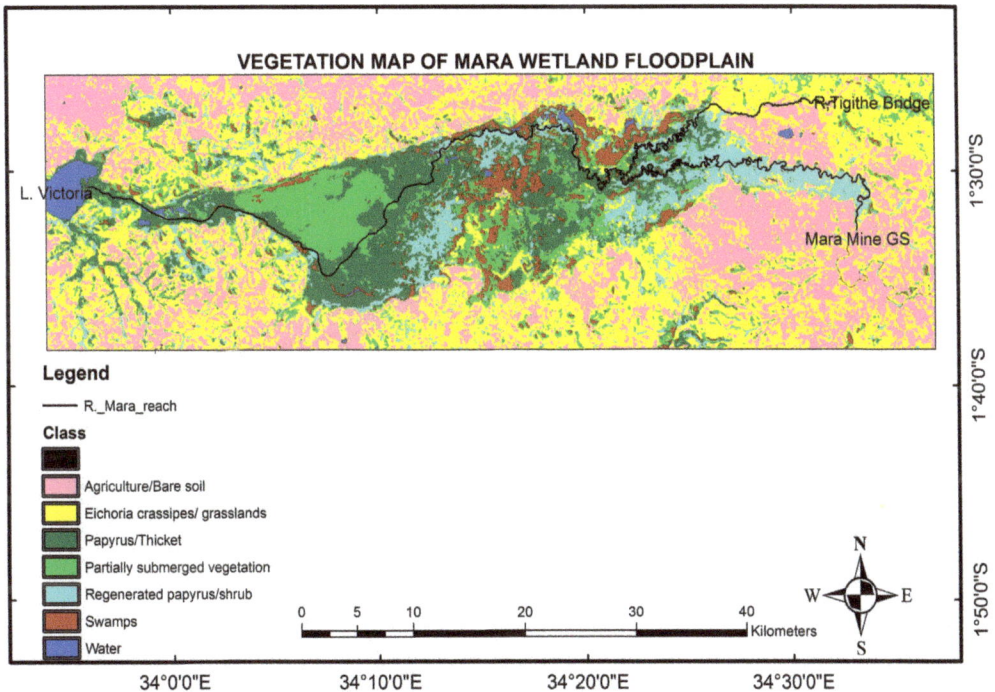

Figure 9. Vegetation map of Mara wetland floodplain showing seven classes of vegetation types.

72

Table 4. Supervised Classification Accuracy Assessments.

S/No	Class Name	Reference	Classified	Correct	Producers Accuracy (%)	User Accuracy (%)	Kappa
1	Water	11	9	6	54.6	66.67	0.654
2	Swamps	31	34	21	67.7	61.76	0.574
3	Partially submerged vegetation	61	58	38	62.0	65.52	0.567
4	Papyrus/thicket	75	78	54	72.0	69.23	0.589
5	Regenerated papyrus/shrub	56	52	37	66.0	71.15	0.645
6	Eichorhoea crassipes/Grassland	37	43	29	78.38	67.44	0.628
7	Agriculture/Bareland	29	26	22	75.86	84.62	0.829
	TOTAL	300	300	207	Average		0.624

Overall Classification Accuracy = 69.00%, Kappa Coefficient = 0.624.

4.1. Relative Surface Roughness

Relative surface roughness results shows that densely vegetated areas have higher values than less dense areas. For example densely vegetated papyrus vegetation and savannah grasslands have K_s of the same order. These indicate that the K_s contains vegetation characteristics that can be used for spatial hydraulic roughness parameterization. Figure 10 below shows map of K_s at a scale of 0 to 1.

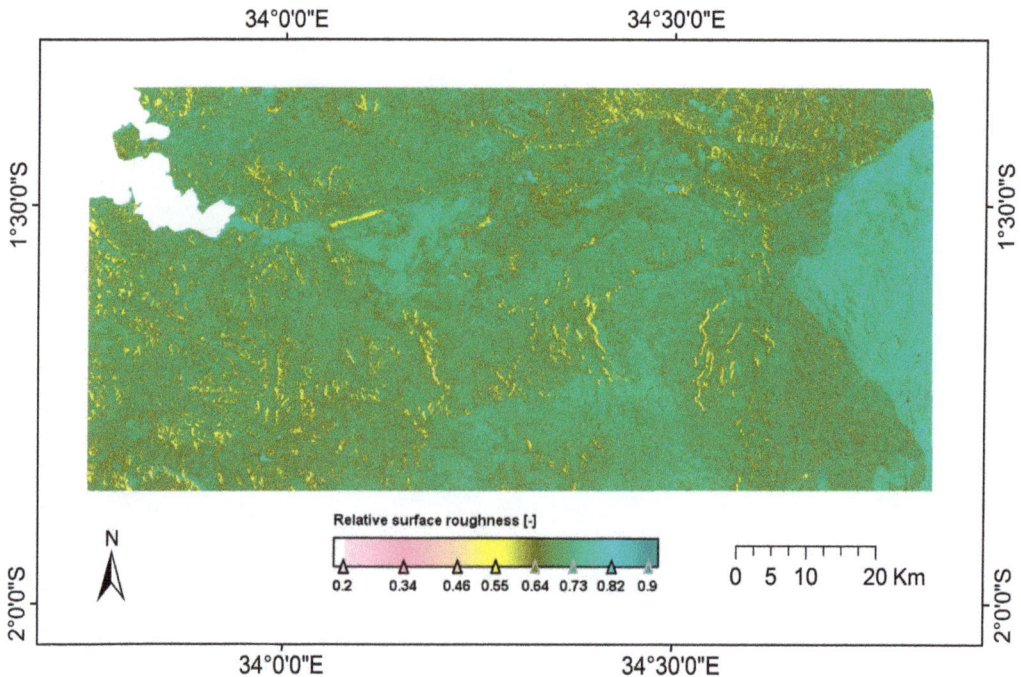

Figure 10. Relative surface roughness (K_s) of the study area on a scale of 0 to 1.

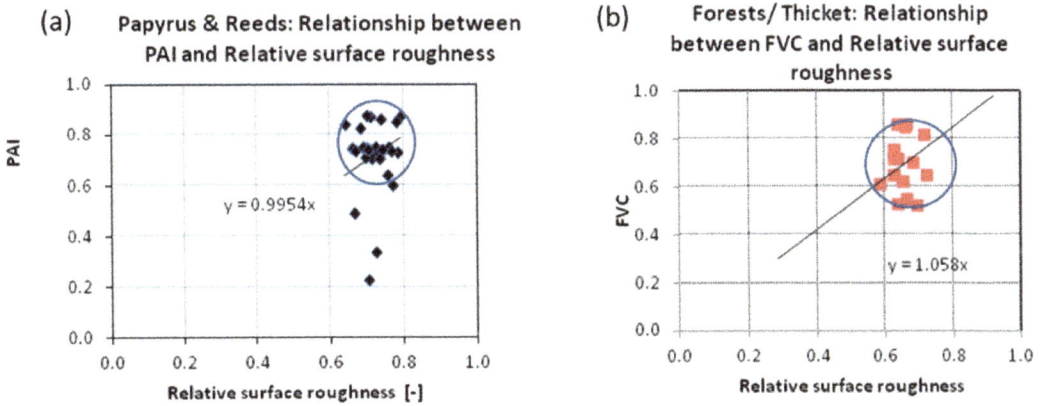

Figure 11. Correlation between Plant Area Index (PAI) and Fractional Vegetation Cover (FVC) against relative surface roughness. The correlation coefficient for plots (**a**), (**b**), are 0.03, 0.08, respectively.

Direct field verification of Manning's roughness values was practically not feasible. Nevertheless, the derived remote sensing based relative surface roughness was correlated to PAI and FVC to check if these vegetation attributes are reflected in the SAR relative surface roughness values. The relative roughness and Manning's roughness were extracted at locations where PAI and FVC were determined on the site. These values were plotted to determine whether they are correlated (Figure 11). Within vegetation sampled, PAI or FVC *versus* relative roughness has an agreement in the shape but week positive linear correlation. These show that relative surface roughness may not be directly related to PAI or FVC (Figure 11a,b). It can support determination of vegetation attributes variations within the vegetation type that is directly related to spatial hydraulic roughness. From Figure 11 both vegetation types have relative surface roughness and Manning's roughness within a range of 0.6–0.8. These results does not prove beyond reasonable doubt that there exists a relationship between PAI or FVC and K_s from the few data points collected.

The plots show good agreement in data spread but low correlation. The correlation between PAI and FVC against relative surface roughness was obtained to be 0.03 and 0.08, respectively, so it is not possible to relate K_s to these parameters, which are easy to define in the field. The major limitation is that the backscatter is defined by the surface and the whole volume of the vegetation canopy, whilst the effective hydraulic roughness on the forested floodplains is defined by the lower part of the vegetation, since water levels are usually under the crown base height.

4.2. Roughness Map

There are overlaps of relative surface roughness ranges between land cover classes. The minimum relative roughness in this case is between 0.09 in swamps to 0.432 in water bodies (with waves), other classes fall within this range. The maximum relative surface roughness for all the classes lies between 0.89 for swamps and 0.92 for open water bodies. The standard deviation of relative surface roughness was observed to be below 0.06 for all classes except on swampy and grassland areas with standard deviation of the relative roughness values of 0.68 and 0.38, respectively. The average relative surface roughness values for all the seven classes were between 0.6 and 0.78. The lowest mean was recorded in swampy areas and highest mean in water bodies. Table 5 below shows minimum, average, maximum K_s and calculated average Manning's coefficient per vegetation class.

Table 5. Relative surface roughness and calculated average Manning's for different vegetation classes.

No	Vegetation Class	$(K_s)_{min,c}$ (-)	$(K_s)_{ave}$ (-)	$(K_s)_{max,c}$ (-)	Calculated Average Manning's Coefficient (m$^{-1/3}$/s)
1	Water	0.43	0.78	0.92	0.07
2	Swamps	0.09	0.68	0.89	0.27
3	Partially submerged vegetation	0.20	0.69	0.89	0.39
4	Papyrus/Thicket	0.18	0.71	0.92	0.62
5	Regenerated papyrus/Shrub	0.41	0.70	0.89	0.32
6	Eichorrhoea Crassipes/Grassland	0.20	0.69	0.91	0.27
7	Agriculture/bare soil	0.27	0.70	0.90	0.24

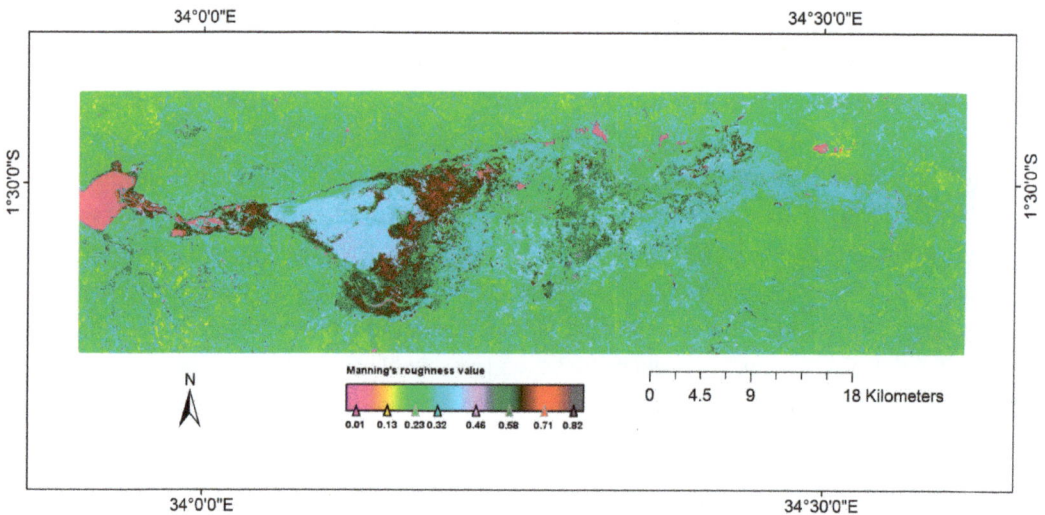

Figure 12. Spatial Manning's roughness coefficient (m$^{-1/3}$/s) derived from Landsat TM based vegetation map and relative surface roughness from SAR imagery.

The spatial roughness map was produced using Equation (4). The result in Figure 12 indicates that water bodies have low values while densely vegetated papyrus vegetation has higher values. This can give a qualitative indication that the approach used in determination of the spatial hydraulic roughness is appropriate.

4.2. Hydraulic Modeling Results

Calibration and sensitivity analysis of single Manning's roughness on model performance was evaluated based on Nash—Sutcliffe efficiency criterion (E) and Root mean square error (RMSE).

The model was run for a range of single values of floodplain friction from between 0.2 m$^{-1/3}$/s and 0.85 m$^{-1/3}$/s at increment of 0.05 m$^{-1/3}$/s (around 14 simulations in total). The results are shown in Table 6 and Figure 13 shows the model performance during the flood event from 21 April to 30 June 2012. The results indicate that at the internal gauging station 2, the single floodplain friction value for optimum performance is 0.65 m$^{-1/3}$/s, which yield 0.92 and 0.02 m for E and RSME, respectively.

Table 6. Model performance parameter as per different single friction roughness parameterizations.

Run No.	1	2	3	4	5	6	7	8	9	10	11	12	13	14
N (m$^{-1/3}$/s)	0.2	0.25	0.3	0.35	0.4	0.45	0.5	0.55	0.6	0.65	0.7	0.75	0.8	0.85
E	0.22	0.28	0.43	0.50	0.56	0.65	0.74	0.84	0.89	0.92	0.84	0.74	0.60	0.50
RMSE (m)	0.35	0.25	0.19	0.16	0.13	0.092	0.08	0.062	0.03	0.02	0.031	0.04	0.075	0.12

The modeling results due to hydraulic roughness parameterizations were evaluated based on the simulated hydrographs for (1) optimum floodplain roughness (n_p) with channel roughness $n_c = 0.03$ m$^{-1/3}$/s(SWL1); (2) calibrated values for both floodplain roughness(n_p) and channel roughness n_c (SWL2); and (3) calibrated channel roughness n_c and spatial Manning's coefficients as derived with aid of relative surface roughness(SWL3). The emphasis was put in the capturing of flood events when the flows are spread in the floodplain. The results of three simulations for the six-month period of from 30 January 2004 to 30 June 2012 including a major flood event are shown Figure 14 below. A bank-full level at internal gauging station 5 is provided in the figure below for reference purposes of flood wave characterization above which flow extends to the floodplain. The results discussed include results from partially calibrated model (SWL1), Fully calibrated model (SWL2) and simulation using spatial Manning's coefficients as derived with aid of relative surface roughness (SWL3).

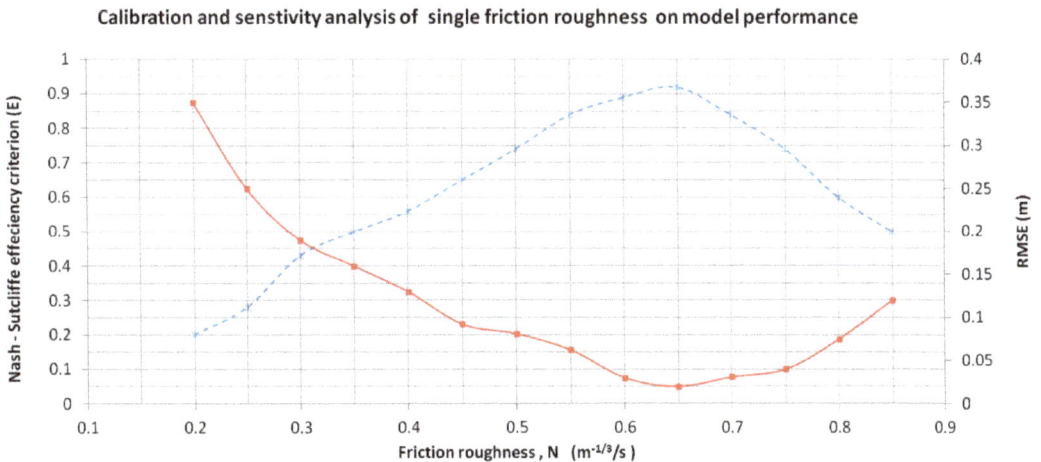

Calibration and senstivity analysis of single friction roughness on model performance

Figure 13. Calibration and sensitivity analysis of single friction roughness on model performance (SWL1). Friction roughness (n), Root Mean Square Error (RMSE) for observed and simulated water level hydrographs.

The resulting water stages are shown in Figure 14. OWL is the observed water level at the gauging station. SWL1 is simulated water levels with optimum floodplain roughness (n_p) with channel roughness $n_c = 0.03$ m$^{-1/3}$/s, SWL2 is simulated water levels with calibrated values for both floodplain roughness $n_p = 0.65$ m$^{-1/3}$/s and channel roughness $n_c = 0.021$ m$^{-1/3}$/s and SWL3 is simulated water levels with calibrated channel roughness n_c and spatial Manning's coefficients as derived with aid of relative surface roughness.

The model performance for the three simulations results is presented in Table. 7. The overall performance for scenario 3 for the whole simulated period is superior as it is proved by the better performance statistics of E = 0.97 compared to calibrated model E = 0.94. The use of spatial roughness parameterization improved the model performance in comparison to the calibrated model but also improved model performance significantly in capturing low flows stages. For this case it can be seen that spatial roughness parameterization can improve flood wave progression modeling without calibrating the Manning's roughness value for the whole floodplain. Scenarios 2 and 3 slightly overestimate the water levels, but the errors are higher in capturing the peaks in scenario 1. In scenario 2 and 3, water stages at low flows are overestimated this may be explained by uncertainty of accuracy of topographic data and downstream boundary condition. At high flood events the hydrographs are captured well at high flows for SWL1 and SWL2 simulations. During flood events or periods the calibrated model and that parametrized using spartial roughness derived from relative roughness performed equally. The results in Figure 13 also indicate that the model is sensitive to river channel roughness parameter.

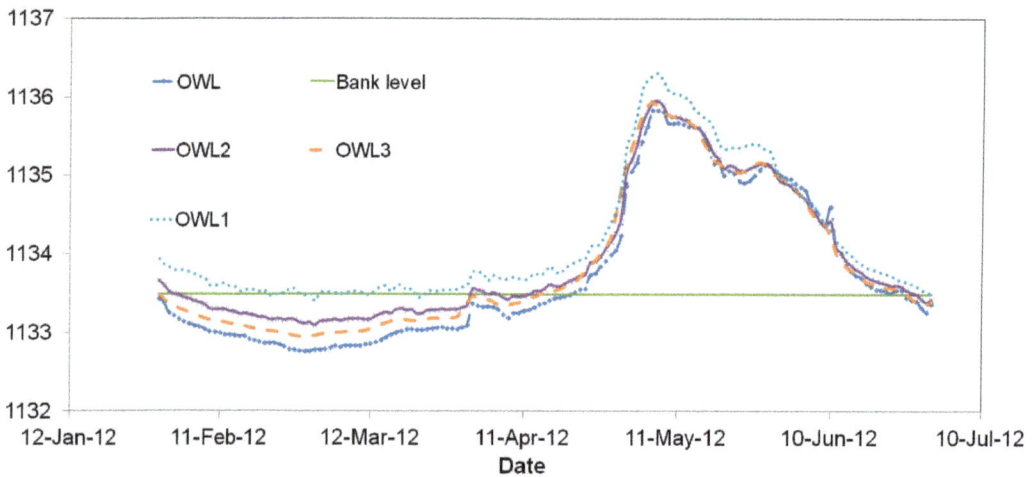

Figure 14. Observed and simulated hydrographs for three simulations SWL1, SWL2 and SWL3.

During low flows all the simulated water levels are higher than the observed water levels. This may be due to the simplified parameterization of the river channel in the model among others. During flooding the water inundates the floodplain, where the parameterization of the resistance against flow in the vegetated riparian areas is becoming important in simulating the water levels.

Table 7. Model performance parameter as per different Manning's roughness parameterizations. SWL1 is simulated water levels with optimum floodplain roughness (n_p) with channel roughness $n_c = 0.03$ m$^{-1/3}$/s, SWL2 is simulated water levels with calibrated values for both floodplain roughness (n_p) and channel roughness n_c and SWL3 is simulated water levels with calibrated channel roughness (n_c) and spatial Manning's coefficients as derived with aid of relative surface roughness The values outside the bracket represent the overall performance for the period of 21 April 2004 to 30 June 2012, whilst the ones in the brackets represent model evaluation during flood event from 21 April 2004 to 30 June 2012.

Performance Criteria	Scenario 1 (SWL1)	Scenario2 (SWL2)	Scenario 3 (SWL3)
Nash—Sutcliffe efficiency criterion (E)	0.75 (0.85)	0.95(0.98)	0.97 (0.96)
Index of agreement (d)	0.93 (0.97)	0.98(0.99)	0.99 (0.99)
Bias (%)	0.036 (0.022)	0.016(0.006)	0.009 (0.005)
STEYX (m)	0.185 (0.188)	0.085(0.086)	0.106 (0.149)
RMSE (m)	0.22 (0.05)	0.047(0.006)	0.023 (0.11)
Coefficient of Determination (R^2)	0.95 (0.95)	0.99(0.99)	0.98 (0.97)

5. Conclusions

The overall performance of scenario SWL1 is simulated water levels with optimum floodplain roughness ($n_p = 0.65$) with channel roughness $n_c = 0.03$ m$^{-1/3}$/s was characterized with E = 0.75 and $R^2 = 0.95$, which was improved in SWL2 during full model calibration with calibrated values for both floodplain roughness $n_p = 0.65$ m$^{-1/3}$/s and channel roughness $n_c = 0.021$ m$^{-1/3}$/s. In SWL2 simulations the model performance was observed to increase to E = 0.95 and $R^2 = 0.99$. When spatially distributed Manning values derived from SAR relative surface values were parametrized in the model, the model also performed well and yielding E = 0.97 and $R^2 = 0.98$. Improved model performance using spatial roughness shows that spatial roughness parameterization can support flood modeling and provide better flood wave simulation over the inundated riparian areas equally as calibrated models. Important water level differences were observed between the three roughness coefficient parameterizations. Nevertheless, there was no major difference in the timing of the flood peak. The calibration of Manning's roughness value (Figure 13) that optimum value for friction roughness is 0.65 m$^{-1//3}$/s which model performance of E = 0.92 and RMSE = 0.02 m. Using spatial roughness for model parameterization also yielded values E = 0.85 and RMSE = 0.05 m. This shows that spatial parameterization of roughness using SAR properties may improve the model performance equally as calibrated model. These results are limited to the model setup characteristics that include grid size, river channel presentation, and stability criteria. Further sources of error can be found in the inherent inaccuracies in the satellite image processing steps. First of all, the unavoidable inaccuracies of the optical satellite image classification. This contributes to the observed overlaps between the relative roughness ranges of the different vegetation types. Speckle in SAR images was suppressed by a median filtering, but this cannot completely remove this noise, which results in an unknown level of inaccuracy in the calculated relative surface roughness values.

The reasons for the observed discrepancies may be as complex as the floodplain itself, which will need careful further research. First of all, the (in) accuracy of the topographical data and the model mesh resolution need to be investigated. In addition, topography interacts with spatially-distributed friction in a complex manner in the timings of the flood wave. In some studies, adaptive time-step of storage cell models have been shown to increase sensitivity to floodplain friction [72]. For the above results the model performance was good when calibrated and spatial roughness were implemented in the model. The added advantage of spatial parametrization is on the simulation time when calibration requires more simulations. The model needs further validation of the spatial distribution of the simulated inundation using satellite image based actual flood maps. The spatial inundation extent during peak flood event needs to be validated using satellite imagery. In addition, topography and spatially-distributed friction interacted with the flow in a complex manner, which could not be fully parameterized all over the floodplain, but our study made an important step in understanding and predicting flood inundation dynamics in riparian areas. Further research is still required to investigate if complex hydrodynamic models (e.g., DELFT 2D/3D, MIKE 21, SOBEK) utilizing full Saint Vennant's equations could perform better than FLO-2D in data scarce areas.

Usefulness of SAR for spatial hydraulic parameterization for hydraulic modeling has been shown in this paper. Temporal variation analysis of backscatter can support spatial hydraulic roughness parameterization of finite volume conservative hydrodynamic models. The use of spatially-distributed friction (Manning's) coefficients as presented here in detail have not been shown in previous studies. Spatially distributed friction coefficients have important effects on hydrodynamic model performance affecting predicted flood depths. These results show that quantification of spatially distributed Manning's roughness can equally improves the performance of hydraulic models as calibrated model for large flood plains. These models are computationally expensive and cannot be automatically calibrated for large floodplains. As a summary, we can state that our study made a step in understanding and predicting flood inundation dynamics in riparian areas based on SAR images, those can provide information even under thick cloud cover. With the upcoming availability of new remote sensing data with higher revisiting frequencies, e.g., Senitel-1 with its SAR sensor and Sentinel-2 with its high-resolution optical data, a unique opportunity opens in monitoring riparian ecosystems for supporting flood modeling with more accurate data for the parameterization of hydraulic resistances.

Acknowledgments

This work was primarily supported by the European Space Agency (ESA) ALCANTARA project AO/1-7102/12- A15, which supported remote sensed data acquisition, field visit and one year fellowship opportunity to conduct this study at ITC, University of Twente, The Netherlands. The author would like to acknowledge the support of the Mara *flows* Project (UPaRF UNESCO-IHE) and the University of Dar es Salaam who initiated this study and provided PhD fellowship support to study at the University of Dar es Salaam. Special thanks to TIGER project No. 24, the RADARSAT-2 Science and Operational Applications Research and Development Program ("SOAR Program") No. 5126 of the Canadian Space Agency for their support in SAR data aquisation. Special thanks to FLO-2D Software *Inc*—Arizona, USA for hydrodynamic model used in this study. Thanks for Lake

Victoria basin Office for providing the meteorological data and support during fieldwork. Authors express their special thanks to Jonathan Mwania for his help in field data collection.

Author Contributions

All authors contributed extensively to the work presented in this paper. Specific contributions include development of the modeling concept and use of SAR data, provision of data and data acquisition capacity (Preksedis Ndomba, Felix Mtalo, Zoltán Vekerdy); fieldwork, model development, data pre-processing, analysis and preparation manuscript and figures (Joseph Mtamba, Rogier van der Velde, Preksedis Ndomba, Zoltán Vekerdy). Final proof reading of the manuscript was done by Rogier van der Velde, Zoltán Vekerdy, and Felix Mtalo).

Conflicts of Interest

The authors declare no conflict of interest.

References

1. Junk W.; Barley P.B.; Sparks R.E. The flood pulse concept in river flood plain systems. *Can. J. Fish. Aquat. Sci.* **1989**, *106*, 110–127.
2. Poff, N.L.; Allan, J.D.; Bain, M.B.; Karr, J.R.; Prestegaard, K.L.; Richter, B.; Sparks, R.; Stromberget, J. The natural flow regime: A new paradigm for riverine conservation and restoration. *BioScience* **1997**, *47*, 769–784.
3. Postel, S.L.; Richter, B. *Rivers for Life: Managing Water for People and Nature*; Island Press: Washington, DC, USA, 2003.
4. Bunn S.E.; Arthington A.H. Basic principles and ecological consequences of altered flow regimes for aquatic biodiversity. *Environ. Manag.* **2002**, *30*, 492–507.
5. Casas, A.; Lane, S.N.; Yu, D.; Benito, G. A method of parameterising roughness and topographic sub-grid scale effects in hydraulic modelling from LiDAR data. *Hydrol. Earth Syst. Sci.* **2010**, *14*, 1567–1579.
6. Bates, P.; Anderson, M.G.; Walling, D.E.; Simm, D. Modelling floodplain flows using a two-dimensional finite element model. *Earth Surf. Proc. Land* **1992**, *17*, 575–588.
7. Stoesser, T.; Wilson, C.A.M.E.; Bates, P.; Dittrich, A. Application of a 3D numerical model to a river with vegetated floodplains. *J. Hydroinf.* **2003**, *5*, 99–112.
8. O'Brien, J.S. *FLO-2D Reference Manual*; FLO-2D Inc.: Nutrioso, AZ, USA, 2009.
9. Bates, P.D.; de Roo, A.P.J. A simple raster-based model for flood inundation simulation. *J. Hydrol.* **2000**, *236*, 54–77.
10. Horritt M.S.; Bates P.D. Evaluation of 1D and 2D numerical models for predicting river flood inundation. *J. Hydrol.* **2002**, *268*, 87–99.
11. Forzieri, G.; Degetto, M.; Righetti,M.; Castelli, F.; Preti, F. Satellite multispectral data for improved floodplain roughness modelling. *J. Hydrol.* **2011**, *407*, 41–57.

12. Tarekegn, T.H.; Haile, A.H.; Rientjes, T.; Reggiani, P.; Alkema, D. Assessment of an ASTER-generated DEM for 2D hydrodynamic flood modelling. *Int. J. Appl. Earth Observ. Geoinf.* **2010**, *12*, 457–465.

13. Rutherfurd, I.D.; Hoang, T.; Prosser, I.P; Abernethy, B.; Jayasuriya, N. The impact of gully networks on the time-to-peak and size of flood hydrographs. In Proceedings of 23rd Hydrology and Water Resources Symposium, Hobart, TAS, Australia, 21–24 May 1996; Institution of Engineers, Australia NCP 96-05: Canberra, NSW, Australia, 1996; pp. 397–402.

14. Anderson, B.G.; Rutherfurd, I.D.; Western, A.W. An analysis of the influence of riparian vegetation on the propagation of flood waves. *Environ. Model. Softw.* **2006**, *21*, 1290–1296.

15. Chow, V.T. *Open Channel Hydraulics*; McGraw-Hill: New York, NY, USA, 1959.

16. Arcement, G.J. Jr.; Schneider, V.R. *Guide for Selecting Manning's Roughness Coefficients for Natural Channels and Flood Plains*; Water-Supply Paper 2339; US Geological Survey: Washington, DC, USA, 1989.

17. Baptist, M.J.; van den Bosch, L.V.; Dijkstra, J.T.; Kapinga, S. Modelling the effects of vegetation on flow and morphology in rivers. *Large Rivers* **2005**, *15*, 339–357.

18. Straatsma, M.W.; Middelkoop, H. Extracting structural characteristics of herbaceous floodplain vegetation under leaf-off conditions using airborne laser scanner data. *Int. J. Remote Sens.* **2007**, *28*, 2447–2467.

19. Dudley, S.J.; Bonham, C.D.; Abt, S.R.; Fischenich, J.G. Comparison of methods for measuring woody riparian vegetation density. *J. Arid Environ.* **1998**, *38*, 77–86.

20. Forzieri, G.; Guarnieri, L.; Vivoni, E.R.; Castelli, F.; Preti, F. Spectral-ALS data fusion for different roughness parameterizations of forested floodplains. *River Res. Appl.* **2010**, *27*, 826–840.

21. Forzieri, G.; Moser, G.; Vivoni, E.R.; Castelli, F. Riparian vegetation mapping for hydraulic roughness estimation using very high resolution remote sensing and LiDAR data fusion. *J. Hydraul. Eng.* **2010**, *11*, 855–867.

22. Mathieu, R.; Naidoo, L.; Cho, M.A.; Leblon, B.; Main, R.; Wessels, K; Asner, G.P; Buckley, J.; Aardt, J.V.; Erasmus, B.F.N.; *et al.* Toward structural assessment of semi-arid African savannahs and woodlands: The potential of multitemporal polarimetric RADARSAT-2 fine beam images. *Remote Sens. Environ.* **2013**, *138*, 215–231.

23. Straatsma, M.W.; Baptist, M.J. Floodplain roughness parameterization using airborne laser scanning and spectral remote sensing. *Remote Sens. Environ.* **2008**, *112*, 1062–1080.

24. Forzieri, G.; Castelli, F.; Preti, F. Advances in remote sensing of hydraulic roughness. *Int. J. Remote Sens.* **2011**, *33*, 630–654.

25. Nagler, P.L.; Glenn, E.P.; Huete, A.R. Assessment of spectral vegetation indices for riparian vegetation in the Colorado River Delta, Mexico. *J. Arid Environ.* **2001**, *49*, 91–110.

26. Townsend, P.A.; Walsh, J. Remote sensing of forested wetlands: Application of multitemporal and multispectral satellite imagery to determine plant community composition and structure in southeastern USA. *Plant Ecol.* **2001**, *157*, 129–149.

27. Congalton, R.G.; Birch, K.; Jones, R.; Schriever, J. Evaluating remotely sensed techniques for mapping riparian vegetation. *Comput. Electron. Agric.* **2002**, *37*, 113–126.

28. Wilson, M.D.; Atkinson, P.M. The use of remotely sense data to derive floodplain friction coefficients for flood inundation modelling. *Hydrolog. Process.* **2007**, *21*, 3576–3586.

29. Makkeasorn, A.; Chang, N.-B.; Li, J. Seasonal change detection of riparian zones with remote sensing images and genetic programming in a semi-arid watershed. *J. Environ. Manag.* **2009**, *902*, 1069–1080.

30. Akasheh, O.Z.; Neale, C.M.U.; Jayanthi, H. Detailed mapping of riparian vegetation in the middle Rio Grande River using high resolution multi-spectral airborne remote sensing. *J. Arid Environ.* **2008**, *72*, 1734–1744.

31. Mason, D.; Cobby, D.M.; Horritt, M.; Bates, P. Floodplain friction parameterization in two-dimensional river flood models using vegetation heights from airborne scanning laser altimetry. *Hydrol. Processes* **2003**, *17*, 1711–1732.

32. Van der Velde, R.; Su, Z.; Ma, Y. Impact of soil moisture dynamics on ASAR $\sigma°$ signatures and its spatial variability observed over the Tibetan Plateau. *Sensors* **2008**, *8*, 5479–5491.

33. Van der Velde, R.; Su, Z.; van Oevelen, P.; Wen, J.; Ma, Y.; Salama, M.S. Soil moisture mapping over the central part of the Tibetan Plateau using a series of ASAR WS images. *Remote Sens. Environ.* **2012**, *120*, 175–187.

34. Kuenzer, C.; Guo, H.; Huth, J.; Leinenkugel, P.; Li, X.; Dech, S. Flood mapping and flood dynamics of the Mekong Delta: ENVISAT-ASAR-WSM based time series analyses. *Remote Sens.* **2013**, *5*, 687–715.

35. Schumann, G.; di Baldassarre, G.; Bates P.D. The utility of space-borne radar to render flood inundation maps based on multi-algorithm ensembles. *IEEE Trans. Geosci. Remote Sens.* **2009**, *47*, 2801–2807.

36. Schumann, G.J.-P.; Bates P.D.; Horritt, M.S.; Matgen, P.; Pappenberger, F. Progress in integration of remote sensing derived flood extent and stage data and hydraulic models. *Rev. Geophys.* **2009**, doi:10.1029/2008RG000274.

37. Ferrazzoli, P.; Paloscia, S.; Pampaloni, P.; Schiavon, G.; Sigismondi, S.; Solamini, D. The potential of multifrequency polarimetric SAR in assessing agriculture and arboreous biomass. *IEEE Trans. Geosci. Remote Sens.* **1997**, *35*, 5–15.

38. Alberga, V.; Satalino, G.; Staykova, D.K. Comparison of polarimetric SAR observables in terms of classification performance. *Int. J. Remote Sens.* **2008**, *29*, 4129–4150.

39. Cloude, R.S.; Pottier, E. An entrophy based classification scheme for land applications of polarimetric SAR. *IEEE Trans. Geosci. Remote Sens.* **1997**, *35*, 68–78.

40. Lucas, R.M.; Cronin, N.; Lee, A.; Moghaddam, M.; Witte, C.; Tickle, P. Empirical relationships between AIRSAR backscatter and LiDAR-derived forest biomass, Queensland, Australia. *Remote Sens. Environ.* **2006**, *100*, 407–425.

41. Englhart, S.; Keuck, V.; Siegert, F. Aboveground biomass retrieval in tropical forests—The potential of combined X- and L-band SAR data use. *Remote Sens. Environ.* **2011**, *115*, 1260–1271.

42. Garestier, F.; Dubois-Fernandez, P.C.; Guyon, D.; le Toan, T. Forest biophysical parameter estimation using L- and P-band polarimetric SAR data. *IEEE Trans. Geosci. Remote Sens.* **2009**, *47*, 3379–3388.

43. Lewis, A.J.; Henderson, F.M. Radar fundamentals: The geoscience perspective. In *Principles and Applications of Imaging Radar: Manual of Remote Sensing*, Third ed.; Henderson, F.M., Lewis, A.J., Eds.; John Wiley & Sons: New York, NY, USA, 1998; Volume 2, pp. 131–181.

44. Verhoest, N.E.; Lievens, H.; Wagner, W.; Álvarez-Mozos, J.; Moran, M.S.; Mattia, F. On the soil roughness parameterization problem in soil moisture retrieval of bare surfaces from synthetic aperture radar. *Sensors* **2008**, *8*, 4213–4248.

45. Ranson K.J.; Sun, G. Mapping biomass of a northern forest using multifrequency SAR data. *IEEE Trans. Geosci. Remote Sens.* **1994**, *32*, 388–396.

46. Mattia, F.; Toan, L.T.; Souyris, J.-C.; Carolis, G.D.; Floury, N.; Posa, F. The effect of surface roughness on multifrequency polarimetric SAR data. *IEEE Trans. Geosci. Remote Sens.* **1997**, *35*, 954–966.

47. Mtamba, J.O.D.; Ndomba, P.M.; Mtalo, F.; Crosato, A. Hydraulic study of flood rating curve development in the Lower Mara Basin. In Proceedings of 4th International Multidisciplinary Conference on Hydrology and Ecology (HydroEco 2013), Rennes, France, 13–16 May 2013.

48. McClain, M.; Subalusky, A.; Anderson, E.; Dessu, S.; Melesse, A.; Ndomba, P.; Mtamba, J.; Tamatamah, R.; Mligo, C. Comparing flow regime, channel hydraulics and biological communities to infer flow-ecology relationships in the Mara River of Kenya and Tanzania. *Hydrol. Sci. J.* **2013**, doi:10.1080/02626667.2013.853121.

49. Breda, N.J.J. Ground-based measurements of leaf area index: A review of methods, instruments and current controversies. *J. Exp. Bot.* **2003**, *54*, 2403–2417.

50. Jonckheere, I.; Fleck, S.; Nackaerts, K.; Muys, B.; Coppin, P.; Weiss, M.; Baret, F. Review of methods for *in situ* leaf area index determination. Part I. Theories, sensors and hemispherical photography. *Agric. For. Meteorol.* **2004**, *121*, 19–35.

51. Weiss, M.; Baret, F.; Smith, G.J.; Jonckheere, I.; Coppin, P. Review of methods for *in situ* leaf area index (LAI) determination Part II. Estimation of LAI, errors and sampling. *Agric. For. Meteorol.* **2004**, *121*, 37–53.

52. Garrigues, S.; Shabanov, N.V.; Swanson, K.; Morisette, J.T.; Baret, F.; Myneni, R.B. Intercomparison and sensitivity analysis of leaf area index retrievals from LAI-2000, AccuPAR, and digital hemispherical photography over croplands. *Agric. For. Meteorol.* **2008**, *148*, 1193–1209.

53. Congalton, R.; Green, K. *Assessing the Accuracy of Remotely Sensed Data: Principles and Practices*, 2nd ed.; Lewis Publishers: Boca Raton, FL, USA, 1999.

54. Maghsoudi, Y.; Collins, M.J.; Leckie, D.G. On the use of feature selection for classifying multitemporal Radarsat-1 images for forest mapping. *IEEE Geosci. Remote Sens. Lett.* **2011**, *8*, 904–908.

55. Bruzzone, L.; Marconcini, M.; Wegmüller, M.; Wiesmann, A. An advanced system for the automatic classification of multitemporal SAR images. *IEEE Trans. Geosci. Remote Sens.* **2004**, *42*, 1321–1334.

56. Petryk, S.; Bosmajian, G.B. Analysis of flow through vegetation. *J. Hydraul. Div.* **1975**, *101*, 871–884.

57. Ogilvy, J.A.; Foster, J.R. Rough surfaces: Gausian or exponential statistics. *Phys. Rev. D Appl. Phys.* **1989**, *22*, 1243–1251.

58. Geman, S.; Geman, D. Stochastic relaxation, Gibbs distributions, and the Bayesian restoration of images. *IEEE Trans. Pattern Anal. Mach. Intell.* **1984**, *6*, 721–741.

59. Jeng, F; Woods, J. Compound Gauss-Markov random fields for image estimation. *IEEE Trans. Signal Process.* **1991**, *39*, 683–697.

60. Zhu, S.C.; Wu, Y.; Mumford, D. Filters, random fields and maximum entropy (FRAME): Towards a unified theory for texture modelling. *Int. J. Comput. Vis.* **1998**, *27*, 107–126.

61. Freeman, A.; Durden, S. A three-component scattering model for polarimetric SAR data. *IEEE Trans. Geosci. Remote Sens.* **1998**, *36*, 963–973.

62. Winkler, G. *Image Analysis, Random Fields and Markov Chain Monte Carlo Methods: A Mathematical Introduction*, 2nd ed.; Springer-Verlag Berlin Heidelberg: New York, NY, USA, 2003.

63. Kasetkasem, T.; Rakwatin, P.; Sirisommai, R.; Eiumnoh, A. A joint land cover mapping and image registration algorithm based on a markov random field model. *Remote Sens.* **2013**, *5*, 5089–5121.

64. Löw, A. Coupled Modelling Of Land Surface Microwave Interactions Using ENVISAT ASAR Data. Ph.D. Thesis, Ludwig Maximilian University of Munich, Munich, Germany, 2004.

65. Bates, P.D. Remote sensing and flood inundation modelling. *Hydrol. Proc.* **2004**, *18*, 2593–2597.

66. Wang, P. Applying two dimension kalman filtering for digital terrain modeling. In IAPRS, Proceedings of ISPRS Commission IV Symposium, GIS Between Visions and Applications, Stuttgart, Germany, 7–10 September 1998; Fritsch, D., Englich, M., Sester, M., Eds.; 32, pp. 649–656. Available online: http://www.isprs.org/proceedings/XXXII/part4/wang71.pdf (Accessed on 20 March 2013).

67. Gallant, J.C.; Hutchinson, M.F. Producing Digital Elevation Models with Uncertainty Estimates Using a Multi-scale Kalman Filter. Available online: http://www.spatial-accuracy.org/book/export/html/183 (accessed on 14 May 2013).

68. Krause, P.; Boyle, D.P.; Base, F. Comparision of different efficiency criteria for hydrological model assessment. *Adv. Geosci.* **2005**, *5*, 89–97.

69. Hassanzadeh, Y. Hydraulics of Sediment Transport, Hydrodynamics—Theory and Model. Available online: http://www.intechopen.com/books/hydrodynamics-theory-and-model/hydraulics-of-sediment-transport (accessed on 25 December 2014).

70. Legates, D.R.; McCabe, G.J., Jr. Evaluating the use of "goodness-of-fit" measures in hydrologic and hydroclimatic model validation. *Water Resour. Res.* **1999**, *35*, 233–241.

71. Corcoran, J.; Knight, J.; Brisco, B.; Kaya, S.; Cull, A.; Murnaghan, K. The integration of optical, topographic, and radar data for wetland mapping in northern Minnesota. *Can. J. Remote Sens.* **2011**, *37*, 564–582.

72. Hunter, N.M.; Bates, P.D.; Horritt, M.S.; Wilson, M.D. Simple spatially-distributed models for predicting flood inundation: A review. *Geomorphology* **2007**, *90*, 208–225.

Hydrological Impacts of Urbanization of Two Catchments in Harare, Zimbabwe

Webster Gumindoga, Tom Rientjes, Munyaradzi Davis Shekede, Donald Tendayi Rwasoka, Innocent Nhapi and Alemseged Tamiru Haile

Abstract: By increased rural-urban migration in many African countries, the assessment of changes in catchment hydrologic responses due to urbanization is critical for water resource planning and management. This paper assesses hydrological impacts of urbanization on two medium-sized Zimbabwean catchments (Mukuvisi and Marimba) for which changes in land cover by urbanization were determined through Landsat Thematic Mapper (TM) images for the years 1986, 1994 and 2008. Impact assessments were done through hydrological modeling by a topographically driven rainfall-runoff model (TOPMODEL). A satellite remote sensing based ASTER 30 metre Digital Elevation Model (DEM) was used to compute the Topographic Index distribution, which is a key input to the model. Results of land cover classification indicated that urban areas increased by more than 600 % in the Mukuvisi catchment and by more than 200 % in the Marimba catchment between 1986 and 2008. Woodlands decreased by more than 40% with a greater decrease in Marimba than Mukuvisi catchment. Simulations using TOPMODEL in Marimba and Mukuvisi catchments indicated streamflow increases of 84.8 % and 73.6 %, respectively, from 1980 to 2010. These increases coincided with decreases in woodlands and increases in urban areas for the same period. The use of satellite remote sensing data to observe urbanization trends in semi-arid catchments and to represent catchment land surface characteristics proved to be effective for rainfall-runoff modeling. Findings of this study are of relevance for many African cities, which are experiencing rapid urbanization but often lack planning and design.

Reprinted from *Remote Sens.* Cite as: Gumindoga, W.; Rientjes, T.; Shekede, M.D.; Rwasoka, D.T.; Nhapi, I.; Haile, A.T. Hydrological Impacts of Urbanization of Two Catchments in Harare, Zimbabwe. Remote Sens. 2014, 6, 12544-12574.

1. Introduction

Understanding the impacts of land conversion and land cover changes on the hydrological cycle has become a global concern in view of the increasing urban populations [1,2]. Studies by [3,4] concluded that the effects of land conversions on river flows are of major interest to water resource managers and hydrologists as they plan, manage and develop water resources. [5] observed that increases in impervious areas through urbanization may result in the following hydrological impacts (i) reduced interception by tree canopies; (ii) reduced infiltration; (iii) increased surface runoff; (iv) increased flow velocities in urban areas due to decreased surface roughness and (v) increased peak flow discharges. Similarly, [6] noted that conversion of natural catchments to peri-urban or urban areas affect many processes of the hydrological cycle, such as interception, infiltration, evaporation and streamflow by runoff processes. However, the magnitude of impacts of urbanization on hydrological processes is commonly not well known especially in large parts of Africa. Furthermore, conclusive studies on the implications of urbanization on closure of the water balance and availability of water resources are limited [5,7–9]. Yet knowledge on the effects of urbanization on the hydrology of catchments is critical for water resources management in most water-scarce areas, such as those in Africa.

Studies that have assessed impacts of urbanization have adopted different approaches. For instance, [10] studied the effects of suburban developments on runoff generation using hydrograph analysis techniques. They showed that with increased suburban development there was an accelerated recession phase and increased peak flows. Similarly, [11] analyzed hydrograph characteristics at an annual scale for a 38-year runoff record to determine the effects of urbanization on streamflow. The study showed that the annual runoff coefficient of the urban stream (Peachtree Creek) was not significantly greater than that of the less-urbanized watersheds. However, the storm recession period of the urban stream was one to two days less than that of the other streams. [12] applied a water budget and meteorological approach to assess the effects of urbanization on catchment evapotranspiration (ET). The study showed significant decreases in catchment ET that were linked to increases in urban and residential areas. In a different study, [13] assessed the impacts of urbanization of river flow frequencies by a controlled experimental modeling approach using the model MIKE-SHE and the 1D hydrodynamic river model MIKE-11. The study showed that the frequency of low flows decreased with increasing urban expansion and that the frequency of average and high-flow events increased with increasing urbanization. Recently, there have been attempts to incorporate remote sensing data in hydrological models to enhance understanding of the effects of urbanization on the hydrological cycle. [14] applied a coupled distributed Hydrologic Engineering Center's Hydrologic Modeling System (HEC-HMS) for runoff simulations with the integrated Markov Chain and Cellular Automata model (CA-Markov model) for development of future land use scenario maps. Landsat and CBERS satellite data were used. The results showed that increases in annual runoff volume, daily peak flows and flood volume between the years 1988–2009 could be related to urbanization. These hydrological variables were projected to further increase with increasing urbanization. These studies have encouraged incorporation of land use change information in distributed hydrologic models. However, the assessment of urbanization on hydrological impacts of catchments remains complicated due the spatial heterogeneity of the land surface in urban areas (see [5,7]). In this regard, satellite remote sensing provides an opportunity to assess and track changes in land cover over selected space and time domains thereby serving as an important input in impact assessments and modeling studies. Despite the importance of remote sensing in providing land cover maps which are critical inputs to hydrological models, there are often inconsistencies that may arise from image misclassifications or registration errors [15]. It is therefore important to correct for such inconsistencies through assessment of land use and land cover (LULC) spatial and temporal patterns [15,16] and or through accuracy assessment of classified images [17].

This study relied on satellite remote sensing data to represent land cover and elevation characteristics as inputs for the topographically driven TOPMODEL, which served to simulate the relationship between rainfall and runoff. TOPMODEL is a semi-distributed, mass conservative model which relies on a simple representation of basin characteristics and hydrologic processes [18] as compared to fully distributed and data demanding models like MIKE SHE [19]. The semi-distributed form of TOPMODEL makes full use of elevation data which is freely available through the Advanced Spaceborne Thermal Emission and Reflection Radiometer (ASTER) or the Shuttle Radar Topography Mission (SRTM) Digital Elevation Models (DEMs). TOPMODEL requires a small number of topographic and land surface based parameters and makes optimized parameter values physically meaningful [20]. Furthermore, in its setup, the model can adapt to a specific catchment and specific modeling purposes [21]. However, TOPMODEL mostly has applications in natural catchments [22–26] with only few applications in urban catchments [6,27]. Latter applications have characterized urbanized land cover by introducing impervious surfaces with very low percolation and surface infiltration rates [6,9,23,28] which resulted in increased and more rapid runoff responses. Despite these efforts, the use of TOPMODEL approach in an urban setting as shown in [6] indicated that the ISBA-TOPMODEL simulations underestimated total streamflow during dry periods whereas it overestimates streamflow during rain events and wet weather conditions. In the study by [29], modifications of TOPMODEL (TOPURBAN v.1 and v.2) were tested for urbanized watersheds by altering the topographic index and the mechanism to generate surface runoff but detailed descriptions on the processing of data including remote sensing data were missing. In fact, several studies [9–11,13,14] that have characterized urbanized land cover types for hydrological assessments have failed to adequately capture relevant spatial information of historical land surfaces in urban catchments. In that regard, this study determined historical changes in land cover and incorporated topographical attributes through ASTER DEM hydro-processing approaches as a first step towards assessing impacts of urbanization on hydrology.

Within the African context, the assessment of the impacts of urbanization on streamflow is important for water development and management. Urbanization in Africa is common due to urban migration resulting in increases in paved and built-up areas in the urban setting. According to [30], Africa is one of the hotspots of serious urban growth and will continue to be so for the next four decades. It is projected that the population of African cities will increase by 0.9 billion by 2050. In Zimbabwe, the population of Harare has grown from 1.8 million in 2002 to 2.1 million in 2012 [31]. As a consequence, the demand for land for housing increased and peri-urban and rural areas have been converted to urban areas. In addition to exterior sprawling, densification is a strategy also being applied to grow the city of Harare. Densification promotes the growth of the city through the construction of buildings on lands previously left as open spaces thus increasing the extent of paved and build-up areas. Densification and sprawling have had the concomitant effect of intensifying urbanization. In Harare City, Marimba and Mukuvisi catchments are two catchments that are experiencing rapid urbanization as characterized by rapid growth and densification. Both catchments constitute the greater part of the built-up environment of the city. These catchments are the most urbanized in Harare City and therefore were selected for this study. The urban areas are characterized by middle-to-low income housing, office complexes and industrial areas. Hydrology and water related studies in and around the Marimba and Mukuvisi catchments have mainly focused on water quality and pollution, evapotranspiration and urban drainage [32–34]. Only few studies have focused on hydrology and quantification of water resources [35,36], but within the broader context of the Upper Manyame catchment. Detailed studies on hydrological impacts of urbanization of the catchments are unknown to the authors. Objectives of this study are to: (1) assess trends in rainfall and streamflow; (2) assess changes in land cover in the Marimba and Mukuvisi catchments; and (3) assess hydrological impacts of urbanization and land conversion by rainfall-runoff model simulations.

In Section 2 descriptions of the study area and available data are given including satellite data. Methods used in this study are described in Section 3. Findings of the study are presented and discussed in Section 4. Section 5 gives the conclusions and an outline of the recommendations.

2. Study Area and Data

2.1. Description of Study Area

For this study the Marimba and Mukuvisi River catchments in Harare City are selected which are tributary catchments of the Upper Manyame basin in Zimbabwe (Figure 1). Mukuvisi catchment has an area of 223.1 km^2 and a longest flow path of 44.7 km whereas Marimba catchment has an area of 220.5 km^2 and a longest flow path of 38.6 km. Both catchments have similar elevation ranges between 1350 m and 1550 m above mean sea level. The soil is primarily sandy clay loam. The mean annual rainfall for the period 2000–2010 is 810 mm/yr whereas potential evapotranspiration is around 1600 mm/yr. These two catchments were selected for this study since both are characterized by a rapid increase of built-up area and urbanization. The dominant residential housing in both catchments is high to medium density houses with limited space for gardens. The population density in the catchments is around 2.540 people/km^2 according to [31]. However, there are low-density areas in the northern and eastern parts of the catchments that have spacious gardens. Trees are also kept within the residential stands.

2.2. Hydro-Meteorological Data

For this study, daily streamflow data for the years 1970 to 2008 for the gauging stations of Marimba and Mukuvisi catchments (C22 and C24, respectively) were made available by the Zimbabwe National Water Authority. Time series of daily meteorological data including rainfall data were acquired from the Meteorological Office of Zimbabwe. Potential Evapotranspiration at daily time step is estimated from the meteorological data using the FAO-Penman Monteith method as outlined in [37].

The relation between rainfall and streamflow was assessed as part of data screening. For rainfall time series data from Airport, Belvedere and Kutsaga rain stations was used whereas for streamflow time series data from gauging stations C22 and C24, for Mukuvisi and Marimba catchments, respectively, was used (see Figure 1). Correlation between time series suggests dependency of streamflow on rainfall in both catchments (Table 1).

Figure 1. Mukuvisi and Marimba catchments in Zimbabwe showing elevation and locations of meteorological stations and streamflow gauging stations.

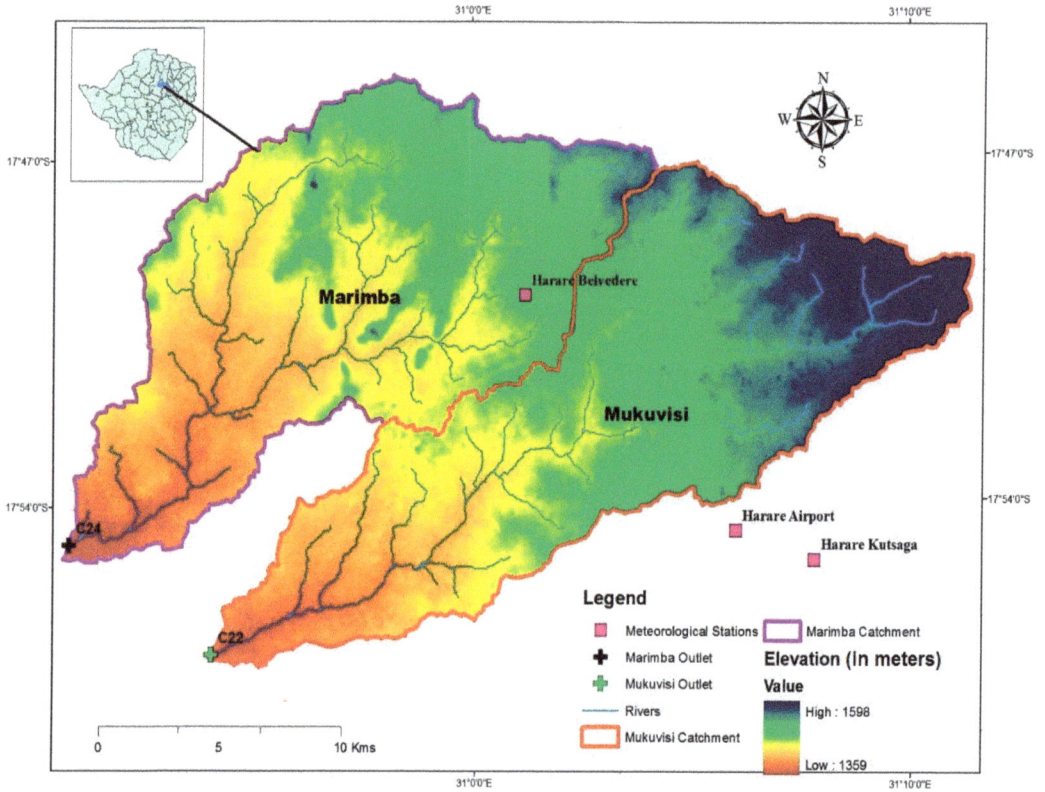

Table 1. The significant relationship ($p < 0.05$) between rainfall and streamflow in the study area. All p values were equal to 0.000.

Rain Station	Streamflow Gauging Station	Correlation R
Airport	C22	0.675
Airport	C24	0.649
Belvedere	C22	0.651
Belvedere	C24	0.656
Kutsaga	C22	0.667
Kutsaga	C24	0.642

2.3. Satellite Data

For estimation of the topographic index an ASTER DEM (30 m resolution) covering the study area was retrieved from the Global ASTER GDEM. For land cover change assessments we used land cover images from Landsat satellites, which were processed in ILWIS open source software. Landsat TM images analyzed in the study (path 170 and row 72) were downloaded from the United States Geological Survey (USGS) Global Visualization Viewer (GLOVIS) for the years 1986, 1994 and

2008, all in the dry month of August. Google Earth imagery of the study area was used to assess the imperviousness of the two catchments.

3. Methodology

3.1. Land Cover by Remote Sensing

Table 2 shows the dates of image acquisition, spatial resolution as well as the bands used for land cover change analysis. The false color composites (5, 4, 3) were used in the classification process because of their ability to enhance image interpretation that ultimately facilitates differentiation of land cover types, such as: grass, woodland, cropped area, aquatic weeds and bare surfaces which are critical for assessing changes in land cover as a result of urbanization.

Table 2. Description of imagery used for land cover classification.

Sensor	Date of Acquisition	Spatial Resolution	Bands Used	Cloud Cover
Landsat 5 TM	31 August 1986	30 meters	5, 4, 3	0
Landsat 5 TM	21 August 1994	30 meters	5, 4, 3	0
Landsat 5 TM	10 August 2008	30 meters	5, 4, 3	10 *

* Although the overall scene had 10% cloud cover, the study site had less than 5% cloud cover.

Prior to image classification, all the images were georeferenced to the Universal Transverse Mercator zone 36 south projection based on the WGS84 datum. A minimum of 15 ground control points were used during image registration. The nearest neighbour resampling method was used for image registration and a root mean square error less than 0.2 pixels (~6 m) was obtained. To ensure comparability of the images across the years, digital numbers were converted to radiance and from this to a dimensionless top-of-atmosphere (ρTOA) reflectance:

$$\rho TOA = \pi \cdot L\lambda \cdot d2/ESUN\lambda \cdot Cos\theta s \qquad (1)$$

where:

$L\lambda$ = the spectral radiance at the sensor
d = the Earth-sun distance in astronomical units
$ESUN\lambda$ = the mean solar exo-atmospheric irradiance for each band and
$Cos\theta s$ = the solar zenith angle in degrees (Irish 1998).

The sensor calibration information, such as solar zenith angle and earth-sun distance, was extracted from the header file of the imagery [38]. Once converted to ρTOA reflectance, bands 5, 4 and 3 representing the Short-wave Infrared, Near Infra-Red and the Red bands of Landsat, respectively, were combined in the ILWIS Geographic Information System (GIS) to create a color composite, which enhanced visualization before image classification. The images were then classified using the maximum likelihood classification algorithm in a GIS environment based on the six land cover classes in the study area (Table 3). The maximum likelihood classification algorithm is based on the probability that a pixel belongs to a particular class and thus a pixel is assigned to a predefined set of classes to which it has the highest probability of belonging to [17]. The maximum likelihood was used for the study because of its robustness [39]. Image classification was important in the determination of land cover, an important determinant of streamflow generation in hydrological catchments. Table 3 provides a description of the classes used in this study.

Table 3. Description of the land cover classes used in the study.

Class	Description
Aquatic weeds	Area under aquatic weeds
Cropped Field	Area under crops
Grassland	Area predominantly covered with grass for a significant part of the year
Urban	Area covered with bare surfaces that have been cleared for urban developments, impervious surfaces, such as roads and buildings
Water	Area occupied by water, such as rivers and wetlands
Woodland	Area covered with sparse to dense woody species, such as shrubs, bushes and trees. Miombo species dominate this cover.

Results of classification were assessed for their accuracy using Kappa statistic. The Kappa statistic was based on 1720 ground control points for Marimba catchment and 985 points for Mukuvisi catchment. The ground control points were taken from high resolution Google Earth imagery and aerial photographs for the dates which coincided with the Landsat imagery acquisition dates. The selection of these points was based on the relative proportion of each land cover type derived from visual interpretation of the image. Table 4 shows the average number of these ground control points used for accuracy assessment in the study.

Table 4. Average number of points used for validating the classified landcover images for Marimba and Mukuvisi Catchment.

Land Cover	Marimba	Mukuvisi
Aquatic weed	42	*
Urban	356	322
Cropped Field	128	104
Grassland	678	292
Water	375	89
Woodland	141	178

* No aquatic weeds were observed in Mukuvisi catchment. Aquatic weeds were mainly observed in Lake Chivero.

After image classification, overlay analysis was performed on the 1986, 1994 and 2008 images to assess land conversion and urbanization in the study area. The result of the overlay analysis is a confusion matrix, which shows land cover and land cover conversions for both catchments for respective years.

3.2. Trend Analysis and Hydro-Meteorological Time Series

Trend analysis for rainfall and streamflow time series is critical to assess if changes in streamflow could be related to changes in rainfall as possibly caused by climate change. In this study, the Mann-Kendall test was used to test whether, statistically, significant trends at monthly and annual basc (1954–2006) could be identified. We tested for significant levels at $p < 0.05$. The Mann-Kendal (MK) test, also known as the as the Kendall's tau statistic [40,41], is a rank-based non-parametric statistical test that is commonly applied for trend detection [40,42,43]. The test compares the relative magnitudes of sample data rather than the data values themselves [44]. The tau statistic, τ, reads:

$$\tau = \frac{2S}{n(n-1)} \tag{2}$$

where:

$$S = \sum_{i-1}^{n-1} \sum_{j=i+1}^{n} sign(Y_{j\_}X_i) \tag{3}$$

$$sign(Y_{j\_}X_i) = \begin{pmatrix} 1 \ if \ (Y_{j\_}X_i) > 0 \\ 0 \ if (Y_{j\_}X_i) = 0 \\ -1 \ if (Y_{j\_}X_i) < 0 \end{pmatrix} \tag{4}$$

The quantity S in Equation (2) shows the number of concordant pairs minus the number of discordant pairs. A high positive value of S is an indicator of an increasing trend, whereas a low negative value indicates a decreasing trend.

3.3. Hydrological Modeling

For this study a TOPMODEL code was developed at Faculty of Geo-information Science and Earth Observation (ITC), University of Twente, for application in a semi-distributed fashion. The code was developed in the IDL programming language and is a conversion of FORTRAN (viz. FORmula TRANslator) version of TOPMODEL in [45]. This version was selected to allow for infiltration excess overland flow simulation by urbanization and land conversion. This was implemented by means of the Green and Ampt equation [46]. Table 5 shows the most relevant model parameters and the Green and Ampt parameters [47–49], which were obtained from literature and linked to soil texture and soil compactness in the study area. The imperviousness of the two catchments was obtained through visualization of freely available Google Earth imagery of the study area and the texture was assessed using soil maps of the study area.

The Green and Ampt approach relies on physically based equations and serves to estimate infiltration rates from a maximum to minimum rate [46]. This study adopts the power function formulation with power $n = 1$ and 2 developed by [50]. In this approach only the linear and exponential forms are considered and serves to allow a faster decay of infiltration and more rapid generation of runoff [50].

TOPMODEL is a mass conservative rainfall-runoff model based on the variable contributing area concept. Predominant factors affecting the formation of runoff are (1) the topographic index; (2) the overland flow and channel network and (3) negative exponential function which links transmissivity of the soil with the vertical distance from the land surface by means of a scaling parameter m [51]. Full details of the governing equations and the rationale behind the model structure are available in [44,51–53].

Table 5. TOPMODEL parameters.

Parameter	Description	Equation
m (m)	Scaling parameter of the exponential transmissivity function which is a function of local storage deficit or depth to the water table [51]. Value range 0.01–1.0 m.	$T = T_o e^{-s_i/m}$
T_o (m²/h)	Transmissivity of the soil profile at full saturation. Value range 0.01–2.25 m²/h	$T = T_o e^{-s_i/m}$
t_d (h)	Time delay constant for routing unsaturated flow. Value range 0.01–24 h [52].	$q_v = \dfrac{S_{uz}}{S_i t_d}$
CHV (m/h)	Channel and overland flow routing velocity. Ranges vary with specific catchment	$t_d = \sum\limits_{i=1}^{N} \dfrac{x_i}{CHV\ \tan \beta_i}$
RV (m/h)	Channel flow inside catchment (vary with specific catchment)	
SR_{max} (m)	The root zone available water storage capacity. Value range 0–0.3 m	$E_a = E_p (1 - SRZ / SR_{max})$
Q_b (m/h)	Initial stream discharge to represent baseflow. Used in recession curve analysis (function of rainfall-runoff relationship in the specific area).	$\dfrac{dQ_b}{dt} = \dfrac{Q_b}{AS_m} \dfrac{dQ_b}{d\delta}$
$SR0$ (m)	Initial value of root zone deficit, also called $SRinit$. Value range 0.001–0.1 m	
$INFEX$ (–)	An infiltration flag set to 1 to include infiltration excess calculations, otherwise 0.	[46,50,53]
K_o (m/hr)	Surface value of the saturated hydraulic conductivity (Ks)	[46,50,53]
Ψ_f (m)	Effective suction head for the calculation of infiltration excess flow	[46,47,53]
θ (–)	Water content change across the wetting front (Beven, 1984).	[46,50,53]

3.3.1. The Topographic Index

TOPMODEL is mathematically and parametrically simple and relies on the processing of digital terrain data to calculate the topographic index distribution function of the catchment. For estimation of the topographic index an ASTER DEM was processed in the ILWIS GIS software. Local depressions were removed and local slopes and drainage patterns were defined. The topographic index (TPI) combines the (local) topographic slope and the specific runoff contributing area α as critical input to model simulations. TPI serves to predict local variations in water table [54,55] as the main driver to generate runoff. The topographic index (TPI) reads:

$$TPI = Ln \left(\frac{\alpha}{tan\beta} \right)$$

(5)

where:

Ln = the natural logarithm
α = specific runoff contributing area
tanβ = the average outflow gradient from the DEM grid element.

3.3.2. Overland Flow and Channel Network Routing

To simulate the flow travel time, TOPMODEL uses a simple scheme called a delay approach [56]. Fractional area and its distance from the outlet are required as well as channel velocity, which is a constant across the catchment. The model computes the time it takes for a water particle to travel from each fractional area to contribute to the catchment outlet. Then for each area, contributions are defined and accumulated for the calculation time steps [44]. We note that the rainwater networks for the urban part of the catchments are integrated into the natural streams and rivers and therefore the mechanism of surface runoff generation and flow routing is maintained. Selecting a DEM grid element as catchment outlet, a distance map was produced showing the shortest distance to the catchment outlet for any DEM element. The distance map was then sliced into thirteen distance classes of equal size for surface flow routing [57].

The Transmissivity Profile

The scaling parameter *m*, is also known as a decay parameter that controls the decrease of transmissivity, T_o, with depth from the land surface when full saturation is considered. Following [58], recession curve analysis of streamflow data was performed to estimate the scaling parameter. A larger value of *m* increases infiltration whereas a smaller value decreases infiltration and thus *m* directly affects simulation results.

Land Cover Parameterization

To evaluate impacts of urbanization and land cover change on hydrological processes, scaling parameter (*m*), soil transmissivity (T_o), root zone available water capacity (SR_{max}) and saturated hydraulic conductivity (K_s) were used. These parameters were selected because they have been found to be the most sensitive TOPMODEL parameters in literature [18,49,58,59]. Also parameter values are affected by soil characteristics and vegetation cover and therefore are of relevance in land cover change impacts assessments.

SR_{max} was selected since it represents maximum root zone storage, which directly affects actual evapotranspiration from the root zone (Table 5). Moreover, net precipitation in excess to SR_{max} causes runoff generation by overland flow. To simulate effects of urbanization in TOPMODEL, using the Green and Ampt infiltration excess approach four additional parameters (Table 4) are required. In residential areas, with compacted soils, the concept adopted from [49] provides a number of infiltration decay methods to increase flexibility in matching the increased incidence of infiltration excess runoff. The Green and Ampt parameters remain constant (*i.e.*, frozen) during a model simulation run. The spatial variability and distribution of hydraulic conductivity (K_s) is represented in the model setup by specifying K_s values for different land covers.

For comparison of streamflow for the 10-year simulation periods, which enclose the 1986, 2004 and 2008 images, m, SR_{max} and K_s were changed for both catchments to mimic the variation and change in land cover by urbanization. The approach for the land cover change impact assessment in this study is that for different years of image analysis different land cover types apply and thus distribution of hydraulic conductivity change as well. In a semi-distributed fashion, values for the parameters are weighted based on area size by each land use and subsequently averaged for the whole catchment. The initial estimates of the parameters were extracted from literature addressing parameterization of land cover in TOPMODEL [25,26,49,51,59].

In order to implement the subsurface storage, each land cover type was allowed to have its specific mean catchment deficit (S_{LUi}). The average specific mean catchment deficit (S_i) was obtained by area weighted averaging. The recharge rate from each land cover type was areally-weighted and summed before updating S_i at each time step. The calibrated and validated model parameter set and the meteorological forcing date for the period enclosing 2008 was applied to 10-year periods enclosing the 1986 and 1994 land cover images. In this approach, hydrological impacts by urbanization and land conversion were made explicit since only effects by land cover changes (*i.e.*, urbanization) are considered.

Interception and Evaporation

For the estimation of rainfall interception in this study, an interception technique adopted from the agro-hydrological model Soil-Water-Atmosphere-Plant (SWAP) [60,61] was adopted. Interception is assumed not to contribute to infiltration or runoff production and therefore an interception depth is subtracted from the rainfall before infiltration and runoff production are estimated. Interception loss was therefore estimated from Leaf Area Index (*LAI*) values, which were calculated from the Soil Adjusted Vegetation Index (SAVI) [62,63] using the 1986, 1994 and 2008 Landsat images. *SAVI* reads:

$$SAVI = \frac{NIR - R}{NIR + R + L}(1 + L) \tag{6}$$

where:

NIR = near-infrared reflectance;
R = red reflectance;
L = soil adjustment factor, most often defined as 0.5 for intermediate vegetation.

LAI is defined as the ratio of the total area of all leaves on a plant to the ground area covered by the plant. The *LAI* was computed from the *SAVI* map as follows:

$$LAI = \frac{SAVI - C1}{C2} \tag{7}$$

where:

$C1$, $C2$ = empirical constants.

Literature values for *SAVI* constants and their ranges are summarized (after [64]).

In order to determine changes in evapotranspiration due to land cover changes, evapotranspiration from each specific vegetation type or crop evapotranspiration (*ETc*) was calculated using the crop coefficient approach (*Kc*) according to Allen *et al.* [37]. In the crop coefficient approach, crop evapotranspiration is calculated by multiplying the reference evapotranspiration (*ET$_o$*) by the *Kc* values as follows:

$$ET_c = K_c ET_o \tag{8}$$

where:

ETc = crop evapotranspiration (mm·d^{-1});
ETo = reference crop evapotranspiration (mm·d^{-1});
Kc = crop coefficient.

For assessing impacts of urbanization on runoff, the hydrographs obtained in the periods 1980–1990, 1990–2000 and 2000–2010 were compared visually. Also percentage changes in accumulated and yearly maximum streamflow amongst the 3 periods were compared. In addition, flow duration curves were used to evaluate changes in the flow regimes for both catchments for the above specified periods.

3.3.3. Model Calibration and Validation

Before the model was applied for land cover change impact assessment, it was initialized, calibrated and validated. Initialization or warming of the model was for the period October 2000–September 2001, which makes up a hydrological year. Selection of periods for calibration and validation were based on the split sample test approach. For calibration, the period October 2001–September 2007 was selected whereas for validation the period 2007–2010 was selected. Results of calibration and validation were evaluated graphically by comparing observed and simulated streamflow hydrographs and numerically by the Nash-Sutcliffe Efficiency (NS) and Relative Volume Error (RVE) objective functions. For NS, values between 0.6 and 0.8 commonly indicate that the model performs fair (0.6) to good (0.8). Values between 0.8 and 0.9 indicate that the model performs very well and values between 0.9 and 1.0 indicate that the model performs extremely well [65]. RVE can vary between +∞ and −∞ with optimum value of 0. A RVE value of 0 indicates that there is no difference between simulated and observed streamflow volume. A RVE between +5% to −5% indicates that a model performs well whereas a RVE between +5% and +10% and −5% and −10% indicates a model with fair performance [65,66]. We note that interpreting model performance indices is not trivial and refer to recent studies by [67–70].

Calibration in this study was done through an iterative process in which the model parameters were manually adjusted to optimize model performance. Initial parameter values were set by considering values from literature [21,53,58,71–75]. The first step in calibration aimed at simulation of baseflow in the dry season after which calibration aimed at higher streamflows and the hydrographs in general. Secondly, parameters were calibrated so that simulated and observed recession periods matched. Lastly, parameters *m*, *SR$_{max}$*, and *T$_o$* were tuned until the rising limb of the simulated hydrograph and timing of the peak flow matched to counter parts of the observed hydrograph. By optimizing the *SR$_{max}$* parameter, the timing of the peak flow could be improved since a higher value of *SR$_{max}$* results in a model response that cause better fit of the rising limb. The model was validated for the period October 2007–2010.

96

4. Results and Discussion

4.1. Trend Analysis

4.1.1. Rainfall

Results of the Mann-Kendall test performed on rainfall data from the three meteorological stations showed that in some months there are statistically significant changes in long-term monthly rainfall (see Table 6). Specifically, statistics for the Airport rainfall station indicated a downward trend for all the months except January, March, June and August. January and March marks the midway of the rainfall season in Zimbabwe while June and August marks the middle of the dry season. Analyses further indicate a statistically significant downward trend for the months of April, May and September. April and May mark the end of the rainfall season in Zimbabwe while September marks the end of the dry season. For Kutsaga rainfall station, there was an overall decreasing trend in rainfall for seven months of the year with May being the only month that showed a statistically significant decreasing trend.

Belvedere station showed an increasing trend for eight months of the year and experienced a decreasing trend for four months of the year but the trends were not significant. For greater part of the rainfall season, which covers the period November till February, no significant trends were detected in the three stations. There is a general decreasing trend in both annual and monthly rainfall for most of the months at Kutsaga station. However, these trends are statistically not significant except for the month of May. Furthermore, findings show that although there are negative trends in annual rainfall, trends are not statistically significant. Since most of the decreases in rainfall have been observed in the dry season, these changes are likely to have minimal effect on streamflow as they contribute little to runoff production.

4.1.2. Streamflow

Table 7 illustrates annual and monthly streamflow trends based on data from Marimba and Mukuvisi gauging stations. Trend analysis results of annual and monthly streamflow between 1970 and 2006 showed that there was a significant increase in streamflow generated in the catchment as observed at Marimba and Mukuvisi gauging stations.

Table 6. The Kendall test statistic (tau) for trend analysis at annual and monthly base for rainfall. Statistically significant trends ($p < 0.05$) are shown in bold.

Month	Airport			Belvedere			Kutsaga		
	tau	z-Score	p-Value	tau	z-Score	p-Value	tau	z-Score	P-Value
January	0.015	0.118	(0.906)	0.111	0.94	(0.347)	0.006	0.041	(0.967)
February	−0.015	−0.118	(0.906)	0.013	0.095	(0.924)	−0.013	−0.095	(0.924)
March	0.089	0.759	(0.44)	0.162	1.399	(0.162)	0.127	1.076	(0.282)
April	−0.258	−2.237	**(0.025)**	−0.191	−1.649	(0.099)	−0.222	−1.894	(0.058)
May	−0.24	−2.156	**(0.031)**	−0.206	−1.859	(0.063)	−0.262	−2.293	**(0.022)**
June	0.074	0.723	(0.45)	0.083	0.853	(0.394)	0.098	0.949	(0.343)
July	−0.137	−1.352	(0.177)	−0.062	−0.591	(0.554)	−0.068	−0.621	(0.535)
August	0.106	1.076	(0.282)	0.133	1.32	(0.187)	0.108	1.092	(0.275)
September	−0.267	−2.366	**(0.018)**	0.016	0.123	(0.902)	−0.208	−1.829	(0.067)
October	−0.01	−0.068	(0.946)	−0.111	−0.94	(0.347)	−0.04	−0.327	(0.744)
November	−0.01	−1.839	(0.066)	0.089	0.749	(0.454)	−0.14	−1.185	(0.236)
December	−0.01	0.041	(0.967)	0.749	0.454	(0.089)	0.029	0.231	(0.819)
Annual	−0.107	−0.916	(0.36)	−0.051	−0.422	(0.673)	−0.14	−1.185	(0.236)

Table 7. The Kendall test statistic (tau) for trend analysis for annual and monthly streamflow measured from 1970 to 2008 at Marimba and Mukuvisi gauging stations. Statistically significant trends ($p < 0.05$) are shown in bold.

Month	Mukuvisi Gauging Station			Marimba Gauging Station		
	tau	z-Score	p-Value	Tau	Z-Score	p-Value
January	0.276	2.393	**(0.017)**	0.203	1.785	(0.074)
February	0.165	1.426	(0.154)	0.124	1.081	(0.280)
March	0.255	2.21	**(0.027)**	0.195	1.71	(0.087)
April	0.264	2.289	**(0.022)**	0.283	2.49	**(0.013)**
May	0.354	3.074	**(0.002)**	0.42	3.696	**(0.000)**
June	0.554	4.813	**(0.000)**	0.539	4.752	**(0.000)**
July	0.435	3.78	**(0.000)**	0.607	5.356	**(0.000)**
August	0.572	4.97	**(0.000)**	0.63	5.558	**(0.000)**
September	0.614	5.337	**(0.000)**	0.631	5.558	**(0.000)**
October	0.628	5.454	**(0.000)**	0.578	5.093	**(0.000)**
November	0.444	3.858	**(0.000)**	0.35	3.08	**(0.002)**
December	0.336	2.917	**(0.004)**	0.337	2.97	**(0.003)**
Annual	0.357	3.099	**(0.001)**	0.295	2.59	**(0.009)**

Streamflow of Mukuvisi catchment showed a significant positive trend ($p < 0.05$) for all months of the year except for February in which streamflow has increased (Table 6). For Marimba catchment, streamflow does not show any significant trend between January and March indicating that the streamflow did not notably change for these months. Analysis of streamflow measured at Marimba and Mukuvisi gauging stations indicate that mean monthly streamflow increased from 7.55 m³ and 4.51 m³ in 1970 to 35.01 m³ and 25.18 m³ in 2006, respectively which suggest large changes in mean monthly streamflow during a period covering nearly four decades.

4.2. Topographic Index

Figure 2 (left panels) shows the spatial variation of elevation for the Marimba and Mukuvisi catchments. Elevation in both catchments range between 1300 m and 1600 m and, as such, indicates little variation. The dominant flow direction in both catchments is south west (middle panels). Regions of higher topographic index (>20) for both catchments are found along rivers and along gentle slopes (right panels). The upstream areas that represent low topographic index are called runoff contributing areas. Comparatively the low lying areas which showed a high topographic index represent zones of saturation [21,76]. In this work, the topography of the two catchments, which is critical for hydrological simulation, is represented by use of a satellite derived ASTER DEM.

4.3. Land Cover Changes

Land cover for Marimba and Mukuvisi catchments was analyzed for the years 1986, 1994 and 2008 for which satellite images were available. Results of the accuracy assessment based on Kappa statistics show that the accuracy levels were above 0.92 for all years (Table 8). A Kappa statistic of more than 92% indicate that there is almost an agreement between land cover indicated by the classified images and ground control points relative to the agreement that can be expected by chance [77,78]. As such, the classified images well represent the land cover in the catchments and the results are suitable for further use.

Figure 2. The spatial variation of topographic derivatives, such as elevation (**left pannels**), local flow direction (**middle panels**) and yopographic index (**right panels**).

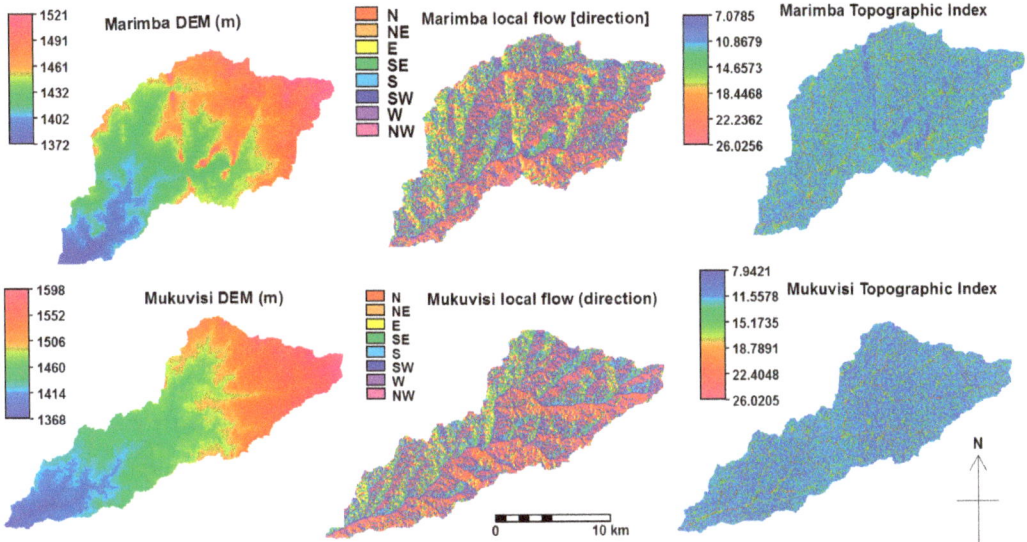

Table 8. Kappa statistics for the classified images of the Marimba and Mukuvisi catchments for 1986, 1994 and 2008.

Catchment	Year	Kappa Statistic
	1986	0.981
Marimba	1994	0.983
	2008	0.922
	1986	0.995
Mukuvisi	1994	0.978
	2008	0.93

4.3.1. Marimba Catchment

Marked changes in land cover were observed in Marimba catchment between 1986 and 2008 (Table 9 and Figure 3). The urban area increased from 34.62 km^2 in 1986 to 40.15 km^2 in 1994 and subsequently increased to 71.95 km^2 in 2008. The increase in urban areas was larger for the period 1994–2008 compared to the period 1986–1994. Results also showed that the area covered by woodlands in Marimba catchment decreased from 99.94 km^2 in 1986 to 57.26 km^2 in 2008. The decrease in woodland area was larger between 1986 and 1994 (25.98 km^2) compared to the decrease in area between 1994 and 2008 (16.70 km^2). Grasslands showed an increase from 1986 to 1994 and a decrease in 1994 and 2008 thus changes were not consistent over the 22-year period. The changes in grasslands can be attributed to conversion of grassland into other land cover types, such as woodland (e.g., through reforestation) as well as clearance for urban developments (see discussion on land cover conversions in Marimba catchment). The aquatic weeds and the water class remained relatively unchanged over the study period.

Table 9. Land conversions in square kilometers for Marimba catchment for the period 1986–2008.

Land Cover	1986	1994	2008
Aquatic weeds	0.11	0.09	0.11
Urban	34.62	40.15	71.95
Cropped Field	9.24	3.25	2.70
Grassland	76.41	102.95	88.38
Water	0.25	0.18	0.17
Woodland	99.94	73.96	57.26
Total	220.57	220.57	220.57

The land cover conversions in Marimba catchment for the years 1986, 1994 and 2008 were determined using overlay analysis in a GIS. Results from land cover analysis show that significant proportions of grasslands (13.75 km^2) and woodlands (6.82 km^2) were converted to urban area between 1986 and 1994. Although there were some conversions from grasslands to woodland and *vice versa*, there was a pronounced decrease of 31.22 km^2 in woodlands over the same period. The second period (1994–2008) was characterized by considerable conversion of grasslands (31.58 km^2) and woodlands (18.56 km^2) to urban area. Despite some conversions from urban to grasslands (13.74 km^2) and to woodlands (4.73 km^2) there is a net gain of the urban area from the two classes.

Figure 3. Results of Land cover classification in Marimba (**top panels**) and Mukuvisi (**bottom panels**) catchments for 1986, 1994 and 2008.

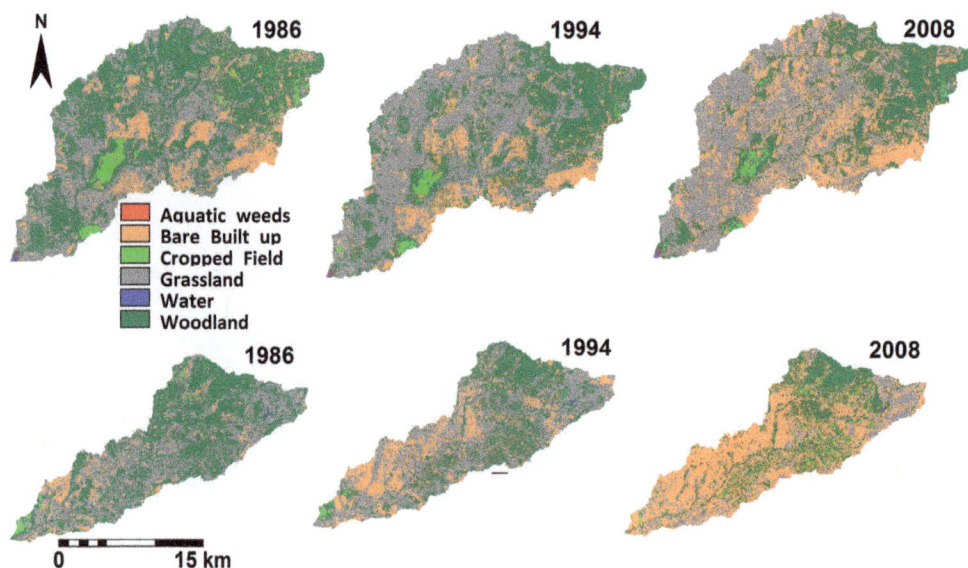

4.3.2. Mukuvisi Catchment

For Mukuvisi catchment findings indicated that the urban area increased from 21.43 km² to 56.31 km² between 1986 and 1994 (Table 10). The same land cover class increased substantially to 135.04 km² in 2008. Cropped fields and grasslands varied during the same period. In contrast, woodlands decreased considerably from 100.78 km² in 1986 to 65.19 km² in 1994 and subsequently decreased to 52.12 km² in 2008. Thus, woodlands decreased by nearly 50% whereas the urban area increased by more than 500% over the twenty-two year period.

Table 10. Land conversions in the Mukuvisi catchment between 1986 and 2008 (Area is in square kilometers).

Land Cover	1986	1994	2008
Urban	21.43	56.31	135.04
Cropped Field	3.45	1.24	5.67
Grassland	97.15	100.21	30.12
Water	0.38	0.24	0.24
Woodland	100.78	65.19	52.12
Total	223.20	223.20	223.20

An analysis of land cover in the Mukuvisi catchment showed that between 1986 and 1994, 27.63 km² of grassland was converted to urban area. About 14.45 km² of woodland was also converted to urban area. Table 11 provides a summary of land cover conversions that occurred in the catchment.

Table 11. Land conversions in Mukuvisi catchment between 1986 and 1994 (Area is in square kilometers).

1986	1994				
Land Cover Type	Urban	Cropped Field	Grassland	Water	Woodland
Urban	13.82	0.11	6.16	0.00	1.33
Cropped Field	0.37	0.80	0.51	0.00	1.77
Grassland	27.63	0.18	54.03	0.02	15.29
Water	0.04	0.00	0.08	0.16	0.11
Woodland	14.45	0.15	39.42	0.07	46.69

The period 1994 to 2008 experienced a conversion of 64.87 km² of grassland to urban area whereas 23.11 km² of woodland was converted to the urban class (Table 12).

Table 12. Land conversions in Mukuvisi catchment between 1994 and 2008 (Area is in square kilometers).

1994	2008				
Land Cover Type	Urban	Cropped Field	Grassland	Water	Woodland
Urban	46.11	1.04	4.96	0.02	4.18
Cropped Field	0.88	0.15	0.08	0.00	0.14
Grassland	64.87	0.99	20.50	0.05	13.79
Water	0.07	0.01	0.01	0.12	0.03
Woodland	23.11	3.48	4.57	0.05	33.97

102

Findings for grassland indicate that 13.79 km² was converted to woodland during the same period whereas about 4.57 km² of woodland was converted to grassland. When combined this indicates a decrease of woodland.

4.4. Model Calibration and Validation

Figures 4 and 5 show results of model calibration, whereas Table 13 shows the optimized model parameter values. Generally, the model was able to successfully reproduce the peak flows and the baseflow throughout the years 2000–2010. However, for Mukuvisi catchment there are overestimations of simulated streamflow throughout the simulation period. For Marimba and Mukuvisi catchments Nash-Sutcliffe (NS) efficiencies of 0.79 and 0.70 were obtained, respectively, suggesting a fair model performance. A Relative Volume Error (RVE) of 6% and 5.2% was obtained for Marimba catchment and Mukuvisi catchment, respectively and indicate that the total streamflow is somewhat overestimated. However, RVE values were in the range of −10% to 10% [66] which, by itself, suggests a fair model performance in terms of representing the catchment water balance. [22] and [79] assert that the proper characterization of topography plays an important role in runoff generation and thus the obtained NS and RVE objective function values indicate that the topographic-index distribution function for both catchments are adequate.

For Marimba and Mukuvisi catchments, for the validation period (2009–2010) the model reproduced the observed streamflow hydrographs quite well but mostly overestimated peak flow discharges. The NS efficiency for the validation period for Marimba and Mukuvisi catchments were 0.74 and 0.65, respectively. The RVE for Marimba and Mukuvisi catchments were 7.4% and 10%, respectively, which also indicates that the model overestimated the streamflow volume.

Figure 4. Model calibration results Marimba catchment (October 2001–September 2007).

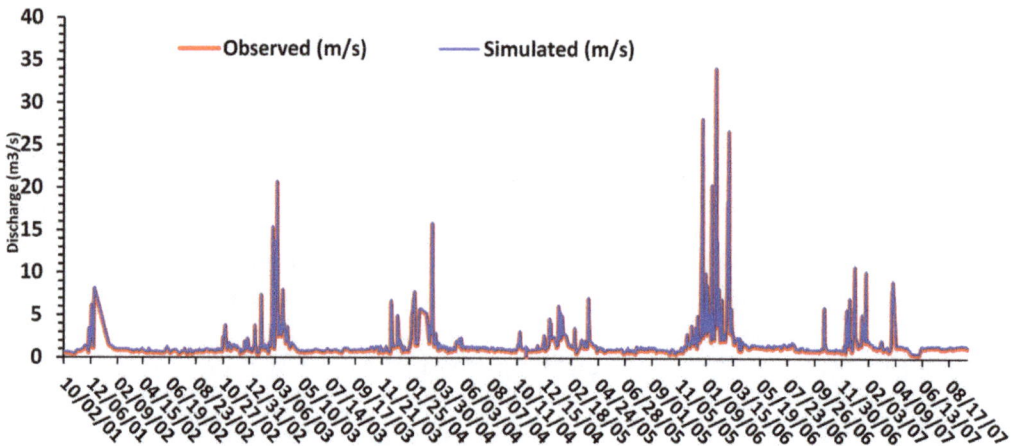

Figure 5. Model calibration results Mukuvisi catchment (October 2001–September 2007).

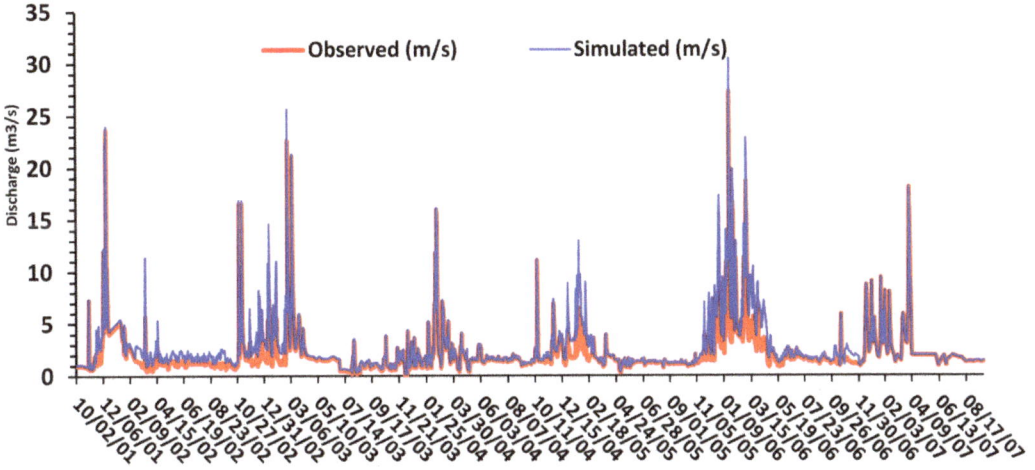

Table 13. Optimized parameter values used in the model for Marimba and Mukuvisi catchments.

Parameter	m (m)	T_o (m²/h)	Td (h)	CHV (m/h)	RV (m/h)	SRMAX (m)	Q0 (m/day)	SR0 (m)
Marimba	0.045	5	20	3600	1700	0.045	0.000286	0.001
Mukuvisi	0.035	5	22	3500	1500	0.035	0.000329	0.002

Figure 6. Observed and simulated streamflow for Marimba catchment (2000–2010).

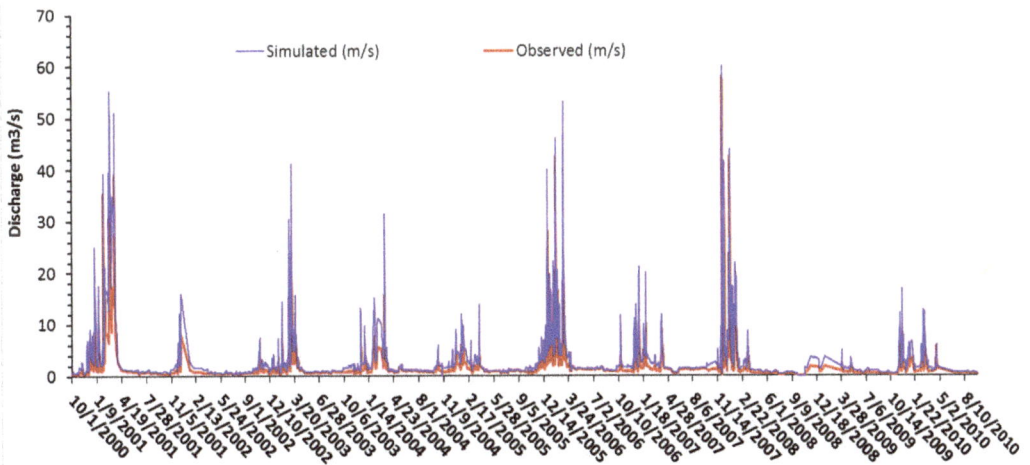

4.5. Simulation Results under Land Conversions

For assessment of hydrologic impacts, the optimized parameter sets were applied to the historic time periods 1980–1990 and 1990–2000. For these periods, classified land covers served as TOPMODEL input to make hydrologic impacts and effects of land cover changes explicit. In the procedure the optimized parameters for each land cover for the period 2000–2010 are selected and applied to the land cover of the historic time period. In this procedure, changes in land cover areas are represented by a re-distribution of optimized model parameters across the catchments. The premise for model impact assessments is that any difference in the streamflow hydrograph characteristics is a direct result of the changed spatial distribution of model parameters by the different degrees of urbanization. It is noted that rainfall and ET for the 2000–2010 period is used for the 1990–2000 and 1980–1990 periods and therefore model forcing remained unchanged. In such procedure the simulated streamflow hydrographs for the period 2000–2010 also may serve as reference to simulation results of the historic period to assess hydrological impacts. Results of streamflow simulations are shown in Figures 6 and 7. It is shown that peak flows and baseflows in both catchments were well represented.

Figure 7. Observed and simulated streamflow for Mukuvisi catchment (2000–2010).

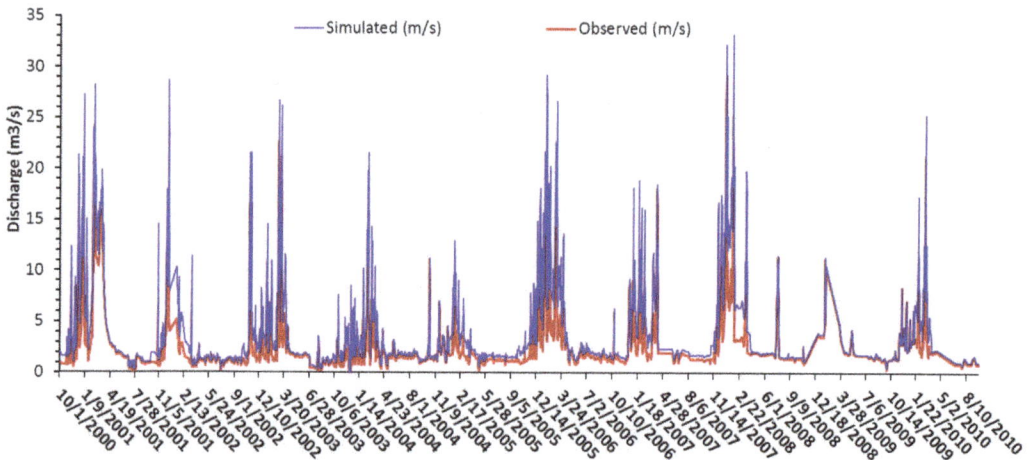

A comparison of simulated streamflow hydrograph for all three periods indicates that runoff behavior has changed in both Marimba and Mukuvisi catchments (Figures 8 and 9). Streamflow hydrographs for the simulation period showed that during the period 2000–2010 more runoff was generated than in the periods 1990–2000 and 1980–1990. The peak flows of the 2000–2010 period were higher than the peak flows of the other two periods. The baseflow of the period 1980–1990 was higher than the baseflow of the period 1990–2000, with the period 2000–2010 being the lowest. The average yearly streamflow for Marimba and Mukuvisi in Table 14 shows that streamflow increased between 1980–1990 and 1990–2000 in both catchments. For the period 1980–2010 there was a notable increase in streamflow by 46% and 45%, respectively for the Marimba and Mukuvisi catchments. Table 14 also shows an increase in the yearly highest streamflow for the same period, which coincided with an increased loss of forest area. The increase in streamflow simulated by TOPMODEL for the three consecutive periods (see Table 14) suggests that the changes in vegetation and soil permeability through urbanization is the main cause for changes in the streamflow. Results of model simulations for the period 1980–2000 in Marimba and Mukuvisi catchments indicated relative increases in mean annual streamflow by 8.5% and 8.4% respectively. For the 20 year period (1990–2010) increases as large as 34.6% and 33.3% were experienced in Marimba and Mukuvisi catchments, respectively, and suggest a progressive and accelerated impact. For the entire 30-year period (1980–2010), increases of mean annual streamflow were as large as 45.9% and 44.5% for Marimba and Mukuvisi catchments, respectively. The highest streamflow values observed during the simulation period were used to assess the impacts of land conversions. Table 14 shows the same trend of increase as experienced in the mean annual streamflow from 1980 to 2010. Findings indicated that urbanization resulted in enlarged areas of reduced infiltration potential thus causing more frequent rapid-runoff responses but also increasing streamflow discharges. The baseflow of the period 2000–2010 in the two catchments is lower than for the period 1980–1990 presumably because of reduced infiltration. We note that reduced forest area caused a reduction in baseflow as forest soils often are characterized by relatively high infiltration whereas root zones are relatively deep and hold and store water. From a modeling point of view, findings suggest the ability of TOPMODEL to add to insights on the changes in hydrological system behavior due to urbanization and land conversion as observed by satellite remote sensing.

Figure 8. Comparison of streamflow hydrographs in Marimba catchment for the three periods.

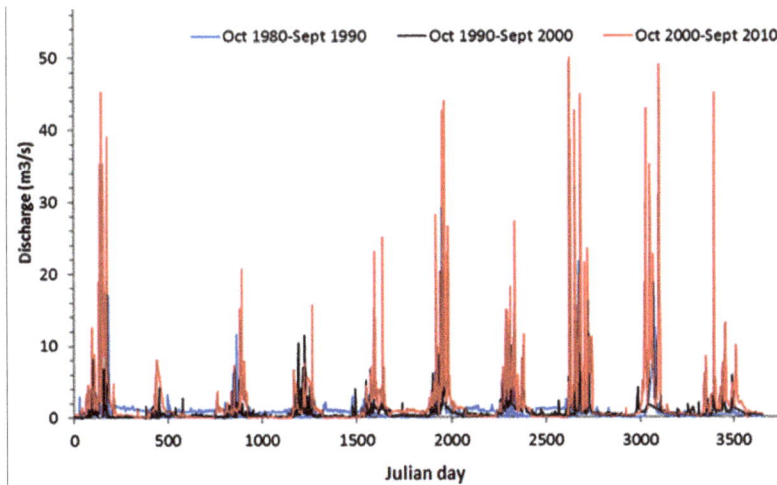

Figure 9. Comparison of streamflow hydrographs in Mukuvisi catchment for the three periods.

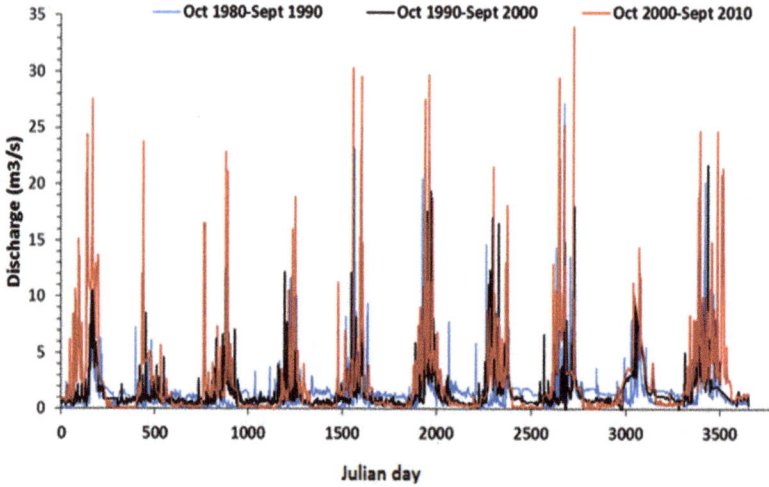

Table 14. Comparison of simulated Mean Annual Streamflow (Q_{mean}) and Yearly Highest Streamflow (Q_{hst}) for respective periods.

Period	Catchment	Q_{mean} (m³/s)	Q_{hst} (m³/s)
1980–1990	Marimba	491.2	16.2
	Mukuvisi	529.8	21.9
1990–2000	Marimba	532.7	29.4
	Mukuvisi	574.1	27.2
2000–2010	Marimba	716.8	50.0
	Mukuvisi	771.0	34.0
		% Change in Q_{mean}	**% Change in Q_{hst}**
1980–2000	Marimba	8.5	81.6
	Mukuvisi	8.4	55.6
1990–2010	Marimba	34.6	70.1
	Mukuvisi	33.3	24.9
		% Change in Qmean	**% Change in Qhst**
1980–2010	Marimba	45.9	208.8
	Mukuvisi	44.5	55.6

Figures 10 and 11 show the flow duration curves for Marimba and Mukuvisi catchments which were used to assess changes in the flow regimes. The flow duration curves show the relationship between the magnitude of streamflow discharges and % number flow discharges are exceeded or equaled. Inter-comparison of the curves and their shifts serves to assess hydrological impacts due to land conversions. Low flows (<1 m³/s) in the Marimba catchment were exceeded 20% of the times for the period 1980–1990, some 22% for the period 1990–2000 and 54% for the period 2000–2010. For Mukuvisi catchment, flow discharge of 1 m³/s was exceeded 58% of the times and decreased to 45% for the period 1990–2000 and to 60% for the period 2000–2010. Furthermore, streamflow discharge of 10 m³/s was equaled or exceeded <1% of the times in both the periods 1980–1990 and 1990–2000 as compared to 7% in the period 2000–2010 for Marimba catchment. For Mukuvisi catchment, stream flow discharge of 10 m^3/s was exceeded 5% in the period 2000–2010. The slope of all the flow duration curves in the period 2000–2010 for higher streamflow is much steeper as compared to the other periods for both catchments. This is an indication of the effect of urbanization which causes higher streamflow responses during the rainy season [80]. For streamflow discharges higher than 30 m^3/s there is no percentage exceedance for the 1980–1990 and 1990–2000 period and are alike conclusions found in [81] in a Tanzanian catchment where changes in flow duration curves were attributed to effects of land conversions.

The flow duration curve for the period 2000–2010 indicates that hydrological impacts are more pronounced compared to earlier periods and particularly applies for higher streamflow discharges. These findings well match to results of land cover classification, which indicate accelerated urbanization for the last (2000–2010) time period.

Figure 10. Flow duration curves in the Marimba catchment for the periods 1980–1990, 1990–2000 and 2000–2010.

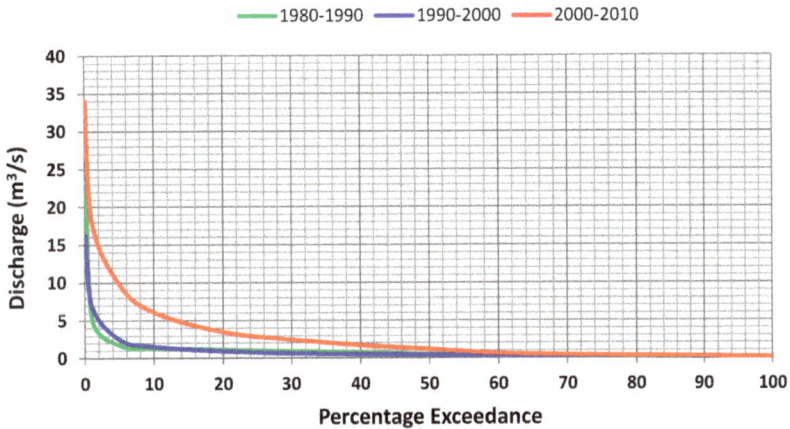

Figure 11. Flow duration curves in the Mukuvisi catchment for the periods 1980–1990, 1990–2000 and 2000–2010.

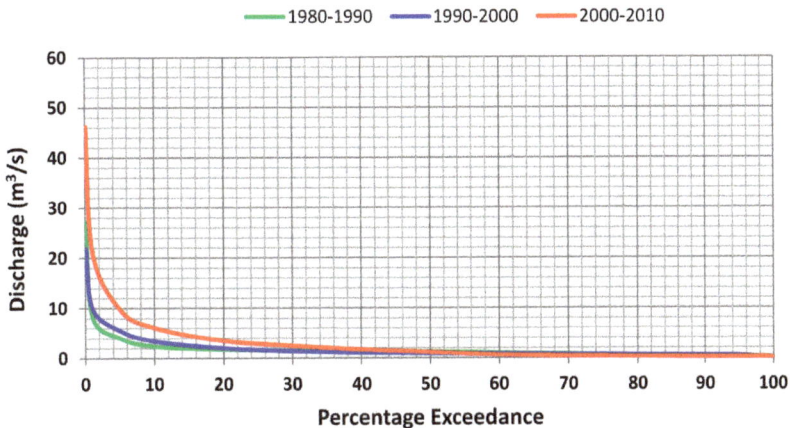

5. Conclusions and Recommendations

Results of satellite image classification for land cover change assessment in Mukuvisi and Marimba catchments in the city of Harare have shown that the urban area increased by more than 500% in the Mukuvisi catchment and by more than 200% in the Marimba catchment between 1986 and 2008. Woodlands decreased by more than 40% over the same period in the two catchments with a larger decrease in Marimba than in Mukuvisi. Findings on land conversion showed increased conversion of grasslands and woodlands to urban area over the past decades. This accelerated urbanization suggests that several land cover types have been converted to impervious surfaces over the past few decades.

Statistical analysis on rainfall and streamflow time series indicated a significant decreasing trend ($p < 0.05$) for rainfall and significant increasing trend ($p < 0.05$) for streamflow. The increasing trends in streamflow could be attributed to the increase in low permeability land surfaces in the two catchments. Results of streamflow modeling for Marimba and Mukuvisi catchments indicated that the mean annual streamflow increased by 46% and 45%, respectively from 1980 to 2010. These increases coincided with the decrease in forest area and an increase in urban area over the same period. As such, findings of this study indicate clear impacts by urbanization in the two catchments. The observed streamflow increases due to land conversions in this study are relatively high compared to other studies (e.g., [82]) which have shown that a 10% increase in imperviousness, results in an increase in the range of 9.8% to 10.2% in annual mean streamflow. A significant impact of urbanization on hydrological regimes is the increase of impervious surfaces, which cause increased streamflow volumes due to the reduction in soil infiltration capacity. As such, urbanized surfaces are likely to generate more runoff than areas, which are densely covered with vegetation especially woodlands. Also, the increase in paved and roofed surfaces reduces the area over which precipitation can infiltrate the soil and results in increased overland flow which, by itself, contributes to quick runoff and streamflow. It can be concluded that clearance of woodlands through urbanization has significantly altered the streamflow regimes in both catchments. These opposing signals in rainfall and streamflow trends signify that the increase in low permeable land surfaces as a result of urbanization probably is the main cause for the streamflow increases.

This study further demonstrated that a widely accepted rainfall-runoff modeling approach can be extended beyond its basic purpose of predicting local variations in water table utilizing the topographic index. To simulate impacts of land use change in this study, land surface parameterization for the rainfall-runoff model was successfully carried out through quantifying the topographic indices, land cover and vegetation indices for urbanization impact assessment. Parameterization served to estimate interception loss, evapotranspiration loss and infiltration excess overland flow by means of the Green and Ampt approach. An approach was applied that used State of the Art GIS and satellite imagery to represent land cover for the years 1986, 1994 and 2008, respectively. For this study, TOPMODEL was run for periods of 10 years which enfolded the dates the satellite images were acquired. This study therefore provided insights into the hydrologic cycle and its regime when a natural or peri-urban catchment undergoes urbanization. Results can be used in the broader spectrum of integrated water resources management and are consistent with observations by [83].

Finally, the study provided insights into hydrologic impacts by an increase in built-up areas and paved surfaces as a result of the urbanization of natural or peri-urban catchments. The findings of this study are highly relevant to many African countries, which are facing accelerated rural-urban migration over the past decades. The latter has been shown in several demographic surveys across Africa with many catchments undergoing rapid urbanization. For instance in Nairobi, Kenya [84] showed the rapid encroachment of urban areas using satellite imagery but hydrological impact assessments are still lacking. Similarly, [85] indicated a rapid increase in urban settlement between 1990 and 2000 in Port Elizabeth, South Africa. These results have important implications on water resources management in Africa, where a number of countries are undergoing rapid urbanization.

The authors recommend that besides field measurements to verify model parameters, future work must apply hydrologic models with a clear physical base, such as the Representative Elementary Watershed model (see [86,87]) to allow better evaluation of the impacts of land cover changes and rainfall distributions on the hydrologic regime. In addition, changes in actual evapotranspiration as caused by urbanization must be assessed spatially. Future work also should integrate climate change impacts with impacts of land conversions on streamflow since both have feedback impacts. Therefore, studying these impacts will greatly benefit the water managers in decision-making. In addition, by urbanization, unmonitored wastewater disposal into urban streams have impacts on streamflow and this is scheduled for future work.

Acknowledgments

The authors would want to thank Upper Manyame Sub-catchment Council (UMSCC) for supporting this research. Scott Peckham is greatly acknowledged for the IDL TOPMODEL version which was adapted in this study. Special mention is given to METI and NASA for the ASTER GDEM and USGlovis for the Landsat images.

Author Contributions

W.G. had the original idea for runoff simulation under land conversion in Harare city. He was responsible for the TOPMODEL simulations. T.R. was responsible for the research approach and conceptualization and prepared the TOPMODEL code. He also made large contributions to the manuscript write-up. M.D.S. was responsible for land cover classification and trend analysis for rainfall and runoff. D.T.R. was responsible for reviewing relevant literature, writing up the introduction section and analyzing the simulated flows to produce the flow duration curves. I.N. was responsible for synthesizing the introduction and conclusions and relating findings to previous work in the Upper Manyame catchment. A.T.H. assisted in interpretation of modeling results and land cover classification.

Conflicts of Interest

The authors declare no conflict of interest.

References

1. Dube, T.; Gumindoga, W.; Chawira, M. Detection of landcover changes based on traditional remote sensing image classification techniques, around Lake Mutirikwi, Zimbabwe. *Afr. J. Aquat. Sci.* **2014**, *39*, 89–95.

2. Wijesekara, G.N.; Gupta, A.; Valeo, C.; Hasbani, J.G.; Qiao, Y.; Delaney, P.; Marceau, D.J. Assessing the impact of future land-use changes on hydrological processes in the Elbow River watershed in southern Alberta, Canada. *J. Hydrol.* **2012**, *412–413*, 220–232.

3. Jothityangkoon, C.; Hirunteeyakul, C.; Boonrawd, K.; Sivapalan, M. Assessing the impact of climate and land use changes on extreme floods in a large tropical catchment. *J. Hydrol.* **2013**, *490*, 88–105.

4. Meenu, R.; Rehana, S.; Mujumdar, P.P. Assessment of hydrologic impacts of climate change in Tunga-Bhadra river basin, India with HEC-HMS and SDSM. *Hydrol. Process.* **2013**, *27*, 1572–1589.

5. Dams, J.; Dujardin, J.; Reggers, R.; Bashir, I.; Canters, F.; Batelaan, O. Mapping impervious surface change from remote sensing for hydrological modeling. *J. Hydrol.* **2013**, *485*, 84–95.

6. Furusho, C.; Chancibault, K.; Andrieu, H. Adapting the coupled hydrological model ISBA-TOPMODEL to the long-term hydrological cycles of suburban rivers: Evaluation and sensitivity analysis. *J. Hydrol.* **2013**, *485*, 139–147.

7. Ackerman, D.; Stein, E.D. Estimating the variability and confidence of land use and imperviousness relationships at a regional scale1. *JAWRA J. Am. Water Resour. Assoc.* **2008**, *44*, 996–1008.

8. Bach, M.; Ostrowski, M. Analysis of intensively used catchments based on integrated modelling. *J. Hydrol.* **2013**, *485*, 148–161.

9. Braud, I.; Breil, P.; Thollet, F.; Lagouy, M.; Branger, F.; Jacqueminet, C.; Kermadi, S.; Michel, K. Evidence of the impact of urbanization on the hydrological regime of a medium-sized periurban catchment in France. *J. Hydrol.* **2013**, *485*, 5–23.

10. Burns, D.; Vitvar, T.; McDonnell, J.; Hassett, J.; Duncan, J.; Kendall, C. Effects of suburban development on runoff generation in the Croton River basin, New York, USA. *J. Hydrol.* **2005**, *311*, 266–281.

11. Rose, S.; Peters, N.E. Effects of urbanization on streamflow in the Atlanta area (Georgia, USA): A comparative hydrological approach. *Hydrol. Process.* **2001**, *15*, 1441–1457.

12. Dow, C.L.; DeWalle, D.R. Trends in evaporation and Bowen Ratio on urbanizing watersheds in eastern United States. *Water Resour. Res.* **2000**, *36*, 1835–1843.

13. Chu, M.L.; Knouft, J.H.; Ghulam, A.; Guzman, J.A.; Pan, Z. Impacts of urbanization on river flow frequency: A controlled experimental modeling-based evaluation approach. *J. Hydrol.* **2013**, *495*, 1–12.

14. Du, J.; Qian, L.; Rui, H.; Zuo, T.; Zheng, D.; Xu, Y.; Xu, C.Y. Assessing the effects of urbanization on annual runoff and flood events using an integrated hydrological modeling system for Qinhuai River basin, China. *J. Hydrol.* **2012**, *464–465*, 127–139.

15. Verbeiren, B.; van de Voorde, T.; Canters, F.; Binard, M.; Cornet, Y.; Batelaan, O. Assessing urbanisation effects on rainfall-runoff using a remote sensing supported modelling strategy. *Int. J. Appl. Earth Observ. Geoinf.* **2013**, *21*, 92–102.

16. Wang, D.; Gong, J.; Chen, L.; Zhang, L.; Song, Y.; Yue, Y. Spatio-temporal pattern analysis of land use/cover change trajectories in Xihe watershed. *Int. J. Appl. Earth Observ. Geoinf.* **2012**, *14*, 12–21.

17. Lillesand, T.M.; Kiefer, R.W. *Remote Sensing and Digital Image Interpretation*, 4th ed.; Wiley: New York, NY, USA, 2000; p. 724.

18. Beven, K. TOPMODEL: A critique. *Hydrol. Process.* **1997**, *11*, 1069–1085.

19. Refsgaard, J.C.; Storm, B.; Refsgaard, A. Validation and applicability of distributed hydrological models. In *Modelling and Management of Sustainable Basin-Scale Water Resource Systems*; IAHS Publ.: Oxfordshire, UK, 1995; No. 231, pp. 387–397.

20. Sorooshian, S.; Gupta, V. Model calibration. In *Computer Models of Watershed Hydrology*; Singh, V.P., Ed.; Water Resources Management Publications: Highlands Ranch, CO, USA, 1995; pp. 23–67.

21. Quinn, P.F.; Beven, K.J.; Lamb, R. The Ln(a/TanB) Index: How to calculate it and how to use it within the Topmodel framework. *Hydrol. Process.* **1995**, *9*, 161–182.

22. Beven, K.; Freer, J. A dynamic TOPMODEL. *Hydrol. Process.* **2000**, *15*, 1993–2011.

23. Beven, K.J. *Rainfall-Runoff Modelling: The Primer*; John Wiley & Sons: Lancaster, UK, 2001.

24. Gumindoga, W.; Rwasoka, D.T.; Murwira, A. Simulation of streamflow using TOPMODEL in the Upper Save River catchment of Zimbabwe. *Phys. Chem. Earth.* **2011**, *36*, 806–813.

25. Gunter, A.; Uhlenbrook, S.; Sibert, J.; Leibundgut, C. Multi-criterial validation of TOPMODEL in a mountaneous catchment. *Hydrol. Process.* **1999**, *13*, 1603–1620.

26. Huang, B.; Jiang, B. AVTOP: A full integration of TOPMODEL into GIS. *Environ. Model. Softw.* **2002**, *17*, 261–268.

27. Nourani, V.; Roughani, A.; Gebremichael, M. TOPMODEL capability for rainfall-runoff modeling of the Ammameh watershed at different time scales using different terrain algorithms. *J. Urban Environ. Eng.* **2011**, *5*, 1–14.

28. Braud, I.; Fletcher, T.D.; Andrieu, H. Hydrology of peri-urban catchments: Processes and modelling. *J. Hydrol.* **2013**, *485*, 1–4.

29. Valeoa, C.; Moinb, S.M.A. Variable source area modelling in urbanizing watersheds. *J. Hydrol.* **2000**, *228*, 68–81.

30. United Nations-Department of Economic and Social Affairs, Population Division. *World Population Prospects: The 2012 Revision, Highlights and Advance Tables*; Working Paper No. ESA/P/WP.228; United Nations: New York, NY, USA, 2013.

31. ZIMSTATS. *Census 2012 Preliminary Report*; ZIMSTATS: Harare, Zimbabwe, 2012; p. 123.

32. Gumbo, B. Re-engineering the urban drainage system for resource recovery and protection of drinking water supplies. In Proceedings of the 2nd WARFSA/WaterNet Symposium: Integrated Water Resources Management: Theory, Practice, Cases, Cape Town, South Africa, 30–31 October 2001.

33. Hranova, R.; Gumbo, B.; Kaseke, E.; Klein, J.; Nhapi, I.; van der Zaag, P. The challenge of integrated water resources management in the Chivero basin, Zimbabwe. In Proceedings of the 2nd WARFSA/WaterNet Symposium: Integrated Water Resources Management: Theory, Practice, Cases, Cape Town, South Africa, 30–31 October 2001.

34. Rwasoka, D.T.; Gumindoga, W.; Gwenzi, J. Estimation of actual evapotranspiration using the Surface Energy Balance System (SEBS) algorithm over land surfaces. *J. Phys. Chem. Earth* **2011**, *36*, 736–746.

35. Japanese International Co-Operation Agency (JICA). *The Study of Water Pollution Control in Upper Manyame River Basin in the Republic of Zimbabwe*; MLGRUD: Harare, Japan; Nippon Jogeduido Sekkei Co. Ltd.: Tokyo, Japan; Nippon Koei Co. Ltd.: Tokyo, Japan, 1996.

36. Nhapi, I.; Zvikomborero, Z.; Siebel, M.A.; Gijzen, H.J. Assessment of the major water and nutrient flows in the Chivero catchment area, Zimbabwe. *Phys. Chem. Earth* **2002**, *27*, 783–792.

37. Allen, R.G.; Pereira, L.S.; Raes, D.; Smith, M. Crop evapotranspiration-guidelines for computing crop water requirements. In *FAO Irrigation and Drainage Paper 56*; FAO: Rome, Italy, 1998.

38. Tebbs, E.J.; Remedios, J.J.; Harper, D.M. Remote sensing of chlorophyll-a as a measure of cyanobacterial biomass in Lake Bogoria, a hypertrophic, Saline–Alkaline, Flamingo lake, using Landsat ETM+. *Remote Sens. Environ.* **2013**, *135*, 92–106.

39. Lu, D.; Weng, Q. A survey of image classification methods and techniques for improving classification performance. *Int. J. Remote Sens.* **2007**, *28*, 823–870.

40. Zheng, H.; Zhang, L.; Liu, C.; Shao, Q.; Fukushima, Y. Changes in streamflow regime in headwater catchments of the Yellow River basin since the 1950s. *Hydrol. Process.* **2007**, *21*, 886–893.

41. Rientjes, T.H.M.; Haile, A.T.; Kebede, E.; Mannaerts, C.M.M.; Habib, E.; Steenhuis, T.S. Changes in land cover, rainfall and streamflow in Upper Gilgel Abbay catchment, Blue Nile basin—Ethiopia. *Hydrol. Earth Syst. Sci.* **2011**, *15*, 1979–1989.

42. Yue, S.; Pilon, P.; Cavadias, G. Power of the Mann-Kendall and Spearman's rho tests for detecting monotonic trends in hydrological series. *J. Hydrol.* **2002**, *259*, 254–271.

43. Kahya, E.; Kalayci, S. Trend analysis of streamflow in Turkey. *J. Hydrol.* **2004**, *289*, 128–144.

44. Beven, K.J.; Kirkby, M.J. A physically based, variable contributing area model of basin hydrology. *Hydrol. Sci. Bull. Sci. Hydrol.* **1979**, *24*, 43–69.

45. Gilbert, R.O. *Statistical Methods for Environmental Pollution Monitoring*; Wiley: New York, NY, USA, 1987.

46. Green, W.H.; Ampt, G. Studies of soil physics, Part I—The flow of air and water through soils. *J. Agric. Sci.* **1911**, *4*, 1–24.

47. Barry, D.A.; Parlange, J.Y.; Li, L.; Jeng, D.S.; Crapper, M. Green-Ampt approximations. *Adv. Water Resour.* **2005**, *28*, 1003–1009.

48. Swartzendruber, D. Derivation of a two-term infiltration equation from the Green-Ampt model. *J. Hydrol.* **2000**, *236*, 247–251.

49. Van den Putte, A.; Govers, G.; Leys, A.; Langhans, C.; Clymans, W.; Diels, J. Estimating the parameters of the Green-Ampt infiltration equation from rainfall simulation data: Why simpler is better. *J. Hydrol.* **2013**, *476*, 332–344.

50. Wang, J.; Endreny, T.A.; Hassett, J.M. Power function decay of hydraulic conductivity for a TOPMODEL-based infiltration routine. *Hydrol. Process.* **2006**, *20*, 3825–3834.

51. Beven, K.; Lamb, R.; Quinn, P.; Romanowicz, R.; Freer, J. TOPMODEL. In *Computer Models of Watershed Hydrology*; Sing, V.P., Ed.; Water Resources Publications: Colorado Springs, CO, USA, 1995; pp. 627–668.

52. Peters, N.E.; Freer, J.; Beven, K. Modelling hydrologic responses in a small forested catchment (Panola Mountain, Georgia, USA): A comparison of the original and a new dynamic TOPMODEL. *Hydrol. Process.* **2003**, *17*, 345–362.

53. Beven, K.J. Infiltration into a class of vertically non-uniform soils. *Hydrol. Sci.* **1984**, *29*, 425–434.

54. Wolock, D.M.; McCabe, G.J., Jr. Comparison of single and multiple flow direction algorithms for computing topographic parameters in TOPMODEL. *Water Resour. Res.* **1995**, *31*, 1315–1324.

55. Ibbitt, R.; Woods, R. Re-scaling the topographic index to improve the representation of physical processes in catchment models. *J. Hydrol.* **2004**, *293*, 205–218.

56. Fedak, R. Effect of Spatial Scale on Hydrologic Modeling in a Headwater Catchment. Master's Thesis, Virginia Polytechnic Institute and State University, Blacksburg, VA, USA, 1999.

57. Gumindoga, W. Hydrologic Impacts of Landuse Change on the Hydrology of Upper Gilgel Abbay Basin—Application of TOPMODEL Concept. Master's Thesis, University of Twente, Enschede, The Netherlands, 2010; p. 99.

58. Ambroise, B.; Beven, K.; Freer, J. Toward a generalization of the TOPMODEL concepts: Topographic indices of hydrological similarity. *Water Resour. Res.* **1996**, *32*, 2135–2145.

59. Kim, S.; Delleur, J.W. Sensitivity analysis of extended TOPMODEL for agricultural watersheds equipped with tile drains. *Hydrol. Process.* **1997**, *11*, 1243–1261.

60. Kroes, J.G.; van Dam, J.C. *Reference Manual SWAP (version 3.0.3)*; Alterra Green World Research; Alterra-Report 773; Waggeningen UR: Wageningen, The Netherlands, 2003.

61. Van Dam, J.C. *Field-Scale Water Flow and Solute Transport: SWAP Model Concepts, Parameter Estimation and Case Studies*; Wageningen Institute for Environment and Climate Research: Wageningen, The Netherlands, 2000.

62. Huete, A.R. A Soil-Adjusted Vegetation Index (SAVI). *Remote Sens. Environ.* **1988**, *25*, 295–309.

63. Van Leeuwen, W.J.D.; Huete, A.R.; Walthall, C.L.; Prince, S.D.; Bégué, A.; Roujean, J.L. Deconvolution of remotely sensed spectral mixtures for retrieval of LAI, fAPAR and soil brightness. *J. Hydrol.* **1997**, *188–189*, 697–724.

64. Parodi, G.N. *AHVRR Hydrological Analysis System Algorithms and Theory—Version 1.3. AHAS User Guide*; WRES-ITC: Enschede, The Netherlands, 2000.

65. Nash, J.E.; Sutcliffe, J.V. River flow forecasting through conceptual models. Part I: A discussion of principles. *J. Hydrol.* **1970**, *10*, 282–290.

66. Janssen, P.H.M.; Heuberger, P.S.C. Calibration of process-oriented models. *Ecol. Model.* **1995**, *83*, 55–66.

67. De Vos, N.J.; Rientjes, T.H.M. Multi-objective performance comparison of an artificial neural network and a conceptual rainfall-runoff model. *Hydrol. Sci. J.* **2007**, *52*, 397–413.

68. De Vos, N.J. and Rientjes, T.H.M. Multi-objective training of artificial neural networks for rainfall-runoff modeling. *Water Resour. Res.* **2008**, *44*, W08434.

69. Deckers, D.E.H.; Booij, M.; Rientjes, T.M.; Krol, M. Catchment variability and parameter estimation in multi-objective regionalisation of a rainfall-runoff model. *Water Resour. Manag.* **2010**, *24*, 3961–3985.

70. Rientjes, T.H.M.; Perera, B.U.J.; Haile, A.T.; Reggiani, P.; Muthuwatta, L.P. Regionalisation for lake level simulation—The case of Lake Tana in the Upper Blue Nile, Ethiopia. *Hydrol. Earth Syst. Sci.* **2011**, *15*, 1167–1183.

71. Beven, K.J.; Wood, E.F. Catchment geomorphology and the dynamics of runoff contributing areas. *J. Hydrol.* **1983**, *65*, 139–158.

72. Fisher, J.; Beven, K.J. Modelling of streamflow at Slapton Wood using TOPMODEL within an uncertainty estimation framework. *Field Stud.* **1996**, *8*, 577–584.

73. Lamb, R. Distributed Hydrological Prediction Using Generalised TOPMODEL Concepts. Ph.D. Thesis, Lancaster University, Lancaster, UK, 1996.

74. Molicová, H.; Grimaldi, M.; Bonell, M.; Hubert, P. Using TOPMODEL towards identifying and modelling the hydrological patterns within a headwater, humid, tropical catchment. *Hydrol. Process.* **1997**, *11*, 1169–1196.

75. Saulnier, G.-M.; Beven, K.; Obled, C. Including spatially variable effective soil depths in TOPMODEL. *J. Hydrol.* **1997**, *202*, 158–172.

76. Quinn, P.F.; Beven, K.J. Spatial and temporal predictions of soil moisture dynamics, runoff, variable source areas and evapotranspiration for Plynlimon, Mid-Wales. *Hydrol. Process.* **1993**, *7*, 425–448.

77. Doswell, C.A.; Davies-Jones, R.; Keller, D.L. On summary measures of skill in rare event forecasting based on contingency tables. *Weather Forecast.* **1990**, *5*, 576–585.

78. Landis, J.R.; Koch, G.G. The measurement of observer agreement for categorical data. *Biometrics* **1977**, *33*, 159–74.

79. Wolock, D.M.; Price, C.V. Effects of digital elevation model map scale and data resolution on a topography-based hydrologic model. *Water Resour. Res.* **1994**, *30*, 3041–3052.

80. Brown, A.E. Predicting the Effect of Forest Cover Changes on Flow Duration Curves. Ph.D. Thesis, University of Melbourne, Melbourne, VIC, Australia, 2008.

81. Kashaigili, J.J.; Majaliwa, A.M. Implications of land use and land cover changes on hydrological regimes of the Malagarasi river, Tanzania. *J. Agric. Sci. Appl.* **2013**, *2*, 45–50.

82. Bhaduri, B.; Minner, M.; Tatalovich, S.; Harbor, J. Long-term hydrologic impact of urbanization: A tale of two models. *J. Water Resour. Plan. Manag.* **2001**, *127*, 13–19.

83. Fohrer, N.; Haverkamp, S.; Eckhardt, K.; Frede, H.G. Hydrologic response to land use changes on the catchment scale. *Phys. Chem. Earth Part B Hydrol. Ocean. Atmos.* **2001**, *26*, 577–582.

84. Mundia, C.N.; Aniya, M. Analysis of land use/cover changes and urban expansion of Nairobi city using remote sensing and GIS. *Int. J. Remote Sens.* **2005**, *26*, 2831–2849.

85. Odindi, J.; Mhangara, P.; Kakembo, V. Remote sensing land-cover change in Port Elizabeth during South Africa's democratic transition. *South Afr. J. Sci.* **2012**, *108*, 60–66.

86. Reggiani, P.; Rientjes, T.H.M. Flux parameterization in the Representative Elementary Watershed (REW) approach: Application to a natural basin. *Water Resour. Res.* **2005**, *41*, doi:10.1029/2004WR003693.

87. Reggiani, P.; Rientjes, T.H.M. Closing horizontal groundwater fluxes with pipe network analysis: An application of the REW approach to an aquifer. *Environ. Model. Softw.* **2010**, *25*, 1702–1712.

Effect of Bias Correction of Satellite-Rainfall Estimates on Runoff Simulations at the Source of the Upper Blue Nile

Emad Habib, Alemseged Tamiru Haile, Nazmus Sazib, Yu Zhang and Tom Rientjes

Abstract: Results of numerous evaluation studies indicated that satellite-rainfall products are contaminated with significant systematic and random errors. Therefore, such products may require refinement and correction before being used for hydrologic applications. In the present study, we explore a rainfall-runoff modeling application using the Climate Prediction Center-MORPHing (CMORPH) satellite rainfall product. The study area is the Gilgel Abbay catchment situated at the source basin of the Upper Blue Nile basin in Ethiopia, Eastern Africa. Rain gauge networks in such area are typically sparse. We examine different bias correction schemes applied locally to the CMORPH product. These schemes vary in the degree to which spatial and temporal variability in the CMORPH bias fields are accounted for. Three schemes are tested: space and time-invariant, time-variant and spatially invariant, and space and time variant. Bias-corrected CMORPH products were used to calibrate and drive the Hydrologiska Byråns Vattenbalansavdelning (HBV) rainfall-runoff model. Applying the space and time-fixed bias correction scheme resulted in slight improvement of the CMORPH-driven runoff simulations, but in some instances caused deterioration. Accounting for temporal variation in the bias reduced the rainfall bias by up to 50%. Additional improvements were observed when both the spatial and temporal variability in the bias was accounted for. The rainfall bias was found to have a pronounced effect on model calibration. The calibrated model parameters changed significantly when using rainfall input from gauges alone, uncorrected, and bias-corrected CMORPH estimates. Changes of up to 81% were obtained for model parameters controlling the stream flow volume.

Reprinted from *Remote Sens.* Cite as: Habib, E.; Haile, A.T.; Sazib, N.; Zhang, Y.; Rientjes, T. Effect of Bias Correction of Satellite-Rainfall Estimates on Runoff Simulations at the Source of the Upper Blue Nile. *Remote Sens.* **2014**, *6*, 6688-6708.

1. Introduction

Any rainfall-runoff modeling requires accurate rainfall data as model input. However, accurate rainfall information in many world regions is hampered by limitations of ground-based observational networks. Rain gauge networks often have inadequate coverage and density, represent only point scale estimates and suffer from problems relating to data quality and inconsistency [1,2]. Alternative to *in situ* network data are satellite rainfall estimates (SREs), which potentially can be a viable alternative. However, SREs are known to suffer from sampling and estimation inaccuracies, which are manifested in the form of systematic (bias) and random errors [3–7]. Though a number of studies report on usage of SREs for runoff and soil moisture simulations [6,8,9], aspects of accuracy and representativeness of SREs for hydrologic modeling are not well investigated.

In the present study, we focus on bias correction of a particular SRE—the Climate Prediction Center (CPC) Morphing technique (CMORPH; [10]), and the effect of bias correction on hydrologic simulations for the Gilgel Abbay catchment, Lake Tana basin, Ethiopia. CMORPH is considered in this study owing to its relatively high space-time resolutions (30 min, 8 km). In this study CMORPH estimates are accumulated to a daily resolution, which matches the rain gauge sampling interval and the time step of the rainfall-runoff model used herein. A number of studies investigated the accuracy of CMORPH products across a range of space-time scales. Examples include seasonal or daily estimates at $0.25° \times 0.25°$ spatial resolution [11–13], three-hourly estimates at $0.25° \times 0.25°$ [14,15], and one-hourly estimates at 8 km × 8 km [1,5]. Results from these studies suggest that CMORPH estimates have significant systematic biases but also that estimates have random errors. Smith *et al.* [16] stated that CMORPH biases might be due to a diurnal sampling bias, tuning of the instrument or the rainfall algorithm, or unusual surface or atmospheric properties, which the algorithm does not correctly interpret. A Kalman filter approach has been recently adopted by the CMORPH developers to optimally integrate satellite-based estimates with rain gauge observations [17,18]. In the scientific literature, some evidence is presented that CMORPH bias exhibits spatio-temporal variation. For instance, Haile *et al.* [1] show that the total bias and its different components exhibit spatial variation in the Gilgel Abbay catchment, which is also selected for the present study. The authors concluded that over mountain areas CMORPH bias mostly is affected by missed rainfall detection. Particularly for lower elevated areas bias is affected by missed rainfall, false rainfall and differences in hit-rainfall estimates. For the Nile basin area, Habib *et al.* [11] showed that CMORPH bias (and other SREs) is largely affected by topography and latitude. The same study showed that CMORPH bias in the wet and dry seasons can be quite different.

The aforementioned studies indicate that it is crucial to reduce the systematic and random errors in SREs before products can be used in hydrologic and water resources applications. Methodologies for bias correction are developed in multi-sensor, radar-gauge approaches [19–21] and triggered applications in satellite remote sensing. Examples are on monthly-based bias correction [22], disaggregation of bias at daily scale to hourly scale [23], and merging satellite and gauge data by means of a non-parametric kernel smoother [24]. Vila *et al.* [25] compared five merging schemes: additive bias correction, ratio bias correction, gauge-to-satellite monthly correction factors, and a combined scheme. The authors concluded that the combined scheme, which considers both additive and multiplicative bias, outperformed the others. Other studies of bias-correction to satellite rainfall products are reported in Hong *et al.* [26], Chiang *et al.* [27], Tobin and Bennett [28], and Tian *et al.* [29]. These studies suggest that selection of bias correction scheme should depend on the desired level of accuracy and assumptions to represent spatial and temporal rainfall characteristics. However, selection also depends on the data requirements, computational expenses, and, more importantly, the hydrologic application the bias-adjusted product is used for. Overall, bias-corrected satellite rainfall products are expected to better match station records compared to satellite only products even in complex terrain [30] and as such correction should improve hydrological applications by improved rainfall representation. However, results by hydrological applications are not consistent and require further assessment.

Artan *et al.* [31] showed that negative bias of a satellite rainfall product in rainfall-runoff modeling could result in deterioration of modeling results. They also showed that a hydrologic model requires recalibration when satellite rainfall data is used to replace use of *in-situ* rainfall data. Zeweldi *et al.* [32] reported increased performance of a rainfall-runoff model when the model was calibrated using satellite data than when it was calibrated using rain gauge data. However, calibration could result in parameter values that are unrealistic and beyond limits as the model attempts to compensate for the large errors in rainfall input. Behrangi *et al.* [33] found that bias-adjustment of satellite-based precipitation products is critical and can yield substantial improvement in capturing both the streamflow pattern and magnitude at six-hourly and monthly time scales. Yong *et al.* [34] showed an improvement in the performance of a rainfall-runoff model after applying gauge-based bias correction to satellite-only rainfall products. However, Bitew and Gebremichael [8] reported improved model performance when using a satellite-only product (TRMM 3B42RT) as compared to a satellite-gauge bias corrected product (TRMM 3B42). Such results could be partly attributed to the fairly poor quality and/or lack of spatial representativeness of the sparse gauges that were used for the bias correction.

In the present study, the focus is mainly on analysis of the spatial and temporal variability of bias in CMORPH 30-min, 8 km × 8 km satellite-based rainfall product and on identifying the critical aspects of such variability from a hydrologic perspective in rainfall-runoff modeling. Results of such analyses are of relevance to product users to guide efforts for product adjustment before being used in further applications [35,36] and also for product developers to identify future needs for algorithmic enhancements. The specific objectives of this study are (i) to assess three bias correction schemes to the CMORPH 30-min, 8 km × 8 km product to adjust for spatial and temporal biases; and (ii) to assess how rainfall-runoff model calibration results are affected when bias-corrected CMORPH data is used instead of uncorrected and *in situ* rainfall data. The area of study is the Gilgel Abbay catchment in Ethiopia. The paper is structured as follows. Section 2 describes the data sets, the bias correction schemes, the hydrologic model, and the calibration approach. Section 3 presents results and discusses the findings and Section 4 concludes on the study.

2. Data and Methods

2.1. Study Setting

The study area is Gilgel Abbay catchment, which is the largest contributor to Lake Tana [37], the source of Upper Blue Nile River in Ethiopia. In the present study, the focus is on the gauged part of Gilgel Abbay for which daily time series of streamflow have been available since the 1970s (Figure 1). This part of the watershed is situated between latitudes of 10°56′N–11°22′N and longitudes of 36°49′E–37°24′E. It covers an area of about 1655 km^2 with predominantly agricultural land cover and with clay to clay-loam as the prevailing soil type. The seasonal rainfall distribution of Gilgel Abbay is affected mainly by the location of the Intertropical Convergence Zone (ITCZ) with a rainy season, which coincides with the summer in the northern hemisphere (June–August). At short time scales (daily and sub-daily), rainfall distribution in this watershed is affected by orographic factors and the presence of Lake Tana [38,39]. The lowlands of Gigel Abbay receive more intense and short lasted rainfall as compared to its highlands [40].

2.2. CMORPH and Local Gauge Data

The satellite-rainfall product used in this study is the National Oceanic and Atmospheric Administration's (NOAA) Climate Prediction Center (CPC) morphing technique (CMORPH) [10]. CMORPH combines rainfall estimates from multiple passive microwave (PMW) sensors, which include the Advanced Microwave Sounding Unit (AMSU-B), the Special Sensor Microwave Imager (SSM/I), the TRMM Microwave Imager (TMI), and the Advanced Microwave Scanning Radiometer—Earth Observing System (AMSR-E), respectively. To fill the time and space gap in the combined PMW based rainfall estimates, the algorithm used cloud motion vectors derived from spatial lag correlation of successive geostationary satellite IR images. These vectors are used to propagate the PMW based rainfall features for time periods between two successive PMW overpasses. The shape and intensity of the rainfall patterns is then morphed through linear interpolation using weights that are obtained from forward advection (previous to most current PMW overpass) and backward advection (most current to previous overpass) of rainfall features. The main advantage of CMORPH is its near real-time global coverage at relatively fine temporal and spatial scales (as fine as 30-min and 8 km × 8 km), which makes it a desirable candidate for hydrologic applications. In the current analysis, the 30-min, 8 km × 8 km CMORPH estimates are aggregated to daily time step to be consistent with rain gauge observation interval. In addition, streamflow time series are at daily time interval so aggregation results in optimal use of available satellite data to represent and to correct rainfall in respective time and space domains for stream flow modeling. Based on availability of rain gauge and streamflow data in the watershed, the current analysis period covers two years from January 2003 to December 2004.

Data is obtained from 10 rain gauges in the watershed (Figure 1) which are operated by the Ethiopian Meteorological Agency (EMA). Daily rainfall from these gauges has been evaluated and used in [37,41–43] and serve as reference to evaluate satellite rainfall estimates. For the study period, the average annual rainfall over Gilgel Abbay was about 1700 mm. The average daily rainfall rate exceeded 10 mm/day for 20% of the time and reached as high as 35 mm/day. The gauge network (Figure 1) is sparse and stations are unevenly distributed over the watershed. As a result, the "true" or real world rainfall distribution may not be well represented by the network, as rainfall varies over the area following topographic variation [1,38,39].

Meteorological observations at three stations in the watershed (Dangila, Adet, and Bahir Dar; Figure 1) were used to estimate monthly potential evapotranspiration (PET) using the Penman-Montheith method [44], which were then used as input to the hydrologic model.

Streamflow daily data is available for the upper part of Gilgel Abbay, which is gauged at Wotet Abbay, a small town near Bahir Dar. Consistency of streamflow daily time series was checked by visual inspection of concurrent rainfall and streamflow plots. We noticed that the baseflow record showed an abrupt increase in late 2005. Local people near the gauging site stated that the gauging station was moved by hundreds of meters downstream of the original site towards the end of 2005 due to road construction. However, this was not confirmed by officials in the Ministry of Water

Energy. As a result we limited our analysis period to 2003 and 2004 for which CMORPH data is available and for which we have confidence in the quality of stream flow data.

Figure 1. Study site showing the location of the Gilgel Abbay catchment and its eight sub-catchments within the Nile Basin. The locations of the ten (10) rain gauge stations and the streamflow gauge are indicated. Note that the unit of terrain elevation is meters. (**a**) Geographic location of study area; (**b**) Terrain elevation and rain gauge stations.

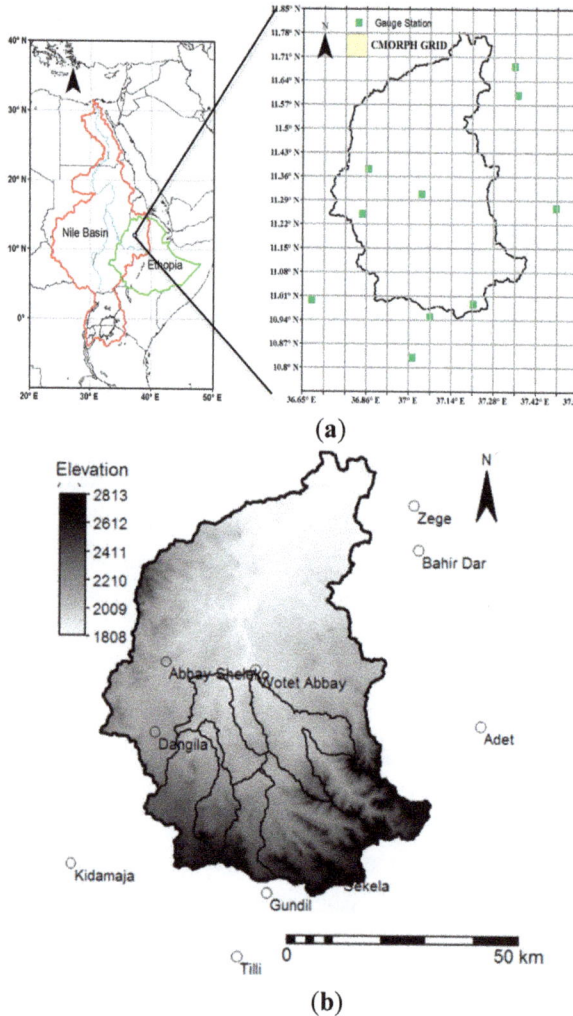

(a)

(b)

2.3. Bias Formulation and Estimation

Satellite-based rainfall estimates exhibit large systematic and random errors. The systematic errors (*i.e.*, bias) persist when the estimates are aggregated over time and, hence, may cause large uncertainties in hydrologic modeling. In addition, models could augment or suppress rainfall biases to larger or smaller streamflow based on the response mode of the model. Therefore, bias in rainfall products should be assessed and corrected before satellite rainfall products can be used in hydrologic applications.

In the current study, we estimate and correct the bias in CMORPH estimates as follows. For a selected day (d) and gauge (i), the multiplicative daily bias factor (BF) at a certain CMORPH pixel with a collocated gauge can be formulated as follows Equation (1):

$$BF_{TSV} = \frac{\sum_{t=d}^{t=d-l} S(i,t)}{\sum_{t=d}^{t=d-l} G(i,t)} \tag{1}$$

where G and S represent daily gauge and CMORPH rainfall estimates, respectively, i refers to gauge location, t refers to a Julian day number; and l is length of a time window for bias calculation. The subscript "TSV" stands for "Time-Space Variable" since the bias in this formulation is estimated for a particular location and a particular day. Based on some preliminary analysis by the authors on rainfall distributions in the study area, a fixed time window of $l = 7$ days was selected to allow for adequate rainfall accumulation for bias calculation while still accounting for temporal variability in BF. The BF factor was calculated for a certain day only when a minimum of five rainy days were recorded within the preceding seven-day window with a minimum rainfall accumulation depth of 5 mm, otherwise no bias is estimated (i.e., assigned a value of 1). We chose five rainy days with minimum accumulation depth of 5 mm to ensure stability of the bias factors and avoid exaggerated values as a result of dividing large satellite estimates by small gauge values. We evaluated sensitivity of BF to different window lengths of 3, 7, and 10 days. We noticed that BF shows relatively lower sensitive during the wet season compared to the dry season. In the wet season, BF shows high variation and becomes highly erratic when the window length is reduced to three days as a result of small accumulation period. In the wet season, BF exceeds 2.0 for a three-day window length but is mostly well below 2.0 for a seven- and 10-day window length.

It is noted that Equation (1) ignores errors introduced by using a single gauge to represent rainfall amounts at the scale of the CMORPH pixel. Haile et al. [1] showed that the error variance of the gauge representativeness error in Gilgel Abbay could contribute as much as 30%–52% to the total variance of CMORPH-gauge rainfall hourly differences. Using the current dataset, we found that the spatial correlation of rainfall in the study area, at 8 km separation distance, increases from 0.55 at a daily scale to 0.91 for a seven-day accumulation scale. Therefore, it is reasonable to assume that the gauge representativeness error will be much smaller at a seven-day time window than that of the hourly base, and, thus, we proceed with using single-gauge observations as a reference for the bias estimation using Equation (1).

2.4. Schemes for Bias Correction

In the current analysis we test three schemes for bias correction:

(i) The first one allows for correcting the bias at a pixel based (i.e., space variable) and at a daily scale (i.e., time varying), and is based on the using the BF_{TSV} factor estimated from Equation (1). To apply a correction that accounts for spatial and temporal variability in the CMORPH bias, the pixel-based daily BF_{TSV} factors were spatially interpolated using the inverse distance weight (IDW) method to yield a spatial and temporally varying field of BFs that cover the entire study area. We followed the approach of Haile *et al.* (2009) [38] in the same study area who showed good interpolation results by IWD. The CMOPRH daily rainfall fields were then multiplied by the BF_{TSV} bias fields for the respective time windows to result in a new set of CMORPH estimates that as such are bias-corrected in a temporally and spatially varying scheme. This procedure is similar to the local-bias correction algorithm developed by Seo and Breidenbach [19], which is adopted in the operational version of the National Weather System-Multisensor Precipitation Estimation (NWS-MPE) system. The use of Equation (1) applies a bias correction factor that varies in space and time domains. We refer to this formulation as time and space variant (TSV) bias correction. To assess the implications for ignoring or for accounting of variability of bias, two more bias estimation and correction schemes were tested:

(ii) Time and space fixed (TSF) bias correction: in this formulation the bias is obtained by using gauge and CMORPH estimates over the entire domain and over the total duration of the sample Equation (2):

$$BF_{TSF} = \frac{\sum_{t=1}^{t=T} \sum_{i=1}^{i=n} S(i,t)}{\sum_{t=1}^{t=T} \sum_{i=1}^{i=n} G(i,t)} \qquad (2)$$

where n is the total number of gauges within the entire domain of the study and T is the full duration of the study period. The bias correction in this case is applied by multiplying the CMORPH estimates by the bias factor, BF_{TSF}, to result in a new set of CMORPH estimates that are bias-corrected in a spatially and temporally-lumped scheme.

(iii) Time variable (TV) bias correction: in this formulation the BF is spatially lumped over the entire domain but is still estimated for each daily time step Equation (3):

$$BF_{TV} = \frac{\sum_{t=d}^{t=d-l} \sum_{i=1}^{i=n} S(i,t)}{\sum_{t=d}^{t=d-l} \sum_{i=1}^{i=n} G(i,t)} \qquad (3)$$

The bias correction in this case is applied by multiplying each daily CMORPH field by the daily bias factor, BF_{TV}, to result in a new set of CMORPH estimates that are bias-corrected in a spatially-lumped but temporally-varying scheme.

2.5. Hydrologiska Byråns Vattenbalansavdelning (HBV-96) Hydrologic Model

In the present study, the Hydrologiska Byråns Vattenbalansavdelning (HBV-96) rainfall-runoff model [45] is used to perform the rainfall-runoff analysis using rainfall estimates by the three correction schemes. The HBV-96 model has been extensively evaluated for different regions on the globe [46–50] including Gilgel Abbay catchment [37,41,43]. HBV-96 can be classified as a conceptual model that relies on water balance equations to simulate runoff and mass exchanges across a set of surface and subsurface zones. Inputs to the model include rainfall, temperature, potential evapotranspiration, and percentage of forested and non-forested catchment areas.

The storage based HBV-96 has four routines that include (i) a precipitation accounting routine; (ii) a soil moisture routine; (iii) a quick runoff routine; and (iv) a base flow routine. The approach is characterised by three stores, which are the soil moisture reservoir, the upper zone store and the lower zone store. From the upper zone store quick runoff is simulated from the lower zone store base flow runoff is simulated. Routing of streamflow is optional and can be de-activated in model simulations. The precipitation accounting routine partitions precipitation into rainfall and snow based on a threshold value (TT). Precipitation is in the form of snow in case the actual temperature (T) is lower than TT. In the Gilgel Abbay area, the temperature is much higher than common values for TT and, as such, precipitation only is in the form of rainfall.

The soil moisture routine controls the formation of direct and indirect runoff. Direct runoff occurs when the simulated soil moisture (SM) in the soil moisture reservoir exceeds the maximum storage capacity as represented by field capacity (FC). Otherwise, rainfall infiltrates (IN) the soil moisture reservoir to add to the actual storage and to add to the flow of water to the upper zone store (indirect runoff).

Indirect runoff (R) is defined as follows:

$$R = IN \left(\frac{SM}{FC} \right)^{BETA} \tag{4}$$

This equation indicates that indirect runoff increases with increasing soil moisture storage (SM) but it reduces to zero when infiltration ceases. BETA is a parameter accounting for the non-linearity of indirect runoff from the soil layer.

Evapotranspiration losses are calculated from the soil moisture reservoir. Actual evapotranspiration (E_a) is highest (i.e., reaches its potential value (E_p)) when SM reaches or exceeds a certain ratio of FC. The ratio, denoted as LP, is used as a calibration parameter. Otherwise, E_a declines linearly as a function of soil moisture deficit represented by SM/FC:

$$E_a = E_p \qquad if \ SM \geq (LP \cdot FC) \tag{5}$$

$$E_a = \frac{SM}{LP \cdot FC} \cdot E_p \qquad if \ SM < (LP \cdot FC) \tag{6}$$

Percolation (PERC) to the lower zone store occurs when water is available in the upper zone store. PERC is treated as a time-invariant process with a fixed value throughout the simulation period. Capillary transport is estimated as a function of soil moisture deficit (FC-SM) and a maximum value for capillary flow (CFLUX):

$$C_f = CFLUX \cdot \left(\frac{FC - SM}{FC} \right) \tag{7}$$

Quick runoff (Q_q) and slow (base) flow (Q_s) are defined as follows:

$$Q_q = K_q UZ^{(1+ALPHA)} \tag{8}$$

$$Q_s = K_s LZ \tag{9}$$

where UZ is the actual storage in the upper zone store, ALFA is a measure for the non-linearity of flow, K_q is a recession coefficient for quick runoff, LZ is the actual storage in the lower zone store and K_s is a recession coefficient for base flow. According to this formulation, the model has 8 parameters that can be used for model optimization and calibration, namely: *FC, BETA, LP, ALPHA, Kq, Ks, PERC,* and *CFLUX*.

Considering the large catchment area and significant topographic variation, and to make use of the spatially distributed data from CMORPH, the catchment has been partitioned in eight sub-catchments (Figure 1). The eight sub-catchments have size of 76, 121, 150, 165, 240, 242, 245, and 414 km^2.

2.6. Model Calibration and Evaluation

The model was calibrated using four sets of rainfall data based on gauge observations, and three variants of bias corrected CMORPH estimates. In both cases, rainfall data were aggregated from their original spatial resolutions (gauge-point, or CMORPH pixel) to the scale of each sub-catchment.

The following two metrics were used to assess performance of the HBV-96 rainfall-runoff model: Nash-Sutcliffe (*NS*) efficiency, which provides a measure of random differences between simulated and observed streamflows, and Q_{Bias}, which measures systematic differences (bias) in the simulated streamflow volumes:

$$NS = 1 - \frac{\sum_{i=1}^{n}(Q_{sim,i} - Q_{obs,i})^2}{\sum_{i=1}^{n}(Q_{obs,i} - \overline{Q_{obs}})^2} \tag{10}$$

$$Q_{Bias} = \left(\frac{\sum_{i=1}^{n}Q_{sim,i}}{\sum_{i=1}^{n}Q_{obs,i}}\right) \tag{11}$$

where Q_{sim} and Q_{obs} represent simulated and observed daily flows, respectively, at a certain day i, and n represents the number of days in the sample. The over-bar symbol denotes the mean statistical operation. The values of *NS*, which is dimensionless, can range between $-\infty$ and 1, where a value of 1 indicates a perfect fit. Similarly, a Q_{Bias} value of 1 reflects bias-free streamflow simulations whereas streamflow overestimation and underestimation are reflected by bias values that are larger or smaller than 1, respectively.

A Monte-Carlo procedure was used to calibrate the HBV-96 model. In this procedure, prior ranges of the eight calibration parameters are selected based on the parameter value ranges specified by Rientjes *et al.* [37] who applied the HBV-96 model in a regionalization study in the (entire) Lake Tana basin area. In that particular study 60,000 parameters sets are generated randomly assuming a uniform distribution of parameter values within the specified, posterior, value ranges. The HBV-96 model was run for each parameters set and the corresponding objective function values (*NS* and Q_{Bias}) are calculated. Following Rientjes *et al.* [37], the optimum parameters set is selected as the average value of the 25 parameter sets that are ranked highest in terms of the *NS* values. It is noted that a similar approach is followed in this study when calibrating the model in case CMORPH rainfall data is used as model input.

3. Results

3.1. Evaluation of CMORPH Estimates

Before presenting the results of the different bias correction schemes, we first examine the temporal and spatial variability in the CMORPH bias field, as reflected by BF_{TSV} (Figure 2). The lowest, highest and mean values of BF_{TSV} for a seven-day moving window are shown in the figure. These values are summarized based on BF_{TSV} values calculated at the ten rain gauge stations within the study area. For each calendar day the minimum, maximum and mean are shown for the ensemble of network stations. The difference between the lowest and highest values shows the extent of the variation of the bias across the 10 stations in the study area. The mean values show pronounced seasonal variations and have different patterns throughout the two years. In general, CMORPH reports smaller rainfall amounts than gauge observations from mid-June to mid-August 2003, but reports larger rainfall amounts towards the end of the rainy season of 2003. This pattern is not shown in 2004, where positive and negative biases in CMORPH show lower variation in time. Overall, these results indicate that the bias in the CMORPH product exhibits pronounced variability in space and time over the study area. Possibly, this could be related to variations in rain generation mechanisms [11] but further investigations are needed for confirmation.

3.2. Results on Rainfall Bias Correction

To examine the implications of space and time variability presented above (Figure 2), we applied the three bias correction schemes described above (Equation (1)), time-space fixed (TSF), time variable (TV), and time-space variable (TSV), to the CMORPH product. We first assess how the different bias correction schemes impact the catchment-average rainfall at a monthly scale. Table 1 shows a summary of the ratios of CMORPH monthly catchment-average rainfall amounts (before and after correction) to the corresponding gauge amounts. Without bias correction, CMORPH mostly underestimated monthly rainfall by up to 37%. In 2003, the deviation of CMORPH during the wettest months (June-August) is reduced when TSF, TV, or TSV corrections are used. TSV correction shows noticeable change (up to 0.19) than TSF correction (only 0.01) in July 2003. In 2004, TSF correction leads to deterioration of the monthly agreement probably since their temporal variation is too pronounced to be ignored. The importance of accounting for temporal variation in CMORPH bias is illustrated again by the fact that TV and TSV corrections reduced the bias in 2004 by up to 0.07 and 0.07–0.14, respectively.

Figure 2. Mean, minimum and maximum of CMORPH daily bias factors (BF_{TSV}, Equation (1)) evaluated for the ensemble of ten network stations.

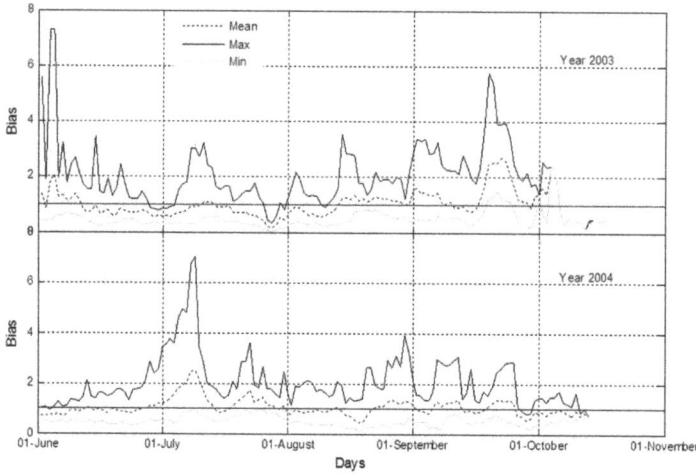

Table 1. Ratios of monthly rainfall amounts of Climate Prediction Center-MORPHing (CMORPH) (without and with three bias correction schemes) to the corresponding gauge amounts.

Year	Rainfall Product	June	July	August	September	October
2003	*CMORPH*	0.69	0.63	0.88	1.25	0.74
	CMORPH TSF	0.71	0.64	0.9	1.28	0.76
	CMORPH TV	0.74	0.82	0.94	0.9	0.63
	CMORPH TSV	0.87	0.81	0.99	0.95	0.63
2004	*CMORPH*	0.83	1.15	0.8	0.99	0.84
	CMORPH TSF	0.79	1.09	0.76	0.94	0.8
	CMORPH TV	0.9	0.87	0.87	0.93	0.93
	CMORPH TSV	0.95	0.86	0.87	0.96	0.98

3.3. Model Parameter Optimization Using Different Rainfall Inputs

We calibrated the HBV-96 model using gauge and CMORPH rainfall inputs for the year 2003–2004 (Table 2). Calibration based on the rain gauge network data serves as a reference for further assessments on effectiveness of bias-correction to the CMORPH estimates. The model was recalibrated using the satellite rainfall fields to examine how the model parameters are affected by bias in the rainfall input. In theory, each rainfall input represents a different model forcing and may result in different parameter values and model performance level as measured by *NS* and Q_{bias}. Independent calibration of the HBV-96 model for different rainfall inputs resulted in satisfactory model performance (*NS* = ~0.8 and Q_{Bias} = ~0.9). All optimum parameter values obtained using the correction schemes are within the allowable value ranges. The values of the optimized model parameters are inter-compared and percent change of each parameter value is shown with respect to the reference case. To allow comparison of parameter values over a common scale, changes are calculated after normalizing the parameter values using their allowable minimum and maximum values, which are set equal for all simulations. We note that the results of parameter optimization are affected by the rainfall input as shown by the percentage errors in Table 2. In particular, parameters (*FC, Beta* and *LP*) which control the volume of the simulated hydrograph showed large changes of up to 81% compared to the parameters using the reference gauge data sets. There is also a significant change in the quick recession coefficient (K_q), whereas those that control groundwater contributions (K_s, *PERC* and *CFLUX*) are less affected.

Table 2. Calibrated values of the Hydrologiska Byråns Vattenbalansavdelning (HBV) model parameters using gauge and bias corrected CMORPH. Numbers in brackets represent percent changes in each parameter value (after normalizing with the allowable minimum and maximum range) with respect to the gauge-driven reference case. The last two rows show the *NS* and Q_{Bias} values.

Parameter	Unit	Minimum	Maximum	Gauge	CMORPH with Bias Correction		
					Time-Space Fixed (TSF)	Space Fixed and Time Variable (TV)	Time-Space Variable (TSV)
FC	mm	100	800	373	186 (−68)	177 (−72)	185 (−69)
BETA	--	1	4	1.351	1.599 (71)	1.562 (60)	1.625 (78)
LP	--	0.1	1	0.544	0.888 (77)	0.905 (81)	0.775 (52)
ALPHA	--	0.1	3	0.271	0.242 (−17)	0.236 (−20)	0.269 (−1)
K_q	day^{-1}	0.0005	0.15	0.073	0.035 (−52)	0.050 (−32)	0.038 (−48)
K_s	day^{-1}	0.0005	0.15	0.087	0.086 (−1)	0.083 (−5)	0.074 (−15)
PERC	mm·day^{-1}	0.1	2.5	1.348	1.422 (6)	1.208 (−11)	1.339 (−1)
CFLUX	mm	0.0005	2.0	0.886	0.898 (1)	0.805 (−9)	0.892 (1)
NS	--	--	--	0.8256	0.703	0.8038	0.8177
Q_{Bias}	--	--	--	0.995	0.982	0.988	0.982

3.4. Effects of Rainfall Bias Corrections on Streamflow Simulations

Next, we applied the calibrated parameter set of the reference case to simulate streamflow using the uncorrected and bias-corrected CMORPH rainfall estimates. We chose to use the reference set of parameters to in all model simulations (gauge and CMORPH) to isolate the effect of rainfall bias from other sources of model uncertainty (e.g., parameter uncertainty). Figure 3 shows streamflow hydrographs for the reference case and for uncorrected CMORPH estimates. At the beginning and middle of the 2003 wet season, differences between CMORPH gauge rainfall amounts are negative for most of the time. For this reason, lower streamflow discharges are simulated. Towards the end of the rainy season of 2003, positive rainfall differences start to appear but this did not cause higher streamflow discharges than observed. It appears that the excess rainfall was stored in the different model stores instead of generating runoff. Relatively large positive differences in rainfall dominate towards the beginning of the wet season of 2004. This rainfall difference resulted in higher streamflow simulations than observed. The rainfall difference becomes negative throughout August 2004, but it did not produce negative streamflow biases, as the model stores had the excess rainwater

128

from previous time steps stored. Overall, these results indicate that CMORPH bias propagates into model simulations. The bias correction affects moisture conditions which yielded lower (excess) runoff than observed.

Figure 3. (**a**) Comparison of daily catchment-average gauge and uncorrected CMORPH rainfall. (**b**) differences in daily rainfall estimates between gauge and CMORPH. (**c**) and (**d**) observed and simulated stream flow hydrographs for the year June 2003–December 2004 based on rainfall inputs from gauges and uncorrected CMORPH.

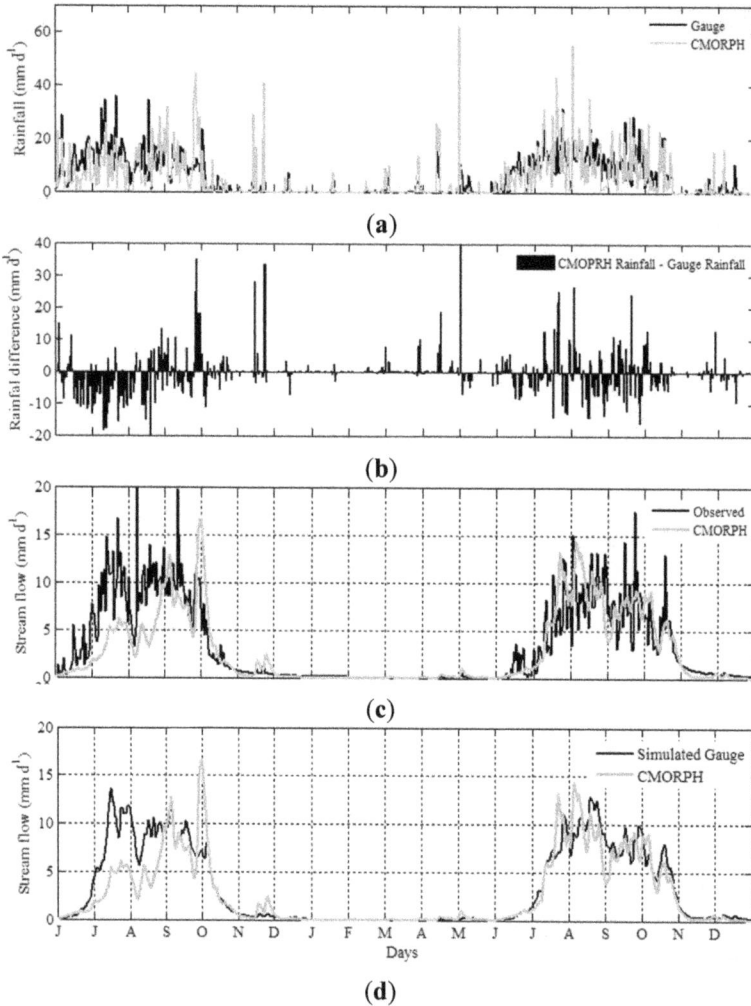

(a)

(b)

(c)

(d)

Next, we evaluated the effect of bias correction in CMORPH on HBV-96 model simulations (Figure 4). Here we show the streamflow hydrograph that is simulated using TSV bias-corrected CMORPH data, which produced the best result. Despite the applied bias correction, there are large differences (>10 mm·d^{-1}) between catchment-average daily rainfall estimates obtained from TSV and gauges. However, it is apparent that the large CMORPH bias in 2003 was substantially reduced. As a result, the patterns and volumes of the observed hydrographs were better captured when using the bias-corrected CMORPH estimates than the uncorrected ones. Some observed peak flows were better captured as a result of correcting for the rainfall bias. The improvements are particularly substantial for the 2003 hydrographs where the uncorrected CMORPH had large negative bias. We note that use of bias corrected rainfall data has some advantages over gauge only data; however, some aspects of observed hydrograph were better captured by using gauge only data (e.g., the second half of July 2003). The simulated hydrographs based on both rainfall inputs show smaller fluctuation than the observed hydrograph. Such mismatches could be caused by deficiencies in the HBV-96 model structure, poor rainfall representation by the low density of the rain gauge network, or errors in streamflow observations, among others.

Figure 4. (**a**) Comparison of daily catchment-average gauge rainfall and the corresponding TSV bias-corrected CMORPH. (**b**) differences in rainfall estimates between gauge and TSV bias-corrected CMORPH. (**c**) and (**d**) observed and simulated stream flow hydrographs for the year June 2003–December 2004 based on rainfall inputs from gauges and TSV bias-corrected CMORPH.

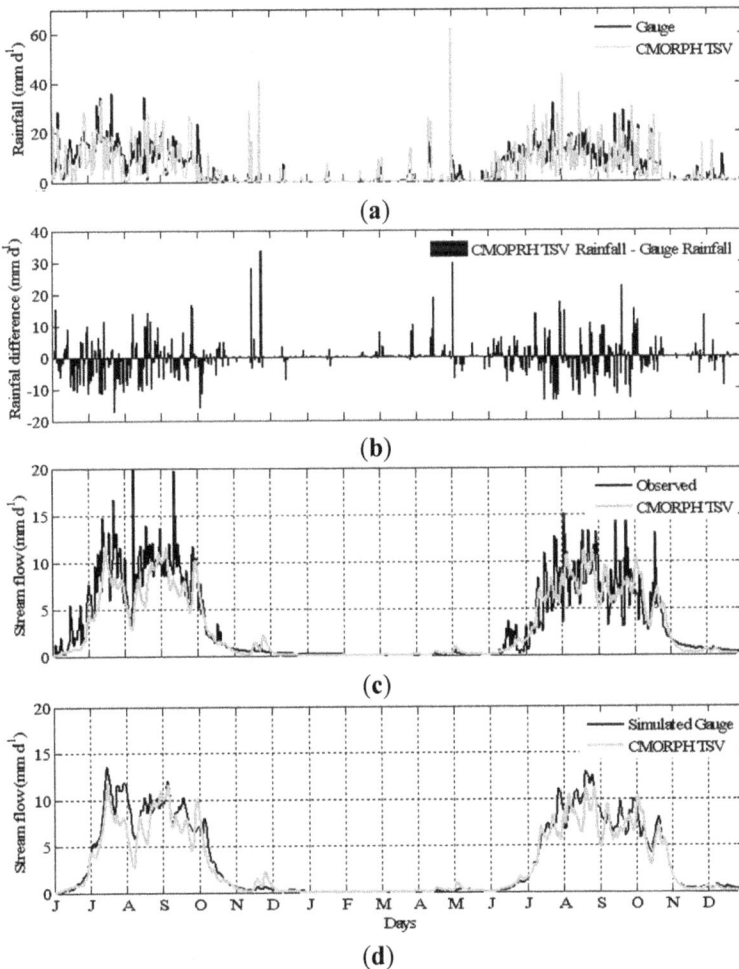

130

Next, we quantify the propagation of CMORPH rainfall biases in model simulations (Table 3). To separate the effect of rainfall errors from effects of model parameter uncertainty on model performance, gauge-based model simulations served as reference to compare the model performance objective functions. In June–August 2003, CMORPH rainfall amounts are smaller than those from gauges by −18% (CMORPH bias ratio = 0.82), which causes difference in streamflow volume of −27% (streamflow bias = 0.73) and increased for the same period of 2004 (from −5% to −27%). This rainfall-to-streamflow bias amplification persists for all of the CMORPH rainfall inputs but its extent became smaller when the temporal variability of the bias is considered (*i.e.*, when using time-varying bias correction, BF_{TSV}). For example, for CMORPH TSV rainfall input, the bias increased from −13% to only −17% in 2003 while it increased from −8% to 20% in 2004. The observed changes of rainfall-to-streamflow biases are probably due to the non-linearity in the rainfall-runoff relation and subsequent runoff generation in the HBV-96 model. For instance, small bias in rainfall input can propagate to result in larger streamflow bias when the catchment is wet than when it is dry.

Table 3. Ratios of catchment-average seasonal rainfall amounts of CMORPH (without and with three bias correction schemes) to the corresponding gauge amounts. The corresponding values for streamflow biases (Q_{Bias}) are also presented. The *NS* efficiency values for the streamflow simulations are shown between brackets.

Year	Performance Measure	CMORPH	CMORPH (TSF)	CMORPH (TV)	CMORPH (TSV)
June–October 2003	*Rainfall Ratio*	0.818	0.819	0.806	0.869
	Streamflow Q_{Bias}	0.734 *(0.19)*	0.762 *(0.21)*	0.764 *(0.71)*	0.831 *(0.79)*
June–October 2004	*Rainfall Ratio*	0.947	0.904	0.898	0.917
	Streamflow Q_{Bias}	0.726 *(0.73)*	0.727 *(0.73)*	0.792 *(0.79)*	0.804 *(0.80)*

Streamflow bias, Q_{Bias} (Equation (9)), obtained using the uncorrected as well as the bias-corrected CMORPH rainfall inputs are shown in Figure 5 (see also Table 3). As compared to gauge-based simulations, the uncorrected and bias-corrected CMORPH data resulted in consistently smaller streamflow in the rainy season (June–August) of 2003 but larger streamflow towards the end of the rainy season. Note that this pattern has some resemblance to that of the rainfall biases (Figure 2). However, the temporal pattern of the streamflow biases in both 2003 and 2004 are smoother than those of the rainfall inputs. This possibly is a result of the filtering effect of the runoff model as it converts highly variable rainfall input to streamflow. The significantly large rainfall bias in October of 2004 is translated to a smaller streamflow bias probably as the model became relatively dry and therefore did not convert the excess rainfall input into surface runoff. CMORPH-based streamflow in 2003 is mostly 0.25 to 0.5 times the gauge observations showing underestimation though this streamflow differences became much smaller in 2004. Overall, these differences were reduced when the bias-corrected CMORPH rainfall amounts served as model inputs. An exception is that TSF, which is obtained using a space-time constant correction factor, only slightly altered the streamflow bias. For most parts of the wet season, the streamflow bias significantly decreased when time variable bias correction is applied. Accounting for both spatial and temporal variation of the CMORPH bias factor further reduced the streamflow bias.

Figure 5. Streamflow bias (Q_{Bias}, Equation (9)) of streamflow simulations driven by different CMORPH rainfall inputs. Q_{Bias} was calculated using a moving window of past 7 days. Streamflow simulations driven by gauge observations served as the reference.

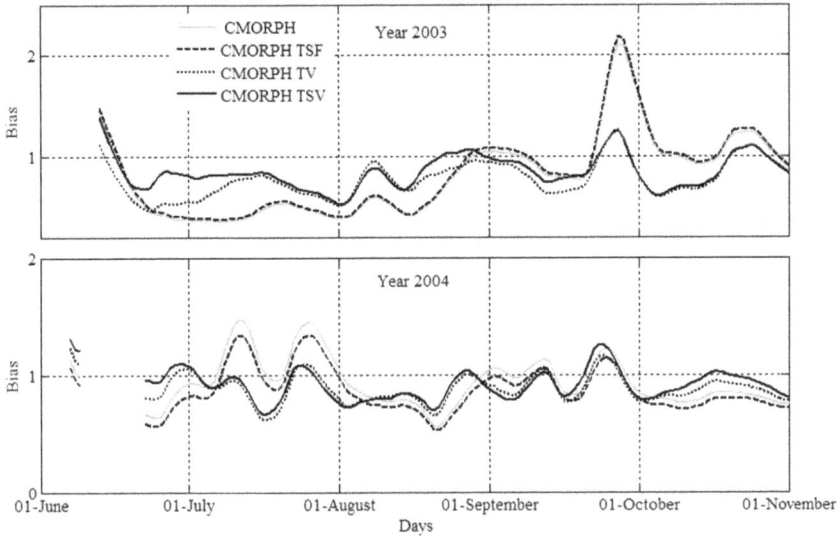

4. Conclusions

Various studies have indicated that satellite rainfall products are contaminated with systematic and random errors. However, not much has been done to illustrate how these products can be made applicable for various purposes by reducing their errors. In this study, the effect of rainfall bias correction in a high-resolution satellite-based product (CMORPH) on the performance of a rainfall-runoff model was assessed. The study is unique as we assess the importance of space and time aspects of CMORPH bias for rainfall-runoff modeling in a data scarce catchment. Our findings contribute to efforts that aim towards enhancing the real-world applicability of satellite rainfall products. The study site is the Gilgel Abbay catchment at the source of the Blue Nile in Eastern Africa—an example of many world regions that can benefit from satellite-based rainfall products for resource assessments and monitoring. Results and conclusions of the present study are summarized below.

CMORPH has large rainfall biases (-2.3 mm·d^{-1} on average and by up to ±30 mm·d^{-1}) with spatial as well as inter-annual and intra-annual variations in Gilgel Abbay catchment. Such biases could be related not only to rain generation mechanisms but also to the sampling and retrieval errors of satellite products [51]. We have showed through cross validation that it is not always the case that gauge-only, or satellite-only estimates, outperform one another. This suggests rainfall estimation can benefit from combined use of satellite and rain gauge data.

We applied three bias correction schemes to correct the bias CMORPH estimates. Space-time fixed, time variable and space-time variable bias factors were estimated to correct CMORPH estimates. One of the significant achievement of analysis indicated that the most important aspect of the CMORPH bias is its temporal variation as accounting for it substantially reduced the rainfall

132

bias. In some instances, it was not possible to noticeably reduce the catchment-average bias. Particularly, absence of gauges in the middle and south-east parts of the catchment is presumed to have contributed to the observed mismatches between bias-corrected CMORPH and gauge rainfall amounts.

An application of the HBV hydrologic model indicated that the model should be calibrated independently for satellite-only or satellite-gauge rainfall data in order to achieve high model performance. However, we observed that the calibration procedure compensated for rainfall input errors by changing the optimum values of model parameters as rainfall input changes. In particular, parameters that control the volume of the simulated hydrograph showed largest sensitivity to the different rainfall inputs. However, it was noted that the optimal parameter values stayed within the physically allowable ranges.

HBV better captured observed hydrograph patterns and volume when bias-corrected CMORPH estimates were used instead of the uncorrected estimates. We observed that the runoff model translates small rainfall errors to larger streamflow errors. The magnitudes of such error amplifications became smaller for bias-corrected satellite data than those for the satellite-only data. In the present study, −18% bias of CMORPH rainfall inputs is translated to −27% streamflow bias. The bias amplification was reduced (from −13% rainfall bias to only −17% streamflow bias) when space and time varying bias corrections were applied to the CMORPH rainfall input. Accounting for the temporal variability of CMORPH bias has the largest influence on model simulations and should be taken into account. The error propagation is found to depend, not only on errors in rainfall inputs, but also on the accumulated rainfall which affect the actual moisture and water storage in the model stores. Future studies should assess comparative advantages of various bias correction algorithms that account for the temporal aspects of bias and that received applications in climate change and radar rainfall studies. There is also a need to devote efforts towards operationalizing the bias correction algorithms.

Acknowledgment

Support for this study was provided by the National Science Foundation (NSF) through Award Number OISE-0914618 to the University of Louisiana at Lafayette, and by the the NASA EPSCoR/BoR DART2 program and the NASA Grant to Louisiana Space (LaSPACE) Consortium. Support provided by the University of Louisiana at Lafayette Computational and Visualization Enterprise (CAVE) Consortium is acknowledged. Support is also received from the CGIAR Research Program on Water, Land and Ecosystems (WLE), which is led by International Water Management Institute (IWMI).

Author Contributions

Emad Habib and Alemseged Tamiru Haile had the original idea for the study with all co-authors carried out the design. Alemseged Tamiru Habib was responsible for mentoring and follow-up of study participants. Alemseged Tamiru Haile and Nazmus Sazib were responsible for data preparation and all authors carried out the analyses. Alemseged Tamiru Haile, and Emad Habib drafted the manuscript, which was revised by Tom Rientjes, Yu Zhang, and the other authors. Tom Rientjes provided the modeling tools. All authors read and approved the final manuscript.

Conflicts of Interest

The authors declare no conflict of interest.

References

1. Haile, A.T.; Habib, E.; Rientjes, T.H.M. Evaluation of the Climate Prediction Center (CPC) morphing technique (CMORPH) rainfall product on hourly time scales over the source of the Blue Nile River. *Hydrol. Process.* **2013**, *27, 1829–1839*.
2. Kondragunta, C.; Shrestha, K. Automated real-time operational rain gauge quality controls in NWS hydrologic operations. In Proceedings of the 20th AMS Conference on Hydrology, Atlanta, GA, USA, 29 January–2 February 2006.
3. AghaKouchak, A.; Mehran, A.; Norouzi, H.; Behrangi, A. Systematic and random error components in satellite precipitation data sets. *Geophys. Res. Lett.* **2012**, *39*, doi:10.1029/2012GL051592.
4. Dinku, T.; Ceccato, P.; Grover-Kopec, E.; Lemma, M.; Connor, S.J.; Ropelewski, C.F. Validation of satellite rainfall products over East Africa's complex topography. *Int. J. Remote Sens.* **2007**, *28*, 1503–1526.
5. Habib, E.; Haile, A.T.; Tian, Y.; Joyce, R. Evaluation of the high-resolution CMORPH satellite-rainfall product using dense rain gauge observations and radar-based estimates. *J. Hydrometeorol.* **2012**, *13*, 1784–1798.
6. Yilmaz, K.K.; Hogue, T.S.; Hsu, K.; Sorooshian, S.; Gupta, H.V.; Wagener, T. Intercomparison of rain gauge, radar, and satellite-based precipitation estimates with emphasis on hydrologic forecasting. *J. Hydrometeorol.* **2005**, *6*, 497–517.
7. Zhang, Y.; Seo, D-J.; Kitzmiller, D.; Lee, H.; Kuligowski, R.J.; Kim, D.; Kondragunta, C.R. Comparative strengths of SCaMPR satellite QPEs with and without TRMM ingest *vs.* gridded gauge-only analyses. *J. Hydrometeorol.* **2013**, *14*, 153–170.
8. Bitew, M.M.; Gebremichael, M. Assessment of satellite rainfall products for streamflow simulation in medium watersheds of the Ethiopian highlands. *Hydrol. Earth Syst. Sci.* **2011**, *15*, 1147–1155.
9. Bitew, M.M.; Gebremichael, M. Evaluation of satellite rainfall products through hydrologic simulation in a fully distributed hydrologic model. *Water Resour. Res.* **2011**, *47*, doi:10.1029/2010WR009917.
10. Joyce, R.J.; Janowiak, J.E.; Arkin, P.A.; Xie, P.P. CMORPH: A method that produces global precipitation estimates from passive microwave and infrared data at high spatial and temporal resolution. *J. Hydrometeorol.* **2004**, *5*, 487–503.
11. Habib, E.; ElSaadani, M.; Haile, A.T. Climatology-focused evaluation of CMORPH and TMPA satellite rainfall products over the Nile Basin. *J. Appl. Meteorol. Climatol.* **2012**, *51*, 2105–2121.

12. Ebert, E.E.; Janowiak, J.E.; Kidd, C. Comparison of near-real-time rainfall estimates from satellite observations and numerical models. *Bull. Am. Meteorol. Soc.* **2007**, *88*, 47–64.

13. Pereira, F.A.J.; Carbone, R.E.; Janowiak, J.E.; Arkin, P.; Joyce, R.; Hallak, R.; Ramos, C.G.M. Satellite rainfall estimates over South America—Possible applicability to the water management of large watersheds. *J. Am. Water Resour. Assoc.* **2010**, *46*, 344–360.

14. Anagnostou, E.N.; Maggioni, V.; Nikolopoulos, E.I.; Meskele, T.; Hossain, F.; Papadopoulos, A. Benchmarking high-resolution global satellite rainfall products to radar and rain-gauge rainfall estimates. *IEEE Trans. Geosci. Remote. Sens.* **2010**, *48*, 1667–1683.

15. Sapiano, M.R.P.; Arkin, P.A. An Intercomparison and validation of high-resolution satellite precipitation estimates with 3-hourly gauge data. *J. Hydrometeorol.* **2009**, *10*, 149–166.

16. Smith, T.M.; Arkin, P.A.; Bates, J.J.; Huffman, G.J. Estimating bias of satellite-based precipitation estimates. *J. Hydrometeorol.* **2006**, *7*, 841–856.

17. CMORPH Improvements: A Kalman Filter Approach to Blend Various Satellite Rainfall Estimate Inputs and Rain Gauge Data Integration. Available online: http://adsabs.harvard.edu/abs/2009EGUGA.11.9810J (accessed on 25 September 2013).

18. Joyce, R.J.; Xie, P.; Janowiak, J.E. Kalman filter based CMORPH. *J. Hydrometeorol.* **2011**, *12*, 1547–1563.

19. Seo, D.-J.; Breidenbach, J. Real-time correction of spatially nonuniform bias in radar rainfall data using rain gauge measurements. *J. Hydrometeorol.* **2002**, *3*, 93–111.

20. Seo, D.-J.; Briedenbach, J.P.; Johnson, E.R. Real-time estimation of mean field bias in radar rainfall data. *J. Hydrol.* **1999**, *223*, 131–147.

21. Zhang, J.; NOAA/NSSL; Norman, O.K.; Howard, K.; Vasiloff, S.; Langston, C.; Kaney, B.; Arthur, A.; van Cooten, S.; Kelleher, K.; *et al.* National Mosaic and QPE (NMQ) system—Description, results and future plan. In Proceedings of the 34th Conference on Radar Meteor, Williamsburg, VA, USA, 6 October 2009.

22. Huffman, G.J.; Adler, R.F.; Bolvin, D.T.; Gu, G.J.; Nelkin, E.J.; Bowman, K.P.; Hong, Y.; Stocker, E.F.; Wolff, D.B. The TRMM multisatellite precipitation analysis (TMPA). Quasi-global, multiyear, combined-sensor precipitation estimates at fine scales. *J. Hydrometeorol.* **2007**, *8*, 38–55.

23. Boushaki, F.I.; Hsu, K.-L.; Sorooshian, S.; Park, G.-H.; Mahani, S.; Shi, W. Bias adjustment of satellite precipitation estimation using ground-based measurement: A case study evaluation over the southwestern United States. *J. Hydrometeorol.* **2009**, *10*, 1231–1242.

24. Li, M.; Shao, Q. An improved statistical approach to merge satellite rainfall estimates and raingauge data. *J. Hydrol.* **2010**, *385*, 51–64.

25. Vila, D.A.; de Goncalves, L.G.G.; Toll, D.L.; Rozante, J. Statistical evaluation of combined daily gauge observations and rainfall satellite estimates over continental South America. *J. Hydrometeorol.* **2009**, *10*, 533–543.

26. Hong, Y.; Hsu, K.; Moradkhani, H.; Sorooshian, S. Uncertainty quantification of satellite precipitation estimation and Monte Carlo assessment of the error propagation into hydrologic response. *Water Resour. Res.* **2006**, *42*, doi:10.1029/2005WR004398.

27. Chiang, Y.-M.; Hsu, K.-L.; Chang, F.-J.; Hong, Y.; Sorooshian, S. Merging multiple precipitation sources for flash flood forecasting. *J. Hydrol.* **2007**, *340*, 183–196.

28. Tobin, K.J.; Bennett, M.E. Adjusting satellite precipitation data to facilitate hydrologic modeling. *J. Hydrometeorol.* **2010**, *11*, 966–978.

29. Tian, Y.; Peters-Lidard, C.D.; Eylander, J.B. Real-time bias reduction for satellite-based precipitation estimates. *J. Hydrometeorol.* **2010**, *11*, 1275–1285.

30. Krakauer, N.Y.; Pradhanang, S.M.; Lakhankar, T.; Jha, A.K. Evaluating satellite products for precipitation estimation in mountain regions: A case study for Nepal. *Remote Sens.* **2013**, *5*, 4107–4123.

31. Artan, G.; Gadain, H.; Smith, J.L.; Asante, K.; Bandaragoda, C.J.; Verdin, J.P. Adequacy of satellite derived rainfall data for stream flow modeling. *Nat. Hazards* **2007**, *43*, 167–185.

32. Zeweldi, D.A.; Gebremichael, M.; Downer, C.W. On CMORPH rainfall for stream flow simulation in a small, Hortonian watershed. *J. Hydrometeorol.* **2011**, *12*, 456–466.

33. Behrangi, A.; Khakbaz, B.; Jaw, T.C.; AghaKouchak, A.; Hsu, K.; Sorooshian, S. Hydrologic evaluation of satellite precipitation products over a mid-size basin. *J. Hydrol.* **2011**, *397*, 225–237.

34. Yong, B.; Ren, L.-L.; Hong, Y.; Wang, J.-H.; Gourley, J.J.; Jiang, S.-H.; Chen, X.; Wang, W. Hydrologic evaluation of Multisatellite Precipitation Analysis standard precipitation products in basins beyond its inclined latitude band: A case study in Laohahe basin, China. *Water Resour. Res.* **2010**, *46*, doi:10.1029/2009WR008965.

35. Sorooshian, S.; AghaKouchak, A.; Arkin, P.; Eylander, J.; Foufoula-Georgiou, E.; Harmon, R.; Hendrickx, J.; Imam, B.; Kuligowski, R.; Skahill, B.; *et al.* Advanced concepts of remote sensing of precipitation at multiple scales. *Bull. Am. Meteorol. Soc.* **2011**, *92*, 1353–1357.

36. Gebremichael, M.; Anagnostou, E.N.; Bitew, M. Critical steps for continuing advancement of satellite rainfall applications for surface hydrology in the Nile River basin. *J. Am. Water Resour. Assoc.* **2010**, *46*, 361–366.

37. Rientjes, T.H.M.; Perera, B.U.J.; Haile, A.T.; Reggiani, P.; Muthuwatta, L.P. Regionalisation for lake level simulation—The case of Lake Tana in the Upper Blue Nile, Ethiopia. *Hydrol. Earth Syst. Sci.* **2011**, *15*, 1167–1183.

38. Haile, A.T.; Rientjes, T.H.M.; Gieske, A.S.M.; Gebremichael, M. Rainfall variability over mountainous and adjacent lake areas: The case of Lake Tana basin at the source of the Blue Nile River. *J. Appl. Meteorol. Climatol.* **2009**, *48*, 1696–1717.

39. Rientjes, T.H.M.; Haile, A.T.; Ayele, A.F. Diurnal rainfall variability over the Upper Blue Nile: A remote sensing based approach. *Int. J. Appl. Earth Obs. Geoinf.* **2013**, *21*, 311–325

40. Haile, A.T.; Rientjes, T.H.M.; Habib, E.; Jetten, V.; Gebremichael, M. Rain event properties at the source of the Blue Nile River. *Hydrol. Earth Syst. Sci.* **2011**, *15*, 1023–1034.

41. Abdo, K.S.; Fiseha, B.M.; Rientjes, T.H.M.; Gieske, A.S.M.; Haile, A.T. Assessment of climate change impacts on the hydrology of Gilgel Abbay catchment in Lake Tana basin, Ethiopia. *Hydrol. Process.* **2009**, *23*, 3661–3669.

42. Rientjes, T.H.M.; Haile, A.T.; Mannaerts, C.M.M.; Kebede, E.; Habib, E. Changes in land cover and stream flows in Gilgel Abbay catchment, Upper Blue Nile basin—Ethiopia. *Hydrol. Earth Syst. Sci. Discuss.* **2011**, *7*, 9567–9598.

43. Wale, A.; Rientjes, T.H.M.; Gieske, A.S.M.; Getachew, H.A. Ungauged catchment contributions to Lake Tana's water balance. *Hydrol. Process.* **2009**, *23*, 3682–3693.

44. Allen, R.G.; Pereira, L.S.; Raes, D.; Smith, M. *Crop Evapotranspiration: Guidelines for Computing Crop Water Requirements*; Irrigation and Drainage Paper; United Nations Food and Agriculture Organization: Rome, Italy, 1998; p. 300.

45. Lindström, G.; Johansson, B.; Persson, M.; Gardelin, M.; Bergström, S. Development and test of the distributed HBV-96 hydrological model. *J. Hydrol.* **1997**, *201*, 272–288.

46. Booij, M.J. Impact of climate change on river flooding assessed with different spatial model resolutions. *J. Hydrol.* **2005**, *303*, 176–198.

47. Deckers, D.L.E.H.; Booij, M.J.; Rientjes, T.H.M.; Krol, M.S. Catchment variability and parameter estimation in multi-objective regionalisation of a rainfall-runoff model. *Water Resour. Manag.* **2010**, *24*, 3961–3985.

48. Merz, R.; Blöschl, G. Regionalisation of catchment model parameters. *J. Hydrol.* **2004**, *287*, 95–123.

49. Rientjes, T.H.M.; Muthuwatta, L.P.; Bos, M.G.; Booij, M.J.; Bhatti, H.A. Multi-variable calibration of a semi-distributed hydrological model using streamflow data and satellite based evapotranspiration. *J. Hydrol.* **2013**, *505*, 276–290.

50. Seibert, J. Estimation of parameter uncertainty in the HBV model. *Nord. Hydrol.* **1997**, *28*, 4–5.

51. Gebremichael, M.; Krajewski, W.F. Characterization of the temporal sampling error inspace-time-averaged rainfall estimates from satellites. *J. Geophys. Res.* **2004**, *109*, doi:10.1029/2004JD004509.

Applicability of Multi-Frequency Passive Microwave Observations and Data Assimilation Methods for Improving Numerical Weather Forecasting in Niger, Africa

Mohamed Rasmy, Toshio Koike and Xin Li

Abstract: The development of satellite-based forecasting systems is one of the few affordable solutions for developing regions (e.g., West Africa) that cannot afford ground-based observation networks. Although low-frequency passive microwave data have been used extensively for land surface monitoring, the use of high-frequency passive microwave data that contain cloud information is very limited over land because of strong heterogeneous land surface emissions. The Coupled Atmosphere and Land Data Assimilation System (CALDAS)was developed by merging soil moisture information estimated from low-frequency data with corresponding high-frequency data to estimate cloud information and, thus, improve weather forecasting over Niger, West Africa. The results showed that the assimilated soil moisture and cloud distributions were reasonably comparable to satellite retrievals of soil moisture and cloud observations. However, assimilating soil moisture alone within a mesoscale model produced only marginal improvements in the forecast, whereas the assimilation of both soil moisture and cloud distributions improved the simulation of temperature and humidity profiles. Rainfall forecasts from CALDAS also correlated well with satellite retrievals. This indicates the potential use of CALDAS as a reliable forecasting tool for developing regions. Further developments of CALDAS and the inclusion of data from several other sensors will be researched in future studies.

Reprinted from *Remote Sens*. Cite as: Rasmy, M.; Koike, T.; Li, X. Applicability of Multi-Frequency Passive Microwave Observations and Data Assimilation Methods for Improving Numerical Weather Forecasting in Niger, Africa. *Remote Sens* **2014**, *6*, 5306–5324.

Nomenclature

AMSR	Advanced Microwave Scanning Radiometer
AMSR-E	AMSR on Earth Observing System
AMSR-2	AMSR-2nd generation on GCOM-W1 satellite
NCEP	National Centers for Environmental Prediction
LPRM	NASA Land Parameter Retrieval Model
ARPS	The Advanced Regional Prediction System
CALDAS	Coupled Atmosphere and Land Data Assimilation System
EnKF	Ensemble Kalman Filter
CMDAS	Cloud Microphysics Data Assimilation System
LDAS	Land Data Assimilation Sysatem
LDAS-A	LDAS coupled with an Atmospheric model
NWP	Numerical Weather Prediction
IR data	InfraRed data

SiB2	Simple Biosphere model version-2
TRMM	Tropical Rainfall Measuring Mission

1. Introduction

Africa is a continent that is vulnerable to the effects of global warming and climate change/variability, which will increase the incidences of abnormal weather, such as frequent flooding and prolonged drought [1]. Numerical Weather Prediction (NWP) in West Africa is still underdeveloped because of several problems (e.g., lack of data, inadequate forecast system, difficulties in simulating land/atmospheric processes, and lack of forecast verification exercises). NWP is mainly associated with the initial-boundary-values problem; *i.e.*, given an estimate of the initial state of the atmosphere and land surface, the model forecasts the evolutions of both. A greater number of high quality observations that more fully represent the complete nature of the initial atmosphere and land surface would greatly improve the forecasting capabilities of NWP models.Therefore, the lack of reliable in situ data with which to initialize the NWP model is the most challenging problem regarding forecasting in underdeveloped regions.

The capability of remote sensing to contribute to the prediction of the Earth's weather and climatic systems has been demonstrated in many scenarios. Although many studies have been based on the use of satellite data in West Africa, only a few have considered the use of satellite data for NWP modeling and validation [2]. Accordingly, this research investigates the use of passive microwave satellite observations with a mesoscale model in West Africa. In microwave frequencies, the dielectric property of materials allows the quantitative estimation of moisture quantities such as soil moisture, vegetation water, snow water content, and cloud condensate [3–6]. Thus, the potential for the use of space-borne microwave observations in NWP has been advanced considerably by the launch of several sophisticated platforms (e.g., Terra, Aqua, and the Global Change Observation Mission).

Furthermore, low-frequency microwave observations provide information on soil moisture that is very useful for enhancing NWP model forecasting through improved initialization [7,8]. However, the few studies that have been performed on the assimilation of soil moisture observations into NWP models have shown several limitations (e.g., retrieval and preprocessing of soil moisture products, Cumulative Distribution Function matching, difficulties in defining an observation operator that changes spatially and temporarily, and time constraints for the use of near-real-time applications). Thus, they recommended the application of assimilation schemes that are more advanced [9,10]. To overcome such problems, an on-line system (*i.e.*, Land Data Assimilation System coupled with Atmospheric model (LDAS-A)) was developed to directly assimilate lower frequency microwave radiance from the Advanced Microwave Scanning Radiometer (AMSR) for updating realistic soil moisture content in a mesoscale model and the results were presented in [11]. However, because of limited satellite observations (e.g., AMSR-E and AMSR-2, twice daily maximum), the improved land surface conditions often suffered from significant errors and drift due to biases in the predicted forcing (e.g., rainfall and solar radiation) that misguided the subsequent forecast [11,12]. Thus, this particular problem rendered the on-line land data assimilation ineffective in NWP models.

Clouds directly affect surface forcing (*i.e.*, downward radiation and rainfall), and thus they affect the estimation of the Earth's surface water and energy budgets. Similar to microwave sounding observations (e.g., Advanced Microwave Sounding Unit (AMSU)), passive microwave images (e.g., AMSR-E and AMSR-2) also contain information on cloud fields at higher frequencies (e.g., 89 GHz). It is also noteworthy to mention the Global Precipitation Measurement mission, which is the first coordinated international satellite network to provide near-real-time observations of radar and microwave images across a range of frequencies (10–183 GHz) every 3 h anywhere on the globe. However, the increasing abundance of data arising from several platforms has not been used well within NWP models. In fact, the 89-GHz frequency is more sensitive to cloud information compared with other channels of the AMSR-E, but it also contains information of land surface emission. Obtaining atmospheric information from combined land-atmosphere signals is very challenging because the strong heterogeneous land surface emissions mask the very weak atmospheric signals. When land surface emissions are estimated as accurately as possible, then the weak signal of the cloud fields from the combined signals can be converted into realistic representations of the cloud fields for the NWP models.

Consequently, the Coupled Atmosphere and Land Data Assimilation System (CALDAS) was developed as an improvement of the LDAS-A. CALDAS assimilates lower frequency data of AMSR-E/AMSR-2 to improve the representation of land surface conditions and it merges them with higher frequency data of AMSR-E/AMSR-2 to improve the representation of cloud conditions over land surfaces. CALDAS results from the Tibetan Plateau have shown that simultaneous assimilation of land and cloud conditions improved the biases in land surface states, cloud representation, and forcing to land surface models, which enhanced land-atmosphere interactions [12].

As a continuation from our previous study [12], this work addresses the application of CALDAS as a weather forecasting system, simulating short-term weather conditions over a mesoscale domain of Niger in West Africa. Model performance is investigated using available ground-based data and satellite observations.

2. Dataset, Models and Method

2.1. Dataset

In this research, data from the National Centers for Environmental Prediction (NCEP) were used for model simulations, and AMSR-E brightness temperature data were used for soil moisture and cloud data assimilation. The other data sets listed under this section were used for the analysis and validation of the model outputs.

2.1.1. Initial and Boundary Conditions

The initial and lateral boundary conditions were derived from the NCEP data. These NCEP-FNL (Final) operational global analysis data are available as a $1° \times 1°$ grid at 6-hourly intervals. To obtain the corresponding initial and boundary conditions required to run the mesoscale model, analysis data available at 26 pressure levels were used. Variables (e.g., pressure, geopotential height, temperature, relative humidity, and u and v winds) required for the mesoscale model simulations were interpolated from the pressure levels.

2.1.2. AMSR-E Brightness Temperature Data

AMSR-E is a total power passive microwave radiometer that measures horizontally and vertically polarized brightness temperatures at separate frequencies of 6.925, 10.65, 18.7, 23.8, 36.5, and 89.0 GHz. Its individual measurements have spatial resolutions varying from ~0.05° at 89.0 GHz to ~0.5° at 6.925 GHz. Because of spatial resolution differences, we used the nearest neighbor interpolation method to resample the data to fit the horizontal resolution of our model (~0.05°). We used vertical polarizations of 6.925 and 10.65 GHz to retrieve soil moisture information, because the atmosphere can be considered transparent at these frequencies. The 89-GHz band of the AMSR-E shows high sensitivity to cloud water content and precipitation compared with other channels. Consequently, the 89-GHz vertical polarization was used to assimilate cloud information over the land surface. The motivation for using vertical polarization is that it is less sensitive than horizontal polarization to land surface roughness and vegetation. The calibrated brightness temperature (Level 1B) data were obtained from the Japan Aerospace Exploration Agency (JAXA).

2.1.3. *In Situ* Data

In situ data were obtained from the African Monsoon Multidisciplinary Analysis database. Atmospheric profile observations were retrieved from radiosondes (Vaisala radiosonde RS92) launched four times daily from the Niamey (2.17°E, 13.48°N) station in June 2006. The associated soil conditions were either missing or unsuitable for validation during the simulation period. Consequently, comparisons were performed using satellite-derived soil moisture products.

2.1.4. Satellite-Derived Soil Moisture Products

JAXA's soil moisture products are based on AMSR-E data by applying the 10–36 GHz algorithm for simultaneous retrieval of the soil moisture and vegetation water content from two indices: the polarization index and index of soil wetness [13]. These global moisture products are available twice daily at 0.1° spatial resolution.

National Aeronautics and Space Administration (NASA) soil moisture products were derived from the AMSR-E data and Land Parameter Retrieval Model (LPRM). The LPRM uses dual polarized channels (6.925 or 10.65 GHz) for the retrieval of both surface soil moisture and vegetation water content [14]. These global moisture products are also available twice daily at 0.25° spatial resolution.

2.1.5. Satellite Cloud and Rainfall Products

The global full-resolution (~0.04°) infrared (IR) data, merged from several geostationary satellites (~11 micron channels), and available from NASA at 30-min intervals, were used as cloud observations. Lower brightness temperatures in the IR images indicate sufficiently thick clouds whose cloud tops radiate at the atmospheric temperature at higher altitudes.

Rainfall data derived from the Tropical Rainfall Measuring Mission (TRMM) were used for model validation. These gridded estimates (3B42) are available with 3-hourly temporal resolution and $0.25° \times 0.25°$ spatial resolution in a global belt extending across latitudes from 50°S–50°N.

2.2. Models

CALDAS (Figure 1) has three subsystems: (1) a land-atmosphere coupled mesoscale model (ARPS-SiB2); (2) a Land Data Assimilation System (LDAS); and (3) a Cloud Microphysics Data Assimilation System (CMDAS). Here, the combinations of the ARPS-SiB2-LDAS and ARPS-SiB2-LDAS-CMDAS models are named LDAS-A and CALDAS, respectively. All three models (*i.e.*, ARPS-SiB2, LDAS-A, and CALDAS) shared the same atmosphere-land couple model (*i.e.*, ARPS-SiB2) and same settings of physics and parameterizations. ARPS-SiB2 model simulation was a simple dynamical downscaling of GCM analysis data. The only difference between ARPS-SiB2 and LDAS-A simulations is that LDAS-A used additional satellite observed soil moisture contents compared to ARPS-SiB2. Similarly, the only difference between LDAS-A and CALDAS simulations is that CALDAS used additional satellite observed cloud data. Detailed information and formulations on LDAS-A and CALDAS are given in [11,12], respectively; however, the following sections provide a brief explanation of their formulation.

Figure 1. Schematic diagram of the Coupled Atmosphere and Land Data Assimilation System (CALDAS).

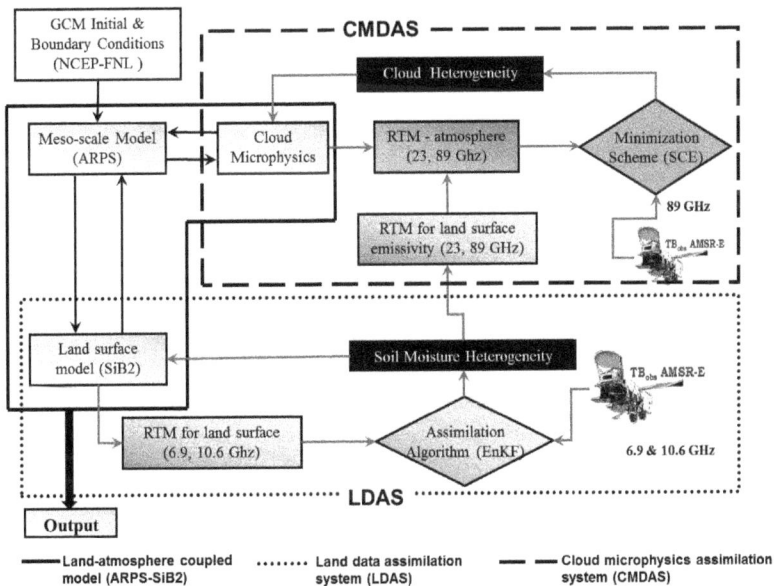

The Advanced Regional Prediction System (ARPS) is a comprehensive regional- to storm-scale prediction model, and its atmospheric prediction component is a three-dimensional nonhydrostatic compressible model [15]. The LDAS consists of the Simple Biosphere model version 2 (SiB2) [16], which functions as a model operator for LDAS as well as the land surface scheme for ARPS, together with a physics-based radiative transfer model as the observation operator, and an Ensemble Kalman Filter (EnKF) [17] as the assimilation algorithm.

A land surface radiative transfer algorithm has been developed based on the large contrast between the dielectric constants of dry soil (~4) and water (~80) for lower microwave frequencies (i.e., 1–15 GHz). By neglecting atmospheric and rainfall effects and by assuming that the reflection at the surface is much less than the radiation from the surface and vegetation layers at lower frequencies, the brightness temperature T_b at the satellite level is given by [18]:

$$T_b = (1 - R_p) * T_s * \exp(-\tau_c) + (1 - \omega_c) * T_c * (1 - \exp(-\tau_c)) \tag{1}$$

here, R_p is the surface reflectivity and T_s is the surface physical temperature (K), ω_c is the single-scattering albedo of the canopy, τ_c is the vegetation optical thickness and T_c is the canopy temperature (K).

In the field of hydrology, the EnKF has been applied to soil moisture estimation and it has been found to perform well against the variational assimilation method [19,20]. An overview of the EnKF for soil moisture assimilation is given below.

Consider X=$[w_1, w_2, w_3]^T$ as a state variable and the first estimate, for which w_1, w_2 and w_3 are the soil moisture contents of the surface, root and deep soil layers, respectively. The first estimate is used to create an ensemble of size (N) by adding pseudo-random noise with known statistics. By dropping the time notation, each member of state variable X_i is given by

$$X_i = \bar{X} + e_i \qquad e_i(i = 1, 2....N) \sim N(0, P) \tag{2}$$

where e_i is the random error vector of each member obtained from a multivariate Gaussian distribution with zero mean and error covariance matrix P, and $\bar{X}$ is the expectation of the first estimate X. In the forecast step, the forecast state member X_i^f is determined from the nearest analysis state member X_i^a according to

$$X_i^f = M(X_i^a) + u_i \qquad u_i \sim N(0, Q) \tag{3}$$

where M is the model operator and u_i is the model error vector of each member, obtained from a multivariate Gaussian distribution with zero mean and error covariance matrix Q.

In the analysis step, the AMSR-E observation data are perturbed by adding a random observation error and each member of the analyzed state variable X_i^a is updated as

$$X_i^a = X_i^f + K((Y_o + v_i) - H(X_i^f)) \qquad v_i \sim N(0, R) \tag{4}$$

where K is a Kalman gain matrix, H is the observation operator, Y_o is the observation, R is the observation error covariance and v_i is a random error vector of the observation with zero mean and covariance matrix R.

CMDAS was developed for use over only sea surfaces to improve the atmospheric moisture fields by assimilating higher-frequency AMSR-E observations [21]. To apply CMDAS over a land surface, land surface emissivity derived from the assimilated soil moisture was used as a boundary condition. To estimate the effects of atmospheric absorption, emission, and scattering on the upwelling radiation at 89 GHz, the 4-stream fast model [22] was used, and the Shuffled Complex Evolution technique [23] was adopted as a minimization scheme.

Compared with land data assimilation (50 members), cloud data assimilation is computationally very expensive because of the several hundred iterations of cloud parameters that are required to minimize the cost function. Nevertheless, the CALDAS model is enabled on parallel computing technology to satisfy the increasingly high-performance computing requirements of operational NWP models.

2.3. Method

1. As shown in Figure 1, the land-atmosphere mesoscale model (ARPS-SiB2) was established using initial and boundary conditions from NCEP-FNL data.
2. The ARPS model was integrated for a predefined period (10 min) and the calculated atmospheric forcing data transferred to the SiB2 model.
3. At the beginning of the SiB2 integration, the ensemble (50 members) of soil moisture profiles was generated. SiB2 was executed independently for each ensemble member of the soil moisture profile, retaining the same model parameters and atmospheric forcing. At the end of the SiB2 calculation, the mean values of the updated soil state and fluxes were computed and fed back to the ARPS model as the lower boundary conditions of the atmospheric model. Then, the ARPS-SiB2 model was integrated forward in time.
4. At times, when AMSR-E observations were available, the brightness temperatures at 6.9 and 10.65 GHz were perturbed to produce an ensemble of observations with prescribed statistics. The SiB2-driven ensemble of soil moisture profiles, surface temperature, and canopy temperature were used to obtain the simulated brightness temperatures using the forward microwave radiative transfer model. The EnKF calculated the assimilated soil moisture profiles using simulated and observed brightness temperatures, as shown in Equation (4). In the case of soil moisture assimilation (no cloud assimilation), the updated soil state and fluxes were fed back to the ARPS model and the ARPS-SiB2 model was integrated forward in time.
5. In the case of cloud data assimilation, CMDAS was activated as soon as the LDAS completed the soil moisture assimilations. The control variables (profiles of temperature, specific humidity, pressure, air density, mixing ratio of cloud water, rain water, hail, snow, and cloud ice) were obtained from ARPS as an initial state to run the model operator (Lin's ice microphysics [24]). The 4-stream fast model calculated the modeled brightness temperatures for 89 GHz at the satellite level by considering the land surface as the lowest boundary. Land surface emissivity was calculated using assimilated soil moisture content. The Shuffle Complex Evolution scheme was used to estimate the assimilated cloud parameters (*i.e.*, cloud liquid

water, snow, and rain) by minimizing the cost function calculated between the modeled and observed brightness temperatures. Then, the updated soil state and fluxes were fed back to the ARPS-SiB2 model.

6. Finally, with the reinitialized land surface and atmospheric conditions, the ARPS-SiB2 model was integrated forward in time to predict the land and atmospheric evolution until the next AMSR-E observations were available. The results from the ARPS-Sib2 model were recorded at 30-min intervals.

3. Experiment Descriptions

The performance of CALDAS was assessed based on three simulations: (1) the ARPS-SiB2 (hereafter ARPS) run, where a one-way nesting procedure employed the land-atmosphere mesoscale model without any assimilation; (2) the LDAS-A run, in which an ARPS run was accompanied by sequential land data (soil moisture) assimilation; and (3) the CALDAS run, in which the LDAS-A run was accompanied by the cloud microphysics data assimilation.

Study Domain and Model Configuration

Two model domains, domain-1 ($\sim$0.23° horizontal resolution) and domain-2 ($\sim$0.05° horizontal resolution) were established (Figure 2). Domain-1 was used to downscale the initial and lateral boundary conditions derived from NCEP-FNL by ARPS to domain-2. Then, ARPS, LDAS-A , and CALDAS were applied to domain-2 independently. The physical parameterization options were configured with a 1.5-order turbulent kinetic-energy-based closure scheme for sub-grid-scale turbulent mixing, latitude-dependent Coriolis parameters, Kain and Fritsch cumulus parameterization [25] (only for domain-1), and Lin ice microphysics.

Because integrated values of moisture fields were assimilated by CALDAS, the assimilated values were distributed with predefined profiles. Cloud Liquid Water Content (CLWC) was assumed to have a parabolic distribution (single-layer cloud) with zero values both above the top and below the bottom of the cloud layers. The lower and upper bounds of integrated CLWC were set to 0 and 1.5 kg·m^{-2}, respectively.

In this study, the assimilation of the 23-GHz channel for water vapor was not performed because such assimilation was found to cause errors (*i.e.*, underestimation) in the assimilated water vapor profile. This underestimation in turn caused assimilated clouds to evaporate quickly to compensate for the water vapor deficiency (not shown). To avoid this particular problem, air was assumed saturated only within the cloud layer. However, more accurate information on the water vapor profiles would improve the development of convective systems in the model and thus, enhance rainfall prediction. Therefore, we are investigating the use of water vapor profile information within the CALDAS system, obtained from the AMSU and Atmospheric Infrared Sounder (AIRS).

The distributions of rain and snow followed a skewed profile. They began to form at the cloud tops, grew to their maximum at the cloud bottoms, and subsequently decreased because of evaporation and the breakup of the raindrops or snowflakes. Further information on the model and the setup can be referenced from [12].

Figure 2. Mesoscale model domains.

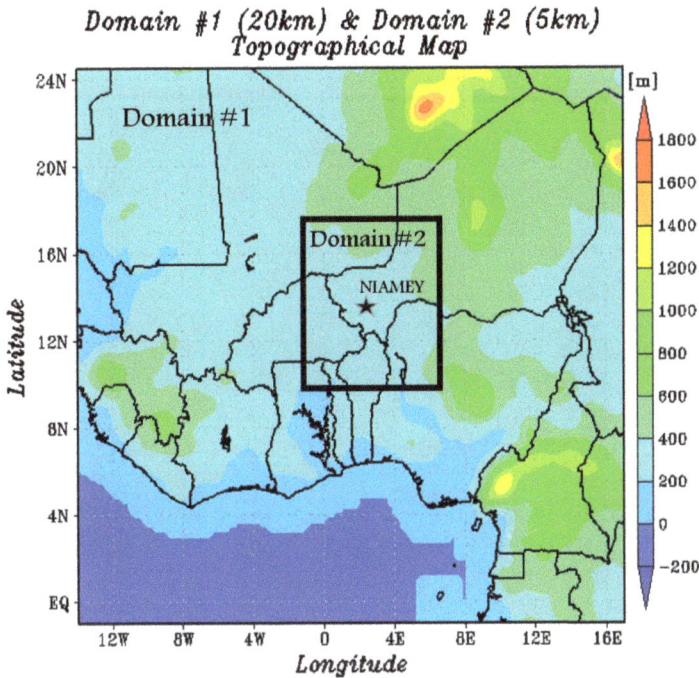

4. Results and Discussions

This study investigated the applicability of multi-frequency passive microwave observations for improving weather forecasting. Simulations were performed during a cloudy period starting from 6 June 2006 to 7 June 2006. The AMSR-E observed full coverage of domain-2 at 0140 UTC on 6 June 2006, which captured the cloud activities over the domain. The results from domain-2 are discussed in the following sections.

4.1. Distribution of Surface Soil Moisture

To examine the reliability of assimilated land surface emission, model simulations of soil moisture content were compared with AMSR-E brightness temperatures at 6.9 GHz and satellite-derived soil moisture products immediately after land data assimilation. Although a linear relationship cannot be assumed between the surface soil moisture content and brightness temperatures at 6.9 GHz, visual classification can be performed based on lower brightness temperatures being related to higher soil moisture content. Figure 3a,b represent the spatial distributions of simulated surface soil moisture for both the ARPS and the CALDAS models, and Figure 3c represents the distribution of brightness temperatures at 6.9 GHz. The magnitude of the soil moisture content simulated by ARPS (derived from NCEP-FNL) was higher in most of the model grids (particularly in the south) and produced completely different spatial distributions when compared with the observed brightness temperatures. Conversely, the distribution of CALDAS soil moisture content was reasonably comparable with that of brightness temperature at 6.9 GHz, and the clusters of dry and wet regions observed at 6.9 GHz were well defined in the CALDAS simulation.

Figure 3. Spatial distribution of volumetric surface soil moisture (m³/m³) at 0200 UTC on 6 June 2006; **(a)** ARPS; **(b)** CALDAS; **(c)** dvanced Microwave Scanning Radiometer (AMSR)-E brightness temperature (K) observed at 6.9 GHz; **(d)** Japan Aerospace Exploration Agency (JAXA) product; and **(e)** National Aeronautics and Space Administration (NASA) product, respectively.

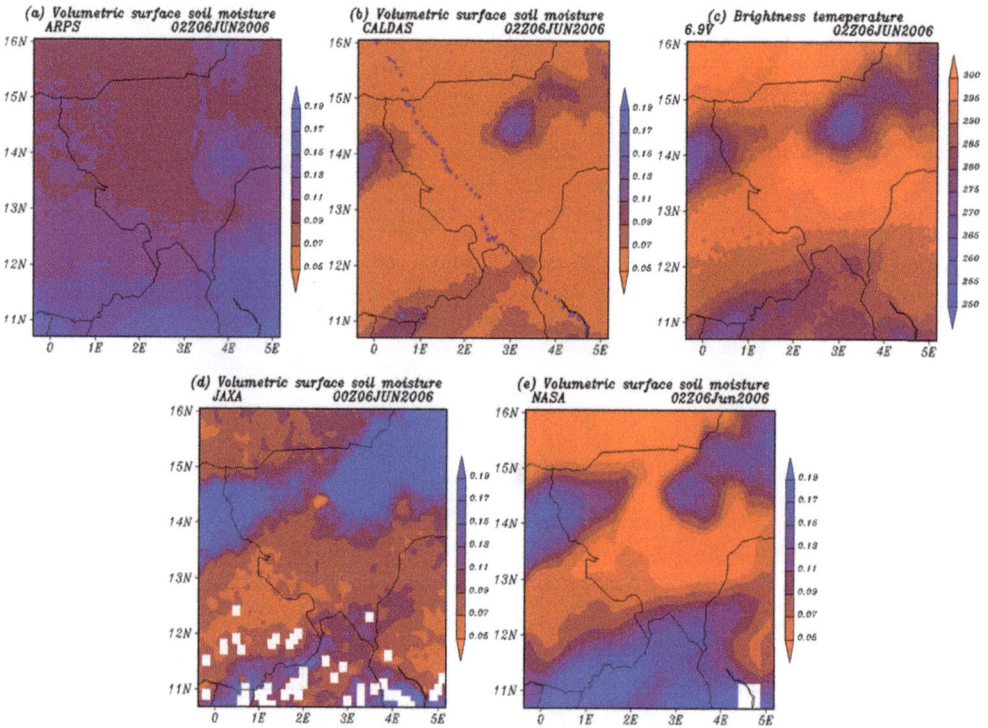

Quantitative validation of assimilated soil moisture content was not possible owing to lack of *in situ* soil moisture data during this period. Therefore, JAXA and NASA (c-band) soil moisture products were obtained for independent verification of the assimilated soil moisture information. As shown in Figure 3d,e, the distribution of CALDAS assimilated soil moisture was comparable and correlated reasonably well with the products of JAXA (Pearson product-moment correlation coefficient R = 0.6) and NASA (R = 0.5). It is also worth noting that the absolute values of soil moisture obtained from the JAXA and NASA products differed greatly, even though they both used the same satellite information. A previous study [26] on the validation of the JAXA and NASA satellite-derived soil moisture products reported significant differences between them and that neither provided reliable estimates for all the conditions represented by the four watershed sites in the United States. The significant differences between both products in our domain also indicated that further coordinated validation studies are necessary to understand and resolve the problems associated with the retrieval algorithms for African regions. However, as CALDAS performed better than ARPS and spatially, its soil moisture correlated well with the satellite-derived soil moisture products, the improved soil moisture content from CALDAS (using the lower frequencies) was used to calculate land surface emission at higher frequencies to facilitate the integration of cloud microphysics data by assimilating AMSR-E higher-frequency observations over land surfaces.

4.2. Comparisons of Cloud Condensate With Satellite IR Observations

The quantitative validation of individual or integrated atmospheric moisture variables (*i.e.*, cloud liquid water, ice, snow, and hail) is still difficult, even at point locations, because of the unavailability of the necessary data sets. Therefore, to investigate the retrieval capability of the land and atmosphere coupled data assimilation processes and to evaluate the reliability of the assimilated cloud related parameters, qualitative comparisons were performed of the spatial distributions between vertically integrated condensate (summation of liquid and solid phases) and IR satellite data. The results were investigated ~1 h after the assimilation (*i.e.*, from 0300 UTC on 6 June 2006).

As shown in the first column of Figure 4, the IR observation reflected very active cloud activity over the model domain, which was successfully introduced by the assimilation of passive microwave observations in CALDAS. The assimilated cloud distributions from CALDAS were coherent and compared well with the observed IR cloud cell distributions. Conversely, the ARPS and LDAS-A models predicted almost no cloud activity at this time. In addition, hourly evolutions of cloud activity (each column) showed that the seeding of cloud information using the 89-GHz observation within the model enhanced the forecasting capability of the model. Thus, the clouds predicted by CALDAS correlated well with IR observations 6 h after the assimilation, whereas ARPS and LDAS-A showed much lower cloud activity during this period. The identical results of cloud condensate obtained from ARPS and LDAS-A during the model forecasting showed that the assimilation of soil moisture did not improve the cloud simulation, and therefore, the reliable initialization of cloud information is shown to be crucial for improving weather forecasting during this particular period.

148

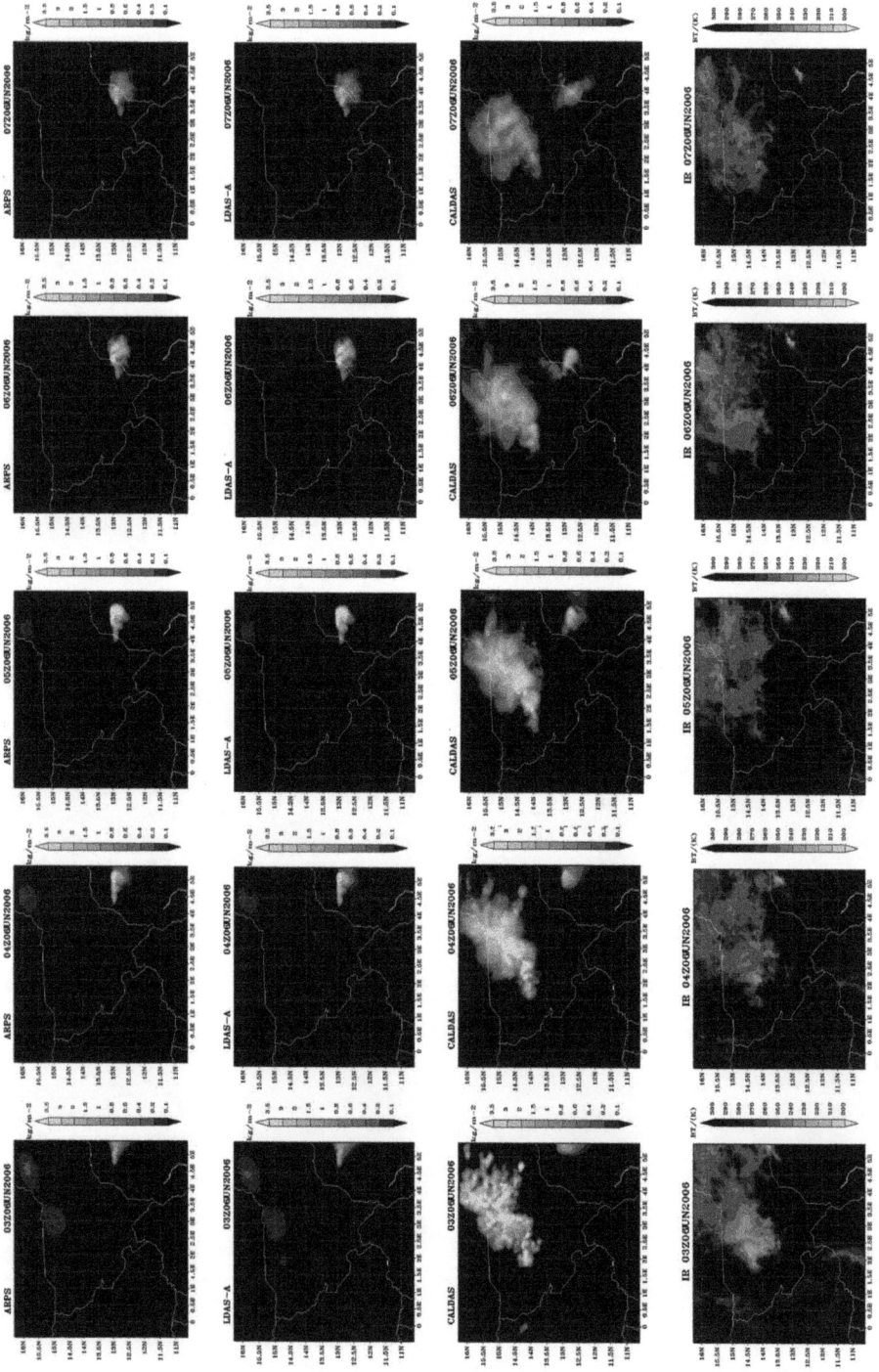

Figure 4. Hourly variation of integrated condensate (kg/m²) and IR data (K); first row: ARPS, second row: Land Data Assimilation System coupled with Atmospheric model (LDAS-A), third row: CALDAS, and fourth row: infrared (IR) brightness temperature.

Figure 5 depicts the diurnal variation in spatial correlations calculated between cloud condensate simulated by the models and IR observations for 6 June 2006. To eliminate the shift between the AMSR-E and IR observations, the IR data and the model's cloud condensate were scaled up to 0.25°. Negative values in the spatial correlation indicate mismatches between the IR data and model simulations of cloud positions, whereas positive values indicate coherence between the IR observations and model simulations. As shown, the spatial correlations derived from ARPS and LDAS-A were very similar and showed mostly negative or very low spatial correlations with the IR observations. However, the CALDAS results showed a rapid increase (+0.49) in the spatial correlation at 0300 UTC. Two hours after the assimilation, the correlation reached a maximum of +0.52. This result indicates that CALDAS was reinitialized with improved cloud distributions (which coincided well with the IR data) and that the assimilation of cloud condensate enhanced the simulated cloud activities over the model domain. Moreover, the correlation of CALDAS was larger than ARPS and LDAS-A for about 6 h from the assimilation before decreasing. Therefore, CALDAS, as a multi-frequency assimilation system, has the potential to improve cloud information over land surfaces. However, quantitative information on each atmospheric moisture variable and microphysical property must be investigated in the future, using reliable data sets, to assess the full capabilities of the CALDAS model.

Figure 5. Hourly variation of spatial correlations calculated from model simulated cloud top temperatures and IR cloud top temperatures for 6 June 2006.

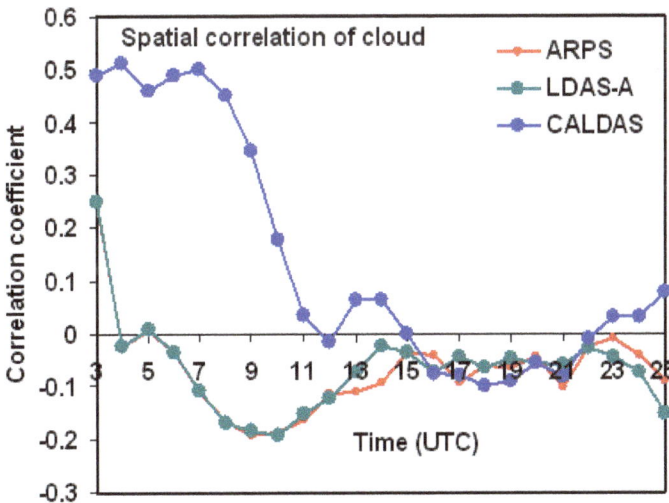

4.3. Evaluation of Land–Atmosphere Interactions

Improving the representations of land surface conditions and cloud will improve the simulations of surface (energy and water) budgets and turbulent heat fluxes, which will eventually improve the representation of land-atmosphere interactions in the model. Radiosonde observations of temperature and humidity in the planetary boundary layer, which reflect average turbulent fluxes over the observation domain, are commonly used to determine the degree to which the land affects the atmosphere, and vice versa. Radiosonde soundings obtained at Niamey station were used to investigate the model performance in simulating the land-atmosphere feedback mechanism over the model domain.

Figure 6a,b compare the observed potential temperature and specific humidity soundings with the simulated soundings from ARPS, LDAS-A, and CALDAS at 1030 UTC (1130 Local Time (LT)) on 6 June 2006. During this time, the land was heated by the sun and turbulent mixing was created close to the land surface (from surface to ~875 hPa), as shown by the observed temperature soundings. ARPS underestimated the potential temperature sounding (Mean Bias Error (MBE) = −0.52 K, Root Mean Square Error (RMSE) = 2.00 K) and overestimated the humidity sounding (MBE = 1.85 g/kg, RMSE = 2.92 g/kg) from the surface to ~700 hPa. In the case of LDAS-A, the assimilation of surface soil moisture resulted in marginal improvements in both the temperature (MBE = −0.48 K, RMSE = 1.94 K) and the humidity (MBE = 1.82 g/kg, RMSE = 2.88 g/kg) profiles, whereas improvements were seen in CALDAS for the humidity profile (MBE = 1.32 g/kg, RMSE = 2.01 g/kg).

Figure 6. Comparison of observed and model soundings at 1030 UTC on 6 June 2006; (**a**) potential temperature (K) and (**b**) specific humidity (g/kg).

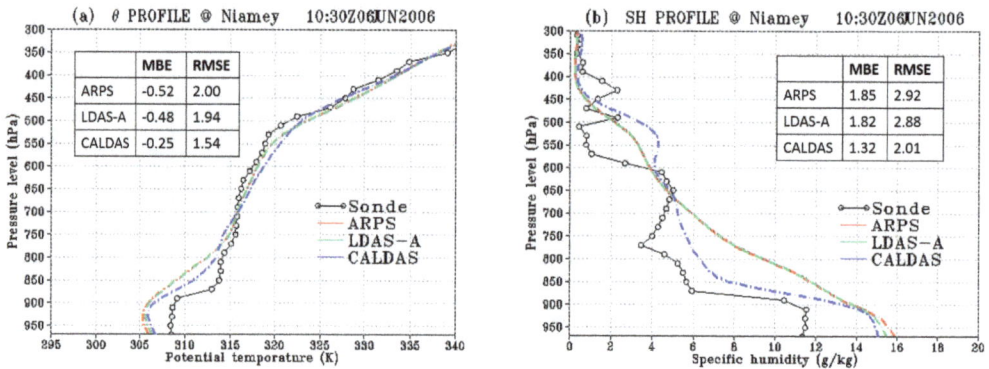

Figure 7 compares the observed soundings with the simulated soundings at 2230 UTC (2330 LT) on the same day. The observed atmospheric conditions at this (night) time, as shown by the temperature profile, were characterized by statically stable air with neutrally stratified layers. In the case of the ARPS and LDAS-A models, profiles with higher humidity and lower temperature were simulated. Further investigation of the model results indicated that these two models produced higher soil moisture during this period (not shown). However, the observed trend was captured well by CALDAS and it compared well with the observed soundings of potential temperature (MBE = 0.26 K, RMSE = 1.02 K) and specific humidity (MBE = 1.06 g/kg, RMSE = 1.31 g/kg). The improvement extended from the surface to the middle atmospheric layers (~650 hPa) in the CALDAS simulation. These results show that the synchronized land and atmosphere re-initialization resulted in better prediction of the atmospheric sounding (*i.e.*, land-atmosphere interactions) during the forecast period of the model.

Figure 7. Same as Figure 6 but for 2230 UTC on 6 June 2006; (**a**) potential temperature (K) and (**b**) specific humidity (g/kg).

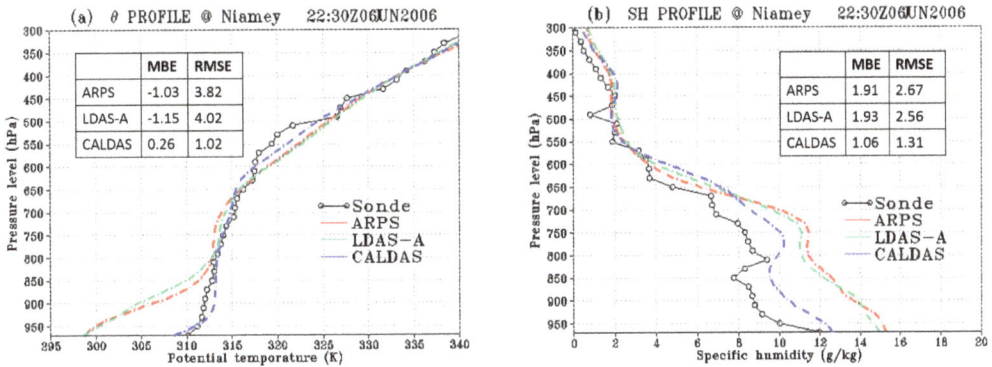

(a) θ PROFILE @ Niamey   22:30Z06JUN2006

	MBE	RMSE
ARPS	-1.03	3.82
LDAS-A	-1.15	4.02
CALDAS	0.26	1.02

Sonde
ARPS
LDAS-A
CALDAS

(b) SH PROFILE @ Niamey   22:30Z06JUN2006

	MBE	RMSE
ARPS	1.91	2.67
LDAS-A	1.93	2.56
CALDAS	1.06	1.31

Sonde
ARPS
LDAS-A
CALDAS

4.4. Evaluation oF Rainfall Forecast

Forecasting precipitation remains very challenging, and the skill and accuracy of quantitative forecasts of precipitation are generally low, even over well-instrumented regions. Figure 8 compares the 6-h (from 03 UTC–09 UTC) accumulated rainfall forecast by ARPS and CALDAS, and estimated by the TRMM satellite on 6 June 2006. As shown in Figure 8a, the location of the rainfall from ARPS was completely different from the location where rain was actually observed by TRMM (Figure 8c). In contrast, the assimilation of the 89-GHz information for cloud data produced rainfall at the same locations as observed by TRMM. Although the rainfall amounts predicted by CALDAS did not correspond completely with the TRMM retrievals, the locations of the heavy rainfall were better predicted by CALDAS. In the case of ARPS, the spatial correlation of rainfall with TRMM was very low (R = 0.01), whereas in the case of CALDAS, the spatial correlation of rainfall with TRMM was improved significantly (R = 0.47).

The quantification of model accuracy, errors, and uncertainties is crucial in determining the reliability of model forecasts, but it is a very challenging task without accurate observations in this study region. In this research, we utilized satellite retrievals (*i.e.*, soil moisture, cloud distribution, and rainfall) because of the unavailability of ground-based observations (except radiosonde data). However, these products are subjected to significant uncertainties due to bias of retrieval algorithms (e.g., soil moisture measurements obtained from JAXA and NASA differed greatly, even though data from the same sensor were used for the retrievals), few calibration and verification practices, low spatial resolutions, and the indirect nature of satellite measurements, which included uncertainties from various factors (e.g., estimation through cloud top reflectance, thermal radiance, and penetration depth of microwave signals). Consequently, our present discussions on validating model outputs with satellite retrievals are based mainly on statistical correlations, because the retrievals capture spatial distribution well, even though the absolute values differ among the similar products used. Conversely, further coordinated efforts are required to organize more intensive and comprehensive field observations to calibrate and assess the uncertainties of the retrievals, as well as to validate model outputs quantitatively.

Figure 8. Comparison of 6 hours (from 03UTC to 09UTC) accumulated rainfall (mm) obtained from (**a**) ARPS; (**b**) CALDAS; and (**c**) TRMM, respectively.

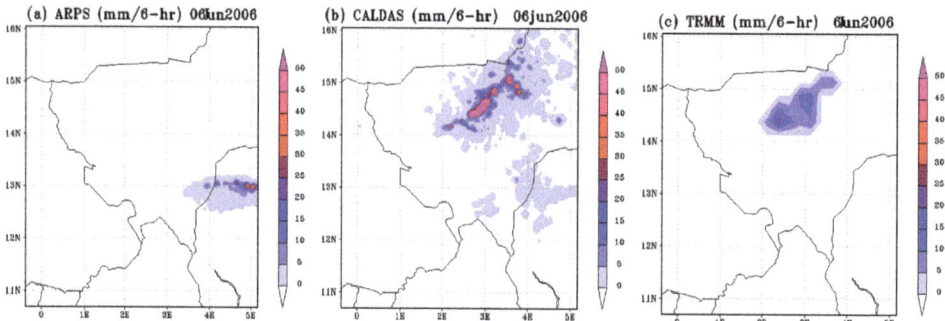

5. Conclusions

This research investigated the applicability of space-borne multi-frequency passive microwave observations for improving weather forecasting by retrieving soil moisture and cloud information using two data assimilation systems (*i.e.*, LDAS-A and CALDAS) over a mesoscale domain of Niger, Africa.

The results showed that the assimilated soil moisture was reasonably comparable with satellite retrievals obtained from NASA and JAXA, whereas the no-assimilation case produced higher soil moisture contents and completely different spatial distributions compared with the assimilated case. However, assimilating soil moisture alone produced only marginal improvements in the model forecasts of atmospheric profiles (*i.e.*, potential temperature and specific humidity).

The assimilated cloud distributions from CALDAS were coherent and they compared well with the observed IR cloud cell distributions. The seeding of cloud information from high-frequency passive microwave radiance enhanced the self-development of simulated cloud activities over the model domain. Consequently, simulated cloud activities exhibited positive correlations for a period of about 12 h following the assimilation, whereas the no-assimilation case demonstrated mostly negative correlations with the IR observations during this period.

The simulated soundings of both potential temperature and specific humidity from CALDAS showed significant improvements from the surface to the middle atmospheric layers ($\sim$650), which resulted from improvements in the representations of land and cloud conditions in the model. Furthermore, the rainfall predicted by CALDAS correlated better with the TRMM retrievals than the no-cloud assimilation cases. In particular, the locations of the heavy rainfall were better predicted by CALDAS, whereas the no-assimilation case produced rainfall in completely different locations compared with the TRMM rainfall retrievals.

These results are encouraging in terms of producing reliable land and atmospheric states as well as regional forecasts, especially in poorly gauged or un-gauged regions, because CALDAS requires only satellite data as input. The consideration of using other satellite information such as the Global Precipitation Measurement mission, AMSU, and AIRS within the system will be researched in future studies to improve the model's robustness and performance further.

Acknowledgments

This research was carried out as part of the Research Program on Climate Change Adaptation (RECCA), verification experiment for AMSR-E/AMSR-2 (GCOM-W), and Data Integration and Analysis System (DIAS) project (2011–2015). The authors would like to thank the Japan Aerospace Exploration Agency (JAXA) and African Monsoon Multidisciplinary Analysis (AMMA) for providing data for this research. The authors are grateful to the Editors and anonymous reviewers for their thoughtful and constructive comments, which improved the quality of the paper.

Author Contributions

Mohamed Rasmy coupled the satellite data assimilation systems within a mesoscale model, performed numerical simulations, analyzed satellite retrievals and model results, and wrote the manuscript. Xin Li contributed significantly for developing the data assimilation components and provided constructive comments and suggestions for coupling the components within a mesoscale model. Toshio Koike supervised the research work, outlined the design of data assimilation systems, and assisted with the system developments as well as the manuscript writing.

Conflicts of Interest

The authors declare no conflict of interest.

References

1. Intergovernmental Panel on Climate Change. *Fourth Assessment Report: Climate Change 2007: The AR4 Synthesis Report*; Intergovernmental Panel on Climate Change: Geneva, Switherland, 2007.
2. Sohne, N.; Chaboureau, J.; Guichard, F. Verification of cloud cover forecast with satellite observation over West Africa. *Mon. Weath. Rev.* **2008**, *136*, 4421–4434.
3. Jackson, T.J.; Schmugge, T. Vegetation effects on the microwave emission of soils. *Remote Sens. Environ.* **1991**, *36*, 203–212.

4. Jackson, T.; le Vine, D.M.; Hsu, A.Y.; Oldak, A.; Starks, P.J.; Swift, C.T.; Isham, J.D.; Haken, M. Soil moisture mapping at regional scales using microwave radiometry: The Southern Great Plains Hydrology Experiment. *IEEE Trans. Geosci. Remote Sens.* **1997**, *37*, 2136–2151.

5. Paloscia, S.; Pampaloni, P. Microwave polarization index for monitoring vegetation growth. *IEEE Trans. Geosci. Remote Sens.* **1988**, *26*, 617–621.

6. Wentz, F.J.; Meissner, T. *Supplement 1 Algorithm Theoretical Basis Document: AMSR-E Ocean Algorithms*; NASA: Santa Rosa, CA, USA, 2007.

7. Koster, R.; Suarez, M.; Higgins, R.; van den Dool, H. Observational evidence that soil moisture variations affect precipitation. *Geophys. Res. Lett.* **2003**, *30*, doi:10.1029/2002GL016571.

8. Koster, R.; Suarez, M.; Liu, P.; Jambor, U.; Berg, A.; Kistler, M.; Reichle, R.; Rodell, M.; Famiglietti, J. Realistic initialization of land surface states: impacts on subseasonal forecast skill. *J. Hydrometeor.* **2004**, *5*, 1049–1063.

9. Scipal, K.; Drusch, M.; Wagner, W. Assimilation of a ERS scatterometer derived soil moisture index in the ECMWF numerical weather prediction system. *Adv. Water Resour.* **2008**, *31*, 1101–1112.

10. Draper, C.S.; Mahfouf J.-F.; Walker, J. P. An EKF assimilation of AMSR-E soil moisture into the ISBA land surface scheme. *J. Geophys. Res.* **2009**, *114*, doi:10.1029/2008JD011650.

11. Rasmy, M.; Koike, T.; Boussetta, S.; Lu, H.; Li, X. Development of a satellite land data assimilation system coupled with a mesoscale model in the Tibetan Plateau. *IEEE Trans. Geosci. Remote Sensing* **2011**, *49*, 2847–2862.

12. Rasmy, M.; Koike, T.; Kuria, D.; Mirza, C.; Li, X.; Yang, K. Development of the coupled atmosphere and land data assimilation system (CALDAS) and its application over the Tibetan Plateau. *IEEE Trans. Geosci. Remote Sens.* **2012**, *50*, 4227–4242.

13. Fujii, H.; Koike, T.; Imaoka, K. Improvement of the AMSR-E Algorithm for soil moisture estimation by introducing a fractional vegetation coverage dataset derived from MODIS data. *J. Remote Sens. Soc. Jpn.* **2009**, *29*, 282–292.

14. Jeu, R.; Wagner, W.; Holmes, T.; Dolman, A.; Giesen, N.; Friesen, J. Global soil moisture patterns observed by space borne microwave radiometers and scatterometers. *Surv. Geophys.* **2008**, *29*, 399–420.

15. Xue, M.; Droegemeier, K.K.; Wong, V.; Shapiro, A.; Brewster, K.; Carr, F.; Weber, D.; Liu, Y.; Wang, D. The Advanced Regional Prediction System (ARPS)— A multi-scale nonhydrostatic atmospheric simulation and prediction tool. Part II: Model physics and applications. *Meteorol. Atmos. Phys.* **2001**, *76*, 143–166.

16. Sellers, P.J.; Los, S.; Tucker, C.; Justice, C.; Dazlich, D.; Collatz, G. A revised land surface parameterization (SiB2) for atmospheric GCMs, Part II: The generation of global fields of terrestrial biophysical parameters from satellite data. *J. Clim.* **1996**, *9*, 706–737.

17. Evensen, G. The ensemble Kalman filter: Theoretical formulation and practical implementation. *Ocean Dyn.* **2003**, *53*, 343–367.

18. Fujii, H.; Koike, T. Development of a TRMM/TMI algorithm for precipitation in the Tibetan Plateau by considering effects of land surface emissivity. *J. Meteorol. Soc. Jpn.* **2001**, *79*, 475–483.

19. Reichle, R.H.; McLaughlin, D.B. Hydrologic data assimilation with the ensemble Kalman filter. *Mon. Weath. Rev.* **2002**, *130*, 103–114.

20. Huang, C.; Li, X.; Lu, L. Experiments of one-dimensional soil moisture assimilation system based on ensemble Kalman filter. *Remote Sens. Environ.* **2008**, *112*, 889–900.

21. Mirza, C.R.; Koike, T.; Yang, K.; Graf, T. Retrieval of Atmospheric integrated water vapor and cloud liquid water content over the ocean from satellite data using the 1-D-Var ice cloud microphysics data assimilation system (IMDAS). *IEEE Trans. Geosci. Remote Sens.* **2008**, *46*, 119–129.

22. Liu, G. A fast and accurate model for microwave radiance calculations. *J. Meteorol. Soc. Jpn.* **1998**, *76*, 335–343.

23. Duan, Q.; Sorooshian, S.; Gupta, V. Effective and efficient global optimization for conceptual rainfall-runoff models. *Water Resour. Res.* **1992**, *28*, 1015–1031.

24. Lin, Y.; Farley, R.; Orville, H. Bulk parameterization of the snow field in a cloud model. *J. Clim. Appl. Meteorol.* **1983**, *22*, 1065–1089.

25. Kain, J.; Fritsch, J. Convective Parameterization for mesoscale models: The kain-fritsch scheme. *Meteoro. Monogr. Amer. Meteor. Soc.* **1993**, *24*, 165–170.

26. Jackson, T.J.; Cosh, M.H.; Bindlish, R.; Starks, P.J.; Bosch, D.D.; Seyfried, M.; Goodrich, D.C.; Moran, M.S.; Du, J. Validation of advanced microwave scanning radiometer soil moisture products. *IEEE Tran. Geosci. Remote Sens.* **2010**, *48*, 4256–4272.

Chapter 3:
Evapotranspiration

A Life-Size and Near Real-Time Test of Irrigation Scheduling with a Sentinel-2 Like Time Series (SPOT4-Take5) in Morocco

Michel Le Page, Jihad Toumi, Saïd Khabba, Olivier Hagolle, Adrien Tavernier, M. Hakim Kharrou, Salah Er-Raki, Mireille Huc, Mohamed Kasbani, Abdelilah El Moutamanni, Mohamed Yousfi and Lionel Jarlan

Abstract: This paper describes the setting and results of a real-time experiment of irrigation scheduling by a time series of optical satellite images under real conditions, which was carried out on durum wheat in the Haouz plain (Marrakech, Morocco), during the 2012/13 agricultural season. For the purpose of this experiment, the irrigation of a reference plot was driven by the farmer according to, mainly empirical, irrigation scheduling while test plot irrigations were being managed following the FAO-56 method, driven by remote sensing. Images were issued from the SPOT4 (Take5) data set, which aimed at delivering image time series at a decametric resolution with less than five-day satellite overpass similar to the time series ESA Sentinel-2 satellites will produce in the coming years. With a Root Mean Square Error (RMSE) of 0.91mm per day, the comparison between daily actual evapotranspiration measured by eddy-covariance and the simulated one is satisfactory, but even better at a five-day integration (0.59mm per day). Finally, despite a chaotic beginning of the experiment—the experimental plot had not been irrigated to get rid of a slaking crust, which prevented good emergence—our plot caught up and yielded almost the same grain crop with 14% less irrigation water. This experiment opens up interesting opportunities for operational scheduling of flooding irrigation sectors that dominate in the semi-arid Mediterranean area.

Reprinted from *Remote Sens.* Cite as: Le Page, M.; Toumi, J.; Khabba, S.; Hagolle, O.; Tavernier, A.; Kharrou, M.H.; Er-Raki, S.; Huc, M.; Kasbani, M.; El Moutamanni, A.; Yousfi, M.; Jarlan, L. A Life-Size and Near Real-Time Test of Irrigation Scheduling with a Sentinel-2 Like Time Series (SPOT4-Take5) in Morocco. *Remote Sens.* **2014**, *6*, 11182-11203.

1. Introduction

In the southern Mediterranean region, as well as in other arid and semi-arid areas in the world, water withdrawals have significantly increased over the last decades [1]. In Africa, 12.5 million hectares out of the 210 hectares grown are under fully- or partly-controlled irrigation (about 6% of croplands) according to [2]. In the southern Mediterranean region, this relatively small surface absorbs 86% of water withdrawals. Localized irrigation, like drip irrigation or micro-sprinkler, represents less than 5% of all irrigated areas, while surface and sprinkler irrigations account for 78% and 17%, respectively. Although the move to drip irrigation is strongly encouraged through subsidies in several countries, flooding and sprinkler will certainly remain the dominant techniques in the future, like in Europe [3], but will eventually shift to more sprinkler as pressurized systems appear.

In the arid and semi-arid areas, where water is one of the most limiting factors to the improvement of agricultural production [4], the challenge is to increase agricultural production with limited water resources, thus by improving irrigation management for a more efficient and productive use of irrigation water. Furthermore, the southern Mediterranean region is pointed out as a hotspot of climate change [5] and the part of water allowed to agriculture will be reduced from 90% in 1995 down to 70% by 2020 [6]. Also, in Africa, crop development may be sub-optimal because of nitrogen stress [7,8].

According to [9], irrigation scheduling is the decision-making process of the farmer related to "when" to irrigate and "how much" water to apply to a crop but, as Pereira noticed, *"research has made available a large number of irrigation scheduling tools including procedures to compute crop water requirement, to simulate soil water balance, to estimate the impact of water deficits on yields, and to simulate the economic returns of irrigation. However, irrigation scheduling is not yet utilized by the majority of farmers"*.

A sophisticated, yet easy to use, assessment of crop development and advice for irrigation scheduling could, then be an interesting tool both at the farming and district scales, especially in a context of surface or sprinkler irrigation and under-optimal crop because of water and/or nitrogen stress. Spatial remote sensing imagery can help to fill the gap between research and implementation of irrigation scheduling.

Depending on the wavelength of the sensor acquiring Remote Sensing Imagery, different biophysical variables may be derived [10]. Several vegetation indices have been made up (SAVI, EVI, *etc.*) [11], but the Normalized Difference Vegetation Index (NDVI) is considered a common denominator amongst several satellite sensors. Other parameters can also be successfully retrieved from visible remote sensing like the Leaf Area Index (LAI), the Fraction Cover (f_c) and albedo.

Several reviews of current modeling of Evapotranspiration and Water Balance with Remote Sensing imagery are available [12–17]. They agreed to group methods into four main classes of increasing complexity: (1) empirical direct methods whereby a relationship links Net Radiation to the difference of Air and Surface Temperature; (2) vegetation indices or inference methods, whereby a crop coefficient inferred from remote sensing modulates a Reference Evapotranspiration locally measured; (3) the Residual Method of Surface Energy Balance combines empirical relationships to physical models in order to assess the energy budget components, minus the latent Heat Flux which is determined as the residue of the other fluxes; and (4) the mechanistic approach based on so called Soil Vegetation Atmosphere Transfer models, whereby all the components of the energy and water budgets are computed.

In this experiment, the second method is used with the FAO-56 dual-crop method under water stress conditions described in [18] and its extension [19]. The crop coefficients and fraction covers are inferred from a high temporal and spatial resolution data set acquired and processed in near real-time (a couple of days). To achieve this, the experiment relied on the SPOT4 (Take5) experiment [20] which aimed to provide a Sentinel-2 like data set from February to June 2013. In fact, the two Sentinel-2 satellites will provide high-resolution images of all land areas, every fifth day. This experiment also benefited from the low cloud cover observed in semi-arid areas, which enabled getting a high frequency of cloud-free acquisitions.

This experiment was not carried out in a fully-controlled experimental environment, but on the contrary, on a plot located within an irrigation scheme running as usual. As the scheme distributes irrigation water arranged by rotational turns, the main constraint is then to anticipate the irrigation event at least three days in advance. Such a system is also subject to various contingencies like element ruptures for example. Hopefully, the only contingency during the experiment was a three-day closure of all irrigation events for accounting and recovery of payments. To resolve the timing constraints, the data were gathered in this way: meteorological data from our local station were sent by telemetry, irrigation data from the Agricultural Office were communicated through phone calls, and images through the Internet.

This paper is organized as follows. In Section 2, we provide a description of the study area, the *in-situ* measurements and satellite imagery, and the specialized FAO-56 method. Section 3 presents the results of irrigation scheduling by satellite imagery, which was carried out on the durum wheat. Section 4 discusses the strengths and weaknesses of the proposed method, and suggests some perspectives.

2. Data and Methods Used

2.1. Study Site and Experimental Plots

The Haouz plain is a sedimentary plain bordered to the North by the Jbilet Mountains and to the South by the High-Atlas Mountains. The plain is crossed from South to North by several *wadis*, which are the tributaries of two large collectors, the Tensift and Oum er Rbia, which gave their name to their respective watershed.

The plain climate is semi-arid, with average annual rainfall of 250 mm concentrated between November and April, whilst evaporative demand according to the reference evapotranspiration (ET_0) is about 1600 mm per year [21].

The agricultural sector is managed by the *Office Régional de Mise en valeur Agricole du Haouz*, or ORMVAH. About two thirds of the 473,000 ha of croplands are irrigated, and half of them (144,600 ha) are irrigated from dams.

The experimental plots are located in the R3 irrigation scheme (Figure 1), which encompasses about 3000 ha of irrigated land, composed of about 745 individual fields. R3 scheme is supplied with water from the Hassan dam through the Rocade main canal, which also provides other irrigated areas and the drinking water of Marrakech City. Irrigation water is supplied to the fields via open primary, secondary and tertiary concrete canals that are well-maintained. *Neyrpic modules à masque*—undershot gate devices—provide control of flow rates by ±5% [22]. According to [22], the allocation of water is determined in three steps. Initially, the seasonal distribution of water is negotiated between the major water stakeholders depending on the availability of water in reservoirs. At the beginning of the irrigation season, the Water User Associations (WUA) are informed about the seasonal water allocated to their sector. A second negotiation phase between WUAs and ORMVAH leads to the monthly and seasonal requirements according to the expected cropping patterns. Thus, the number of irrigation cycles and water depths is fixed in proportion to farm size. During the season, the WUAs adjust the irrigation scheduling by choosing the start date of each irrigation cycle, usually based on simple observations of weather or by visual assessment of crop and soil water status.

Figure 1. The picture shows the area covered with the SPOT4-Take5 images (background is the first image of the Time Series taken on 31 January 2013), and the location of R3 scheme (green), 40 km east of Marrakech, Morocco. The scheme is fed by the Rocade Canal (dark blue).

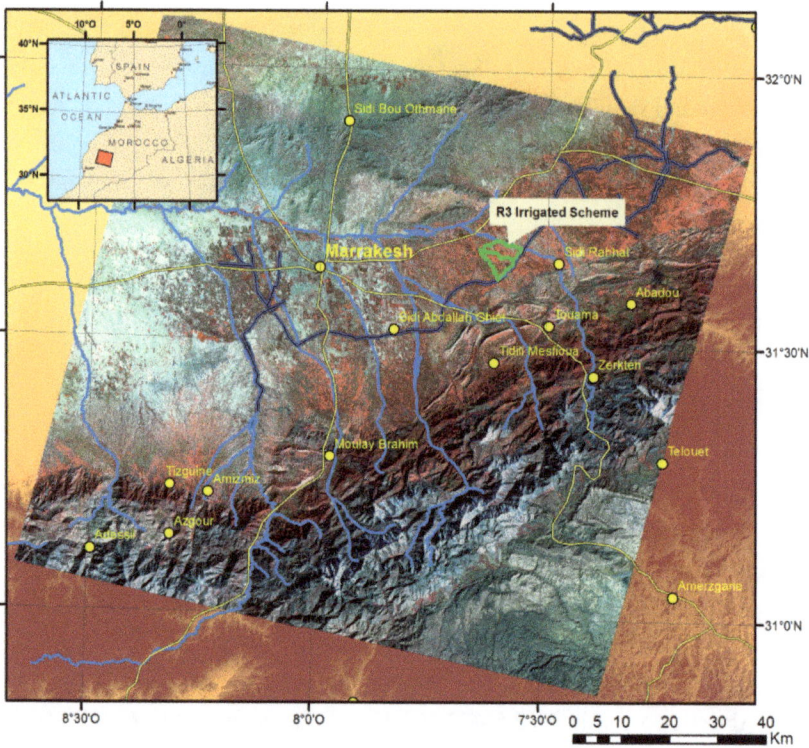

The dominant soil-type is xerosol, developed on colluvial materials of the High Atlas mountains, resulting in homogeneous deep soils (generally more than 1 m), with fine clay to loamy texture. The main crops are cereals (mostly wheat) adding up to 80% of the area on average, the remaining area being planted in olive trees or covered by fallows and villages [22].

Two plots were selected at the proximity of a primary canal head (Figure 2) in order not to disrupt the scheduling of the irrigation scheme and to ensure water availability to the experiment. The first plot, hereafter "experimental", covers an area of 3.7 hectares and irrigation was driven by our model. The second plot (3.3 hectares) is irrigated "as usual" by the farmer, and it is hereafter referred to as "reference". The four soil samplings conducted at 20 cm and 40 cm depth showed a texture of 36% clay and 20% sand. The pedotransfer function [23] is taken from [24] to obtain a field capacity of $0.32 \ \text{m}^3/\text{m}^3$ and a wilting point of $0.18 \ \text{m}^3/\text{m}^3$.

Figure 2. Setup at the south of the scheme: The experimental plot (yellow) is located north of the reference plot (pink) at a distance of about 1km from the main canal. A flux tower (green circle) was installed at the center of each plot (E stands for "experimental", and R, for "Reference"). The meteorological station (green square) is located 2 km west of the plots.

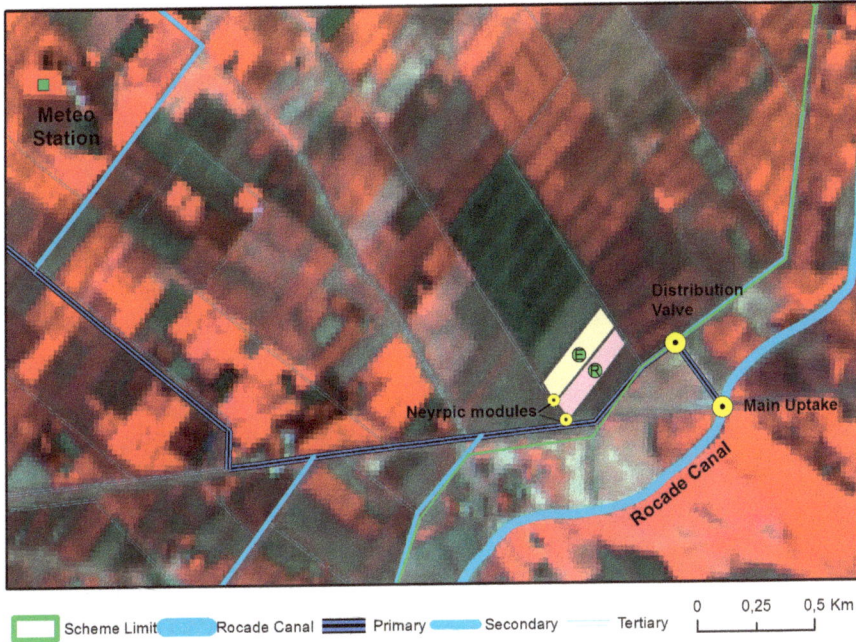

2.2. Technical Itinerary

After a proper soil preparation (deep pre-plowing in the summer and soil refinement), durum wheat (Saragola) was sown on 24 December 2012, at a rate of 200 kg/ha. We will further refer to the dates following sowing as DaS (Date after Sowing). Fertilizer was applied at sowing, consisting of 200 kg/ha of Diammonium Phosphate (DAP), and then at the beginning of the grain-filling phase on DaS 106 with 100 kg/ha of Ammonitrate. Weeds were controlled with specific chemical applications on DaS 51 and 108. Harvest took place on DaS 171 (12 June 2013).

Both plots are surface-irrigated. The irrigation technique is Border Irrigation [25]. This technique is widely used in the so-called Modern Irrigation areas of the Haouz plain where the plots have been leveled. A quaternary channel dug by the farmer orthogonally to the distribution canal allows diverting water flow to each strip of land within the plot. Water then travels down the slope while fully filling the soil, thus preventing the control of applied water for each irrigation event. The total length of each border is equal to the width of the plot (about 100 meters) and the width of the border is about 8 m. It takes about half an hour to irrigate one border with two operators, and about 5 to 6 h per hectare. Nevertheless, as described in [25], this technique may not ensure an adequate filling of the root zone, in particular because of poor land grading, wrong stream size, or stopping the inflow at the wrong time. Also according to [26], in the study area, deep percolation was evaluated between

164

26% and 31% of water supplied (rainfall and irrigation), depending on the stage of crop development.

Because of the clay soil texture, a slaking crust occurs when soil moisture content is below field capacity. In our case, the two plots were irrigated immediately after sowing to ensure germination. Since there was no rain after this period, a second irrigation was necessary after a few days in order to ease wheat emergence (which was realized for the farmer plot). Unfortunately, due to a misunderstanding with the farmer, the test plot did not receive this second irrigation on time.

Also, previous experiments [7,24] on winter wheat on this irrigation scheme had already shown that rooting depth does not exceed 60 to 80 centimeters with this irrigation technique.

2.3. The Time Series of Remote Sensing Images

This experiment was mainly driven by the use Sentinel-2 like data thanks to the SPOT4 (Take5) experiment [20]. The SPOT4 (Take5) experiment consisted in lowering the altitude of SPOT4 orbit by a few kilometers, putting it on a five-day repeat cycle orbit, in order to be able to observe a limited number of sites every fifth day under constant viewing angles. Therefore, the achieved repetitiveness is the same as the one provided by Sentinel-2 mission with two satellites, and the experiment was aimed at helping users prepare their methods and applications to get ready for Sentinel-2 data availability in 2015.

Finally, 45 sites were chosen, among them the Moroccan site in which resides the Haouz plain (Figure 2), and 28 successive images were acquired every fifth day in 2013, from 31 January to 15 June. Data were processed in near real-time to level 2A using the MultiSensor Atmospheric Correction and Cloud Screening software (MACCS) [27,28]. Level 2A data is ortho-rectified images expressed in surface reflectance after atmospheric correction, provided with a cloud/cloud shadow mask.

In total, 18 scenes (64%) were cloud-free for our two plots. As displayed on Figure 3, there were six gaps: two gaps with one missing image (10 days) and four gaps with two missing images (15 days). As the Take5 experiment started one month after sowing, and the success of orbit altitude change was not assured, a SPOT5 programming was also performed, with one image every fifteenth days from December to April. Like all other satellites of the SPOT family, the SPOT5 has the capability to make off-track pointing. Thus the major time interval between two images was only 25 days. SPOT5 atmospheric corrections were carried out with SMAC [29] using the data of the Saada Photometer, 50 km west of the plots.

Figure 3. Time series of satellite images Spot4-Take5, Spot5 and Landsat8 during the 2012/13 season. Landsat8 scenes were not used during the experiment and are only indicated as reference. The gaps in the SPOT4 series are only due to cloud cover.

The NDVI (Equation (1)) based on red (R) and infra-red (NIR) reflectances was finally calculated for each image and averaged for each plot, eliminating edge pixels. This choice is important. First, the border irrigation technique does not allow modifying the applied quantity of irrigation, resulting in the fact that the plot level is the most adequate scale of management for irrigation. Second, although it is very likely that a variability of soil and irrigation exists within the plot, we believe that we cannot afford such accuracy on an operational level, so the water budget may be inaccurate at the pixel level. Third, the algorithm is simplified and computing time is reduced.

$$\text{NDVI} = (NIR - R)/(NIR + R) \tag{1}$$

2.4. The FAO-56 Method Driven with Remotely Sensed NDVI

The FAO-56 method [18,19] is used to estimate crop water requirements based on the concepts of the reference evapotranspiration ET_0 (defining atmospheric demand), and crop coefficients introduced to regulate demand with regards to the specific response of particular crops. The definition of ET_0 is associated with assumptions on key biophysical variables of the system: leaf LAI, surface resistance, albedo, canopy height. Although several alternative methods may be used to approximate ET_0, the recommended method is the Penman-Monteith equation [30].

In optimal agronomical conditions (no disease, water, salt or nutrient stress), the crop evapotranspiration is called ET_c. The crop coefficient K_c is the intermediary between ET_0 and ET_c; it takes into account the physical and physiological differences between reference and studied crops. $ET_{c\,adj}$ is the evapotranspiration under non-standard conditions Equation (2). In our study, we only deal with water stress introducing the stress coefficient K_s varying from 0 to 1:

$$ET_{c\,adj} = K_s \times ET_c = K_s \times K_c \times ET_0 \tag{2}$$

The "Dual Crop Coefficient" approach includes a soil water budget module. This is a procedure to predict the impact of specific wetting events that is better suited for irrigation management at the scale. The crop coefficient is split into a basal crop coefficient (K_{cb}) for crop transpiration and a coefficient K_e for evaporation as in Equation (3) [18].

$$ET_{c\,adj} = (K_s \times K_{cb} + K_e) \times ET_0 \tag{3}$$

In this main equation, K_s is explicitly separated while the reduction coefficient of evaporation K_r has been previously introduced into K_e Equation (4):

$$K_e = \min[K_r(K_{c\,max} - K_{cb}); f_{ew} \times K_{c\,max}] \tag{4}$$

where $K_{c\,max}$ reflects the natural constraints of available energy. It is the upper limit of evapotranspiration for every cropped surface and ranges from 1.05 to 1.3 depending on wind, humidity and crop height. In the spatialized version of the model, a single steady value of 1.15 is used. The fraction of soil that is both exposed and wetted (f_{ew}) is equal to 1 for bare soil, and equals to the soil not covered by vegetation ($1 - f_c$) as crop develops.

The review from [31] shows that early works like [32] have found that NDVI was highly correlated with LAI and f_c, and that crop coefficient K_c derived from NDVI were independent of time parameters. The study of [33] showed that the basal crop coefficient K_{cb} can also be derived from a linear relationship of vegetation indices, but pointed out that NDVI saturates at a LAI of about 3.0. The SAVI index [34] that saturates at higher LAI (about 5.0) and that is also less sensitive to changes in soil brightness and moisture is then more appropriate. The linear relationships between f_c and NDVI have shown to be very reliable. For example [35] and [36] found $r^2 > 0.9$.

In relatively recent years, progresses have been made on the estimation of the temporal evolution of K_c from remote sensing measurements of vegetation indices (VIs). Some authors have in fact suggested that relationships between crop coefficient and VIs are linear [32], but others have found non-linear relationships [37]. These relationships have been studied for several crops and recently for potato [38], cotton [39], sugar beet [40], maize [41], grapes [42,43] and citrus orchard [44].

For the winter wheat grown in the Haouz plain, we used the relationships established by [24] for NDVI- f_c and [21] for NDVI- K_{cb}. They determined the NDVI- K_{cb} empirically using eddy-covariance and ET0 measurements and only keeping the data of unstressed and fully covered days (Equation (5)). The NDVI-f_c relationship was empirically established between fraction cover measured by hemispherical photography and NDVI retrieved with a portable radiometer (Equation (6)). Note that, for the latter, the $NDVI_{min}$ parameter has been replaced with a static value of 0.14. Fraction cover has been maintained to its higher value until harvest time;

$$K_{cb} = 1.64 \times NDVI - 0.23 \tag{5}$$

$$f_c = 1.18 \times NDVI - 0.16 \tag{6}$$

A bucket-type water budget is calculated, where a surface bucket is the water source for evaporation, another one for transpiration of the rooting zone and a deeper one for deep infiltration with a simple parameterization to simulate eventual capillary rises. The depth of the evaporative layer is set to 11 cm, with a Readily Evaporable Water threshold set to 10 mm. The second layer varies from 11 to 60 cm at maximum rooting depth in linear relationship to the fraction cover. The total column is set to 150 cm, so that the depth of the third layer is computable by subtracting the second layer.

The Total Water Storage Capacity (TAW) of each layer is the one between field capacity and wilting point (0.32 m^3/m^3 and 0.18 m^3/m^3 respectively as mentioned earlier). For each layer, a threshold p separates the readily accessible water (RAW) from the one accessed with difficulty [18]. A threshold of 0.55 for winter wheat is set after Table 22 of FAO-56. After reaching this threshold, the layer in question begins to reduce its water losses to ET. Reduction coefficients K_s and K_r are calculated with the same formalism for the water balance of the root layer and upper layer, respectively: If there is less depletion (Di) than the threshold, there is no reduction and then $K_s = 1$. If the depletion exceeds the threshold, the stress coefficient is computed as the ratio between available water for the process and the less accessible water (Equation (7)).

$$K_s = \frac{TAW - Di}{(1 - p) \times TAW} \ if \ Di > RAW \tag{7}$$

The rooting depth (Zr) is calculated at each time step with a linear relationship to f_c (Equation (8)) where the Zrmax and Zrmin are the maximum and minimum rooting depth and $f_c max$ is the maximum fraction cover of the crop. According to the previous studies mentioned above, Zrmin and Zrmax were set to 0.15 and 0.6 meter respectively

$$Zr = Zrmin + \frac{f_c}{f_c \max} \times (Zrmax - Zrmin) \tag{8}$$

Depletion of each layer was updated at the end of each day by subtracting rainfall and irrigation, and adding computed $ET_{c\,adj}$ and eventual deep percolation. Capillary rise was set to null for this experiment, and as the plots have been graded with a gentle slope for border irrigation, we assumed that runoff to and from the plot was negligible.

As no significant rainfall had occurred during the previous months, the initial soil water content was set to 15%. The model also implements the partition of wetted area by irrigation or rainfall as described in [19], for which the fraction wet was set to 100%.

2.5. Meteorological Data and In-Situ Measurements

Although ORMVAH is well-equipped with four agro-meteorological stations over the Haouz plain, it was decided to put a full meteorological station equipped with telemetry for two reasons: ORMVAH stations are located more than 40 km away from the study site and meteorological data must be available in near real-time. Conventional climatic data were collected over a site located 2 km away from the experimental plots (square in Figure 2). A 2-meter tower was equipped to measure the air temperature and relative humidity using CS215 from Campbell Sc.; wind speed and direction using a Young 5103 anemometer; incoming solar radiation using an Apogee CS300, and finally rainfall was measured using a tipping bucket automatic rain gauge (ARG100, Campbell Scientific Inc., North Logan, UT, USA). The measured values were aggregated to a 30 min time interval, and then transmitted with a GPRS modem connected from the CR200 data logger.

The station was installed over an alfalfa crop maintained in the range 10–15 cm height during the whole experiment. It was considered appropriate to calculate Reference Evapotranspiration following the Penman-Monteith equation. ET_0 appeared to be consistent with the surrounding stations, corroborating that the surface resistance of this crop was compatible with the reference grass described in the FAO-56 method. A total of 99 mm of rain and 604 mm of ET_0 were recorded during the cropping season (Figure 4). Actual applied irrigation inputs of both plots were provided by ORMVAH, based on the opening time of the irrigation canals. As stated earlier, the flow of water into the plot is well controlled thus the irrigation water depth can easily be inferred.

Two towers were installed in the center of each plot at a height of 2 m (green circle in Figure 2). They provided the elements to calculate the four principal components of energy balance. An Eddy covariance instrument provided high-frequency (20 Hz) measurements of three-dimensional air velocity and temperature (CSAT3, Campbell Scientific Inc.); and synchronous water vapor concentration measured with high-speed hygrometers (KH2O, Campbell Scientific Inc.). Soil fluxes were measured with heat flux plates (HFT3-L, Campbell Scientific Inc.), and finally the four net radiation components were measured with a radiometer (CNR4, Kipp & Zonen, Delft, Netherlands). Data was averaged at a 30 min time interval and checked for energy balance closure with the Bowen ratio method with an r^2 of 0.85 (slope of 1.1) for the experimental plot and 0.92 (slope of 1.06) for the reference plot. Although not being applied here, it may be noted that other energy balance closure techniques exist. In particular, the study of [45] showed that differences of +10% to −20% of yearly cumulated ET compared to the Bowen Ratio technique may appear. Actual evapotranspiration measured with this system will be further referred to as ET_{EC}.

Figure 4. Time evolution of Rainfall and Reference Evapotranspiration (ET₀) during the experimental setup.

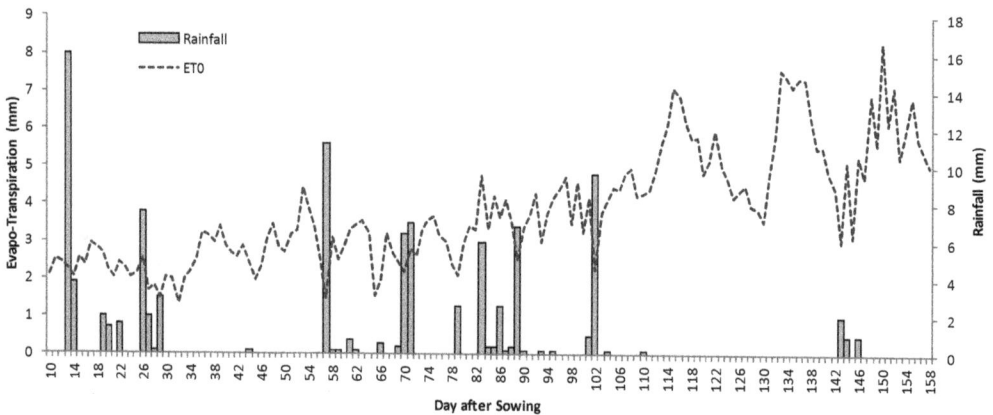

At harvest time, above-ground biomass and grain yield measurements were carried out by counting the total number of straw bales (dry biomass) and the number of grain sacks. Mean plot yields were finally computed by multiplying those accounts by the average weight of sacks and bales obtained through weight sampling.

2.6. Kcb Extrapolation

Changes in NDVI over a five-day period (return period proposed by SPOT 4 (Take 5) or Sentinel 2) are rather small. However, cloud cover increases this interval, and an extrapolation between the latest clear satellite image and the current day is required. Tables 11 and 17 from [18] were used to set up future inflexion points, choosing the California Desert Winter Wheat crop with a cycle length of 160 days. For the FAO-56 approach, the inflexion points are obtained combining the duration of three development stages (initial = 20 days, development = 50 days, mid-season = 60 days and late season = 30 days, for the selected crop), and the basal crop coefficients at initial, mid and late season (0.15, 1.1, 0.25 for the selected crop). The FAO-56 inflexion points were combined with the existing points obtained from remote sensing to calculate a full season by linear interpolation.

2.7. Climatic Extrapolation or Forecast

A forecast of climatic data is needed to help the irrigation decision-making process. The ASP weather model, which is a hybrid-mass coordinate primitive equation hydrostatic model with equations expressed in flux form [46] has been running over the Moroccan domain at 10km resolution during the whole experiment. The rainfall output of the model was scrutinized and compared to freely available model output from the Web. Smallest rain forecasts were simply ignored. Nevertheless, ET₀ forecasts were extracted from climatology because of technical difficulties. We will discuss the importance of these data in the last part of this paper.

2.8. Irrigation Decision-Making Process

As mentioned above, the irrigation scheduling of the reference and the experimental plots were managed separately. The reference plot was managed "as usual" by the farmer, meaning that he had to take not only into account this plot, but also the hundreds of hectares he is growing, and to manage its workforce.

The experimental plot was managed according the FAO-56 method guided by remote sensing NDVI, but taking into account that network management and labor mobilization generally takes a minimum of three days. The triggering of irrigation may thus have been underoptimal in a modeling point of view, as the RAW may not have been fully depleted. It was also decided to let the farmer irrigate by himself the experimental plot before vegetation emergence was noticeable on satellite images (DaS 27 with a NDVI of 0.17), and for the last irrigation.

The irrigation water depths actually applied are dictated by the irrigation technique. They are between 50 and 60 mm.

3. Results

3.1. NDVI and Basal Crop Coefficient

The time series of NDVI for both plots is shown on Figure 5. The first three dates are calculated from SPOT5, the remaining dates are obtained from SPOT4 (Take5). The reference plot shows a lower NDVI during the growth period. The two plots reach the same level of NDVI around DaS 100 (23 March). The jumps in both curves may be related either to soil moisture conditions as wetting events are likely to impact the soil albedo [47], either to a possible increase of aerosols thickness in particular on hot and windy days.

On Figure 6 showing results of the linear conversion from NDVI to Kcb, it can be seen that Kcb is limited to Kcmax (1.15), and that the extrapolator was quite efficient. The reception of new images could correct the Kcb trajectory, especially on DaS 63 when the Kcb curve began inflecting, and at DaS 133 at the early start of senescence.

3.2. ET Estimates

Figure 7 shows the comparison between ETc_{adj} and ET_{EC} for the experimental plot. With RMSE's of 0.91 mm per day and 0.95 mm per day, and r^2 values of 0.75 and 0.67 for the experimental and reference plots, respectively, the comparison shows a fair yet not excellent fit for both fields. These error ranges are quite comparable to previous studies [21,24,48]. The score is particularly low at the beginning of the season, in particular because of lower evaporation during DaS 23–29 and 32–49 (Figure 7). A post-experiment calibration procedure showed that the use of the capillary rise equations would have improved the overall RMSE by 0.09 mm per day (0.28 mm per day for the first 40 days), and that a full calibration of ET_{Cadj} and ET_{EC} would have helped gain an extra 0.07 mm per day. It can also be remembered that Eddy-covariance measurements are prone to a certain amount of error [49]. In addition, an error analysis at a five-day time interval shows RMSE of 0.59 mm per day and r^2 of 0.9 (0.46 and 0.91 with corrected diffusion) for the experimental plot. The gain in RMSE obtained by this temporal integration shows that if the technique does not perform very well on a daily time step, the results are improved over a longer period of time. With regards to irrigation, this information may be interpreted in this way: during the colder period of growth (December–February), average ET was 2.4 mm per day, the irrigation scheduling may be shifted up to one day per every five days, while during the warmer period (average ET of 4.8 mm per day), it goes down to one day per every 10 days. It suggests that the estimates of evapotranspiration are enough accurate for the purpose of irrigation scheduling.

Figure 5. Evolution of NDVI for the two plots: Reference (Ref) and Experimental (Exp). Note that the first three points come from SPOT5 and others from SPOT4. The DaS of each input image has been labeled.

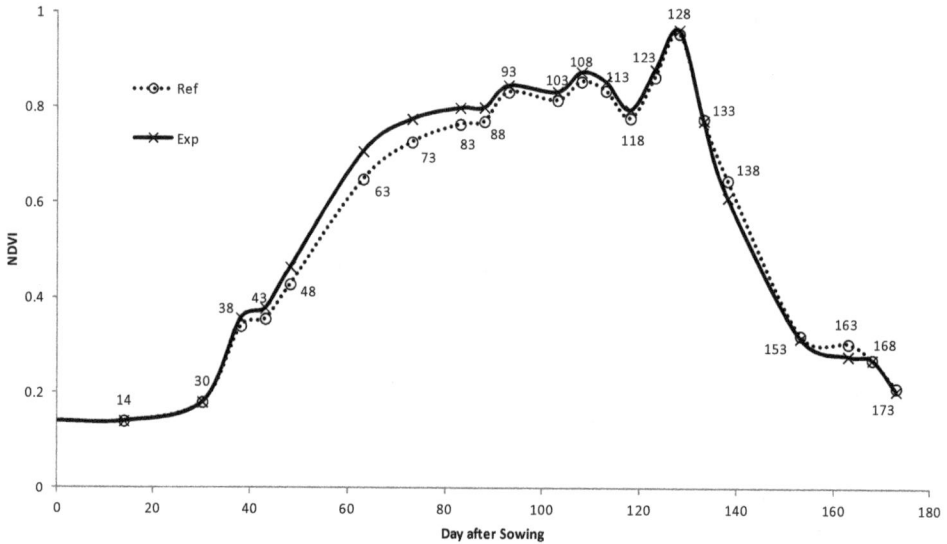

3.3. Irrigation Decisions

As shown on Figures 8 and 9, eleven irrigations events (640.8 mm) were done on the reference and nine on the experimental plot (562.9 mm). They were scheduled fairly differently. As explained earlier, the irrigation decisions were left to the farmer before noticing the emergence of vegetation on NDVI images for both plots.

The farmer decided its first irrigation on DaS 17 with 92 mm (note that a rainfall of 16mm was registered this same day). It is the highest amount in single irrigation during the whole season.

Figure 6. Kc$_b$ derived from NDVI (dashed line) and evolution of extrapolations as new images are arriving (dark to light gray) for the experimental plot. Each time a new image is received (dot and DaS small label), the Kc$_b$ trajectory is corrected. The three inflexion points extracted from the FAO-56 tables are indicated by a star and a bold label.

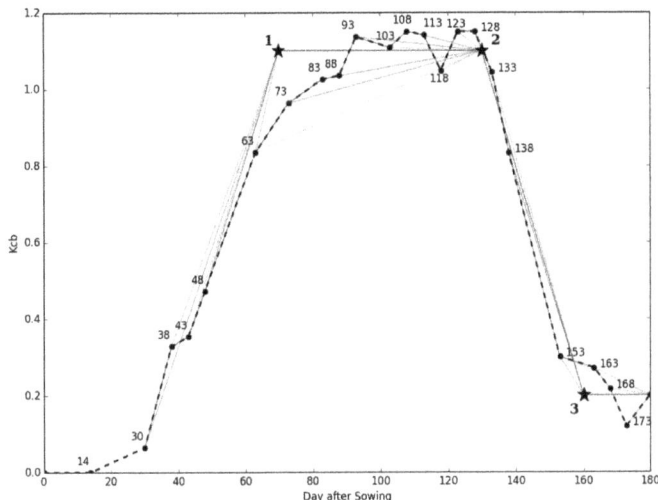

Figure 7. Comparison between the daily values of estimated ET_{adj} by the FAO-56 model and the measured one (ET_{EC}) by the Eddy-covariance system for the experimental plot. Error indicators are MSE (Mean Square Error), RMSE (Root Mean Square Error), MBE (Mean Bias Error), MAE (Mean Absolute Error) and MAPE (Mean Absolute Percentage Error).

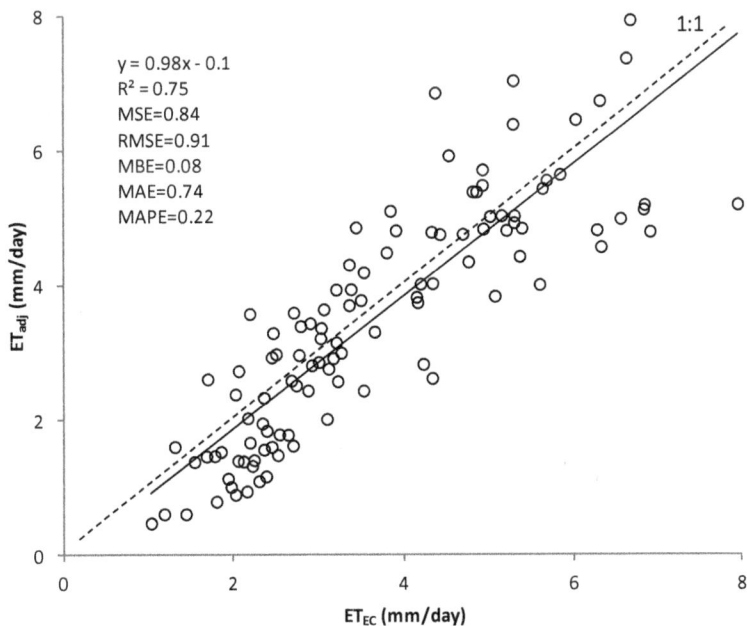

Figure 8. Forcing (ET$_0$, Rainfall and Irrigation) of the Water Balance and comparison of measured Evapotranspiration (ET_EC) and Calculated Evapotranspiration (ETadj) for the experimental plot.

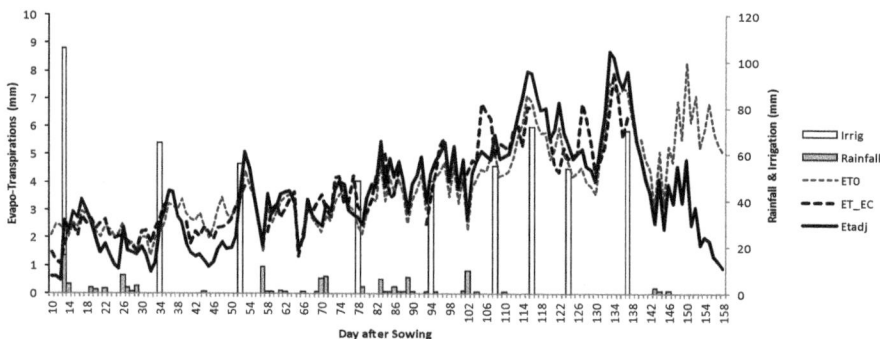

Figure 9. Comparison of irrigation events between the two plots. Time intervals between events are indicated in the same figure. NDVI is for the experimental plot.

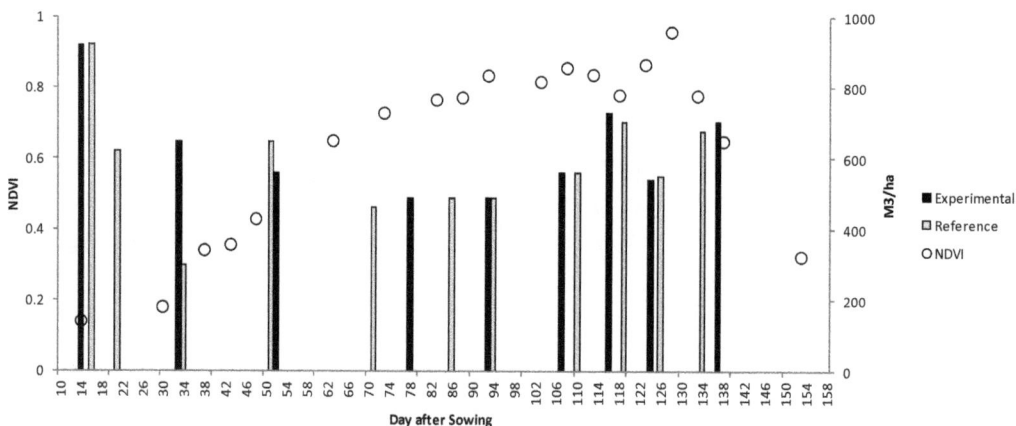

The farmer decided a second irrigation on DaS 22 and before plant emergence to prevent the appearance of soil crusting. As our water balance model is only efficient after the emergence of the crop, both plots should have been watered the same way. But, although there was no plant emergence, misunderstandings with the farmer led him not to irrigate the experimental plot simultaneously with the reference one. Unfortunately, the soil crust actually appeared on the experimental plot and imperiled the wheat development for the rest of the season (see discussion below), by limiting the number of tiller.

The two following irrigations were done at the same time on DaS 34 and 52.

The next irrigation schedule was significantly different. Indeed, the farmer decided to irrigate on DaS 71, while the experimental plot had been irrigated a week later (DaS 79). It may be noticed that the changing rainfall forecast had moved the date of this irrigation from DaS 75 to 79, and that subsequent measurement of rain and ET$_0$ showed that this irrigation could have been delayed until DaS 81.

The farmer did irrigate on DaS 87 as well. The next irrigation was carried out on the same day (DaS 94 and 95). For the next irrigation, the farmer waited three more days than we did. It is possible that the farmer had overestimated the amount of rain on DaS 101–103.

The third remaining irrigation events were quite similar in terms of quantity and timing of each plot, but with a three-day interval. The decision to make the last irrigation for both plots was left to the farmer.

3.4. Final Budget and Yields

At the end of the experiment, the final irrigation budget of the experimental plot (563 m³/ha) is lower than the reference plot (640.8 m³/ha). The biomass of the reference plot is 24% higher than the experimental plot and the grain yield is slightly higher (+6% for the reference plot). The almost 14% of extra water used on the reference plot is due to the fact that the farmer used the remaining of irrigation of DaS34 to irrigate its plot and that he did an unnecessary irrigation during the period of Das 71–94.

Table 1. Actual irrigation events of both plots and comparison of irrigation events three to eight to the water balance.

	Reference			Experimental				
#	Dates (DaS)	Quantity (mm)	#	Dates (DaS)	Quantity (mm)	Water Balance	Absolute Difference (WB-Exp)	Percentage (WB-Exp)
1	9 January (17)	92	1	7 January (14)	91.8	-	-	-
2	14 January (22)	62.1	-	-	-	-	-	-
3	26 January (34)	30	2	26 January (34)	64.8	-	-	-
4	13 February (52)	64.8	3	14 February (53)	56	38	18	32
5	4 March (71)	46	4	12 March (79)	48.6	56	−7.4	−15
6	20 March (87)	48.6	-	-	-	-	-	-
7	27 March (94)	48.6	5	28 March (95)	48.6	49	-0.4	-1
8	13 April (111)	56	6	10 April (108)	56	53	3	5
9	22 April (120)	70.2	7	19 April (117)	72.9	47	25.9	36
10	29 April (127)	55	8	27 April (125)	54	48.9	5.1	9
11	7 May (134)	67.5	9	10 May (137)	70.2	-	-	-
Total Irrigation		640.8			562.9			
Total with Rainfall		739.8			661.9			

Table 1 shows two things: first, a comparison of quantities applied to both plots and second, a comparison of actually applied quantities for the experimental plot to the quantities recommended by the water balance model (the water needed to fill-up the two upper buckets referring to the surface and the root zones respectively), for irrigation events three to eight. We chose to compare only those events because during the initial phase (no vegetation cover), the irrigation scheduling was handled by the farmer and the decision of the last irrigation was also left to the farmer. The comparison to the water budget shows a surplus of 13% for irrigation events three to eight, so the difference is not too large. Nevertheless, the amount of water applied to the first two irrigations is much more important than the budget proposed water. This difference is directly related to the irrigation technique itself, which requires a minimum amount of water to irrigate the entire field, especially on dry and plowed soils at this time of the year.

4. Discussion and Perspective

4.1. Strengths and Weaknesses

A time series of optical high resolution images, very close to what the Sentinel-2 product will look like, was used to compute a daily water balance in near real-time for a wheat field in Morocco, and irrigation was scheduled according to the model prediction. Calculations were performed at the plot level because it is not possible to irrigate a subpart of the plot and also to cut calculation time. A good knowledge of local parameters (soil and rooting depths), realistic input values of ET_0 and the wetting events and a dense time series of high-resolution remote sensing images allow the FAO-56 model to perform properly. By contrast, with Landsat 8 (launched on 11 February 2013 with the first image of our area available on 19 April and a return period of 16 days), a gap of 47 days (two missing images in a row) occurred between the 5 May and 22 June images (Figure 3). In our case, the high frequency of SPOT4 (Take5) allowed a consistent, high repetitiveness data set.

This simple model provides good advice in the irrigation decision-making process, yet several aspects may affect the water balance. Namely, (a) the farmer's expertise remains crucial; (b) calibration should be adjustable; (c) extrapolation is important to preclude satellite image gaps; (d) the meteorological forecasting especially for rainfall may also be important, but mainly to make the decision of postponing an irrigation event and (e) a time window for irrigation should be provided to the farmer:

(a) The FAO-56 method is dedicated to the estimation of evapotranspiration and to assist irrigation scheduling. Anything that falls out of this area must be addressed by the farmer's expertise. In our case, the appearance of a slaking crust greatly disadvantaged the tillering of the experimental plot, and the crop could never recover. This is the main reason for the much lower biomass obtained on the experimental plot. By contrast, the reference crop may have suffered some stress hampering grain filling, leading to a similar amount of grain yield in both plots. In this case, the water budget guided by remote sensing was more efficient than the farmer's management.

(b) Although predefined calibrations may be assigned for typical crops and soil parameters, they should be adjustable by the farmer, so that the model could reflect local measurements (e.g., Watermarks) or his know-how of irrigation scheduling (delaying or advancing irrigation triggering).

(c) The typical description of crops through crop coefficients includes four inflexion points; hence, it can be assumed that four perfectly-timed images would be sufficient to describe the temporal evolution of the crop. However, at a regional level, the spatial variation of sowing dates is important. It increases this theoretical number to a daily image. In addition, cloud coverage forces the need for extrapolation.

(d) Although it was not demonstrated here, the prevailing forecast data is the rainfall amount. An expected rainfall of 20 mm may boost the root layer for the next few days. But, if the actual rainfall is much lower than the forecast, an irrigation decision must be made rapidly, and still a couple days ahead of the irrigation event because of the technical constraint at the scheme scale. In semi-arid areas, the irrigation decision system should take into account the uncertainty of rainfall forecast, including ignoring the forecasts of small rainfall.

(e) In the introduction, we said that an irrigation scheme may suffer contingencies, and it is also clear that a farmer has to deal with his equipment and workforce availability. On the other hand, the FAO-56 model suggests a single best date of irrigation. To overcome the problem, it would be nice to provide an irrigation window to the farmer, which would also take into account the preceding remark.

Interestingly enough, during tillering and stem extension the NDVI of the experimental plot was slightly higher than the reference plot, when actually there were about 20% fewer stems on average. This may be explained by the fact that at higher density of leaves, higher shade occurs within the canopy, thus radiation interception is higher [50]. In fact, due to the limited penetration of the visible signal in the canopy, the three dimensional complexity of a crop canopy can hardly be seen by optical remote sensing imagery; hence NDVI remains a broad indicator of crop development, in particular for biomass.

The soils of the Haouz plain are poor and the climate is semi-arid. The yield depends on irrigation, but above all on fertilization. The water productivity indices in the region are generally low (1.63 m^3/kg for 2002/03 and 1.49 m^3/kg for 2003/04 [7]. In our study, water productivity is 1.34 m^3/kg against 1.52 m^3/kg for the reference plot (calculated only on irrigation water), which represents a gain of 0.18 m^3/kg. This conclusion goes into the way of increasing water productivity which is one of the main concerns of the National Agricultural Plan of Morocco [51].

4.2. Conclusions and Perspective

This experiment demonstrated how remote sensing time series could be used in an almost operational way for irrigation scheduling at the plot scale in a semi-arid environment. In this particular case, this irrigation advice showed to be a possible way to improve performance while saving on water, even without changing the irrigation technique. High-resolution and high-frequency satellite optical images, like the ones that will be available with Sentinel-2, may be of great help for irrigation scheduling at the plot scale. The simulated Sentinel-2 data set provided by the SPOT4 (Take5) experiment proved to be perfectly fitted for this experiment.

Additionally to high temporal and spatial resolution reflectance images, a better assessment of surface soil moisture and root soil moisture can be achieved through assimilation techniques of high temporal and spatial resolution microwave data on the one hand, and thermal data on the other hand [52]. Active microwave in the L and C bands have proven to be particularly interesting for the estimation of soil moisture in the first centimeters [53] for bare soils. With a revisit time of one to three days and a ground resolution of 5×20 m, Sentinel-1 data would be a good candidate to help a reflectance-based crop coefficient approach in the early part of the season. This real-life experiment was carried out over a single plot. Other works show the potentiality of the approach for a whole irrigated sector, introducing a new Irrigation Priority Index [54] and an optimization scheme [55]. But the change of scale is not at all straightforward. Initialization will be difficult, in particular because dealing with a large number of plots would imply to automate several steps like crop detection and its corresponding Kcb profile for interpolation. At running time, actual irrigations, which is crucial input appears to be extremely difficult to collect. Again, SAR data with high revisit time may be a solution to retrieve irrigation dates and then force a refill of the two upper layers of the water budget. Finally, since the data of land cover, soil type and irrigation are the main limitations to operational service, we are currently developing a Web Service called SAT-IRR whereby a stripped-down interface allows the user to set up their plot, their crop and introduce their own irrigations. The server is in charge of gathering the remaining weather and satellite data from remote sources, and to compute a daily water balance with the herein described technique.

Acknowledgments

This research was conducted within the International Joint Laboratory TREMA (Télédétection et Ressources en Eau en Méditerranée semi-Aride, http://trema.ucam.ac.ma). The authors would like to thank the MISTRALS/SICMED program (financed projects METASIM and CHAMO), the project ANR AMETHYST—ANR-12-TMED-0006-01, the project RS/2011/09 financed by the CNRST, Morocco, CESBIO and the IRD for their financial support. CNES is acknowledged for the fast treatment and availability of the SPOT4 (Take5) images, and Astrium for the cheap access to SPOT5 images through ISIS #691. The staff of the Regional Office of Agricultural Development (ORMVAH) is also kindly acknowledged for their assistance during the course of this experiment. We thank the farmer, Mohammed Tarbaoui, who agreed to take part in this game, and without whom the experiment would not have been possible. The authors would finally like to thank the reviewers and co-editor who greatly helped us to improve this manuscript.

Author Contributions

Le Page Michel, Lionel Jarlan, Saïd Khabba and Salah Er-Raki designed the study, developed the methodology. Le Page Michel performed the analysis, and wrote the manuscript. Lionel Jarlan, Saïd Khabba and Salah Er-Raki reviewed the manuscript. M. Hakim Kharrou, Salah Er-Raki, Saïd Khabba and Mohamed Kasbani have coordinated the work between the different partners and the farmer. Jihad Toumi, M. Hakim Kharrou, Adrien Tavernier, Abdelilah El Moutamanni, Mohamed Kasbani, Mohamed Yousfi, Abdelilah El Moutamanni and Le Page Michel collected and processed the field data. Olivier Hagolle, Mireille Huc and Le Page Michel processed the satellite data.

Conflicts of Interest

The Authors declare that there is no conflict of interest.

References

1. Blinda, M. *More Efficient Water Use in the Mediterranean*; Water Effi. Plan Bleu: Valbonne, France, 2012.
2. Frenken, K. *L'irrigation en Afrique en Chiffres*; FAO: Rome, Italy, 2005.
3. Dwyer, J.; Baldock, D.; Caraveli, H.; Petersen, J.E.; Sumpsi-Vinas, J.; Varela-Ortega, C. The Environmental Impacts of Irrigation in the European Union. Available online: http://ec.europa.eu/environment/agriculture/pdf/irrigation.pdf (accessed on 7 November 2014).
4. Oweis, T.; Pala, M.; Ryan, J. Stabilizing rainfed wheat yields with supplemental irrigation and nitrogen in a Mediterranean climate. *Agro. Jour.* **1998**, *90*, 672–681.
5. IPCC. Intergovernmental Panel on Climate Change. Available online: http://www.ipcc.ch/ (accessed on 7 November 2014).
6. Pinstrup-Andersen, P.; Pandya-Lorch, R. World food needs toward 2020. *Am. J. Agric. Econ.* **1997**, *79*, 1465–1466.
7. Hadria, R.; Khabba, S.; Lahrouni, A.; Duchemin, B.; Chehbouni, A.; Ouzine, L.; Carriou, J. Calibration and validation of the shoot growth module of STICS crop model: Application to manage water irrigation in the Haouz plain, Marrakech plain. *Arab. J. Sci. Eng.* **2007**, *31*, 87–101.
8. Bationo, A.; Hartemink, A.; Lungu, O.; Naimi, M.; Okoth, P.; Smaling, E.; Thiombiano, L.; Waswa, B. *Improving Soil Fertility Recommendations in Africa Using the Decision Support System for Agrotechnology Transfer (DSSAT)*; Kihara, J., Fatondji, D., Jones, J.W., Hoogenboom, G., Tabo, R., Bationo, A. Eds.; Springer Netherlands: Dordrecht, The Netherlands, 2012.
9. Pereira, L.S. Higher performance through combined improvements in irrigation methods and scheduling: A discussion. *Agric. Water Manag.* **1999**, *40*, 53–169.
10. Weiss, M.; Baret, F. Evaluation of canopy biophysical variable retrieval performances from the accumulation of large swath satellite data. *Remote Sens. Environ.* **1999**, *70*, 293–306.
11. Leprieur, C.; Kerr, Y.; Mastorchio, S.; Meunier, C.J. Monitoring vegetation cover across semi-arid regions: Comparison of remote observations from various scales. *Int. J. Remote Sens.* **2000**, *21*, 281–300.
12. Olioso, A.; Chauki, H.; Courault, D.; Wigneron, J. Estimation of evapotranspiration and photosynthesis by assimilation of remote sensing data into SVAT models. *Remote Sens. Environ.* **1999**, *68*, 341–356.
13. Courault, D.; Seguin, B.; Olioso, A. Review on estimation of evapotranspiration from remote sensing data: From empirical to numerical modeling approaches. *Irrig. Drain. Syst.* **2005**, *19*, 223–249.
14. Calcagno, G.; Mendicino, G.; Monacelli, G.; Senatore, A.; Versace, P. Distributed estimation of actual avpotranspiration through remote sensing techniques. In *Method and Tools for Drought Analysis and Management*; Springer: Dordrecht, The Netherlands, 2007; pp. 125–147.

15. Gowda, P.H.; Chavez, J.L.; Colaizzi, P.D.; Evett, S.R.; Howell, T.A.; Tolk, J.A. ET mapping for agricultural water management: present status and challenges. *Irrig. Sci.* **2007**, *26*, 223–237.

16. Li, Z.-L.; Tang, R.; Wan, Z.; Bi, Y.; Zhou, C.; Tang, B.; Yan, G.; Zhang, X. A review of current methodologies for regional evapotranspiration estimation from remotely sensed data. *Sensors* **2009**, *9*, 3801–3853.

17. Chirouze, J.; Boulet, G.; Jarlan, L.; Fieuzal, R.; Rodriguez, J.C.; Ezzahar, J.; Er-Raki, S.; Bigeard, G.; Merlin, O.; Garatuza, J.; *et al.* Inter-comparison of four remote sensing based surface energy balance methods to retrieve surface evapotranspiration and water stress of irrigated fields in semi-arid climate. *Hydrol. Earth Syst. Sci. Discuss* **2013**, *10*, 895–963.

18. Allen, R.; Pereira, L.; Raes, D.; Smith, M. *FAO Irrigation and Drainage N.56: Guidelines for Computing Crop Water Requirements*; FAO: Rome, Italy, 1998.

19. Allen, R.; Pereira, L.; Smith, M.; Raes, D.; Wright, J. FAO-56 dual crop coefficient method for estimating evaporation from soil and application extensions. *J. Irrig. Drain. Eng.* **2005**, *131*, 2–13.

20. Hagolle, O.; Huc, M.; Dedieu, G.; Sylvander, S.; Houpert, L.; Leroy, M.; Clesse, D.; Daniaud, F.; Arino, O.; Koetz, B.; *et al.* SPOT4 (Take5): Time series over 45 sites to prepare Sentinel-2 applications and methods. In Proceedings of the ESA's Living Planet Symposium, Edinburgh, UK, 11 September 2013.

21. Duchemin, B.; Hadria, R.; Er-Raki, S.; Boulet, G.; Maisongrande, P.; Chehbouni, A.; Escadafal, R.; Ezzahar, J.; Hoedjes, J.C.B.; Kharrou, M.H.; *et al.* Monitoring wheat phenology and irrigation in Central Morocco: On the use of relationships between evapotranspiration, crops coefficients, leaf area index and remotely-sensed vegetation indices. *Agric. Water Manag.* **2006**, *79*, 1–27.

22. Kharrou, M.H.; Le Page, M.; Chehbouni, A.; Simonneaux, V.; Er-Raki, S.; Jarlan, L.; Ouzine, L.; Khabba, S.; Chehbouni, A. Assessment of equity and adequacy of water delivery in irrigation systems using remote sensing-based indicators in semi-arid region, Morocco. *Water Resour. Manag.* **2013**, *27*, 4697–4714.

23. Wosten, J.H.M.; Lilly, A.; Nemes, A.; Le Bas, C. Development and use of a database of hydraulic properties of European soils. *Geoderma* **1999**, *90*, 169–218.

24. Er-Raki, S.; Chehbouni, A.; Guemouria, N.; Duchemin, B.; Ezzahar, J.; Hadria, R. Combining FAO-56 model and ground-based remote sensing to estimate water consumptions of wheat crops in a semi-arid region. *Agric. Water Manag.* **2007**, *87*, 41–54.

25. Brouwer, C.; Prins, K.; Kay, M.; Heibloem, M. *Irrigation Water Management: Irrigation Methods, Training Manual N°5*; FAO: Rome, Italy, 1990.

26. Khabba, S.; Jarlan, L.; Er-Raki, S.; Le Page, M.; Ezzahar, J.; Boulet, G.; Simonneaux, V.; Kharrou, M.H.; Hanich, L.; Chehbouni, G. The SudMed program and the Joint International Laboratory TREMA: A decade of water transfer study in the Soil-Plant-Atmosphere system over irrigated crops in semi-arid area. *Proced. Environ. Sci.* **2013**, *19*, 524–533.

27. Hagolle, O.; Dedieu, G.; Mougenot, B.; Debaecker, V.; Duchemin, B.; Meygret, A. Correction of aerosol effects on multi-temporal images acquired with constant viewing angles: Application to Formosat-2 images. *Remote Sens. Environ.* **2008**, *112*, 1689–1701.

28. Hagolle, O.; Huc, M.; Villa Pascual, D.; Dedieu, G. A multi-temporal method for cloud detection, applied to FORMOSAT-2, VENμS, LANDSAT and SENTINEL-2 images. *Remote Sens. Environ.* **2010**, *114*, 1747–1755.

29. Rahman, H.; Dedieu, G. SMAC: A simplified method for the atmospheric correction of satellite measurements in the solar spectrum. *Int. J. Remote Sens.* **1994**, *15*, 123–143.

30. Monteith, J.L. Evaporation and environment. *Symp. Soc. Exp. Biol.* **1965**, 19, 205–234.

31. Glenn, E.P.; Neale, C.M.U.; Hunsaker, D.J.; Nagler, P.L. Vegetation index-based crop coefficients to estimate evapotranspiration by remote sensing in agricultural and natural ecosystems. *Hydrol. Process.* **2011**, *25*, 4050–4062.

32. Neale, C.M.U.; Bausch, W.C.; Heerman, D.F. Development of reflectance-based crop coefficients for corn. *Trans. ASAE* **1989**, *32*, 1891–1899.

33. Choudhury, B.J.; Ahmed, N.U.; Idso, S.B.; Reginato, R.J.; Daughtry, C.S.T. Relations between evaporation coeflqcients and vegetation indices studied by model simulations. *Remote Sens. Environ.* **1994**, *50*, 1–17.

34. Huete, A. A Soil-Adjusted Vegetation Index (SAVI). *Remote Sens. Environ.* **1988**, *25*, 295–309.

35. Wittich, K.P.; Hansing, O. Area-averaged vegetative cover fraction estimated from satellite data. *Int. J. Biometeorol.* **1995**, *38*, 209–215.

36. Trout, T.J.; Johnson, L.F.; Gartung, J. Remote sensing of canopy cover in horticultural crops. *HortScience* **2008**, *43*, 333–337.

37. Hunsaker, D.J.; Pinter, P.J.; Kimball, B.A. Wheat basal crop coefficients determined by normalized difference vegetation index. *Irrig. Sci.* **2005**, *24*, 1–14.

38. Jayanthi, H.; Neale, C.M.U.; Wright, J.L. Development and validation of canopy reflectance-based crop coefficient for potato. *Agric. Water Manag.* **2007**, *88*, 235–246.

39. Garatuza-Payan, J.; Watts, C.J. The use of remote sensing for estimating ET of irrigated wheat and cotton in Northwest Mexico. *Irrig. Drain. Syst.* **2005**, *19*, 301–320.

40. Gonzalez-Dugo, M.P.; Mateos, L. Spectral vegetation indices for benchmarking water productivity of irrigated cotton and sugarbeet crops. *Agric. Water Manag.* **2008**, *95*, 48–58.

41. Li, S.; Kang, S.; Li, F.; Zhang, L. Evapotranspiration and crop coefficient of spring maize with plastic mulch using eddy covariance in northwest China. *Agric. Water Manag.* **2008**, *95*, 1214–1222.

42. Campos, I.; Neale, C.M.U.; Calera, A.; Balbontín, C.; González-Piqueras, J. Assessing satellite-based basal crop coefficients for irrigated grapes (*Vitis vinifera* L.). *Agric. Water Manag.* **2010**, *98*, 45–54.

43. Er-Raki, S.; Rodriguez, J.C.; Garatuza, J.P.; Watts, C.; Chehbouni, G. Determination of crop evapotranspiration of table grapes in a semi-arid region of Northwest Mexico using multi-spectral vegetation index. *Agric. Water Manag.* **2013**, *122*, 12–19.

44. Consoli, S.; D'Urso, G.; Toscano, A. Remote sensing to estimate ET-fluxes and the performance of an irrigation district in southern Italy. *Agric. Water Manag.* **2006**, *81*, 295–314.

45. Teixeira, A.H.; Bastiaanssen, W.G.M. Five methods to interpret field measurements of energy fluxes over a micro-sprinkler-irrigated mango orchard. *Irrig. Sci.* **2012**, *30*, 13–28.

46. Boone, A. The Atmosphere Surface Forecast Prediction. 2012. Available online: http://aaron.boone.free.fr/weather.html. (accessed on 1 January 2012).

47. Lobell, D.B.; Asner, G.P. Moisture effects on soil reflectance. *Soil Sci. Soc. Am. J.* **2002**, *66*, 722–727.

48. Kharrou, M.H.; Er-Raki, S.; Chehbouni, A.; Duchemin, B.; Simonneaux, V.; Le Page, M.; Ouzine, L.; Jarlan, L. Water use efficiency and yield of winter wheat under different irrigation regimes in a semi-arid region. *Agric. Sci.* **2011**, *2*, 273–282.

49. Allen, R.G.; Pereira, L.S.; Howell, T.A.; Jensen, M.E. Evapotranspiration information reporting: I. Factors governing measurement accuracy. *Agric. Water Manag.* **2011**, *98*, 899–920.

50. O'Connell, M.G.; O'Leary, G.J.; Whitfield, D.M.; Connor, D.J. Interception of photosynthetically active radiation and radiation-use efficiency of wheat, field pea and mustard in a semi-arid environment. *Field Crop. Res.* **2004**, *85*, 111–124.

51. Schilling, J.; Freier, K.P.; Hertig, E.; Scheffran, J. Climate change, vulnerability and adaptation in North Africa with focus on Morocco. *Agric. Ecosyst. Environ.* **2012**, *156*, 12–26.

52. Li, F.; Crow, W.T.; Kustas, W.P. Towards the estimation root-zone soil moisture via the simultaneous assimilation of thermal and microwave soil moisture retrievals. *Adv. Water Resour.* **2010**, *33*, 201–214.

53. Kornelsen, K.C.; Coulibaly, P. Advances in soil moisture retrieval from synthetic aperture radar and hydrological applications. *J. Hydrol.* **2013**, *476*, 460–489.

54. Belaqziz, S.; Khabba, S.; Er-Raki, S.; Jarlan, L.; Le Page, M.; Kharrou, M.H.; El Adnani, M.; Chehbouni, A. A new irrigation priority index based on remote sensing data for assessing the networks irrigation scheduling. *Agric. Water Manag.* **2013**, *119*, 1–9.

55. Belaqziz, S.; Mangiarotti, S.; Le Page, M.; Khabba, S.; Er-Raki, S.; Agouti, T.; Drapeau, L.; Kharrou, M.H.; El Adnani, M.; Jarlan, L. Irrigation scheduling of a classical gravity network based on the Covariance Matrix Adaptation—Evolutionary Strategy algorithm. *Comput. Electron. Agric.* **2014**, *102*, 64–72.

Monitoring of Irrigation Schemes by Remote Sensing: Phenology *versus* Retrieval of Biophysical Variables

Nadia Akdim, Silvia Maria Alfieri, Adnane Habib, Abdeloihab Choukri, Elijah Cheruiyot, Kamal Labbassiand Massimo Menenti

Abstract: The appraisal of crop water requirements (CWR) is crucial for the management of water resources, especially in arid and semi-arid regions where irrigation represents the largest consumer of water, such as the Doukkala area, western Morocco. Simple and (semi) empirical approaches have been applied to estimate CWR: the first one is called K_c-NDVI method, based on the correlation between the Normalized Difference Vegetation Index (NDVI) and the crop coefficient (K_c); the second one is the analytical approach based on the direct application of the Penman-Monteith equation with reflectance-based estimates of canopy biophysical variables, such as surface albedo (r), leaf area index (LAI) and crop height (h_c). A time series of high spatial resolution RapidEye (REIS), SPOT4 (HRVIR1) and Landsat 8 (OLI) images acquired during the 2012/2013 agricultural season has been used to assess the spatial and temporal variability of crop evapotranspiration ET_c and biophysical variables. The validation using the dual crop coefficient approach (K_{cb}) showed that the satellite-based estimates of daily ET_c were in good agreement with ground-based ET_c, *i.e.*, $R^2 = 0.75$ and RMSE = 0.79 *versus* $R^2 = 0.73$ and RMSE = 0.89 for the K_c-NDVI, respective of the analytical approach. The assessment of irrigation performance in terms of adequacy between water requirements and allocations showed that CWR were much larger than allocated surface water for the entire area, with this difference being small at the beginning of the growing season. Even smaller differences were observed between surface water allocations and Irrigation Water Requirements (IWR) throughout the irrigation season. Finally, surface water allocations were rather close to Net Irrigation Water Requirements (NIWR).

Reprinted from *Remote Sens.* Cite as: Akdim, N.; Alfieri, S.M.; Habib, A.; Choukri, A.; Cheruiyot, E.; Menenti, K.L.M. Monitoring of Irrigation Schemes by Remote Sensing: Phenology *versus* Retrieval of Biophysical Variables. *Remote Sens.* **2014**, *6*, 5815-5851.

1. Introduction

The interest for the assessment of irrigation performance using satellite data developed in the late 1980s due to growing consensus on the difficulty of collecting the required ground data continuously, reliably and in a consistent way across all major irrigation schemes worldwide [1–4]. Since in a large part of irrigated lands water is allocated proportionally to irrigated area, work was initially focused on the relation between allocated water and irrigated area, observable with multispectral satellite data [5]. Later on, other aspects of irrigation water management were evaluated, like crop water requirements [6–12], actual consumptive water use [13–18], water productivity [17,19–22] and water and salinity stress [23–29]. On a higher level of abstraction, irrigation performance may be evaluated for different objectives such as equity, adequacy, or effectiveness [30–32]. In this study, we focus on the evaluation of adequacy by relating water allocation to water requirement.

The most common and practical approach used for estimating crop water requirements (CWR) is the FAO-56 method (the Food and agricultural Organization of United States (FAO) Irrigation and Drainage Paper No. 56) [33], based on the combination of reference evapotranspiration ET_0 and crop coefficients (K_c) to determine crop evapotranspiration (ET_c) under unrestricted water availability. In the majority of the studies, the K_c values are obtained by the single crop coefficient approach, where crop transpiration and soil evaporation are combined into a single K_c coefficient. Infrequently, the dual crop coefficient (k_{cb}) approach is used, where the effects of crop transpiration and soil evaporation are determined independently [34–38]. The FAO-56 method is based on the use of crop specific parameters. While this is not an issue for on-farm evaluations of CWR, it becomes rather challenging when the objective is to monitor CWR of large irrigation schemes. In this paper, we have adopted the definitions CWR = ET_c and irrigation water requirement IWR = ET_c − P, where P is precipitation.

Frequent mapping of crop types is, in principle, feasible using multispectral and multi-temporal satellite data [39–47], but accurate classification requires, in most cases, ground reference data and analysis of images acquired at multiple dates [48]. This makes timely availability of crop maps rather unlikely, thus reducing the timeliness and the reliability of a CWR monitoring service based on satellite data.

Remote sensing methods which do not require knowledge of crop type to determine CWR have been developed taking advantage of the strong physical relationship between the spectral response of cropped surfaces and the corresponding values of CWR and K_c. Examples of these approaches can be found in [11,48–51]. To such end, empirical relationships have been found to retrieve the value of K_c or K_{cb} from simple vegetation indexes, i.e., NDVI (Normalized difference Vegetation Index) [52].

Several indices have been proposed as alternatives to NDVI to estimate Kc, such as the Perpendicular Vegetation Index PVI [53], the Soil-Adjusted Vegetation Index SAVI [54–57], the Weighted Difference Vegetation Index WDVI [58] and the Global Environment Monitoring Index GEMI [59]. These indices have been formulated in order to reduce the influence of perturbing effects such as the soil background or the atmospheric influence, which may alter significantly the reflectance of vegetated surfaces.

A different approach is by using directly the Penman-Monteith equation either to determine a generic K_c (crop identification not required) or ET_c [6,10,60,61]. This approach is based on retrieving from satellite data the crop properties which determine ET_c, i.e., crop height (h_c), surface albedo (α), and leaf area index (LAI). The crop height influences the aerodynamic resistance (r_a) term of the FAO Penman-Monteith equation and the turbulent transfer of vapor from the crop into the atmosphere. The r_a term appears twice in the full form of the FAO Penman-Monteith equation. The surface albedo of the crop-soil surface influences the net radiation at the surface, Rn, which is the primary forcing factor of the transpiration and evaporation processes. The surface albedo (α) is affected by the fraction of ground covered by vegetation and by the soil surface wetness. The canopy resistance of the crop to vapor transfer is affected by leaf area (number of stomata), leaf age and condition, and leaf-level stomatal control. The fraction of exposed soil also affects K_c.

In this study, we have applied and evaluated two methods to determine CWR, *i.e.*, the one based on the correlation between K_c and NDVI (K_c-NDVI method) and the one using directly the Penman-Monteith equation (analytical method). We have applied both methods to evaluate the adequacy of water allocation in the Doukkala irrigation scheme.

In the Doukkala area (Western Morocco), water demand has significantly increased over the last decades while fresh water resources are becoming increasingly scarce. This is mainly due to the combined effect of climate change, persistent drought and the increase of water demand related to increase in irrigated area, urbanization and recreational projects. This shows the necessity to use available water resources as effectively as possible in order to avoid or at least mitigate the consequences of recurring droughts. This is particularly important for water management in agricultural areas, where irrigation represents the biggest water consumer.

This paper is organized as follows. After the description of the study area and the data collected (Section 3), we describe in detail our implementation of the two methods (Section 4), followed by the presentation of results, including the evaluation against *in-situ* observations, and concluding with global interpretation of our results.

2. Study Area

The Doukkala region lies in western Morocco, between 32°15′N and 33°15′N latitude and 7° 55′W and 9°20′W longitude in the downstream portion of the hydraulic basin of Oum Er_Rabia (Figure 1). Geo-morphologically speaking, Doukkala is divided into three parts: the coastal area, the Sahel and the plain. The latter extends over an area of 3500 km² and it is located at about 120–130 m above sea level [62], with favorable conditions for agricultural development as regards arable land and soil fertility.

The irrigated area of Doukkala is among the largest (96,000 ha) and earliest developed areas in Morocco, remarkable for its size and strategic importance for national production, specially sugar beet (38%) and commercialized milk (20%). The important crops grown in the study area include wheat, corn, sugar beet, and alfalfa [63].

The climate of Doukkala is typically semi-arid to temperate and mild in winter whereas in summer it is generally warm and dry, with a large inter and intra-annual variability of rainfall, which amounts to 316 mm/year on average (1964–2009). Reference evapotranspiration (ET_0) is 1434 mm/year on average 2000–2008.

The irrigation is performed by means of different techniques with different efficiency (surface, sprinkler and drip irrigation) with a rotational interval of 15 days. Surface irrigation is the dominant system. Water allocation is calculated on the basis of current irrigated area, as determined on the basis of requests by farmers for each rotational interval. The gross irrigation water depth allocated to farmers is 864 m³/ha by rotational interval, equivalent to 1728 mm/month, which is largely sufficient to meet the water needs of the dominant crops. The water resources mobilized for irrigation come mainly from the dam Al Massira, a major water storage structure in the watershed of Oum Er-Rbia with a capacity of approximately 2760 × 10⁶ m³.

184

Figure 1. Location map of Doukkala region.

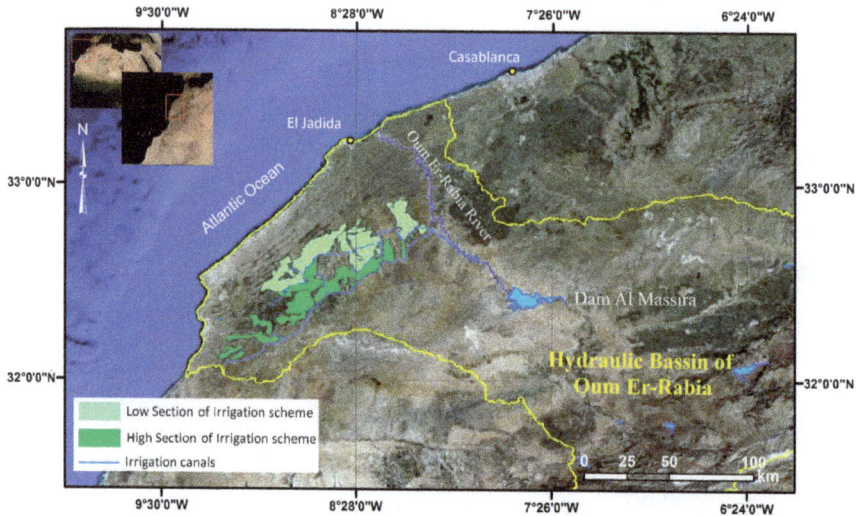

The Doukkala irrigation scheme is divided into the High and Low Sections. We focused on the Low Section, which contains three main districts: Faregh, Sidi Bennour and Zemamra, respectively, from the East to the West. Each district is divided into a number of Centers of Irrigation Management (CGR) irrigated with different irrigation systems (Table 1). The Doukkala irrigation scheme is managed by the Regional Office of Agricultural Development in Doukkala (ORMVAD).

Table 1. Irrigation Systems in different Centers of Irrigation Management (CGRs) in three districts of the Doukkala Irrigation Scheme (ORMVAD-personal communication).

District	CGR	Area (Ha)	Irrigation System
SidiBennour	330	5305.25	Surface Irrigation
	331	3520.06	
	333	4293.39	
	335	3202.49	
	336	4197.1	
	337	3112.55	
	338	1738.92	
	338 sprinkler	1905.3	Sprinkler Irrigation
Zemamra	320	2995.18	Sprinkler Irrigation
	321	5327.4	
	322	3243.68	
	324	4565.71	
	325	3122.27	
Faregh	312	4840.24	Drip Irrigation
	332	1490.46	Surface Irrigation
	310	4468.60	
	311	5021.25	

3. Data Collection

3.1. Satellite Data

To meet the combined requirements of high spatial resolution and frequent imaging to monitor crop development and water requirements, we have combined image data acquired by multiples sensors at high and very high spatial resolution: (Landsat 8 (30 m), Spot 4 HRVIR1 (20 m) and Rapideye (5 m)) during the irrigation season 2012/2013 (Table 2). All satellite images acquired were corrected geometrically with the following system of coordinate: UTM, WGS-84, zone 29.

An unprecedented time series consisting in Spot4 data (level 2A) with a five-day revisit interval has been acquired and analyzed. The Level 2A product is in-band surface reflectance corrected from atmospheric effects using the SMAC model [64].

3.2. Meteorological and Water Flow Data

Daily meteorological data (*i.e.*, Air temperature (T_a), relative humidity (RH), solar radiation (R_s), wind speed (U), precipitation (R) and reference evapotranspiration ET_0) on the same dates of the satellite observations were collected at the Zemamra and Khmiss Mettouh meteorological stations and provided by ORMVAD. The observations at Zemamra were used for the district of Zemamra and the district of Sidi Bennour and the observations of Khmiss Mettouh were used for the district of Faregh.

Data on monthly water allocation by CGR were provided by ORMVAD. The water allocation is calculated monthly for each tertiary and secondary irrigation unit on the basis of requests submitted by farmers for each rotational interval. The irrigated area of each farm is multiplied by the irrigation module (*i.e.*, 864 m^3/ha) to determine the duration of water delivery to each farm and the total volume to be delivered. These farm-level irrigation water depths were added up to obtain the monthly water allocation by the CGRs we have used in our assessments on the adequacy of water allocations.

3.3. Data In-Situ

Periodical field campaigns were carried out with multiple objectives: to explore the structure of the irrigation system over the area as well as the irrigation systems (surface irrigation, sprinkler irrigation and drip irrigation); to collect phenological data on the dominant crops and to collect data on fractional cover and crop height.

The fractional cover (f_c) and crop height (h_c) of the dominant crops (wheat, sugar beet, alfalfa, berseem and corn) were measured in 22 pilot plots in the Zemamra and Sidi Bennour districts on 17–19 December, 27–28 February, 3–4 April, 23–24 May, and 20 July in the growing season 2012/2013. During the May and July surveys, 14 additional plots in the Faregh district were sampled.

Table 2. Overview of image data characteristics and coverage of the study area.

SENSOR	DATE	Area	SPECTRAL RESOLUTION (μm)	Spatial Resolution	ORBIT
SPOT4-HRVIR1	From January to June 2013	FAREGH	XS1: 0.500–0.590 XS2: 0.610–0.680 XS3: 0.790–0.890 SWIR (HRVIR): 1.530–1.750	20 m	Altitude:832 km revisit: 5 days
RapidEye-REIS	10 December 2012 8 February 2013	ZEMAMRA SIDIBENNOUR	Blue: 0.440–0.510 Green: 0.520–0.590 Red: 0.630–0.685 Red-Edge: 0.690–0.730 NIR: 0.760–0.850	5 m	Altitude:630 km revisit: Daily (off-nadir); 5.5 days (at nadir)
Landsat 8-OLI	19 April 2013 26 April 2013 13 June 2013 29 June 2013 15 July 2013	SIDIBENNOUR ZEMAMRA	Coastal/Aerosol: 0.433–0.453 Blue: 0.450–0.515 Green: 0.525–0.600 Red: 0.630–0.680 NearInfrared: 0.845–0.885 SWIR: 1.560–1.660 SWIR: 2.100–2.300 Cirrus: 1.360–1.390 Panchromatic: 0.500–0.680	30 m 15 m	Altitude:705 km revisit: 16 days

For each plot (max. 1 ha), we measured h_c in several points (5–6 points) with a graduated stick, and used the mean value of h_c as representative of the plot. In the same plot, we estimated f_c at 5–6 locations using an approximate target of 1 m^2, and took the mean value of f_c as representative of the plot. Then, we calculated *in-situ* crop evapotranspiration ET$_c$ using the ground measurement of h_c and f_c combined with meteorological data (see Section 4.3.2 Dual crop coefficient approach). For each plot where we had such *in-situ* ET$_c$, we obtained the corresponding satellite ET$_c$ by sampling image data as described in Section 5.3.3.

In some cases, the difference between the dates of satellite data acquisition and field work date was significant. In these cases, we interpolated the temporal observations of h_c and f_c linearly to obtain estimates on the dates of acquisition of the satellite data. This gave 80 pairs of ground and satellite observations throughout the growing season.

4. Methods

4.1. Work Flow

Crop evapotranspiration ET_c is the basic information for the evaluation of crop water requirements and irrigation management. In this work, we estimated ET_c with two different methods: FAO-56 (k_c—NDVI) and analytical approach. After the pre-processing of satellite images, ET_c was estimated for the entire study area and the 2012–2013 growing season and validated using *in-situ* observations (Figure 2). The FAO-56 method is the most widely used method to compute ET_c and is based on the estimation of the so-called "crop coefficient" K_c, defined as the ratio of total evapotranspiration ET_c by reference evapotranspiration ET_0. We evaluated our RS estimates of ET_c using the FAO-56 dual approach to calculate the basal crop coefficient k_{cb} and the evaporation coefficient k_e. with the f_c and h_c observed in the field during the above mentioned surveys.

Figure 2. Workflow of the methodology applied.

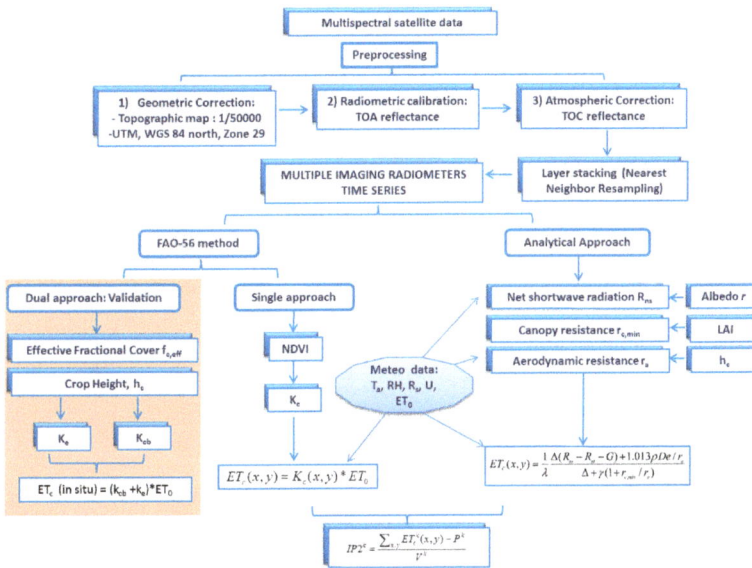

The analytical approach is based on the direct application of the Penman-Monteith equation. The required vegetation properties are surface albedo (r), Leaf Area Index (LAI) and the crop height(h_c) that are obtained from the processing of E.O data. LAI is used to compute canopy resistance ($r_{c,min}$),,crop height to calculate the aerodynamic resistance (r_a) and surface albedo is used to calculate the net shortwave radiation (R_{ns}).

4.2. Pre-Processing

The study area is rather flat at an elevation of 150 m ± 15 m, thus we did not carry out an additional geometric correction taking into account the topography.

The radiometric calibration of satellite images was achieved in two steps:

Firstly, by conversion of DN values into radiance; the following Equation (1) was used for conversion:

$$L_\lambda = (X/A) + B \tag{1}$$

where, L is Top Of Atmosphere (TOA) band spectral radiance observed by the satellite. X is a Digital Number. A is the absolute calibration gain for relevant spectral band and B is the absolute calibration bias for relevant spectral band.

Secondly, the atmospheric correction was performed on Landsat 8 and RapidEye data using the FLAASH model that incorporates the MODTRAN 4 model, and the input parameters used in this study are presented in Table 3. The model calculates the Top Of Canopy (TOC) in-band Lambertian reflectance ρ_p by means of a radiance-to-reflectance conversion, using the Equation (2):

$$\rho_p = \frac{\pi \cdot L_\lambda \cdot d^2}{E_\lambda^0 \cdot \cos\theta_s} \tag{2}$$

where:

L_λ: TOA in-band spectral radiance observed by the satellite; θ_s: Solar zenith angle; E_λ^0 is the mean in-band solar exo-atmospheric constant; d is the Earth-Sun distance at sensor's aperture, in astronomical units; $d = 1 - 0.01674 \cos(0.9856 (JD-4))$, where JD is Julian Day.

Table 3. MODTRAN input parameters used in this study.

PARAMETER	VALUE	Date
Model Atmosphere	Mid Latitude Summer	10 December 2012
		8 February 2013
		19 April 2013
		26 April 2013
		13 June 2013
	Tropical	29 June 2013
		15 July 2013
Aerosol Model	Rural	
Aerosol Retrieval	None	
Visibility	40 km (Default)	
Ground Altitude Above Sea Level	150 m	

For each band (VIS, NIR and SWIR), a time series of the pre-processed multispectral reflectance data was constructed using the nearest neighbor resampling to match the spatial resolution of the master image (UTM-WGS84). This layer stack was then exported as a multilayer GeoTIFF file, which is easily read and analyzed by Matlab.

To evaluate the consistency of the multi-sensor reflectance data (Table 2), we have simulated [65] the Top of Canopy (TOC) spectral reflectance (400–2400 nm with 1 nm spectral resolution) for a very heterogeneous soil vegetation scene with widespread irrigation. We constructed a data set including 60,000 samples. For each sample, the 2101 spectral bands were convolved with the spectral response functions of the three sensors OLI (Landsat8), HRVIR1 (Spot4) and REIS

(RapidEye) to simulate the TOC reflectance observed by each sensor. The red and NIR reflectances were used to calculate the NDVI and WDVI for each sensor and for each sample. The OLI sensor was taken as the reference and the other two sensors (REIS and HRVIR1) were compared with OLI.

For NDVI, the RMSE values were respectively 0.0469, 0.0328 and 0.02 for the pairs OLI-HRVIR1, OLI-REIS and HRVIR1-REIS, while for WDVI the RMSE values were 0.0288, 0.0343 and 0.006 for the same sensor combinations. Accordingly, we neglected these differences and concluded that OLI, HRVIR1 and REIS gave consistent observations of WDVI and NDVI.

4.3. Application of FAO-56 Model

In 1998, FAO proposed the FAO-56 Penman-Monteith reference evapotranspiration (ET_0) for irrigation scheduling [33]. This method has been widely used because it gives satisfactory results under various climate conditions across the world [66–69].

According to this model, two parameters are required to estimate the ET_c: the crop coefficient kc and ET_0. Crop coefficient curves provide simple, reproducible means to estimate ET_c from weather-based ET_0 values. ET_0 is defined as the evapotranspiration of a reference grass, completely covering the soil, well-watered and actively growing under optimal agronomic conditions.

In FAO-56, two approaches to determine k_c are presented: the single crop coefficient approach (k_c), which we have applied with remote sensing data, and the dual crop coefficient approach (k_{cb}), which we applied only with ground measurement.

4.3.1. Single Crop Coefficient Approach: k_c-NDVI Method

In the single crop coefficient (K_c), the effect of crop transpiration and soil evaporation is combined into a single K_c. The remotely sensed spectral reflectance data can be used to estimate K_c, because both K_c and spectral vegetation indices are correlated to leaf area index and fractional ground cover [70]. The simplest approach to derive K_c from remotely sensed data uses a linear relationship between K_c and *NDVI* (Normalized Difference Vegetation Index). *NDVI* is obtained from red (R) and near infrared (*NIR*) reflectance (0.6–0.7 μm and 0.7–1.3 μm, respectively), which are present in most imaging radiometers (Equation (3)).

$$NDVI = \frac{NIR - R}{NIR + R}$$ (3)

This approach was introduced by [71] and used and validated in further case studies by [72–75]. The theoretical basis has been established by [70]. This approach is one of the most promising ones for operational applications [8].

We have further simplified this concept, following [49,50] by using their K_c-NDVI relationship (Equation (4)). This is a relationship between the maximum NDVI (set as 0.8) and the maximum K_c (1.2 at effective full cover) and the minimum (bare soil) NDVI (0.16) and bare soil K_c (0.4), respectively. These values are valid for NDVI calculated from in-band surface reflectance and they are not crop-dependent.

$$K_c = 1.25 \times NDVI + 0.2 \qquad (4)$$

4.3.2. Dual Crop Coefficient Approach

The dual crop coefficient approach of FAO-56 is intended to improve daily estimates of ET_c by considering separately the contribution of soil evaporation (k_e) and crop transpiration (k_{cb}). The dual method utilizes "basal" crop coefficients (k_{cb}) representing ET of a crop interspersed with dry soil, where:

$$k_c = k_{cb} + k_e \qquad (5)$$

As crops grow, the crop height and the leaf area change, and due to the differences in evapotranspiration during the various growth stages, the k_c for a given crop will vary over each period. Following the FAO-56 approach (page 187–189) [33], growth season of the crop is divided into four distinct growth stages: initial, crop development, mid-season stage and late season. The k_{cb} mid-season can be estimated from simple field observations and measurement of fractional cover (f_c) and crop height (h_c):

$$K_{cb\text{-mid}} = K_{c\text{-min}} + \left(K_{cb\text{-full}} - K_{c\text{-min}}\right)\left(\min\left(1, \ 2f_c, \left(f_{c\text{-eff}}\right)^{\frac{1}{1+hc}}\right)\right) \qquad (6)$$

where:
$K_{cb\text{-mid}}$ is the estimated basal K_{cb} during the mid-season when plant density and/or leaf area are lower than for full cover conditions;
$K_{cb\text{-full}}$ is the estimated basal K_{cb} during the mid-season (at peak plant size or height);
$K_{c\text{-min}}$ is the minimum K_c for bare soil (in the presence of vegetation) ($K_{cmin} \approx 0.15$–0.20),
f_c is the observed fraction of soil surface that is covered by vegetation as observed from nadir [0.01–1], $f_{c\text{-eff}}$ is the effective fraction of soil surface covered or shaded by vegetation [0.01–1].
h_c is the plant height (m).
In Equation (6), we have applied the $K_{c\text{-min}}$ and $K_{c\text{-full}}$ values given by FAO-56 approach [33] (Table 17, page 137).
The soil evaporation coefficient (K_e) can be estimated from $(1 - f_c)$ using an empirical relationship given by [76] and applied in [15] for irrigated wheat field in Morocco:

$$K_e = K_{e\max} \times (1 - f_c) \qquad (7)$$

It gives the maximum soil evaporation coefficient ($K_{e\,\max}$) when the soil is bare ($f_c \approx 0$) and $= 0$ when the vegetation attains full cover ($f_c \approx 1$). We adopted $K_{emax} = 0.25$ according to [15].
The choice of this value is not random but is based on the frequency and quantity of water supply and the rate of the reference evapotranspiration ET_0. We have estimated $K_{e\,\max} = K_{c\,ini}$ at low f_c, taking into account the frequency of irrigation and ET_0. In our case, the frequency of irrigation is 15 days during the growing season and the mean value of ET_0 is 3.5–4 mm per day.
Finally, we have applied the procedure described above to calculate ET_c for all reference plots and dates using our ground measurements of fractional cover (f_c) and crop height (h_c) and FAO-56

guide [33] (Table 17, page 137). We have used this ground based ET_c data set to evaluate the remote sensing estimates of ET_c obtained with the K_c-NDVI and the analytical methods described below.

4.4. Analytical Approach

Analytical approach is based on the direct application of the Penman-Monteith Equation (8) using crop characteristics estimated from satellite images, in analogy to the direct calculation proposed by FAO [10]:

$$ET_C = \frac{1}{\lambda}\frac{\Delta(R_{ns} - R_{nl} - G) + 1.013\rho De / r_a}{\Delta + \gamma(1 + r_{c,min} / r_a)} \tag{8}$$

where:
λ is the latent heat of vaporization [MT/kg]; R_{ns} is the net short wave radiation (MJ/m²·d); R_{nl} is the net long wave radiation (MJ/m²·d); G is the soil heat flux (KJ/m²·s), D_e is the vapor pressure deficit of the air (KPa); ρ is the mean air density at constant pressure (kg/m³); γ is the psychometric constant (KPa/°C); Δ is the slope of the saturation vapor pressure temperature relationship (KPa/°C); $r_{c,min}$ and r_a are the minimum surface (in the absence of water stress) respective of the aerodynamic resistance.

Note that Equation (8) gives an estimate of K_c by dividing ET_c by ET_0.

In Equation (8), we have a radiative and an aerodynamic term. The former is the net shortwave radiation, R_{ns}, while the latter accounts for turbulent transport of heat and vapor.

The crop resistances $r_{c,min}$ and aerodynamic resistance r_a require the knowledge of canopy geometrical characteristics.

The surface albedo (r), the Leaf Area Index (LAI) and crop height (h_c) can be integrated in the Equation (8) as follows:

$$R_{ns} = (1 - r) \times R_s \tag{9}$$

$$r_{c,min} = \frac{100}{0.5 \times LAI} \tag{10}$$

$$r_a = \frac{\ln(\frac{Z_U - (2/3)h_c}{0.123h_c})\ln(\frac{Z_h - (2/3)h_c}{0.0123h_c})}{0.168U_z} \tag{11}$$

where: R_s is the total incoming solar radiation (MJ/m²·d); Z_U and Z_h are the measurement heights for wind and humidity, respectively (m); h_c is the crop height (m) and U is wind speed at height z (m/s).

To determine the aerodynamic resistance (r_a) with Equation (11), we have used the Equation (4) page 20 in FAO-56 [33] where we have set the zero plane displacement height ($d = 2/3\ h_c$), the roughness length for momentum $Z_{0m} = 0.123\ h_c$ and the roughness length for heat $Z_{0h} = 0.0123\ h_c$, since we are dealing with full homogeneous vegetation canopies.

The surface albedo (r) is the spectrally integrated hemispherical solar reflectance and is the driving variable of the radiation budget of a surface. The estimation of (r) can be done using measurements of the reflected solar radiance $K^{\uparrow}(\vartheta, \Phi, \lambda)$ (Wm^{-2}·sr^{-1}) at a wavelength λ (nm) and can be expressed as a function of viewing zenith, ϑ, and azimuth, Φ angles, respectively (Equation (12)). However, the current sensor capabilities impose several simplifications. In the first instance, the observed surface is considered as Lambertian. In this case, the dependence of $K^{\uparrow}$ on ϑ and Φ will be neglected and r can be estimated from any direction of observation, by means of the Equation (13), using the reflectance corrected values for atmospheric effects, ρ_λ and the weighting coefficient W_λ [77]

The weighting coefficients calculated from the extraterrestrial solar irradiance E$^\circ_\lambda$ for each band for RapidEye (REIS), Landsat8 (OLI) and SPOT4 (HRVIR1) used in our study area are summarized in Table 4.

$$r = \int_{300}^{2500} \frac{\left[\int_{0}^{2\pi} \int_{0}^{\pi/2} K^{\uparrow}(\vartheta, \Phi, \lambda) \cos\vartheta \sin\vartheta \, d\vartheta \, d\Phi \right]}{K^{\downarrow}(\lambda)} d\lambda \tag{12}$$

$$r = \sum_\lambda W_\lambda \rho_\lambda \tag{13}$$

Table 4. Weighting coefficients for the calculation of albedo α by using Equation (13) for different sensors.

Sensor	Spectral Band (μm)	Weighting Coefficient W_λ
RapidEye (REIS)	Blue: 0.440–0.510	0.2455
	Green: 0.520–0.590	0.2989
	Red: 0.630–0.685	0.1973
	NIR: 0.760–0.850	0.2583
Landsat 8 (OLI)	Blue: 0.450–0.515	0.2935
	Green: 0.525–0.600	0.2738
	Red: 0.630–0.680	0.233
	NIR: 0.845–0.885	0.1554
	SWIR: 1.560–1.660	0.0322
	SWIR: 2.100–2.300	0.0121
Spot 4 (HRVIR1)	XS1: 0.500–0.590	0.3925
	XS2: 0.610–0.680	0.3339
	XS3: 0.790–0.890	0.224
	SWIR:1.530–1.750	0.0496

The Leaf Area Index (LAI) quantifies the amount of foliage area per unit ground surface area, and is an important structural property of vegetation canopies [58]. It is a crucial variable controlling many biological and physical processes associated with vegetation on the earth's surface, such as photosynthesis, respiration, transpiration, carbon and nutrient cycle, and rainfall interception. In an operational context, the estimation of LAI from measurements of spectral reflectance has been mostly based on the (semi) empirical relationships between this parameter and vegetation indices.

The Weighted Difference Vegetation Index (WDVI) (Equation (14)) has the advantage to reduce to a great extent the influence of soil background on the spectral signal [58], by means of the factor C (Equation (15)). The soil line slope (C) represents a linear relationship between red and NIR reflectance of bare soil, and accounts for the effects of the soil background on the vegetation index, and depends on soil type, soil texture and soil moisture.

$$WDVI = r_{ir} - C \cdot r_r \tag{14}$$

$$C = r_{s,ir} / r_{s,r} \tag{15}$$

where r_{ir} is the total measured NIR reflectance, r_r is the total measured red reflectance; The $r_{s,ir}$ is the NIR reflectance of the bare soil, and $r_{s,r}$ is the red reflectance of bare soil.

Once WDVI and WDVI$_\infty$ (representing the asymptotically limiting value for $WDVI$ when LAI tends to infinity) are determined, a light extinction coefficient α^* has to be estimated in order to determine the LAI through the Equation (16):

$$LAI = (-1/\alpha^*)\ln(1 - WDVI/WDVI_\infty) \tag{16}$$

where α^* represents the light extinction through the vegetation canopy, while it is dependent on crop geometry and solar zenith angle. We have used the average value ($\alpha^* = 0.37$) established by [78] from field measurements of LAI and WDVI for different crops.

An accurate estimation of crop height (h_c) using spectral reflectance data is quite difficult. Several studies were conducted in this framework using airborne laser altimeter [79–81], and other studies, *i.e.*, [82] using logarithmic relationships between h_c and different vegetation indices (SAVI, WDVI, NDVI, TVI, *etc.*). The same author evaluated these relationships against field measurements of alfalfa and grass. Since h_c can be estimated indirectly from the canopy roughness length (Z_{0m}), [83] Brutsaert proposed a formula using (Z_{0m}) (Equation (17)). Several relationships between NDVI-Z_{0m} have been proposed by [84–86]. We have tested some of these equations and finally we have chosen the Equation (18) [84]. The values of the C2 and C3 coefficients have been determined by comparing estimates with Equation (18) and the C2, C3 values given by [84]. The values $C2 = (-5.2)$ and $C3 = 5.3$ gave the best agreement with our observations.

$$h_c = Z_{0m} / 0.123 \tag{17}$$

$$Z_{0m} = \exp(C2 + C3 \times NDVI) \tag{18}$$

4.5. Irrigation Performance Indicators

Several researchers have demonstrated the use of satellite remote sensing derived information in conjunction with canal flow data for the evaluation of irrigation command [87–89]. A considerable amount of work has been undertaken in the past 30 years to develop a framework for irrigation performance assessment, related to equity, adequacy, reliability, productivity and

194

sustainability [5,20,30–32,90–93]. A list of irrigation performance indicators that can be quantified by use of remote sensing has been proposed by [20].

In our study on the Doukkala irrigation scheme, irrigation performance was assessed on the basis of adequacy of irrigation water allocations by CGR. For a real assessment of irrigation performance, the precipitation should be taken into account by using Irrigation Water Requirement (IWR), *i.e.*, CWR—Precipitation. Subsequently the value of performance indicator 2 (IP2) is determined for all CGRs [5]:

$$IP2^k = \frac{\sum_{x,y} ET_c^k(x, y) - P^k}{V^k} \tag{19}$$

where: ET_c is the total crop evapotranspiration (m³); k is the reference unit (in our case is the CGR), P^k is the total precipitation in the reference unit k (m³), and V^k is the volume received at reference unit k (m³).

In this study, we calculated directly ET_c (x, y) using the k_c-NDVI and the analytical methods, which can then be integrated over the area of interest and compared with irrigation volumes to determine *IP2* with Equation (19). In Equation (19), ET_c (x, y) in (m³) is obtained by multiplying *ET*c in (mm) by the area of the pixel (x, y).

5. Results and Discussions

5.1. Retrieval of Crop Bio-Physical Variables

The bio-physical variables required for ET_c estimation (the surface albedo (r), Leaf area index (LAI) and crop height (h_c)) are derived by applying the equations described in Section 4.3, to the time series constructed with the images listed in Table 2.

5.1.1. The Surface Albedo *r*

The surface albedo *r* depends on the sun elevation and zenith and azimuth view angles. The effect of sun elevation on the surface reflectance has been quantified by [77]. Since we used multi-sensor satellite data to construct albedo time series, the differences in terms of viewing angle and sun elevation were taken into account in the atmospheric correction and by considering the surface observed as lambertian. However, the differences in the spectral (number and width of bands) and spatial properties of the sensors used affected the estimated *r*.

To assess the impact on our retrievals of surface albedo (r) due to the differences in spectral and spatial properties of the sensors, we compared (r) estimated by means of different sensors in the same date and area. This was possible within the Faregh district on 26 April 2013 by comparing HRVIR1 (SPOT4) and OLI (Landsat8) *r*-estimates.

We chose 50 random pixels and then performed a linear correlation of OLI *vs.* HRVIR1 *r*-estimates (Figure 3). The RMSE = 0.0135 and the correlation coefficient $R^2 = 0.768$ indicate a good agreement of the estimated albedo with two sensors. The values of coefficient (0.77) and the offset (0.036) suggest that the residual error is not negligible. To evaluate whether this might be due to collocation errors of our data points, we repeated the comparison using larger samples.

Figure 3. Scatter plot of the estimated (*r*) by HRVIR1 (High Resolution in Visible and Infrared) *vs.* OLI (Operational Land Imager); Faregh district, 26 April 2013.

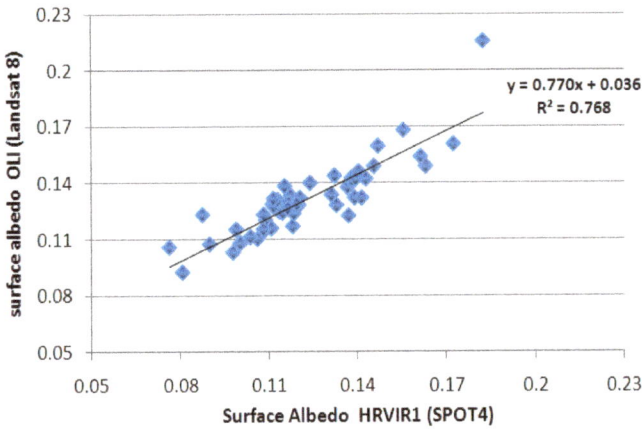

We selected 10 independent and heterogeneous samples (20 pixels × 20 lines) of (*r*) estimated with each sensor (HRVIR1 and OLI). The analysis of the two populations of samples is presented in Figure 4. OLI data gave higher mean values of albedo than HRVIR1, with a lower spatial variability. The average of the 10 values of albedo estimated with HRVIR1 is 0.12 against 0.13 estimated with OLI: this slight difference is mainly due to the contribution of the blue band (not sampled by HRVIR1) to the OLI albedo.

Figure 4. Surface albedo (*r*) estimated with overlapping HRVIR1 (SPOT4) and OLI (Landsat8) data: distribution for the area of overlap (**left**), mean and standard deviation (σ) for 10 samples of 20 pixels × 20 lines (**right**); Faregh district, 26 April 2013.

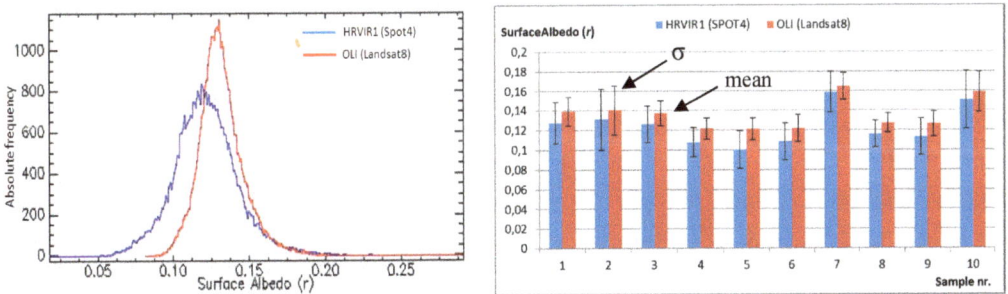

We have evaluated the significance of the differences in the mean values of (*r*) estimated with the two sensors by calculating the mean and standard deviation value of (*r*) over all samples and applied a *t*-test (Student test) for the case of two variables with different variances. Taking into account the sample size and the standard deviations, the *t*-test (*t*) confirmed that the difference between these two estimates of albedo is not significant with 0.18 bilateral where t is the value of the *t* statistic, and p is the threshold of significance). The *t*-test gave very similar results for all samples except sample 5. The difference between the mean OLI-albedo and the mean

HRVIR1-albedo was 0.01: this gives a difference in ET$_c$ with 0.01 mm/day in winter and 0.05 mm/day in summer, which are both negligible.

Generally, in the entire study area, the surface albedo is low in winter, with a high frequency of values arranged between 0.1 and 0.15, and larger in summer with a high frequency of values between 0.14 and 0.23 (Figure 5).

Figure 5. Comparison between the spatial distribution (absolute frequency) of surface albedo estimated by SPOT4 (HRVIR1) in winter and summer (Faregh District).

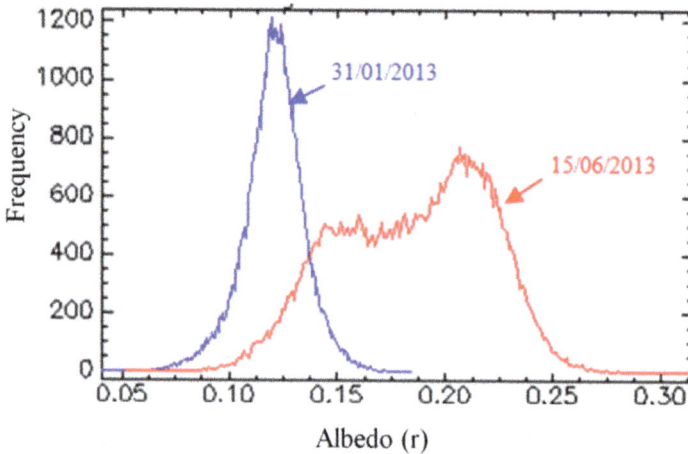

5.1.2. The Leaf Area Index (LAI)

For the estimation of LAI using Equation (16), we have to determine three parameters α*, the coefficient C (Equation (15)) and WDVI∞. In our application, we applied α* = 0.37 estimated by [78].

The C values were estimated by fitting a soil line to the scatter plot (Figure 6) of red *versus* NIR reflectance for all images of each study area (see Table 1). To determine the soil line, at first the dataset of each time series (study areas) was divided into multiple 0.002 intervals of red reflectance. Thus, within each interval the minimum value of infrared was selected [94]. To determine the soil line slope, a linear regression model was applied to the resulting r$_{ir}$ vs. r$_{ir}$ (minimum) subset taking into account only r$_{ir}$ values less than 0.4 (bare soil). The slope of the soil line gives the value of C (Equation (15)) for each area: C = 1.20 (Sidi Bennour), C = 1.02 (Zemamra) and C = 1.25 (Faregh).

The values of WDVI∞ were calculated from the WDVI time series for each study area. In each image, the mean WDVI (WDVIm) and the standard deviation (σ) were calculated. The WDVI∞ of each image was estimated as WDVIm + 3σ to filter out outliers. Finally, the WDVI∞ for each zone was calculated as the mean value (over all images) of WDVIm + 3σ. These values were equal to 0.46, 0.4 and 0.51 in Sidi Bennour, Zemamra and Faregh, respectively.

Figure 6. Scatter plot of minimum NIR *vs.* red reflectance and estimated soil line; Sidi Bennour district, see Table 1 for acquisition dates.

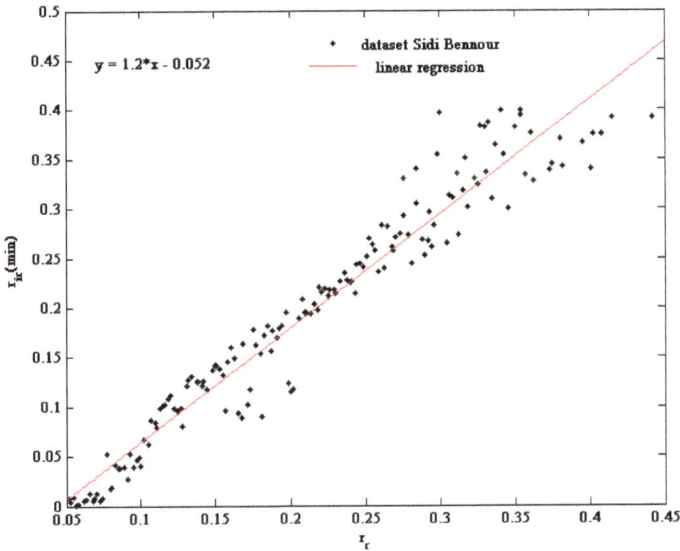

The time series of the maps of the variables, *i.e.*, LAI in Figure 7, clearly show the pattern of land cover and its temporal evolution. In the Sidi Bennour district, LAI values were small and quite variable in space on 15 December (mean = 1.78; σ = 1.97) when some crops are at the beginning of their development stage (sugar beet and alfalfa) and others are going to be sown (wheat) while sugar beet, wheat and alfalfa reach the maximum vegetative development in February (mean = 2.98; σ =2.11). In July (mean = 1.28; σ =1.36), LAI values are lower because of the smaller cultivated area.

LAI maps present sparse outliers that are generally isolated except for some plots where the high LAI values are further extended in space (see arrow in Figure 7c). Outliers in LAI maps are mostly due to saturation effect [95], *i.e.*, the received radiance at the satellite exceeds the maximum value that can be measured by the sensor. This occurs in general for the NIR band. In the Faregh area, saturated pixels (filtered out in the generation of the SPOT 4 data product) were surrounded by pixels with very high, although not saturated, NIR reflectance (Blooming effect [96,97]). This case gives very high WDVI values and therefore very high LAI values (Figure 8).

5.1.3. Soil Moisture and Radiation Control: LAI *vs.* Albedo

The evolution over time of LAI and albedo is correlated (Figure 9). During spring and summer, the soil surface becomes drier and crops reach maturation and are harvested, leading to an increase in albedo and a decrease in LAI. However, we notice that both albedo and LAI increase in autumn and winter because of initial wetting by precipitation and crop development at the beginning of the growing season.

Figure 7. Spatial and temporal variability of LAI in Sidi Bennour (**a**) (LAI inset in December (**b**); February (**c**) and July (**d**)).

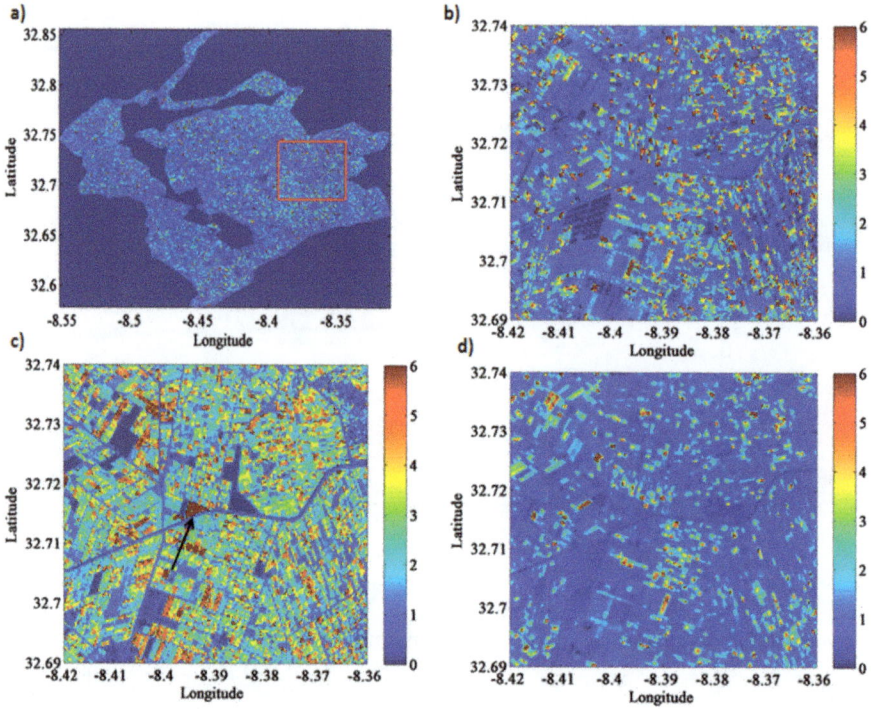

Figure 8. (**a**) HRVIR1 saturated pixels: (**a**) Red band (negatives values), (**b**) NIR band (red color maximum value, yellow color blooming) and (**c**) Blooming effect on LAI values; Faregh district on 10 June 2013.

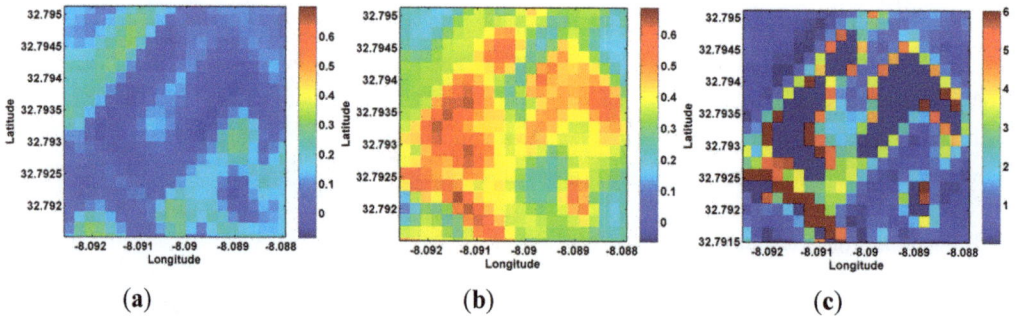

(**a**)   (**b**)   (**c**)

Figure 9. Temporal profile of LAI (Blue) and Albedo (green) in the Sidi Bennour (**a**) and Zemamra (**b**) districts.; both LAI and albedo are mean values over each district.

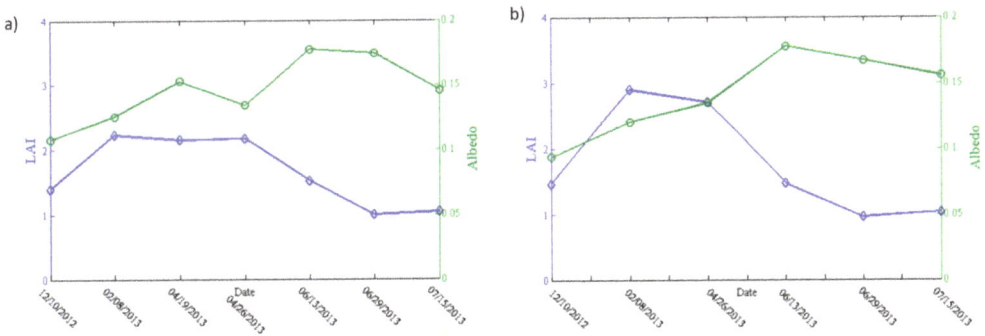

5.1.4. The Crop Height h_c

To estimate the aerodynamic properties of a vegetation canopy, we need the zero plane displacement height (d), the roughness length for momentum, z_{0m} and heat, z_{0h}, transport. When the surface is uniformly covered by vegetation, these properties are simply related to the crop height (h_c):

We analyzed the spatial variability of the estimated h_c in winter (December 2012) and summer (July 2013) (Figure 10). Our own field observations (see Section 3.3) and local knowledge (ORMVAD personal communication) indicate that dominant crops in our study area (sugar beet, wheat, maize and fourage) have a maximum h_c of approximately 1.2 m for wheat in winter, 2 m for maize in summer and 0.6 m for sugar beet. For the perennial crops, such as trees, we expect some high values in a few plots (max $h_c = 3$ m).

In winter, we notice a significant dominance of h_c values between 0.1 and 0.5 m, with a lower frequency of values ranging between 2 and 3 m. In summer, a high frequency of small values of h_c was noted since winter crops have just been harvested and summer crops are at the beginning of the development stage (maize, fruits and vegetables). The spatial variability of h_c is larger in winter than in summer.

The high values of h_c in winter could be due to an over estimation depending on the relationship used to derive Z_{0m} (see Equation (18)). When crops are completing the vegetative development, their chlorophyll content starts to increase even though their h_c remains constant. During this phase, the high values of NDVI give high values of h_c estimated by Equation (18). The effect of over-estimated h_c values on ET_c estimation was evaluated by a sensitivity analysis described in the following Section 5.2.

5.2. Sensitivity Analysis of ET_c to Bio-Physical Variables

The ET_c calculated with the analytical method (Equation (8)) depends explicitly on r, LAI and h_c, while the k_c-NDVI does not depend directly on any of these variables, since ET_0 is calculated using constant values, *i.e.*, $r = 0.23$, LAI $= 2.88$ and $h_c = 0.12$. The overall dependence of ET_c on crop conditions is accounted for by the value of k_c and its evolution over time. The spatial variability of ET_c calculated with the analytical method can, therefore, be different from ET_c calculated with the K_c-NDVI method, because of this different sensitivity to land surface conditions.

Figure 10. Spatial variability (absolute frequency) of h_c in winter (10 December 2012) and summer (15 July 2013) in Sidi Bennour.

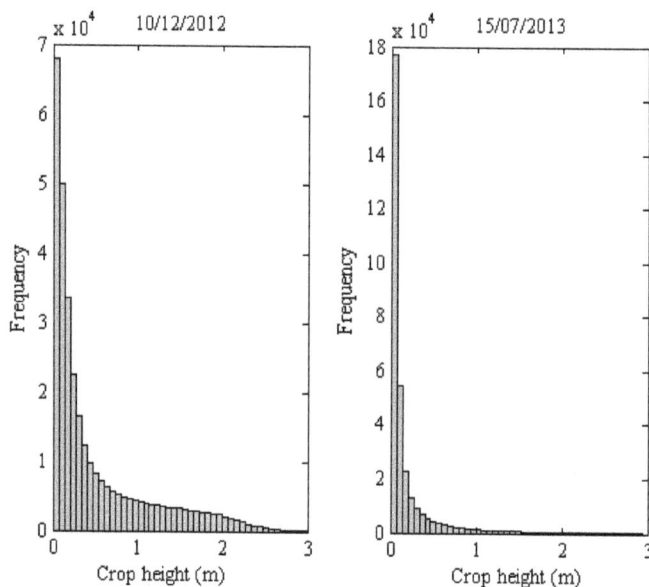

5.2.1. ET_c (Analytical) *versus* (r-LAI)

Figure 11 shows ET_c *vs. r* for different LAI values. Here, ET_c was calculated using meteorological data observed on 10 December 2012 and 15 July 2013, *i.e.*, for low and high values of R_n. The value of h_c was set to 0.40 m. These variables affect directly the values of ET_c calculated by means of the analytical approach (see the points in blue color). The relationship between ET_c and r can be very well approximated by a linear function, *i.e.*, ET_c (analytical approach) decreases with increasing value r, and increases with increasing value of LAI. The sensitivity of ET_c to LAI is higher than to r. The impact of r on ET_c (analytical approach) is slightly more pronounced in summer with higher solar irradiance.

To assess the sensitivity of ET_c calculated from K_c-NDVI method to LAI, we have derived NDVI using the following formula [98]:

$$NDVI = 0.0653 \ln (LAI) + 0.5872 \tag{20}$$

We notice that ET_c (K_c-NDVI) increases slightly with increasing value of LAI in both winter and summer, showing a small deviation and sensitivity to this geometric and structural variable. ET_c (K_c-NDVI) does not depend on r. The difference between ET_c estimated with the two methods, however, becomes greater with increasing r, and even more pronounced in summer with increasing solar radiation. In this context, we analyzed the temporal and spatial variability of ET_c estimated with the two methods at different scales in the following section.

Figure 11. Relationship between crop evapotranspiration ET_c estimated with the analytical approach (blue) as function of surface albedo r for different values of LAI and crop height $h_c = 0.4$ m; values of ET_c calculated with K_c-NDVI approach (black) for two values of $NDVI_{min}$, $NDVI_{max}$ calculated with Equation (20) for the values LAI_{min}, LAI_{max} shown in this figure; (**a**) December 2012 and (**b**) June 2013.

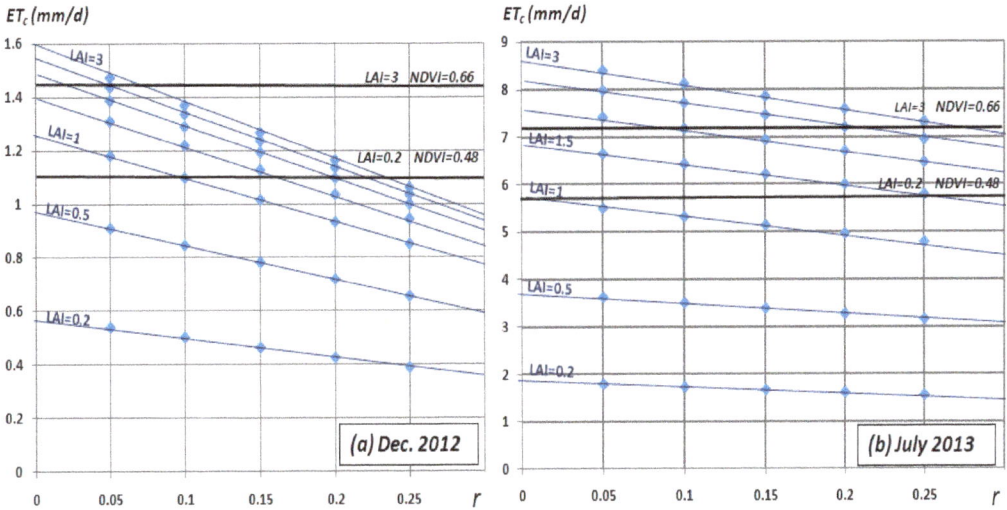

5.2.2. ET_c (Analytical) *versus* (r-h_c)

We have assessed the sensitivity of ET_c to h_c and r under winter conditions (10 December 2012) and summer conditions (15 July 2013). In both cases, the value of LAI was set to 2, the surface albedo was varying from 0.05–0.25 and h_c from 0.05–3 (Figure 12).

In general, we notice that h_c is not a critical variable for the estimation of ET_c under both winter and summer conditions. As illustrated in Figure 12a, ET_c hardly depends on h_c, and increases very little with increasing h_c. The increase is higher in summer when high values of the vapor pressure deficit (D_e) occur (Figure 12b). According to Equation (11), the direct effect of h_c is negligible because it appears in both the numerator and denominator of the argument of the logarithm.

The variability of ET_c as function of albedo r is lower. Assuming a value of h_c equal to 0.4 m, this assumption determines an error of ET_c not larger than 0.2 mm during winter, and not more than 1 mm under summer conditions. The increase of ET_c that corresponds to the decrease in r is slightly more pronounced in summer conditions than in winter because of the higher solar irradiance in summer.

202

Figure 12. Relationship between crop evapotranspiration ET$_c$ (analytical method) and the surface albedo r for different values of h$_c$ and LAI = 2, in December 2012 (**a**) and June 2013 (**b**).

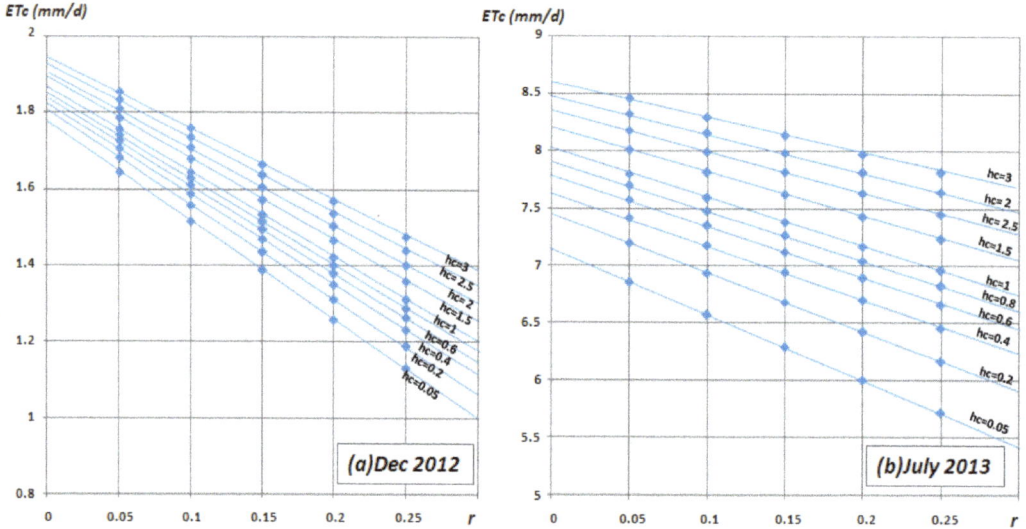

5.3. Estimation of Crop Evapotranspiration ET$_c$: kc-NDVI vs. Analytical Approach

5.3.1. Temporal Variability

Figure 13 shows the temporal evolution of daily ET$_c$ estimated by the k$_c$-NDVI method and the analytical approach in the Sidi Bennour, Zemamra and Faregh irrigation districts during the growing season of 2012/2013. ET$_c$ estimated by the two methods has rather similar evolution while the difference increases significantly from winter to summer (Figure 13), thereby supporting the hypothesis raised previously in Section 5.2, about the combined effect of LAI and r in ET$_c$ values in summer. The values of ET$_c$ (analytical approach) are slightly higher than ET$_c$ values calculated by K$_c$-NDVI method. The radiative term of Equation (9) increases with decreasing albedo. The albedo used in the K$_c$-NDVI method is constant (r = 0.23) and higher than the mean value of albedo observed in our study, which explains the higher ET$_c$ values obtained with the analytical method. In another study [11], it has been shown that the k$_c$-NDVI approach without a local calibration produces an average over-estimation of ET$_c$ of 17% in the case of corn and 19% for alfalfa.

In the Faregh district, both methods show a high value of ET$_c$ at the beginning of the growing season. This is mainly due to the values of solar radiation Rs, relative humidity RH, air temperature Ta and especially the vapor pressure deficit D$_e$ observed on the day of acquisition of the satellite data.

In general, the temporal evolution of ET$_c$ reflects the dominant crop development in the study area. In Sidi Bennour and Zemamra, the dip in June represents the transition between winter crops (wheat and sugar beet) and summer crops (Maize).

Figure 13. Comparison between daily ET_c estimates with the K_c-NDVI method and the analytical approach in the Zemamra, Sidi Bennour and Faregh districts.

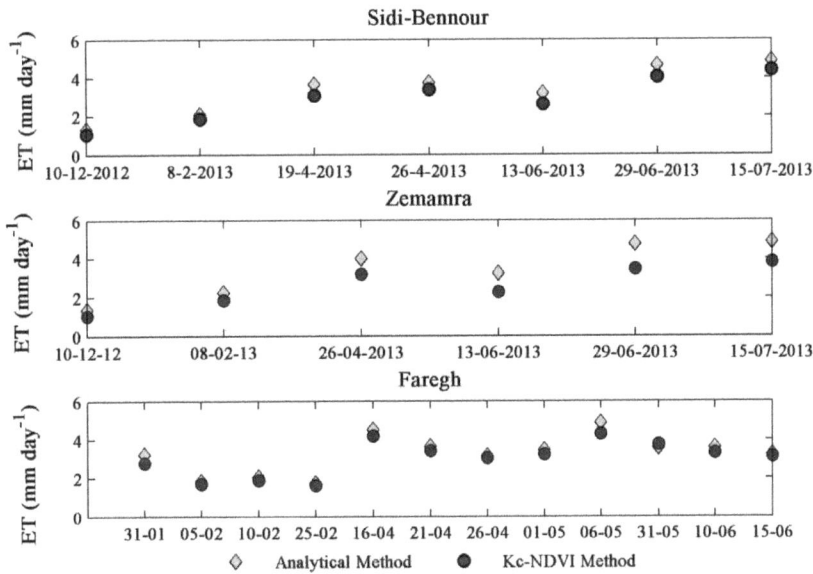

5.3.2. Spatial Variability

Since the mean value of ET_c estimated by the two methods is similar, we evaluated the spatial distribution of ET_c at different spatial scales by calculating statistics (Table 5) of ET_c with samples of 20 pixels × 20 pixels with respect to 200 pixels × 200 pixels.

The mean values of ET_c are quite similar for the two samples while the standard deviation (*i.e.*, the spatial variability) of ET_c estimated by means of the analytical approach is significantly larger than ET_c estimated with the k_c-NDVI method. This applies to all months and depends on the combined effect of surface albedo, LAI and crop height in the analytical method.

Table 5. Statistical variability of the daily ET_c values at different spatial scales.

		Kc-NDVI Approach		Analytical Approach	
		Mean	Standard Deviation	Mean	Standard Deviation
	December	1.22	0.25	1.47	0.55
20 × 20	April	4.06	0.74	4.08	0.90
	June	4.22	0.76	4.24	0.92
	December	1.26	0.36	1.55	0.74
200 × 200	April	4.58	0.72	4.78	1.08
	June	4.43	1.08	4.61	1.31

The spatial variability of ET_c must be taken into account when evaluating remote sensing estimates against point observations (see next section).

5.3.3. Validation

The primary variable of interest to compute the performance indicator (Equation (19)) is the maximum evapotranspiration ET_c. Accordingly, we compared our satellite-based estimates of ET_c, by the k_c-NDVI and analytical methods, with values calculated by means of the dual crop coefficient approach (k_{cb}), using our ground observation of f_c and h_c. We used the k_{cb} method as reference, because it is the most accurate for partial canopies and it takes explicitly into account vegetation fractional cover and crop height. It should be noted that the three methods are completely independent except for the use of the solar irradiance and the vapor pressure deficit.

We noted in some cases anomalous values in the satellite-based estimates of ET_c. We identified outliers in two ways: by filtering out estimates deviating more than 2.5σ and more than 2σ from the mean value of the difference. The latter filter leaves out about 18% of data points for either method, while the former about 13%.

The K_c-NDVI method gave a better agreement with the reference ground based ET_c with RMSE = 0.86 mm/d and RMSE = 0.79 mm/d, when applying the 2.5σ–2σ filter (Figure 14(Left)). Contrariwise, the analytical method gave a RMSE = 0.99 mm/d and RMSE = 0.89 mm/d when applying the same filters (2.5σ and 2σ) (Figure 14(Right)).

The analytical approach gave slightly higher RMSE than the K_c-NDVI method, namely it was 13% higher when filtering values greater than 2.5σ and 11% when applying the 2σ threshold to identify outliers. In a previous statistical analysis, the two methods provided rather similar mean values both at the regional scale (Figure 13) and for the different sample sizes (20×20) and (200×200) (Table 5). However, since in the study area the typical plot size is 1 ha, to perform the validation of satellites based estimates using ground observations, we used 3 pixel $\times$ 3 pixel samples to extract the ET_c values from the maps obtained with Landsat 8 data. We carried out a statistical analysis of ET_c estimated with the K_c-NDVI method and the analytical approach at the plot scale using this sampling scheme. The ΔET_c samples were analyzed separately for each acquisition date (Figure 15). The standard deviation of ΔET_c gives an indication of how likely the agreement is between the ground measurements and satellites estimates. The time series of σET_c was evaluated for each sample (from A–M) (Figure 16).

As expected, we noticed that ΔET_c is significantly larger at the plot scale (0.28 mm·d^{-1} in December and 1.16 mm·d^{-1} in June) than at the regional scale (0.25 mm in December and 0.02 mm in July (Table 5)), especially under summer conditions with high values of deficit vapor pressure (D_e) and solar radiation (R_s). The spatial variability of ET_c within (1 ha) was very large especially for the samples B, C, F, H, K, L and M. At such locations, a large difference should be expected between *in-situ* and remote sensing observations. The analytical approach captures the spatial variability of ET_c better than the Kc-NDVI method.

Figure 14. *In-situ* ET$_c$ estimated with the dual k$_c$ approach and ET$_c$ estimated with the analytical approach (**a**), the K$_c$-NDVI method (**b**), using ($\Delta ET_c - \overline{\Delta ET_c} < 2.5\sigma$) (**c**), and for ($\Delta ET_c - \overline{\Delta ET_c} < 2\sigma$) (**d**).

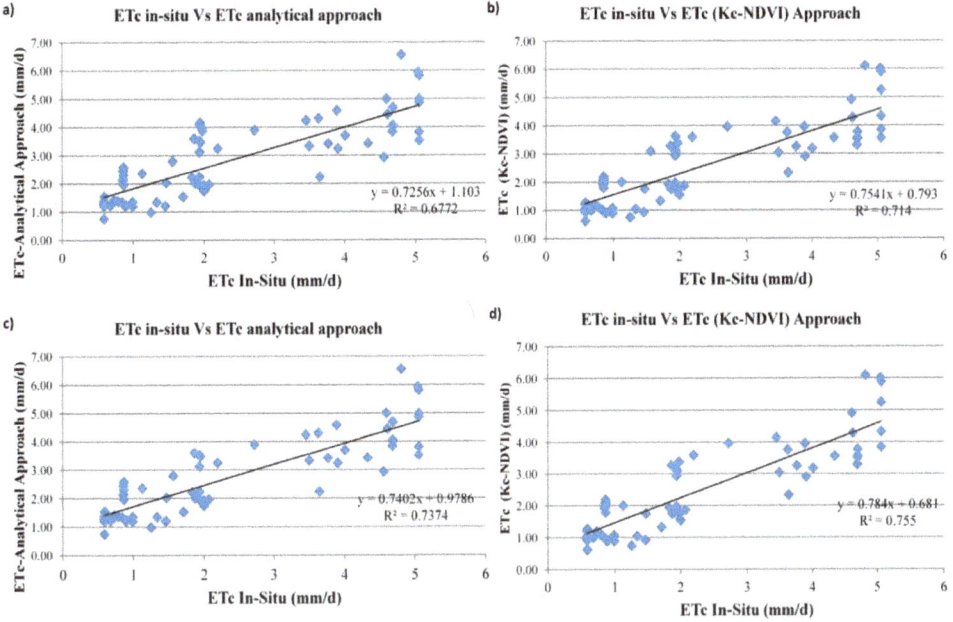

Figure 15. Temporal analysis of ΔET$_c$ between estimates with the K$_c$-NDVI method and the analytical approach for 10 (1ha) samples (3 pixels × 3 pixels) during the 2012–2013 growing season.

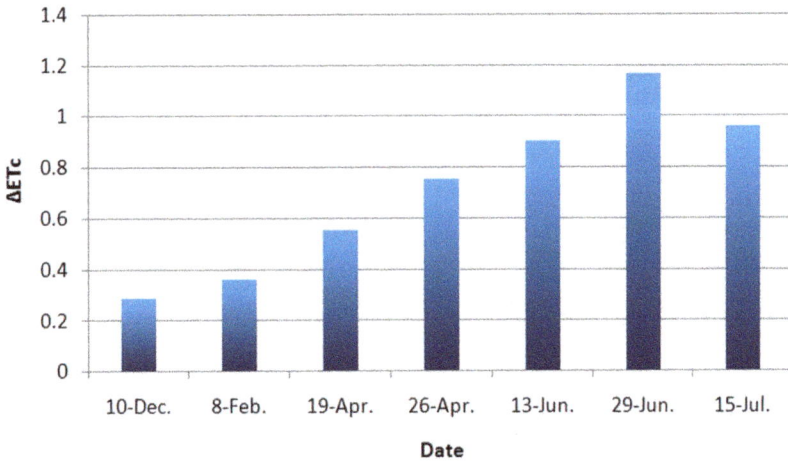

Figure 16. Statistical analysis of ETc standard deviation in 10 samples (3 × 3).

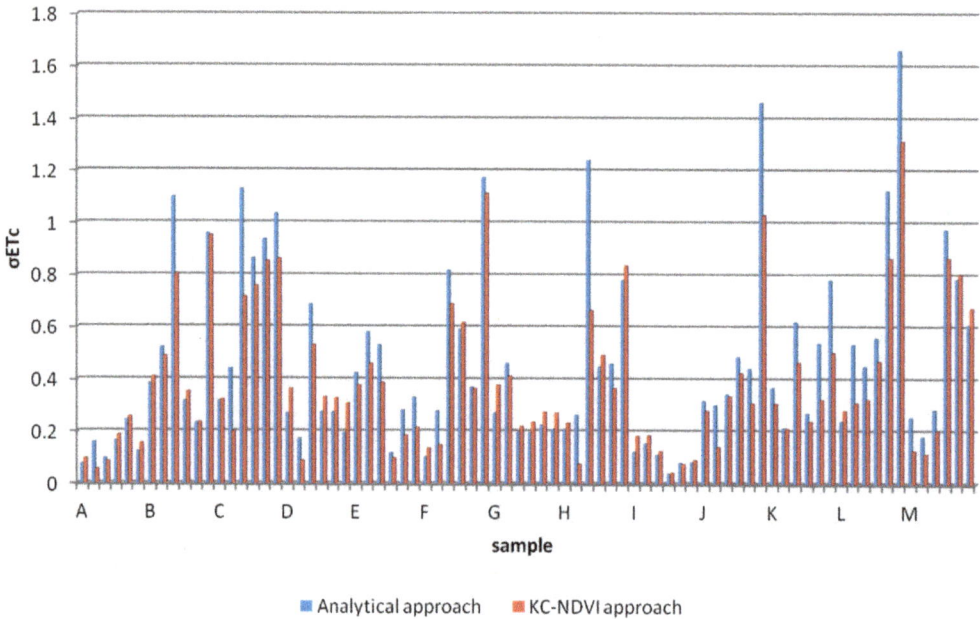

■ Analytical approach   ■ KC-NDVI approach

5.4. Irrigation Performance Indicator

5.4.1. CWR and IWR *versus* Water Allocation

We collected the monthly surface irrigation water volumes allocated in the Low Section of the Doukkala irrigation scheme, in each district and each CGR (Table 1) (ORMVAD-personal communication). Monthly values of the performance indicator IP2 were calculated for each CGR and for each district, since data on water allocation were provided at this temporal and spatial aggregation. To obtain the monthly IP2 values, the remote sensing pixel-wise estimates of K_c available on specific days have been interpolated linearly to obtain monthly ET_c values by multiplying the interpolated daily K_c value by the daily ET_0.

Less irrigation water is allocated in some cases when precipitations are significant. Irrigation Water Requirement (IWR) has been obtained by subtracting precipitation from the mean value of CWR (mm/month) estimated using the analytical approach. The comparison between CWR, IWR and water allocation (mm/month) for different districts and CGR, irrigated with different irrigation systems, is shown in Figure 17.

CWRs were larger than water allocation for both the entire irrigation scheme and the irrigation units (CGR). The irrigation water deficit was low at the beginning of the growing season (December–February), and larger at the end of the season (June and July). Except for Faregh district, the mismatch between CWR and allocations is significant for all months. Water allocation was roughly constant throughout the year, irrespective of the increasing water requirements during summer (for Maize). In the district of Sidi Bennour and Zemamra, it should be taken into account that precipitation (IWR) slightly reduces the mismatch between water allocation and demand, especially in winter when in some cases, the irrigation water deficit is converted into excess, *i.e.*, February for a number of CGR.

Unfortunately, the gaps in satellite data time series (January, March, May) in Sidi Bennour and Zemamra did not allow capturing fully the temporal evolution of water allocation and requirements. The total values of pixel-wise interpolated ET_c in January, February and March were respectively 194 mm and 204 mm against 80.5 mm and 117.19 mm of irrigation water depth for Sidi Bennour and Zemamra district, and a total of 130.5 mm of rainfall for each district. This implies that IWR = 64 mm (Sidi Bennour) and IWR = 74 mm (Zemamra), *i.e.*, water allocation was adequate. We concluded that from January to March, which represents the critical stage and the months of maximum development of the dominant crops *i.e.*, sugar beet and wheat, the contribution of precipitation to meet CWR was significant.

In the Faregh district, the precipitation does not reduce the mismatch between requirement and allocations in all cases, at the district and CGR scale.

The ratio between IWR and water allocation (IP2) allows us to assess and understand the adequacy of water allocation in the entire area, in different locations of the primary canal of irrigation.

5.4.2. Spatial Distribution of IP2

The ratio between IWR volumes and surface irrigation water volumes gives the value of IP2. We have calculated IP2 for each CGR and district. An example of the spatial distribution of IP2 in February 2013, for the Sidi Bennour and Zemamra districts and the individual CGRs is shown in Figure 18.

We noticed that the irrigation performance is not uniform over the whole study area. In some cases, the IP2 was lower than 1, which means that water allocation exceeded irrigation water requirements, *i.e.*, CGR 338 in Sidi Bennour and CGR 322, 325 in Zemamra. To some extent, this excess is necessary to compensate for water losses and it remains to be evaluated whether a fraction of it could be used for supplemental amount of irrigation water volumes in other CGR where crops suffer from a significant water deficit, such as CGR 330, 333, 336, 337, 320, and 321.

The irrigation performance indicator IP2 was used to assess the spatial pattern of adequacy between water consumption and allocation in the head–tail reaches of the primary irrigation canal (Figure 19). The water allocations show for some dates that no irrigation was provided to farmers in a given CGR. Accordingly, we have evaluated the IP2 head–tail end pattern only for February, April and June of the 2013 growing season.

Large differences were found in irrigation performance between the head, the middle and the end of the system.

In February 2013, the mean IP2 was approximately 1.06 in the beginning and the middle of the primary irrigation system which means a properly performing irrigation, but was around 1.33 at the end of the system. Although water is reported to be sufficiently available during the main season, it can be concluded that there are significant differences in adequacy towards the tail-end of the system (CGR 322, 321, 320 and 325).

Figure 17. Comparison between the temporal variability of CWR, IWR mean values and water allocation (mm/month) for different districts with an example of a CGR.

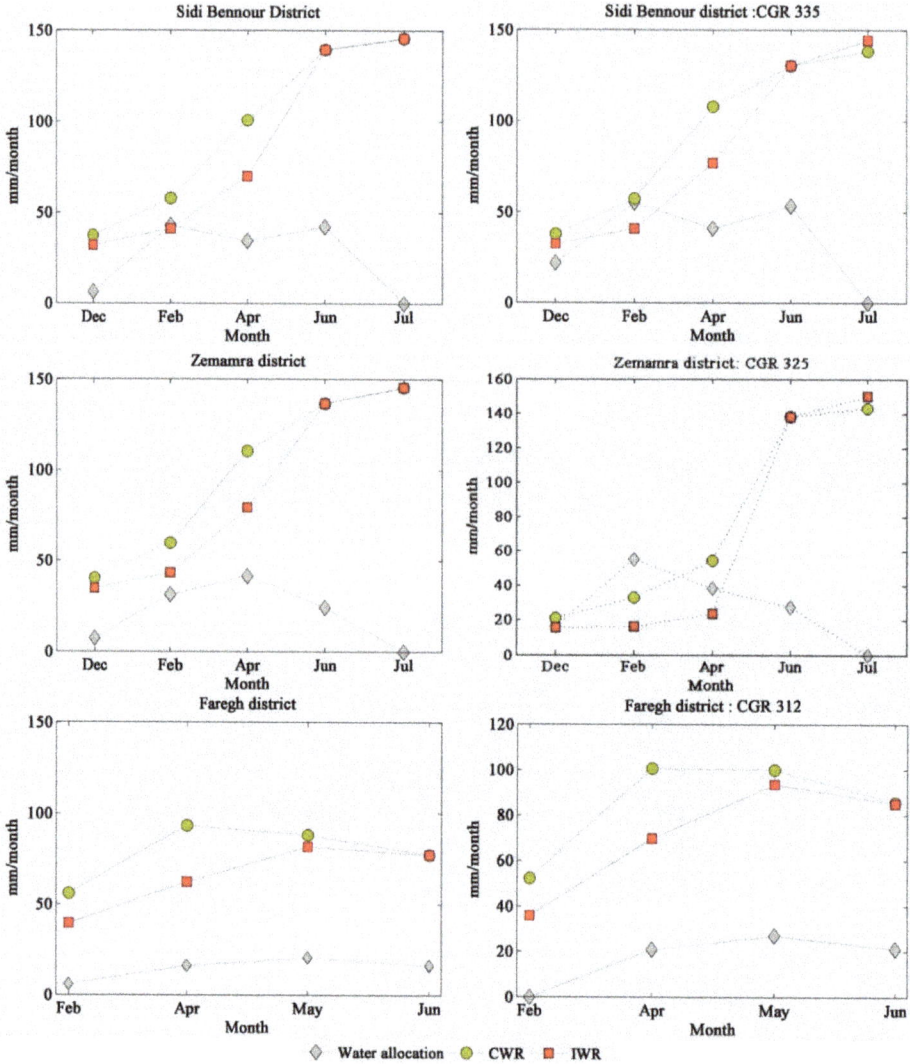

In April and July 2013, the adequacy was lower at the beginning and the end of the system and better in the middle of the system, *i.e.*, no clear head–tail end pattern in irrigation performance was observed.

These two results can be explained by the continuous control of water flows in the irrigation system, which apparently was less effective in offsetting head–tail end patterns in February 2013. On the other hand, it should be noted that the irrigation performance was lower in April 2013, when water allocations were about half the water requirement.

Figure 18. Distribution of IP2 per CGR for both Sidi Bennour and Zemamra districts in February 2013 using the background of RapidEye image.

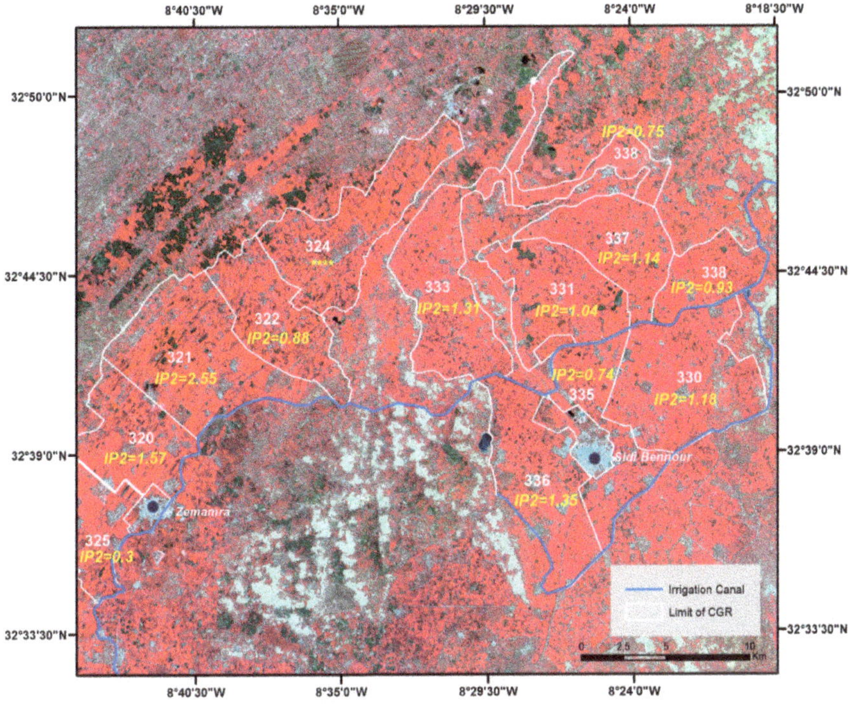

Figure 19. Head–tail end patterns of IP2 in the irrigated perimeter of Doukkala.

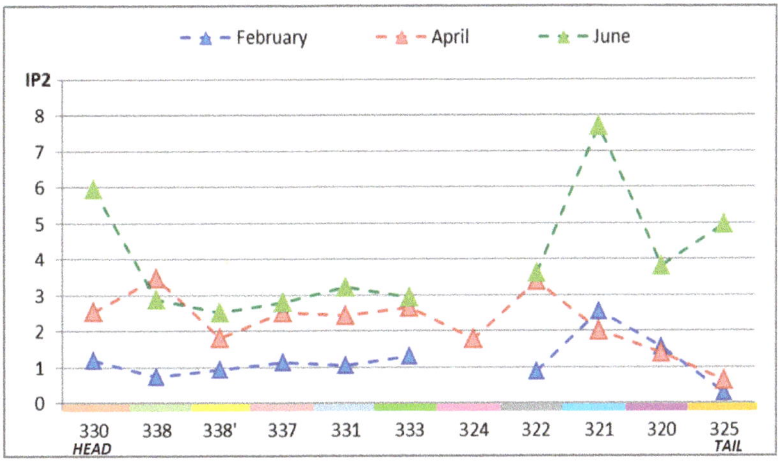

6. Application in Irrigation Water Management

In this study, we have demonstrated the potential of using satellite remote sensing as a practical tool for CWR estimation for improved understanding of water use in major irrigation schemes such as the Doukkala. Repetitive multispectral and high resolution imaging of this agricultural area was used to provide a precise and quantitative evaluation of the crop water needs during different irrigation periods during the growing season of 2012/2013.

In practice, the information provided by remote sensing could be used for irrigation water management in two ways: pixel-wise CWR data or aggregated CWR data by CGR or district.

The pixel-wise CWR data provide a reference for better precision in quasi-real time scheduling of irrigation water depth. The primary users of this information are farmers and the operators of the tertiary canals. As shown in Figure 14b, the difference between reference ground based and estimated CWR (using the analytical approach) was RMSE = 0.86 when applying the 2.5σ filter. In this case, the pixel-wise CWR data will present useful information for precise irrigation scheduling, when the spatial variability of CWR in the plot scale is higher than RMSE.

The CWR data aggregated by CGR and district provide a reference to adjust water allocation. The primary user of this information is the water management body, in our case ORMVAD, at the different management levels involved in planning and operation of water distribution. In general, it is necessary to take into account the difference between CWR, irrigation water requirement (IWR), and net irrigation water requirement (NIWR) in order to determine water allocation.

The assessment of the irrigation performance can only be done by simultaneously assessing the CWR, IWR, and more precisely using NIWR, which is the quantity of water necessary for crop growth, taking into account the rainfall. Information on irrigation efficiency is necessary to be able to estimate IWR given NIWR. The water balance in the soil–plant–atmosphere continuum can be described by models such as the Soil Water Atmosphere Plant (SWAP) model to estimate the NIWR by parameterizing root water uptake as a function of soil pressure head and soil water deficit. We have estimated NIWR by adding the soil water deficit on all dates of irrigations, where the latter are determined by maintaining crop transpiration at the potential rate. An estimation of the monthly NIWR (mm) for the dominant crops in the study area (wheat, sugar beet and alfalfa) using the SWAP model [99,100] for the growing season of 2000/2001 gave a mean value of 54.7 $mm\cdot m^{-1}$ [101].

As illustrated in Figure 17, the CWR is significantly larger than water allocation for the entire study area with 20–30 $mm\cdot m^{-1}$ of mismatch in winter for both the Sidi Bennour and the Zemamra districts. In summer, the CWR becomes much larger than water supply by 90–145 $mm\cdot m^{-1}$ in June and July, respectively. In the Faregh district, CWR is much higher than irrigation water depth, *i.e.*, around 50 $mm\cdot m^{-1}$ in winter and 70 $mm\cdot m^{-1}$ in summer.

This mismatch between requirement and allocation is improved for the entire study area when taking into consideration rainfall by means of IWR. For example, in February 2013, water allocation was almost equal to IWR in Sidi Bennour, while being just 10 mm lower than IWR in Zemamra. In the same month, water allocation exceeded IWR by 38 mm in CGR 325 and by 14 mm in CGR 335. In this case, water allocations were adequate.

In general, under summer conditions with an absence of rainfall, the mismatch between requirement and allocation remains high. However, the NIWR (54.7 $mm\cdot m^{-1}$) is rather close to water allocation in winter and adequacy is reasonable in summer with the mismatch decreasing from 90–145 $mm\cdot m^{-1}$ to 30–50 $mm\cdot m^{-1}$. We cannot conclude, however, that water allocations can meet NIWR since we should take into account conveyance and operational irrigation water losses from the secondary canal to the plot. Bos, M.G. *et al* (1974) [102] evaluated over 250 irrigation schemes worldwide and estimated irrigation water losses at 50%. Taking into account water losses, net (on farm) water allocation would still be lower than NIWR in winter and much lower in summer.

Spatially speaking, and as shown in Figure 18, the adequacy of water allocation could be improved by reducing the water excess in some CGR and by using it in others where a deficit has been assessed. Likewise, the temporal distribution of water allocation could be improved by reducing water allocation at the beginning of the growing season and increasing it in summer.

7. Conclusions

The study confirmed that crop water requirement (CWR) can be estimated with satisfactory accuracy using a generic algorithm, which does not require prior classification of crops. The appraisal of irrigation performance in terms of adequacy between requirements and water allocation at both the district and CGR (Centers of Irrigation Management) level documented a significant mismatch of requirements and allocations. Taking rainfall into account, the difference between requirements and water supply becomes acceptable in winter, but the irrigation water deficit increases in summer (90–145 mm·m^{-1}). The mismatch in both winter and summer becomes even lower when the net irrigation water requirement (NIWR) is taken as a reference.

This was achieved by constructing a time series of multi-spectral satellite image data with different spatial, temporal and spectral resolutions (SPOT4 HRVIR1, Landsat8 (OLI) and RapidEye (REIS)), and implementing (semi-) empirical algorithms to assess phenology (K_c-NDVI method) and retrieve canopy biophysical parameters (analytical approach), such as surface albedo (r), leaf area index (LAI), and crop height (h_c). The spatial distribution of K_c, r, LAI and h_c was used in conjunction with ground-based meteorological data for mapping maximum crop evapo-transpiration (ET_c). These methods are fast, robust and easily applicable to large data sets and thus suitable for operational purposes.

Calera, A.B. *et al* (2005) [49] recommended the use of K_c-NDVI method to estimate CWR, and mentioned that the methods based on the retrieval of canopy biophysical variables are very complicated and give similar final results to the K_c-NDVI method. In our case, we have assessed the difference between the two methods, and we have concluded that they give rather similar mean values of ET_c but the analytical approach captures much larger spatial variability which is very useful for precision irrigation scheduling using pixel-wise CWR data.

In general, both spatially and temporally, the adequacy of water allocation to requirements could be improved by judicious management of irrigation, *i.e.*, by reducing the water excess in some CGR/date and using it in others in deficit.

Acknowledgments

This study was supported by ESA in the framework of the ALCANTARA program. The authors thank the ORMVAD ("Office Regional de Mise en Valeur Agricole Des Doukkala", El Jadida, Morocco) for its technical help and for access to the field sites. The authors are indebted with

Alijafar Mousivand for simulating and providing the synthetic hyperspectral reflectance data used in this study.

Author Contributions

Nadia Akdim was responsible for the study and the write up of the manuscript with contributions by Silvia Maria Alfieri and Massimo Menenti. Field data were acquired by Nadia Akdim, Adnane Habib, Abdeloihab Choukri and Kamal Labbassi. Field and satellite data were processed by Nadia Akdim, Silvia Maria Alfieri and Elijah Cheruiyot; and analyzed by Nadia Akdim, Silvia Maria Alfieri, Kamal Labbassi and Massimo Menenti.

Conflicts of Interest

The authors declare no conflict of interest.

References

1. Lahlou, O.; Vidal, A. Remote Sensing and Management of Large Irrigation Projects. In *Options Mediterraneennes. Serie A, Seminaires Mediterraneens, No. 4*; Seminaires Mediterraneens: Montpellier, France, 1991; pp. 131–138.
2. Vidal, A.; Sagardoy, J.A. Use of remote sensing techniques in irrigation and drainage. *Water Rep.* **1995**, *4*, 173–178.
3. Molden, D.J. Accounting for Water Use and Productivity. Available online: http://www.iwmi.cgiar.org/Publications/SWIM_Papers/PDFs/SWIM01.PDF (accessed on 18 February 2014).
4. Molden, D.; Sakthivadivel, R. Water accounting to assess use and productivity of water. *Int. J. Water Resour. Dev.* **1999**, *15*, 55–71.
5. Menenti, M.; Visser, T.N.M.; Morabito, J.A.; Drovandi, A. Appraisal of Irrigation Performance with Satellite Data and Georeferenced Information. In *Irrigation Theory and Practice*; Rydzewsky, J.R., Ward, K., Eds.; Pentech Press: London, UK, 1989; pp. 785, 801.
6. Azzali, S. High and low resolution satellite images to monitor agricultural land. *Rep. Winand Star. Cent. Wagening.* **1990**, *61*, 41–56.
7. Moran, M.S.; Jackson, R.D. Assessing the spatial distribution of evapotranspiration using remotely sensed inputs. *J. Environ. Qual.* **1991**, *20*, 725–737.
8. Moran, M.S.; Inoue, Y.; Barnes, E.M. Opportunities and limitations for image-based remote sensing in precision crop management. *Remote Sens. Environ.* **1997**, *61*, 319–346.
9. Kustas, W.P.; Perry, E.M.; Doraiswamy, P.C.; Moran, M.S. Using satellite remote sensing to extrapolate evapotranspiration estimates in time and space over a semiarid rangeland basin. *Remote Sens. Environ.* **1994**, *49*, 275–286.
10. D'Urso, G.; Menenti, M. Section of Remote Sensing for agriculture, Forestry, and Natural Resources. In *Mapping Crop Coefficients in Irrigated Areas from Landsat TM Images*; International Society for Optics and Photonics: Paris, France, 1995; pp. 41–47.

11. D'Urso, G. Operative Approaches to Determine Crop Water Requirements from Earth Observation Data: Methodologies and Applications. In Proceedings of Earth Observation for Vegetation Monitoring and Water Management, Naples, Italy, 10–11 November 2005; pp. 14–25.

12. Vuolo, F.; D'Urso, G.; Richter, K.; Prueger, J.; Kustas, W. Physically-Based Methods for the Estimation of Crop Water Requirements from EO Optical Data, 2008. In Proceedings of Geosciences and Remote Sensing Symposium, IGARSS 2008, Boston, Massachusetts, USA, 8–11 July 2008; pp. IV-275–IV-278.

13. Bastiaanssen, W.G.M.; Molden, D.J.; Makin, I.W. Remote sensing for irrigated agriculture: Examples from research and possible applications. *Agric. Water Manag.* **2000**, *46*, 137–155.

14. Er-Raki, S.; Chehbouni, A.; Guemouria, N.; Duchemin, B.; Ezzahar, J.; Hadria, R. Combining FAO-56 model and ground-based remote sensing to estimate water consumptions of wheat crops in a semi-arid region. *Agric. Water Manag.* **2007**, *87*, 41–54.

15. Er-Raki, S.; Chehbouni, A.; Duchemin, B. Combining satellite remote sensing data with the FAO-56 dual approach for water use mapping in irrigated wheat fields of a semi-arid region. *Remote Sens.* **2010**, *2*, 375–387.

16. Yang, Y.; Yang, Y.; Liu, D.; Nordblom, T.; Wu, B.; Yan, N. Regional water balance based on remotely sensed evapotranspiration and irrigation: An assessment of the Haihe Plain, China. *Remote Sens.* **2014**, *6*, 2514–2533.

17. Teixeira, A.H.D.C. Determining regional actual evapotranspiration of irrigated crops and natural vegetation in the São Francisco river basin (Brazil) using remote sensing and penman-monteith equation. *Remote Sens.* **2010**, *2*, 1287–1319.

18. Alexandridis, T.K.; Cherif, I.; Chemin, Y.; Silleos, G.N.; Stavrinos, E.; Zalidis, G.C. Integrated methodology for estimating water use in mediterranean agricultural areas. *Remote Sens.* **2009**, *1*, 445–465.

19. Bastiaanssen, W.G.M.; Thiruvengadachari, S.; Sakthivadivel, R.; Molden, D.J. Satellite remote sensing for estimating productivities of land and water. *Int. J. Water Resour. Dev.* **1999**, *15*, 181–194.

20. Bastiaanssen, W.G.M.; Bos, M.G. Irrigation performance indicators based on remotely sensed data: A review of literature. *Irrig. Drain. Syst.* **1999**, *13*, 291–311.

21. Zwart, S.J.; Bastiaanssen, W.G.M. Review of measured crop water productivity values for irrigated wheat, rice, cotton and maize. *Agric. Water Manag.* **2004**, *69*, 115–133.

22. Teixeira, A.D.C.; Bassoi, L.H. Crop water productivity in semi-arid regions: From field to large scales. *Ann. Arid Zone* **2009**, *48*, 1–13.

23. Jackson, R.D.; Idao, S.B.; Reginato, R.J.; Pinter, P.J. Remotely Sensed Crop Temperatures and Reflectances as Inputs to Irrigtion Scheduling. Available online: http://cedb.asce.org/cgi/WWWdisplay.cgi?31647 (accessed on 18 February 2014).

24. Jackson, R.D.; Idso, S.B.; Reginato, R.J.; Pinter, P.J., Jr. Canopy temperature as a crop water stress indicator. *Water Resour. Res.* **1981**, *17*, 1133–1138.

214

25. Jackson, R.D. Section of Remote Sensing: Critical Review if Technology. In *Remote Sensing of Vegetation Characteristics for Farm Management*; International Society for Optics and Photonics: Arlington, Washington, DC, USA, 1984; pp. 81–97.
26. Makin, I.W. *Applications of Remotely Sensed, Multi-Spectral Data in Monitoring Saline Soils*; Irrigation and Power Research Institute (IPRI): Punjab, Amritsar, India, 1986.
27. Ben-Asher, J.; Charach, C.; Zemel, A. Infiltration and water extraction from trickle irrigation source: The effective hemisphere model. *Soil. Sci. Soc. Am. J.* **1986**, *50*, 882–887.
28. Moran, M.S.; Clarke, T.R.; Inoue, Y.; Vidal, A. Estimating crop water deficit using the relation between surface-air temperature and spectral vegetation index. *Remote Sens. Environ.* **1994**, *49*, 246–263.
29. Er-Raki, S.; Chehbouni, A.; Hoedjes, J.; Ezzahar, J.; Duchemin, B.; Jacob, F. Improvement of FAO-56 method for olive orchards through sequential assimilation of thermal infrared-based estimates of ET. *Agric. Water Manag.* **2008**, *95*, 309–321.
30. Wolters, W. *Influences on the Efficiency of Irrigation Water Use*; International Institute for Land Reclamation and Improvement: Wageningen, the Netherlands, 1992.
31. Murray-Rust, H.; Snellen, W.B. *Irrigation System Performance Assessment and Diagnosis*; Irrigation Water Management Institute: Colombo, Sri Lanka, 1993.
32. Bos, M.G.; Wolters, W. Influences of Irrigation on Drainage. In *Drainage Principles and Applications*, Ritzema, H.P., Ed.; International Institute for Land Reclamation and Improvement: Wageningen, the Netherlands, 1994; pp. 513–531.
33. Allen, R.G.; Pereira, L.S.; Raes, D.; Smith, M. Crop Evapotranspiration-Guidelines for Computing Crop Water Requirements-FAO Irrigation and Drainage Paper 56. Available online: http://www.kimberly.uidaho.edu/water/fao56/fao56.pdf (accessed on 18 February 2014).
34. López-Urrea, R.; Martín de Santa Olalla, F.; Montoro, A.; López-Fuster, P. Single and dual crop coefficients and water requirements for onion (*Allium cepa* L.) under semiarid conditions. *Agric. Water Manag.* **2009**, *96*, 1031–1036.
35. Casa, R.; Russell, G.; Lo Cascio, B. Estimation of evapotranspiration from a field of linseed in central Italy. *Agric. For. Meteorol.* **2000**, *104*, 289–301.
36. Benli, B.; Kodal, S.; Ilbeyi, A.; Ustun, H. Determination of evapotranspiration and basal crop coefficient of alfalfa with a weighing lysimeter. *Agric. Water Manag.* **2006**, *81*, 358–370.
37. Paço, T.A.; Ferreira, M.I.; Conceição, N. Peach orchard evapotranspiration in a sandy soil: Comparison between eddy covariance measurements and estimates by the FAO 56 approach. *Agric. Water Manag.* **2006**, *85*, 305–313.
38. Er-Raki, S.; Chehbouni, A.; Guemouria, N.; Ezzahar, J.; Khabba, S.; Boulet, G.; Hanich, L. Citrus orchard evapotranspiration: Comparison between eddy covariance measurements and the FAO-56 approach estimates. *Plant Biosyst.* **2009**, *143*, 201–208.
39. Johnson, L.F.; Trout, T.J. Satellite NDVI assisted monitoring of vegetable crop evapotranspiration in California's San Joaquin valley. *Remote Sens.* **2012**, *4*, 439–455.
40. De Bie, C.A.J.M.; Khan, M.R.; Smakhtin, V.U.; Venus, V.; Weir, M.J.C.; Smaling, E.M.A. Analysis of multi-temporal SPOT NDVI images for small-scale land-use mapping. *Int. J. Remote Sens.* **2011**, *32*, 6673–6693.

41. Benhadj, I.; Simonneaux, V.; Maisongrande, P.; Khabba, S.; Chehbouni, A. Combined Use of NDVI Time Courses at Low and High Spatial Resolution to Estimate Land Cover and Crop Evapotranspiration in Semi-Arid Areas. In Proceedings of International Workshop on the Analysis of Multi-temporal Remote Sensing Images, 2007, MultiTemp 2007, Leuven, Belgium, 18–20 July 2007; pp. 1–6.

42. Thenkabail, P.S.; Wu, Z. An Automated Cropland Classification Algorithm (ACCA) for Tajikistan by combining Landsat, MODIS, and secondary data. *Remote Sens.* **2012**, *4*, 2890–2918.

43. Amri, R.; Zribi, M.; Lili-Chabaane, Z.; Duchemin, B.; Gruhier, C.; Chehbouni, A. Analysis of vegetation behavior in a North African semi-arid region, using SPOT-VEGETATION NDVI data. *Remote Sens.* **2011**, *3*, 2568–2590.

44. Atzberger, C.; Rembold, F. Mapping the spatial distribution of winter crops at sub-pixel level using AVHRR NDVI time series and neural nets. *Remote Sens.* **2013**, *5*, 1335–1354.

45. Gumma, M.K.; Thenkabail, P.S.; Hideto, F.; Nelson, A.; Dheeravath, V.; Busia, D.; Rala, A. Mapping irrigated areas of Ghana using fusion of 30 m and 250 m resolution remote-sensing data. *Remote Sens.* **2011**, *3*, 816–835.

46. González-Dugo, M.P.; Escuin, S.; Cano, F.; Cifuentes, V.; Padilla, F.L.M.; Tirado, J.L.; Oyonarte, N.; Fernández, P.; Mateos, L. Monitoring evapotranspiration of irrigated crops using crop coefficients derived from time series of satellite images. II. Application on basin scale. *Agric. Water Manag.* **2013**, *125*, 92–104.

47. Conrad, C.; Fritsch, S.; Zeidler, J.; Rücker, G.; Dech, S. Per-field irrigated crop classification in arid central Asia using SPOT and ASTER data. *Remote Sens.* **2010**, *2*, 1035–1056.

48. Azzali, S.; Menenti, M. Mapping isogrowth zones on continental scale using temporal Fourier analysis of AVHRR-NDVI data. *Int. J. Appl. Earth Obs. Geoinform.* **1999**, *1*, 9–20.

49. Calera, A.B.; Jochum, A.M.; García, A.C.; Rodríguez, A.M.; Fuster, P.L. Irrigation management from space: Towards user-friendly products. *Irrig. Drain. Syst.* **2005**, *19*, 337–353.

50. D'Urso, G. Current status and perspectives for the estimation of crop water requirements from Earth Observation. *Ital. J. Agron.* **2010**, *5*, 107–120.

51. Rocha, J.; Perdigão, A.; Melo, R.; Henriques, C. Managing Water in Agriculture through Remote Sensing Applications. In Proceedings of 30th EARSeL Symposium on Remote Sensing for Science, Education, and Natural and Cultural Heritage, Paris, France, 31 May–3 June 2010; pp. 223–230.

52. Rouse, J.W.; Haas, R.H.; Schell, J.A.; Deering, D.W.; Harlan, J.C. *Monitoring the Vernal Advancement and Retrogradation (Greenwave Effect) of Natural Vegetation*; Texas Agricultural and Mecanical University, Remote Sensing Center: Texas, TX, USA, 1974.

53. Richardson, A.J.; Weigand, C.L. Distinguishing vegetation from soil background information. *Photogramm. Eng. Remote Sens.* **1977**, *43*, 1541–1552.

54. Huete, A.R. A soil-adjusted vegetation index (SAVI). *Remote Sens. Environ.* **1988**, *25*, 295–309.

216

55. Ray, S.S.; Dadhwal, V.K. Estimation of crop evapotranspiration of irrigation command area using remote sensing and GIS. *Agric. Water Manag.* **2001**, *49*, 239–249.

56. Garatuza-Payan, J.; Tamayo, A.; Watts, C.; Rodríguez, J.C. Estimating Large Area Wheat Evapotranspiration from Remote Sensing Data. In Proceedings of IEEE International Geoscience and Remote Sensing Symposium, 2003, IGARSS '03, Toulouse, France, 21–25 July 2003, pp. 380–382.

57. Gontia, N.K.; Tiwari, K.N. Estimation of crop coefficient and evapotranspiration of wheat (*Triticum aestivum*) in an irrigation command using remote sensing and GIS. *Water Resour. Manag.* **2010**, *24*, 1399–1414.

58. Clevers, J. Application of a weighted infrared-red vegetation index for estimating leaf area index by correcting for soil moisture. *Remote Sens. Environ.* **1989**, *29*, 25–37.

59. Pinty, B.; Verstraete, M.M. GEMI: A non-linear index to monitor global vegetation from satellites. *Vegetatio* **1992**, *101*, 15–20.

60. Menenti, M.; Bastiaanssen, W.; Van Eick, D.; Abd el Karim, M.A. Linear relationships between surface reflectance and temperature and their application to map actual evaporation of groundwater. *Adv. Space Res.* **1989**, *9*, 165–176.

61. D'Urso, G.; Menenti, M.; Santini, A. Regional application of one-dimensional water flow models for irrigation management. *Agric. Water Manag.* **1999**, *40*, 291–302.

62. Ferré, M.; Ruhard, J.P. Les bassins des Abda-Doukkala et du Sahel de Azemmour à Safi. *Notes Mém. Serv. Géol.* **1975**, *23*, 261–297.

63. Guemimi, A. Plan D'action D'économie de L'eau Dans le Périmètre Des Doukkala. In Proceedings of La Modernisation de L'agriculture Irriguée. Actes du Séminaire Euro-Méditerranéen, Thème 1: Aspects Techniques de la Modernisation des Systèmes Irriguès, Rabat, Maroc, 19–23 April 2004; pp. 1–10.

64. Rahman, H.; Dedieu, G. SMAC: A simplified method for the atmospheric correction of satellite measurements in the solar spectrum. *Int. J. Remote Sens.* **1994**, *15*, 123–143.

65. Mousivand, A.; Menenti, M.; Gorte, B.; Verhoef, W. Global sensitivity analysis of the spectral radiance of a soil-vegetation system. *Remote Sens. Environ.* **2014**, *145*, 131–144.

66. Smith, M. *CROPWAT: A Computer Program for Irrigation Planning and Management*; Food and Agriculture Organization (FAO): Rome, Italy, 1992; Volume 46.

67. Kashyap, P.S.; Panda, R.K. Evaluation of evapotranspiration estimation methods and development of crop-coefficients for potato crop in a sub-humid region. *Agric. Water Manag.* **2001**, *50*, 9–25.

68. Allen, R.G. Using the FAO-56 dual crop coefficient method over an irrigated region as part of an evapotranspiration intercomparison study. *J. Hydrol.* **2000**, *229*, 27–41.

69. Suleiman, A.A.; Tojo Soler, C.M.; Hoogenboom, G. Evaluation of FAO-56 crop coefficient procedures for deficit irrigation management of cotton in a humid climate. *Agric. Water Manag.* **2007**, *91*, 33–42.

70. Choudhury, B.J.; Ahmed, N.U.; Idso, S.B.; Reginato, R.J.; Daughtry, C.S.T. Relations between evaporation coefficients and vegetation indices studied by model simulations. *Remote Sens. Environ.* **1994**, *50*, 1–17.

71. Heilman, J.L.; Heilman, W.E.; Moore, D.G. Evaluating the crop coefficient using spectral reflectance. *Agron. J.* **1982**, *74*, 967–971.

72. Bausch, W.C.; Neale, C.M.U. Spectral inputs improve corn crop coefficients and irrigation scheduling. *Trans. ASAE* **1989**, 32, 1901–1908.

73. Neale, C.M.U.; Bausch, W.C.; Heermann, D.F. Development of reflectance-based crop coefficients for corn. *Trans. ASAE* **1989**, 32, 1891–1900.

74. Bausch, W.C. Remote sensing of crop coefficients for improving the irrigation scheduling of corn. *Agric. Water Manag.* **1995**, *27*, 55–68.

75. Bausch, W.C. Soil background effects on reflectance-based crop coefficients for corn. *Remote Sens. Environ.* **1993**, *46*, 213–222.

76. Simonneaux, V.; Duchemin, B.; Helson, D.; Er-Raki, S.; Olioso, A.; Chehbouni, A.G. The use of high-resolution image time series for crop classification and evapotranspiration estimate over an irrigated area in central Morocco. *Int. J. Remote Sens.* **2008**, *29*, 95–116.

77. Menenti, M.; Bastiaanssen, W.G.M.; van Eick, D. Determination of surface hemispherical reflectance with Thematic Mapper data. *Remote Sens. Environ.* **1989**, *28*, 327–337.

78. Vuolo, F.; Neugebauer, N.; Bolognesi, S.; Atzberger, C.; D'Urso, G. Estimation of leaf area index using DEIMOS-1 data: Application and transferability of a semi-empirical relationship between two agricultural areas. *Remote Sens.* **2013**, *5*, 1274–1291.

79. Menenti, M.; Ritchie, J.C. Estimation of effective aerodynamic roughness of Walnut Gulch watershed with laser altimeter measurements. *Water Resour. Res.* **1994**, *30*, 1329–1337.

80. Ritchie, J.C.; Menenti, M.; Weltz, M.A. Measurements of land surface features using an airborne laser altimeter: The HAPEX-Sahel experiment. *Int. J. Remote Sens.* **1996**, *17*, 3705–3724.

81. De Vries, A.C.; Kustas, W.P.; Ritchie, J.C.; Klaassen, W.; Menenti, M.; Rango, A.; Prueger, J.H. Effective aerodynamic roughness estimated from airborne laser altimeter measurements of surface features. *Int. J. Remote Sens.* **2003**, *24*, 1545–1558.

82. Payero, J.O.; Neale, C.M.U.; Wright, J.L. Comparison of eleven vegetation indices for estimating plant height of alfalfa and grass. *Appl. Eng. Agric.* **2004**, *20*, 385–393.

83. Brutsaert, W. *Evaporation into the Atmosphere: Theory, History, and Applications*; Reidel Dordrecht: Dordrecht, The Netherlands, 1982.

84. Moran, M.S.; Jackson, R.D.; Hart, G.F.; Slater, P.N.; Bartell, R.J.; Biggar, S.F.; Gellman, D.I.; Santer, R.P. Obtaining surface reflectance factors from atmospheric and view angle corrected SPOT-1 HRV data. *Remote Sens. Environ.* **1990**, *32*, 203–214.

85. Su, Z.; Schmugge, T.; Kustas, W.P.; Massman, W.J. An evaluation of two models for estimation of the roughness height for heat transfer between the land surface and the atmosphere. *J. Appl. Meteorol.* **2001**, *40*, 1933–1951.

86. Jacob, F.; Olioso, A.; Gu, X.F.; Su, Z.; Seguin, B. Mapping surface fluxes using airborne visible, near infrared, thermal infrared remote sensing data and a spatialized surface energy balance model. *Agronomie* **2002**, *22*, 669–680.

87. Perry, C.J. The IWMI water resources paradigm—Definitions and implications. *Agric. Water Manag.* **1999**, *40*, 45–50.

88. Rao, P.S. Review of Selected Literature on Indicators of Irrigation Performance. Available online: http://publications.iwmi.org/pdf/H_13467i.pdf (accessed on 18 February 2014).

89. Nageswara Rao, P.P.; Mohankumar, A. Cropland inventory in the command area of Krishnarajasagar project using satellite data. *Int. J. Remote Sens.* **1994**, *15*, 1295–1305.

90. Menenti, M.; Azzali, S.; d'Urso, G. Management of Irrigation Schemes in Arid Countries. In *Use of Remote Sensing Techniques in Irrigation and Drainage: Proceedings of the Expert Consultation*; Food and Agricultural Organization (FAO): Montpellier, France, 1993; pp. 2–4.

91. Bastiaanssen, W.G.M.; Pelgrum, H.; Droogers, P.; de Bruin, H.A.R.; Menenti, M. Area-average estimates of evaporation, wetness indicators and top soil moisture during two golden days in EFEDA. *Agric. For. Meteorol.* **1997**, *87*, 119–137.

92. Bastiaanssen, W.G.M.; Menenti, M.; Feddes, R.A.; Holtslag, A.A.M. A remote sensing surface energy balance algorithm for land (SEBAL). 1. Formulation. *J. Hydrol.* **1998**, *212*, 198–212.

93. Bos, M.G. Performance indicators for irrigation and drainage. *Irrig. Drain. Syst.* **1997**, *11*, 119–137.

94. Xu, D.; Guo, X. A study of soil line simulation from landsat images in mixed grassland. *Remote Sens.* **2013**, *5*, 4533–4550.

95. Duong, N.D.; Thoa, K.; Hoan, N.T. Some Advanced Techniques for SPOT 4 Xi Data Handling. Available online: http://www.geoinfo.com.vn/userfiles/file/cac%20cong%20trinh/15.pdf (accessed on 18 February 2014).

96. Tomasko, M.G.; Doose, L.R.; Smith, P.H.; West, R.A.; Soderblom, L.A.; Combes, M.; Bézard, B.; Coustenis, A.; de Bergh, C.; Lellouch, E. The Descent Imager/Spectral Radiometer aboard Huygens. Available online: http://www.rssd.esa.int/SB/HUYGENS/docs/SP1177/tomask_1.pdf (accessed on 18 February 2014).

97. Szafranek, I.; Amir, O.; Calahora, Z.; Adin, A.; Cohen, D. Blooming effects in indium antimode focal plane arrays. *Proc. SPIE* **1997**, *3061*, 633–639.

98. Walker, D.; Epstein, H.; Jia, G.; Balser, A.; Copass, C.; Edwards, E.; Gould, W.; Hollingsworth, J.; Knudson, J.; Maier, H. Phytomass, LAI, and NDVI in northern Alaska: Relationships to summer warmth, soil pH, plant functional types, and extrapolation to the circumpolar Arctic. *J. Geophys. Res. Atmos.* **2003**, *108*, 1984–2012.

99. Feddes, R.A.; Kowalik, P.J.; Zaradny, H. Simulation Monographs. In *Simulation of Field Water Use and Crop Yield*; Centre for Agricultural Publishing and Documentation (PUDOC): Wageningen, The Netherlands, 1978; p. 188.

100. Van Dam, J.; Huygen, J.; Wesseling, J.; Feddes, R.; Kabat, P.; van Walsum, P.; Groenendijk, P.; van Diepen, C. Theory of SWAP version 2.0; Simulation of water flow, solute transport and plant growth in the soil-water-atmosphere-plant environment. *Tech. Doc.* **1997**, *45*, 970–973.

101. TIGER. Avalable online: http://www.tiger.esa.int/PDF/news_45/20.pdf (accessed on 18 February 2014).

102. Bos, M.G.; Nugteren, J. *On Irrigation Efficiencies*; International Institute for Land Reclamation and Improvement: Wageningen, The Netherlands, 1974; Volume 19.

Earth Observation Based Assessment of the Water Production and Water Consumption of Nile Basin Agro-Ecosystems

Wim G.M. Bastiaanssen, Poolad Karimi, Lisa-Maria Rebelo, Zheng Duan, Gabriel Senay, Lal Muthuwatte and Vladimir Smakhtin

Abstract: The increasing competition for water resources requires a better understanding of flows, fluxes, stocks, and the services and benefits related to water consumption. This paper explains how public domain Earth Observation data based on Moderate Resolution Imaging Spectroradiometer (MODIS), Second Generation Meteosat (MSG), Tropical Rainfall Measurement Mission (TRMM) and various altimeter measurements can be used to estimate net water production (rainfall (P) > evapotranspiration (ET)) and net water consumption (ET > P) of Nile Basin agro-ecosystems. Rainfall data from TRMM and the Famine Early Warning System Network (FEWS-NET) RainFall Estimates (RFE) products were used in conjunction with actual evapotranspiration from the Operational Simplified Surface Energy Balance (SSEBop) and ETLook models. Water flows laterally between net water production and net water consumption areas as a result of runoff and withdrawals. This lateral flow between the 15 sub-basins of the Nile was estimated, and partitioned into stream flow and non-stream flow using the discharge data. A series of essential water metrics necessary for successful integrated water management are explained and computed. Net water withdrawal estimates (natural and humanly instigated) were assumed to be the difference between net rainfall (P_{net}) and actual evapotranspiration (ET) and some first estimates of withdrawals—without flow meters—are provided. Groundwater-dependent ecosystems withdraw large volumes of groundwater, which exceed water withdrawals for the irrigation sector. There is a strong need for the development of more open-access Earth Observation databases, especially for information related to actual ET. The fluxes, flows and storage changes presented form the basis for a global framework to describe monthly and annual water accounts in ungauged river basins.

Reprinted from *Remote Sens.* Cite as: Bastiaanssen, W.G.M.; Karimi, P.; Rebelo, L.-M.; Duan, Z.; Senay, G.; Muthuwatte, L.; Smakhtin, V. Earth Observation Based Assessment of the Water Production and Water Consumption of Nile Basin Agro-Ecosystems. *Remote Sens.* **2014**, *6*, 10306-10334.

1. Introduction

Water is becoming an increasingly scarce resource worldwide as a result of economic and demographic development pressures. While agriculture is generally assumed to be the largest consumer of water in Africa and Asia, future increases in food production will be critical to ensure human wellbeing in both these regions and globally. Projections indicate that producing enough food to meet the demands of a global population of 9.1 billion people by 2050 require levels of food production in 2007 to be increased by approximately 60%, and doubled in sub-Saharan Africa and parts of South and East Asia ([1,2]). In order to achieve this level of increase on a sustainable basis, it is critical that new strategies and approaches are employed to increase agricultural productivity, while sustaining ecological systems and the services they provide.

Water scarcity, defined as of the threat to people's livelihoods due to lack of access to safe and affordable water for drinking, sanitation, and food production (Rijsberman, [3]), is increasing in many regions along with salinization and pollution of rivers and water bodies and degradation of water-related ecosystems (FAO, [4]). Mobilizing the necessary water resources to increase food production will require informed decisions within the water sector and in related sectors (Thenkabail et al., [5]; Rebelo et al., [6]). It is estimated that annual agricultural water use will need to increase from approximately 7,100 km^3 globally to between 8,500 and 11,000 km^3 in order to meet projected food requirements in 2050 (de Fraiture et al., [7]). Projections suggest that over the next several decades the population of people whose livelihoods will be affected by water scarcity will rise to two-thirds of the world population ([8–12]).

In order to effectively manage resources, decision makers require information on how much water is available and how much is being used and consumed for various purposes, as well as an understanding of how water availability will change under future scenarios (e.g., Droogers et al., [13]). Increases in water scarcity are obvious from fast declining groundwater reservoirs, degraded wetlands and terrestrial lands and associated ecosystem services, and increased vulnerability to hazards such as droughts and floods. Most water is consumed through natural evaporative processes, but a portion is also consumed through anthropogenic influences (e.g., actual evapotranspiration from reservoirs, irrigation, aquaculture, domestic use, plantations, greenhouses, etc.). Closed basins occur where the total actual evapotranspiration (ET) has exceeded the gross inflow from rainfall and interbasin transfers. Flows in streams and aquifers could be enhanced by appropriate reductions in ET (e.g., [14]) to meet reserved flows for environmental purposes (e.g., Smakhtin et al., [15]), navigation needs, or reserved flows for downstream commitments. A reduction in consumption means that less water is evaporated or exported in products or flows to sinks. This reduction is also known as a real water saving (Seckler, [16]) because water resources will remain physically in the basin for a longer period of time. Upstream *versus* downstream water availability is an extremely sensitive issue in trans-boundary river basins such as the Nile River. A reduction in withdrawals leaves more water in rivers, lakes, reservoirs and aquifers, which is beneficial if most of the available water is utilized downstream. If most of the available water is utilizable, then a reduced withdrawal is a lost opportunity to boost local agronomies, industries and environments, provided environmental flow commitments are met (which is basically included in the definition of utilizable flows). Access to data that underpins information for decision making is largely erratic in vast basins such as the Nile. The availability of hydrological data and information related to water management is limited to local agencies. While many attempts are made through, for example, the Nile Basin Initiative (NBI) and the Food and Agricultural Organization of the United Nations FAO-Nile agricultural water management program, it remains a challenge to describe the key hydrological and catchment-scale processes in a geographically and spatially distributed manner. Water management information such as withdrawals, stocks, waste water return flows and groundwater well yields are kept by the individual water use sectors. Most of the classical hydrological research (e.g., Shahin, [17], Sutcliffe and Parks, [18]) is based on rainfall gauges and streamflow measurements, and quantification of the

ET process is not given much attention. Actual ET has however a major impact on streamflow, and is paramount for quantifying ecosystem services. A recent overview of several open-access rainfall products was published by Serrat-Capdevilla *et al.* ([19]).

The objectives of this paper are to demonstrate (i) how Earth observation data from multi-platform satellites can contribute to the generation of an open-access data set that provides insights into the major water flows and fluxes of the Nile River basin and (ii) the need for a sound system of water metrics and the use of a consistent, comprehensive framework, in order to enhance the understanding of river basins by stakeholders who make decisions on the retention, allocation and release of scarce water resources. The volume of renewable water produced by various agro-ecosystems is quantified, and it is demonstrated where, and how much water of the Nile is consumed, all on the basis of Earth observations.

2. Materials and Methods

The River Nile is fed by two main river systems: the White Nile, with its sources in the Equatorial Lake Plateau (Burundi, Rwanda, Tanzania, Kenya, Democratic Republic Congo and Uganda), and the Blue Nile, with its sources in the Ethiopian highlands. The sources are located in humid regions, with an average rainfall of over 1000 mm per year. The arid region starts in Sudan and extends into northern Ethiopia and Egypt. The Nile basin comprises 15 sub-basins (see Figure 1). The network of hydro-meteorological gauging stations in the Nile basin is meager. *In situ* data are generally not easily accessible. Earth observation measurements have the potential to complement the lack of available and accessible *in situ* measurements. Earth observation data are available on-line; they are based on true measurements made at particular moments (t) and for a specific geographic location (x,y). By sequencing repetitive measurements, a time series can be created. Flags can be used in the data sets to indicate or identify the quality of data. Data set methods to interpret and convert raw satellite measurements into hydrological variables are described in downloadable manuals, conference proceedings (e.g., Neale and Cosh, [20]) and journal special issues (e.g., Batelaan *et al.*, [21]). Following (Nagesh Kumar and Reshmidevi, [22]; Karimi *et al.*, [23]), it is argued that the remote sensing science has progressed sufficiently to ensure certain levels of accuracy. Many products now claim to have an accuracy better than 90–95 percent.

The most important Earth observation data set for hydrology and water management is rainfall. In this case study, the open-access rainfall product from the Tropical Rainfall Measurement Mission (TRMM) version 7-product 3B43 is explored. The Famine Early Warning System Network FEWS-NET Satellite Rainfall Estimate [24], processed and archived by the Famine Early Warning System Network, has been utilized as well. The period of consideration lapses from 2005 to 2010. Only monthly rainfall data sets have been created in this study. The two selected rainfall products were both subjected to first-level calibrations, using rain gauges, by the provisional agencies. The official rain gauges of the Nile basin registered with the World Meteorological Organization (WMO) have been verified, and it seems that most of them have substantial data gaps. These gaps make an independent verification of TRMM and RFE cumbersome, and in the absence of independent verification, it was decided to use the data products as they are. The average values of the two products were computed.

222

Figure 1. Location of sub-basins of the Nile River system delineated by the Nile
Basin Initiative.

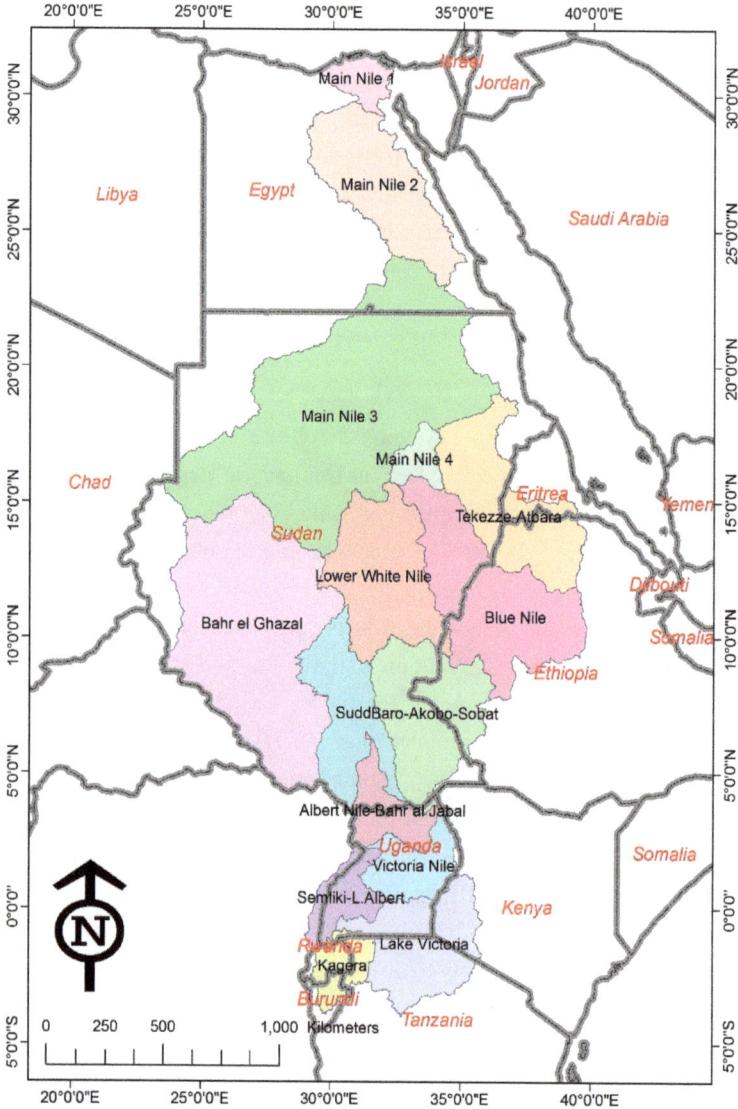

The second largest hydrological flow path is actual ET. While open-access data sets for rainfall
have a long heritage and are rapidly increasing, the generation of similar data sets for actual ET is in
its infancy. Several leading institutes have started to develop global-scale ET maps; however, they
are currently in the process of verifying the maps before opening new ET data portals. Open-access
actual ET data sets are currently provided by a few institutes only, and include the Land Satellite
Application Facility (LandSAF) under aegis of Eumetsat using the Tiled ECMWF Surface Scheme
for Exchange over Land (TESSEL) algorithm (Van den Hurk *et al.*, [25]; Belsamo *et al.*, [26]). The
ET data are computed with the Second Generation Meteosat (MSG) data having a spatial resolution
of 3 km at the equator and extending to 5 km for European cover. The Global Evaporation Monitoring

Amsterdam Model (GLEAM) is published by Mirales *et al.* ([27]) at a spatial resolution of 17 km, and while these data are not available online, the data were acquired for this study from the authors on request. The standard MODIS 16 ET product (Mu *et al.*, [28]) has an open-access status at 1 km spatial resolution, although the University of Montana requires a form of registration for use of the product. Several ET comparison studies have concluded that MOD16 is not favorable (e.g., Kim *et al.*, [29]; Trambauer *et al.*, [30]). Guerschman *et al.* [31] from the Commonwealth Scientific and Industrial Research Organisation (CSIRO) and the National University of Australia have developed the global coverage CRMSET model. Anderson *et al.* [32] have applied their Alexi energy balance model to the global scale. These two latter data sets are expected to become open-access within the next year. Bastiaanssen *et al.* [33] developed the two-layer ETLook energy balance algorithm and computed ET for the Nile basin for the years 2007 (complete basin), and 2005 and 2010 (equatorial Nile only). The best ETLook pixel dimension currently is 250 m. Because the ETLook model output is not routinely available, the retrieval of multiple year ET data for the current study was based on the Operational Simplified Surface Energy Balance (SSEBop) model, recently developed and tested by U.S. Geological Survey Earth Resources Observation and Science (EROS) Center (Senay *et al.*, [34]). SSEBop has a pixel dimension of 1 km and computes the surface energy balance on the basis of thermal infrared measurements by the MODIS satellite. It is an example of a model in the category of potential ET-soil moisture reduction methods. The SSEBop model output has a high operational potential and the data will be compared with the ETLook model, which was used successfully in earlier studies in the Nile basin and provides a good reference (Karimi *et al.*, [35]). SSEBop and ETLook both utilize climatic data from weather stations and or atmospheric models. A complicating factor with most ET algorithms is that they utilize satellite images during clear sky days only. Thermal images are taken during cloud-free days when wet surfaces are unlikely to occur. During rainfall events, leaves, soil and paved surfaces get wet for a short period, and this interception process is not properly considered by most remote sensing-based surface energy balance models. The water film on the surface induces a temporary extra net radiation value, that contributes to a quicker drying of the wet surface; this is also known as the positive drying feedback effect (Dickinson, [36]). This enhanced drying process is a result of lower albedo ($\Delta\alpha{\sim}0.05$) and lower surface temperatures ($\Delta T{\sim}2K$). Except for GLEAM, none of the other ET algorithms includes this process of temporary enhanced interception energy and ET. This aspect needs to be considered when compiling the water balance for each land use class, as it can be expected that existing remote sensing algorithms have a systematic underestimation of actual ET during wet events.

Africover is a land cover data set that was compiled for East Africa, with data available for all African countries, except Ethiopia (e.g., Di Gregorio and Jansen, [37]). The purpose of Africover is to produce the geographic information required for decision making, planning and natural resources management in African countries. The original Africover data set contains more than 500 land use classes. The land use classification system developed for Water Accounting (WA) by Karimi *et al.* [38] was used in this paper and the Africover classes have been converted into 60 individual WA+ classes that can be regrouped again into four categories that are relevant for water management: (i) protected land use (PLU), (ii) utilized land use (ULU), (iii) modified land use (MLU) and (iv) managed water use (MWU). While PLU is legally protected, ULU has a light

224

utilization and several features of natural ecosystems. MLU and especially MWU are controlled by the human desire to boost capital growth and improve the quality of living.

Access to monthly and annual values of rainfall and ET by land use class, opens an innovative pathway to start describing rainfall excess, and hence lateral water flows in an ungauged river basin system. Land use that fulfills P > ET will generate surface runoff, interflow, drainage, groundwater recharge, seepage and base flow. Besides P-ET being the main driver for flows in streams and rivers, P-ET also impacts the conditions affecting aquifer recharge. Agro-ecosystems where P > ET are referred to as net producers of water and are typically present in the forested upstream end of river basins. Such excess water moves downgradient in a given tributary to be used by other agro-ecosystems. Land use classes fulfilling ET > P are net water consumers. Agro-ecosystems that are net water consumers have an incremental ET that cannot be attributed to rainfall only, but also to other water sources with a natural origin, such as groundwater seepage, shallow water tables, interflow or inundations during annual wet seasons with high river flow levels. The incremental ET can also be caused by anthropogenic factors such as withdrawals for irrigation, drinking water supply, sanitation, industries, water retention by reservoirs, *etc.* Ahmad *et al.* ([39]) and van Eekelen *et al.* ([40]) for instance demonstrated that incremental ET (ET_{increm}) can be inferred from net precipitation (ET_{Precip}) and the total ET:

$$ET_{total} = ET_{Precip} + ET_{increm}$$

Net precipitation (ET_{Precip}) can be determined from gross precipitation (P_{gross}) using a certain efficiency factor ω (e.g., Schreiber, [41]; Budyko *et al.*, [42]; Gerrits *et al.*, [43]):

$$ET_{Precip} = \omega\, P_{gross}$$

Estimates of ET_{increm} represent the net withdrawal and it is a first step in the estimation of actual gross withdrawals. ET_{increm} is referred to as a net value because substantial portions of the gross withdrawals return to the system. Information on both gross and net withdrawals is difficult to obtain from the national departments within the Nile Basin, and is the source of political concern. The next section shows some first estimates of the net withdrawals that can be made on the basis of using Earth observation data without flow meter data. This is new information related to Nile basin abstractions that has not yet been published.

Information on the annual changes in storage (ΔS) is needed to link P and ET to flows. The total annual storage changes include (i) soil moisture, (ii) surface water and (iii) groundwater. Surface water changes were determined from altimeter observations over six major lakes and reservoirs (Lake Victoria, Lake Kyoga, Lake Kwania, Lake Tana, Lake Roseires and Lake Nasser). The altimeters used are Topex/Poseidon for rivers, and ERS-1 & 2, Envisat, Jason-1 and Geosat Follow-On (GFO) data for lakes. The Hydrology from Space website (http://www.legos.obs-mip.fr/soa/hydrologie/hydroweb/) was consulted (Cretaux *et al.*, [44]). Duan and Bastiaanssen [45] undertook a comparative analysis of several Earth observation-based altimeter products and concluded that space-borne altimeters provide water level fluctuations in 4.6–13.1% of *in situ* measurements. Space-borne altimeters are essential for the estimation of storage changes in large open water bodies of trans-boundary river systems. The changes in soil moisture were taken from the Global Land Data Assimilation System (GLDAS) database (Syed *et al.*, [46]), and they are derived from simulated soil moisture profiles in the vadose zone using a numerical land surface model. The changes in groundwater were measured on the basis of Gravity Recovery And Climate Experiment (GRACE) data, after corrections of soil moisture and surface water changes. GRACE estimates the total change of terrasphere storage from modifications of the gravitational forces. These three individual data sets (GRACE, GLDAS and altimeters) together will give an estimate of the total ΔS and the breakdown in each component. It is beyond the scope of this paper to discuss all scientific aspects of storage changes and gravity, and we refer to the GRACE studies for storage changes provided by Syed *et al.* [46], Klees *et al.* [47] and Rodell *et al.* [48].

3. Results

3.1. Rainfall

The difference in annual rainfall from TRMM and RFE is presented in Figure 2. The bar chart demonstrates that TRMM gives for most years a slightly higher rainfall over the Nile basin than RFE. The average value for the period 2006–2010 is 647 mm for TRMM and 600 mm for RFE. This is a difference of 47 mm or 8%. While this difference seems reasonable, the consequence of 8% difference at an average rainfall volume of 2013 km^3 is a potential error of 161 km^3/yr, which is two times the historic Nile flow at Dongola. This example demonstrates the need for a high accuracy in rainfall products. Because the official WMO rain gauge network is limited in number, and both TRMM and RFE have advantages and disadvantages, it is difficult to favor one of the rainfall products without in-depth research. For pragmatic reasons, the pixel values of the two rainfall products were linearly averaged. Ensemble rainfall products based on Earth observation data will likely lead the way forward to obtain reliable rainfall data layers. The new Climate Hazard Group IR Precipitation Station (CHIRPS) rainfall product is an example of an ensemble product based on various interpolation schemes to create spatially continuous grids from raw point data based on climatology, satellite measurements and ground precipitation observations from a variety of sources (Funk *et al.*, [49]).

The average rainfall of the two products for the period 2005–2010 is 624 mm/yr (2013 km^3/yr for a basin area of 3,229,038 km^2). The FAO-Nile report (Hilhorst *et al.*, [4]) gives an average rainfall volume of 2008 km^3/yr for a basin area of 3,170,418 km^2, which computes to 633 mm/yr, and which is close to the value used here (deviation is +1.4%). The FAO irrigation potential study mentions a rainfall of 615 mm/yr (deviation is −1.4%) for an area of 3,112,369 km^2 or 1914 km^3/yr (FAO, 1997). Kirby *et al.* estimated the Nile basin rainfall volume to be 2,043 km^3/yr (627 mm/yr; deviation is 0.5%) using data from the Climate Research Unit at the University of East Anglia (CRU TS 2.10) covering the period 1901–2002. Karimi *et al.* [50] summarized the water balance of the Nile, and estimated the total rainfall for the wet year 2007 as 2045 km^3/yr, which is plausible for an above-average rainfall volume. Hence, the estimates appear to correspond closely.

The consistency between the TRMM and RFE rainfall products for all 15 sub-basins is shown in Figure 3. The scatterplot demonstrates that the overall agreement is acceptable, but that large differences in local rainfall occur (RMSE is 153 mm/yr). The largest differences in absolute rainfall amounts occur over the equatorial Nile zone. The rainfall rates over Lake Victoria and the Victoria

Nile sub-basins are frequently more than 250 mm/yr different: the difference of 250 mm/yr is unlikely. Rainfall estimates of the Blue Nile sub-basin also seem to have unreasonable differences. It is therefore concluded that more research is needed to validate local rainfall products from Earth observation data. The success of calibration will increase if the data are exposed to a downscaling procedure first. Examples of calibrating downscaled rainfall products are provided in Duan and Bastiaanssen ([51]) and Hunink *et al.* ([52]).

Figure 2. Annual rainfall of the Nile as estimated by the Tropical Rainfall Measurement Mission (TRMM) and RainFall Estimates (RFE) open-access rainfall products.

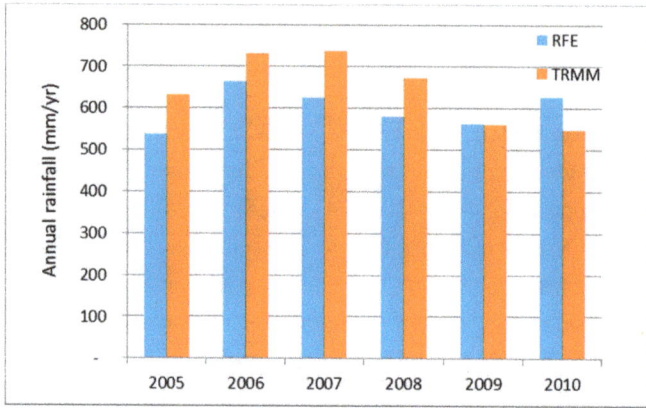

Figure 3. Annual rainfall data from TRMM 3B43 and FEWS-NET RFE presented by sub-basin (n = 15) for the period 2005–2010 (n = 6).

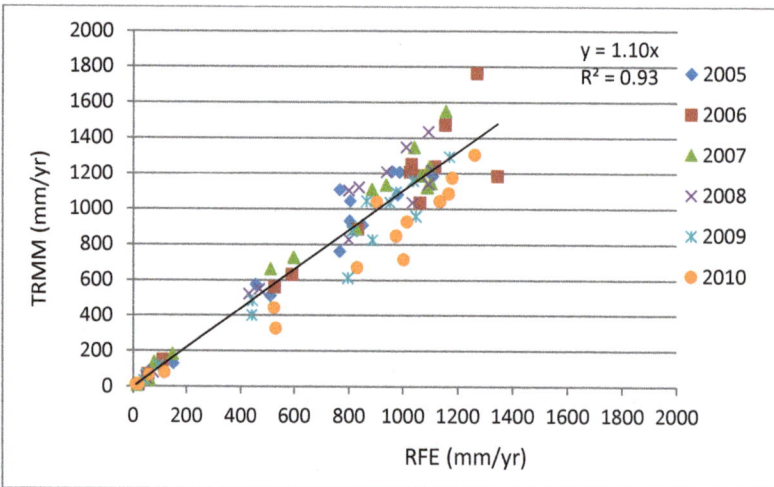

The rainfall regimes differ significantly across the different sub-basins, and rainfall evaluation should actually focus on the sub-basins only (see Table 1). The humid tropical Kagera, Lake Victoria, Semiliki and Lake Albert and Victoria Nile basins all show substantial amounts of rainfall throughout the year, with their peaks occurring during March and April (P > 100 mm/month). The central part of the Nile basin receives rainfall during June, July and August and is relatively dry during the period November to March. The area downstream of the Blue Nile sub-basin receives substantially lower rainfall amounts. Their access to water resources depends entirely on the rainfall surplus (P-ET) from the upstream basins.

Table 1. Monthly rainfall from the combined TRMM and RFE products averaged for the period 2005–2010 by sub-basin.

Sub-Basin	Area (km²)	Jan	Feb	Mar	Apr	May	Jun	Jul	Aug	Sep	Oct	Nov	Dec	Total (mm)
Main Nile 1	39.896	9	10	5	4	2	0	0	0	0	3	2	6	40
Main Nile 2	199.564	2	2	2	2	2	0	0	0	0	0	1	2	13
Main Nile 3	743.913	1	0	0	1	2	3	21	31	8	1	0	0	70
Tekezze-Atbara	231.492	3	2	9	20	31	58	150	155	74	19	3	2	524
Main Nile 4	35.338	0	0	0	2	7	5	31	51	19	4	0	0	121
Blue Nile	307.262	6	11	22	36	87	138	231	225	140	63	14	6	978
Lower White Nile	237.429	1	1	1	13	31	68	131	129	83	54	4	0	517
Bahr el Ghazal	549.714	4	6	19	47	78	107	149	176	124	70	14	3	795
Sudd	167.354	8	12	32	76	118	135	148	160	137	103	32	7	968
Baro-Akobo-Sobat	230.368	23	22	50	94	130	119	140	140	134	111	62	27	1051
Albert Nile-Bahr al Jabal	80.432	18	41	79	126	143	98	119	132	140	114	75	36	1121
Victoria Nile	86.192	39	65	111	142	134	68	92	105	131	123	98	58	1166
Semliki-L.Albert	70.646	55	90	121	109	99	50	54	93	122	120	118	64	1095
Lake Victoria	191.317	104	105	163	167	136	53	45	58	88	101	140	125	1285
Kagera	58.115	98	133	142	128	86	19	14	26	57	74	122	89	986

3.2. Actual Evapotranspiration

The application of the SSEBop model to the African continent is relatively new. Since 2012, ET anomaly products have been operationally served for various regions of the world at http://earlywarning.usgs.gov/fews/. While the U.S. Geological Survey (USGS) Earth Resources Observation and Science (EROS) Center is in the process of preparing an operational ET service (absolute magnitudes in addition to the already served anomalies) for Africa in line with the RFE rainfall products, the ET data have not been formally released, and are available only for special ET validation and water balance studies, such as presented in this paper, in Alemu *et al.* ([53]) and Senay *et al.* ([54]). SSEBop has been validated for locations in the US using flux towers (Senay *et al.*, [34]; Velpuri *et al.*, [55], Singh *et al.*, [56]), water balances (Senay *et al.*, [57]; Velpuri *et al.*, [55]) and lysimetric observations in the semi-arid Texas High Plains (Gowda *et al.*, [58]; Senay *et al.*, [59]). In addition, Senay *et al.* ([57]) described the comparison of ET layers between the earlier version of the algorithm (SSEB) and the more detailed output data computed with the Mapping EvapoTranspiration at high Resolution and Internalized Calibration (METRIC) model (Allen *et al.*, [60]). METRIC is the American version of the Surface Energy Balance Algorithm for Land (SEBAL) model (Bastiaanssen *et al.*, [61]). The METRIC comparisons for the State of Idaho, US, revealed a strong correlation ($r^2 = 0.90$) for elevations less than 2000 m. The thermodynamic modelling of the land surface fluxes in METRIC is known to be substantially more advanced compared to SSEBop. The SSEBop algorithm is an improved version of the SSEB, which is capable of handling a range of elevations (Velpuri *et al.*, [55]). Furthermore, the simplification of the parameterization of the surface energy balance in SSEBop is justifiable and required in order to cover the African continent in an operational mode. Comparison studies for Africa will provide more insights into the absolute differences and agreements, and this paper contributes to that.

Karimi *et al.* ([35,50]) report results of the ETLook model developed by Bastiaanssen *et al.* ([33]) for water balance and water accounts of the Nile basin during 2007. The total ET was estimated to be 2014 km^3/yr during the 2007 above-average rainfall year. The availability of the 1 km × 1 km ETLook data set for 2007, as well as of the equatorial Nile for the years 2005 and 2010 (Immerzeel *et al.*, [62]), provided a great opportunity for comparison with SSEBop output data. A bias correction factor of 1.12 appears to be required. This bias can be related to hot and cold pixel computations with SSEBop that have been strongly simplified as compared to the original work of Bastiaanssen *et al.* [63] related to the selection of surface temperature end members. There is a single overall calibration coefficient in SSEBop that represents elements of the crop coefficient approach of FAO and the maximum values of evaporative fraction that are not necessarily 1.0 due to certain small values of sensible heat flux (H) over wet surfaces. This coefficient is ideal for bias corrections. The overall scatter in the spatial data of ET is higher ($R^2 = 0.65$) than observed for the open-access rainfall products ($R^2 = 0.93$). The larger scatter implies that the spatial variability of ET patterns in ETLook and SSEB are larger than for rainfall. This difference can partially be ascribed to the higher spatial resolution (1 km × 1 km) of the ET product that generates more spatial contrast. Ground truth data are necessary to independently validate the ET products from Earth observations. This validation can best be done with the help of flux towers at a few locations, and with accurate water balances of paired catchments.

After this bias correction of 1.12, the longer term average actual ET for the period 2005–2010 was computed with SSEBop as 1863 km^3/yr. This is the actual ET rate that occurs due to radiative and advective energy. The additional energy due to net radiation increment of wet leaves is not included. After inclusion of the additional "interception energy", the longer term total actual ET value totaled 1987 km^3/yr, which is 6.7% more than the ET under dry surface conditions. The term "total ET" is used to express the inclusion of this extra interception energy. The lowest total ET of 1851 km^3/yr occurred during the year 2010, while 2007 had the highest total ET volume (2142 km^3/yr). The coefficient of variation of the 6 years analyzed is 6.2% only, and this indicates a temporary stable situation of actual ET for the whole Nile basin. The temporal stability can be ascribed to the large area of desert with negligible ET values.

The monthly ET layer for each sub-basin is presented in Table 2. These values represent a mix of agro-ecosystems. The sub-basin with the highest ET per unit of land is the Sudd (1209 mm/yr) followed by the Albert Nile—Bahr Al Jabal (1144 mm/yr). Both sub-basins host extensive wetlands and tropical forests, and have a rich biodiversity. Mohamed *et al.* ([64]) estimated the ET rates of a 38,600-km^2-wide Sudd area to be 1636 mm/yr. The areas prone to floods were included in their analysis. The central part of the Sudd (shown in Figure 4) displays actual ET rates between 1500 and 2000 mm/yr, similar to the ET values reported by Mohamed *et al.* ([64]). The spatial coverage of the entire Sudd sub-basin is 167,354 km^2, which is substantially larger than the area used by

Mohamed *et al.* ([64]), and includes the pastures surrounding the wetlands. This difference in area provides a logical explanation for the lower ET value shown in Table 2 (1209 mm/yr). The lowest ET by sub-basin occurs in northern Sudan and upper Egypt where the majority of the agro-ecosystem is desert (74 mm/yr). The impact of irrigation on the sub-basin averaged ET is clear when considering the sub-basin average ET value for the Main Nile 1 sub-basin (472 mm/yr). This value is roughly an average value for a landscape consisting of double cropping systems in the Nile delta (ET of 1000–1500 mm/yr) and desert land (ET of 10–100 mm/yr).

Table 2. Monthly actual evapotranspiration (including interception) from the calibrated SSEBop model (mm/month). The update is based on a bias factor and a correction term for "interception energy" during periods when the land surface contains a water film and net radiation is temporarily enhanced.

Sub-Basin	Area (km²)	Jan	Feb	Mar	Apr	May	Jun	Jul	Aug	Sep	Oct	Nov	Dec	Total (mm)
Main Nile 1	39,896	13	21	36	49	46	60	76	75	45	24	15	11	472
Main Nile 2	199,564	3	4	6	8	8	9	10	10	7	4	3	2	74
Main Nile 3	743,913	1	1	1	1	3	10	28	30	11	5	3	1	97
Tekezze-Atbara	231,492	19	15	15	16	23	53	79	79	70	44	23	18	453
Main Nile 4	35,338	2	2	3	3	7	18	51	61	25	5	1	1	180
Blue Nile	307,262	45	38	37	39	52	72	89	87	90	83	57	47	737
Lower White Nile	237,429	35	29	25	22	34	62	95	100	80	64	38	34	617
Bahr el Ghazal	549,714	51	48	53	50	68	81	93	95	88	80	62	53	823
Sudd	167,354	97	87	86	79	102	105	104	103	109	119	114	103	1,209
Baro-Akobo-Sobat	230,368	87	70	65	70	82	82	86	82	92	106	99	91	1,012
Albert Nile-Bahr al Jabal	80,432	85	72	88	87	105	100	94	93	105	114	113	88	1,144
Victoria Nile	86,192	93	77	90	89	91	85	82	84	90	101	99	86	1,067
Semliki-L.Albert	70,646	92	82	90	86	86	83	78	78	84	88	90	87	1,023
Lake Victoria	191,317	96	90	102	95	89	82	80	79	83	86	88	85	1,056
Kagera	58,115	77	67	82	83	82	73	68	69	68	71	79	71	891

Hence, the average total ET for the Nile basin for the period 2005–2010 is 1987 km³/yr. The FAO-Nile program estimated the total ET to be 1991 km³/yr (Hilhorst *et al.*, [4]), being a difference of less than 0.2%. Their methodology is based on a classical water balance calculation for sub-basins. Actual ET over rainfed areas is assumed to be equal to the reference ET$_0$ when there is enough water stored in the soil to allow actual ET fluxes to be equal to ET$_0$. During dry periods, a soil moisture reduction term is used. The land use map for irrigated areas was derived from AQUASTAT, and it was assumed that actual ET equaled ET$_0$ after correction with a crop coefficient that reflects pristine growing conditions, without any moisture or salinity stress, following Allen *et al.* ([65]). Open water and wetland evaporation was estimated from reference evapotranspiration and calibrated through closure of the water balance. Unfortunately, the boundaries of the sub-basins reported by Hilhorst *et al.* ([4]) are not identical to the official boundaries of the sub-basins outlined by the Nile Basin Initiative and used in this paper. It is therefore more useful to compare ET expressed as a rate per unit of land, and not by volume. Except for Bahr el Ghazal and Sudd, the differences are within

15% (see Table 3) and the RMSE is 295 mm/yr. The overall correlation of $R^2 = 0.92$ between the FAO-Nile ET by basin and the aggregated ET value on the basis of 1 km × 1 km pixels is encouraging, despite the differences between the geographical coverage.

Figure 4. Distribution of the annually accumulated actual evapotranspiration across the Nile basin averaged for the period 2005–2010. The pixel size is 1 km. The data are based on SSEBop being calibrated against ETLook and adjusted for interception energy.

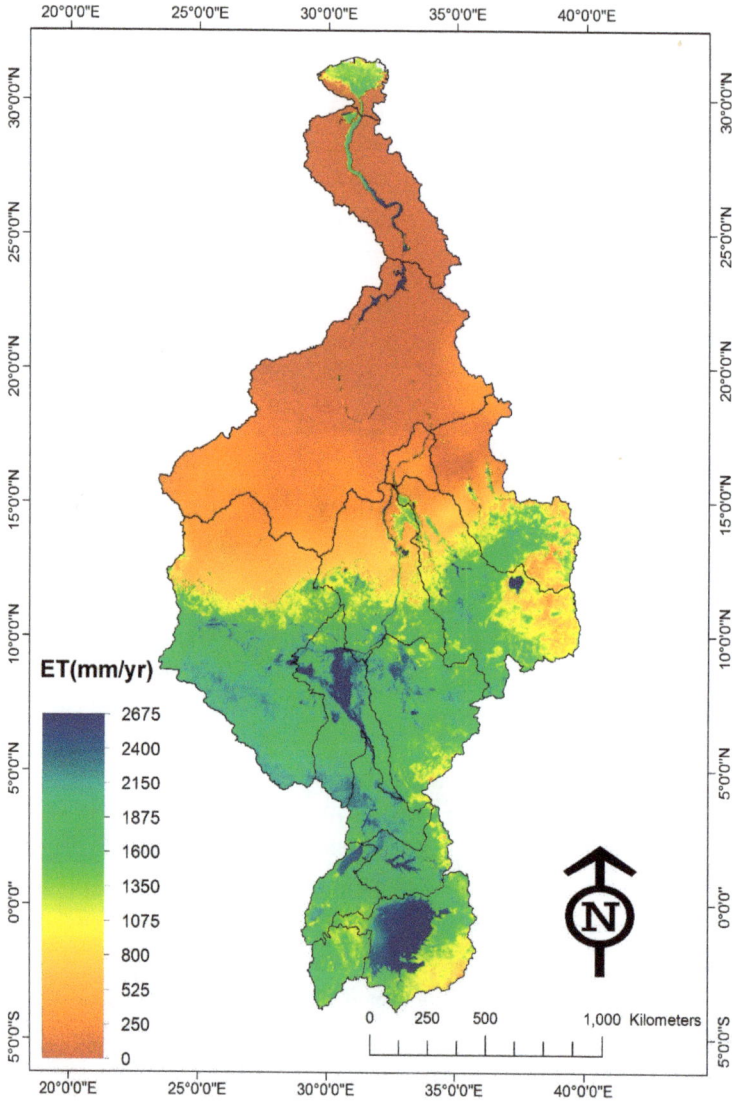

Kirby *et al.* ([66]) estimated the ET (referred to as "water use") to be 1828 km³/yr, with the majority of the water being evapotranspired by grassland (937 km³/yr), followed by woodland and other (563 km³/yr), rainfed agriculture (264 km³/yr), and irrigated agriculture (65 km³/yr). Remarkably, they did not report on the evaporation from open water bodies, and that is one of the reasons that their total value of 1828 km³/yr is lower than the 1987 km³/yr found in this study. The Kirby et al. model calculates actual ET based on potential ET and water availability. Potential ET acts as an upper boundary for actual ET. Storage of water in soil and on the surface is calculated using a soil moisture bookkeeping procedure and water balance equation. Sub-basins identified as Main Nile 1 and 2 are the irrigated areas in Egypt. The average ET of these sub-basins for the period 2005–2010 is 35 km³/yr, and this number compares well with the 35 km³/yr of ET in agriculture mentioned in the National Water Resources Plan for Egypt (MWI, [67]). In addition, there is ET from surface evaporation (2.5 km³/yr), fallow land 0.9 km³/yr, fish ponds (0.2 km³/yr), and municipal and industrial use (2.5 km³/yr). Some of these national-scale non-agricultural ET values occur, however, outside the Nile basin. So, this ET number officially endorsed by the Egyptian authorities agrees well with the Earth observation estimates.

Table 3. Comparison of longer term ET volumes (km³/yr) and fluxes (mm/yr) estimated by FAO-Nile (1960–1990) and the adjusted USGS EROS SSEBop model (2005–2010). The deviation is based on the ET flux values.

FAO-Nile				Adjusted SSEBop Model				
Description	Area (km²)	ET (km³/yr)	ET (mm/yr)	Description	Area (km²)	ET (km³/yr)	ET (mm/yr)	Dev (%)
Main Nile d/s Atbara	877,866	108.8	124	Main Nile 1,2,3	98,3375	105.7	107	13.3
Atbara	237,044	94.1	397	Tekezze-Atbara	231,492	104.8	453	−14.0
Main Nile d/s Khartoum	34,523	7.3	211	Main Nile 4	35,338	6.4	180	14.7
Blue Nile	308,198	266.0	863	Blue Nile	307,262	226.4	737	14.6
White Nile	260,943	144.5	554	Lower White Nile	237,429	146.4	617	−11.4
Bahr el Ghazal & el Arab	606,428	454.1	749	Bahr el Ghazal - Sudd	717,069	654.7	913	−21.9
Pibor-Akabo-Sobat	246,779	223.8	907	Baro-Akobo-Sobat	230,369	233.2	1012	−11.6
Bahr el Jebel	136,400	163.1	1,196	Albert Nile-Bahr - al Jabal	80,433	92.0	1144	4.3
Kyoga-Albert	197,253	221.6	1,124	Victoria Nile Semliki - L.Albert	156,839	164.2	1047	6.8
Lake Victoria basin	264,985	307.5	1,160	Lake VictoriaKagera	249,433	253.8	1018	12.3
Total and average	3,170,419	1990.8	628		3,229,039	1987.6	616	

3.3. Water Balance

The long-term average flows in the River Nile are measured and published in several sources. Based on data from Sutcliffe and Parks [18] and others, river flows at various points in the Nile basin were compiled before by Dai and Trenberth [68], Awulachew et al. [69], El-Shabraway [70] and Johnston [71]. The water balance for every sub-basin can also be computed from the Earth observation-based P, ET and ΔS data. It is a unique situation to have spatially distributed ET data. In combination with flow data, the ET data will facilitate the understanding of the water balance of the Nile.

Congruency between P, ET and ΔS values can be obtained from the mass balance. The inflow and outflow across sub-basins can be determined by accounting for P, ET and ΔS. Stream discharge gauges were used to measure flows in the main river courses. Outflow of sub-basins does not follow the main river course only. Inter-basin transfer also occurs through floods, smaller ungauged streams, groundwater flow and irrigation canals. The large flood plain in southern Sudan, stretching from the west (Bahr al Ghazal), across the center (Sudd) to the east (Baro Akabo Sobat), transfers large amounts of water between sub-basins when the area is submerged. Ungauged streams traverse the hilly topography and undulating terrain of the Bahr al Ghazal and Victoria Nile. The Gezira irrigation system in Sudan is located in the lower White Nile sub-basin, and receives water released by the Roseires reservoir in the Blue Nile sub-basin. Irrigation is thus an essential inter-basin transfer process. Groundwater flow occurs in areas with strong groundwater recharge. A map with annual recharge rates of 100–400 mm/yr in the Nile basin was provided by MacAllister et al. [72] based on earlier work of Bonsor et al. [73]. The greatest recharge occurs in the Blue Nile, Sudd and Bahr al Ghazal. These areas are conspicuous in Figure 5. Areas with P-ET > 700 mm/yr occur in the Blue Nile sub-basin at the highlands of Ethiopia and Kenya. Values of P-ET also seem to be systematically higher in the southwestern part of Lake Victoria than at the eastern side of the lake. All the blue areas with P > ET in Figure 5 are net water producing areas. Besides the typical water towers at higher elevations, there are also vast areas in north Sudan that produce a thin layer of water (P-ET > 50–100 mm/yr), which volume-wise, contributes substantially to groundwater recharge and river flow (commonly via the groundwater flow system and baseflow).

Excess water from the production areas are conveyed via streams and aquifers to the lower parts of the basin. The largest water consumers (i.e., most negative P-ET pixel values) are (i) lakes and reservoirs, (ii) wetlands and (iii) irrigated areas. The flooded area of the Sudd appears dominantly in Figure 5 due to a high percentage of permanent open water bodies and swamps. The wetland areas in the Bahr al Ghazal and Sobat also appear to be vast, although flood duration might be shorter than in the Sudd; due to shorter flood seasons, the ET rates in Figure 4 do not exceed 1400 mm/yr. The irrigation systems in Sudan are clearly visible, but P-ET is higher in absolute terms than for Egypt where the role of rainfall is basically excluded. The Kenana irrigation system located at the right bank of the White Nile near the town of Rabak is an exception, with rates of P-ET < 1000 mm/yr due to commercial monoculture sugarcane plantations. Egypt and the Sudd wetlands have large areas where the ET exceeds P by more than 1000 mm/yr, hence a substantial net withdrawal must occur.

Congruency of P, ET and ΔS was verified by comparing accumulated P-ET values from upstream to downstream with outflow measured at discharge gauges. The accumulation of P-ET is zero at the water divide and increases when going downstream along the river course trajectory. The accumulation of P-ET reverses after passing flood plains and other points of substantial withdrawals. The closure errors between the measured flow at discharge gauges and P-ET for each sub-basin are ascribed to ungauged inter-basin transfer processes discussed before. Maps of groundwater recharge were used to estimate whether the inter-basin-transfer is of surface or groundwater origin. The results are provided in Table 4. It is estimated that the lower White Nile is a net receiver of surface and groundwater from floods and groundwater seepage in the southern part of the sub-basin. Bahr al Ghazal and the Sudd receive inter-basin transfer during the flood season. Lake Victoria and the Victoria Nile discharge excess water through ungauged streams and by means of deep percolation towards the flood plain areas in southern Sudan. Inter-basin transfer in Egypt happens by leakage towards the Nubian Sandstone and Mohra aquifers, among others (Shahin, [17]; Elsawwaf et al., [74]). Surface water resources are withdrawn and transferred to newly reclaimed desert areas outside the Nile basin such as in Sinai, North Coast, Toshka and Oweinat. The net result of all these exchanges is that 5 km³/yr of groundwater leaves the Nile basin by inter-basin transfer in addition to a surface water amount of 2 km³/yr. The average outflow to the Mediterranean Sea through the Nile branches and drainage canals of the Nile delta is estimated to be 14 km³/yr. The Egyptian National Water Resources Plan reports 12 km³/yr (drainage: 11.7 km³/yr; rivers 0.2 km³/yr). Faures *et al.* [75] estimated that 12.5 km³/yr drains to the sea and lakes, 1.2 km³/yr directly via the rivers. The agreement between the estimated outflow in the main river course of each sub-basin and the flow measurements is presented in Figure 6. The RMSE is 3 km³/yr without any bias for the 15 discharge stations and the correlation is excellent ($R^2 = 0.98$).

Figure 5. Water production areas (P > ET) and net water consumption areas (ET > P) for the Nile basin over the period 2005–2010. Lateral transport of water occurs from positive to negative areas, also non-conventionally via floods and groundwater flow.

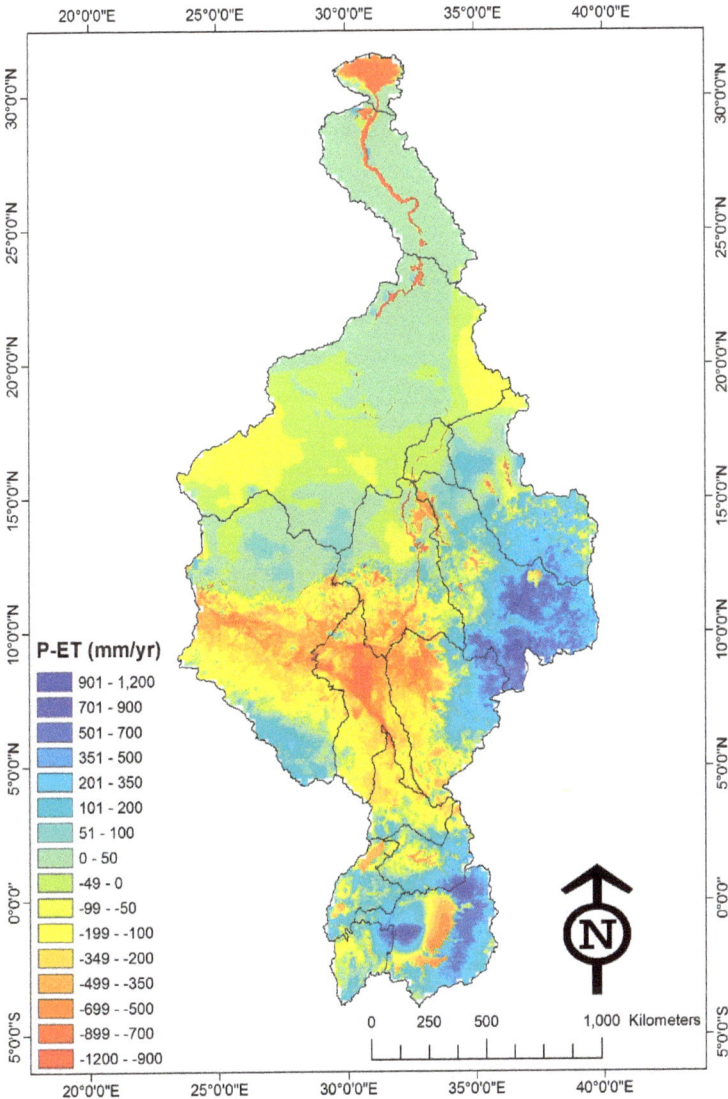

Table 4. Annual water balance by sub-basin, averaged for the period 2005–2010. Inter-basin transfer is estimated from closure of the water balance of each sub-basin.

SB Name	Inflow (km³/yr)	P (km³/yr)	ET+I (km³/yr)	Net GW interbasin (km³/yr)	Net SW interbasin (km³/yr)	ΔS (km³/yr)	Outflow (km³/yr)
Main Nile 1	36	2	19	4	1	−0.09	14
Main Nile 2	55	3	16	4	1	−0.22	36
Main Nile 3	79	51	71	4	1	0.10	55
Tekezze-Atbara	0	121	105	1	2	1.19	12
Main Nile 4	87	4	6	4	2	−0.07	79
Blue Nile	0	299	237	5	6	1.54	50
Lower White Nile	25	122	141	−11	−7	−0.25	25
Bahr el Ghazal	0	435	446	−3	−10	1.00	1
Sudd	35	162	201	−9	−6	−1.25	12
Baro-Akobo-Sobat	0	242	232	−1	0	−1.17	13
Albert Nile-Bahr al Jabal	33	90	91	−2	−1	−0.08	35
Victoria Nile	28	100	93	3	5	−0.56	28
Semliki-L.Albert	0	78	72	0	0	0.43	5
Lake Victoria	5	246	208	6	8	2.00	28
Kagera	0	57	52	0	0	0.78	5
NILE 2005 to 2010		2013	1987	5	2	3	14

The water balance check shows that P, ET and ΔS spatial data from different types of satellites are consistent with the classical hydrological observations. The minor differences can be ascribed to ungauged lateral transfer of water outside the main river courses, apart from the fact that the reporting period of the discharge stations and the satellite data are not identical. Discharge measurement stations are not always located at the boundary between adjoining basins. For instance, the peak flow in Dongola is generally known to be 84 km³/yr, but Dongola is located in the central part of the Main Nile 3 sub-basin. The peak flow computed from P-ET is 87 km³/yr and occurs upstream of the Merowe dam.

Considering this good agreement with flows and the plausibility of non-conventional inter-basin transfers, which cannot be neglected, it can be concluded that the P-ET values are highly realistic. Figure 5 is, to the authors' knowledge, the first map of water producing and water consuming areas published for the Nile basin, and due to the acceptable values for inter-basin border flow, this information can be used for water accounting. Some examples of estimated drainage and withdrawals by agro-ecosystems are discussed in the next section.

Figure 6. Relationship between estimated outflow through the main outlet of each sub-basin for the period 2005–2010 and the longer term averaged measured river discharge.

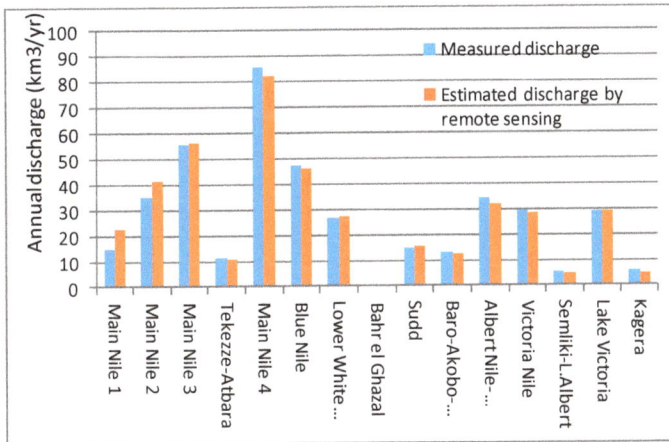

3.4. Net Water Producers and Consumers

The FAO-Africover Land Use-Land Cover (LULC) product was used as a basis for identifying the agro-ecosystems with net production and net consumption of water. The results of the LULC map and the link to rainfall and ET are provided in Table 5. The complete data set for 43 LULC classes is provided in Appendix 1. The net water production comprises all non-consumed and non-utilized water resources and is defined simply as P-ET. It is the total drainage of all non-consumed water, assuming zero changes in storage. In cases where ET > P prevails, an effective rainfall coefficient of 70% ($\omega = 0.7$) has been used arbitrarily, hence the surface runoff and drainage production is 30% of P. In reality, the ω-factor is climate, soil and land use dependent (e.g., Dastane, [76]), and more research is required to improve the estimation of ω across heterogeneous landscapes. The remaining ET that cannot be met from P is referred to as the incremental ET due to surface water or groundwater withdrawals. This extra water consumption is feasible due to lateral transfer of water in the basin and can happen by manmade infrastructure or through natural groundwater flows.

This analysis reveals that plantations are, in a relative sense, the major sources of exploited water in the River Nile (P-ET: 300 mm/yr). Rainfed crops follow with a water layer of P-ET being 181 mm/yr. Plantations and rainfed crops are both Modified Land Use. It is a positive signal that these forms of land use generated by humans are the source of lateral flow. Because the rainfed crop area is five times larger than for forest plantations, crop plantations generate 68.7 km³/yr excess water (runoff, drainage, recharge), in addition to providing staple food. This is an excellent example of contributing simultaneous to food security and ecosystem services (Rebelo and McCartney, 2012 [77]). The majority of these plantations can be found in the equatorial Nile region.

Table 5. Water production and consumption by land use-land cover class in the entire Nile basin for the period 2005–2010. Production was computed as P-ET when P > ET. For cases with ET > P, production was computed as 0.3 P. Net withdrawal was computed as ET-0.7 P.

WA+ Code	LULC	Area	ET	ET	P	P	P-ET	Production	Net Withdrawals
		(km²)	(mm/yr)	(km³/yr)	(mm/yr)	(km³/yr)	(mm/yr)	(km³/yr)	(km³/yr)
MLU2	Rainfed crops	380,180	749	284.6	929	353.3	181	68.7	0
MLU1	Plantations	74,806	850	63.6	1150	86.0	300	22.4	0
ULU4	Savannah	892,666	846	755.3	870	776.2	23	20.9	0
ULU8	Pastures	441,240	423	186.9	442	194.9	18	8.1	0
MWU17	Wetlands	10,057	1040	10.5	1092	11.0	52	0.5	0
MWU11	Urban areas	58	769	0.0	815	0.0	46	0.0	0
ULU19	Sinks	987	297	0.3	146	0.1	−150	0.0	−0,2
MLU12	Urban areas	4740	473	2.2	142	0.7	−330	0.2	−1,8
PLU4	Bare land	196,017	91	17.9	64	12.6	−27	3.8	−9,1
MWU6	Reservoirs	6310	1566	9.9	48	0.3	−1518	0.1	−9,7
ULU11	Bare land	679,835	62	42.1	52	35.4	−10	10.6	−17,3
MWU1	Irrigated crops	54,733	812	44.5	282	15.5	−530	4.6	−33,6
ULU16	Rivers and natural lakes	89,489	1445	129.3	1335	119.4	−110	35.8	−45,7
ULU10	Wetlands	112,648	1206	135.9	960	108.1	−247	32.4	−60,2
ULU1	Forests	285,271	1067	304.4	1053	300.5	−14	90.2	−94,1
Total		3,229,039		1987.3		2014.1		298	−272

A layer of 1581 and 530 mm/yr, respectively, is evaporated in excess of rainfall with respect to reservoirs and irrigated crops. As pointed out before, the source of this incremental ET is not only surface water, but can also be groundwater that seeps out in flood plain areas and in shallow water table areas. The largest volumetric water consumer is forest. In addition to rainfall, forests consume 94 km³/yr extra, most likely due to deep rooting and use of groundwater interflows. Forests in the Nile must be distinct groundwater-dependent ecosystems. The total forest ET is 304.4 km³/yr, indicating that forest in the Nile basin are substantial water consumers (304/1987 × 100%=15%). The gross rainfall is 300.5 km³/yr, and all this rainfall water is certainly not available for forest ET, especially during heavy storms when surface runoff from forested mountains is large. There is no runoff from forests during the dry season, and the fact that ET remains high during the dry season implies that groundwater must be tapped into. The $\omega = 0.7$ correction factor is used for forests also. More research is needed on local forest hydrology processes to estimate which part of the rainfall is stored in the root zone and subsequently available for root uptake during dry years, and which part of the rainfall runs off. Van Eekelen et al. ([40]) found for the Incomati basin that the net withdrawals by forests were also substantially higher than irrigated crops, which supports the findings for the Nile basin. Wetlands in the Nile basin have a net withdrawal of −60.2 km³/yr, which reduces the flow of water to downstream areas. The incremental ET from open water bodies (i.e., rivers and natural lakes; see Appendix 1) totals 45.7 km³/yr (Table 5). This total ET value for natural bodies of water should not be confused with the reservoirs that have an incremental ET of 9.7 km³/yr. Hence, the natural lakes and rivers consume more water than the manmade lakes.

The irrigated area of 54,733 km² is in agreement with earlier estimates of 55,360 km² and 49,010 km² made by Awulachew *et al.* [78] and Bastiaanssen and Perry [79], respectively. FAO [80] estimated the total irrigation-equipped area in the basin at 50,790 km². Hence, there are some acceptable differences that can be attributed to the type of data survey, the period under consideration, and definitions of irrigated areas. Table 5 indicates that the net withdrawal by the irrigation sector is 33.6 km³/yr. Bastiaanssen and Perry [79] estimated the total net withdrawal to be 31.2 km³/yr for the year 2007; their value is remarkably similar to the value derived in this study, given that different data sources were used, and 2007 was a wet year with lower irrigation water requirements.

Awulachew *et al.* [78] estimated the current annual irrigation requirement—or gross withdrawal—to be 65.9 km³/yr, which implies an average irrigation efficiency of 50% at a net withdrawal of 33.6 km³/yr, which is a reasonable number for the current flood irrigation practices in the Nile basin. Hilhorst *et al.* [4] report a basin-wide average irrigation efficiency of 49%. The FAO estimates of gross withdrawal were 67.7 km³/yr in 1997 and are thus close to Awulachew *et al.* [78].

4. Discussion

The growing water scarcity in developed and developing countries prompts water professionals to deal more accurately with the management of water resources. Hydro-meteorological stations of rainfall, weather and stream discharge provide point measurements. The data are not publicly accessible, and many stations are dysfunctional. Actual ET is a key component of the water balance, but cannot be measured easily using *in situ* devices. There are only three flux towers in the Nile basin that measure ET flux. Access to spatially distributed P, ET and ΔS data from Earth observations has a number of advantages including: (i) providing actual ET data that cannot be measured *in situ*, (ii) providing information on all agro-ecosystems in Prediction of Ungauged Basins PUBs, (iii) the data are open-access, (iv) the data are scientifically verified by the providing agencies who have strict data protocols, (v) the source of the measurements is provided and can be inspected and verified in case of disputes, and (vi) the data are archived indefinitely. The contribution of Earth observation data for improved management of water resources in Africa will increase substantially if, in addition to rainfall products, competing ET products are made freely accessible. The SSEBop and Alexi data sets would be a valuable addition to the existing open-access MOD16 and LandSAF products for Africa. The provision of 30 m EEFlux data in the near future will be another important contribution (Allen *et al.*, [81]; Morton *et al.*, [82]). The new version of the Alexi model developed by Anderson *et al.* ([32]) has a spatial resolution of 1 km, and is also based on thermal infrared radiation. Alexi has been applied to the Nile basin in some recent National Aeronautics and Space Administration (NASA) studies (http://svs.gsfc.nasa.gov/vis/a000000/a004000/a004044/).

This paper has demonstrated that P-ET information is of strategic importance for describing the redistribution of water resources in basins without flow meters. Agro-ecosystems that produce water will need to be managed carefully, to ensure that they continue to provide this ecosystem service, and that they sustain rainfall (van der Ent *et al.*, [83]). Spatially discretized information on net water withdrawals or incremental ET is also fundamental for discussions on redistribution of water among competing user groups. While classical statements concerning withdrawals are typically limited to irrigated crops, this study demonstrates that groundwater-dependent ecosystems extract twice as much by utilizing groundwater. In situations of droughts or dry years, it is especially relevant to prioritize water allocations in terms of the most urgent and important user needs.

Although not discussed in detail, it was found that satellite information on storage changes is fundamental, but not yet reliable. The GRACE data are only applicable for areas with minimum dimensions of 300 km in the x and y direction. Changes in sub-soil moisture are not really measureable with multi-spectral radiometers from space. While claims are made that sub-surface moisture is a reflection of top-soil moisture processes (e.g., De Lange et al., [84]), this supposition cannot be supported in soil physical terms because complex 3-dimensional interactions between moisture in the soil and the vertical and horizontal water fluxes are at play. Better Earth observations systems are needed to compute sub-surface storage changes, especially for monthly-scale applications.

Every remote sensing algorithm has its own uncertainty and limitations. The accuracy is expressed by the deviation between two values pertaining to a certain space and time domain, which is often referred to as an error. The uncertainty is expressed by the range of likely values and the RMSE is a good proxy for describing that range. The rainfall product from TRMM and RFE is validated by the provisional agencies, and is estimated to have an error of 0.5 to 1.4 % at the basin scale when comparing with independent sources. The RMSE values at sub-basin scale is 153 mm/yr ($R^2 = 0.93$), hence locally there is more uncertainty of rainfall. The actual ET product deviates 0.2% at basin scale from the only reliable alternative data source based on earlier FAO work. At the spatial scale of a sub-basin, the RMSE is 295 mm/yr ($R^2 = 0.92$). This implies that local values of actual ET are more uncertain than for rainfall. The congruency between rainfall and actual ET is excellent because the estimated river flows at 15 discharge stations have a RMSE of 3,0 km^3/yr only, and the mean discharge is 100% accurate. We therefore believe that the water balance presented at the annual time scale for all sub-basin level is highly accurate. Locally, the uncertainty will be substantially larger, especially when dealing with monthly time steps. This is natural to any water balance analysis.

Bastiaanssen ([85]) and Karimi et al. ([38]) recommended a water accounting system based on Earth observation data (see also www.wateraccounting.org). This paper confirms that some fundamental processes can indeed be derived from satellite measurements, even for vast basins with many ungauged catchments such as the Nile. Water accounting with the use of public domain Earth observation data is feasible if such data can be coupled to a local hydrological model that recognizes essential processes such as drainage, irrigation, recharge, baseflow and lateral groundwater flow. This is an attractive possibility next to the more complex accounting systems proposed by System of Environmental-Economic Accounting for Water SEEAW (United Nations, [86]), Australia (Australian Bureau Statistics, [87]) and Europe (Crouzet et al., [88]).

5. Conclusions

This paper is the first to describe the water balances of the Nile Basin and its 15 major sub-basins, based on Earth Observation data. Excess water flows from the water towers in the mountains and forests in the equatorial Nile belt and the Ethiopian Highlands are quantified. The lateral transport of water throughout the ungauged basin as a result of floods, groundwater movement, small ungauged streams, and withdrawals—planned for irrigation and unplanned for groundwater-dependent ecosystems—are quantified.

Open-access data portals for rainfall and ET are fundamental in making water resources assessment reporting feasible. While there are more than 15 operational rainfall products available, open-access ET layers are currently provided only by the LandSAF product (*i.e.*, TESSEL) and MOD16. ET data layers from GLEAM, SSEBop, Alexi, CRMSET and EEFlux will likely become available within the next year. New high frequency and high resolution spatial data measured by Proba-V, Sentinel and Landsat-8 series satellites will enrich the current satellite data archives and enable refinement of ET and land use products. While sufficient materials have been used to demonstrate the overall accuracy of the results (the RMSE's and R2 are acceptable, the discharge is congruent with P and ET, the irrigation withdrawals match), further independent scientific ground verification of ET fluxes and rainfall is required in these type of vast and transboundary river basins. Ideally, this should be done for small and instrumented catchments, through the inclusion of more eddy covariance towers and rainfall measurements using recent advances in technologies such as acoustic disdrometers (van de Giesen *et al.*, [89]) or the attenuation of microwave signals between telecommunication towers (e.g., Overeem *et al.*, [90]).

The fluxes, flows and stocks presented in the paper can be used to determine monthly and annual water accounts. The analysis of the Nile basin indicates that most water is produced in rainfed crop areas (69 km^3/yr) followed by areas of forest plantations (22 km^3/yr). The four agro-ecosystems with the highest net withdrawals are forests (94 km^3/yr), wetlands (60 km^3/yr), open water areas (46 km^3/yr) and irrigated crops (34 km^3/yr). Because the net withdrawal by irrigated crops could be validated against third sources, and the total water balance for every sub-basin is closed, all class averaged net withdrawals are believed to be accurate.

Acknowledgement

The authors acknowledge the help and data received from the U.S. Geological Survey Earth Resources Observation and Science Center (SSEBop model) and eLEAF Competence Center (ETLook model). This work has been undertaken as part of the CGIAR Research Program on Water, Land and Ecosystems.

Author Contributions

All authors contributed extensively to this manuscript.

Conflicts of Interest

The authors declare no conflict of interest.

240

Appendix

Table A1. Comprehensive list of rainfall and ET by land use-land cover class. The classes are coded by the standard defined by Water Accounting Plus (WA+).

WA+ Code	Description	LULC	Area (km²)	ET (mm/yr)	ET (km³/yr)	P (mm/yr)	P (km³/yr)	P-ET (mm)	P-ET (km³/yr)
PLU4	Sand dunes	Bare land	196017	91.4	17.9	64.1	12.6	−27.3	−5.3
ULU1	Closed trees with closed to open shrubs	Forests	9802	1182.2	11.6	1290.3	12.6	108.1	1.1
ULU2	Closed multilayered trees (broadleaved evergreen)	Forests	7109	1154.6	8.2	1286.4	9.1	131.7	0.9
ULU3	Open trees with open shrubs	Forests	268360	1060.7	284.7	1038.7	278.7	−22.1	−5.9
ULU4	Closed shrubs	Savannah	116480	1068.7	124.5	1096.5	127.7	27.8	3.2
ULU5	Open general shrubs with closed to open herbaceous	Savannah	342789	845.0	289.7	866.8	297.1	21.8	7.5
ULU6	Closed shrubs with sparse trees	Savannah	312953	693.8	217.1	738.6	231.1	44.8	14.0
ULU7	Closed low trees with closed to open shrubs	Savannah	120445	1030.2	124.1	998.2	120.2	−32.0	−3.9
ULU8	Sparse herbaceous	Pastures	419102	405.8	170.1	415.6	174.2	9.8	4.1
ULU9	Closed to very open grassland	Pastures	22138	758.0	16.8	936.4	20.7	178.4	4.0
ULU10	River bank	Wetlands	535	751.8	0.4	193.1	0.1	−558.8	−0.3
ULU11	Bare soil stony (deep soil)	Bare land	117213	33.1	3.9	30.7	3.6	−2.3	−0.3
ULU12	Bare soil stony under reclamation	Bare land	23530	386.9	9.1	384.2	9.0	−2.7	−0.1
ULU14	Bare rock with a thin sand layer	Bare land	396767	51.9	20.6	39.6	15.7	−12.2	−4.9
ULU16	River	Open water	4423	967.7	4.3	409.8	1.8	−557.9	−2.5
ULU17	Natural lakes	Open water	85066	1469.3	125.0	1382.9	117.6	−86.4	−7.4
ULU18	Post Flooding Herbaceous Crop, Medium Fields	Wetlands	25093	1111.3	27.9	938.4	23.5	−172.9	−4.3
ULU19	Salt fields	Sinks	987	297.0	0.3	146.5	0.1	−150.5	−0.1
ULU20	Closed medium herbaceous on permanently flooded land - brackish water	Wetlands	112	787.4	0.1	74.6	0.0	−712.8	−0.1
ULU21	Bare soil	Bare land	142325	59.8	8.5	49.5	7.0	−10.3	−1.5
ULU24	Open general woody with closed to open herbaceous on temporarily flooded land	Wetlands	77743	1231.0	95.7	980.5	76.2	−250.5	−19.5
ULU25	Closed trees on permanently flooded land - fresh water	Wetlands	9166	1286.6	11.8	894.6	8.2	−392.0	−3.6
MLU1	Forest Plantation	Plantations	74806	850.0	63.6	1150.0	86.0	300.0	22.4
MLU2	Rainfed Tree Crop	Rainfed crops	189204	869.4	164.5	1142.8	216.2	273.5	51.7
MLU4	Rainfed Herbaceous Crop	Rainfed crops	8952	752.4	6.7	350.7	3.1	−401.8	−3.6

Table A1. *Cont.*

WA+ Code	Description	LULC	Area (km²)	ET (mm/yr)	ET (km³/yr)	P (mm/yr)	P (km³/yr)	P-ET (mm)	P-ET (km³/yr)
MLU5	Rainfed Herbaceous Crop	Rainfed crops	107661	541.4	58.3	618.1	66.5	76.7	8.3
MLU6	Rainfed Shrub Crop/orchard	Rainfed crops	2634	904.1	2.4	1103.2	2.9	199.1	0.5
MLU7	Rainfed Herbaceous Crop	Rainfed crops	70564	732.2	51.7	895.9	63.2	163.7	11.6
MLU8	Rainfed Shrub Crop	Rainfed crops	1165	866.1	1.0	1077.7	1.3	211.6	0.2
MLU12	Dumps / deposits	Urban areas	3	177.7	0.0	147.7	0.0	−30.1	0.0
MLU14	Airport	Urban areas	94	401.9	0.0	360.5	0.0	−41.4	0.0
MLU16	Urban areas	Urban areas	3490	503.3	1.8	60.3	0.2	−443.0	−1.5
MLU17	Rural settlements	Urban areas	1153	386.0	0.4	372.4	0.4	−13.6	0.0
MWU1	Irrigated Herbaceous Crop	Irrigated crops	12983	744.3	9.7	287.8	3.7	−456.5	−5.9
MWU2	Irrigated Herbaceous − Cereal	Irrigated crops	9275	835.5	7.7	351.2	3.3	−484.3	−4.5
MWU2	Irrigated Herbaceous Crop (1 add. Crop) Large to Medium Fields – Maize, Clover	Irrigated crops	5621	839.2	4.7	41.7	0.2	−797.5	−4.5
MWU3	Irrigated Orchard, Small Fields - Citrus spp.	Irrigated crops	17698	822.3	14.6	428.3	7.6	−394.0	−7.0
MWU4	Irrigated Herbaceous Crop (1 add. Crop) Small Fields	Irrigated crops	6449	921.0	5.9	68.8	0.4	−852.2	−5.5
MWU5	Irrigated Forest Plantation - Eucalyptus	Irrigated crops	2707	682.2	1.8	75.7	0.2	−606.5	−1.6
MWU6	Artificial Lakes or Reservoirs	Open water	12	1156.3	0.0	461.9	0.0	−694.4	0.0
MWU8	Snow	Open water	5918	1642.6	9.7	46.1	0.3	−1596.5	−9.4
MWU10	Fish Pond	Open water	381	391.3	0.1	66.2	0.0	−325.1	−0.1
MWU11	Refugee camp	Urban areas	33	765.7	0.0	821.9	0.0	56.2	0.0
MWU15	Urban Areas Vegetated	Urban areas	25	772.8	0.0	806.5	0.0	33.7	0.0
MWU17	Open trees with closed to open herbaceous on temporarily flooded land	Wetlands	5443	984.4	5.4	1101.4	6.0	117.0	0.6
MWU18	Very open trees with closed to open shrubs on temporarily flooded land - fresh water	Wetlands	4614	1105.2	5.1	1080.4	5.0	−24.7	−0.1
Total			3229039		1987.3		2014.1		26.8

References

1. Rockstrom, J.; Steffen, W.; Noone, K.; Persson, A.; Chapin, F.S.; Lambin, E.F.; Lenton, T.M.; Scheffer, M.; Folke, C.; Schellnhuber, H.J.; *et al.* A safe operating space for humanity. *Nature* **2009**, *461*, 472–475.
2. Alexandratos, N.; Bruinsma, J. *World Agriculture towards 2030/2050: The 2012 Revision*; FAO: Rome, Italy, 2012.
3. Rijsberman, F.R. Water scarcity: Fact or fiction? *Agric. Water Manage.* **2006**, *80*, 5–22.
4. Hilhorst, B.; Burke, J.; Hoogeveen, J.; Fremken, K.; Faures, J.-M.; Gross, D. *Information Products for Nile Basin Water Resources Management*; FAO: Rome, Italy, 2011; p. 130.

5. Thenkabail, P.S.; Knox, J.W.; Ozdogan, M.; Gumma, M.K.; Congalton, R.G.; Wu, Z.; Milesi, C.; Finkral, A.; Marshall, M.; Mariotto, I.; *et al.* Assessing future risks to agricultural productivity, water resources and food security—How can remote sensing help? *Photogramm. Eng. Remote Sens.* **2012**, *78*, 773–782.

6. Rebelo, L.-M.; Johnston, R.; Karimi, P.; McCornick, P.G. Determining the dynamics of agricultural water use: Cases from Asia and Africa. *J. Contemporary Water Res. Education* **2014**, *153*, 79–90.

7. De Fraiture, C.; Molden, D.; Wichelns, D. Investing in water for food, ecosystems, and livelihoods: An overview of the comprehensive assessment of water management in agriculture. *Agric. Water Manage.* **2010**, *97*, 495–501.

8. Alcamo, J.; Döll, P.; Kaspar, F.; Siebert, S. *Global Change and Global Scenarios of Water Use and Availability: An Application of Watergap1.0*; Center for Environmental Systems Research (CESR), University of Kassel: Kassel, Germany, 1997.

9. Seckler, D.; Amarasinghe, U.; Molden, D.; Silva, R.d.; Barker, R. *World Water Demand and Supply, 1990 to 2025: Scenarios and Issues*; International Water Management Institute: Colombo, Sri Lanka, 1998.

10. Shiklomanov, I.A. Appraisal and assessment of world water resources. *Water Int.* **2000**, *25*, 11–32.

11. Wallace, J.S. Increasing agricultural water use efficiency to meet future food production. *Agric. Ecosyst. Environ.* **2000**, *82*, 105–119.

12. Vörösmarty, C.J.; Green, P.; Salisbury, J.; Lammers, R.B. Global water resources: Vulnerability from climate change and population growth. *Science* **2000**, *289*, 284–288.

13. Droogers, P.; Immerzeel, W.W.; Terink, W.; Hoogeveen, J.; Bierkens, M.F.P.; van Beek, L.P.H.; Debele, B. Water resources trends in middle east and north africa towards 2050. *Hydrol. Earth Syst. Sci. Discuss.* **2012**, *16*, 3101–3114.

14. Bosch, J.M.; Hewlett, J.D. A review of catchment experiments to determine the effect of vegetation changes on water yield and evapotranspiration. *J. Hydrol.* **1982**, *55*, 3–23.

15. Smakhtin, V.; Revenga, C.; Döll, P. *Taking into Account Environmental Water Requirements in Global-Scale Water Resources Assessments*; Comprehensive Assessment Secretariat: Colombo, Sri Lanka, 2004.

16. Seckler, D. *The New Era of Water Resources Management: From "Dry" to "Wet" Water Savings*; International Irrigation Management Institute: Colombo, Sri Lanka, 1996.

17. Shahin, M. *Assessment of Groundwater Resources in Egypt*; UNESCO-IHE: Delft, The Netherlands, 1991.

18. Sutcliffe, J.V.; Parks, Y.P. *The Hydrology of the Nile*; IAHS Press, Institute of Hydrology: Wallingford, UK, 1999; p. 179.

19. Serrat-Capdevila, A.; Valdes, J.B.; Stakhiv, E.Z. Water management applications for satellite precipitation products: Synthesis and recommendations. *JAWRA J. Am. Water Resour. Assoc.* **2014**, *50*, 509–525.

20. Neal, C.M.U.; Cosh, M.H. *Remote Sensing and Hydrology*; IAHS: Wallingford, UK, 2012.

21. Batelaan, O. Special Issue: Hydrological Remote Sensing. http://www.mdpi.com/journal/remotesensing/special_issues/hydrological (accessed on 15 April 2014).

22. Nagesh Kumar, D.; Reshmidevi, T.V. Remote sensing applications in water resources. *J. Indian Inst. Sci.* **2013**, *93*, 163–187.

23. Karimi, P.; Bastiaanssen, W.G.M. Spatial evapotranspiration, rainfall and land use data in water accounting—Part 1: Review of the accuracy of the remote sensing. *Hydrol. Earth Syst. Sci. Discuss.* **2014**, *11*, 1073–1123.

24. Xie, P.; Arkin, P.A. Global precipitation: A 17-year monthly analysis based on gauge observations, satellite estimates, and numerical model outputs. *Bull. Amer. Meteor. Soc.* **1997**, *78*, 2539–2558.

25. Van den Hurk, B.J.J.M.; Viterbo, P.; Beljaars, A.C.M.; Betts, A.K. *Offline Validation of the Era40 Surface Scheme*; ECMWF: Reading, UK, 2000; p. 43.

26. Balsamo, G.; Beljaars, A.; Scipal, K.; Viterbo, P.; van den Hurk, B.; Hirschi, M.; Betts, A.K. A revised hydrology for the ecmwf model: Verification from field site to terrestrial water storage and impact in the integrated forecast system. *J. Hydrometeorol.* **2009**, *10*, 623–643.

27. Miralles, D.G.; Holmes, T.R.H.; De Jeu, R.A.M.; Gash, J.H.; Meesters, A.G.C.A.; Dolman, A.J. Global land-surface evaporation estimated from satellite-based observations. *Hydrol. Earth Syst. Sci.* **2011**, *15*, 453–469.

28. Mu, Q.; Heinsch, F.A.; Zhao, M.; Running, S.W. Development of a global evapotranspiration algorithm based on modis and global meteorology data. *Remote Sens. Environ.* **2007**, *111*, 519–536.

29. Kim, H.W.; Hwang, K.; Mu, Q.; Lee, S.O.; Choi, M. Validation of modis 16 global terrestrial evapotranspiration products in various climates and land cover types in Asia. *KSCE J. Civ. Eng.* **2012**, *16*, 229–238.

30. Trambauer, P.; Dutra, E.; Maskey, S.; Werner, M.; Pappenberger, F.; van Beek, L.P.H.; Uhlenbrook, S. Comparison of different evaporation estimates over the african continent. *Hydrol. Earth Syst. Sci.* **2014**, *18*, 193–212.

31. Guerschman, J.P.; Van Dijk, A.I.J.M.; Mattersdorf, G.; Beringer, J.; Hutley, L.B.; Leuning, R.; Pipunic, R.C.; Sherman, B.S. Scaling of potential evapotranspiration with modis data reproduces flux observations and catchment water balance observations across australia. *J. Hydrol.* **2009**, *369*, 107–119.

32. Anderson, M.C.; Kustas, W.P.; Norman, J.M.; Hain, C.R.; Mecikalski, J.R.; Schultz, L.; González-Dugo, M.P.; Cammalleri, C.; d'Urso, G.; Pimstein, A.; *et al.* Mapping daily evapotranspiration at field to continental scales using geostationary and polar orbiting satellite imagery. *Hydrol. Earth Syst. Sci.* **2011**, *15*, 223–239.

33. Bastiaanssen, W.G.M.; Cheema, M.J.M.; Immerzeel, W.W.; Miltenburg, I.J.; Pelgrum, H. Surface energy balance and actual evapotranspiration of the transboundary indus basin estimated from satellite measurements and the etlook model. *Water Resour. Res.* **2012**, *48*, W11512.

34. Senay, G.B.; Bohms, S.; Singh, R.K.; Gowda, P.H.; Velpuri, N.M.; Alemu, H.; Verdin, J.P. Operational evapotranspiration mapping using remote sensing and weather datasets: A new parameterization for the sseb approach. *JAWRA J. Am. Water Resour. Assoc.* **2013**, *49*, 577–591.

35. Karimi, P.; Molden, D.; Bastiaanssen, W.G.M.; Cai, X. Water accounting to assess use and productivity of water—Evolution of the concept and new frontiers. In *Water Accounting: International Approaches to Policy and Decision-Making*; Chalmers, K.; Godfrey, J., Eds.; Edgar Elger: Cheltenham, UK, 2012; pp. 76–88.

36. Dickinson, R.E. Land surface processes and climate surface albedos and energy-balance. In *Advances in Geophysics*; Academic Press, Inc.: New York, NY, USA, 1983; Vol. 25, pp. 305–353.

37. Di Gregorio, A.; Jansen, L.J.M. *Land Cover Classification System (LCCS), Classification Concepts and User Manual*; FAO: Rome, Italy, 2000; pp. 80–99.

38. Karimi, P.; Bastiaanssen, W.G.M.; Molden, D. Water accounting plus (WA+)—A water accounting procedure for complex river basins based on satellite measurements. *Hydrol. Earth Syst. Sci.* **2013**, *17*, 2459–2472.

39. Ahmad, M.-U.-D.; Bastiaanssen, W.G.M.; Feddes, R.A. A new technique to estimate net groundwater use across large irrigated areas by combining remote sensing and water balance approaches, Rechna Doab, Pakistan. *Hydrogeol. J.* **2005**, *13*, 653–664.

40. Van Eekelen, M.; Bastiaanssen, W.G.M.; Jarmain, C.; Jackson, B.; Ferreira, F.; Zaag, P.V.d.; Okello, A.S.; Bosch, J.H.; Dye, P.; Bastidas-Obando, E.; *et al.* A novel approach to estimate direct and indirect water withdrawals from satellite measurements: A case study from the incomati basin. *Agric. Ecosyst. Environ.* **2014**, submitted.

41. Schreiber, P. Uber die beziehungen zwischen dem niederschlag und der wasserfuhrung der flube in mitteleuropa. *Z. Meteorol.* **1904**, *21*, 441–452.

42. Budyko, M.I. *Climate and Life*; Academic: Orlando, FL, USA, 1974.

43. Gerrits, A.M.J.; Savenije, H.H.G.; Veling, E.J.M.; Pfister, L. Analytical derivation of the budyko curve based on rainfall characteristics and a simple evaporation model. *Water Resour. Res.* **2009**, *45*, W04403.

44. Crétaux, J.F.; Jelinski, W.; Calmant, S.; Kouraev, A.; Vuglinski, V.; Bergé-Nguyen, M.; Gennero, M.C.; Nino, F.; Abarca Del Rio, R.; Cazenave, A.; *et al.* Sols: A lake database to monitor in the near real time water level and storage variations from remote sensing data. *Adv. Space Res.* **2011**, *47*, 1497–1507.

45. Duan, Z.; Bastiaanssen, W.G.M. Estimating water volume variations in lakes and reservoirs from four operational satellite altimetry databases and satellite imagery data. *Remote Sens. Environ.* **2013**, *134*, 403–416.

46. Syed, T.H.; Famiglietti, J.S.; Rodell, M.; Chen, J.; Wilson, C.R. Analysis of terrestrial water storage changes from GRACE and GLDAS. *Water Resour. Res.* **2008**, *44*, W02433.

47. Klees, R.; Liu, X.; Wittwer, T.; Gunter, B.C.; Revtova, E.A.; Tenzer, R.; Ditmar, P.; Winsemius, H.C.; Savenije, H.H.G. A comparison of global and regional grace models for land hydrology. *Sur. Geophys.* **2008**, *29*, 335–359.

48. Rodell, M.; Velicogna, I.; Famiglietti, J.S. Satellite-based estimates of groundwater depletion in India. *Nature* **2009**, *460*, 999–1002.

49. Funk, C.C.; Peterson, P.J.; Landsfeld, M.F.; Pedreros, D.H.; Verdin, J.P.; Rowland, J.D.; Romero, B.E.; Husak, G.J.; Michaelsen, J.C.; Verdin, A.P. *A Quasi-Global Precipitation Time Series for Drought Monitoring*; U.S. Geological Survey Data Series 832 ed.; 2014.

50. Karimi, P.; Molden, D.; Notenbaert, A.; Peden, D. Nile basin farming systems and productivity. In *The Nile River Basin, Water, Agriculture, Governance and Livelihoods*; Chalmers, K., Godfrey, J., Eds.; Edgar Elger: Cheltenham, UK, 2012; pp. 133–153.

51. Duan, Z.; Bastiaanssen, W.G.M. First results from version 7 TRMM 3b43 precipitation product in combination with a new downscaling–calibration procedure. *Remote Sens. Environ.* **2013**, *131*, 1–13.

52. Hunink, J.E.; Immerzeel, W.W.; Droogers, P. A high-resolution precipitation 2-step mapping procedure (hip2p): Development and application to a tropical mountainous area. *Remote Sens. Environ.* **2014**, *140*, 179–188.

53. Alemu, H.; Senay, G.B.; Kaptue, A.T.; Kovalskyy, V. Evapotranspiration variability and its association with vegetation dynamics in the Nile Basin, 2002–2011. *Remote Sens.* **2014**, *6*, 5885–5908.

54. Senay, G.B.; Gowda, P.H.; Bohms, S.; Howell, T.A.; Friedrichs, M.; Marek, T.H.; Verdin, J.P. Evaluating the ssebop approach for evapotranspiration mapping with landsat data using lysimetric observations in the semi-arid texas high plains. *Hydrol. Earth Syst. Sci. Discuss.* **2014**, *11*, 723–756.

55. Velpuri, N.M.; Senay, G.B.; Singh, R.K.; Bohms, S.; Verdin, J.P. A comprehensive evaluation of two MODIS evapotranspiration products over the conterminous united states: Using point and gridded fluxnet and water balance ET. *Remote Sens. Environ.* **2013**, *139*, 35–49.

56. Singh, R.; Senay, G.; Velpuri, N.; Bohms, S.; Scott, R.; Verdin, J. Actual evapotranspiration (water use) assessment of the colorado river basin at the landsat resolution using the operational simplified surface energy balance model. *Remote Sens.* **2014**, *6*, 233–256.

57. Senay, G.B.; Budde, M.E.; Verdin, J.P. Enhancing the simplified surface energy balance (SSEB) approach for estimating landscape ET: Validation with the METRIC model. *Agric. Water Manage.* **2011**, *98*, 606–618.

58. Gowda, P.H.; Senay, G.B.; Colaizzi, P.D.; Howell, T.A. Lysimeter validation of the simplified surface energy balance (SSEB) approach for estimating actual ET. *Trans. ASABE* **2009**, *25*, 665–669.

59. Senay, G.B.; Velpuri, M.N.; Bohms, S.; Demissie, Y.; Gebremichael, M. Understanding the hydrologic sources and sinks in the Nile Basin using multi-source climate and remote sensing data sets. *Water Resour. Res.* **2014**, submitted.

60. Allen, R.; Tasumi, M.; Trezza, R. Satellite-based energy balance for mapping evapotranspiration with internalized calibration (METRIC)—model. *J. Irrig. . Drain. Eng.* **2007**, *133*, 380–394.

61. Bastiaanssen, W.G.M.; Menenti, M.; Feddes, R.A.; Holtslag, A.A.M. A remote sensing surface energy balance algorithm for land (SEBAL). 1. Formulation. *J. Hydrol.* **1998**, *212–213*, 198–212.

62. Immerzeel, W.W.; Droogers, P.; Terink, W.; Hoogeveen, J.; Hellegers, P.J.J.G.; Bierkens, M.F.P.; Beek, L.P.H.v. *Middle-East and Northern Africa, Water Outlook*; FutureWater: Wageningen, The Netherlands, 2011; p. 136.

63. Bastiaanssen, W.G.M.; Hoekman, D.H.; Roebeling, R.A. *A Methodology for the Assessment of Surface Resistance and Soil Water Storage Variability at Mesoscale Based on Remote Sensing Measurements: A Case Study with Hapex-Efeda Data*; International Association of Hydrological Sciences: Wallingford, UK, 1994.

64. Mohamed, Y.A.; Bastiaanssen, W.G.M.; Savenije, H.H.G. Spatial variability of evaporation and moisture storage in the swamps of the upper nile studied by remote sensing techniques. *J. Hydrol.* **2004**, *289*, 145–164.

65. Allen, R.G.; Pereira, L.S.; Raes, D.; Smith, M. *Crop Evapotranspiration—Guidelines for Computing Crop Water Requirements*; FAO: Rome, Italy, 1998.

66. Kirby, M.; Eastham, J.; Mainuddin, M. *Water-Use Accounts in CPWF Basins: Simple Water-Use Accounting of the Nile Basin*; The CGIAR Challenge Program on Water and Food: Colombo, Sri Lanka, 2010.

67. NWRP Project. *National Water Resources Plan 2017*; Arab Republic of Egypt, Ministry of Water Resources and Irrigation: Cairo, Egypt, 2005.

68. Dai, A.; Trenberth, K.E. New estimates of continental discharge and oceanic freshwater transport. In Proceedings of JP1.11 AMS Symposium on Observing and Understanding the Variability of Water in Weather and Climate, Long Beach, CA, USA, 9–13 February 2003.

69. Awulachew, S.R., Lisa-Maria; Molden, David. The nile basin: Tapping the unmet agricultural potential of nile waters. *Water Int.* **2010**, *35*, 623–654.

70. El-Shabrawy, G.M. Lake Nasser-Nubia. In *The Nile: Origin, Environments, Limnology and Human Use*; Dumont, H.J., Ed.; Springer: Dordrecht, The Netherlands, 2009; pp. 125–156.

71. Johnston, R. Availability of water for agriculture in the nile basin. In *The Nile River Basin, Water, Agriculture, Governance and Livelihoods*; Awulachew, S.B., Smakhtin, V., Molden, D., Peden, D., Eds.; Routledge - Earthscan: Abingdon, UK, 2012; pp. 61–83.

72. MacAlister, C.; Pavelic, P.; Tindimugaya, C.; Ayenew, T.; Ibrahim, M.E.; Meguid, M.A. Overview of groundwater in the nile river basin. In *The Nile River Basin, Water, Agriculture, Governance and Livelihoods*; Awulachew, S.B., Smakhtin, V., Molden, D.; Peden, D., Eds.; Routledge - Earthscan: Abingdon, UK, 2012; pp 186–211.

73. Bonsor, H.C.; Mansour, M.M.; MacDonald, A.M.; Hughes, A.G.; Hipkin, R.G.; Bedada, T. Interpretation of grace data of the nile basin using a groundwater recharge model. *Hydrol. Earth Syst. Sci. Discuss.* **2010**, *7*, 4501–4533.

74. Elsawwaf, M.; Feyen, J.; Batelaan, O.; Bakr, M. Groundwater–surface water interaction in lake Nasser, Southern Egypt. *Hydrol. Processes* **2014**, *28*, 414–430.

75. Faures, J.-M.; Svendsen, M.; Turral, H. Reinventing irrigation. In *Water for Food, Water for Life, a Comprehensive Assessment of Water in Agriculture*; Molden, D.J., Ed.; IWMI and Earthscan: London, UK, 2007; pp. 353–394.

76. Dastane, N.G. *Effective Rainfall in Irrigated Agriculture*; FAO: Rome, Italy, 1974.

77. Rebelo, L.-M.; McCartney, M. Wetlands of the Nile Basin, distributions, functions and contribution to livelihoods. In *The Nile River Basin: Water, Agriculture, Governance and Livelihoods*; Awulachew, S.B., Smakhtin, V., Molden, D., Peden, D., Eds.; Routledge - Earthscan: Abingdon, UK, 2012; pp. 212–228.

78. Awulachew, S.B.; Demissie, S.S.; Hagos, F.; Erkossa, T.; Peden, D. Water management intervention analysis in the nile basin. In *The Nile River Basin: Water, Agriculture, Governance and Livelihoods*; Awulachew, S.B., Smakhtin, V., Molden, D., Peden, D., Eds.; Routledge - Earthscan: Abingdon, UK, 2012; pp. 292–316.

79. Bastiaanssen, W.; Perry, C. *Agricultural Water Use and Water Productivity in the Large Scale Irrigation (LSI) Schemes of the Nile Basin*; Nile Basin Initiative (NBI): Entebbe, Uganda, 2009.

80. FAO. *Assessment of the Irrigation Potential of the Nile Basin*; Land and Water Division, FAO: Rome, Italy, 1997; p. 41.

81. Allen, R.G.; Burnett, B.; Kramber, W.; Huntington, J.; Kjaersgaard, J.; Kilic, A.; Kelly, C.; Trezza, R. Automated calibration of the metric-landsat evapotranspiration process. *JAWRA J. Am. Water Resour. Assoc.* **2013**, *49*, 563–576.

82. Morton, C.G.; Huntington, J.L.; Pohll, G.M.; Allen, R.G.; McGwire, K.C.; Bassett, S.D. Assessing calibration uncertainty and automation for estimating evapotranspiration from agricultural areas using metric. *JAWRA J. Am. Water Resour. Assoc.* **2013**, *49*, 549–562.

83. Van der Ent, R.J.; Savenije, H.H.G.; Schaefli, B.; Steele-Dunne, S.C. Origin and fate of atmospheric moisture over continents. *Water Resour. Res.* **2010**, *46*, W09525.

84. De Lange, R.; Beck, R.; van de Giesen, N.; Friesen, J.; de Wit, A.; Wagner, W. Scatterometer-derived soil moisture calibrated for soil texture with a one-dimensional water-flow model. *IEEE Trans. Geosci. Remote Sens.* **2008**, *46*, 4041–4049.

85. Bastiaanssen, W.G.M. *Water Accountants: De Nieuwe Generatie Waterbeheercontroleurs, Intreerede*; Delft University of Technology: Delft, The Netherlands, 2009.

86. UN. *System of Environmental Economic Accounting for Water*; United Nations: Geneva, Switzerland, 2007.

87. ABS. *Water Account Australia 2000-01*; Australian Bureau of Statistics: Canberra, ACT, Australia, 2004.

88. Crouzet, P.; Gall, G.L.; Campling, P.; Basso, M.; Weber, J.L.; Gomez, O.; Kurnik, B. *Results and Lessons from Implementing the Water Assets Accounts in the Eea Area, from Concept to Production*; European Environmental Agency (EEA): Copenhagen, Denmark, 2013.

89. Van de Giesen, N.; Hut, R.; Selker, J. The Trans-African Hydro-Meteorological Observatory (TAHMO). *Wiley Interdisciplinary Reviews: Water* **2014**, *1*, 341–348.

90. Overeem, A.; Leijnse, H.; Uijlenhoet, R. Country-wide rainfall maps from cellular communication networks. *Proc. Natl. Acad. Sci. USA* **2013**, *110*, 2741–2745.

Application of a Remote Sensing Method for Estimating Monthly Blue Water Evapotranspiration in Irrigated Agriculture

Mireia Romaguera, Maarten S. Krol, Mhd. Suhyb Salama, Zhongbo Su and Arjen Y. Hoekstra

Abstract: In this paper we show the potential of combining actual evapotranspiration (ET_{actual}) series obtained from remote sensing and land surface modelling, to monitor community practice in irrigation at a monthly scale. This study estimates blue water evapotranspiration (ET_b) in irrigated agriculture in two study areas: the Horn of Africa (2010–2012) and the province of Sichuan (China) (2001–2010). Both areas were affected by a drought event during the period of analysis, but are different in terms of water control and storage infrastructure. The monthly ET_b results were separated by water source—surface water, groundwater or conjunctive use—based on the Global Irrigated Area Map and were analyzed per country/province. The preliminary results show that the temporal signature of the total ET_b allows seasonal patterns to be distinguished within a year and inter-annual ET_b dynamics. In Ethiopia, ET_b decreased during the dry year, which suggests that less irrigation water was applied. Moreover, an increase of groundwater use was observed at the expense of surface water use. In Sichuan province, ET_b in the dry year was of similar magnitude to the previous years or increased, especially in the month of August, which points to a higher amount of irrigation water used. This could be explained by the existence of infrastructure for water storage and water availability, in particular surface water. The application presented in this paper is innovative and has the potential to assess the existence of irrigation, the source of irrigation water, the duration and variability in time, at pixel and country scales, and is especially useful to monitor irrigation practice during periods of drought.

Reprinted from *Remote Sens*. Cite as: Romaguera, M.; Krol, M.S.; Salama, M.S.; Su, Z.; Hoekstra, A.Y. Application of a Remote Sensing Method for Estimating Monthly Blue Water Evapotranspiration in Irrigated Agriculture. *Remote Sens*. **2014**, *6*, 10033-10050.

1. Introduction

The assessment of water use is crucial in a changing environment in which water is an essential but scarce resource. From a water management perspective, an accurate evaluation of the irrigation water used in agriculture is of high importance. The AQUASTAT database [1] shows a wide range of values on water withdrawal for irrigation, with values ranging for example from 0.6% of total national water withdrawal in the Netherlands to 60% or 85% in Spain and Tanzania, respectively.

Crop water use or evapotranspiration (ET_{actual}) has traditionally been separated into a "green" and "blue" component, referring to the origin of the used water: precipitation or irrigation water, respectively. Early studies estimated blue and/or green water use at country, continental or global levels [2–5]. Later studies made global estimates of consumptive water use for a number of specific crops per country [6–9]. At a global scale and higher spatial resolution, Alcamo *et al.* [10]

estimated blue water withdrawal and Döll and Siebert [11] the irrigation water requirements. More recently, a few studies estimated global green and blue water consumption in crop production at spatial resolutions of 30 and 5 arc minutes [12–20].

The aforementioned approaches used hydrological models with the objective of estimating actual evapotranspiration from croplands per crop type, distinguishing between blue and green ET_{actual}. However, the input used and the type of output produced, differed. The results were calculated and presented at different spatial resolutions and covered different time periods. The inputs of the methods were national statistics, reports, climatic databases and crop-related maps. The spatial and temporal resolutions of the source data were coarse in some cases, especially where extracted from statistical databases, implying in some cases the use of disaggregation techniques.

Bearing this in mind, remote sensing techniques may improve the estimates of blue and green water use since they provide global coverage, varied temporal and spatial resolution and broad spectral information. This allows characterizing the physical processes and monitoring crops in appropriate space and time scales. In this context, Romaguera et al. [21,22] included the use of remote sensing data and proposed a methodology to estimate blue water evapotranspiration (ET_b) that could benefit from the remote sensing advantages. This method allows the estimation of ET_b at different time scales, i.e., hourly, daily, monthly and yearly, which is supposedly an improvement with respect to the existing static maps for monitoring irrigation practice. At regional scale, other works used remote sensing to evaluate irrigation performance [23–25].

Moreover, in recent years, several studies have approached the problem of global irrigation mapping, using national statistical data as input [26,27] or making use of spectral and temporal remote sensing data to perform classifications and obtain irrigated areas [28,29]. These methods provide information about areas equipped for irrigation, about crop dominance and irrigation source, and about existence or absence of irrigation, but none of the methods quantifies the actual amount of water received by the crops through irrigation, or blue water. In particular, the source of irrigation water was determined by Thenkabail et al. [29,30] in their Global Irrigated Area Map (GIAM), where irrigated areas were classified as a function of three sources of irrigation supply: surface water, groundwater, and conjunctive use (due to usage of stored rain water).

The objective of this paper is to apply the remote sensing method by Romaguera et al. [21,22] and obtain ET_b values at relevant time scales for water management purposes, that is at monthly and country/province scale, as well as to show preliminary results and the potential of exploiting these data when combined with the source of irrigation water, from the aforementioned GIAM map. The regions and period of study are the Horn of Africa (period 2010–2012) and the Chinese province of Sichuan (period 2001–2010), both affected by a drought event during the period of study, but with differences in terms of water control and storage infrastructure.

Section 2 describes the method and datasets used in this paper and Section 3 the selected study areas. Section 4 includes ET_b time series per source of irrigation water in the study areas and a sensitivity analysis. Section 5 discusses relevant aspects of the application tackled in this research and finally the conclusions of this work are summarized.

2. Method and Data

The method to estimate ET_b used in this paper is described in Romaguera *et al.* [21,22]. It is based on the calculation of the differences in actual evapotranspiration (ET_{actual}) given by remotely sensed ET_{actual} data (RS–ET in the following) and the Global Land Data Assimilation System (GLDAS) ET_{actual} model simulations (GLDAS–ET in the following). The former included the effect of irrigation where relevant, whereas irrigation was not incorporated in GLDAS simulations. A bias between the two datasets is calculated in rain-fed croplands, where no irrigation is supplied, and then used to correct the whole dataset, obtaining ET_b as:

$$ET_b = \Delta ET - bias \tag{1}$$

where ΔET is the difference between RS–ET and GLDAS–ET and *bias* is this difference calculated only in rain-fed croplands. The idea behind this formulation relies on the fact that GLDAS–ET products do not account for extra water supply in form of irrigation in the land surface model [31], whereas RS–ET are based on the energy balance and therefore are able to observe full ET_{actual} from croplands, including all sources of water. Therefore, the difference between the two datasets provides information about the water used in the form of irrigation.

2.1. Bias Estimation

Since the two datasets present systematic discrepancies, rain-fed croplands were used to calculate a reference bias to correct for this effect and isolate the differences due to irrigation practices. The GlobCover land cover map (version 2.3) [32] was used to identify rain-fed croplands.

Previous literature showed temporal and spatial variations of this bias [21,33]. For example, in Europe the bias amplitude changed through the year roughly resembling a positive concave curve. The maximum amplitude value reached up to 3 mm/day and occurred in the months of spring and summer in northern latitudes [21]. In that paper, the spatial variability of the bias was taken into account by performing a classification of the study area and calculating the spatial mean bias per class and per month. Normalized Difference Vegetation Index (NDVI) and satellite observation angle were the input parameters for the classification. The validity of the bias curves obtained was carried out by analyzing their representativeness in bigger areas, providing satisfactory results in majority classes.

The classification scheme was improved in recent literature [22] by testing different classification approaches and proposing a new set of input parameters. This allowed to obtain a better differentiation of the bias curves, reduced the standard deviation of the data and captured the expected variability of the maximum bias.

Therefore, following Romaguera *et al.* [22], in the present work a yearly classification of every study area was carried out with the k-means algorithm and using the following parameters as inputs: a yearly climatic indicator (CI) based on net radiation and precipitation, the maximum value of monthly ET_{actual} along the year (ET_{mmax}), the month where the ET_{mmax} occurs (t_ET_{mmax}) and the maximum NDVI ($NDVI_{max}$) in the year of interest. The optimal number of classes was calculated using a scattering distance (SD) quality index [34].

For every year and area, a classification was generated and biases per month were obtained by spatially averaging the bias obtained in rain-fed croplands per class. Finally, Equation (1) was used in the study areas to calculate the total ET$_b$ per month and the GIAM map to assign the source of irrigation water per pixel.

2.2. Data

Table 1 describes the main characteristics of the datasets used in the present work which are detailed in the following paragraphs.

Remote sensing ET$_{actual}$ estimates were obtained from two sources: the Meteosat Second Generation products provided by the Land Surface Analysis–Satellite Applications Facility (LSA–SAF) [35] for the region of Africa (period 2010–2012) and the dataset produced by Chen *et al.* [36,37] over China during the years 2001 till 2010. The periods of study and areas were (partially) determined by the availability of data at the moment of writing this paper. The inclusion of the region of China allowed the analysis of a longer time series of data, which was limited in the Meteosat products over Africa, and also allowed the estimation of ET$_b$ in a region with more extensive irrigation practices and infrastructure, which is China.

Table 1. Specification of the datasets used in the present work.

Data	Source	Spatial Coverage	Spatial Resolution	Temporal Resolution	Details
ET$_{actual}$	LSA–SAF *	MSG disk **	3 km at nadir	daily	Availability of data: Europe: Jan. 2007–present The rest: Sept. 2009–present Used for the study area in Africa
	Chen *et al.* [36,37] (SEBS model)	China	0.1°	monthly	Availability of data: Years 2001–2010 Used for the study area in China
	GLDAS (Noah model)	Global	0.25° (~30 km at equator)	monthly	Availability of data: March 2000–present
Land Cover	MERIS	Global	300 m	Static	GlobCover map calculated in year 2009
R$_n$, P	GLDAS (Noah model)	Global	0.25° (~30 km at equator)	monthly	Availability of data: February 2000–present
NDVI	AVHRR	Africa	1 km	monthly	Generated by IGBP Period: April 1992–March 1993 Used for the study area in Africa
	SPOT–VEG	Global	1 km	monthly	Used for the study area in China
Irrigation source	GIAM	Global	10 km	Static	Data: Type of irrigation Primary data used: —AVHRR from 1997–1999 —TOA NDVI from 1982–2000

* List of acronyms: LSA–SAF (Land Surface Analysis–Satellite Applications Facility); MSG (Meteosat Second Generation); GLDAS (Global Land Data Assimilation System); SEBS (Surface Energy Balance System); MERIS (Medium Resolution Imaging Spectrometer); R$_n$ (Net Radiation); P (Precipitation); Normalized Difference Vegetation Index (NDVI); AVHRR (Advanced Very High Resolution Radiometer); IGBP (International Geosphere–Biosphere Programme Data); SPOT–VEG (Satellite Pour l'Observation de la Terre–Vegetation); GIAM (Global Irrigated Area Map); TOA (Top Of Atmosphere); ** Meteosat disk covers latitudes between −60° and +60° and longitudes between −60° to +60°.

The MSG ET$_{actual}$ model is a simplified Soil–Vegetation–Atmosphere Transfer (SVAT) scheme that uses as input a combination of remote sensed data and atmospheric model outputs. The inputs based on remote sensing are LSA–SAF products of albedo, and downwelling short and longwave radiation fluxes [35,38].The dataset from Chen *et al.* [37] is based on the Surface Energy Balance System (SEBS) [39], which uses multi-sensor remote sensing based NDVI, albedo, surface emissivity and temperature.

Simulated ET$_{actual}$ data with the Noah model [40] were acquired from the Global Land Data Assimilation System (GLDAS) [41]. The Noah land surface model is a 1D column model that describes the physical processes of the soil, vegetation and snowpack. The inputs of this model are satellite and ground-based observational data. The calculation of the latent (LE) and sensible (H) heat flux start from potential LE (LE$_p$), based on the soil moisture, atmosphere states, and vegetation characteristics. Constrains to LE$_p$ are applied resulting in the actual LE and ET$_{actual}$.

The GlobCover land cover map (version 2.3) [32] was used to identify rain-fed croplands. This map is based on classification techniques which use the surface reflectance observed by the Medium Resolution Imaging Spectrometer (MERIS).

The inputs for the classification of the study areas were obtained from the following sources. Net radiation (R$_n$) (as a sum of longwave and shortwave radiation) and precipitation (P) (as a sum of rainfall and snowfall rate) were also taken from the GLDAS dataset. These were used to calculate the climatic indicator as the ratio LP/R$_n$, where L(J/kg) is the latent heat of vaporization, P (mm) is the annual precipitation and R$_n$ (W/m^2) is the annual net radiation. The monthly ET$_{actual}$ used for the classification was taken from GLDAS. Data on NDVI was obtained from the Advanced Very High Resolution Radiometer (AVHRR) delivered by the Deutsches Zentrum für Luft- und Raumfahrt (DLR) and from the Satellite Pour l'Observation de la Terre (SPOT–Vegetation). These NDVI sources were selected as inputs for the classification because they are the ones used for the RS–ET estimations, and their values may influence the differences/biases between RS–ET and model simulations.

The Global Irrigated Area Map by Thenkabail *et al.* [30] was used to identify the source of irrigation, *i.e.*, surface water, groundwater or conjunctive use. This map shows global irrigated areas and classifies them depending on the type of irrigation. The "surface water" (SW) class includes major and medium irrigation from surface water based on large and medium dams. The "groundwater" (GW) class describes minor irrigation from groundwater, small reservoirs and tanks. The "conjunctive use" (CU) class comprises predominately minor irrigation from groundwater, small reservoirs and tanks, but with some mix of surface water irrigation from major reservoirs. This map was generated using classification techniques whose input data were remote sensing based reflectivity, NDVI, rainfall, tree cover and elevation, combined with ground data and Google Earth imagery.

From a technical point of view the inputs were resampled to a common grid and projection, and the resolution of remote sensing data was chosen to calculate the ET$_b$ results, that is 0.030 and 0.1 degree for the Horn of Africa and the Chinese region respectively. The separation of SW, GW and CU was carried out at the resolution of the GIAM map, which is 10 km. The temporal resolution of a month was chosen in this analysis. In order to homogenize the data, daily ET_{actual} values from MSG were monthly aggregated.

3. Study Areas

Based on the availability of remote sensing data, two study areas, both affected by a drought event during the period of study, but with differences in terms of water control/storage infrastructure were selected. First, the Horn of Africa was affected by a drought in the year 2011 [42,43]. In particular, Ethiopia is considered a water scarce country. Despite the abundance of water in some parts of the country (central, western and southwestern parts), the distribution and availability of water is erratic both in space and time due to the lack of water control/storage infrastructures [44]. Strategies have been implemented at national level to improve in this direction, like the Irrigation and Drainage Project [45].

Secondly, China is a country with abundant water resources where dams and reservoirs are numerous, built for hydropower generation, flood control, irrigation and drought mitigation. In particular, in the province of Sichuan we can find the Dujiangyan irrigation project [46], a more than 2000 year old system that was developed to prevent flood and nowadays is crucial in draining off flood water, irrigating farms and provide water resources for more than 50 cities in the province. This region suffered a severe drought in 2006 [47,48].

Figure 1 shows the GIAM map, location and size of the study areas. For the sake of comparison, the neighboring countries/provinces were included in the study area, which computed a total of 1,680,000 and 875,000 km^2 in the regions of East Africa and Southwest of China respectively. Based on this map, irrigated areas were scarce in the Horn of Africa, mainly concentrated in the center and middle-north of Ethiopia, middle-west and southeast of Kenya and in the coastal areas of south Somalia. In Sichuan province, irrigated areas were abundant in the eastern part and they were scattered in Yunnan province.

Figure 1. Global Irrigated Area Map (GIAM) map in the regions of study (**a**) Horn of Africa and (**b**) Southwest of China, where SW, GW and CU stand for surface water, groundwater and conjunctive use, respectively.

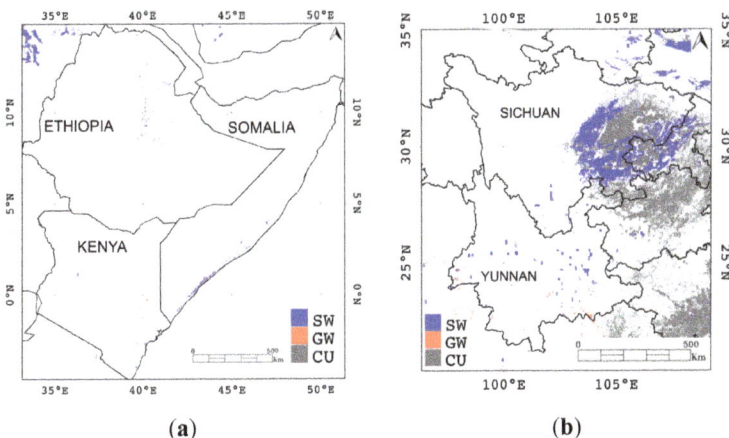

(**a**)   (**b**)

4. Results

4.1. Bias Curves

The spatial distribution of the bias was obtained monthly for every study area (not shown here). These computed a total of 36 images in the Horn of Africa, and 120 in the Chinese area, for the time periods analyzed (three and 10 years respectively). After the classification of the study areas, the monthly bias value was obtained per class by averaging the monthly biases in rain-fed croplands.

Figure 2 shows the inter-annual variability of the resulting bias curves. The yearly classification of the study areas provided the following number of classes: six (for 2010 and 2011) and eight (for 2012) in the Horn of Africa; six (for 2001, 2003, 2004, 2006), seven (for 2002, 2005, 2007, 2008, 2009) and eight (for 2010) in the Chinese area. In general, largest biases and similar patterns over the years were found in the Southwest of China, with amplitudes between −80 and 80 mm/month. The largest biases were found for the years 2005 and 2010, and the lowest biases for 2009.

The bias curves found in the Horn of Africa show no clear pattern over the years for some classes, which may be explained by the relatively low number of rain-fed pixels used to obtain them. This is the case of classes 1 and 2 in 2010 and class 1 in 2011, where the number of pixels used is one or two orders of magnitude lower than the rest of the classes. Moreover, in some classes the absence of a clear centered peak as observed in China is related to the incoming solar radiation patterns at these latitudes (between 5°S and 15°N). At the equator, maximum radiation values are found at the equinoxes (March and September) and a single maximum is developed with increasing latitude. In this paper, the biases were calculated as spatial averages per class, therefore the combination of values from different latitudes may partially explain the fluctuations of the curves.

For all classes, the magnitude of the biases in the African region is relatively modest compared to the ones found in the region of China. As a reference, we provide the average monthly ET_{actual} in both regions for the year 2010 obtained from the GLDAS–ET data set. The value was computed over all land pixels shown in Figure 1. The average monthly ET_{actual} ranged from 25–60 mm/month and from 25–120 mm/month in the African and Chinese regions respectively. These differences can partly explain the magnitude of the amplitudes found in the bias curves.

4.2. Monthly ET_b and Source of Irrigation

This section contains preliminary results of the application of the ET_b method in the study areas. Monthly ET_b was calculated for the Horn of Africa for the period 2010 till 2012 and for the Southwest of China for the years 2001–2010. Monthly ET_b values were extracted from the pixels labeled by GIAM as irrigated and assigned to the corresponding source of irrigation (SW, GW, CU). Pixel values were converted to volumes (Mm^3/month) by using the pixel area and then aggregated per country/province. Figure 3 shows the first results in the study areas where the monthly values of precipitation aggregated over the evaluated pixels are also included.

Figure 2. Spatial mean bias per class for (**a**) Horn of Africa and (**b**) Southwest of China study areas, obtained in rain-fed croplands as the difference of remotely sensed ET$_{actual}$ data (RS–ET) and Global Land Data Assimilation System ET$_{actual}$ model simulations (GLDAS–ET). (Note that the discrete ET$_b$ monthly values are connected to ease visualization).

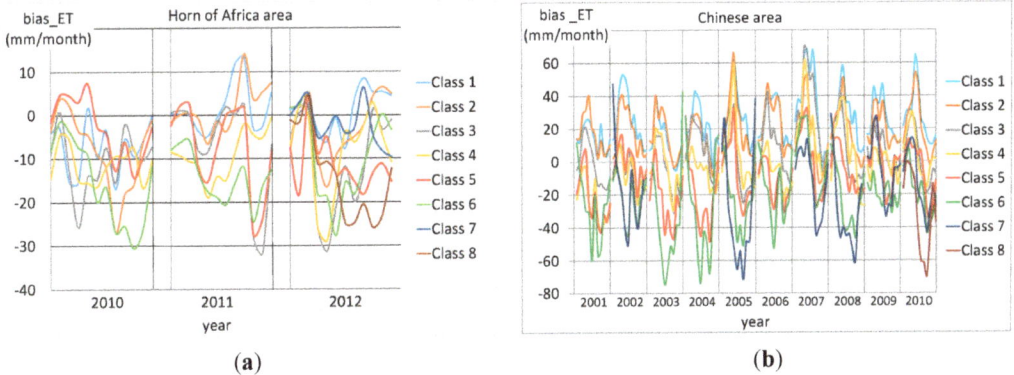

(**a**) (**b**)

The temporal signature of ET$_b$ allows seasonal patterns to be distinguished within a year and also inter-annual ET$_b$ dynamics, especially in the long series of ET$_b$ obtained in the provinces of China. The ET$_b$ pattern in Yunnan province was found to be relatively regular, contrary to what was observed in Sichuan, with some ET$_b$ peaks in the years 2006 and 2007 and lower general values in 2009 and 2010.

Precipitation showed a significant decrease in the year 2011 in Ethiopia and in the year 2006 in Sichuan province. This corresponds to drought periods as explained in Section 3.

In Ethiopia, a general decrease of ET$_b$ was observed in 2011, which points at a lower amount of irrigation water used. In particular, total ET$_b$ was estimated to decrease from 21 Mm^3/month in the wet year 2010 to 10 Mm^3/month in the dry year 2011. Moreover, in this period an increase of groundwater use at the expense of surface water use was observed, which is consistent with the report by Hendrix [49]. Despite the existence of the drought, national crop production did not appear to be significantly affected as reported by the Food and Agriculture Organization of the United Nations [50]. This might be explained by the fact that the drought mainly affected the east and south of the country and the majority of croplands use rain-fed production systems and are located in the other part of the country [51].

A decrease of ET$_b$ in half of the year 2011 was also observed in Kenya. The precipitation values in this period were only slightly lower than for other years. The drought in Kenya affected the northeastern regions of the country and therefore there is no significant effect on the precipitation in irrigated areas. The study of longer time series of data would allow inter-annual variability, trends and anomalies to be analyzed with a better statistical representation, and therefore have a better interpretation of these patterns.

Figure 3. Monthly ET$_b$ per source of irrigation water (surface water, SW; groundwater, GW; conjunctive use, CU) in irrigated areas of (**a**) Ethiopia and (**b**) Kenya (years 2010–2012) and the Chinese provinces of (**c**) Sichuan and (**d**) Yunnan (years 2001–2010). The figure also shows monthly precipitation (Note that the discrete ET$_b$ monthly values are connected to ease visualization).

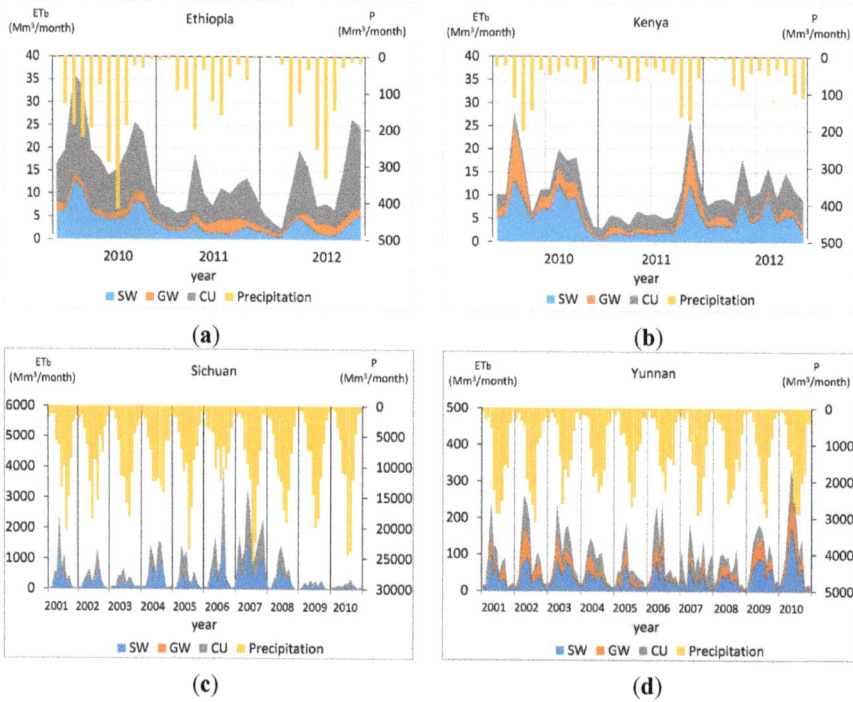

(**a**) (**b**)

(**c**) (**d**)

In Sichuan province, the values of ET$_b$ in the dry year were of similar magnitude to the previous years or increased, especially in the month of August, which points to a higher amount of irrigation water used. In particular, ET$_b$ was estimated at 200 Mm^3/month in the wet year 2005 and 400 Mm^3/month in the dry year 2006. The National Bureau of Statistics of China [52] reported that total water resources in Sichuan decreased by 26% in 2006 with respect to the average of other reported years (2004–2012), but still with a high value of 187 billion m^3. Moreover, the grain production in 2006 was only 10% lower than in year 2005. These two facts suggest that water was still available for irrigation and it was used when precipitation decreased.

In order to better interpret the results obtained, Figure 4 shows the input RS–ET and GLDAS–ET values in August 2006, where a peak of ET$_b$ is found in Sichuan. In this study area the range of RS–ET values was double the ones given by the land surface model in GLDAS–ET, with values up to 330 mm/month. In particular, a hot spot was found in Sichuan province near the border with Chongqing province with low values of GLDAS–ET. There is a high density of irrigated agriculture in this area (see Figure 1), so that the aggregated results per province are highly influenced by these values. Figure 4 also includes the temporal series of these two ET$_{actual}$ estimates in a pixel of the hot spot, where the significant decrease of GLDAS–ET outputs in the year 2006 can be observed. Due to the lack of precipitation, the ET$_{actual}$ model outputs given by the land surface model are lower.

Figure 4. Monthly ET$_{actual}$ in the study area of Southwest of China in August 2006 obtained from (**a**) the remote sensing estimates with the SEBS method and (**b**) GLDAS data; and (**c**) 10 years of monthly ET$_{actual}$ in the identified hot spot (30°34′N, 105°19′E). (Note that the discrete monthly values are connected to ease visualization).

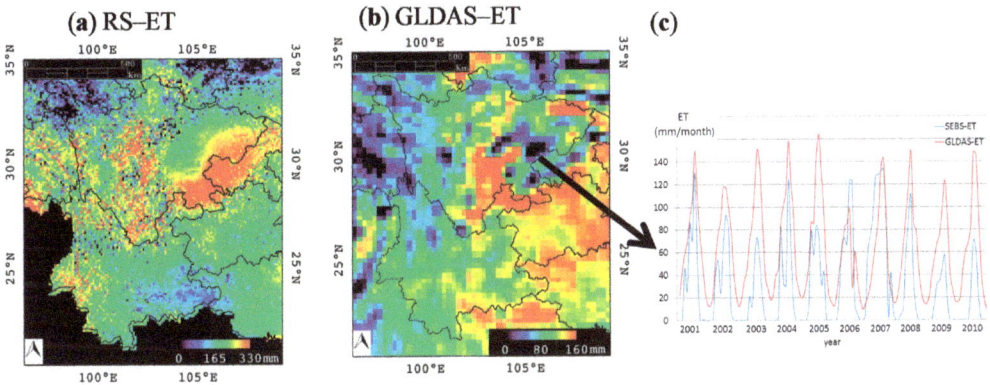

In Yunnan province, in which there was no significant dry year during the period of analysis, the total ET$_b$ curves show relatively regular patterns and values ten times smaller than in Sichuan. The use of the three sources of irrigation water is observed in this province with a major use of surface water.

In general, the preliminary results shown in Figure 3 also reveal features that could not be explained, like the ET$_b$ peaks in Sichuan in 2007 or low ET$_b$ values in 2009 and 2010. Although further research is needed to fully understand the patterns, this paper exemplifies the potential exploitation of the temporal dimension of ET$_b$, combined with the source of irrigation water, which could be useful for water management purposes.

The analysis of data in longer periods of time showed an advantage when interpreting and better understanding the ET$_b$ patterns. Bearing this in mind, the following section about sensitivity was elaborated using the case study of Sichuan province (years 2001–2010).

4.3. Sensitivity to Bias Curve Assignment

Since the principal aspect of the ET$_b$ method used is the definition of the bias curves, this section analyzes whether the ET$_b$ estimates are sensitive to the bias assignment. Figure 5 shows the monthly ET$_b$ results obtained in the province of Sichuan in the irrigated pixels as indicated in Figure 1. Four cases were considered depending on the bias assigned per month: (i) maximum of all classes; (ii) minimum of all classes; (iii) bias assigned based on the classification and (iv) mean bias calculated in all rain-fed croplands when no classes are considered.

Figure 5. ET$_b$ obtained in the province of Sichuan using the maximum, minimum, assigned-per-class and mean bias.

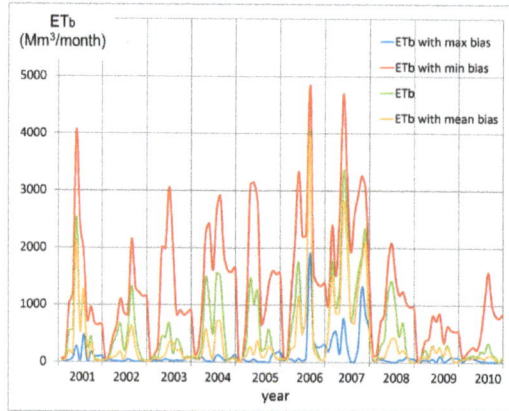

All four cases show maximum values of ET$_b$ in the years 2006 and 2007. However, inter-annual variability was found to be sensitive to the selection of the bias. ET$_b$ presented low monthly values in most of the study period when using the maximum bias, whereas higher values and relatively regular patterns were obtained with the minimum bias. The curves obtained with the mean and assigned-per-class bias showed intermediate values, with ET$_b$ in general lower in the former case. In this context, Romaguera *et al.* [22] showed that the bias estimation was improved when using different classes instead of a single mean bias obtained for all rain-fed pixels. Therefore, despite the possible ET$_b$ similarities between these two cases, the classification approach is preferred to evaluate ET$_b$.

5. Discussion

This paper illustrates preliminary results of the potential of using a remote sensing based method for obtaining time series of blue water evapotranspiration and combining it with the source of irrigation to monitor irrigation practices. The details and drawbacks of the models and data used were discussed in Romaguera *et al.* [21,22] and Thenkabail *et al.* [30].

The outputs produced in this paper need to be understood as preliminary examples of application. A better understanding of the ET$_{actual}$ inputs used would be required in order to obtain concluding outcomes. Regarding the bias, Section 4.3 showed how the ET$_b$ estimates were sensitive to the bias assignment.

Accuracies, Errors and Uncertainties

The uncertainties in ET$_{actual}$ estimation from remote sensing and the land surface modelling played an important role in the total ET$_b$ uncertainty. Kalma *et al.* [53] showed that remote sensing data provided typically relative errors of 15%–30% in ET$_{actual}$ estimation. In the case of the GLDAS products, the ET$_{actual}$ accuracy was not sufficiently evaluated in the literature, although some estimates exist. Fang *et al.* [54] reported the uncertainty in GLDAS–ET estimates by continent as equivalent heights of water based on 1979–2007 outputs from the four models included in the system. The climatology values of ET$_{actual}$ were 550 mm/year in Africa and 430 mm/year in Asia, with an uncertainty of ±60 mm/year in both cases. Besides, the definition of the bias curves has a standard deviation associated to the spatial averaging of the values per class. Despite the lack of detailed information about the GLDAS–ET accuracies, the aforementioned quantities were used (not shown here) to obtain the contributions of these three aspects to the total uncertainty by using the first order Taylor series expansion, where the covariance terms were neglected (inputs are independent) and linearity was assumed. A typical daily ET$_{actual}$ rate of 5 mm, a 30% in error of RS–ET, an average uncertainty of 5 mm/month in GLDAS–ET, and a bias curve in the Sichuan province were assumed. It was found that the error in RS–ET was the major contributor (50%–95%), modulated by the error of the bias which oscillated in time from around 5%–50%. The contribution of the GLDAS–ET inaccuracy was negligible. Increasing daily ET$_{actual}$ rates resulted in higher relative contribution of RS–ET, as expected, while decreasing the role of the bias, and being insignificant, the GLDAS–ET impact. Decreasing daily ET$_{actual}$ rates resulted in higher relative contribution of GLDAS–ET, with a maximum of 20% when a low value of daily ET$_{actual}$ was considered (0.1 mm/day).

These values served as an indication of the relative importance of RS–ET, GLDAS–ET and the bias, to the total uncertainty of ET$_b$. In the case of irrigated areas, ET$_{actual}$ values are expected to be high, and therefore the role of the bias accuracy is less significant. However a better estimate of the GLDAS–ET uncertainty is required to properly quantify the different contributions.

Moreover, the accuracy of the static GlobCover and GIAM maps may decrease in time. These are used in the method to define rain-fed areas and to assign the type of irrigation respectively. Therefore, they are also a source of uncertainty.

The results in this paper were aggregated per country/province, which may be appropriate for regional planning purposes. However, specific spatial features may be lost in big areas due to the aggregation process, such as multiple cropping practices. Therefore, analysis at different spatial scales is recommended when examining particular features. Besides, the spatial resolution of the input data may be a limitation in heterogeneous areas, and therefore scaling techniques [55,56] are advised for understanding the sub-pixel variability.

In order to obtain concluding results about the application shown in this paper, long time series of data are desired to be able to properly analyze trends, and possible anomalies in the climatology. From the point of view of the land surface models, global data can be obtained for long time periods, from the year 1970 until the present for the Noah model in GLDAS. However, remote sensing ET$_{actual}$ outputs are more limited in time and space and depend a lot on the geometry of observation, technical characteristics, and lifetime of the sensors on board the satellites. In this context, Mu *et al.* [57] provided global ET$_{actual}$ products every eight days at 1 km resolution between the years 2001–2010. Their algorithm is based on the Penman-Monteith equation using daily meteorological reanalysis data and 8-day remotely sensed vegetation property dynamics from the Moderate–Resolution Imaging Spectroradiometer (MODIS) as inputs.

In general terms, the interpretation of the results regarding irrigation practices bears an uncertainty related to the multiple situations that can be found in reality. Water availability and decisions taken by the farmers to irrigate or not and how much, are factors that influence the results. However, in the face of an extreme event like a drought, the results obtained in the case studies of the present paper indicated the possibility of identifying and explaining the episode in terms of irrigated water.

Finally, compared with the existing literature about ET_b given by Liu and Yang [16] and Mekonnen and Hoekstra [19], the method applied in this paper is innovative in two aspects: first it uses physically based remote sensing data instead of statistical data, and second it provides a better temporal resolution, more suitable for water management applications. Moreover, from an implementation point of view the method has a reasonably straightforward application procedure.

6. Conclusions

This paper illustrates the potential of using remote sensing and simulated actual evapotranspiration (ET_{actual}) time series combined with an existing "type of irrigation" map, to monitor irrigation practice. It provides new tools to obtain monthly blue evapotranspiration (ET_b) and shows the application in two relevant study areas: the Horn of Africa and the Chinese province of Sichuan, both affected by a drought event during the periods of analysis, but with differences in terms of water control and storage infrastructure. Further, monthly ET_b are subdivided into the source of irrigation water: surface water, groundwater and conjunctive use, which relates to the availability of water resources.

The preliminary results show seasonal and inter-annual patterns in ET_b. In the face of an extreme event like a drought, changes in ET_b (*i.e.*, irrigated water) can be identified, as well as the relative use of different sources of irrigation water. In Ethiopia, total ET_b is estimated to decrease from 21 Mm^3/month in the wet year 2010 to 10 Mm^3/month in the dry year 2011, while ET_b from groundwater increased; in Sichuan ET_b is estimated at 200 Mm^3/month in the wet year 2005 and 400 Mm^3/month in the dry year 2006; these very different patterns of drought response, as found for the two locations, are qualitatively consistent with the literature. However, further research is needed to fully understand the whole of the temporal patterns found.

The research also reveals methodological and data limitations. The results in Sichuan are found to be dependent on the bias assignment required in the method. Moreover, particular spatial ET_{actual} patterns are encountered in the input data. Finally, the use of longer time series of data for better interpretation of the results is recommended.

The application shown in this paper is innovative compared to similar literature in two aspects: first it uses physically based remote sensing data instead of statistical data, and second it provides a better temporal resolution, more suitable for water management applications. This paper constitutes a starting point for global temporal ET_b analysis, applying an innovative remote sensing based approach and further research will contribute to the achievement of more concluding and operative results. In the field of water management, the approach has potential to assess the existence of irrigation, the source of irrigation water, the duration and variability in time, at pixel and country scales, and could be especially useful to monitor irrigation practice during periods of drought.

Acknowledgments

The authors wish to thank the Land Surface Analysis Satellite Applications Facility (LSA–SAF) of the European Organisation for the Exploitation of Meteorological Satellites (EUMETSAT), as well as the European Space Agency (ESA) and the ESA GlobCover Project, led by MEDIAS–France, for providing the products used in this paper. The GLDAS data used in this study were acquired as part of the mission of NASA's Earth Science Division and archived and distributed by the Goddard Earth Sciences (GES) Data and Information Services Center (DISC). The authors also wish to thank the SPOT/VEGETATION data center (www.vgt.vito.be) for the NDVI imagery and Prasad Thenkabail for providing the Global Irrigated Area Map.

Author Contributions

This work was carried out in the framework of the PhD research of the first author M. Romaguera. The coauthors are the supervisors linked to this project.

Conflicts of Interest

The authors declare no conflict of interest.

References

1. FAO: Aquastat Database—Food and Agriculture Organization of the United Nations. Rome, Italy. Available online: http://faostat.Fao.Org/site/544/default.Aspx (accessed on 1 June 2014).
2. Postel, S.L.; Daily, G.C.; Ehrlich, P.R. Human appropriation of renewable fresh water. *Science* **1996**, *271*, 785–788.
3. Seckler, D.; Amarasinghe, U.; Molden, D.J.; de Silva, R.; Barker, R. *World Water Demand and Supply, 1990–2025: Scenarios and Issues*; IWMI: Colombo, Sri Lanka, 1998.
4. Rockstrom, J.; Gordon, L.; Falkenmark, M.; Folke, C.; Engvall, M. Linkages among water vapor flows, food production, and terrestrial ecosystem services. *Conserv. Ecol.* **1999**, *3*, 5.
5. Shiklomanov, I.A.; Rodda, J.C. *World Water Resources at the Beginning of the Twenty-First Century*; Cambridge University Press: Cambridge, UK, 2003.
6. Hoekstra, A.Y.; Hung, P.Q. *A Quantification of Virtual Water Flows between Nations in Relation to International Crop Trade*; UNESCO–IHE: Delft, The Netherlands, 2002.
7. Chapagain, A.K.; Hoekstra, A.Y. *Water Footprints of Nations*; UNESCO–IHE: Delft, The Netherlands, 2004.
8. Hoekstra, A.Y.; Chapagain, A.K. Water footprints of nations: Water use by people as a function of their consumption pattern. *Water Resour. Manag.* **2007**, *21*, 35–48.
9. Hoekstra, A.Y.; Chapagain, A.K. *Globalization of Water. Sharing the Planet's Freshwater Resources*; Blackwell Publishing: Oxford, UK, 2008; pp. 1–208.
10. Alcamo, J.; Florke, M.; Marker, M. Future long-term changes in global water resources driven by socio-economic and climatic changes. *Hydrol. Sci. J. J. Sci. Hydrol.* **2007**, *52*, 247–275.

11. Döll, P.; Siebert, S. Global modeling of irrigation water requirements. *Water Resour. Res.* **2002**, *38*, doi:10.1029/2001WR000355.

12. Rost, S.; Gerten, D.; Bondeau, A.; Lucht, W.; Rohwer, J.; Schaphoff, S. Agricultural green and blue water consumption and its influence on the global water system. *Water Resour. Res.* **2008**, *44*, doi:10.1029/2007WR006331.

13. Siebert, S.; Döll, P. *The Global Crop Water Model (GCWM): Documentation and First Results for Irrigated Crops*; Institute of Physical Geography, University of Frankfurt (Main): Frankfurt, Germany, 2008.

14. Siebert, S.; Döll, P. Quantifying blue and green virtual water contents in global crop production as well as potential production losses without irrigation. *J. Hydrol.* **2010**, *384*, 198–217.

15. Liu, J.G.; Zehnder, A.J.B.; Yang, H. Global consumptive water use for crop production: The importance of green water and virtual water. *Water Resour. Res.* **2009**, *45*, doi:10.1029/2007WR006051.

16. Liu, J.G.; Yang, H. Spatially explicit assessment of global consumptive water uses in cropland: Green and blue water. *J. Hydrol.* **2010**, *384*, 187–197.

17. Hanasaki, N.; Inuzuka, T.; Kanae, S.; Oki, T. An estimation of global virtual water flow and sources of water withdrawal for major crops and livestock products using a global hydrological model. *J. Hydrol.* **2010**, *384*, 232–244.

18. Fader, M.; Gerten, D.; Thammer, M.; Heinke, J.; Lotze-Campen, H.; Lucht, W.; Cramer, W. Internal and external green-blue agricultural water footprints of nations, and related water and land savings through trade. *Hydrol. Earth Syst. Sci.* **2011**, *15*, 1641–1660.

19. Mekonnen, M.M.; Hoekstra, A.Y. The green, blue and grey water footprint of crops and derived crop products. *Hydrol. Earth Syst. Sci.* **2011**, *15*, 1577–1600.

20. Pfister, S.; Bayer, P.; Koehler, A.; Hellweg, S. Environmental impacts of water use in global crop production: Hotspots and trade-offs with land use. *Environ. Sci. Technol.* **2011**, *45*, 5761–5768.

21. Romaguera, M.; Krol, M.S.; Salama, M.S.; Hoekstra, A.Y.; Su, Z. Determining irrigated areas and quantifying blue water use in Europe using remote sensing Meteosat Second Generation (MSG) products and Global Land Data Assimilation System (GLDAS) data. *Photogramm. Eng. Remote Sens.* **2012**, *78*, 861–873.

22. Romaguera, M.; Salama, S.; Krol, M.S.; Hoekstra, A.Y.; Su, Z. Towards the improvement of blue evapotranspiration estimates by combining remote sensing and model simulation. *Remote Sens.* **2014**, *6*, 7026–7049.

23. Bastiaanssen, W.G.M.; Bos, M.G. Irrigation performance indicators based on remotely sensed data: A review of literature. *Irrig. Drain. Syst.* **1999**, *13*, 192–311.

24. Santos, C.; Lorite, I.J.; Tasumi, M.; Allen, R.G.; Fereres, E. Performance assessment of an irrigation scheme using indicators determined with remote sensing techniques. *Irrig. Sci.* **2010**, *28*, 461–477.

25. D'Urso, G.; De Michele, C.; Vuolo, F. Operational irrigation services from remote sensing: The irrigation advisory plan for the campania region, Italy. In *Remote Sensing and Hydrology*; Neale, C.M.U., Cosh, M.H., Eds.; Int. Assoc. Hydrological Sciences: Wallingford, WA, USA, 2012; Volume 352, pp. 419–422.

26. Siebert, S.; Döll, P.; Hoogeveen, J.; Faures, J.M.; Frenken, K.; Feick, S. Development and validation of the global map of irrigation areas. *Hydrol. Earth Syst. Sci.* **2005**, *9*, 535–547.

27. Siebert, S.; Hoogeveen, J.; Frenken, K. *Irrigation in Africa, Europe and Latin America. Update of the Digital Global Map of Irrigation Areas to Version 4*; Institute of Physical Geography, University of Frankfurt (Main): Frankfurt, Germany, 2006.

28. Ozdogan, M.; Gutman, G. A new methodology to map irrigated areas using multi-temporal MODIS and ancillary data: An application example in the continental US. *Remote Sens. Environ.* **2008**, *112*, 3520–3537.

29. Thenkabail, P.S.; Biradar, C.M.; Noojipady, P.; Dheeravath, V.; Li, Y.J.; Velpuri, M.; Gumma, M.; Gangalakunta, O.R.P.; Turral, H.; Cai, X.L.; *et al.* Global irrigated area map (GIAM), derived from remote sensing, for the end of the last millennium. *Int. J. Remote Sens.* **2009**, *30*, 3679–3733.

30. Thenkabail, P.S.; Biradar, C.M.; Noojipady, P.; Dheeravath, V.; Li, Y.J.; Velpuri, M.; Reddy, G.P.O.; Cai, X.L.; Gumma, M.; Turral, H.; *et al. A Global Irrigated Area Map (GIAM) Using Remote Sensing at the End of the Last Millenium*; International Water Management Institute: Colombo, Sri Lanka, 2008; p. 63.

31. Ozdogan, M.; Rodell, M.; Beaudoing, H.K.; Toll, D.L. Simulating the effects of irrigation over the united states in a land surface model based on satellite-derived agricultural data. *J. Hydrometeorol.* **2010**, *11*, 171–184.

32. UCLouvain; ESA. *Globcover 2009. Products Description and Validation Report*; European Space Agency: Paris, France, 2011.

33. LSA–SAF. *LSA–SAF Validation Report. Products LSA–16 (MET), LSA–17 (DMETt)*; Doc. Num. SAF/LAND/RMI/VR/0.6; The EUMETSAT Network of Satellite Application Facilities: Darmstadt, Germany, 2010.

34. Rezaee, M.R.; Lelieveldt, B.P.F.; Reiber, J.H.C. A new cluster validity index for the fuzzy *c*-mean. *Pattern Recognit. Lett.* **1998**, *19*, 237–246.

35. Ghilain, N.; Arboleda, A.; Gellens-Meulenberghs, F. Evapotranspiration modelling at large scale using near-real time MSG SEVIRI derived data. *Hydrol. Earth Syst. Sci.* **2011**, *15*, 771–786.

36. Chen, X. The Plateau Scale Land—Air Interaction and Its Connections to Troposphere and Lower Stratosphere. PhD Thesis, ITC Dissertation 237, Faculty of Geo-Information and Earth Observation (ITC), University of Twente, Enschede, The Netherlands, 2013.

37. Chen, X.; Su, Z.; Ma, Y.; Liu, S.; Xu, Z. Development of a 10 year (2001–2010) 0.1° dataset of land-surface energy balance for mainland China. *Atmos. Chem. Phys. Discuss.* **2014**, *14*, 14471–14518.

38. Gellens-Meulenberghs, F.; Arboleda, A.; Ghilain, N. Towards a continuous monitoring of evapotranspiration based on MSG data. In Proceedings of IAHS Symposium on Remote Sensing for Environmental Monitoring and Change Detection, Perugia, Italy, 2–13 July 2007; IAHS Press: Wallingford, WA, USA.

39. Su, Z. The surface energy balance system (SEBS) for estimation of turbulent heat fluxes. *Hydrol. Earth Syst. Sci.* **2002**, *6*, 85–99.

40. Chen, F.; Mitchell, K.; Schaake, J.; Xue, Y.K.; Pan, H.L.; Koren, V.; Duan, Q.Y.; Ek, M.; Betts, A. Modeling of land surface evaporation by four schemes and comparison with FIFE observations. *J. Geophys. Res. Atmos.* **1996**, *101*, 7251–7268.

41. Rodell, M.; Houser, P.R.; Jambor, U.; Gottschalck, J.; Mitchell, K.; Meng, C.J.; Arsenault, K.; Cosgrove, B.; Radakovich, J.; Bosilovich, M.; *et al.* The Global Land Data Assimilation System. *Bull. Am. Meteorol. Soc.* **2004**, *85*, 381–394.

42. Anderson, W.B.; Zaitchik, B.F.; Hain, C.R.; Anderson, M.C.; Yilmaz, M.T.; Mecikalski, J.; Schultz, L. Towards an integrated soil moisture drought monitor for East Africa. *Hydrol. Earth Syst. Sci.* **2012**, *16*, 2893–2913.

43. Viste, E.; Korecha, D.; Sorteberg, A. Recent drought and precipitation tendencies in Ethiopia. *Theor. Appl. Climatol.* **2013**, *112*, 535–551.

44. Awulachew, S.B.; Yilma, A.D.; Loulseged, M.; Loiskandl, W.; Ayana, M.; Alamirew, T. *Water Resources and Irrigation Development in Ethiopia. IWMI Working Paper 123*; International Water Management Institute (IWMI): Colombo, Sri Lanka, 2007; p. 66.

45. Onimus, F. *Ethiopia—Irrigation and Drainage Project: P092353—Implementation Status Results Report: Sequence 15*; World Bank: Washington, DC, USA, 2014.

46. Zhang, S.H.; Yi, Y.J.; Liu, Y.; Wang, X.K. Hydraulic principles of the 2268-year-old Dujiangyan project in China. *J. Hydraul. Eng. ASCE* **2013**, *139*, 538–546.

47. Dai, Z.J.; Du, J.Z.; Li, J.F.; Li, W.H.; Chen, J.Y. Runoff characteristics of the Changjiang river during 2006: Effect of extreme drought and the impounding of the Three Gorges dam. *Geophys. Res. Lett.* **2008**, *35*, doi:10.1029/2008GL033456.

48. Wang, Y.Q.; Shi, J.C.; Liu, Z.H.; Liu, W.J. Application of Microwave Vegetation Index (MVI) to monitoring drought in Sichuan Province of China. In Proceedings of 2012 First International Conference on Agro-Geoinformatics (Agro-Geoinformatics), Shanghai, China, 2–4 August 2012; pp. 458–463.

49. Hendrix, M. Water in Ethiopia: Drought, disease and death. *Glob. Major. E J.* **2012**, *3*, 110–120.

50. FAOSTAT. Food and Agriculture Organization of the United Nations, FAOSTAT Database. Available online: http://faostat3.Fao.Org/faostat–gateway/go/to/download/q/qc/e (accessed on 27 April 2014).

51. See, L.; McCallum, I.; Fritz, S.; Perger, C.; Kraxner, F.; Obersteiner, M.; Baruah, U.D.; Mili, N.; Kalita, N.R. Mapping cropland in Ethiopia using crowdsourcing. *Int. J. Geosci.* **2013**, *4*, 6–13.

52. National Data Base. National Bureau of Statistics of China. Available online: http://data.Stats.Gov.Cn/index (accessed on 17 April 2014).

53. Kalma, J.D.; McVicar, T.R.; McCabe, M.F. Estimating land surface evaporation: A review of methods using remotely sensed surface temperature data. *Surv. Geophys.* **2008**, *29*, 421–469.

54. Fang, H.; Beaudoing, H.K.; Rodell, M.; Teng, W.L.; Vollmer, B.E. Global Land Data Assimilation System (GLDAS) products, services and applications from NASA Hydrology Data and Information Services Center (HDISC). In Proceedings of the ASPRS 2009 Annual Conference, Baltimore, Maryland, 8–13 March 2009.

55. Anderson, M.C.; Kustas, W.P.; Norman, J.M. Upscaling flux observations from local to continental scales using thermal remote sensing. *Agron. J.* **2007**, *99*, 240-254.

56. Jia, S.; Zhu, W.; Lŭ, A.; Yan, T. A statistical spatial downscaling algorithm of TRMM precipitation based on NDVI and DEM in the Qaidam basin of China. *Remote Sens. Environ.* **2011**, *115*, 3069–3079.

57. Mu, Q.Z.; Zhao, M.S.; Running, S.W. Improvements to a MODIS global terrestrial evapotranspiration algorithm. *Remote Sens. Environ.* **2011**, *115*, 1781–1800.

FAO-56 Dual Model Combined with Multi-Sensor Remote Sensing for Regional Evapotranspiration Estimations

Rim Amri, Mehrez Zribi, Zohra Lili-Chabaane, Camille Szczypta, Jean Christophe Calvet and Gilles Boulet

Abstract: The main goal of this study is to evaluate the potential of the FAO-56 dual technique for the estimation of regional evapotranspiration (ET) and its constituent components (crop transpiration and soil evaporation), for two classes of vegetation (olives trees and cereals) in the semi-arid region of the Kairouan plain in central Tunisia. The proposed approach combines the FAO-56 technique with remote sensing (optical and microwave), not only for vegetation characterization, as proposed in other studies but also for the estimation of soil evaporation, through the use of satellite moisture products. Since it is difficult to use ground flux measurements to validate remotely sensed data at regional scales, comparisons were made with the land surface model ISBA-A-gs which is a physical SVAT (Soil–Vegetation–Atmosphere Transfer) model, an operational tool developed by Météo-France. It is thus shown that good results can be obtained with this relatively simple approach, based on the FAO-56 technique combined with remote sensing, to retrieve temporal variations of ET. The approach proposed for the daily mapping of evapotranspiration at 1 km resolution is approved in two steps, for the period between 1991 and 2007. In an initial step, the ISBA-A-gs soil moisture outputs are compared with ERS/WSC products. Then, the output of the FAO-56 technique is compared with the output generated by the SVAT ISBA-A-gs model.

Reprinted from *Remote Sens.* Cite as: Amri, R.; Zribi, M.; Lili-Chabaane, Z.; Szczypta, C.; Calvet, J.C.; Boulet, G. FAO-56 Dual Model Combined with Multi-Sensor Remote Sensing for Regional Evapotranspiration Estimations. *Remote Sens.* **2014**, *6*, 5387-5406.

1. Introduction

In semi-arid regions, and the Mediterranean basin in particular, agricultural productivity and water resources regularly suffer from serious crises, as a consequence of limited levels of precipitation, combined with the occurrence of long periods of drought, which are typical features of the Mediterranean climate [1]. In this context, the accurate monitoring of vegetation cover and hydric stress can be a valuable tool, especially for areas which rely on rainfed agriculture in water-limited environments. A second challenge for the well-adapted management of this agriculture is to accurately determine the level of evapotranspiration, in order to quantify the soil water stock. In recent years, a variety of physical surface models have been proposed at regional and global scales [2]. The most accurate of these are the SVAT models [3–6].

On the other hand, the initial FAO-56 model is the most commonly used and practical approach for the estimation of crop water requirements and local scale evapotranspiration [7], and is based on a simple combination of a reference evapotranspiration value (ET_0) and crop coefficients. The FAO-56 dual model distinguishes between the respective contributions of plant transpiration (K_{cb})

and soil evaporation (K_e) [8]. In recent years, various attempts have been made to combine the latter model with remote sensing data for operational applications [9–14]. These studies were motivated, in particular, by the need to retrieve vegetation cover dynamics from vegetation indices derived from optical satellite observations [9,10,15,16]. Most applications of this model are related to the study of irrigated areas, for the effective planning and use of irrigation water.

In recent years, sustained scientific activity based on the interpretation of remotely sensed data has made it possible to develop various methodologies for the characterization of the spatio-temporal variability of continental surface parameters (vegetation characteristics and soil moisture), on both local and global scales. In the case of vegetation cover, various indices based on the use of optical data have been proposed for the retrieval of vegetation characteristics (e.g., leaf area index, vegetation fraction, NDVI (Normalized Difference Vegetation Index), *etc.*). The NDVI is the most commonly used index, and is expressed by the ratio: NDVI = (RNIR − RRED)/(RNIR + RRED), where RNIR is the near-infrared (NIR) reflectance and RRED is the red reflectance. This index is sensitive to the presence of green vegetation [17,18], and has been used in various studies dealing with the estimation of the potential photosynthetic activity of vegetation [10,19–22]. As a consequence of its formulation, the NDVI is able to robustly characterize green vegetation, despite varying atmospheric conditions in the red and NIR bands [23,24]. Nevertheless, in the case of high levels of vegetation density, the NDVI can suffer from saturation effects [25], which bias the estimated levels of evapotranspiration. In semi-arid regions, this behavior is rarely observed. Several different methodologies based on the interpretation of microwave sensor data have also been developed for the determination of soil moisture [26–31]. A large number of studies have demonstrated the potential of low-resolution spaceborne (active microwave) scatterometers for land surface characterization, in particular for the estimation of soil moisture [32–37].

In this context, the aim of the present study is to illustrate the ability of the FAO-56 dual approach to estimate evapotranspiration at regional scales, without making use of complex physical surface models requiring large quantities of input data. In an initial step, the effectiveness of the FAO-56 dual approach, which is commonly used for irrigation management [9,10], is evaluated. Then, the combination of this model with data generated by remote sensing is considered, not only for the study of vegetation as proposed in other studies [9–12,14], but also for the purposes of soil moisture analysis. Under such conditions, the spatio-temporal variations of these two fundamental parameters can be taken into account in the transpiration and evaporation estimations.

In Section 2, the studied site and the remotely sensed and ground databases are presented. In Section 3, the FAO dual model is introduced, and the concurrent use of remotely sensed data is discussed. In Section 4, the proposed approach is compared with the outputs of a physical SVAT model: the ISBA-A-gs. Finally, the authors' conclusions are presented in Section 5.

2. Database and Processing

2.1. Studied Site

The studied site is the Kairouan plain, which is situated in central Tunisia (35°–35°45′N; 9°30′–10°15′E) (Figure 1) and is characterized by a semi-arid climate [38]. The average annual rainfall is approximately 300 mm per year, with a rainy season lasting from October to May. The rainfall patterns in this semi-arid area are highly variable in time and space. The mean daily temperature in Kairouan City ranges between a minimum of 10.7 °C in January and a maximum of 28.6 °C in August, with a mean value equal to 19.2 °C. The mean annual potential evapotranspiration (Penman) is close to 1600 mm. The landscape has no relief and land use is dominated by agriculture, with two main types of vegetation cover: annual agriculture and olive trees. A ground campaign carried out over more than 20 test fields revealed that the mean soil texture is composed of 45% sand, 32% clay and 23% loam. Sandy soils are more commonly observed in the areas characterized by olive tree cultivation.

Figure 1. Satellite imagery of the studied area, indicating the locations of the rainfall and climate network stations present on the Kairouan plain.

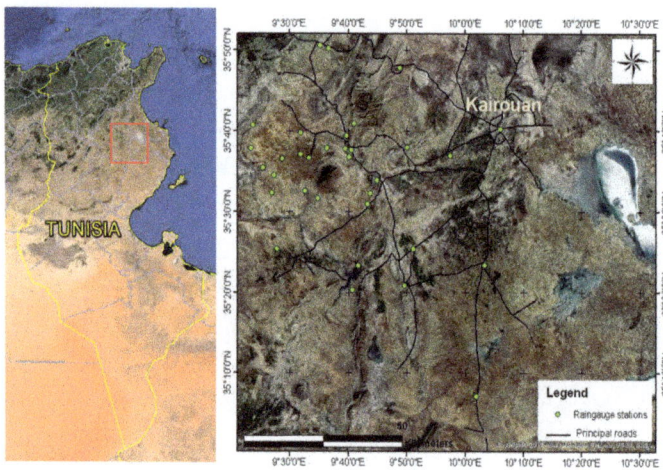

2.2. Satellite Products

2.2.1. ERS/WSC Moisture Products

The scatterometers carried by ESA's dual ERS satellites (launched in 1991 and 1995, the dual ERS satellite mission was finally decommissioned in September 2011) are operated in the C-band (5.3 GHz) in the vertical polarization. Over land, the measured radar backscatter coefficient is sensitive to surface parameters (soil moisture, surface roughness, vegetation characteristics) and the emission characteristics of the radar (incidence angle, polarization and frequency). A change detection approach developed by the Institute of Photogrammetry and Remote Sensing (IPF), Vienna University of Technology (TU-Wien), has been applied to the estimation of soil moisture, based on radar measurements [31,32,39,40]. The proposed moisture products have been validated in different regions of the globe, such as the Canadian Prairies [32], the Iberian Peninsula [35], Western Africa [34,41], France [37] and Australia [42]. The TU-Wien algorithm is based on scaling of the normalized backscattering coefficient to a value corresponding to a 40° incidence angle, lying between the lowest value which occurs during the driest conditions, and the highest value which occurs during the wettest conditions. The retrieved moisture index, referred to as the "surface soil moisture" (SSM), can range between 0% and 100% and represents the water content present in the first 5 cm of soil. Based on the interpretation of cells approximately 50 × 50 km in size, the TU-Wien products have a grid spacing of 25 km, and a temporal resolution of approximately two to three measurements per week. In order to compare SSM values with ground measurements on the same study site [43] or modeled surface moisture values, these products were converted to physical units of $m^3 \cdot m^{-3}$, using the method described by Pellarin *et al.* [36]. The TU-Wien Soil Water Index (SWI), which provides the water content along a 1 m deep profile, is derived from the SSM values measured on different successive dates. These products have already been validated and used in various hydrological studies [29,35–37,44]. Amri *et al.* [43] discussed the validation of ERS and ASCAT products, through the use of continuous thetaprobe recordings of ground soil moisture in the Kairouan plain. These authors revealed a strong correlation between the ground measurements recorded between 2010 and 2011, and satellite products corresponding to the same period, with an RMSE (the root mean square error) equal to 0.06 cm^3/cm^3 for surface moisture and 0.039 cm^3/cm^3 for the SWI.

2.2.2. SPOT-VGT NDVI Products

The 10-day synthesis (S10) products derived from SPOT-VGT (SPOT-VEGETATION) data are available at full resolution (1 km), and include 10-day NDVI data [45]. For these products, top-of-atmosphere corrections were applied using the SMAC (Simplified Method for Atmospheric Corrections) algorithm [46], which corrects for molecular and aerosol scattering, water vapor, ozone and other gas absorption effects. The parameters taken into account in the atmospheric corrections are the aerosol optical depth (AOD), the atmospheric water vapor and ozone, and a Digital Elevation Model used for atmospheric pressure estimation [47]. The water vapor parameter is obtained once every six hours from Météo-France with a 1.5° × 1.5° grid cell resolution. Although the AOD is currently retrieved from B0 data (blue spectral band of SPOT-VGT, 0.43–0.47 µM), combined with the NDVI [47], prior to 2001 it was a static data set which varied as a function of latitude only. Different systematic errors (misregistration of the different channels, calibration of the linear array detectors for each spectral band) are corrected in the final P product, which is re-sampled to a Plate carrée geographic projection. The S10 products are available [48].

2.3. Ground Measurements

2.3.1. Precipitation Data

Precipitation estimations were based on a network of 30 rain gauges, distributed over the entire site (Figure 1). The Inverse Distance Weighting (*IDW*) interpolation algorithm was used to derive daily precipitation maps. This algorithm estimates values at non-sampled points, by computing the weighted average of data observed at nearby points [49,50]. The landscape is mainly flat in the validation areas, and there is no mountainous terrain able to influence the spatial distribution of rainfall. The precipitation time series and the NDVI data set are thus available with a spatial resolution of 1 km.

2.3.2. Meteorological Data

Meteorological data, including air temperature, humidity, wind speed, net radiation and rainfall measurements, have been recorded over the last 20 years by an automatic weather station located in the Kairouan Plain. Daily averaged climatic parameters were computed in order to determine the daily reference evapotranspiration ET_0 (mm/day), in accordance with the FAO-56 Penman-Monteith parameterization [51].

The global radiation was determined from Météosat data [52] retrieved from the SoDa server (Solar radiation Databases for environment [53]), established by the Mines ParisTech graduate school. In this database, global radiation data is available from January 1985 to December 2005, at temporal intervals of one day and a spatial resolution of 20 km.

2.4. Land Use Mapping

Low spatial resolution SPOT Vegetation NDVI images were used to map the land into three characteristic classes: olive trees, annual agriculture and pastures. It is important to note that these classes were labeled: "Olive trees", "Annual Agriculture" corresponding to cereals and "Pastures". Later, for the purposes of estimating regional evapotranspiration, only two principal classes were considered: Olive trees and annual agriculture.

In recent years, several approaches have focused on the disaggregation of low resolution mixed pixels in different land cover classes [54]. A linear mixing theory is generally used, in which it is assumed that the reflectance (respectively NDVI) of a mixed pixel is given by the sum of the mean reflectance (respectively NDVI) values of the different land cover classes within the pixel, weighted by their corresponding fractional cover. The identification of typical NDVI profiles, representative of each of these land cover classes, is the first step in the disaggregation methodology. These pixels are identified by making use of information related to the class composition of each pixel, retrieved from a high-resolution land-cover map.

The general expression is given by Equation (1), and the RMS of this quantity is given by Equation (2):

$$Y_i(t) = \sum_{j=1}^{p} \pi_{ij} \times \rho_j(t) + \varepsilon_i(t) \tag{1}$$

$$RMSE_i = \sqrt{\frac{1}{T} \times \sum_{t=1}^{T} (\varepsilon_i(t))} \tag{2}$$

where: $Y_i(t)$ is the average signal observed at pixel i and time t, and is estimated from the NDVI time series produced by SPOT-VGT; π_{ij} is the area occupied by the jth class in the ith pixel: the unknown term in Equation (1); $\rho_j(t)$ is the signal assigned to the jth class at time t, and is calculated for each class (pure pixels are considered) from the NDVI time series produced by SPOT-VGT; ε_i is the error term; p is the number of classes, T is the number of observations, i is the pixel index, and j is the class index.

The error term defined in Equation (2) is the square of the differences between the NDVI signature assigned to each class, and the NDVI profile observed in the *i*th pixel.

Disaggregation techniques are designed to estimate the proportion (between 0 and 1) of specific classes occurring within each pixel. The result is a certain number of fraction images, each corresponding to the relevant land-cover class. While this information describes the composition of the class, it does not provide any indication as to how the classes are spatially distributed within the pixel. The outcome is thus quite different from that obtained with conventional classification algorithms, in which a single "crisp" land cover map, containing all classes, is produced. Figure 2 shows a land-use map for two classes. The pixels having low proportions (dark blue areas) for all 3 classes of land use correspond to areas covered by water (sebkhas, dams). These areas are masked in all of the maps used in the following analysis.

Figure 2. Land use map for the 2008–2009 agricultural season at 1 km spatial resolution, showing (on a scale ranging from 0–1) the proportion of coverage represented by each of three different classes of vegetation present in this area: (**a**) annual agriculture; (**b**) pastures; (**c**) olive trees.

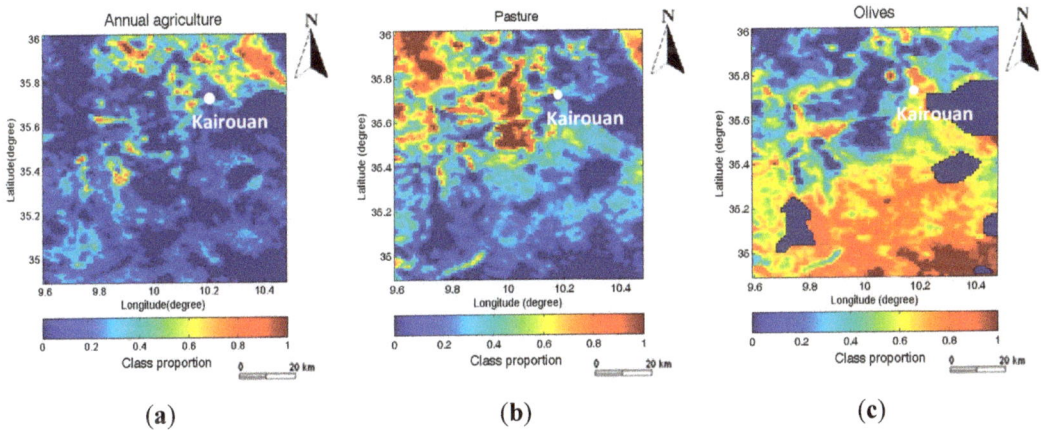

(**a**)   (**b**)   (**c**)

3. Proposed Approach for the Retrieval of Evapotranspiration

3.1. Description of the Basic FAO-56 Model

The algorithm used in the present study is based on the FAO-56 dual crop coefficient model developed by [7], which describes the relationship between crop evapotranspiration under non-standard conditions (ET) and a reference level of evapotranspiration (ET_0). The crop coefficient (K_c) is separated into two components: the basal crop coefficient (K_{cb}) and the soil water evaporation coefficient (K_e):

$$ET = (K_s \cdot K_{cb} + K_e) * ET_0 \tag{3}$$

where ET_0 is estimated at 24 h intervals using the FAO Penman-Monteith equation [7]; K_{cb}: the basal crop coefficient, K_S is the stress coefficient and K_e: the soil water evaporation coefficient.

The daily reference evapotranspiration ET_0 is determined with a spatial resolution of 20 km, allowing ET_0 maps to be derived at daily intervals, for each growing season (from September of one year to August of the following year), with the same resolution as the SoDa data. The cumulative annual ET_0 values are consistent with the observed levels of ET in this region (approximately 1600 mm/year).

3.2. Application with a Dual Vegetation Cover

In the present study, two types of vegetation cover are considered for each pixel: cereals (annual agriculture) and olive trees. For each pixel, the evapotranspiration ET under non-standard conditions is estimated using:

$$ET = \left[Fr_{cereals}\left[\left(f_{c-c}K_{cb-c}K_s + (1-f_{c-c})K_e\right)\right] + Fr_{olive}\left[\left(f_{c-o}K_{cb-o}K_s + (1-f_{c-o})K_e\right)\right]\right] * ET_0 \qquad (4)$$

where: f_c is the fractional cover; Fr is the cover percentage per pixel, for each class, K_S is the stress coefficient and the indices "o" and "c" denote the cereal and olive tree classes, respectively.

The parameters (K_{cb} and K_e) used in Equation (4) are derived from the remotely sensed SPOT-VGT NDVI index, and ERS/WSC soil moisture products, respectively.

K_s describes the effect of water stress on crop transpiration.

$$K_s = \frac{TAW - D_r}{(1-p)TAW} \qquad (5)$$

D_r: root zone depletion (mm). The equation number 86 of the FAO No. 56 guidelines [7] is used to calculate this parameter.

TAW: Total available soil water in the root zone (mm), estimated using the equation number 82 of the FAO No. 56 guidelines [7].

p: fraction of TAW that a crop can extract from the root zone without suffering water stress. This parameter is derived for each class from table 22 of the FAO No. 56 guidelines [7].

3.2.1. Computing the Values of K_{cb} and f_c

Further details of the proposed annual agriculture (cereals) estimations can be found in Er-raki et al. [9]. The calibrations made at the Tensift site in Morocco Er-raki et al. [9] were also applied to the studied site, due to their similarities in terms of climate and cereal yields.

K_{cb} is defined as:

$$K_{cb} = 1.07 \times \left[1 - \left(\frac{NDVI - NDVI_{min}}{NDVI_{max} - NDVI_{min}} \right)^{\frac{0.84}{0.54}} \right] \qquad (6)$$

where $NDVI_{min}$ and $NDVI_{max}$ are the minimum and maximum values of the $NDVI$ associated with bare soil and dense vegetation, respectively. The values retrieved from the SPOT-VGT NDVI index time series, used in the present study, are 0.1 and 0.6.

f_c is the vegetation cover fraction defined by Er-raki et al. [9]:

$$fc = 1.18 * \left(NDVI - NDVI_{min}\right) \qquad (7)$$

In the case of olive trees, the crop coefficients proposed in the FAO Bulletin No. 56 [7] for the estimation of the water requirements of olive trees are not applicable to the present case study, due to the low percentage of coverage corresponding to this culture. In a recent study, Testi, *et al.* [55] established a relationship between the *ET* and the cover fraction, which is used to determine the value of K_{cb} applicable in the present study.

In the case of olive trees, a tree spacing of approximately 20 m and a tree diameter of approximately 4 m area were considered. This leads to a value of *fc* value equal to 8%.

3.2.2. Computation of the Parameter K_e

In recent years, various different approaches have been proposed to relate soil resistance to soil moisture [56–58]. Chanzy and Bruckler [57] proposed an empirical method relating soil evaporation to soil moisture and climate demand, for different types of soil texture. In arid and semi-arid regions, the soil evaporation which occurs after a rainfall event is a process of major importance, whenever the local agriculture is characterized by a low density vegetation cover. When this term is determined accurately, it allows a reliable estimation to be made of the stock of water available for use by the vegetation. In this section, a simple approach is used for the estimation of soil evaporation, which is equal to the ET_0 whenever the surface layer is saturated. A method developed by Merlin et al. [58] was used, allowing soil evaporation to be related to the surface soil moisture (0–5 cm) estimated from radar satellite measurements. The parameter K_e can then be written as:

$$K_e = \left[\frac{1}{2} - \frac{1}{2} \cdot \cos(\pi \cdot \theta_L / \theta_{max}) \right]^P \quad \text{For } \theta_L \le \theta_{max} \tag{8}$$

where θ_L is the soil water content in the soil layer of thickness L, θ_{max} is the soil moisture at saturation, and P is a parameter given by the following expression:

$$P = \left(\frac{1}{2} + A_3 \frac{L - L_1}{L_1} \right) \frac{LE_p}{B_3} \tag{9}$$

In this expression, L_1 is the thinnest layer of soil represented (here 0–5 cm), and A_3 (unit-less) and B_3 ($W \cdot m^{-2}$) are a priori the two best-fit parameters, which depend on the soil's texture and structure.

θ_{max} was estimated from continuous ground thetaprobe measurements, acquired over a period of three years.

θ_L was considered to be equal to the remotely sensed surface soil moisture products produced by the ERS and ASCAT scatterometers. As described above, these products were converted to physical units of $m^3 \cdot m^{-3}$, using the method described by Pellarin et al. [36].

Soil evaporation is assumed to reach its maximum value in the case of saturated soils, with a value defined as being equal to ET_0, which is close to zero for very dry surfaces.

3.3. Description of the ISBA Model Used to Evaluate the FAO Dual Approach

Since it would not be realistic to use local flux measurements made at the field scale to validate the FAO-56 dual approach for applications involving low spatial resolution data ($0.5° \times 0.5°$, *i.e.*, regional scale), the validation used in the present study was based on comparisons made with the operational ISBA model developed by Météo France. This model uses the force-restore method proposed by Deardoff [59] to compute the corresponding variations in soil surface energy and water budget [5]. The model uses three layers to represent the soil's hydrology: the upper surface layer, the root-zone layer and the deep soil layer [60], and water interception storage and snow pack variations are also taken into account [61]. Within each grid, the heterogeneity of infiltration, precipitation, topography and vegetation are accounted for, and the conversion of precipitation into runoff over saturated surfaces is based on the TOPMODE approach [62,63]. Within each grid cell, the heterogeneity of land cover and soil depths is taken into account through the use of a tile approach, in which the cell is divided into a series of sub-grid patches. For each tile in a grid cell, distinct energy and water budgets are calculated. The multiplicative model developed by Jarvis [64] is used to determine the stomatal resistance of the vegetation. ISBA-A-gs is a variant of the ISBA model [5], which takes photosynthesis and its coupling with leaf-level stomatal conductance into account, and in which a biochemical soil–vegetation–atmosphere transfer representation is used to model the diurnal cycle of photosynthesis [65]. Then, the canopy conductance to water vapor is computed by integrating the photosynthesis model over the vegetation canopy, using a one-dimensional radiative transfer model within the vegetation. The canopy conductance is then used in the original ISBA model [66] to calculate plant transpiration. The other components of evapotranspiration (soil evaporation and evaporation of intercepted rain) are simulated in the same manner as in the original ISBA model. In addition to meteorological variables and surface temperature, soil evaporation depends on surface soil moisture and the vegetation coverage fraction. The interception reservoir is assumed to evaporate at the potential rate and depends on the LAI and the vegetation coverage fraction. The ECOCLIMAP look-up tables are used to generate the LAI (Leaf Area Index) inputs used by the ISBA model. ISBA-A-gs also implements a new representation for soil moisture stress, in which two different drought responses can be applied: one is used for herbaceous vegetation [67], and the other for forests [68].

4. ISBA-A-gs Model Inter-Comparison with the FAO-56 Approach

4.1. Analysis of the ISBA-A-gs Soil Moisture Output

In Figure 3, the ERS/WSC soil moistures, validated by Amri *et al.* [43] over the studied site and described in Section 2.2, are compared with the output generated by the ISBA model, and are plotted together with the precipitation time series. Three statistical parameters: RMSE (the root mean square error), R^2 (the coefficient of determination), and the bias are used to compare the ERS and ISBA datasets.

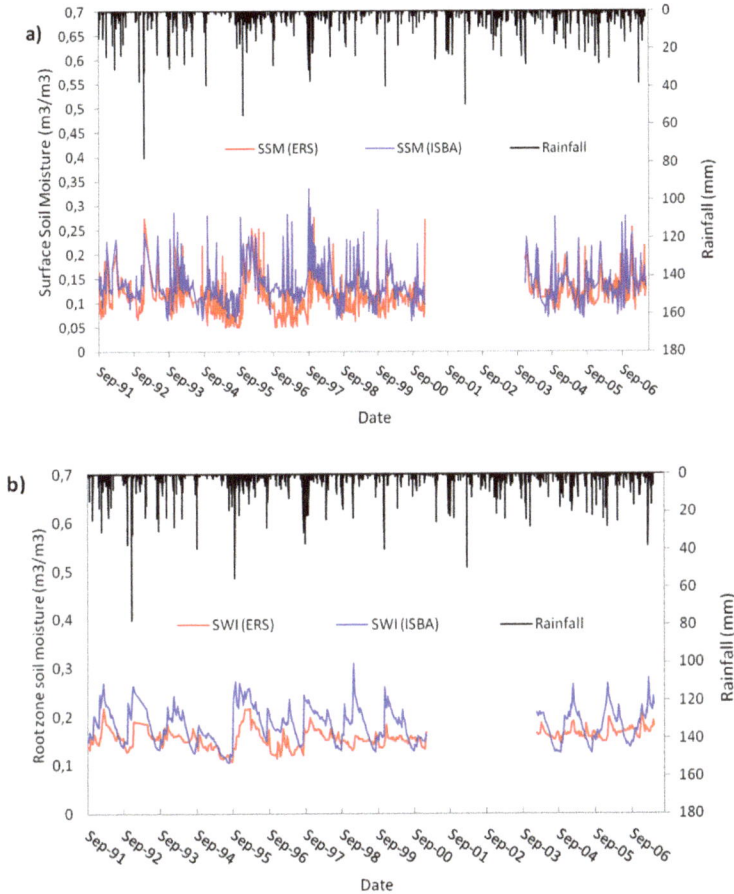

Figure 3. Inter-comparison between ISBA-A-gs soil moisture outputs and ERS/WSC products during the period from 1991–2007 (no data for 2001–2003): (**a**) surface soil moisture (SSM); (**b**) Soil Water Index (SWI) corresponding to the root zone moisture.

The satellite data products are compared with two ISBA-A-gs outputs; the modeled top layer (the first five centimeters of soil) and the soil moisture profiles (down to a depth of 100 cm), for the 16-year period from 1991–2007. It should be noted that the ERS measurements and the corresponding remotely sensed moisture products are not available for the period from 2001–2003. As can be seen in Figure 3a, local variations in the ERS surface moisture products are not completely retrieved by the ISBA outputs. In fact, the latter corresponds to the first 10 centimeters of soil, which are less strongly influenced by atmospheric conditions (rain, wind and solar radiation) than the first five centimeters of soil, which affect the ERS/WSC radar measurements. Despite the strong heterogeneity of the moisture profile in the first centimeters, the statistics of the resulting comparison are good: RMSE = 0.04 $m^3 \cdot m^{-3}$, bias = 0.02, and $R^2 = 0.52$.

In Figure 3b, the monthly SWI ISBA-A-gs outputs (0.5 m depth) are compared with the ERS estimations, showing that these two products have a good degree of coherence. In general, a delay of several days is observed between significant rainfall events and the corresponding increase in water soil content, determined using the ERS measurements.

The statistics of the compared data are good: RMSE equal to 0.03 $m^3 \cdot m^{-3}$, $R^2 = 0.5$, and a low bias equal to 0.008. A change in behavior can be observed in 1997: before this date, less than one data point (determined using the ERS1 scatterometer only) was available per week, from which the algorithm could be developed. After 1997, the ERS1 and ERS2 scatterometers were both able to provide a combined two to three data points per week, for use in the SWI estimations, and there is thus a difference in the product's accuracy between these two periods. Although the SWI itself is not used in the evaporation estimation, comparisons made with this product are used to validate the ISBA-A-gs model over the studied site.

4.2. Inter-Comparison between ISBA-A-gs and FAO-56 Approaches

Figure 4 compares the ISBA and FAO model evapotranspiration simulations over the studied area, for a single ISBA pixel with a spatial resolution corresponding to 0.5° in latitude and 0.5° in longitude, for the periods between 1998 and 2000 and between 2004 and 2005. These two products are compared only on dates for which ERS/WSC-based determinations of the K_e evaporation parameter are available. The two products are found to be in good agreement, and the statistical parameters derived from the FAO simulation are reasonable: RMSE = 0.36 mm/day, with a correlation $R^2 = 0.55$. The discrepancies observed at some points in this figure are related, in particular, to the soil evaporation component, and to the occurrence (or not) of a precipitation event: all of the data points characterized by a strong discrepancy between the two models involve a precipitation event, which was taken into account by one of the models, but not the other. This is mainly due to the input used in the ISBA-Ags model. In the case of the present study, the ISBA model was driven by the ERA-Interim atmospheric forcing, corresponding to the ERA-Interim global ECMWF atmospheric reanalysis, projected onto a 0.5° grid. Since precipitation is underestimated by the latter product [69], the monthly Global Precipitation Climatology Centre (GPCC) precipitation product was used to correct the precipitation bias in the 3-hourly ERA-Interim estimates [70]. The availability of a smaller number of GPCC precipitation observations for northern Africa, than for Europe, thus provides an explanation for the reduced robustness of the Tunisian precipitation series. The retrieved levels of FAO approach are in coherency with other studies realized over other semi-arid regions [9,71].

The FAO model values for the ISBA pixel (0.5° × 0.5°) correspond to the mean value of 1 km pixel estimations. In order to evaluate the scale effect, the FAO model is computed directly at the ISBA pixel scale, by considering the mean NDVI values at this scale, for vegetation fractions and K_{cb} estimations, and for two types of vegetation cover (olive trees and cereals). The intercomparison between FAO model levels, calculated at two different scales, reveals an rms error of 0.2 mm/day and a correlation coefficient equal to 0.85, for the period (1991–2007). This result indicates that the FAO dual model has only a limited scaling effect.

Figure 4. Evapotranspiration (ET) simulated by the ISBA-A-gs model as a function of the ET levels simulated by the FAO-56 model over a single ISBA pixel, during the period from 1991–2007, and on dates when remotely sensed ERS/WSC observations were recorded.

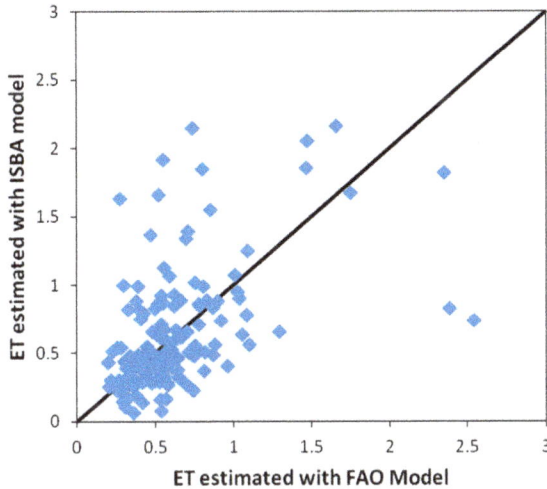

Figure 5a compares the ET simulated by the ISBA-A-gs model with that predicted by the FAO-56 model, for the 1998–1999 agricultural season during which the total precipitation was approximately 280 mm. A good degree of consistency is observed for the results obtained with these two models. However, on some dates, large differences are observed between FAO-56 estimates and ISBA-A-gs outputs, probably as a consequence of the rainfall events taken into account in the ISBA-A-gs model. The statistical parameters derived from the simulation are reasonable: RMSE equal to 0.39 mm/day, correlation strength given by $R^2 = 0.55$, and a low bias equal to 0.009 mm/day.

The second example, illustrated in Figure 5b, corresponds to the 1999–2000 agricultural season, characterized by a relatively dry growing season and a total annual precipitation of approximately 250 mm. During this season, the results given by the two models are found to be more consistent. The FAO-56 model retrieves almost the same trends as the ISBA-A-gs model, for both high and low values of ET. The statistical parameters from the comparison are good: RMSE = 0.25 mm/day, correlation strength $R^2 = 0.61$, and a low bias equal to 0.01. This comparison shows that the models are more consistent during the driest period of the season (from June to August). This is mainly due to the very limited number of rainfall events at this time of the year. Under these conditions, it is easier for the FAO dual model to retrieve the evaporation dynamics corresponding to small temporal variations in moisture, caused for example by a single rainfall event during a period of drought.

278

Figure 5. Inter-comparison between ET outputs from the FAO-56 and ISBA-A-gs models, on dates when remotely sensed ERS/WSC observations were recorded: **(a)** 1998–1999 agricultural season; **(b)** 1999–2000 agricultural season.

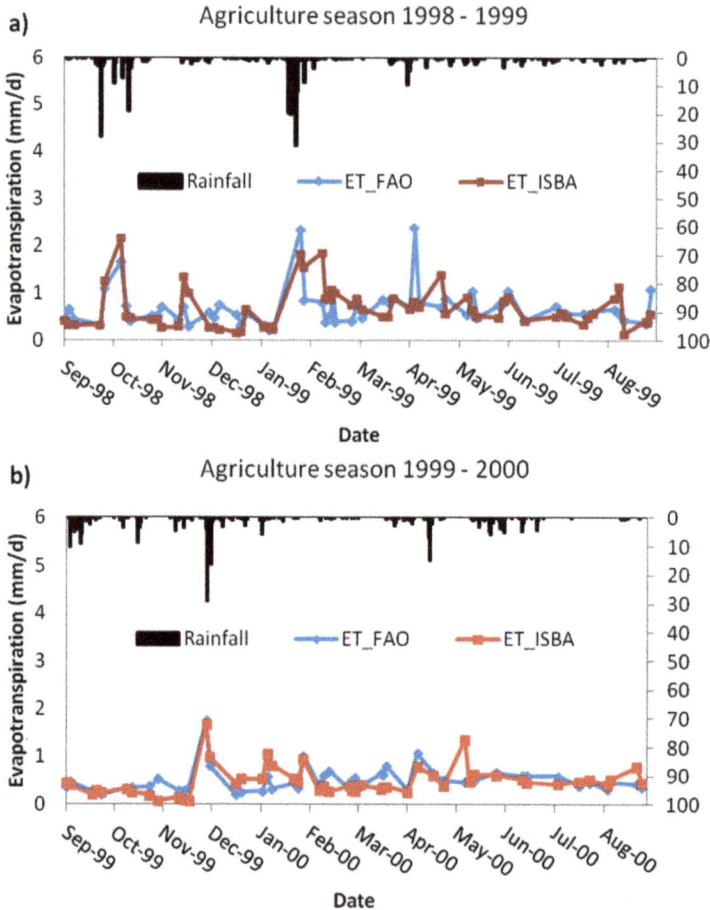

For both of these agricultural seasons, the two models are found to be well correlated with the rainfall events, with a clear increase in the level of evapotranspiration following strong precipitation events. However, some distinct inconsistencies can be observed between the two simulations. As an example, the FAO-56 model retrieves a high level of ET, due to a 10 mm precipitation event in April 1999, which was not detected by the ISBA-A-gs simulations. Conversely, in May 2000, a rise in the level of ET is predicted by the ISBA-A-gs model, whereas this is not predicted by the FAO-56 model. Prior to the May 2000 peak observed in the ISBA-A-gs model, one small (1 mm) rainfall event occurred, which probably had a very small influence on the real level of ET. It is thus likely that the discrepancy observed between the two models results from an exaggeration of this event, at the precipitation input of the ISBA-A-gs model.

Finally, the cumulative seasonal water requirement maps for the 1998–1999, 1999–2000 and 2004–2005 agricultural seasons, estimated by the FAO-56 dual crop coefficient model combined with satellite data, are shown in Figure 6. It should be noted that these maps are well matched with the land-use maps. As an example, during the 1998–1999 agricultural season the water requirements were slightly greater for annual crops (400 mm) than for olive trees (200 mm). The very low ET areas generally correspond to olive trees with a low vegetation cover fraction, as can be seen in Figure 2. All of the areas covered by water (dams, sebkhas) were masked, and were not taken into account in the ET calculations.

By comparing the water requirements of annual crops with the average annual rainfall in the Kairouan plain (300 mm), an initial estimation is made of the hydric deficit, which should be offset through the use of irrigation.

Figure 6. Total annual evapotranspiration maps: (**a**) 1998–1999 agricultural season; (**b**) 1999–2000 agricultural season; (**c**) 2004–2005 agricultural season.

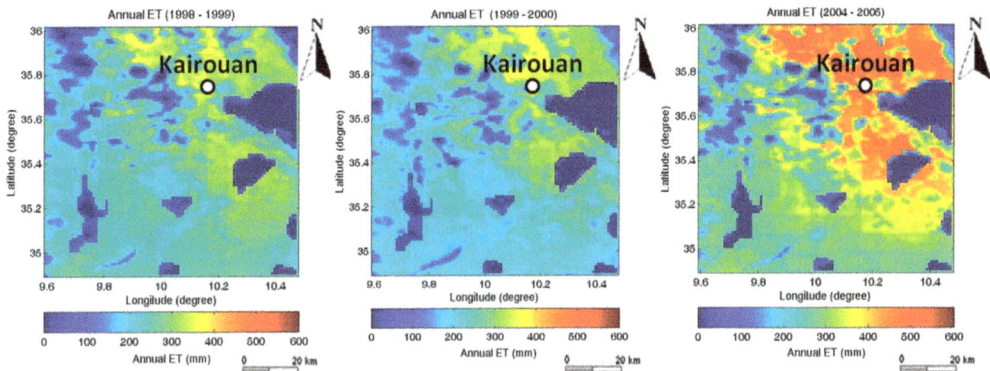

5. Conclusions

The FAO-56 dual approach, which is commonly used for irrigation management, is applied to the simulation of evapotranspiration at the regional scale. This is combined with remotely sensed, multi-sensor data. Two main types of vegetation cover, cereals and olive trees, are considered in this analysis. The vegetation fractions represented by these two classes of vegetation are retrieved by means of a multi-temporal classification of SPOT-VGT time series images. The evapotranspiration is computed over the studied area for two vegetation classes, weighted by their respective vegetation fractions.

The cereal crop coefficient and vegetation fraction estimations are based on NDVI SPOT-VGT data, through the use of an empirical relationship. Soil evaporation is estimated by a simple approach developed by Merlin, *et al.* [58] established between this quantity and surface soil moisture, using ERS/WSC radar products developed by the TU-Wien. Saturated soils are associated with the highest level of evaporation, and the driest soils have approximately zero evaporation.

The ISBA-A-gs SVAT model is compared to the FAO approach, using simulations covering the period between 1991 and 2007. The ISBA soil moisture outputs are validated using ERS/WSC products developed by the TU-Wien. A good degree of coherence is observed for surface moisture, with an RMSE equal to 0.04 $m^3 \cdot m^{-3}$, R^2 equal to 0.52 and a bias equal to 0.002. The soil moisture profiles are also in good agreement, with an RMSE equal to 0.03 $m^3 \cdot m^{-3}$, R^2 equal to 0.5, and a bias equal to 0.008. When the ISBA and FAO models are compared for the same study area, they are found to be strongly coherent. In the case of daily comparisons, an RMSE equal to 0.36 mm/day is found, which is low by comparison with the mean ET values, estimated at approximately 2 mm/day. The soil moisture profiles are well correlated, with R^2 equal to 0.5. These results illustrate the strong potential of this simple approach, in which the FAO-56 model is combined with several satellite observations (optical and microwaves), to retrieve evapotranspiration levels particularly in semi-arid regions.

Acknowledgments

This study was funded by two sources: the French national MISTRALS program, and the ANR AMETHYST "ANR-12-TMED-0006-01". The authors extend their thanks to VITO for kindly providing them with its SPOT-VEGETATION NDVI products, and to the ISIS program for providing them with the SPOT images used in this study. They would also like to thank the Tunisian Ministry of Agriculture for providing the precipitation data used in this study, as well as the technical teams of the IRD, INAT, CTV-Chebika and INGC for their strong collaboration and support with the implementation of ground-truth measurements.

Author Contributions

Rim Amri and Mehrez Zribi proposed modifications and application of FAO-56 model and discussions of results. Gilles Boulet helps on analysis and interpretation of the use of FAO-56 model. Zohra Lili-Chabaane participates to ground measurements and analysis of correlation between results and the site climate. Camille Szczypta and Jean-Christophe Calvet proposed simulations of ISBA-A-gs model and participate to interpretation of comparison between FAO-56 and ISBA-Ags.

Conflicts of Interest

The authors declare no conflict of interest.

References

1. Amri, R.; Zribi, M.; Lili-Chabaane, Z.; Duchemin, B.; Gruhier, C.; Chehbouni, A. Analysis of vegetation behavior in a north African semi-arid region, using SPOT-Vegetation NDVI data. *Remote Sens.* **2011**, *3*, 2568–2590.
2. Mueller, B.; Hirschi, M.; Jimenez, C.; Ciais, P.; Dirmeyer, P.A.; Dolman, A.J.; Fisher, J.B.; Jung, M.; Ludwig, F.; Maignan, F.; *et al.* Benchmark products for land evapotranspiration: LandFlux-EVAL multi-data set synthesis. *Hydrol. Earth Syst. Sci.* **2013**, *17*, 3707–3720.

3. Braud, I.; Dantas Antonio, A.C.; Vauclin, M.; Thony, J.L.; Ruelle, P. A Simple Soil Plant Atmosphere Transfer Model (SisPAT): Development and field verification. *J. Hydrol.* **1995**, *166*, 213–250.

4. Mahfouf, J.-F.; Manzi, O.; Noilhan, J.; Giordani, H.; Déqué, M. The land surface scheme ISBA within the Météo-France climate model ARPEGE. Part I: Implementation and preliminary results. *J. Clim.* **1995**, *8*, 2039–2057.

5. Calvet, J.C.; Noilhan, J.; Roujean, J.L.; Bessemoulin, P.; Cabelguenne, M.; Alioso, A.; Wigneron, J.P. An interactive vegetation SVAT model tested against data from six contrasting sites. *Agric. For. Meteorol.* **1998**, *92*, 92–95.

6. Saux-Picart S.; Ottlé, C.; Perrier, A.; Decharme, B.; Coudert, B.; Zribi, M.; Boulain, N.; Cappelaere, B.; Ramier, D. SEtHyS_Savannah: A multiple source land surface model applied to sahelian landscapes. *Agric. For. Meteorol.* **2009**, *149*, 1421–143.

7. Allen, R.G.; Pereira, L.S.; Raes, D.; Smith, M. Chapter 5. In *Crop Evapotranspiration-Guidelines for Computing Crop Water Requirements, Irrigation and Drain*; Paper No. 56; Food and Agriculture Organization: Rome, Italy, 1998; p. 300.

8. Allen, R.G. Using the FAO-56 dual crop coefficient method over an irrigated region as part of an evapotranspiration intercomparison study. *J. Hydrol.* **2000**, *229*, 27–41.

9. Er-Raki, S.; Chehbouni, G.; Guemouria, N.; Duchemin, B.; Ezzahar, J.; Hadria, R. Combining FAO-56 model and ground-based remote sensing to estimate water consumptions of wheat crops in a semi-arid region. *Agric. Water Manag.* **2007**, *87*, 41–54.

10. Er-Raki, S.; Chehbouni, A.; Boulet, G.; Williams, D.G. Using the dual approach of FAO-56 for partitioning ET into soil and plant components for olive orchards in a semi-arid region. *Agric. Water Manag.* **2010**, *97*, 1769–1778.

11. Belaqziz, S.; Khabba, S.; Er-Raki, S.; Jarlan, L.; le Page, M.; Kharrou, M.H.; El Adnani, M.; Chehbouni, A. A new irrigation priority index based on remote sensing data for assessing the networks irrigation scheduling. *Agric. Water Manag.* **2013**, *119*, 1–9.

12. Campos, I.; Villodre, J.; Carrara, A.; Calera, A. Remote sensing-based soil water balance to estimate Mediterranean holm oak savanna (dehesa) evapotranspiration under water stress conditions. *J. Hydrol.* **2013**, *494*, 1–9.

13. Conrad, M.; Rahmann, M.; Machwitz, M.; Stulina, G.; Paeth, H.; Dech, S. Satellite based calculation of spatially distributed crop water requirements for cotton and wheat cultivation in Fergana Valley, Uzbekistan. *Glob. Planet. Chang.* **2013**, *110*, 88–98.

14. Mateos, L.; González-Dugo, M.P.; Testi, L.; Villalobos, F.J. Monitoring evapotranspiration of irrigated crops using crop coefficients derived from time series of satellite images. I. Method validation. *Agric. Water Manag.* **2013**, *125*, 81–91.

15. González-Piqueras, J. *Crop Evapotranspiration by Means of Remote Sensing Determination of the Crop Coefficient. Regional Scale Application: 08–29 Mancha Oriental Aquifer*; Universitat de València: Valencia, Spain, 2006.

16. Purevdorj, T.; Tateishi, R.; Ishiyama, T.; Honda, Y. Relationships between percent vegetation cover and vegetation indices. *Int. J. Remote Sens.* **1998**, *19*, 3519–3535.

17. Sellers, P.J. Canopy reflectance, photosynthesis and transpiration. *Int. J. Remote Sens.* **1985**, *6*, 1335–1372.

18. Deblonde, G.; Cihlar, J. A multiyear analysis of the relationship between surface environmental variables and NDVI over the Canadian landmass. *Remote Sens. Rev.* **1993**, *7*, 151–177.

19. Delbart, N.; Kergoat, L.; Toan, T.L.; Lhermitte, J.; Picard, G. Determination of phenological dates in boreal regions using normalized difference water index. *Remote Sens. Environ.* **2005**, *97*, 26–38.

20. Myneni, R.B.; Los, S.O.; Asrar, G. Potential gross primary productivity of terrestrial vegetation from 1982 to 1990. *Geophyis. Res. Lett.* **1995**, *22*, 2617–2620.

21. Propastin, P.; Kappas, M. Modeling net ecosystem exchange for grassland in Central Kazakhstan by combining remote sensing and field data. *Remote Sens.* **2009**, *1*, 159–183.

22. Laurila, H.; Karjalainen, M.; Kleemola, J.; Hyyppä, J. Cereal yield modeling in Finland using optical and radar remote sensing. *Remote Sens.* **2010**, *2*, 2185–2239.

23. Fraser, R.S.; Kaufman, Y.J. The relative importance of aerosol scattering and absorption in remote sensing. *IEEE Trans. Geosci. Remote Sens.* **1985**, *23*, 625–633.

24. Holben, B.N.; Kaufaman, Y.J.; Kendall, J.D. NOAA-11 AVHRR visible and near-IR inflight calibration. *Int. J. Remote Sens.* **1990**, *11*, 1511–1519.

25. Simonneaux, V.; Duchemin, B.; Helson, D.; Er-Raki, S.; Olioso, A.; Chehbouni, A.G. The use of high resolution image time series for crop classification and evapotranspiration estimate over an irrigated area in central Morocco. *Int. J. Remote Sens.* **2008**, *29*, 95–116.

26. Jackson, T.J.; Schmugge, J.; Engman, E.T. Remote sensing applications to hydrology: Soil moisture. *Hydrol. Sci. J.* **1996**, *41*, 517–530.

27. Ulaby, F.T.; Dubois, P.C.; van Zyl, J. Radar mapping of surface soil moisture. *J. Hydrol.* **1996**, *184*, 57–84.

28. Paris Anguela, T.; Zribi, M.; Baghdadi, N.; Loumagne, C. Analysis of local variation of soil surface parameters with TerraSAR-X radar data over bare agricultural fields. *IEEE Trans. Geosci. Remote Sens.* **2010**, *48*, 874–881.

29. Calvet, J.C.; Wigneron, J.P.; Walker, J.; Karbou, F.; Chanzy, A.; Albergel, C. Sensitivity of passive microwave observations to soil moisture and vegetation water content: L-Band to W-Band. *IEEE Trans. Geosci. Remote Sens.* **2011**, *49*, 1190–1199.

30. Das, N.N.; Entekhabi, D.; Njoku, E.G. An algorithm for merging SMAP radiometer and radar data for high-resolution soil-moisture retrieval. *IEEE Trans. Geosci. Remote Sens.* **2011**, *49*, 1504–1512.

31. Kolassa, J.; Aires, F.; Polcher, J.; Prigent, C.; Jimenez, C.; Pereira, J.M. Soil moisture retrieval from multi-instrument observations: Information content analysis and retrieval methodology. *J. Geophys. Res.: Atmos.* **2013**, *118*, 4847–4859.

32. Wagner, W.; Noll, J.; Borgeaud, M.; Rott, H. Monitoring soil moisture over the Canadian Prairies with the ERS scatterometer. *IEEE Trans. Geosci. Remote Sens.* **1999**, *37*, 206–216.

33. Zribi, M.; Le Hegarat-Mascle, S.; Ottlé, C.; Kammoun, B.; Guerin, C. Surface soil moisture estimation from the synergistic use of the (multi-incidence and multi-resolution) active microwave ERS Wind Scatterometer and SAR data. *Remote Sens. Environ.* **2003**, *86*, 30–41.

34. Zribi, M.; André, C.; Decharme, B. A method for soil moisture estimation in Western Africa based on the ERS scatterometer. *IEEE Trans. Geosci. Remote Sens.* **2008**, *46*, 438–448.

35. Ceballos, A.; Scipal, K.; Wagner, W.; Martinez-Fernandez, J. Validation of ERS scatterometer-derived soil moisture data in the central part of the Duero Basin, Spain. *Hydrol. Process.* **2005**, *19*, 1549–1566.

36. Pellarin, T.; Calvet, J.C.; Wagner, W. Evaluation of ERS scatterometer soil moisture products over a half-degree region in southwestern France. *Geophys. Res. Lett.* **2006**, *33*, doi:10.1029/2006GL027231.

37. Paris Anguela, T.; Zribi, M.; Hasenauer, S.; Habets, F.; Loumagne, C. Analysis of surface and root-zone soil moisture dynamics with ERS scatterometer and the hydrometeorological model SAFRAN-ISBA-MODCOU at Grand Morin watershed (France). *Hydrol. Earth Syst. Sci.* **2008**, *12*, 1415–1424.

38. Zribi, M.; Chahbi, A.; Shabou, M.; Lili-Chabaane, Z.; Duchemin, B.; Baghdadi, N.; Amri, R.; Chehbouni, A. Soil surface moisture estimation over a semi-arid region using ENVISAT ASAR radar data for soil evaporation evaluation. *Hydrol. Earth Syst. Sci.* **2011**, *15*, 345–358.

39. Naeimi, V.; Bartalis, Z.; Wagner, W. ASCAT soil moisture: An assessment of the data quality and consistency with the ERS scatterometer heritage. *J. Hydrometeorol.* **2008**, doi:10.1175/2008JHM1051.1.

40. Albergel, C.; Dorigo, W.; Balsamo, G.; Muñoz-Sabater, J.; de Rosnay, P.; Isaksen, L.; Brocca, L.; de Jeu, R.; Wagner, W. Monitoring multi-decadal satellite earth observation of soil moisture products through land surface reanalyses. *Remote Sens. Environ.* **2013**, *138*, 77–89.

41. Wagner, W.; Scipal, K. Large scale soil moisture mapping in western Africa using the ERS scatterometer. *IEEE Trans. Geosci. Remote Sens.* **2000**, *38*, 1777–1782.

42. Su, C.; Ryu, D.; Young, R.; Western, A.; Wagner, W. Inter-comparison of microwave satellite soil moisture retrievals over the Murrumbidgee Basin, southeast Australia. *Remote Sens. Environ.* **2013**, *134*, 1–11.

43. Amri, R.; Zribi, M.; Lili-Chabaane, Z.; Wagner, W.; Hauesner, S. Analysis of ASCAT-C band scatterometer estimations derived over a semi-arid region. *IEEE Trans. Geosci. Remote Sens.* **2012**, *50*, 2630–2638.

44. Loew, A.; Stacke, T.; Dorigo, W.; de Jeu, R.; Hagemann, S. Potential and limitations of multidecadal satellite soil moisture observations for selected climate model evaluation studies. *Hydrol. Earth Syst. Sci.* **2013**, *17*, 3523–3542.

45. Holben, B.N. Characteristics of maximum-value composite images from temporal AVHRR data. *Int. J. Remote Sens.* **1986**, *7*, 1417–1434.

46. Rahman, H.; Dedieu, G. SMAC: A simplified method for the atmospheric correction of satellite measurements in the solar spectrum. *Int. J. Remote Sens.* **1994**, *15*, 123–143.

47. Maisongrande, P.; Duchemin, B.; Dedieu, G. VEGETATION/SPOT: An operational mission for the Earth monitoring; Presentation of new standard products. *Int. J. Remote Sens.* **2004**, *25*, 9–14.

48. Teegavarapu, R.; Chandramouli, V. Improved weighting methods, deterministic and stochastic data-driven models for estimation of missing precipitation records. *J. Hydrol.* **2005**, *312*, 191–206.

49. SPOT-VEGETATION Images. Available online: http://free.vgt.vito.be (accessed on 27 January 2014).

50. Shepard, D. A Two Dimensional Interpolation Function for Regularly Spaced Data. In Proceedings of the National Conference of the Association for Computing Machinery, Princeton, NJ, USA, 30 April–2 May 1968; pp. 517–524.

51. Monteith, J.L. *Evaporation and Environment.* In Symposia of the Society for Experimental Biology Number XIX the State and Movement of Water in Living Organisms; Cambridge University Press: Cambridge, UK, 1965; pp. 205–234.

52. Olseth, J.A.; Skartveit, A. Solar irradiance, sunshine duration and daylight illuminance derived from METEOSAT data at some European site. *Theor. Appl. Climatol.* **2001**, *69*, 239–252.

53. Solar Radiation Databases for Environment. Available online: http://www.soda-is.com (accessed on 27 January 2014).

54. Settle, J.J.; Drake, N.A. Linear mixing and the estimation of ground cover proportions. *Int. J. Remote Sens.* **1993**, *14*, 1159–1177.

55. Testi, L.; Villalobos, F.J.; Orgaz, F. Evapotranspiration of a young irrigated olive orchard in southern Spain. *Agric. For. Meteorol.* **2004**, *121*, 1–18.

56. Mahfouf, J.F.; Noilhan, J. Comparative study of various formulations of evaporation from bare soil using *in situ* data. *J. Appl. Meteorol. Climatol.* **1991**, *30*, 1354–1365.

57. Chanzy, A.; Bruckler, L. Significance of soil surface moisture with respect to daily bare soil evaporation. *Water Resour. Res.* **1993**, *29*, 1113–1125.

58. Merlin, O.; Al Bitar, A.; Rivalland, V.; Beziat, P.; Ceschia, E.; Dedieu, G. An analytical model of evaporation efficiency for unsaturated soil surfaces with an arbitrary thickness. *J. Appl. Meteorol. Climatol.* **2011**, *50*, 457–471.

59. Deardorff, J.W. Efficient prediction of ground surface temperature and moisture, with inclusion of a layer of vegetation. *J. Geophys. Res.* **1978**, *83*, 1889–1903.

60. Douville, H.; Royer, J.-F.; Mahfouf, J.-F. A new snow parameterization for the Meteo-France climate model. Part 1: Validation in stand-alone experiments. *Clim. Dyn.* **1995**, *12*, 21–35.

61. Boone, A.; Calvet, J.-C.; Noilhan, J. Inclusion of a third soil layer in a land surface scheme using the force restore method. *J. Appl. Meteorol.* **1999**, *38*, 1611–1630.

62. Beven, K.; Kirkby, M.J. A physically based variable contributing area model of basin hydrology. *Hydrol. Sci. Bull.* **1979**, *24*, 43–69.

63. Decharme, B.; Douville, H.; Boone, A.; Habets, F.; Noilhan, J. Impact of an exponential profile of saturated hydraulic conductivity within the ISBA LSM: Simulations over the Rhône basin. *J. Hydrometeorol.* **2006**, *7*, 61–80.

64. Jarvis, P.G. The interpretation of the variations in the leaf water potential and stomatal conductance found in canopies in the field. *Philos. Trans. R. Soc. Lond.* **1976**, *273*, 593–610.

65. Arora, V.K. Modeling vegetation as a dynamic component in soil-vegetationatmosphere transfer schemes and hydrological models. *Rev. Geophys.* **2002**, *40*, doi:10.1029/2001RG000103.

66. Noilhan, J.; Mahfouf, J.-F. The ISBA land surface parameterization scheme. *Glob. Planet. Chang.* **1996**, *13*, 145–159.

67. Calvet, J.-C. Investigating soil and atmospheric plant water stress using physiological and micrometeorological data. *Agric. For. Meteorol.* **2000**, *103*, 229–247.

68. Calvet, J.-C.; Rivalland, V.; Picon-Cochard, C.; Guehl, J.-M. Modelling forest transpiration and CO_2 fluxes—Response to soil moisture stress. *Agric. For. Meteorol.* **2004**, *124*, 143–156.

69. Szczypta, C.; Calvet, J.-C.; Albergel, C.; Balsamo, G.; Boussetta, S.; Carrer, D.; Lafont, S.; Meurey, C. Verification of the new ECMWF ERA-Interim reanalysis over France. *Hydrol. Earth Syst. Sci.* **2011**, *15*, 647–666.

70. Szczypta, C.; Decharme, B.; Carrer, D.; Calvet, J.-C.; Lafont, S.; Faroux, S.; Somot, S.; Martin, E. Impact of precipitation and land biophysical variables on the simulated discharge of European and Mediterranean rivers. *Hydrol. Earth Syst. Sci.* **2012**, *16*, 3351–3370.

71. Benhadj, I. Observation Spatial de L'Irrigation D'Agrosystèmes Semi-Arides et Gestion Durable de la Ressource en Eau en Plaine de Marrakech. Thèse de Doctorat, Université Paul Sabatier-Toulouse III, Toulouse, France, 2008; Chapter 4, p. 296.

Validation of Global Evapotranspiration Product (MOD16) using Flux Tower Data in the African Savanna, South Africa

Abel Ramoelo, Nobuhle Majozi, Renaud Mathieu, Nebo Jovanovic, Alecia Nickless and Sebinasi Dzikiti

Abstract: Globally, water is an important resource required for the survival of human beings. Water is a scarce resource in the semi-arid environments, including South Africa. In South Africa, several studies have quantified evapotranspiration (ET) in different ecosystems at a local scale. Accurate spatially explicit information on ET is rare in the country mainly due to lack of appropriate tools. In recent years, a remote sensing ET product from the MODerate Resolution Imaging Spectrometer (MOD16) has been developed. However, its accuracy is not known in South African ecosystems. The objective of this study was to validate the MOD16 ET product using data from two eddy covariance flux towers, namely; Skukuza and Malopeni installed in a savanna and woodland ecosystem within the Kruger National Park, South Africa. Eight day cumulative ET data from the flux towers was calculated to coincide with the eight day MOD16 products over a period of 10 years from 2000 to 2010. The Skukuza flux tower results showed inconsistent comparisons with MOD16 ET. The Malopeni site achieved a poorer comparison with MOD16 ET compared to the Skukuza, and due to a shorter measurement period, data validation was performed for 2009 only. The inconsistent comparison of MOD16 and flux tower-based ET can be attributed to, among other things, the parameterization of the Penman-Monteith model, flux tower measurement errors, and flux tower footprint *vs.* MODIS pixel. MOD16 is important for global inference of ET, but for use in South Africa's integrated water management, a locally parameterized and improved product should be developed.

Reprinted from *Remote Sens.* Cite as: Ramoelo, A.; Majozi, N.; Mathieu, R.; Jovanovic, N.; Nickless, A.; Dzikiti, S. Validation of Global Evapotranspiration Product (MOD16) using Flux Tower Data in the African Savanna, South Africa. *Remote Sens.* **2014**, *6*, 7406-7423.

1. Introduction

Globally, water is an important resource required for the daily sustenance and survival of human beings. Water is crucial to facilitate livelihoods and economic growth, e.g., vital for the industrial and agricultural sector. Today, irrigated agriculture is the main fresh water user, accounting for about 70% of the water from lakes, rivers and ground aquifers [1]. In South Africa, almost 50% of the available surface water resource usage is attributed to agricultural activities. In essence, South Africa is a semi-arid environment, with evaporation rates exceeding the rates of precipitation by a considerable margin [2]. Therefore, it is crucial to develop methods or tools to quantify water use and availability (e.g., evapotranspiration, ET) over large spatial scales in order to inform decision makers on sustainable utilization and management of this resource. For example, the program by the South African Department of Water Affairs (DWA) to verify and validate country's water use is critical [2] for water use license purposes, which requires information on ET, water use and river flows. This program is also linked to the legal requirement on the sustainable utilization of water resources outlined in the National Water Act (Act 36 of 1998).

ET is the second most important element of the hydrological cycle after precipitation because it facilitates the continuation of precipitation by replacing the vapor lost through condensation [3]. ET is listed as one of 48 observation priorities of water societal benefit area (Water SBA) by the Group on Earth Observation; see GEO [4,5]. GEO is an intergovernmental organization working to improve availability, access and use of earth observation to benefit society [5]. ET is also crucial for the transportation of minerals and nutrients required for plant growth; creates a beneficial cooling process to plant canopies in many climates; and influences the Earth's energy and water balance because of the direct association with latent heat flux (LE). ET consumes large amounts of energy during the conversion of liquid water to vapor, hence playing an important role in hydrology, agriculture, climatology and meteorology. Accurate estimates of ET contribute to improved quantification of the catchment water balance and in the facilitation of decision making for sustainable water resource management [6–8].

ET is a difficult component to measure, especially in arid and semi-arid regions where the magnitude of the ET flux is relatively smaller than in wetter regions. In these areas, plants are prone to water stress and are adapted accordingly to prolonged dry conditions. ET is highly variable over space and time, depending on landscape heterogeneity, topography, climate, vegetation type, soil properties, management and environmental constraints [7,9]. The conventional point-based ET estimation methods do not capture large spatial scale variability of ET and are very difficult to obtain due to time and cost constraints. These methods include (i) direct measurements with porometry and lysimeters [10]; (ii) atmospheric measurements, including energy balance and micrometeorological techniques like Bowen ratio [11], eddy covariance [12], scintillometry [13], as well as methods based on weather data, for example; used for the calculation of the Penman-Monteith reference grass evapotranspiration (ETo) and crop coefficient [9]; (iii) soil measurements [14] and the application of the soil water balance. The *in situ* estimation of ET using the above techniques was successful in a number of agricultural and natural environments within South Africa, e.g., natural vegetation [15], wetlands [13] and crops [16].

Remote sensing-based techniques have the capability to estimate spatial and temporal variation of ET from catchment to global scales. Several studies reviewed different remote sensing techniques used to estimate ET [17–19]. They are classified into;

- empirical methods that involve the use of statistically-derived relationships between ET and vegetation indices such as the normalized difference vegetation index (NDVI) or the enhanced vegetation index (EVI) [20–25],
- residual methods of surface energy balance (single- and dual-source models) [8,26] which include the Surface Energy Balance Algorithm over Land (SEBAL) [27,28], Surface Energy Balance System (SEBS) [8,29,30] and Mapping EvapoTranspiration at high Resolution with Internalized Calibration (METRIC) [6,31,32],
- physically-based methods that involve the application of the combination of Penman-Monteith [7,33,34] and Priestley-Taylor types of equations [35–39], and

- Data assimilation methods adjoined to the heat diffusion equation [40] and through the radiometric surface temperature sequences [41].

Various remote sensing-based ET estimation algorithms (e.g., SEBS, SEBAL and METRIC) have been partially evaluated in South Africa [42]. Although estimates of net radiation and ET were accurate, the soil and sensible heat fluxes were more complex and challenging to determine. Furthermore, Sun et al. [43] used local flux tower data (Skukuza) to evaluate a remote sensing-based continental ET product at 3 km resolution, which was developed by combining data from the MODIS sensor and SEVIRI sensor onboard the geostationary-orbiting MSG satellite. Although the results were reasonable during the wet season, low correlations were observed during the dry season, due to factors such as, the spatial scale differences between the satellite sensor and flux tower observations. In South Africa, SEBAL was used to provide information on water use efficiency of irrigated crops, including grapes, deciduous fruits, sugarcane, and grain crops [29]. A recent study by Gibson et al. [29], discussed the use of SEBS in South Africa for various agricultural and natural systems, and finally recommended the validation of existing global ET products in South Africa to encourage their use.

Remote sensing-based global estimates of ET have been produced by different algorithms. For example, the MODerate Resolution Imaging Spectroradiometer (MODIS) MOD16 [7,34], and the EUMETSAT MSG ET product [43]. The MOD16 ET product has a spatial resolution of 1 km and is available on an eight-day, monthly and yearly basis. The EUMETSAT MSG ET product is available at 3 km spatial resolution every 30 min or daily. These products have been calibrated and validated mainly in the Northern hemisphere, with sites located in North and South America, Europe, Asia and sometimes Australia. For instance, Cleugh et al. [33] applied the Penman-Monteith algorithm and MODIS data to estimate ET for the Australian continent. Mu et al. [7,34] modified this algorithm and used flux tower data from Ameriflux stations to validate the global MOD16 product. Kim et al. [44] later validated the global product using Asiaflux stations. Jia et al. [36] evaluated spatiotemporal MODIS ET product in the Hai river basin. Inaccuracies, such as over- or underestimation, and no relationship were observed between the flux tower and MOD16 ET estimates in the above publications. Errors or uncertainties are assumed to be caused by misclassification of landcover types from the global MODIS land cover product, scaling from flux tower to landscape, and algorithm limitations. Accurate classification of land cover type is critical for ET estimation. In South Africa, the first attempt to evaluate the MOD16 ET was done by Jovanovic et al. [45], using historical in situ ET measurements. Jovanovic et al. [45] found that the MOD16 method underestimated ET, but the in situ data collected was not sufficient to evaluate remote sensing based products. Most of this in situ ET data were for small fields, not sufficient to cover a 1 km × 1 km pixel. There is a need to evaluate global remote sensing products again at long term monitoring sites in South Africa.

The main objective of this study is to evaluate the MOD16 global ET product using eddy covariance flux tower-derived ET in the savanna and woodland biome in the Kruger National Park, South Africa. Thirty-minute latent heat fluxes (LE) were acquired from the CARBOAFRICA project for Skukuza and Malopeni towers, and converted to daily ET. The flux tower-derived ET was then summed to an eight-day ET corresponding to the MOD16 ET values for comparison.

2. Materials and Methods

The historical flux tower data records from two sites in the Kruger National Park were used. The flux tower data were preprocessed to acquire daily, eight-day and monthly ET. The measured ET from the flux tower was then compared to the modelled MOD16 global ET product. We used basic statistical metrics, including the coefficient of determination, root mean square error and bias, to assess MOD16 ET against measured ET.

2.1. Study Area and Flux Tower Instrument

Two eddy covariance flux tower sites, Skukuza and Malopeni, were selected (Figure 1). From 2000 to 2005, the Skukuza flux tower was equipped with a closed path gas analyser for carbon dioxide and water, but from 2006 this changed to a Licor Li7500 open path gas analyser with a Campbell Scientific CSAT sonic anemometer. The Malopeni flux tower is equipped with a Licor Li7500 gas analyser and a Gill WindMaster Pro sonic anemometer. These flux towers contributed to the CARBOAFRICA network, a project which ran from 2007 to 2010, and was designed to contribute to the quantification, understanding and prediction of the carbon cycle and energy fluxes in Sub-Saharan Africa. Though these two sites are located in Kruger National Park, their localities present an interesting contrast in weather, soil, geology and vegetation types.

Established in 2000, the Skukuza flux tower (25.0197°S, 31.4969°E), lies at 365 m above the sea level, in an area with 547 mm/year of mean annual rainfall, which falls between November and April [46]. The annual temperature ranges between 14.5 and 29.5 °C. This site lies at the boundary of two distinct savanna vegetation types which include broad-leafed *Combretum* savanna and fine-leafed *Acacia* savanna. These contrasting savanna types occur on soils of differing texture, water holding capacity and nutrient levels, and are characterized by different physical structure, physiology and phenology [46]. These two savanna types are typical of the southern Kruger National Park, and placing the flux tower on the ecotone allows for an integrated measurement of the fluxes over these different savanna types.

The Malopeni flux tower (23.8325°S, 31.2145°E) is located in the northern part of KNP approximately 130 km north-west of Skukuza and 12 km from the town of Phalaborwa. The tower was established in 2009 on a site dominated by the broad-leaf *Colophospermum mopane* characteristic of a hot and dry savanna, 384 m above the sea level, with a mean annual rainfall of 472 mm/year [47]. Temperature ranges between 12.4 and 30.5 °C.

Figure 1. Study area map showing an insert of South African and Kruger National Park boundaries, with projected coordinates.

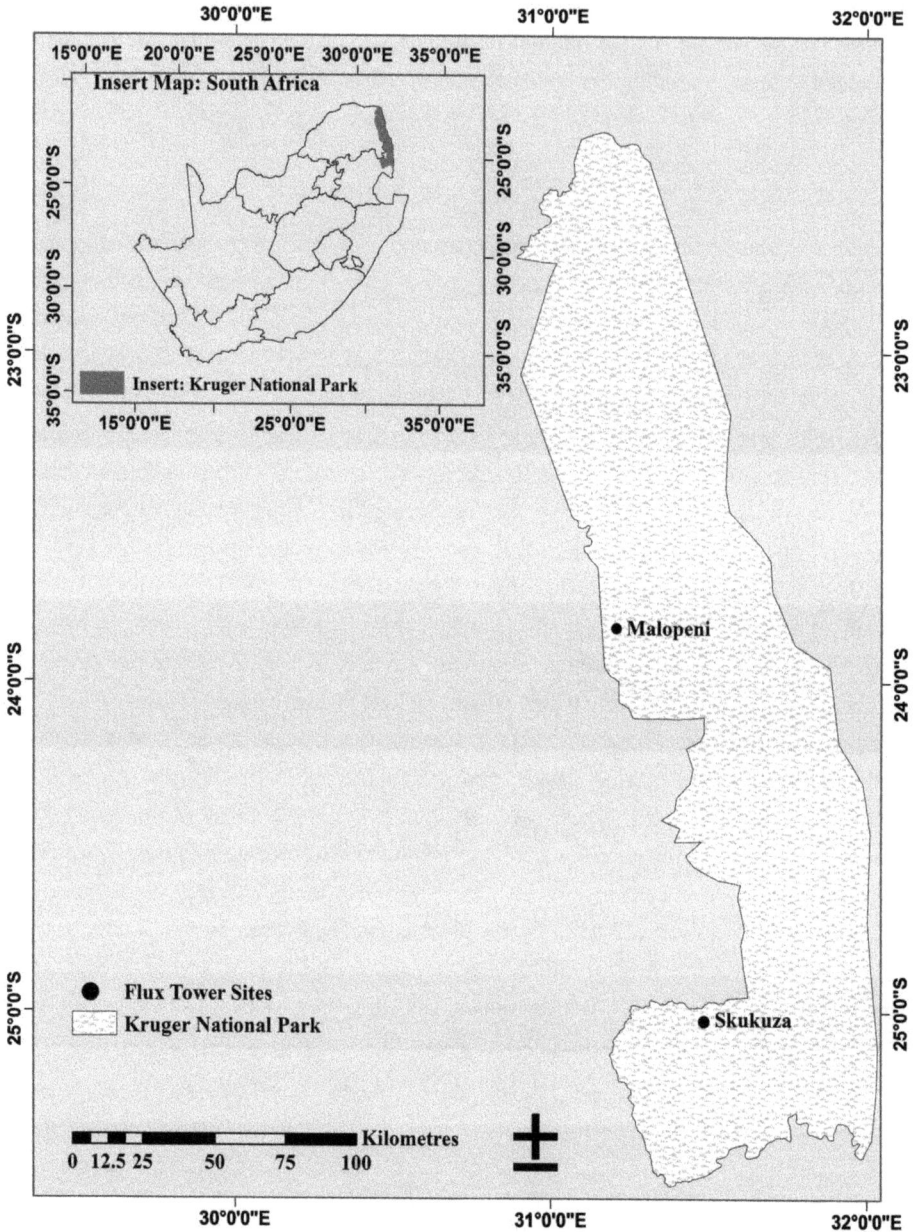

2.2. Flux Tower Data

To evaluate the global 1 km, eight-day MOD16 ET product, eddy covariance LE data from the Skukuza (2000–2010) excluding 2002 and 2006, and Malopeni (2009) flux towers were used. LE for 2002 and 2006 were excluded from the analysis, since the measurement years were incomplete. The 16 m and 7 m measurement height of Skukuza and Malopeni towers, respectively, are adequate to validate the 1 km pixel of MOD16. The size of eddy covariance source area or footprint does not only depend on instrument height [48], but also on the wind direction and velocity, atmospheric stability and the underlying surface conditions [49]. The source area or footprint modelling was not carried out because the location of the flux tower was homogeneous. The LE data observed every 30 min were MODIS-driven estimation of terrestrial latent heat flux in China based on a modified Priestley-Taylor algorithm converted to daily ET using equations presented in Mu *et al.* [34]. In addition, only reliable 30 min measurements were prioritized, exceeding 40 per day. The derived daily ET was further summed over eight days for each year to match the MOD16 ET product. Some of the data were excluded from analysis because of insufficient ET measurements. The number of the 30 min ET measurements per day (over 40) was prioritized in the validation process, to avoid compromising the completeness and reliability of the flux tower data. For further analysis, eight day summations were done to create monthly ET for Malopeni and Skukuza.

2.3. MOD16 Global ET Data

The MOD16 ET product with temporal resolution of eight days and spatial resolution of 1 km, were acquired for free from the University of Montana's Numerical Terradynamic Simulation group (ftp://ftp.ntsg.umt.edu/pub/MODIS/NTSG_Products/MOD16/MOD16A2.105_MERRAGMAO/). The ET values corresponding to Skukuza from 2000–2010 and 2009 for the Malopeni sites were extracted from each pixel of the MOD16 ET images using ArcGIS 10×. The algorithm used to derive the MOD16 ET product was modified by Mu *et al.* [7,34], from Cleugh *et al.* [33]'s Penman-Monteith derived model.

2.4. Rainfall Data

The rainfall data sets were collected from the South African Weather Service (SAWS) for Skukuza and Phalaborwa rainfall stations, for the dates corresponding to the ET data. Skukuza station is approximately 20 km away from the flux tower, while Phalaborwa rainfall station is located approximately 10 km away from the Malopeni flux tower. We used rainfall data to interpret variability in ET in relation with the wetness conditions in the landscape and the response of vegetation to drought and water stress.

2.5. Data Analysis

For assessing the relationship between the MOD16 ET and flux tower derived ET, the coefficient of determination (R^2), root mean square error (*RMSE*) (Equation (1)), bias (Equation (2)) and percent bias (PBias) (Equation (3)) were used. These statistical techniques were commonly used for comparing pairs of variables, e.g., Sun et al. [43]. The R^2 was used to determine the strength of the relationship between the flux tower measured and the MOD16 modelled ET. Bias, on the other hand, is a measure of how a modelled value deviates from the true value, and indicates whether there is under- or overestimation, while the percent bias is a percentage of bias relative to the observed mean.

The RMSE, Bias and PBias were computed using the following equations:

$$RMSE = \sqrt{\frac{\sum (FET\text{-}MET)^2}{N}} \tag{1}$$

$$Bias = \frac{\sum (MET\text{-}FET)}{N} \tag{2}$$

$$PBIAS = \frac{Bias}{(\frac{1}{N})\sum FET} \times 100 \tag{3}$$

where FET is flux tower ET, MET is MOD16 ET and N number of measurements. Bias and RMSE values close to zero signify that the MOD16 ET does not deviate from the true ET value (flux tower), indicating that the MOD16 is deemed accurate, while higher values of these statistic metrics indicate a high level of inaccuracy. A negative value of bias signifies underestimation, while a positive value shows overestimation by the modelled value or MOD16.

3. Results

For the Skukuza site, the results show an inconsistent comparison of the flux tower and MOD16 ET values over a period of time (Table 1; Figure 2). From 2000–2010, excluding 2002 and 2006, the highest correlations were obtained in 2005 and 2007 achieving R^2 of 0.81 and 0.85, respectively. According to the RMSE, 2003 ($R^2 = 0.58$, RMSE = 3.4 mm/8-days) and 2005 ($R^2 = 0.81$, RMSE = 2 mm/8-days) achieved the lowest values which indicate reasonable accuracy of the MOD16 ET product. In the years 2003 and 2005, the relationship is almost 1:1, with 2005 yielding the lowest RMSE, with the second highest R^2. In 2009 and 2010, there is almost a complete set of measurements from the flux tower measurements (36 and 44 out 45 eight-day periods, respectively). In 2009 and 2010, the validation results ($R^2 = 0.78$ and 0.74, respectively) are poorer compared to 2005 and 2007 based on the lower coefficient of determination, as well as the higher RMSE (7.39 and 4.3 mm/8-days). In general, Skukuza results showed that there is an overestimation of MOD16 ET in 2000 and 2004, with a Bias ranging from 2.80 to 3.08 mm/8-days (22.4–33.33% of the observed mean) (Table 1). Whilst, there was evidence of high underestimation of MOD16 ET in 2007, 2008, 2009 and 2010 with Bias ranging from −12.1 to −2.6 mm/eight-days (−55.7 to −16.44% of the observed mean). The 2003 and 2005 results, specifically, yielded low Bias and PBias which confirms the reasonable prediction of MOD16 during these years.

Table 1. Validation of MOD16 products using flux tower based evapotranspiration (ET) from Skukuza and Malopeni site.

Flux Tower	Year	R^2	RMSE (mm/8-days)	Bias (mm/8 day)	PBias (%)	No. of Measurements *
Skukuza	2000	0.26	5.22	3.08	33.33	18
	2001	0.35	3.60	0.58	6.63	26
	2003	0.58	3.40	−0.31	−3.02	32
	2004	0.54	8.00	2.85	22.38	37
	2005	0.81	2.00	−0.24	−3.20	26
	2007	0.85	6.00	−12.11	−55.71	32
	2008	0.36	7.40	−9.47	−51.68	34
	2009	0.78	7.39	−6.46	−29.50	36
	2010	0.74	4.30	−2.57	−16.44	44
Malopeni	2009	0.23	3.00	1.18	21.20	35

* A complete yearly eight days measurements should be 45, as per MOD16 dates.

Figure 2. Eight day validation results of MOD16 ET using Flux tower data (mm/eight-days) in Skukuza (2000–2010). The 1:1 line is depicted by a red color.

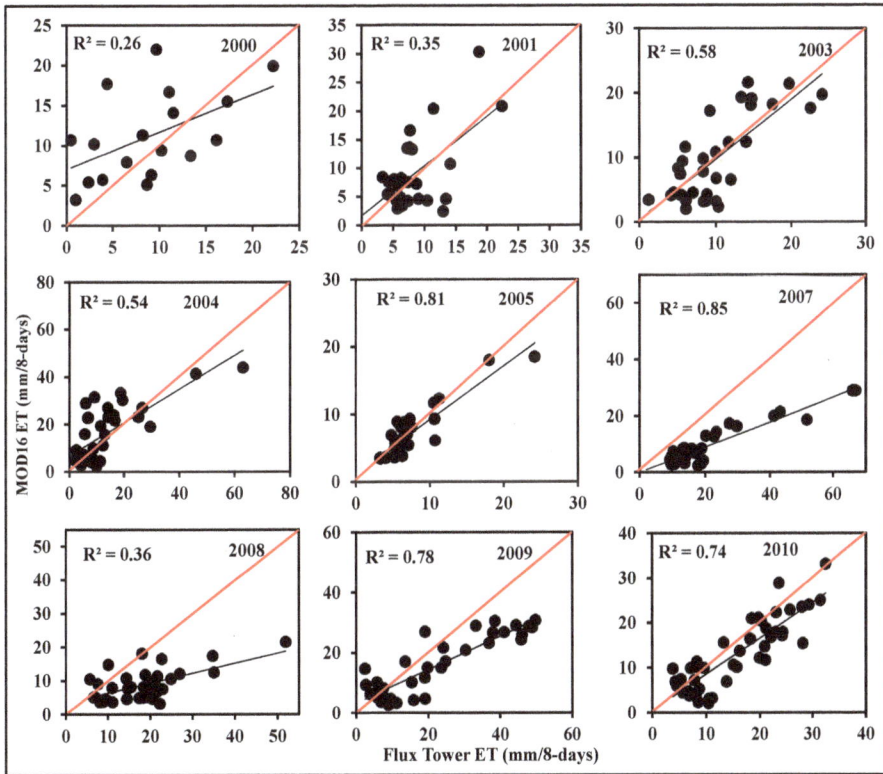

The eight-day and monthly comparison of the ET in Skukuza between January 2000 and December 2010 are shown in Figures 3 and 4. Flux tower ET values are generally higher than MOD16, especially during summer months (December–February), when most of the data gaps occur. During winter season (June–August), when the flux record is more complete, MOD16 and flux tower ET are closely related. This is a confirmation of the results for *Bias* and *PBias* showing systematic underestimation of ET by MODIS ET.

Figure 3. Eight-day time series comparison of MOD16, rainfall and Skukuza flux tower derived ET.

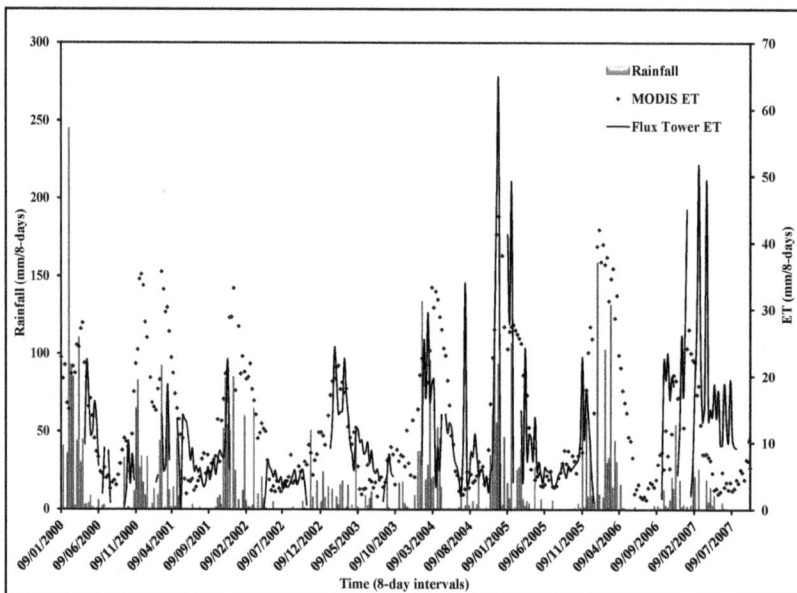

Figure 4. Monthly time series comparison of MOD16 and Skukuza flux tower derived ET.

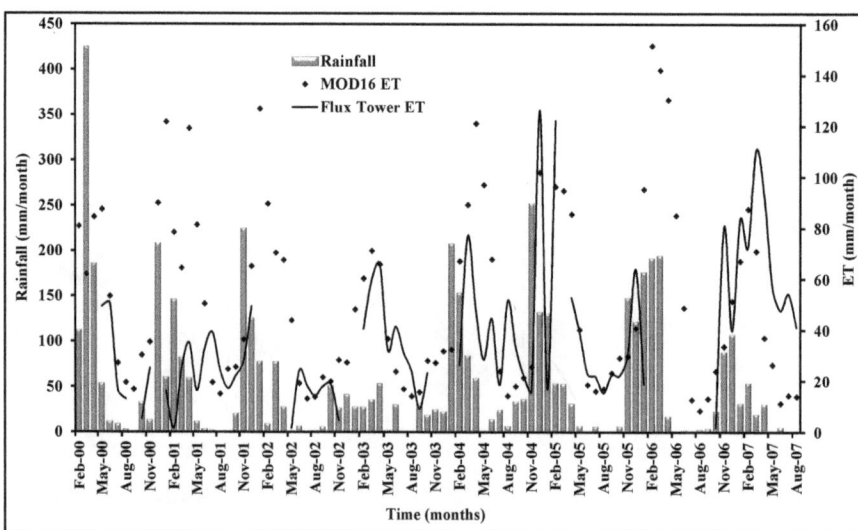

For the Malopeni site, the results show a relatively good *RMSE* (3 mm/8-days) with 35 out of 45 eight-day flux tower measurements (Table 1; Figure 5). Malopeni's validations are similar to those of Skukuza in 2001, 2003 and 2005 in terms of *RMSE*. In terms of the R^2-*value* (0.23), Malopeni validation is the lowest. Generally, the results of Malopeni show that MOD16 ET is overestimated (*Bias* = 1.18, *PBias* = 21% of the observed mean). Figures 6 and 7 show the time series visualization of eight-day and monthly ET. In both the eight-day and monthly time series, MOD16 ET started to increase around August–September while flux tower ET continues to drop.

Figure 5. Validation of MOD16 ET using flux tower data (mm/8-days) in Malopeni (2009). The 1:1 line is depicted by a red color.

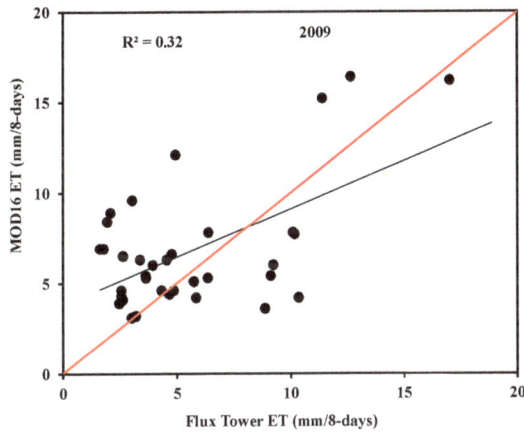

Figure 6. Eight-day time series comparison of MOD16 ET, rainfall and Malopeni flux tower derived ET.

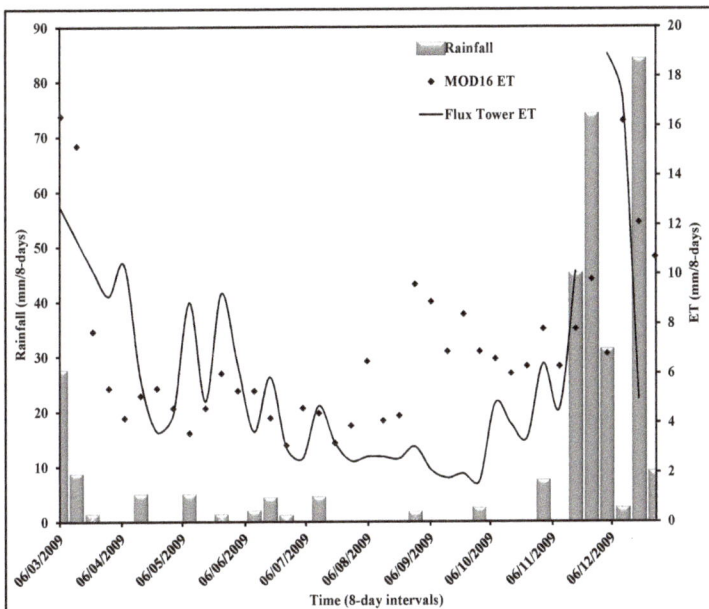

296

Figure 7. Monthly time series comparison of MOD16, rainfall and Malopeni flux tower derived ET.

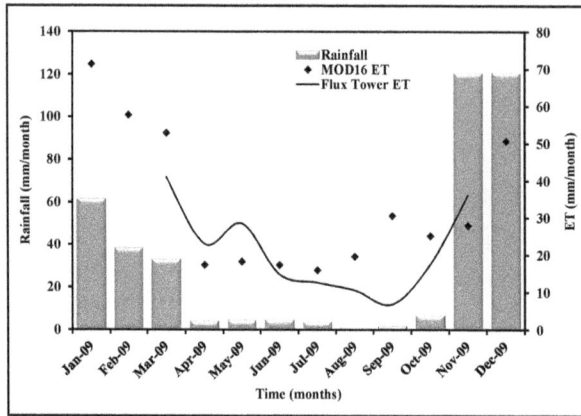

Figure 8 shows all-period data (2000–2010) comparison of eight day and monthly modelled MOD16 and measured flux tower ET for Skukuza as well as monthly for Malopeni in 2009. Results indicate a poor relationship between modelled MOD16 and measured flux tower ET. Eight-day and monthly comparisons yielded R^2 of 0.20 and 0.16, respectively. Monthly comparison for Malopeni obtained R^2 of 0.33, which is relatively higher than the correlation achieved in Skukuza site. This could be a consequence of the number of data points used, only one year for Malopeni and several years for Skukuza used. Generally, the results show that there is a poor relationship between MOD16 and flux tower ET when using all data sets from various dates in each site.

General trends as depicted by Figures 3 and 4 for Skukuza as well as Figures 6 and 7 for Malopeni show that both the flux tower and MODIS ET are to some extent related to rainfall variability (see Table 2). Table 2 shows further analysis of the relationship between rainfall and flux tower measured as well as MOD16 modelled ET for eight days and monthly data. Generally, there is a significant relationship between rainfall and ET. High rainfall peaks are associated with high ET values. In addition, missing flux tower ET values are associated with high rainfall or towards the end of the rainy season, since the measurements from the flux instruments become unreliable when wet.

Table 2. Analysis for the comparison between ET and rainfall.

Sites	Types	Duration	R^2	RMSE(mm/8-days)	* $P < 0.05$
Skukuza	Flux Tower	8day	0.14	8.85	Yes
		Mon	0.01	28.1	No
	MOD16	8day	0.15	9.41	Yes
		Mon	0.16	33.9	Yes
Malopeni	Flux Tower	8day	0.01	5.02	No
		Mon	0.41	8.99	Yes
	MOD16	8day	0.11	4.60	Yes
		Mon	0.32	13.77	Yes

* 95% confidence level, where YES indicates significance; Mon = Monthly.

Figure 8. (**Top left**) indicates all eight day comparisons between MODIS and Flux tower ET from 2000 to 2010, (**top right**) is a monthly comparison for Skukuza from 2000 to 2010 and (**bottom**) is a monthly comparison for Malopeni in 2009.

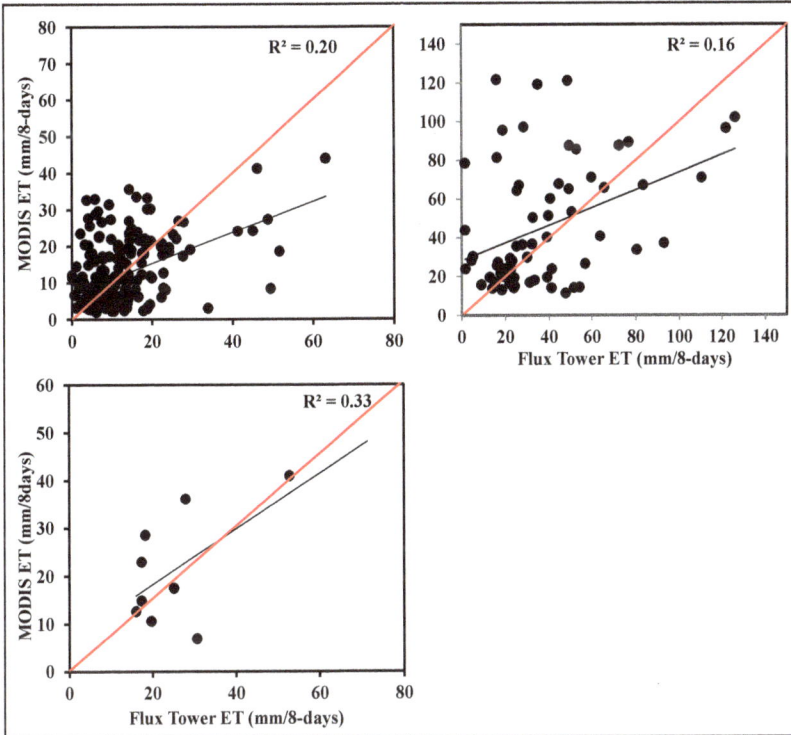

4. Discussion

The paper focused on the evaluation of the MOD16 modelled ET product in South African savannas using flux tower measured ET. Discrepancies between MOD16 ET and flux tower ET can originate from a number of factors, including the parameterization (input data) of the Penman-Monteith model, flux tower measurement error, flux tower footprint *vs.* MODIS pixel size as well as the limitations of the algorithm, most of which were identified by Mu *et al.* [7,34]. The main input data for the MODIS ET modelling include MODIS derived global products such as land cover [50], albedo, leaf area index (LAI), fraction of photosynthetic absorbed radiation (FPAR) as well as meteorological data. These input parameters are coarse scale products, generally poorly or not validated in the semi-arid conditions of South Africa, which are likely to generate significant ET prediction errors. For instance, MODIS global land cover (MOD12Q1) is a relatively coarse product (500 m) which inadequately captures the heterogeneity of savanna ecosystems. Further, the global MODIS based LAI or FPAR products have not been validated in Southern Africa. For generating the LAI product, an inversion of the physically-based model such as PROSAIL is used [51]. However, a backup algorithm based on LAI *vs.* NDVI relationship is used, when the inversion of the physically based model does not provide a solution [51]. In the semi-arid environments like South Africa, it is likely that the LAI is based on the latter approach because of the global parameterization. Therefore, it is crucial that the input data such as land cover, FPAR and LAI are also assessed and validated in the local context, and improved when needed. This exercise will help determine and document error propagation within the MOD16 algorithm and support the development of local parameterization of models for an integrated water management system. Sensitivity analyses are required to identify the variables which influence the ET output the most, and to document the level of agreement between input and output errors.

Uncertainties associated with the flux tower measurements could have also influenced the results. The flux towers have an energy balance closure problem due to the fact that the sum of the net radiation and the ground heat flux is sometimes larger than the sum of the turbulent fluxes of latent and sensible heat [52]. The energy balance closure problem was not corrected in this paper due to the lack of reliable ground heat flux measurement. Flux tower measurements are largely influenced by weather conditions. During rainy and stormy days, flux tower sensors either record abnormal values or simply do not record any data. The missing flux tower measurements affected the cumulative eight-day ET. The advantage of having full eight-day measurements was evident in the 2010 datasets, which provided the best overall relationship between the flux tower and MOD16 ET measurements. The low correlations obtained during the dry season are similarly observed in other regions and probably related to a lag in detecting the plant water stress using remote sensing techniques [53]. The relationship between rainfall with flux tower and MOD16 ET (Table 2) demonstrated that rainfall has a significant influence in the variability of ET, and hence the estimation of ET.

Spatial discrepancy may still exist between the footprint of the flux tower measurements and the MODIS pixel. The height of the sensors on a flux tower [48], wind direction or velocity, atmospheric stability and underlying surface conditions influence the size of the eddy covariance source area [49]. The measurement heights of Skukuza and Malopeni are 16 m and 7 m, respectively, thus the footprints of these towers are 1.6 km and 0.7 km, respectively. The Skukuza flux tower footprint provides a better match to the MODIS pixel size compared to the Malopeni flux tower. In addition, the layout of a single 1 km pixel may not directly match the flux tower footprints and may add further spatial discrepancy between the two. Footprint modelling is a means to reduce the spatial discrepancy between flux tower measurements and MODIS pixel [36], but this was beyond the scope of this study.

Shortcomings associated with the algorithm itself could have influenced the differences between flux tower and MOD16 ET. Mu et al. [34] argued that several physical factors such as micro-climate, plant biophysics for site specific species and landscape heterogeneity influence the soil surface evaporation and plant transpiration processes, which likely affect MOD16 ET estimation accuracy. The MOD16 ET does not account for disturbance history or species composition and stand age [7,34], which could also add further uncertainty. The algorithm makes the assumption that the stomata close during the night, while studies such as Musselman and Minnick [54] have reported stomata opening during the night. This induces underestimation of daily ET because of the bias imposed by night time vegetation transpiration [34].

5. Conclusions

The study evaluated the quality of the MOD16 ET global products. Generally, MOD16 is poor and the accuracy is not consistent over a period of time in selected savanna ecosystem sites. The quantification of errors associated with the MOD16 ET product in the savanna ecosystem presents new findings. The MOD16 product underestimated ET with errors ranging from 2–7 mm/8 days in the Skukuza site and 3 mm/8 days in the Malopeni site. The evaluation of this product and quantified errors has been undertaken exhaustively for the first time in the dry savanna ecosystem, especially in South Africa. Rainfall was found to significantly ($p < 0.05$) influence ET distribution and is associated with the missing data. Several factors could have influenced the inconsistency between MOD16 and flux tower derived ET, including parameterization of the model, scaling from flux tower measurement to a pixel as well as limitations associated with the algorithm used. For further evaluation of MOD16, footprint modelling for the eddy covariance source area should be done to ensure spatial representativeness or to reduce errors associated with scaling from flux tower measurements to a pixel. In addition, the energy balance closure problem should be analyzed, provided that there is reliable soil heat flux data. In future, there is a need to develop locally parameterized models for consistent estimation and mapping of ET in South Africa. It is important to understand existing ET estimation methods in order to improve ET estimation for the South African environment. In addition, future activities should also focus on the improvement of the estimation accuracy of other remote sensing derived input variables such as LAI, albedo and land cover. Accurate and consistent estimation and mapping of ET is crucial for understanding plant or crop water use which is an important component of integrated water resource management.

Acknowledgments

We would like to thank Council for Scientific and Industrial Research for the funding. We are also grateful to CARBOAFRICA project for providing access to the flux tower data. We would also like to thank University of Montana and NASA for the free access to MOD16 Global ET data.

Author Contributions

The manuscript was developed as part of the project entitled "Monitoring of water availability using geo-spatial data and earth observations" funded by the Council for Scientific and Industrial Research led by Nebo Jovanovic, and all the co-authors in this paper are co-investigators in this project. Abel Ramoelo was the main author involved in data collection, data analysis, and paper writing. Nobuhle Majozi contributed to the data collection and paper writing. Nebo Jovanovic and Renaud Mathieu played a supervisory role, involved in the inception of the project idea and contributed to the paper writing. Alecia Nickless was involved in data collection and preprocessing as well as paper writing. Sebinasi Dzikiti contributed to the data analysis and paper writing. Each author contributed immensely to the development of this manuscript.

Conflicts of Interest

The authors declare no conflict of interest.

References

1. Cai, X.; Rosegrant, M.W. Global water demand and supply projections, Part 2: Results and prospects to 2025. *Water Int.* **2002**, *27*, 170–182.
2. Department of Water Affairs and Forestry (DWAF). A Guide to Verifying the Extent of Existing Lawful Water Use. Available online: www.dwaf.gov.za/WAR/documents/VerificationGuide2EdNov06.pdf (accessed on 1 April 2014).
3. Brutsaert, W. *Hydrology: An Introduction*, 4th ed.; Cambridge University Press: New York, NY, USA, 2009.
4. Group on Earth Observation. GEO Task US-09–01a: Critical Earth Observations Priorities Water Societal Benefit Area. Final SBA Report. Available online: http://sbageotask.larc.nasa.gov/water.html (accessed on 22 October 2013).
5. Group on Earth Observations. GEO Task US-09–01a Critical Earth Observation Priorities. Final Report, 2nd ed.; Available online: http://sbageotask.larc.nasa.gov/docpres.html (accessed on 22 October 2013).
6. Allen, R.G.; Tasumi, M.; Morse, A.; Trezza, R.; Wright, J.L.; Bastiaanssen, W.; Robison, C.W. Satellite-based energy balance for mapping evapotranspiration with internalized calibration (METRIC)—Applications. *J. Irrig. Drain Eng.–ASCE.* **2007**, *133*, 395–406.
7. Mu, Q.; Heinsch, F.A.; Zhao, M.; Running, S.W. Development of a global evapotranspiration algorithm based on MODIS and global meteorology data. *Remote Sens. Environ.* **2007**, *111*, 519–536.
8. Su, Z. The surface energy balance system (SEBS) for estimation of turbulent heat fluxes. *Hydrol. Earth Syst. Sci.* **2002**, *6*, 85–100.
9. Allen, R.G.; Pereira, L.S.; Raes, D.; Smith, M. *Crop Evapotranspiration-Guidelines for Computing Crop Water Requirements-FAO Irrigation and Drainage Paper 56*; FAO: Rome, Italy, 1998.
10. Allen, R.G.; Howell, T.A.; Pruitt, W.O.; Walter, I.A.; Jensen, M.E. *Lysimeters for Evapotranspiration and Environmental Measurements*; American Society of Civil Engineers: New York, NY, USA, 1991; p. 456.
11. Bowen, I.S. The ratio of heat losses by conduction and by evaporation from any water surface. *Phys. Rev.* **1926**, *27*, doi:http://dx.doi.org/10.1103/PhysRev.27.779.
12. Monteith, J.; Unsworth, M. *Principles of Environmental Physics*, 2nd ed.; Edward Arnold: London, UK, 1990.
13. Everson, C.S.; Clulow, A.; Mengitsu, M. *Feasibility Study on the Determination of Riparian Evaporation in Non-Perennial Systems*; WRC Report No. TT 424/09; Water Research Commission: Pretoria, South Africa, 2009.
14. Hillel, D. *Introduction to Soil Physics*; Academic press: New York, NY, USA, 1982.

15. Clulow, A.D.; Everson, C.S.; Jarmain, C.; Mengistu, M. *Water-Use of the Dominant Natural Vegetation Types of the Eastern Shores Area, Maputaland*; WRC Report No. 1926/1/12; Water Research Commission: Pretoria, South Africa, 2012.

16. Oelofse, A.; van Averbeke, W. *Nutritional Value and Water Use of African Leafy Vegetables for Improved Livelihoods*; WRC Report No. TT 535/12; Water Research Commission: Pretoria, South Africa, 2012.

17. Li, Z.L.; Tang, R.; Wan, Z.; Bi, Y.; Zhou, C.; Tang, B.; Yan, G.; Zhang, X. A review of current methodologies for regional evapotranspiration estimation from remotely sensed data. *Sensors* **2009**, *9*, 3801–3853.

18. Liang, S.; Wang, K.; Zhang, X.; Wild, M. Review on estimation of land surface radiation and energy budgets from ground measurement, remote sensing and model simulations. *IEEE J. Sel. Top. Appl.* **2010**, *3*, 225–240.

19. Wang, K.; Dickinson, R.E. A review of global terrestrial evapotranspiration: Observation, modeling, climatology, and climatic variability. *Rev. Geophys.* **2012**, *50*, doi:10.1029/2011RG000373.

20. Glenn, E.P.; Huete, A.R.; Nagler, P.L.; Hirschboeck, K.K.; Brown, P. Integrating remote sensing and ground methods to estimate evapotranspiration. *Crit. Rev. Plant Sci.* **2007**, *26*, 139–168.

21. Nagler, P.L.; Cleverly, J.; Glenn, E.; Lampkin, D.; Huete, A.; Wan, Z. Predicting riparian evapotranspiration from MODIS vegetation indices and meteorological data. *Remote Sens. Environ.* **2005**, *94*, 17–30.

22. Nagler, P.L.; Glenn, E.P.; Nguyen, U.; Scott, R.L.; Doody, T. Estimating riparian and agricultural actual evapotranspiration by reference evapotranspiration and MODIS Enhanced Vegetation Index. *Remote Sens.* **2013**, *5*, 3849–3871.

23. Tian, J.; Su, H.; Sun, X.; Chen, S.; He, H.; Zhao, K. Impact of the spatial domain size on the performance of the T_s-VI triangle method in terrestrial evapotranspiration estimation. *Remote Sens.* **2013**. *5*, 1998–2013.

24. Nagler, P.L.; Morino, K.; Murray, R.S.; Osterberg, J.; Glenn, E.P. An empirical algorithm for estimating agricultural and riparian evapotranspiration using MODIS enhanced vegetation index and ground measurements of ET. I. Description of method, *Remote Sens.* **2009**, *1*, 1273–1297.

25. Murray, R.S.; Nagler, P.L.; Morino, K.; Glenn, E.P. An empirical algorithm for estimating agricultural and riparian evapotranspiration using MODIS enhanced vegetation index and ground measurements of ET. II. Application to the Lower Colorado River, U.S. *Remote Sens.* **2009**, *1*, 1125–1138.

26. Bastiaanssen, W.; Menenti, M.; Feddes, R.; Holtslag, A. A remote sensing surface energy balance algorithm for land (SEBAL). 1. Formulation. *J. Hydrol.* **1998**, *212*, 198–212.

27. Ruhoff, A.L.; Paz, A.R.; Collischonn, W.; Aragao, L.E.O.C.; Rocha, H.R.; Malhi, Y.S. A MODIS-based energy balance to estimate evapotranspiration for clear-sky days in Brazilian tropical savannas. *Remote Sens.* **2012**, *4*, 703–725.

28. Alexandridis, T.K.; Cherif, I.; Chemin, Y.; Silleos, G.N.; Stavrinos, E.; Zalidis, G.C. Integrated methodology for estimating water use in Mediterranean agricultural areas. *Remote Sens.* **2009**, *1*, 445–465.

29. Gibson, L.A.; Jarmain, C.; Eckardt, F.E. Estimating evapotranspiration using remote sensing and the surface energy balance system—A South African perspective. *Water SA* **2013**, *39*, 477–484.

30. Jin, X.; Guo, R.; Xia, W. Distribution of actual evapotranspiration over Qaidam Basin, an arid area in China. *Remote Sens.* **2013**, *5*, 6976–6996.

31. Trezza, R.; Allen, R.G.; Tasumi, M. Estimation of actual evapotranspiration along the Middle Rio Grande of New Mexico using MODIS and Landsat imagery with the METRIC Model. *Remote Sens.* **2013**, *5*, 5397–5423.

32. Hankerson, B.; Kjaersgaard, J.; Hay, C. Estimation of evapotranspiration from fields with and without cover crops using remote sensing and *in situ* methods. *Remote Sens.* **2012**, *4*, 3796–3812.

33. Cleugh, H.A.; Leuning, R.; Mu, Q.; Running, S.W. Regional evaporation estimates from flux tower and MODIS satellite data. *Remote Sens. Environ.* **2007**, *106*, 285–304.

34. Mu, Q.; Zhao, M.; Running, S.W. Improvements to a MODIS global terrestrial evapotranspiration algorithm. *Remote Sens. Environ.* **2011**, *115*, 1781–1800.

35. Priestley, C.H.B.; Taylor, R.J. On the assessment of surface heat flux and evaporation using large scale parameters. *Mon. Weather Rev.* **1972**, *100*, 81–92.

36. Jia, Z.; Liu, S.; Xu, Z.; Chen, Y.; Zhu, M. Validation of remote sensed evapotranspiration over the Hai river basin, China. *J. Geophys. Res.: Atmos.* **2012**, *117*, doi:10.1029/2011JD017037.

37. Fisher, J.B.; Tu, K.P.; Baldocchi, D.D. Global estimates of the land-atmosphere water flux based on monthly AVHRR and ISLSCP-II data, validated at 16 FLUXNET sites. *Remote Sens. Environ.* **2008**, *112*, 901–919.

38. Yao, Y.; Liang, S.; Cheng, J.; Liu, S.; Fisher, J.B.; Zhang, X.; Jia, K.; Zhao, X.; Qin, Q.; Zhao, B.; *et al.* MODIS-driven estimation of terrestrial latent heat flux in China based on a modified Priestley-Taylor algorithm. *Agr. Forest Meteorol.* **2013**, *171–172*, 187–202.

39. Bateni, S.M.; Liang, S. Estimating surface energy fluxes usind a dual-source data assimilation approach adjoined to the heat diffusion equation. *J. Geophys. Res.: Atmos.* **2012**, *117*, doi:10.1029/2012JD017618.

40. Yao, Y.; Liang, S. Zhao, S. Zhang, Y.; Qin, Q.; Cheng, Q.; Jia, J.; Xie, X.; Zhang, N.; Liu, M. Validation and application of the modified satellite-based Priestley-Taylor Algorithm for mapping terrestrial evapotranspiration. *Remote Sens.* **2014**, *6*, 880–904.

41. Caparrini, F.; Castelli, F.; Entekhabi, D. Estimation of surface turbulent fluxes through assimilation of radiometric surface temperature sequences. *J. Hydrometeorol.* **2004**, *5*, 145–159.

42. Jarmain, C.; Mengitsu, M.; Jewitt, G.; Kongo, V.; Bastiaanssen, W. *A Methodology for Near-Real Time Spatial Estimation of Evaporation*; WRC Report No. 1751-1-09; Water Research Commission: Pretoria, South Africa, 2009.

43. Sun, Z.; Gebremichael, M.; Ardö, J.; Nickless, A.; Caquet, B.; Merboldh, L.; Kutschi, W. Estimation of daily evapotranspiration over Africa using MODIS/Terra and SEVIRI/MSG data. *Atmos. Res.* **2012**, *112*, 35–44.

44. Kim, H.W.; Hwang, K.; Mu, Q.; Lee, S.O.; Choi, M. Validation of MODIS 16 global terrestrial evapotranspiration products in various climates and land cover types in Asia. *KSCE J. Civ. Eng.* **2012**, *16*, 229–238.

45. Jovanovic, N.; Masiyandima, M.; Naiken, V.; Dzikiti, S.; Gush, M. *Remote Sensing Applications in Water Resources Management—Desktop Validation and Draft Paper*; CSIR Report No. CSIR/NRE/ECOS/IR/2011/0097/A; CSIR: Pretoria, South Africa, 2012.

46. Scholes, R.; Gureja, N.; Giannecchinni, M.; Dovie, D.; Wilson, B.; Davidson, N.; Piggot, K.; McLoughlin, C.; van der Velde, K.; Freeman, A.; *et al.* The environment and vegetation of the flux measurement site near Skukuza, Kruger National Park. *Koedoe* **2001**, *44*, 73–83.

47. Kirton, A.; Scholes, R.J. Site Characterization of the Malopeni Flux Tower Site, Kruger National Park, South Africa. Available online: http://www.carboafrica.net/downs/ws/accra/6-Posters/Malopeni_Site_Characterisation.pdf (accessed on 1 April 2014).

48. Burba, G.; Anderson, D. *A Brief Practical Guide to Eddy Covariance Flux Measurements, Principles and Workflow Examples for Scientific and Industrial Applications*; LI-COR Biosciences: Lincoln, NE, USA, 2010.

49. Liu, S.M.; Xu, Z.W.; Zhu, Z.Z.; Jia, Z.L., Zhu, M.J. Measurements of evapotranspiration from eddy-covariance systems and large aperture scintillometers in the Hai River Basin, China. *J. Hydrol.* **2013**, *22*, 24–38.

50. Friedl, M.A.; McIver, D.K.; Hodges, J.C.F.; Zhang, X.Y.; Muchoney, D.; Strahler, A.H.; Woodcock, C.E.; Gopal, S.; Schneider, A.; Cooper, A.; *et al.* Global land cover mapping from MODIS: Algorithms and early results. *Remote Sens. Environ.* **2002**, *83*, 287–302.

51. Myneni, R.B.; Hoffman, S.; Knyazikhin, Y.; Privette, J.L.; Glassy, J.; Tian, Y.; Wang, Y.; Song, X.; Zhang, Y.; Smith, G.R.; *et al.* Global products of vegetation leaf area and fraction of absorbed PAR from year one of MODIS data. *Remote Sens. Environ.* **2012**, *83*, 214–231.

52. Aubinet, M. Estimates of the annual net carbon and water exchange of forests: The EUROFLUX methodology. *Adv. Ecol. Res.* **2000**, *30*, 113–175.

53. Glenn, E.P.; Neale, M.U.; Doug, J.H.; Nagler, P.L. Vegetation index-based crop coefficients to estimate evapotranspiration by remote sensing in agricultural and natural resources. *Hydrol. Process.* **2011**, *25*, 4050–4062.

54. Musselman, R.C.; Minnick, T.J. Nocturnal stomatal conductance and ambient air quality standards for azone. *Atmos. Environ.* **2000**, *34*, 719–733.

Towards the Improvement of Blue Water Evapotranspiration Estimates by Combining Remote Sensing and Model Simulation

Mireia Romaguera, Mhd. Suhyb Salama, Maarten S. Krol, Arjen Y. Hoekstra and Zhongbo Su

Abstract: The estimation of evapotranspiration of blue water (ET_b) from farmlands, due to irrigation, is crucial to improve water management, especially in regions where water resources are scarce. Large scale ET_b was previously obtained, based on the differences between remote sensing derived actual ET and values simulated from the Global Land Data Assimilation System (GLDAS). In this paper, we improve on the previous approach by enhancing the classification scheme employed so that it represents regions with common hydrometeorological conditions. Bias between the two data sets for reference areas (non-irrigated croplands) were identified per class, and used to adjust the remote sensing products. Different classifiers were compared and evaluated based on the generated bias curves per class and their variability. The results in Europe show that the k-means classifier was better suited to identify the bias curves per class, capturing the dynamic range of these curves and minimizing their variability within each corresponding class. The method was applied in Africa and the classification and bias results were consistent with the findings in Europe. The ET_b results were compared with existing literature and provided differences up to 50 mm/year in Europe, while the comparison in Africa was found to be highly influenced by the assigned cover type and the heterogeneity of the pixel. Although further research is needed to fully understand the ET_b values found, this paper shows a more robust approach to classify and characterize the bias between the two sets of ET data.

Reprinted from *Remote Sens.* Cite as: Romaguera, M.; Salama, M.S.; Krol, M.S.; Hoekstra, A.Y.; Su, Z. Towards the Improvement of Blue Water Evapotranspiration Estimates by Combining Remote Sensing and Model Simulation. *Remote Sens.* **2014**, *6*, 7026-7049.

1. Introduction

Water management in agriculture has been always important, especially in areas where water resources are scarce. In this context, it is relevant to distinguish between the sources of the usage: water supplied by precipitation (called green water) and irrigation (called blue water).

Recent studies obtained blue and green water usage in agriculture at large scale, using data from national agricultural statistics, reports and climatic databases, and making use of hydrological models based on the calculation of actual evapotranspiration (ET_{actual}) [1–4]. Moreover, several studies tackled the problem of retrieving global irrigated areas by using national statistics of areas equipped for irrigation [5] and by statistically analyzing remote sensing products [6].

The potential of using remote sensing data for global studies of green and blue waters use and water footprint estimations is discussed in Romaguera *et al.* [7]. The first approaches to exploit those data at large scale are shown in Romaguera *et al.* [8,9], where the use of the different components of the water cycle is explored, together with the use of land surface models. Other works used remote sensing to evaluate irrigation performance at regional scale [10–12].

In particular, Romaguera et al. [8] obtained large scale blue water evapotranspiration (ET_b), i.e., due to irrigation, based on the differences between remote sensing ET_{actual} obtained from the Meteosat Second Generation (MSG) satellites [13] and ET_{actual} values simulated from the Global Land Data Assimilation System (GLDAS) [14]. In general, it was found that there was a systematic bias between the two datasets in rain-fed pixels and that this difference was variable in time and space. The bias amplitude changed along the year, roughly resembling a positive concave curve. The maximum amplitude value reached up to 3 mm/day and occurred in the months of spring and summer in northern latitudes. The spatial variability of the bias was associated in this paper to vegetation characteristics and the remote sensing observation angle. Romaguera et al. [8] calculated the bias per day and used three parameters to generate a classification map for Europe to discriminate areas with different bias patterns: the maximum value of the NDVI ($NDVI_{max}$), the season where $NDVI_{max}$ occurred, and the viewing zenith angle (VZA) of the sensor on board MSG. Thresholds were assigned to distinguish between classes. Recent work from Romaguera et al. [15] showed that the classification scheme was not sufficient to describe the variability of the bias estimates in the continent of Africa and proposed the inclusion of a climatic indicator in the selection of parameters for the classification.

Similar results in terms of bias between these two datasets were obtained by Ghilain et al. [13] and the validation report from the Land Surface Analysis Satellite Applications Facility (LSA–SAF) [16], who showed that the bias might be explained by the differences in the inputs of incoming solar radiation, the ratio between leaf are index (LAI) and stomatal resistance, and land cover type. Yilmaz et al. [17] identified discrepancies in insolation inputs and analyzed differences in soil moisture, when comparing these data sets in the region of the Nile River basin. Romaguera et al. [8] emphasized that the GLDAS simulations did not account for extra water supply due to irrigation and consequently it was expected that they underestimate ET_{actual} during the cropping season in irrigated areas. Therefore in the aforementioned work, the differences between these two estimates were corrected for the bias in rain–fed croplands, to obtain blue water evapotranspiration according to:

$$ET_b = \Delta ET - bias \tag{1}$$

ΔET is the difference between MSG ET_{actual} (MSG–ET in the following) and GLDAS ET_{actual} (GLDAS–ET in the following), and bias is this difference in ET_{actual} calculated in reference areas, i.e., rain–fed croplands where irrigation practices are not present.

In this paper, we intend to solve the drawbacks of the previous methodology by improving the classification scheme in order to achieve a better spatial representation of the bias in rain–fed croplands, which results in better estimates of ET_b.

First, a better choice of parameters is proposed to represent regions with common hydrometeorological conditions, based on three processes at which ET_{actual} estimation may be affected, namely vegetation characterization, atmosphere/forcing definition, and land–atmosphere interaction. Secondly, an alternative strategy for classification is adopted, based on the use of classifiers, instead of the selection of thresholds published in the aforementioned work. This makes the methodology generic and robust. The suitability of these classifiers is explored via the classification maps and the bias curves obtained per class.

Section 2 explains the selection of parameters for the classification and the properties of the selected classifiers. Next, in Section 3, the datasets used in this paper are detailed. The Results section includes the classification maps, the bias curves and ET_b outputs compared with the original methodology. The application of the method to the region of Africa is shown in Section 5, together with the comparison of the results with existing literature. Finally relevant issues about the proposed improved methodology can be found in the Discussion and Conclusions sections.

2. Method

The objective of this work is to improve the existing methodology for ET_b estimation [8] on two aspects: the selection of parameters for the classification of the study area and the classification method.

2.1. Selection of Parameters for the Classification

The hypothesis here is that the classification output discriminates between areas with different bias curves, *i.e.*, differences along the year of MSG–ET and GLDAS–ET estimates. Therefore, potential variables need to be identified in order to explain the differences in ET_{actual} retrievals. In the present work, a more complete selection of parameters is carried out by accounting for three processes at which ET_{actual} estimation from both sources may differ, namely vegetation characterization, atmosphere/forcing definition and land–atmosphere interaction. In order to account for the vegetation properties, a typical indicator is selected, the NDVI, and in particular its maximum value along the year, $NDVI_{max}$. Precipitation and net radiation are combined into a climatic indicator (CI) to account for the driving forces for ET_{actual}, as follows [18]:

$$CI = \frac{L\,P}{R_n} \qquad (2)$$

where L(J/kg) is the latent heat of vaporization, P (mm) is the annual precipitation and R_n (W/m^2) is the annual net radiation (obtained as the sum of net shortwave (S_n) and net longwave (L_n) radiation). In other words, the LP term is the amount of energy necessary to evaporate the available precipitation P.

The land–atmosphere interaction is included in the selection of parameters by means of the monthly accumulated ET_{actual}, and in particular the maximum value along the year (ET_{mmax}). This aggregated value is chosen in order to reduce relative errors in ET_{actual} estimation. Since the focus of this paper is to obtain the ET bias curves per class, including ET itself in the classification inputs contributes to capture the observed variability. Moreover, as shown in previous literature [8,13], a seasonality was found in this bias. The position of the bias maxima was variable and, therefore, it was reasonable to select t_ET_{mmax} (month when ET monthly maxima occurs) to account for these different patterns.

Being aware that other potential variables might have been included in this selection, such as land surface temperature and albedo, effective precipitation, soil moisture, LAI or topography, the number was limited in order to reduce data redundancy when the parameters were correlated or equivalent in terms of climate and vegetation interaction. Moreover, the use of too many variables in a classification procedure may decrease the classification accuracy [19].

2.2. Classification Methods

Many classification methods exist in the literature and all have their own merits. However, the question of which classification approach is suitable for a specific study is not easy to answer [20]. Romaguera *et al.* [8] used a basic classification method, based on thresholds, as follows: $NDVI_{max}$ higher/lower than 0.4, October to March and April to September periods for the $NDVI_{max}$ to occur, and VZA intervals of $10°$.

In this paper, the use of three different classifiers was explored and discussed. Firstly, a common classification approach was chosen: an unsupervised classification based on k–means. Secondly, a more advanced learning method was selected: an unsupervised classification based on the expectation–maximization algorithm. These two approaches do not use training samples, they work per pixel, they are hard classifiers (*i.e.*, output is a definitive decision) and they do not use spatially neighboring pixel information for the classification, all aspects that are appropriate to this study. Thirdly, an image transform of the selected parameters was carried out by using the principal component analysis in order to reduce data redundancy of correlated bands and concentrate the information contents in the transformed images, exploring the possible clustering.

2.2.1. Unsupervised Classification Based on K–means

An unsupervised classification of the study area was executed in order to cluster pixels based on the k–means statistical technique [21]. This method calculates initial mass means evenly distributed in the data space and then iteratively clusters the pixels into the nearest class using a minimum distance technique. In each iteration, classes' means are recalculated and pixels are reclassified with respect to the new means. All pixels are classified to the nearest class unless a standard deviation or distance threshold is specified, in which case some pixels may be unclassified if they do not meet the selected criteria. This process continues until the number of pixels in each class changes by less than the selected pixel change threshold or the maximum number of iterations is reached. The four-layer input file contained $NDVI_{max}$, ET_{mmax}, t_ET_{mmax} and CI as described in the previous section. A maximum of 100 iterations was fixed to ensure completion of the algorithm, and the default value of 5% was conserved for the pixel change threshold [22].

Neither standard deviation, nor distance thresholds were fixed. In order to select the optimal number of clusters and to evaluate the clustering found by the algorithm, a scattering distance (SD) quality index was calculated according to Rezaee *et al.* [23], which accounts for the intra–cluster and inter–cluster distances as:

$$SD(c) = a\ Scat(c) + Dis(c) \qquad (3)$$

308

where c is the number of clusters, Scat(c) is the average scattering and indicates the average compactness of the clusters (*i.e.*, intra–cluster distance), Dis(c) is the total separation between the c clusters (*i.e.*, an indication of inter–cluster distance), and a is a weighing factor equal to Dis(cmax), where cmax is the maximum number of input clusters.

A small value of Scat(c) indicates a compact cluster. The second term Dis(c) is influenced by the geometry of the cluster centers and increases with the number of clusters. The optimal value for the number of clusters present in the data set is such that minimizes the SD index.

In the present research, the maximum number of clusters for the unsupervised classification was set to 20, assuming a reasonable minimum percentage of pixels per class of 5%. The classification was obtained for 5 up to 10 clusters and SD was calculated. As a result, the number of clusters for which the SD quality index was minimized was found to be 6 in the region of Europe.

2.2.2. Unsupervised Classification Based on the Expectation–Maximization Algorithm (EM)

The Expectation–Maximization algorithm [24] is an iterative procedure that estimates the probabilities of the elements to belong to a certain class, based on the principle of maximum likelihood of unobserved variables in statistical models. The EM iteration alternates between performing an expectation (E) step, which creates a function for the expectation of the log–likelihood evaluated using the current estimate for the parameters, and a maximization (M) step, that computes parameters maximizing the expected log–likelihood found on the E step. These parameter estimates are then used to determine the distribution of the latent variables in the next E step. This classification was performed using the machine learning software WEKA version 3.6.9 (Waikato Environment for Knowledge Analysis) [25], using the implemented Simple EM classifier. This software contains tools and algorithms for the analysis of data and predictive modeling, where the system is trained and can learn from the data and provide classified outputs. EM assigns a probability distribution to each instance which indicates the probability of it belonging to each of the cluster and can decide how many clusters to create by cross validation. In the current research, the maximum number of iterations was set to 100 to ensure completion of the algorithm. Moreover, the software allowed to test the model output by using the 66% of the data as a training set and the rest for testing.

2.2.3. Classification Using Principal Component Analysis (PCA)

The principal component analysis (PCA) [26] consists of a transformation of the input dataset (multilayer file with NDVI$_{max}$, ET$_{mmax}$, t_ET$_{mmax}$ and CI) to produce uncorrelated output bands, segregate noise components, and reduce the dimensionality of data sets. The characteristic matrix (covariance matrix or correlation matrix) of the variables, the eigen values, the eigen vectors (which are the directions of the principal components (PC)), and the coordinates of each data point in the direction of the PC's were calculated. A new set of orthogonal axes was found, which had their origin at the data mean and which were rotated so the data variance was maximized. PC output bands were linear combinations of the original spectral bands and were uncorrelated. The relationships found between the principal components, which led to the clustering of the data are shown in the results section.

3. Data Sets

Table 1 describes the main characteristics of the datasets used in the present work which are detailed in the following subsections.

Table 1. Specifications of the data sets used in the present work.

Data	Source	Spatial Coverage	Spatial Resolution	Temporal Resolution	Details
ET_{actual}	MSG	MSG disk *	3 km at nadir	30'	Availability of data: Europe: January 2007–present The rest: September 2009–present
	GLDAS	Global	0.25° (~30 km at equator)	3 h	Availability of data: February 2000–present
S_n, L_n, P	GLDAS	Global	0.25° (~30 km at equator)	3 h	Availability of data: February 2000–present
Land Cover	MERIS	Global	300 m	Static	GlobCover map calculated in year 2009
NDVI	AVHRR	Europe ——— Africa	1 km	Monthly	Generated by DLR Period: year 1997 ——— Generated by IGBP Period: April 1992–March 1993
Irrigation	Blue WF	Global	5 arcmin (~10 km at equator)	Static	Data: blue WF per year[3]

* Meteosat disk covers latitudes between −60° and +60° and longitudes between −60° to +60°; ** List of acronyms: MSG (Meteosat Second Generation), GLDAS (Global Land Data Assimilation System), MERIS (Medium Resolution Imaging Spectrometer), AVHRR (Advanced Very High Resolution Radiometer), DLR (Deutsches Zentrum für Luft– und Raumfahrt), IGBP (International Geosphere–Biosphere Programme Data), WF (Water Footprint).

From a technical point of view, the combination of data of different spatial resolution, extent and geographical projection was tackled by creating a layer stack where the data were resampled and re–projected to a common output projection. The present work was carried out at the resolution of the MSG–ET products.

3.1. Evapotranspiration and Cover Type Data

Based on Equation (1), the main datasets for obtaining the ET_b were the ET_{actual} products from the MSG satellites provided by the Land Surface Analysis Satellite Applications Facility (LSA–SAF) [13] (MSG–ET) and ET_{actual} from the Global Land Data Assimilation System (GLDAS) generated with the Noah land surface model [27,28] (GLDAS–ET). These datasets are available from the LSA–SAF website (http://landsaf.meteo.pt/) and the NASA Goddard Earth

310

Sciences Data and Information Services Center (GES DISC) (http://disc.sci.gsfc.nasa.gov/hydrology/data-holdings) respectively.

The GlobCover land cover map (ver. 2.3) [29] was used to identify the land cover type, e.g., rain–fed croplands (where the bias was calculated) and bare areas (where ET$_{actual}$ rates are low) (see Figure 1). More detailed information about these datasets can be found in Romaguera *et al.* [8].

Figure 1. Pixel type (only rain–fed croplands, rest of croplands, bare areas, others) from the GlobCover classification map.

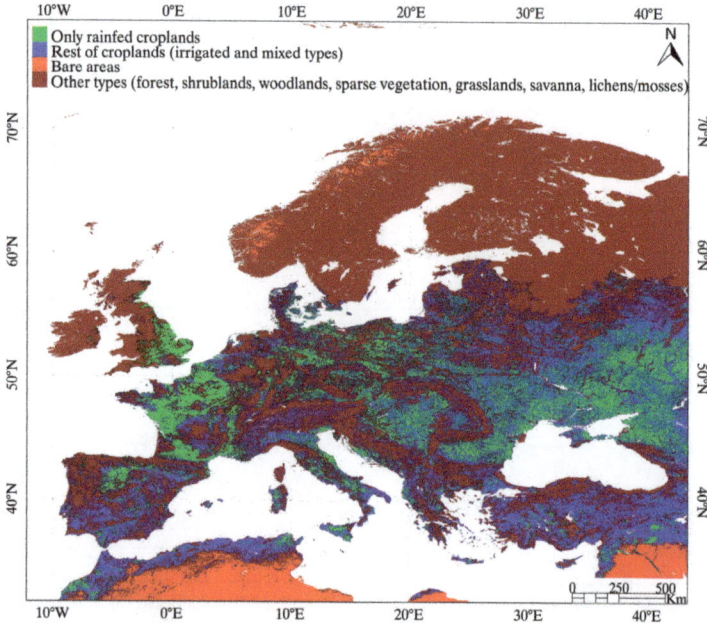

In the research presented, daily MSG–ET values were obtained by temporal integration of the 48 instantaneous values per day, during the year 2010. Linear interpolation in time was used to fill in missing data, due to non–acquisitions. Daily MSG–ET were not considered if missing data occurred during periods of one hour or longer. Daily GLDAS–ET values were obtained by temporal averaging of the eight provided ET$_{actual}$ rates per day. No missing data were found in this dataset.

3.2. Data for the Classification

As explained in previous sections, four parameters were selected for the classification of the study area: (a) NDVI$_{max}$; (b) a climate indicator (CI) based on net radiation, latent heat of vaporization and precipitation; (c) the maximum value of monthly aggregated ET (ET$_{mmax}$); and (d) the month where the ET$_{mmax}$ occurs (t_ET$_{mmax}$). The selected study area was Europe and the year was 2010. Additionally, the classification was also obtained in Africa for testing the method.

In order to ensure consistency of the data, the NDVI$_{max}$ was extracted from the source related to MSG–ET retrieval. That is the ECOCLIMAP database [30], which includes the LAI values that are used in the MSG–ET algorithm. These values are obtained by taking *in situ* maximum and minimum values of LAI and considering Advanced Very High Resolution Radiometer (AVHRR) NDVI series to impose seasonality per class cover. For Europe, the monthly NDVI values generated during the year 1997, by the Deutsches Zentrum für Luft– und Raumfahrt (DLR) [31], are considered, and, for Africa the International Geosphere–Biosphere Programme Data (IGBP) 1 km AVHRR NDVI composites from April 1992 until March 1993 [32]. Although these values may not represent irrigated vegetation in the year of analysis, they influence the MSG–ET retrieval and therefore the difference with GLDAS–ET, which is the focus of this research.

CI was calculated for the year 2010, according to Equation (2), by yearly aggregating R_n and P. Net radiation was obtained as the sum of net shortwave (S_n) and net longwave (L_n) radiation, which were obtained from the GLDAS dataset, together with the precipitation values.

GLDAS–ET values were aggregated monthly and the maximum value was obtained (ET_{mmax}), as well as the month when it occurred (t_ET_{mmax}). Radiation and ET_{actual} values in GLDAS were given as rates every 3 h, so the proper way to calculate the yearly/monthly values was by temporal averaging of the data corrected by the time conversion factor.

The data sets were filtered according to the following criteria. Firstly, bare pixels defined by the GlobCover classification map were masked. These are arid areas, where the estimation of t_ET_{mmax} may be affected by fluctuations of the low values of ET_{actual}. Secondly, coastal pixels with nonrealistic values were also masked. This effect appeared when resampling the data to a common grid and pixel size. Finally, pixels with negative $NDVI_{max}$ value were also masked for the calculations. These were found in the datasets in areas close to water bodies. Figure 2 shows the selected data sets for the classification and the study area. Finally, every dataset was normalized dividing by its maximum and the generated four–layer file was used for the classifications.

3.3. Data for the Test of the Method

The global blue water footprint (WF_b) of crop production estimated by Mekonnen and Hoekstra [3] was used to compare the ET_b outputs produced in Europe and Africa. The water footprint (WF) is defined as the water consumed for crop production, where green and blue stand for precipitation and irrigation water usage. In their method, the computations of crop evapotranspiration were done following Allen *et al.* [33] for the case of crop growth under non–optimal conditions. The model takes into account the daily soil water balance and climatic conditions for each grid cell. Climatic and reference evapotranspiration inputs are averaged for the period 1996–2005 and results are given as average over that time interval. WFs are typically given in units of m^3/ton or mm/year. In the last case, the yield is not considered and therefore WF_b corresponds to total ET_b. This product has global coverage and a spatial resolution of 5 arcmin.

Figure 2. Study area and inputs used in the present paper: (**a**) NDVI$_{max}$; (**b**) Climate indicator; (**c**) ET$_{mmax}$; (**d**) t_ET$_{mmax}$.

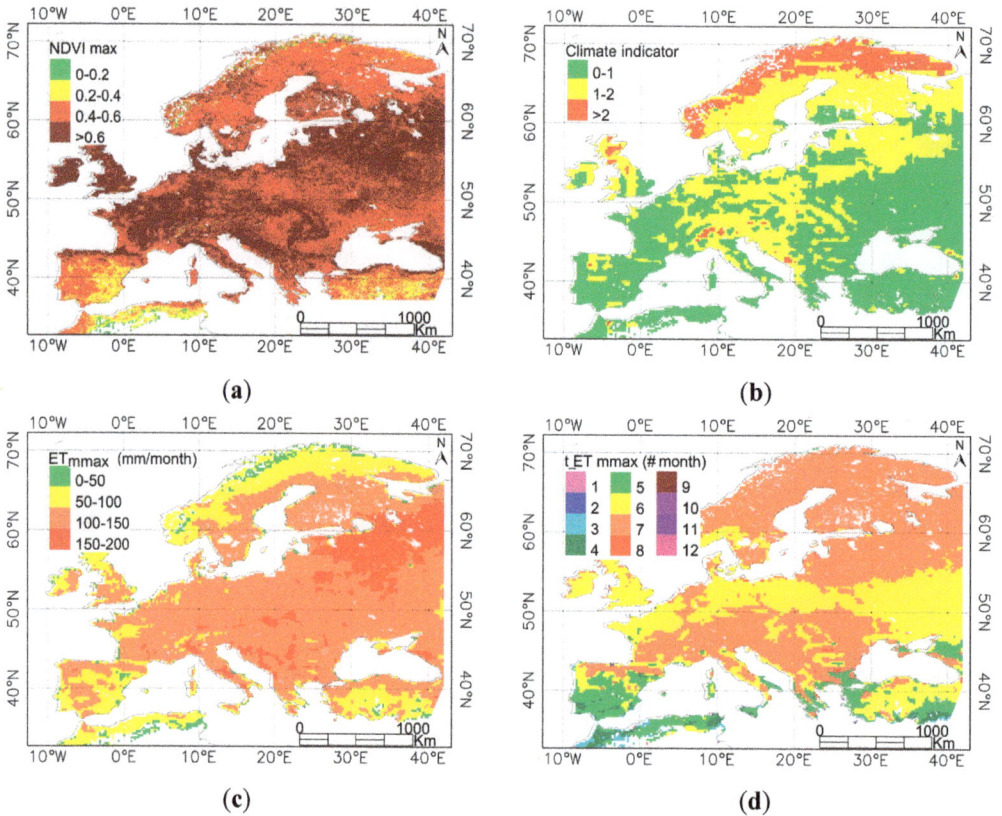

(**a**)

(**b**)

(**c**)

(**d**)

4. Results

This section shows the classification maps generated with the three selected methods. The k–means and EM classifiers have a straightforward application. For the PC approach, the relationship between the first and third PC allowed identifying 11 vertical clusters, whose pixels were assigned to 11 classes (named from 1 to 11, from left to right of the Figure 3). Neither clear relationships nor groupings could be established between the PC1 and the components PC2 and PC4.

4.1. Classification Maps

Figure 4 shows the classification maps obtained using the new set of input parameters and the three classifiers in the region of Europe. The map generated in Romaguera *et al.* [8] (ROM in the following) is also included for the sake of comparison. Information about the abundance of different cover types is shown per class. Three cover types are considered: (i) only rain–fed croplands; (ii) rest of croplands (irrigated and mixed types); and (iii) others (forests, shrublands, woodlands, sparse vegetation, grassland, savanna, lichens/mosses), according to the GlobCover map. Additionally, the number of rain–fed croplands (RC) and the ratio of these over the total croplands (RC/TC) are incorporated.

Figure 3. Scatter plot of first principal component (PC1) *versus* the third (PC3) obtained in Europe from the classification dataset proposed in this paper.

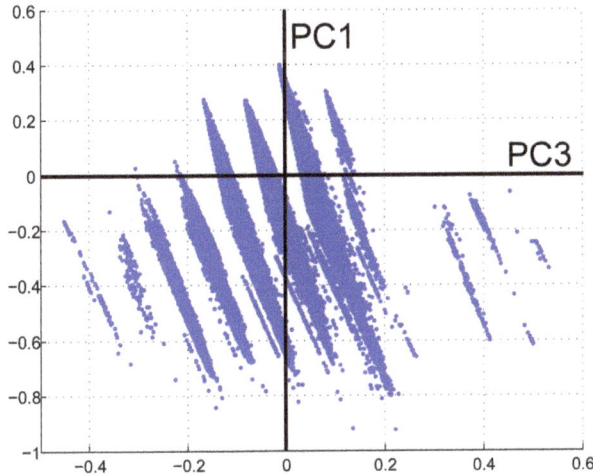

Figure 4. (**a**) Classification maps obtained using the methods ROM, k–means, EM and PCA; (**b**) Abundance of cover type per class (i) only rain–fed croplands; (ii) rest of croplands (irrigated and mixed types); or (iii) others (forests, shrublands, woodlands, sparse vegetation, grassland, savanna, lichens/mosses); (**c**) Total of rain–fed cropland pixels (RC) per class and ratio between RC and the total number of cropland pixels (RC/TC).

(**a**) Classification map

(**b**) Abundance of cover type

(**c**) rain–fed cropland pixels (RC) and ratio with total cropland pixels (TC)

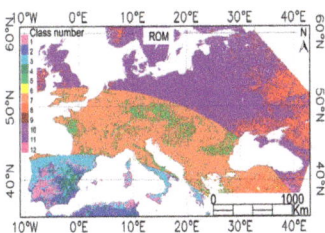

* Only pixels in the map are counted for plotting

Class	#RC	RC/TC (%)
1	750	24
2	302	14
3	1315	33
4	328	21
5	2832	45
6	40	14
7	13,872	40
8	332	13
9	6032	47
10	10	7
11	15,142	40
12	32	7

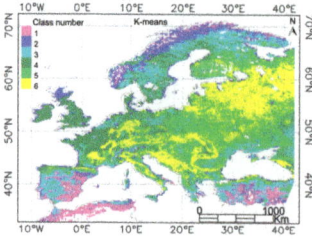

Class	#RC	RC/TC(%)
1	1889	14
2	1654	32
3	6710	34
4	13,181	51
5	17,996	43
6	7880	34

Figure 4. *Cont.*

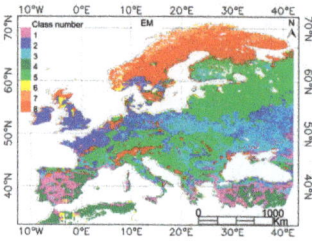

Class	#RC	RC/TC (%)
1	4989	29
2	13,222	50
3	16,876	44
4	2223	16
5	9573	35
6	301	27
7	88	39
8	2038	39

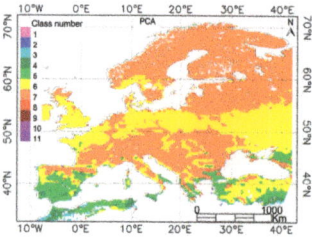

Class	#RC	RC/TC (%)
1	1	3
2	7	11
3	142	8
4	674	16
5	4075	26
6	28,279	44
7	15,948	38
8	152	36
9	8	14
10	24	47
11	0	0

The number of classes generated was 12, 6, 8, and 11 in the ROM, k–means, EM and PCA classifications respectively. The numbers assigned to the classes must be understood as labels and their value is not necessarily related between the different classification outputs. In general, the proposed classifications had a visually more continuous spatial distribution than ROM, which presented the characteristic rings due to the intervals chosen in the VZA criteria. Moreover, some similar grouping can be observed in the generated classifications, such as the areas of Spain and East and center Europe in EM and PCA or Eastern part of Norway in k–means and EM, although this comparison is not easy to evaluate. The distribution of the classes in the k–means and EM classification showed the influence of all four input parameters, whereas in the PCA classification output, the parameter determining the classes was the t_ET_{mmax}. This is associated to the fact that the third component of the PCA captures the variability (after transformation) of the discrete values of t_ET_{mmax}.

Two majority classes were found in the ROM classification, as well as in PCA. That is represented in Figure 4b by the total height of the columns. However, the abundance of the classes in k–means and EM was more balanced, with no significant minority classes in the case of k–means.

The blue color in the graphs indicates the amount of rain–fed croplands per class, which is also indicated in Figure 4c. This number is important since the bias is calculated in this cover type. Therefore in classes 6, 10, 12 from ROM and classes 1, 2, 9, and 10 from PCA, the bias was calculated with relatively few samples. Additionally, class 11 in PCA had no rain–fed pixels and the bias could not be calculated.

Another relevant factor is the relative abundance of rain–fed croplands with respect to the total number of croplands (rain–fed plus rest of croplands) per class. The ET_b is ultimately calculated in all croplands, and therefore this ratio (named RC/TC in the following) is important. The higher RC/TC is, the more representative the bias is for all the pixels in a class, that is in classes 3, 5, 7, 9, 11 from ROM, 2 until 6 in k–means, 2, 3, 5, 7, 8 in EM, and 6, 7, 8, 10 in PCA. These classes had a ratio higher than 30%, as can be observed in Figure 4c.

As a result of this analysis, it was concluded that the k–means and EM classification schemes improved the existing one (ROM) in two aspects: the spatial pattern of the classification map and the increase in number of rain–fed pixels and RC/TC per class, with less minority classes, showing the k–means a better performance. The PCA approach showed weaker changes.

The evaluation of the performance of the classifications is detailed in the Discussion section.

4.2. Bias Curves

The bias values were obtained per pixel at monthly scale. Monthly ET_{actual} values were obtained from the daily MSG–ET and GLDAS–ET estimates during 2010. Due to the lack of some MSG–ET data, and in order to obtain a consistent bias curve, the daily values were only aggregated when both datasets existed. Then, monthly bias values were obtained per pixel and yearly curves were averaged per class, using rain–fed pixels only, for each of the classifications.

Figure 5 shows the bias curves using the three proposed classifiers and ROM. Additionally the mean bias curve using all rain–fed croplands is plotted as a reference. Note that they are discrete monthly values that are connected to facilitate visual interpretation.

In general it was observed that the k–means and EM classifications achieved a better separation of the bias curves, compared with ROM and PCA. The k–means curves (Figure 4a) presented a minimum around the month of June in classes 3, 4, 5, and 6 with a general convex shape and amplitude up to −28 mm/month. A local maximum was found around August. The curves of classes 1 and 2 were concave and corresponded to classes with relatively low number of rain–fed croplands, achieving for class 1 the lowest RC/TC ratio (see Figure 4c).

Classes 1, 2, 3, 5, and 8 in EM also presented a convex bias with maximum amplitude around May–June with values up to −28 mm/month. A local maximum was found around August. Classes 4, 6 and 7 were concave; 4 and 6 corresponded to lowest ratio RC/TC and class 7 to lowest value of rain–fed croplands.

Figure 5. Bias between MSG–ET and GLDAS–ET obtained in rain–fed pixels and averaged per class, using different classifiers. (**a**) k–means; (**b**) EM; (**c**) PCA; (**d**) ROM; (**e**) all rain–fed pixels. Note: Monthly discrete values are connected to facilitate visual interpretation.

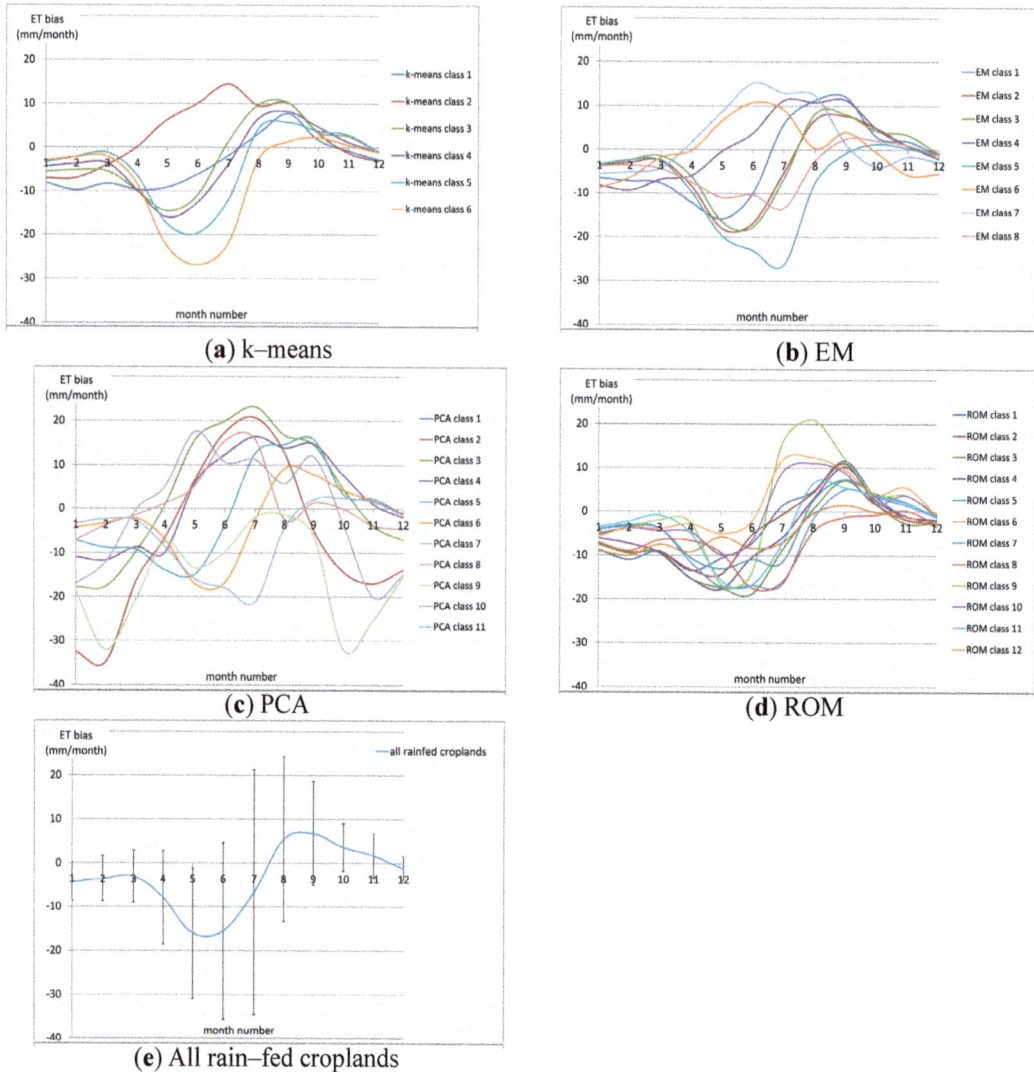

(**a**) k–means

(**b**) EM

(**c**) PCA

(**d**) ROM

(**e**) All rain–fed croplands

In the PCA classification, classes 5, 6, and 7 behaved in a similar way as the convex curves described for k–means and EM. These were the classes with higher number of rain–fed pixels and also high RC/TC. Classes 2, 3, 4, 8 and 10 were concave and presented more irregular bias curves. These were less abundant classes, some of them not even noticeable visually in the figure, in general they had lower ratio RC/TC and relatively low number of pixels. Class number 9 presented an intermediate pattern and classes 1 and 11 were not plotted since they contained only one rain–fed pixel or none.

The biases obtained from the ROM classification were more fuzzy and irregular. Some of the classes (3, 5, 7, 8, 9, and 11) presented a convex shape similar to the aforementioned curves, but the pattern was irregular and it was difficult to distinguish between classes and extract conclusions regarding the amount of rain–fed croplands and RC/TC.

Finally, Figure 5e shows the variability of the bias curve when all rain–fed croplands are averaged and no classes are taken into account. The position of the minimum and local maxima is consistent with what it has been described in this paper, and the amplitude is in general flattened due to the averaging.

In general the k–means and EM approach represented an improvement with respect to the ROM bias results in terms of separation of the bias curves, which means a better discrimination of the classes. Furthermore, differences in the maximum amplitude of the curves were also found. In order to understand them better, Figure 6 shows the comparison of the bias values around the maximum (month six) (represented as dots) together with the standard deviation associated to it (represented as error bars and also in columns). This was obtained for all the classes and for the mean curve where no classes are assigned (Figure 5e).

Figure 6 shows how the standard deviation (σ, in columns and secondary Y axis) changed when adding a classification in the method instead of using a single bias curve averaged for all rain–fed croplands. In general, the diminution of σ was found in majority classes. In the case of the k–means classifier, σ was reduced in classes 4, 5, and 6 and slightly increased in classes 1, 2, and 3. For the EM classifier, σ increased in classes 1, 4, 6, 7, and 8. In general terms, σ increased in the PCA classification, and the values in ROM fluctuated depending on the class. It was also observed that the increase of σ was related with the decrease of number of rain–fed croplands and RC/TC in the class.

In terms of the bias value, the classifiers captured different intervals of variability: between −27 and 10 mm/month in k–means, between −23 and 15 mm/month for EM, between −18 and 20 mm/month for PCA and between −19 and −2 mm/month for ROM. The value of the bias for the single curve was −15 mm/month. Therefore the proposed classifiers captured a higher range of bias with respect to ROM.

Previous literature showed the differences between MSG–ET and GLDAS–ET [13,34]. For the region of Europe, these studies showed the bias relative to the mean MSG–ET, with values ranging from −0.5 to 0.5. The mean MSG–ET in the month of July was also plotted in this literature, with a value of 0.22 mm/h calculated in the range 9–12UTC. If we assume a constant ET$_{actual}$ rate, and 10 h of sun, the aggregated value is 66 mm/month. This value combined with the ±0.5 values of relative bias, produces values of absolute bias up to 33 mm, which is consistent with the intervals found with the proposed classifiers.

As a result of this analysis, the k–means approach showed to improve the existing methodology and perform better than EM and PCA in different aspects related with the bias estimation; first, the ability to differentiate bias curves; second, by reducing the standard deviation of the data when introducing the classes; third, by capturing the expected variability of the maximum bias.

Figure 6. ET bias at month six and standard deviation (error bars and columns) obtained for all the classifiers and classes discussed in this paper. The label "all" refers to the calculation with all rain–fed croplands, where no classes are assigned.

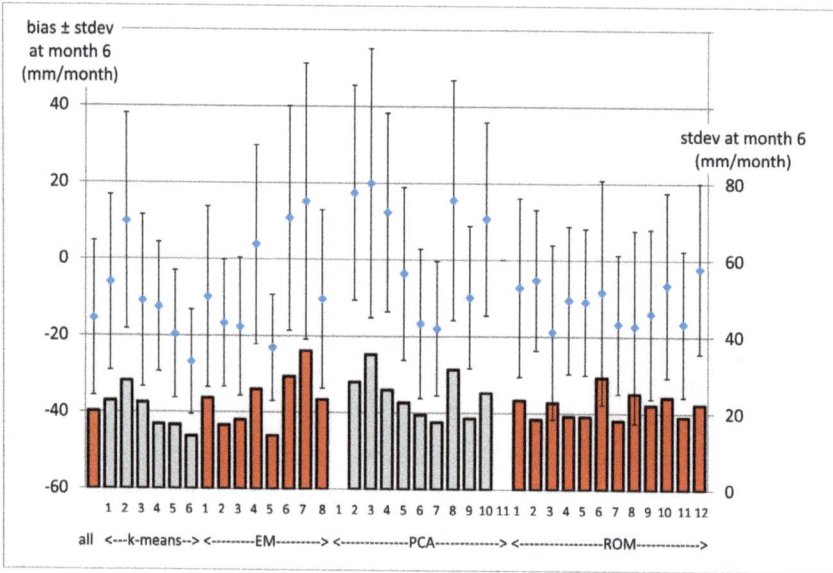

Therefore based on the conclusions extracted from the classification and bias analysis, the k–means is used in the following to estimate ET_b, since it was shown to be the most suitable approach for this research.

4.3. ET_b Estimation

ET_b was obtained in the region of Europe in 2010 following Equation (1), the k–means classification and the bias results. ΔET values were calculated daily when both datasets were available and then accumulated monthly. Monthly bias corrections were undertaken, and the resulting positive ET_b values were aggregated to a yearly scale.

According to the method described in Romaguera *et al.* [8], the GlobCover map was used to mask all cover types except for the rain–fed croplands, irrigated croplands and mixed types that include croplands. Although in that publication a value of 50 mm was suggested as a threshold from which the method was able to detect irrigation, in the current research, no threshold was considered based on the fact that small values of ET_b may be also representative for heterogeneous pixels.

The same procedure was used to obtain ET_b using the classification scheme provided in the literature (ROM). The comparison showed how the outputs changed when using a different classification scheme. ET_b reached differences up to 60 mm/year, being ET_b (k–means) higher in some regions of Ukraine and lower in some regions of Spain, Turkey and coast of France. The spatial distribution of these differences is related with the classification maps, being for example the red areas in Figure 7 associated with class number 2 of the k–means classification that has a convex shaped bias.

Figure 7. Difference between yearly ET_b (ROM) and ET_b (k–means) obtained for the year 2010.

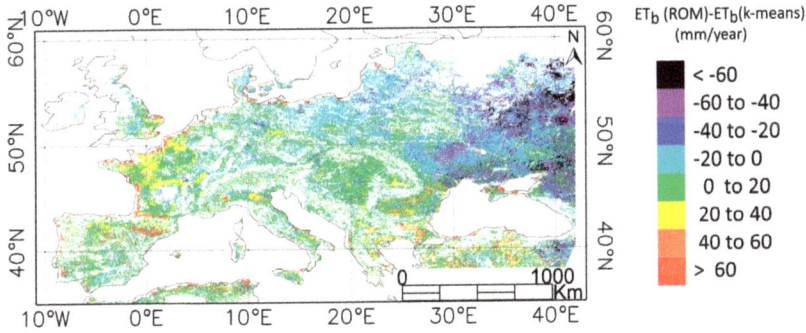

5. Application of the Method

The study area selected to test the method was the whole Meteosat observation disk, which included Europe and Africa, in the year 2010. The analysis was carried out by separating it in three sectors namely Europe, North Africa, and South Africa, like the MSG–ET products delivered by LSA–SAF. These sections are separated at the latitudes of 34°N and the equator approximately.

Following the method explained in the present paper, a classification of every sector was carried out with the k–means algorithm using the proposed input data sets. Pixels labeled as bare areas by the GlobCover map were excluded from the classification. The optimal number of classes was calculated per sector following the procedure explained in Section 2.2.1. The bias curves per class in the rain–fed pixels of North Africa and South Africa are plotted in Figure 8.

Figure 8. Bias curves obtained in rain–fed pixels of the sectors (**a**) North Africa and (**b**) South Africa.

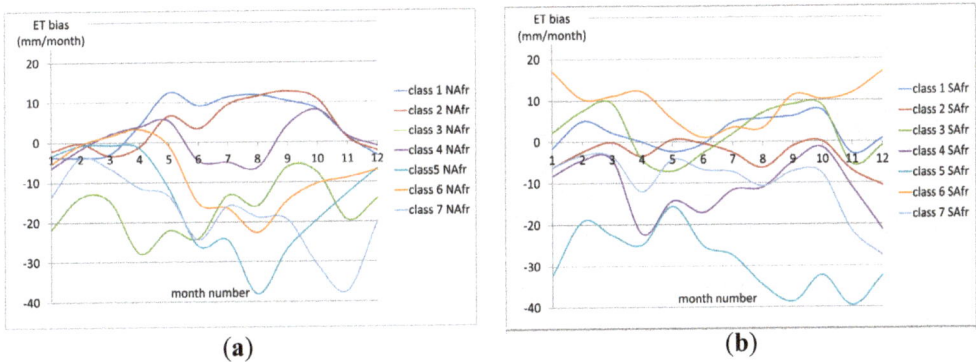

Similarly to the conclusions found in Europe using the k–means approach (Figure 5), the curves present distinguishable patters with different amplitudes and shapes. Classes 1, 4 and 6 in North Africa present more defined convex bias curves which correspond to highest numbers of rain–fed pixels and ratio RC/TC, whereas classes 3 and 7 are represented by the lowest number of rain–fed pixels and ratio RC/RT. In the sector of South Africa, the curves present a shift of about six months with respect to the sectors of Europe and North Africa. This is due to the seasonality patterns of the southern Hemisphere. The most abundant classes in terms of rain–fed pixels were classes 2, 4, and 7. However, due to the relatively low quantity of rain–fed pixels in this sector, the ratio RC/TC is below 10% for all the classes in South Africa.

Equation (1) was used in all the study area to calculate monthly ET_b and the positive values were aggregated to a yearly scale. These were compared with the values of blue water footprint (WF_b) for crop production estimated by Mekonnen and Hoekstra [3]. For consistency, the areas labeled with no irrigation croplands in Mekonnen and Hoekstra [3] were masked (see Figure 9).

Figure 9. Difference between ET_b generated in this paper and the WF_b (mm/year) given by Mekonnen and Hoekstra [3].

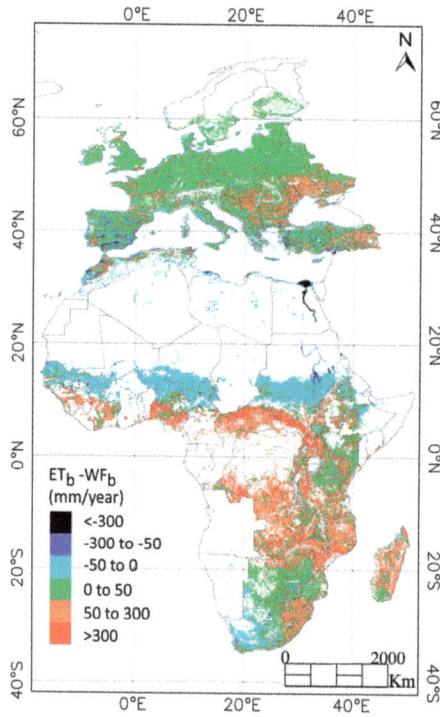

The mean value of the differences found in Europe and Africa was 27 and 62 mm/year with standard deviation of 62 and 142 mm/year, and rmse of 44 and 155 mm/year. No straightforward correlation was found between the two datasets. Differences between the two methods were found to be below 50 mm/year in most of Europe and in some regions of Africa, although in this area the discrepancies (in positive and negative sign) were higher as also shown by the statistical indicators. Several reasons exist to explain the magnitude and sign of the differences.

First, from a methodological point of view, Mekonnen and Hoekstra [3] used the cropmap of Monfreda et al. [35] to correct the calculated ET_{actual} with the percentage of cropland per pixel, and then the irrigated area map from Portmann *et al.* [36] to correct for the area used for irrigation. This resulted in values of annual ET_b lower than the ones generated in this paper, as it can be observed in most of Europe and some regions in Africa indicated with green color.

Higher discrepancies (in orange and red) were found in mid Africa. This can be explained based on the cover type and pixel heterogeneity, since these are areas labeled as forests and shrublands by the GlobCover map. The method by Mekonnen and Hoekstra [3] provided low values of ET_b (below 1 mm/year) in these mixed pixels due to low value of irrigated cropland area. However, the method presented in this paper and the bias estimation were not developed for cover types different than croplands, and therefore the values obtained in these areas are not realistic and should be masked as indicated by Romaguera *et al.* [8]. This is the case also of the forested areas in Spain, Italy, Greece, and Turkey.

The region of Ukraine presented higher values of ET_b compared with the estimates given by Mekonnen and Hoekstra [3]. This can be explained by a combination of factors. Low values of ET_b were obtained by Mekonnen and Hoektra [3] in the majority of this area. Higher values were obtained by the proposed method, which were influenced by the high values of ΔET found in the inputs. Moreover, this region is located in the extremes of the Meteosat observation disk, which might influence the accuracy of the remote sensed ET_{actual} estimates.

Areas painted in cyan were labeled as bare areas by the GlobCover map and were masked in the classification carried out in the present paper. That was done to avoid misclassification caused by the fluctuation of low ET_{actual} rates in these areas. Therefore, the method did not provide ET_b values, whereas Mekonnen and Hoekstra [3] estimated values of ET_b below 1 mm/year. Moreover, other bare areas were also masked in the present paper and they provided significantly high values of ET_b in the estimates of the literature. That was the case of the White Nile in Sudan and some areas in Morocco.

Mekonnen and Hoekstra [3] provided significantly higher values of ET_b in the region of the Nile basin. These are areas where the assigned cropland irrigated area from Portmann *et al.* [36] was higher than 100%, a fact that was related to multiple cropping practices. ET_b estimates were higher than 1000 mm/year in this case. However, the method presented in this paper relied only on the differences between the ET_{actual} inputs from MSG and GLDAS and the bias correction, achieving annual values in this region of 600 mm/year.

Moreover, the discrepancies in some mountainous areas like Mozambique, Ethiopia or north of Italy, may be associated to the effect of the terrain on the radiation. In zones of complex topography, variability in elevation, surface slope and aspect create strong spatial heterogeneity in solar distribution, which determines air temperature, soil temperature, ET, snow melt and land–air exchanges. Therefore, MSG–ET satellite retrievals may need a slope and aspect correction to radiation inputs in areas of significant relief. Some remote sensing methods that include this geometric correction for calculating ET are given by Allen *et al.* [37] and Chen *et al.* [38]. Finally, together with the aforementioned methodological reasons, the differences between the two estimates might be related to the accuracy and parameterization of the input data, and the period of comparison, since the datasets were obtained at different time spans.

322

6. Discussion

6.1. Uncertainty

6.1.1. Performance of the Classifications

Evaluating the performance of the generated classifications is not an easy task, especially due to the fact that a "ground truth classification map" that explains the different classes in terms of the biases between MSG–ET and GLDAS–ET does not exist. However, in order to optimize the performance of the methods, different strategies were adopted. The k–means settings were adjusted to a maximum of 100 iterations in order to ensure completion of the algorithm and the optimal number of clusters was calculated using a quality index [23] that accounts for the intra–cluster and inter–cluster distances, as it is shown in Section 1.2.1.

The EM algorithm was trained with 66% of the data, and tested in the rest of the dataset providing a log likelihood of 7.04 in the test. The overall likelihood is a measure of the "goodness" of the clustering and increases at each iteration of the EM algorithm. The larger this quantity is, the better the model fits the data. In order to interpret this number, the model was applied to the whole dataset, obtaining a log likelihood of 7.6. This value was expected to be higher than using only the test dataset since the inputs included also the training data. However, the two log likelihood values were similar, which served as an indicator of the good performance of the classifier. Moreover, the EM algorithm found the optimal number of clusters for which the log likelihood had a maximum value.

Finally the PC analysis was carried out by using two matrixes, the covariance and the correlation matrixes. The results showed the same clustering using both procedures, where the groups could be visually identified as it is shown in Figure 3. The usefulness of the generated classifications in the application that is presented in this paper (bias estimation and ET$_b$), was discussed in Section 4.2.

6.1.2. Uncertainty of ET$_b$ Estimations

It was pointed out in Romaguera *et al.* [8] that Equation (1) provided some negative values of ET$_b$ which were not physically correct and were due to the uncertainties in the input data and in the bias curves definition. In the present work, these negative values were not used in the calculation of the annual ET$_b$ and in order to evaluate their significance they were aggregated during 2010 and combined with the yearly MSG–ET. The ratio between these two (not shown here) was less than 10% in most of the croplands in the study area. The impact of the negative values was concentrated in areas labeled as "others" (see Figure 1) or in their proximity, like next to the Alps and the Carpathian mountains.

Therefore, it was shown that with the proposed method the non–croplands pixels are more sensitive to errors and, therefore, they should be masked when producing ET$_b$.

6.2. Limitations

This paper focused on the improvement of the classification scheme proposed by Romaguera *et al.* [8]. The analysis presented showed satisfactory results when producing the classification maps and obtaining the bias curves. General limitations and aspects to be taken into consideration when obtaining ET$_b$ are extensively discussed in the aforementioned literature, and are related to the inaccuracies of the inputs, data availability, validation drawbacks and generalization of the method.

Regarding the classification approach presented in this paper, several aspects need to be discussed.

First, the selection of the parameters for the classification is justified in the text. However, additional parameters may have been used and a sensitivity analysis might be carried out to select the optimal set to describe the study area.

Regarding the selection of the classifier, this work showed three strategies, different in concept and complexity. The k–means showed to work better in this research. Nevertheless, having in mind the coarse resolution of the data used, subpixel classifiers may be explored in order to account for mixed pixels and subpixel heterogeneity. In that case, aspects like avoiding over classification and tradeoffs in accuracy, time consumption and computing resources would need to be taken into account. Also, the better performance of the k–means approach, as compared to EM and PCA, was not tested in the application for Africa.

The dimensions of the study area may play an important role when the classifier finds similar pixels apart of each other (e.g., North and South Africa). The bias value may be influenced by the averaging of rain–fed areas in these distant areas. Therefore, a preliminary test on qualitative differences in sub–continental bias patterns is advised to avoid these effects and select reasonable study area sizes.

The GlobCover land cover map was used in several steps in this paper, to mask bare areas in the classification, to identify rain–fed croplands where the bias curves were obtained and finally to filter other land covers for ET$_b$ estimation. The results are therefore influenced by the inaccuracies of this input, since pixels may be misclassified.

From a technical point of view, the rescaling and resampling of the input data to achieve a common spatial resolution may have an impact in the analysis. The nearest neighbor resampling technique was used in this research. In general, impacts on the results are expected due to the heterogeneity of the surface and the question of how representative are the low resolution data disaggregated to a higher scale. Up–downscaling techniques tackle this issue by using the parameters of surface temperature or NDVI [39,40].

The ET$_b$ results compared with existing literature provided differences of 50 mm/year in most of Europe. In Africa, the comparison was highly influenced by the assigned cover type and the heterogeneity of the pixel. Results over regions of high topographic relief point to the need for slope and aspect corrections to radiation inputs to the MSG–ET algorithm. Bearing in mind the advantages of both approaches (literature and present paper), the synergy between them may allow to benefit from the temporal frequency of the remote sensed data and from the better definition of the subpixel heterogeneity.

324

The *in situ* validation of the ET_b estimates produced in this paper is hampered by the scarcity of good quality spatial data on irrigation at regional scale. The availability of irrigation water management information on a detailed scale like farmer fields or for entire river basins is not common. Data to quantify performance indicators are rarely collected. If collected, data frequently is unreliable or not easily accessible [10]. An attempt was made in previous literature to validate the methodology in a corn field in a cropland area in Spain [8]. Difficulties were found due to the unavailability of MSG–ET remote sensing data coincident with the literature, together with the fact that water resources and irrigation are politically critical issues and generally practices are not regulated or *in situ* quality data are difficult to access. Moreover, statistical databases such as AQUASTAT [41] provide static and scattered data. Therefore, the validation of the improved method with *in situ* data remains an open challenge. Although further research is needed to fully understand the ET_b values found, this paper has shown to improve the classification scheme and the estimation of the bias curves between the sets of ET data from MSG and GLDAS. An example of application of this improved method to time series of data in the Horn of Africa and Southwest of China is presented in Romaguera *et al.* [42].

7. Conclusions

This paper provided a more generic and robust methodology to estimate blue water evapotranspiration (ET_b) from remote sensing and simulated ET_actual data, by enhancing the classification scheme employed in the literature. A new selection of input parameters was proposed and the analysis of different classifiers provided the best results for the k–means technique.

The main outcome was the improvement of the definition of the bias between the two ET_actual datasets, *i.e.*, the ability to differentiate bias curves of different classes, reduction of the standard deviation of the data and achievement of the expected variability of the maximum bias.

This paper proposed new tools to evaluate the variability of the biases between remote sensing and simulated ET_actual data. However, the comparison of ET_b in Europe and Africa with existing literature showed the need of further research to fully understand the final ET_b values found.

Acknowledgments

The authors wish to thank the Land Surface Analysis Satellite Applications Facility (LSA–SAF) of the European Organisation for the Exploitation of Meteorological Satellites (EUMETSAT), as well as the European Space Agency (ESA) and the ESA GlobCover Project, led by MEDIAS–France, for providing the products used in this paper. The GLDAS data used in this study were acquired as part of the mission of NASA's Earth Science Division and archived and distributed by the Goddard Earth Sciences (GES) Data and Information Services Center (DISC).

Author Contributions

This work was carried out in the framework of the PhD research of the first author M. Romaguera. The coauthors have contributed as supervisors of this project.

Conflicts of Interest

The authors declare no conflict of interest.

References

1. Hanasaki, N.; Inuzuka, T.; Kanae, S.; Oki, T. An estimation of global virtual water flow and sources of water withdrawal for major crops and livestock products using a global hydrological model. *J. Hydrol.* **2010**, *384*, 232–244.
2. Liu, J.G.; Yang, H. Spatially explicit assessment of global consumptive water uses in cropland: Green and blue water. *J. Hydrol.* **2010**, *384*, 187–197.
3. Mekonnen, M.M.; Hoekstra, A.Y. The green, blue and grey water footprint of crops and derived crop products. *Hydrol. Earth Syst. Sci.* **2011**, *15*, 1577–1600.
4. Siebert, S.; Döll, P. Quantifying blue and green virtual water contents in global crop production as well as potential production losses without irrigation. *J. Hydrol.* **2010**, *384*, 198–217.
5. Siebert, S.; Döll, P.; Hoogeveen, J.; Faures, J.M.; Frenken, K.; Feick, S. Development and validation of the Global Map of Irrigation Areas. *Hydrol. Earth Syst. Sci.* **2005**, *9*, 535–547.
6. Thenkabail, P.S.; Biradar, C.M.; Noojipady, P.; Dheeravath, V.; Li, Y.J.; Velpuri, M.; Gumma, M.; Gangalakunta, O.R.P.; Turral, H.; Cai, X.L.; *et al.* Global Irrigated Area Map (GIAM), derived from remote sensing, for the end of the last millennium. *Int. J. Remote Sens.* **2009**, *30*, 3679–3733.
7. Romaguera, M.; Hoekstra, A.Y.; Su, Z.; Krol, M.S.; Salama, M.S. Potential of using remote sensing techniques for global assessment of water footprint of crops. *Remote Sens.* **2010**, *2*, 1177–1196.
8. Romaguera, M.; Krol, M.S.; Salama, M.S.; Hoekstra, A.Y.; Su, Z. Determining irrigated areas and quantifying blue water use in Europe using remote sensing Meteosat Second Generation (MSG) products and Global Land Data Assimilation System (GLDAS) data. *Photogramm. Eng. Remote Sens.* **2012**, *78*, 861–873.
9. Romaguera, M.; Salama, M.S.; Krol, M.S.; Su, Z.; Hoekstra, A.Y. Remote sensing method for estimating green and blue water footprint. In *Remote Sensing and Hydrology*; Neale, C.M.U., Cosh, M.H., Eds.; The International Association of Hydrological Sciences: Wallingford, UK, 2012; Volume 352, pp. 288–291.
10. Bastiaanssen, W.G.M.; Bos, M.G. Irrigation performance indicators based on remotely sensed data: A review of literature. *Irrig. Drain. Syst.* **1999**, *13*, 192–311.
11. Santos, C.; Lorite, I.J.; Tasumi, M.; Allen, R.G.; Fereres, E. Performance assessment of an irrigation scheme using indicators determined with remote sensing techniques. *Irrig. Sci.* **2010**, *28*, 461–477.
12. D'Urso, G.; de Michele, C.; Vuolo, F. Operational irrigation services from remote sensing: The irrigation advisory plan for the Campania region, Italy. In *Remote Sensing and Hydrology*; Neale, C.M.U., Cosh, M.H., Eds.; The International Association of Hydrological Sciences: Wallingford, UK, 2012; Volume 352, pp. 419–422.

13. Ghilain, N.; Arboleda, A.; Gellens–Meulenberghs, F. Evapotranspiration modelling at large scale using near–real time MSG SEVIRI derived data. *Hydrol. Earth Syst. Sci.* **2011**, *15*, 771–786.

14. Rodell, M.; Houser, P.R.; Jambor, U.; Gottschalck, J.; Mitchell, K.; Meng, C.J.; Arsenault, K.; Cosgrove, B.; Radakovich, J.; Bosilovich, M.; *et al.* The Global Land Data Assimilation System. *Bull. Am. Meteorol. Soc.* **2004**, *85*, 381–394.

15. Romaguera, M.; Salama, S.; Krol, M.S.; Hoekstra, A.Y.; Su, Z. Synergy between remote sensing ET products and simulated data to retrieve irrigated areas and blue ET in Europe and Africa. Poster. Presented at the 5th International Workshop on Catchment Hydrological Modeling and Data Assimilation CAHMDA–V: Catchments in a Changing Climate, Enschede, The Netherlands, 9–13 July 2012; p. 1.

16. LSA–SAF. *LSA–SAF Validation Report. Products LSA–16 (MET), LSA–17 (DMET)*; Doc. Num. SAF/LAND/RMI/VR/0.6; The EUMETSAT Network of Satellite Application Facilities: Darmstadt, Germany, 2010.

17. Yilmaz, M.T.; Anderson, M.C.; Zaitchik, B.; Hain, C.R.; Crow, W.T.; Ozdogan, M.; Chun, J.A.; Evans, J. Comparison of prognostic and diagnostic surface flux modeling approaches over the Nile river basin. *Water Resour. Res.* **2014**, *50*, 386–408.

18. Roerink, G.J.; Menenti, M.; Soepboer, W.; Su, Z. Assessment of climate impact on vegetation dynamics by using remote sensing. *Phys. Chem. Earth* **2003**, *28*, 103–109.

19. Price, K.P.; Guo, X.L.; Stiles, J.M. Optimal Landsat TM band combinations and vegetation indices for discrimination of six grassland types in eastern Kansas. *Int. J. Remote Sens.* **2002**, *23*, 5031–5042.

20. Lu, D.; Weng, Q. A survey of image classification methods and techniques for improving classification performance. *Int. J. Remote Sens.* **2007**, *28*, 823–870.

21. Tou, J.T.; Gonzalez, R.C. *Pattern Recognition Principles*; Addison–Wesley Publishing Company: Reading, MA, USA, 1974.

22. Exelis Visual Information Solutions. *ENVI Tutorial: Classification Methods*; Exelis Visual Information Solutions: Boulder, CO, USA, 2006.

23. Rezaee, M.R.; Lelieveldt, B.P.F.; Reiber, J.H.C. A new cluster validity index for the fuzzy c–mean. *Pattern Recognit. Lett.* **1998**, *19*, 237–246.

24. Dempster, A.P.; Laird, N.M.; Rubin, D.B. Maximum likelihood from incomplete data via the EM algorithm. *J. R. Stat. Soc. B* **1977**, *39*, 1–38.

25. Hall, M.; Frank, E.; Holmes, G.; Pfahringer, B.; Reutemann, P.; Witten, I.H. The WEKA data mining software: An update. *Sigkdd Explor.* **2009**, *11*, 10–18.

26. Richards, J.A.; Jia, X. *Remote Sensing Digital Image Analysis: An Introduction*; Springer–Verlag: Berlin, Germany, 1999; p. 240.

27. Chen, F.; Mitchell, K.; Schaake, J.; Xue, Y.K.; Pan, H.L.; Koren, V.; Duan, Q.Y.; Ek, M.; Betts, A. Modeling of land surface evaporation by four schemes and comparison with FIFE observations. *J. Geophys. Res.: Atmos.* **1996**, *101*, 7251–7268.

28. Koren, V.; Schaake, J.; Mitchell, K.; Duan, Q.Y.; Chen, F.; Baker, J.M. A parameterization of snowpack and frozen ground intended for NCEP weather and climate models. *J. Geophys. Res.: Atmos.* **1999**, *104*, 19569–19585.

29. UCLouvain/ESA. *Globcover 2009. Products Description and Validation Report*; European Space Agency: Paris, France, 2011.

30. Masson, V.; Champeaux, J.L.; Chauvin, F.; Meriguet, C.; Lacaze, R. A global database of land surface parameters at 1–km resolution in meteorological and climate models. *J. Clim.* **2003**, *16*, 1261–1282.

31. Mucher, C.A.; Champeaux, J.L.; Steinnocher, K.T.; Griguolo, S.; Wester, K.; Heunks, C.; Winiwater, W.; Kressler, F.P.; Goutorbe, J.P.; ten Brink, B.; *et al. Development of a Consistent Methodology to Derive Land Cover Information on an European Scale from Remote Sensing for Environmental Modeling*; Center for Geo–Information (CGI) Alterra: Wageningen, The Netherlands, 2001; p. 160.

32. Belward, A.S.; Estes, J.E.; Kline, K.D. The IGBP/DIS global 1–km land–cover data set discover. A project overview. *Photogramm. Eng. Remote Sens.* **1999**, *65*, 1013–1020.

33. Allen, R.; Pereira, L.S.; Raes, D.; Smith, M. *FAO Irrigation and Drainage Paper N. 56: Crop Evapotranspiration*; Food and Agriculture Organization: Rome, Italy, 1998.

34. Gellens–Meulenberghs, F.; Arboleda, A.; Ghilain, N. Towards a continuous monitoring of evapotranspiration based on MSG data. In *IAHS Symposium on Remote Sensing for Environmental Monitoring and Change Detection*; IAHS Press: Wallingford, UK, 2007.

35. Monfreda, C.; Ramankutty, N.; Foley, J.A. Farming the planet: 2. Geographic distribution of crop areas, yields, physiological types, and net primary production in the year 2000. *Glob. Biogeochem. Cycle* **2008**, doi:10.1029/2007GB002947.

36. Portmann, F.T.; Siebert, S.; Döll, P. MIRCA2000–global monthly irrigated and rainfed crop areas around the year 2000: A new high–resolution data set for agricultural and hydrological modeling. *Glob. Biogeochem. Cycle* **2010**, doi:10.1029/2008GB003435.

37. Allen, R.G.; Tasumi, M.; Trezza, R. Satellite–based energy balance for mapping evapotranspiration with internalized calibration (METRIC)–model. *J. Irrig. Drain. Eng.* **2007**, *133*, 380–394.

38. Chen, X.; Su, Z.; Ma, Y.; Yang, K.; Wang, B. Estimation of surface energy fluxes under complex terrain of Mt. Qomolangma over the Tibetan plateau. *Hydrol. Earth Syst. Sci.* **2013**, *17*, 1607–1618.

39. Anderson, M.C.; Kustas, W.P.; Norman, J.M. Upscaling flux observations from local to continental scales using thermal remote sensing. *Agron. J.* **2007**, doi:10.2134/agronj2005.0096S.

40. Jia, S.; Zhu, W.; Lü, A.; Yan, T. A statistical spatial downscaling algorithm of TRMM precipitation based on NDVI and DEM in the Qaidam basin of China. *Remote Sens. Environ.* **2011**, *115*, 3069–3079.

41. FAO. Aquastat Database. Food and Agriculture Organization: Rome, Italy, 2009. Available online: http://www.fao.org/nr/water/aquastat/main/index.stm (accessed on 15 July 2014).

42. Romaguera, M.; Krol, M.S.; Salama, S.; Su, Z.; Hoekstra, A.Y. Application of a remote sensing method for estimating monthly blue water evapotranspiration in irrigated agriculture. *Remote Sens.* **2014**, submitted.

Evaporative Fraction as an Indicator of Moisture Condition and Water Stress Status in Semi-Arid Rangeland Ecosystems

Francesco Nutini, Mirco Boschetti, Gabriele Candiani, Stefano Bocchi and
Pietro Alessandro Brivio

Abstract: Rangeland monitoring services require the capability to investigate vegetation condition and to assess biomass production, especially in areas where local livelihood depends on rangeland status. Remote sensing solutions are strongly recommended, where the systematic acquisition of field data is not feasible and does not guarantee properly describing the spatio-temporal dynamics of wide areas. Recent research on semi-arid rangelands has focused its attention on the evaporative fraction (EF), a key factor to estimate evapotranspiration (ET) in the energy balance (EB) algorithm. EF is strongly linked to the vegetation water status, and works conducted on eddy covariance towers used this parameter to increase the performances of satellite-based biomass estimation. In this work, a method to estimate EF from MODIS products, originally developed for evapotranspiration estimation, is tested and evaluated. Results show that the EF estimation from low spatial resolution over wide semi-arid area is feasible. Estimated EF resulted in being well correlated to field ET measurements, and the spatial patterns of EF maps are in agreement with the well-known climatic and landscape Sahelian features. The preliminary test on rangeland biomass production shows that satellite-retrieved EF as a water availability factor significantly increased the capacity of a remote sensing operational product to detect the variability of the field biomass measurements.

Reprinted from *Remote Sens.* Cite as: Nutini, F.; Boschetti, M.; Candiani, G.; Bocchi, S.; Brivio, P.A. Towards the Improvement of Blue Water Evapotranspiration Estimates by Combining Remote Sensing and Model Simulation. *Remote Sens.* **2014**, *6*, 6300-6323.

1. Introduction

The ecosystem carrying capacity and food security of the West African Sahel relies on annual vegetation production, which is concentrated in a short rainy period of four months, on average, between July to October [1]. The majority of the Sahelian livelihood counts on these wet months to get by in the dry season. The management of existing natural resources by the local population has developed several strategies to cope with climatic difficulties, such as exploiting herd transhumance at the beginning of the dry season [2] or handling the seedling date in the beginning of the rainy period [3].

However, recurrent erratic rainfall or a drought period could affect Sahelian food security, as happened during the great drought of the past century [4] and recent local food crises [5]. Despite several adaptations of the Sahelian population to erratic climate conditions [6], food security still remains a concern, and an accurate estimation of regional yields plays an important role in food security [7].

The awareness of rangeland production in relation to water availability is of major interest for the implementation of operational monitoring systems to support policies aiming at reducing the socio-economic impacts of environmental stresses. As water availability is the main limiting factor for vegetation production, especially where average annual rainfall is lower than 500/600 mm [8,9], the interest to estimate rainfall and soil moisture at the regional scale in relation with biomass production has earned a lot of attention.

Several recent studies analyzed time series of rainfall and vegetation indices highlighting the Sahel as an area where vegetation production is rainfall driven and only locally influenced by human activities [9–13]. Other works in the area compared vegetation production to trends of soil moisture [14] and rain use efficiency [15], identifying water availability as the main driver of vegetation growth and dynamics in the Sahel. A shortwave infrared water stress index (SIWSI) has been proposed as an indicator of vegetation water stress [16], while a combination of thermal data and vegetation index [16] were used to produce qualitative maps of soil moisture along the Senegal River.

Compared to these methods, the estimation of evapotranspiration (ET) at the regional scale could give a more quantitative assessment of vegetation water status. ET is a key component of the water budget, and its estimation at different scales is of outmost importance for water management in agriculture [17] and food security programs [18]. ET can be appropriately measured at the field scale by lysimeters, scintillometers or eddy correlation techniques [18]. However, being highly dynamic in space and time because of complex interactions between soil, vegetation and climate [19], the quantification of its flux at the watershed scale is much more difficult than at a specific site [20].

Traditional methods to estimate ET assume homogeneous vegetation cover and structure, but these conditions are hard to meet for large regions [21]. For studies at regional and continental scales, monitoring models are coupled with remotely sensed data that can cope with the spatial and temporal variability of surface characteristics that affect evapotranspiration processes [18]. Several surface characteristics, such as albedo, vegetation cover, leaf area index and land surface temperature, can be retrieved from satellite observations providing data for ET estimation from space.

Since the launch of Earth Observation satellites with thermal infrared channel, such as Landsat Thematic Mapper, NOAA-AVHRR and Terra/Aqua MODIS, several applications have been developed over near fully agricultural canopy covers and semiarid rangeland basins to estimate instantaneous ET and to scale up such estimations to daily ET.

One of the widely used methods [22] to estimate daily ET is based on the evaporative fraction (EF), which is defined as the ratio between latent heat flux and the total heat leaving the Earth's surface. A strong correlation between the value of EF at midday and the daytime average value has been observed [23,24], and it is often assumed as a constant daytime variable [18,21,25–32].

The EF has a strong link with soil moisture availability [33], which is the limiting factor of latent heat flux [29], and it is essentially controlled by water availability in the root zone [34]. The EF behavior at the landscape scale is correlated to the amount of vegetation cover [35], the timing of rainfall events [36], the successions of wet and dry periods [37], the vapor pressure deficit and vegetation photosynthesis activity [38].

EF has an annual behavior related to rainfall events, with peaks during the rainy season and decreasing when soil is drying [39]. In fact, a work conducted over paddy rice area shows that EF always has values close to one, because soil moisture was almost saturated [38].

Recent works conducted in correspondence with eddy covariance stations in North America [39], the northern Australia savannah [40] and the Sahelian region [41] proposed the EF as an indicator of water stress to correct vegetation production estimation. The results of these studies indicated that the use of field-measured EF values within a light use efficiency (LUE) model allows one to improve the estimate of biomass production.

EF can be derived from satellite data using the NDVI-temperature triangle method [42] or the simplified surface energy balance index (S-SEBI) model [43], following the relationship between albedo and land surface temperature [18]. This last approach found applications with a wide range of remotely sensed data and in different ecosystems.

The accuracy of EF estimated by S-SEBI was demonstrated in comparison with other approaches, both for high resolution ASTER images [44] and low resolution NOAA-imagery [18]. Daily ET values, estimated via the EF approach, were validated at the field scale with flux measurements on cropland [27], as well as at the regional scale over the Iberian Peninsula with Digital Airborne Imaging Spectrometer (DAIS) high resolution data [26].

Outside of Europe, the use of this approach has been reported in the Mediterranean landscapes of Chile [45] and in the cotton crops of Brazil [46], demonstrating the suitability of the method in semi-arid areas.

The aim of this work is to retrieve EF from satellite data in the Sahelian rangeland ecosystem and to evaluate the parameter as a moisture indicator, useful also as a correcting factor in the radiation use efficiency biomass estimation model.

The application of the S-SEBI method requires the presence of wet and very dry surfaces [26,43,47]; these conditions are well satisfied over the West Africa area, thanks to presence of the Sahara Desert and stable, humid ecosystems, such as the Niger Inner Delta and Lake Chad. In particular, the goals of this research are: (i) to set-up an automatic procedure to derive EF maps from MODIS products; (ii) to evaluate EF estimation using *in situ* data and to assess EF maps at the regional scale; and (iii) to evaluate the improvement brought by satellite-retrieved EF to the accuracy of the biomass production model.

2. Study Area

The study area covers 1200 × 2400 km over Niger and Chad. The northern part includes the Sahara Desert, where less than 200 mm of rain falls every year and human presence is almost absent (Figure 1). The central part is located on the Sahelian belt, identified by the isohyets of 200 and 600 mm. This zone is mainly characterized by semi-arid savannah, where pastoralism is the most important livelihood activity, with localized evidence of agricultural activity (<20% of cultivated areas; [48]). The southern part of study area belongs to the Sudanian savannah, characterized by wetter climate (annual rain greater than 600 mm), an intensive farming system and less dependency on rain for vegetation productivity [4,10]. The rainy season of the whole study area is essentially from July to October and slightly longer in southern areas, with almost zero precipitation during the rest of the year.

A number of humanitarian crises have hit this area over recent years; although several are from concurring causes, such as food supply, livestock management, environmental degradation and household coping capabilities, low or erratic rainfall remains the key factor triggering the crisis [4,5,49]. In this region, the population has increased during the past 25 years [50]. The rural population is still growing, contrary to many other parts of the world, leading to heavy pressure on the environment, especially during adverse years.

Figure 1. The study area overlaid on the regional GlobCover (GC) map of Africa [51]; the red star shows the position of the eddy covariance station; the red diamonds represent the field sites; the blue lines indicate the isohyet boundaries of 200–600 mm/year.

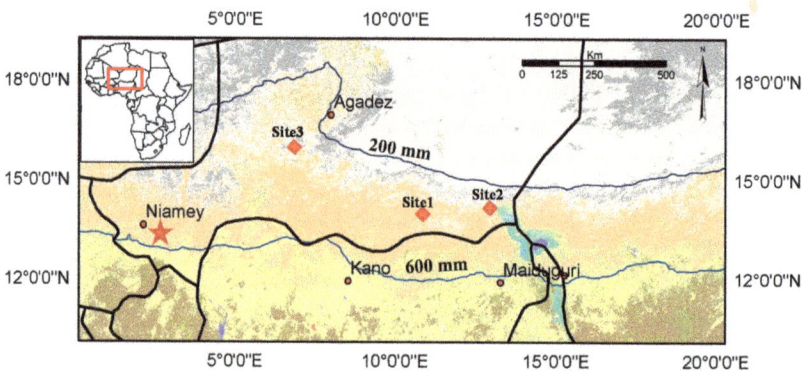

3. Materials

3.1. Earth Observation Data

According to the S-SEBI approach [43], the retrieval of EF from satellite data is based on the relationship between albedo and land surface temperature (LST). Two MODIS products were then considered: MCD43B3, *i.e.*, the hemispherical reflectance (black-sky albedo) 8 days at 1-km spatial resolution, and MOD11A2, *i.e.*, land surface temperature 8 days at 1-km spatial resolution [52]. In order to cover the entire area of interest, two MODIS tiles (h18v07, h19v07) were downloaded for 10 years (2000–2009), summing up to about 450 images per tile.

Other satellite-derived products were used for analysis and evaluation purposes. For the analysis of the EF contribution to biomass production, we used dry matter productivity (DMP) maps [53]. DMP is a satellite-derived product, developed at the Flemish Institute for Technological Research (VITO), that quantifies the daily increase of dry biomass (growth rate) and is expressed as kilograms of dry matter (kg·DM) per hectare per day. The DMP product used in this exercise is a 10-day composite at 1-km spatial resolution covering the period 2000–2009.

Ancillary satellite data consist of rainfall and vegetation maps for the study area. Rainfall estimation (RFE 2.0) is provided by Famine Early Warning Systems Network (FEWS) every 10 days at 8-km spatial resolution [54]. RFE 2.0 is produced by a combination of Meteosat 5 data (satellite infrared data) and daily rain gauge data extracted from the WMO's Global Telecommunication System (GTS) with the additional integration of the two new Special Sensor Microwave/Imager (SSM/I) instruments on-board the Defense Meteorological Satellite Program satellites and the Advanced Microwave Sounding Unit (AMSU). Vegetation maps are represented by Normalized Difference Vegetation Index (NDVI) provided by the SPOT-Vegetation satellite (VGT) sensor every 10-days at 1-km spatial resolution [55]. Finally for the analysis of EF behavior for different vegetation types, the regional GlobCover (GC) map for Africa with 300-m spatial resolution was used [51], which describes land cover classes over the entire study area.

3.2. Field Biomass and Flux Measurements

Field biomass data have been provided by Action Against Hunger (ACF) for three different sites in Niger (Figure 1). Three site were analyzed: the first site is located in a tiger bush area 35 km north of Nigeria border (Site 1, Longitude 10.9, Latitude 13.7); the eastern site is located around Lake Chad (Site 2, Longitude 12.8, Latitude 13.95), while the northern site (Site 3, Longitude 6.8, Latitude 15.8) is located around Agadez, which is the upper limit of pasture activities [56,57]. Biomass measurements were collected following the quick double-sampling technique [58] to calibrate/validate ACF satellite maps of available forage [59]. Overall, 19 annual biomass values/data are available for the period of 2000–2009 (Table 1). These ground samples provide crucial information for the evaluation of EF capability as a water stress factor in biomass estimation.

Table 1. Field data cardinality and average sampled values of the three field biomass measurements.

Site	#Data	Period	AVG (kg/ha)	Max (kg/ha)	Min (kg/ha)	Stand Deviation (kg/ha)
Site 1	6	2003; 2005–2009	963	1,463	342	508
Site 2	8	2000; 2002–2009	371	1,047	0	378
Site 3	5	2001; 2005; 2007–2009	888	1712	326	614

Flux measurements were collected at an eddy covariance tower situated in the Wankama catchment (Figure 1), 60 km east of Niamey, Niger. This site presents the typical Sahelian landscape with sparse savannah and millet fields. Daily data of net radiation (W·m^{-1}) was measured every minute by the tower instruments at a height of 2.5 m; these data are supplied to the user from the CarboAfrica project through FLUXNET measurement network as the average over 30-min periods [60]. This variable is available as a Level 2 product, *i.e.*, not gap-filled, but checked/filtered for out-of-range values or clearly wrong data [60]. The daily latent heat flux data (W·m^{-1}), processed with despiking, double rotation and gap filling following the indications of [61], were obtained from the publication of [62]. Both fluxes are available for the period between June 2005 and June 2007, including the wet season of 2005 and 2006.

4. Method

4.1. Estimation of Evaporative Fraction

The more widely applied method for ET estimation with passive remote sensing is the energy balance equation [18]. The land surface energy balance is the thermo-dynamic equilibrium between turbulent transport processes in the atmosphere and laminar processes in the sub-surface [17]. The basic formulation can be written as:

$$R_n = \lambda E + G_0 + H \tag{1}$$

where Rn is the net radiation, λE is the latent heat flux (λ is the latent heat of vaporization of water and E is evapotranspiration), G_0 is the soil heat flux and H is the sensible heat flux.

Evaporation and transpiration occur simultaneously, and there is no easy way of distinguishing between the two processes: when the crop is small, water is predominately lost by soil evaporation, but once the vegetation completely covers the soil, leaf transpiration becomes the main process [28]. Many satellite-based approaches estimate daily ET, exploiting the EF factor [26,32,44], defined as the ratio between the latent heat flux (λE) and the available energy at the land surface ($R_n - G_0$):

$$EF = \frac{\lambda E}{R_n - G_0} \tag{2}$$

In the present work, the EF estimation is obtained using the albedo-temperature method [43]. This approach allows one to compute the EF for every pixel as the relative distance from two lines, called the dry edge and wet edge, defined through a date-specific albedo-LST relationship (Figure 2). The method's accuracy is dependent on the presence of humid and arid surfaces in the study area.

Figure 3 provides the flow chart of the steps followed for the EF estimation from satellite products, for each available date of satellite products. Albedo and LST data were extracted from the digital numbers (DN) of MCD43B3 (layer 10) and MOD11A2 (layer 1), as indicated by the MODIS product description [52], while information on LST data quality as derived from Layer 2 of the MOD11A2.

Before starting the EF calculation, pixels flagged as "no-data" or "low quality" were masked out and excluded from the analysis.

To perform the EF estimation, the albedo-LST scatterplot is derived for a single date (Figure 2) and analyzed to extract minimum and maximum temperature values for all of the albedo classes identified from statistical analysis [63].

The series of maximum and minimum LST values are used to calculate the date-specific dry and wet edge equation through linear regression:

$$dry\ edge\!: T_H = m_{dry}\alpha_0 + q_{dry} \tag{3}$$

$$wet\ edge\!: T_{\lambda E} = m_{wet}\alpha_0 + q_{wet} \tag{4}$$

where m and q represent the parameters (slope and intercept) of the two regression lines, α_0 represents the albedo, while T is the land surface temperature.

Figure 2. Scatterplot between surface albedo and LST. Blue circles correspond to minimum temperature values for each albedo class, which are used to compute the wet edge (lower limit of the graph) through linear regression. Red circles correspond to the maximum temperature values for each albedo class, which are used to compute the dry edge (upper limit) through linear regression. T$_H$ (maximum temperature) and T$_{\lambda E}$ (minimum temperature) represent the values used in the calculation of the EF for the pixel i.

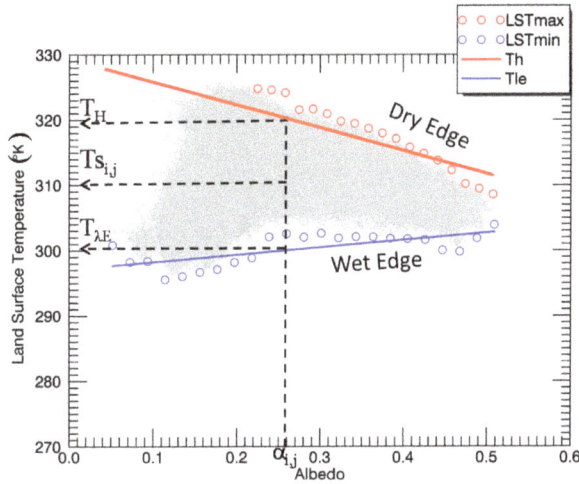

Figure 3. Flowchart for the evaporative fraction estimation from the MODIS products of albedo and land surface temperature.

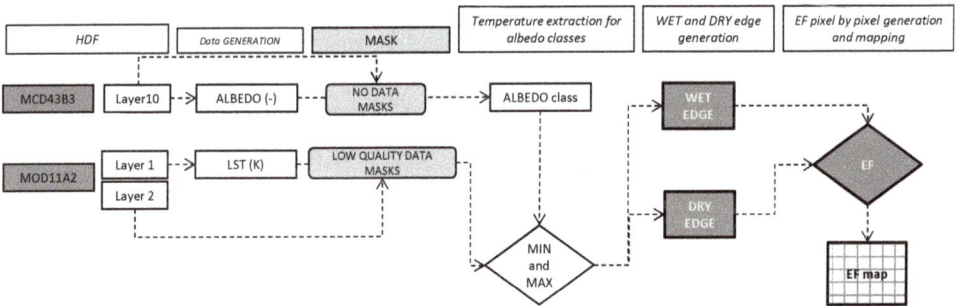

The dry edge was defined considering only pixels in the radiative controlled condition, commonly identified as the maximum temperature data for all of the albedo values greater than the inflection point of the concave temperature-albedo scatterplot [17]. This condition was empirically defined for an albedo value above 0.2, as used also by [44].

Exploiting the dry and wet edge, the EF can be calculated for every pixels i, dividing the difference between T$_H$ and the temperature pixel Ts by the difference between T$_H$ and T$_{\lambda E}$:

$$EF_i = \frac{T_{Hi} - T_{si}}{T_{Hi} - T_{\lambda Ei}} \qquad (5)$$

where T_{si} is the temperature value of the pixel i and T_{Hi} and $T_{\lambda Ei}$ are respectively the maximum and minimum temperature value derived by the dry and wet edge functions for a given albedo value α_i.

The *EF* equation can be rewritten as:

$$EF_i = \frac{(m_{dry}\alpha_i + q_{dry}) - T_{si}}{(m_{dry}\alpha_i + q_{dry}) - (m_{wet}\alpha_i + q_{wet})} \tag{6}$$

this procedure, implemented with an *ad hoc* code in IDL language (Interactive Data Language, version 8.2), was applied to each pixel of the image and on each date available for both MODIS tiles h18v07 and h19v07.

The maps estimated at the same dates were mosaicked, obtaining EF maps of 122 × 2400 km to cover the entire study area.

4.2. Evaluation of the Estimated EF

Since it is well known that *EF* is related to water availability provided by rainfall, particularly in the natural vegetation of semi-arid environments, vegetation growth and land cover [29,35,38,64], we used the RFE, SPOT-VGT NDVI and GlobCover classes to assess the consistency of EF estimation. In particular, average and relative standard deviation (RSD) maps of EF are computed from the 448 8-day EF maps and analyzed for the major land cover classes of the study area, thanks to GlobCover map. This analysis was conducted in order to evaluate the coherence between the EF and the expected behavior over different vegetation covers.

In correspondence with the eddy covariance tower, EF behavior was compared to rainfall events and vegetation growth.

Quantitative evaluation of the reliability of EF estimations as a moisture (water stress) indicator is accomplished using the eddy covariance data from Wankama station.

Due to the different time steps of satellite estimation and flux measurements, the satellite-derived EF was compared with the 8-day average of daily ET corresponding to the MODIS composite. Data from the tower measurements identified as outliers from statistical analysis and EF satellite estimation flagged as low quality were excluded by the analysis. Moreover, thanks to Equation (7), it was possible to compare *in situ* estimation of EF with satellite-derived *EF*:

$$EF_{eddy} = \lambda E_d / R_{nd} \tag{7}$$

where R_{nd} and λE_d are the daily net radiation and the daily latent heat flux, respectively.

4.3. Biomass Estimation

The seasonal cumulative DMP is an indicator of the annual rangeland production [15,65–67], which can be compared to the ground data, that represents the total annual herbaceous production measured in field.

In order to compare satellite data with field samples, EF and DMP were extracted in correspondence with the field sites location in a buffer of 1 km. The 10-day DMP product for the period of July to October, referred to here as JASO, was cumulated to obtain the annual syntheses of dry matter production (DMP^{JASO}).

The EF was exploited as a water availability factor to correct the satellite estimation of vegetation biomass (DMP).

DMP and EF have different time steps (10 and 8 days, respectively); consequently, monthly values were calculated in order to use the water availability/stress factor in the biomass estimation model.

For EF, the monthly average ($\overline{EF}^m$) was computed for every month (m) and every site (s):

$$\overline{EF}_s^m = \frac{\sum_{t=1}^{n} EF_{s,t}^{8D}}{n} \tag{8}$$

where EF^{8D} is the estimated water stress from Equation (6), s is the site and n is the cardinality of the 8-day EF data for each month.

For DMP, the monthly sum (DMP^m) was calculated to represent the total dry biomass produced during every month at each site:

$$DMP_s^m = \sum_{t=1}^{3} DMP_{s,t}^{10D} \tag{9}$$

where DMP^{10D} is the 10-day biomass estimation product, t is the number of DMP data within the month and s is the site.

Monthly $\overline{EF}^m$ and DMP^m values were than integrated and annually cumulated by the following equation for each site:

$$DMP_s^{JASO*} = \sum_{m=1}^{4} DMP_s^m \cdot \overline{EF}_s^m \tag{10}$$

where EF^m and DMP^m are the variable obtained from Equations (8) and (9), s is the site and 4 is the number of months in the JASO period.

Finally, to quantify the improvement of DMP^{JASO}*, the comparison between observed and estimated values were performed and difference-based statistics [68] together with regression analysis and Akaike information criterion (AIC) [69], Equation (11), were conducted.

$$AIC = n \cdot \log(MSE) + 2 \cdot T \tag{11}$$

where n is the number observed/simulated pairs, MSE is the mean square error and T is the number of inputs in the model.

AIC with a lower value indicates whether the increase in the input of a model is compensated for by a significant increase in accuracy.

5. Results and Discussion

5.1. Dry and Wet Edge Statistics

Figure 4 shows the average intercepts (a) and slopes (b) for the calculated dry (empty triangles) and wet (filled circles) edge. Every point represents the average of 10 estimations from 2000 to 2009 together with a bar representing the standard deviation. The dry edge statistics of the slope

338

and intercept are on the second y-axis, to facilitate a comparison with the wet edge statistics. The gray shaded area displays the period when generally no rainfall occurs in the study area.

Figure 4. Eight-day average values of the intercept (**a**) and slope (**b**) obtained from dry and wet edge lines for the 2000–2009 period. Shaded gray areas represent the dry season. Plots show three albedo-LST scatterplots for the year, 2009.

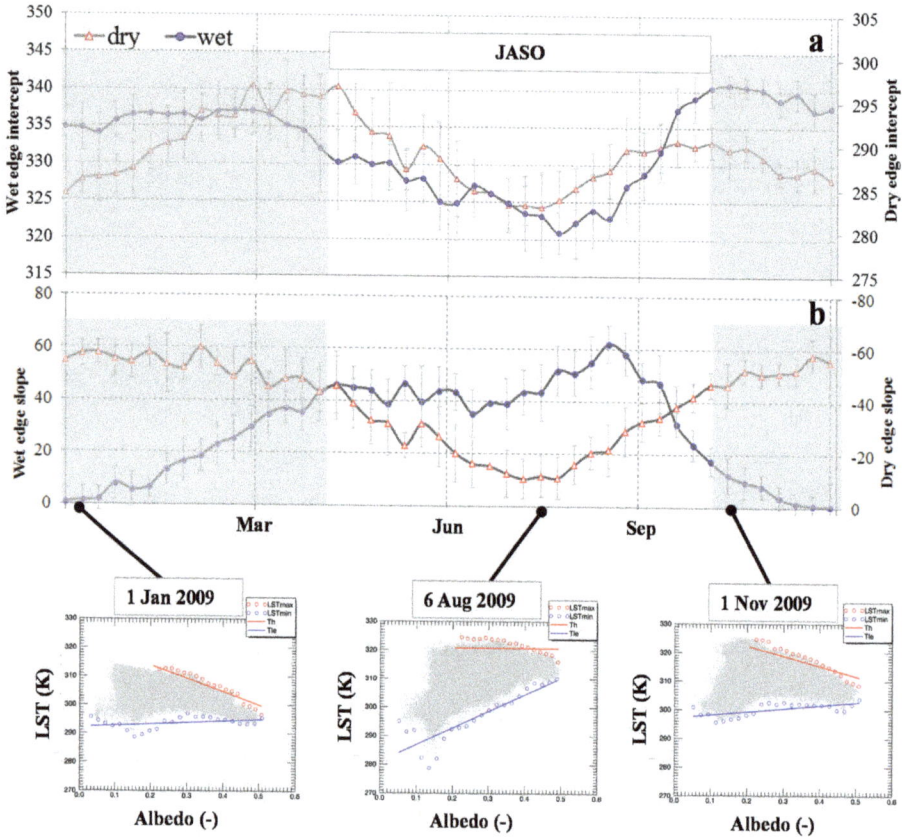

The intercept of the dry edge follows the typical behavior of West African temperature [70], with lower values during the wet season (June–October) and two peaks during the dry season, the former in April and the latter in November.

The Wet edge intercept has lower value in the wet season and stable, higher values during the dry one. The average slope coefficient shows that during the dry season, the wet edge is generally horizontal (values close to zero), while the dry edge has a high negative slope (values down to −60), as shown by a similar analysis conducted in the Mekong Delta [71].

On the contrary, in the rainy period, the dry edge is almost flat, while the wet edge has a strong positive slope (values up to 60). In general, the coefficients of wet and dry edge follow a seasonal behavior driven by rainfall and incoming solar radiation.

Figure 4 shows also three examples of the albedo-LST scatterplots. The first displays the dry season condition with the flat wet edge and the second one the wet condition with the flat dry edge.

The last scatterplot displays an intermediate condition at the end of the rainy season, when the two lines are both oblique and the maximum LST is higher.

The maximum albedo value of 0.6 in the scatterplots highlights the presence of high reflective surfaces [72], which correspond to brighter desert areas. These areas are stable through the season; hence, they are present in every plot.

The areas with lower temperature (below 305 K) and lower albedo (below 0.2) correspond to a permanent humid zone, such as the Lake Chad area and the border of the Niger River.

The permanent presence through the years of these two extreme situations allows one to produce a meaningful scatterplot describing the contrast between dry and wet areas, hence guaranteeing the conditions for the application of the method [17,43].

5.2. Evaluation of EF Spatial Patterns

Figure 5 shows the mean EF map obtained from the estimated 448 maps for the period of 2000–2009 (a), together with the RSD (b). As expected, the mean values vary between zero and one, where zero indicates the hyper-arid condition and one the humid area. The areas with a mean rainfall below 200 mm, belonging to the Sahara Desert, were excluded from the analysis, since these areas are not populated and EF estimation makes sense and is useful only on partially vegetated surfaces. Permanent arid areas ($EF < 0.2$) can be found close to the desert, especially in the Agadez province (Niger) and in the central part of Chad. Both of these areas rely on the "high-risk Sahel's vulnerable zone", where the main livelihood activity is transhumant herding [73].

The hyper-humid area ($EF > 0.7$) can be found in correspondence with permanent water bodies, such as Lake Chad [74], Lake Fitri (central Chad) and Lake Kainji (west Nigeria). Furthermore, the woody hills in Nigeria, characterized by a high level of rainfall (more than 1200 mm/year), are generally well humid. The Sahelian belt is characterized by medium-low average values of *EF* (below 0.5) apart from the river belts: Niger in the west part of study area, Yobe along the Niger-Nigeria border and Chari south of Lake Chad.

The RSD (Figure 5b) is a normalized measure of EF data dispersion. The equation of the RSD is obtained dividing the standard deviation by the mean. The lower percentage indicates a lower variability in the EF time series.

The map highlights areas in red and orange with stable *EF* (RSD < 30%) from 2000–2009. These areas belong to lakes, rivers and wet regions in central Nigeria, which are also the regions characterized by high *EF* average values (Figure 5a). Hence, these well-watered areas have maintained their condition across the years analyzed.

Vice versa, higher variation in EF values (RSD > 50%, blue and light blue) can be found in the northern Sahel (northern-western Niger and central Chad). In particular, in western Niger, the fossil valleys display a stronger variability of EF data compared to the surrounded rangelands, because of the greater water availability thanks to the morphopedological characteristics, as observed in [75]. The high RDS indicates that the EF data of these areas can vary abruptly, thanks to the strong seasonality (intra-annual variability).

Figure 5. The map of the average EF (**a**) and relative standard deviation (**b**) derived from 448 EF eight-day maps (2000–2009). Isohyets were calculated from rainfall estimation (RFE) data for the same period. The hyper-arid areas (<200 mm·year^{-1}) are masked out, and the GlobCover map is in the background.

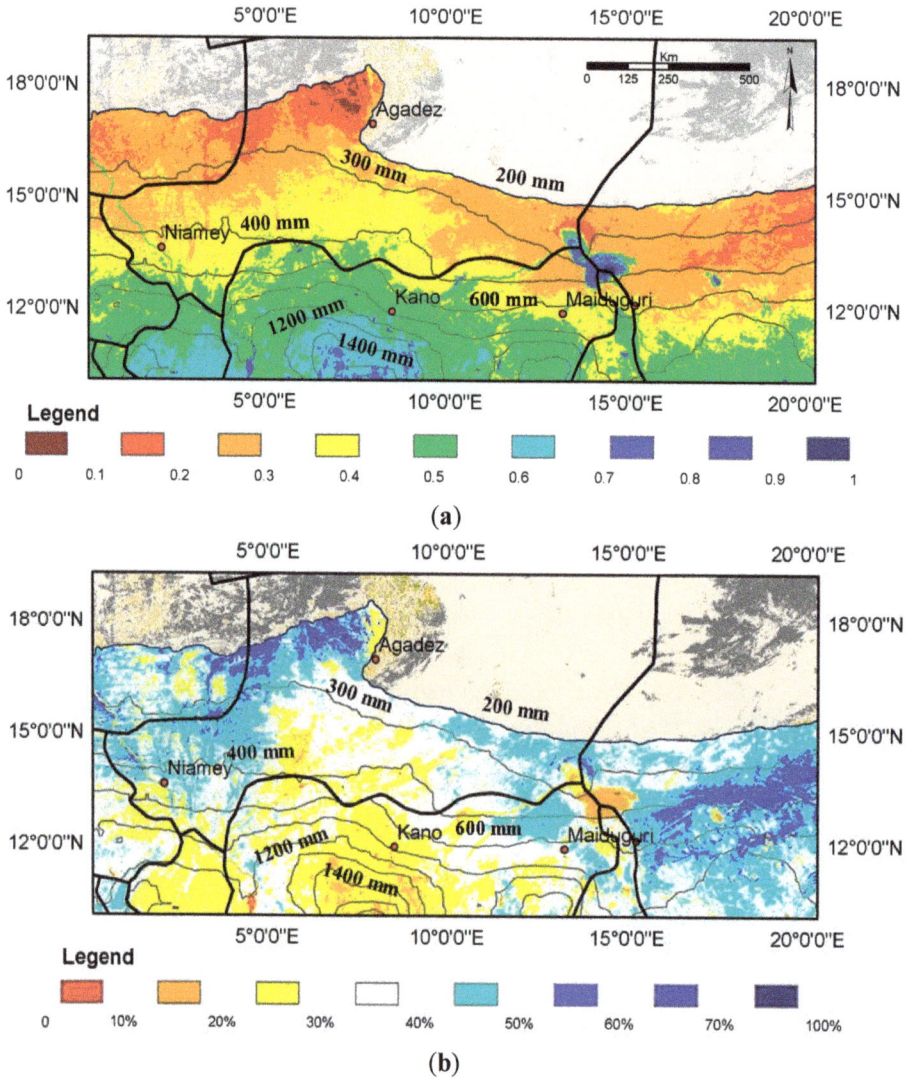

(a)

(b)

The high EF variability of northern areas in Niger, characterized by the small EF average, can be driven by particularly favorable years (inter-annual variability).

The average EF map has been analyzed by GC classes (Figure 6). The GC classes are sorted from the mainly northern classes (GC_200) to the southern (GC_130), except for the classes of wetland (GC_180) and water body (GC_210). The most common classes are the bare areas (GC_200) and grassland savannah (GC_140). These two classes cover 70% of the entire study area.

Figure 6. Percentage of GC classes over the study area (codes and map color are reported) and the statistics of EF data for each LC classes (average (AVG) and relative standard deviation (RSD)). Red and green indicate land cover with a lower of a higher EF average, respectively.

LC (GC)				EF [-]	
Code 1	Code	Description	%	AVG	RSD(%)
	200	Bare areas	16%	0.38	21.91
200	201	Consolidated bare areas (hardpans, gravels, bare rock, stones, boulders)	12%	0.39	21.32
	202	Non-consolidated bare areas (sandy desert)	23%	0.51	17.57
140	140	Closed to open (>15%) herbaceous vegetation (grassland, savannas or lichens/mosses)	11%	0.36	18.01
	144	Closed (>40%) grassland	7%	0.42	16.32
110	110	Mosaic forest or shrubland (50-70%) / grassland (20-50%)	3%	0.44	22.92
30	30	Mosaic vegetation (grassland/shrubland/forest) (50-70%) / cropland (20-50%)	7%	0.46	17.12
20	20	Mosaic cropland (50-70%) / vegetation (grassland/shrubland/forest) (20-50%)	4%	0.53	16.59
10	10	Irrigated croplands	10%	0.60	13.33
60	60	Open (15-40%) broadleaved deciduous forest/woodland (>5m)	0%	0.64	8.39
130	130	Closed to open (>15%) (broadleaved or needleleaved, evergreen or deciduous) shrubland (<5m)	7%	0.58	11.99
180	180	Closed to open (>15%) grassland or woody vegetation on regularly flooded or waterlogged soil - Fresh, brackish or saline water	1%	0.67	18.40
210	210	Water bodies	0%	0.83	11.97

On average, the GC class with the highest EF value (EF = 0.83) is water bodies (GC_210). Among the vegetation classes, only irrigated crops (GC_10), forest (GC_60) and wetland have a mean EF greater than 0.6. The most arid classes (GC_140 and GC_200), with a mean EF lower than 0.4, describe the typical landscape of the northern Sahel [50].

This analysis shows that the spatial patterns of EF data (long-term average) are in agreement with the well-known climatic and landscape features of these areas. A similar analysis conducted in China [21], Europe [26] and Africa [22] demonstrated that EF maps build up spatial and temporal patterns coherent with the presence of different vegetated surfaces, different climatic conditions and different seasonal behavior of vegetation.

5.3. Comparison of Seasonal EF Estimations with Eddy Covariance Data

5.3.1. Temporal Dynamics of the Variables

Figure 7 presents the time series of net radiation, ET and EF measured at the Wankama eddy covariance tower (black lines) and the satellite-derived time series of EF (red dashes), RFE (blue bars), NDVI (green line), albedo (cyan line) and LST (gold line) extracted by the corresponding image pixel for the years 2005 (Figure 7a) and 2006 (Figure 7b). The Wankama eddy tower, placed in millet fields, is characterized by the typical Sahelian behavior of rainfall and vegetation growth [62].

342

Figure 7. From top to bottom, the temporal behavior of daily net radiation; daily evapotranspiration; EF-derived from the eddy covariance tower data at the Wankama site (black lines) together with eight-day EF estimation from MODIS data (red dashes); decadal NDVI-VGT (green line); decadal precipitation (blue bars), eight-day MODIS albedo (gray line) and eight-day MODIS temperature (yellow line) for 2005 (**a**) and 2006 (**b**). Vertical lines represent the start and finish of the JASO period, doy the Day Of the Year.

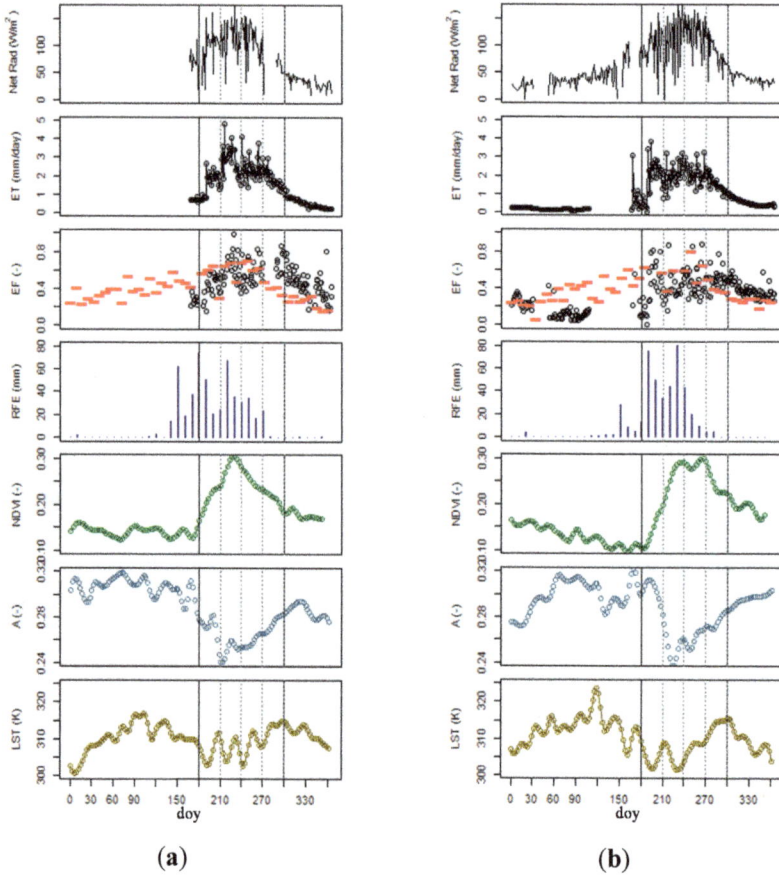

(a)　　　　　　　(b)

Vertical black lines indicate the average Sahelian wet season as being from July to October (JASO). The zero value of the satellite EF estimation, due to cloud contamination or other atmospheric interference in the data, was masked out from this analysis.

Figure 7a shows the time series of remote sensed and measured variables for the year, 2005. The first eddy measurement was recorded in June (doy 160), after the beginning of rain. ET shows the peak (>4 mm/day) in August, as well as net radiation. RFE shows an early start of the rainy season compared to the JASO, with an intense rainfall period of 60 mm in May (doy 155). In total, 524 mm fell in 2005.

The red dashes indicate the eight-day period of satellite EF estimation from Equation (6) together with the *in situ* calculated EF from Equation (7). Both of the EF time series show higher values in the rainy season and drops in correspondence with the low ET value (e.g., doy 210 and 240). MODIS-derived EF decreases smoothly after doy 270, as happens for ET.

The vegetation behavior, highlighted by NDVI, shows the start of the season around doy 200 (19 July), about 50 days after the start of rainfall, because of the necessary time for germination [75]. NDVI has a specular behavior compared to albedo, as expected from the progressive cover of bare soil due to vegetation growth. The last two time series in Figure 7a display albedo and LST data. Both show a high value during the dry season, indicating a warm and bare surface. The rainy season has an average LST of 308 K (~35 °C), lower than the dry season average (311 K), because incoming energy during is exploited by evaporative and transpirative processes.

Figure 7b shows the same variables for 2006. The rainy period is shorter and less abundant compared to the previous one, with 430 mm of total rainfall. The estimated EF reaches a peak of 0.8 at doy 240 (28 August). The main EF drop is visible (doy 210) in correspondence with the drier period of the wet season, between the two main rain events. Both EF and ET rapidly decrease their values at the end of the wet season (doy 270). Higher vegetation growth occurs between August and September, and NDVI shows the presence of vegetation also in November–December (doy 300–360, NDVI ~0.2), even if EF and ET show that the area is completely dry.

The temporal behaviors of field measurements and satellite-derived data for 2005 and 2006 display high variability between dry and wet months. The 2005 wet season had an early start, while 2006 had a very late start, as well as an earlier end. Hence, the two years had different seasonality in terms of rainfall amount and distribution [62]. Among the several satellite-derived variables, estimated EF shows a higher correlation with estimated rainfall (data not shown), as expected from previous field studies [36,37,64,70], and in general, the estimated EF looks to be in accordance with the ET behavior and eddy covariance-derived EF. The temporal behavior of the EF variable is more noisy compared to the time series of other satellite-derived variables and hardly zero also in the absence of rainfall [39], as also displayed by eddy-derived EF.

Both MODIS estimation and eddy EF have comparable values during the wet season (JASO), showing a higher average value for the wetter year, 2005 ($\mu = 0.45$; $\sigma = 0.07$ for satellite and $\mu = 0.51$; $\sigma = 0.13$ for *in situ*, respectively), when compared to the drier 2006 ($\mu = 0.39$; $\sigma = 0.07$ and $\mu = 0.34$; $\sigma = 0.09$ for the satellite and *in situ*, respectively).

5.3.2. Correlation Analysis with ET

In order to evaluate the reliability of EF as a moisture indicator, a correlation analysis has been conducted between satellite *EF* estimation (y-axis) and ET measured by the eddy covariance tower (x-axis) (Figure 8).

The EF resulted in being significantly correlated to ET ($p < 0.001$), with a regressive coefficient of 0.54 pooling together the two years (2005–2006). ANCOVA analysis reveals that single year correlations (2005, $r^2 = 0.62$; 2006, $r^2 = 0.45$) were not significantly different ($p < 0.05$). This correlation is biased by estimated EF in the late, dry season (January–May), when no rain and no vegetation are present, confirming that EF is noisy in the dry season [39].

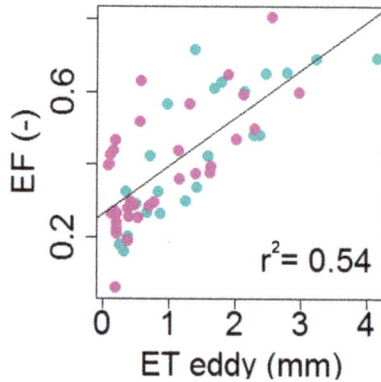

Figure 8. Correlation between estimated EF (y-axis) and measured ET (x-axis) for both years 2005 (gray) and 2006 (purple) ($n = 57$).

As expected the measured net radiation is better correlated with ET ($r^2 = 0.64$), since it represents the climatic driving force of evaporative and transpirative processes. In order to investigate if *EF* can improve the capability to explain the variance of ET, a multiple regression was performed between ET as a dependent variable and two independent variables, the measured Rn and the simulated *EF*.

Result shows that both the explanatory variables significantly contribute to the explanation of ET variability (70% of the total variance). Rn resulted in being more important, explaining about 64% of the total variance ($p < 0.001$), and *EF* significantly improved the model with a further 6% of variance explanation ($p < 0.01$). These results indicate that *EF* estimated with low resolution satellite data is well correlated with the field measured flux and gives a statistically significant contribution to the explanation of ET variability. It is important to remember that the *EF* data is derived by 1-km albedo and LST products; this aspect can strongly limit the comparison with field data acquired on small plots in a heterogeneous environment.

5.3.3. Biomass Estimation Improvements Using *EF* Correction

The results of previous analysis confirmed the validity of *EF* as a moisture indicator supporting the idea of using this satellite estimation as a water stress factor in a radiation use efficiency model. Previous studies, exploiting only field-based *EF*, demonstrated that *EF* can be exploited as a water stress efficiency factor [41].

To assess this contribution, the performance of operational products (DMP^{JASO}) and biomass estimation corrected by *EF* (DMP^{JASO}*) were compared with available annual production data over three test sites in Niger.

In Figure 9a are shown the three sites' specific correlations between the available field data and DMP^{JASO}. The three sites show different correlations: in particular, Site 1 (black dots) presents little correlation ($r^2 = 0.49$, intercept = 1300, slope = 0.3); Site 2 (blue squares) shows an average correlation ($r^2 = 0.51$, intercept = 700, slope = 1.1); and Site 3 (red triangles) has a high correlation ($r^2 = 0.66$, intercept = 400, slope = 0.3). All three sites have the typical Sahelian biomass production [76], ranging from 100 (kg·ha^{-1}), in adverse years, to 20-times higher production in favorable climatic conditions.

Figure 9. The correlation between annual biomass samples and satellite estimation DMP^{JASO} (DMP, dry matter productivity) (**a**), DMP^{JASO}* (**b**) and normalized data (**c**) ($n = 19$). Black dots for Site 1, blue squares for Site 2 and red triangles for Site 3. Black and gray diamonds represent normalized DMP^{JASO} and DMP^{JASO}, respectively. The dotted line indicates the 1:1 line.

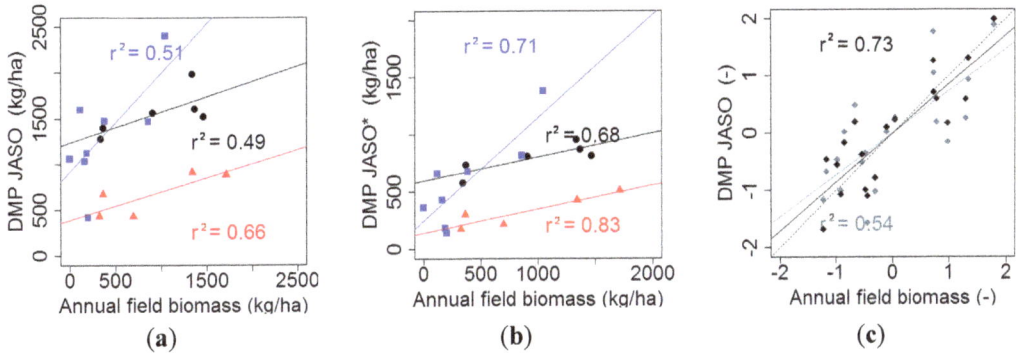

Results demonstrate that the DMP^{JASO} is able to detect the field biomass variability with site-specific, good correlation; however, the analysis of intercept and slope variability across sites indicates that the model is not able to give a robust quantitative biomass estimation. Indeed, the DMP algorithm does not take into account distinct efficiency factors in the conversion of light into biomass among different vegetation types. It should be reminded that, despite the three test sites featuring the same land cover and eco-region, the actual floristic composition and ecological characteristics could be much more different.

In Figure 9b are shown the effect of the EF contribution (DMP^{JASO}*) over the three sites. The plots show a general increase in the capacity of the remote sensing estimation in each site to detect the variability of the field measurements if water stress is taken into account, as indicated by the increasing of regression coefficients. Moreover, in particular, the EF has reduced the overestimation of the model for poorly productive years, as shown by intercepts closer to zero.

Due to the observation of site-specific DMP product performance, in order to directly compare the two biomass estimations, satellite products and field data were normalized for each site. The normalized data allows one to remove the effect of local differences in the relation between satellite outputs and field biomass, visualizing only the overall model capability to detect the field data variance, rather than absolute values. Data were standardized and converted to z-scores by subtracting from each value the site average and then dividing the result by the standard deviation. In the analysis of time series, the z-score is a dimensionless quantity adopted to convert variables with different scales to a common domain [10,77].

In Figure 9c are shown the normalized data, both for DMP^{JASO} (gray dots) and DMP^{JASO}* (black dots). The data close to zero are near the population average, while data values below or above zero indicate a positive or a negative anomaly, respectively. The top right and bottom left corners indicate that years were estimated, and the measured variables' data are in agreement.

The correlation coefficient of the normalized DMP[JASO]* ($r^2 = 0.73$, $p < 0.001$) indicates that there is a significative increase in the capacity of the remote sensing estimation to explain the variance of annual field biomass measurements if water stress (EF) is taken into account.

This result is in accordance with previous work [41], even if the analysis was conducted at a monthly time step using EF derived with MODIS data (LST and albedo), rather than with field-measured EF and MODIS EVI at an eight-day time step.

Finally, AIC was calculated in order to evaluate whether the increase in the input of the model compared to the basic DMP was compensated for by a significant increase in accuracy (a lower AIC value indicates a convenient model improvement) [69]. Despite the correction, proposed to increase the number of inputs in the biomass estimation, the usefulness of the EF approach (DMP[JASO]*) is confirmed by the improved model performance, indicated by the higher correlations shown in Figure 9c and the lower AIC value (106) compared with the one obtained with DMP[JASO] (112).

6. Conclusions

The work exploited an automatic procedure to calculate multitemporal evaporative fraction maps from low resolution albedo and land surface temperature satellite data over Niger and Chad. Up to now, this is the first time that multiyear (2000–2009), eight-day maps of the evaporative fraction were produced from low resolution satellite data and analyzed for the West African Sahel. The adopted methodology, based on previous scientific works and well suited for semi-arid areas, allowed producing maps able to identify patterns of wet and dry condition, which are coherent with the main ecological features related to land cover classes and precipitation regimes.

The satellite estimation of the evaporative fraction, despite the uncertainty related to the 1-km resolution of the data, resulted in being correlated with the measurements of evapotranspiration ($r^2 = 0.54$, $p < 0.001$) acquired for two years (2005–2006) by an eddy flux tower in Niger. The total variance of evapotranspiration is mainly explained by the measured net radiation (64%, $p < 0.001$), while the estimated evaporative fraction significantly improves the model with a further 6% of variance explanation ($p < 0.01$). These results demonstrate that the satellite-derived evaporative fraction is a reliable indicator of moisture, useful for savannah status monitoring.

We further tested the use of the evaporative fraction as a water availability indicator to improve the accuracy of an operational remote sensing product of biomass estimation based on the radiation use efficiency concept. When the satellite-derived evaporative fraction is used as an indicator of water stress in the model, the correlation between annual biomass ground measurements and satellite estimations, for 19 samples over three sites, significantly improves ($r^2 = 0.73$, $p < 0.001$) compared to the performance of the basic satellite product ($r^2 = 0.54$, $p < 0.001$). The appropriate water efficiency term derived from optical and thermal remote sensing data represents an advancement over previous studies conducted using only the evaporative fraction derived by *in situ* eddy covariance data.

These findings are encouraging for the monitoring of biomass over wide savannah areas using a satellite-based approach. Future studies are needed to better parameterize the radiation use efficiency model and to calibrate existing products over different ecosystems, in order to take into account the limiting factors and efficiency in the conversion of light into biomass.

Acknowledgments

This research was partially supported by the Geoland-2 project (Contract No. 218795), which is a collaborative project (2008–2012) funded by the European Union under the seventh Framework Programme (http://www.gmes-geoland.info/) and by Space4Agri project in the framework agreement between Regione Lombardia and National Research Council (Convenzione Operativa No. 18091/RCC, 05/08/2013). We also acknowledge the three anonymous reviewers for the help provided in improving this manuscript.

Author Contributions

Conceived and designed the experiments: Mirco Boschetti, Francesco Nutini, Stefano Bocchi. Performed the experiments: Gabriele Candiani, Francesco Nutini. Analyzed the data: Francesco Nutini, Mirco Boschetti, Pietro Alessandro Brivio. Wrote the paper: Francesco Nutini, Mirco Boschetti, Gabriele Candiani, Pietro Alessandro Brivio.

Conflicts of Interest

The authors declare no conflict of interest.

References

1. Anyamba, A.; Tucker, C. Analysis of Sahelian vegetation dynamics using NOAA-AVHRR NDVI data from 1981–2003. *J. Arid Environ.* **2005**, *63*, 596–614.
2. Zorom, M.; Barbier, B.; Mertz, O.; Servat, E. Diversification and adaptation strategies to climate variability: A farm typology for the Sahel. *Agric. Syst.* **2013**, *116*, 7–15.
3. Soler, C.M.T.; Maman, N.; Zhang, X.; Mason, S.C.; Hoogenboom, G. Determining optimum planting dates for pearl millet for two contrasting environments using a modelling approach. *J. Agric. Sci.* **2008**, *146*, 445–459.
4. Mortimore, M.J.; Adams, W.M. Farmer adaptation, change and "crisis" in the Sahel. *Glob. Environ. Chang.* **2001**, *11*, 49–57.
5. FAO. Available online: www.fao.org/emergencies/crisis/sahel/en (accessed on 14 April 2014).
6. Mertz, O.; Mbow, C.; Maiga, A.; Diallo, D.; Reenberg, A.; Diouf, A.; Barbier, B.; Moussa, I.B.; Zorom, M.; Ouattara, I.; *et al.* Climate factors play a limited role for past adaptation strategies in West Africa. *Ecol. Soc.* **2010**, *15*. Available on line: http://www.ecologyandsociety.org/vol15/iss4/art25/ (accessed on 14 April 2014).
7. Wang, J.; Li, X.; Lu, L.; Fang, F. Estimating near future regional corn yields by integrating multi-source observations into a crop growth model. *Eur. J. Agron.* **2013**, *49*, 126–140.

8. Tucker, C.J.; Vanpraet, C.L.; Sharman, M.J.; van Ittersum, G. Satellite remote sensing of total herbaceous biomass production in the senegalese sahel: 1980–1984. *Remote Sens. Environ.* **1985**, *17*, 233–249.

9. Hein, L.; de Ridder, N.; Hiernaux, P.; Leemans, R.; de Wit, A.; Schaepman, M. Desertification in the Sahel: Towards better accounting for ecosystem dynamics in the interpretation of remote sensing images. *J. Arid Environ.* **2011**, *75*, 1164–1172.

10. Boschetti, M.; Nutini, F.; Brivio, P.A.; Bartholomé, E.; Stroppiana, D.; Hoscilo, A. Identification of environmental anomaly hot spots in West Africa from time series of NDVI and rainfall. *ISPRS J. Photogramm. Remote Sens.* **2013**, *78*, 26–40.

11. Olsson, L.; Eklundh, L.; Ardo, J. A recent greening of the Sahel—Trends, patterns and potential causes. *J. Arid Environ.* **2005**, *63*, 556–566.

12. Seaquist, J.; Hickler, T.; Eklundh, L. Disentangling the effects of climate and people on Sahel vegetation dynamics. *Biogeosciences* **2009**, *6*, 469–477.

13. Herrmann, S.; Anyamba, A.; Tucker, C. Recent trends in vegetation dynamics in the African Sahel and their relationship to climate. *Glob. Environ. Chang.* **2005**, *15*, 394–404.

14. Huber, S.; Fensholt, R.; Rasmussen, K. Water availability as the driver of vegetation dynamics in the African Sahel from 1982 to 2007. *Glob. Planet. Chang.* **2011**, *76*, 186–195.

15. Fensholt, R.; Rasmussen, K. Analysis of trends in the Sahelian "rain-use efficiency" using GIMMS NDVI, RFE and GPCP rainfall data. *Remote Sens. Environ.* **2011**, *115*, 438–451.

16. Sandholt, I.; Rasmussen, K.; Andersen, J. A simple interpretation of the surface temperature/vegetation index space for assessment of surface moisture status. *Remote Sens. Environ.* **2002**, *79*, 213–224.

17. Bastiaanssen, W.G.M.; Menenti, M.; Feddes, R.A.; Holtslag, A.A.M. A remote sensing surface energy balance algorithm for land (SEBAL). 1. Formulation. *J. Hydrol.* **1998**, *212–213*, 198–212.

18. Verstraeten, W.W.; Veroustraete, F.; Feyen, J. Estimating evapotranspiration of European forests from NOAA-imagery at satellite overpass time: Towards an operational processing chain for integrated optical and thermal sensor data products. *Remote Sens. Environ.* **2005**, *96*, 256–276.

19. Allen, R.; Tasumi, M.; Trezza, R. Satellite-based energy balance for mapping evapotranspiration with internalized calibration (METRIC)—Model. *J. Irrig. Drain. Eng.* **2007**, 380–394.

20. Irmak, A.; Ratcliffe, I.; Hubbard, K. Estimation of land surface evapotranspiration with a satellite remote sensing procedure. *Gt. Plains Res.* **2011**, *21*, 73–88.

21. Li, X.; Lu, L.; Yang, W.; Cheng, G. Estimation of evapotranspiration in an arid region by remote sensing—A case study in the middle reaches of the Heihe River Basin. *Int. J. Appl. Earth Obs. Geoinf.* **2012**, *17*, 85–93.

22. Sun, Z.; Gebremichael, M.; Ardö, J.; Nickless, A.; Caquet, B.; Merboldh, L.; Kutschi, W. Estimation of daily evapotranspiration over Africa using MODIS/Terra and SEVIRI/MSG data. *Atmos. Res.* **2012**, *112*, 35–44.

23. Hall, F.; Huemmrich, K. Satellite remote sensing of surface energy balance: Success, failures, and unresolved issues in FIFE. *J. Geophys. Res.* **1992**, *97*, 19061–19089.

24. Cragoa, R.; Brutsaert, W. Daytime evaporation and the self-preservation of the evaporative fraction and the Bowen ratio. *J. Hydrol.* **1996**, *178*, 241–255.

25. Crago, R.D. Conservation and variability of the evaporative fraction during the daytime. *J. Hydrol.* **1996**, *180*, 173–194.

26. Sobrino, J.A.; Gómez, M.; Jiménez-Muñoz, J.C.; Olioso, A. Application of a simple algorithm to estimate daily evapotranspiration from NOAA–AVHRR images for the Iberian Peninsula. *Remote Sens. Environ.* **2007**, *110*, 139–148.

27. Gomez, M.; Olioso, A.; Sobrino, J.; Jacob, F. Retrieval of evapotranspiration over the Alpilles/ReSeDA experimental site using airborne POLDER sensor and a thermal camera. *Remote Sens. Environ.* **2005**, *96*, 399–408.

28. Ciraolo, G.; Minacapilli, M.; Sciortino, M. Stima dell'evapotraspirazione effettiva mediante telerilevamento aereo iperspettrale. *J. Agric. Eng.* **2007**, *38*, 49–60.

29. Gentine, P.; Entekhabi, D.; Chehbouni, A.; Boulet, G.; Duchemin, B. Analysis of evaporative fraction diurnal behaviour. *Agric. For. Meteorol.* **2007**, *143*, 13–29.

30. Venturini, V.; Islam, S.; Rodriguez, L. Estimation of evaporative fraction and evapotranspiration from MODIS products using a complementary based model. *Remote Sens. Environ.* **2008**, *112*, 132–141.

31. Yang, D.; Chen, H.; Lei, H. Analysis of the diurnal pattern of evaporative fraction and its controlling factors over croplands in the Northern China. *J. Integr. Agric.* **2013**, *12*, 1316–1329.

32. Hoedjes, J.C.B.; Chehbouni, A.; Jacob, F.; Ezzahar, J.; Boulet, G. Deriving daily evapotranspiration from remotely sensed instantaneous evaporative fraction over olive orchard in semi-arid Morocco. *J. Hydrol.* **2008**, *354*, 53–64.

33. Bastiaanssen, W.G.M.; Ali, S. A new crop yield forecasting model based on satellite measurements applied across the Indus Basin, Pakistan. *Agric. Ecosyst. Environ.* **2003**, *94*, 321–340.

34. Bastiaanssen, W.G.M.; Pelgrum, H.; Droogers, P.; de Bruin, H.A.R.; Menenti, M. Area-average estimates of evaporation, wetness indicators and top soil moisture during two golden days in EFEDA. *Agric. For. Meteorol.* **1997**, *87*, 119–137.

35. Kustas, W.; Schmugge, T.; Humes, K.; Jackson, T.; Parry, R.; Weltz, M.; Moran, M. Relationships between evaporative fraction and remotely sensed vegetation index and microwave brightness temperature for semiarid rangelands. *J. Appl. Metereol.* **1993**, *32*, 1781–1790.

36. Kurc, S.A.; Small, E.E. Dynamics of evapotranspiration in semiarid grassland and shrubland ecosystems during the summer monsoon season, central New Mexico. *Water Resour. Res.* **2004**, *40*, doi:10.1029/2004WR003068.

37. Guyot, A.; Cohard, J.-M.; Anquetin, S.; Galle, S. Long-term observations of turbulent fluxes over heterogeneous vegetation using scintillometry and additional observations: A contribution to AMMA under Sudano-Sahelian climate. *Agric. For. Meteorol.* **2012**, *154–155*, 84–98.

38. Higuchi, A.; Kondoh, A.; Kishi, S. Relationship among the surface albedo, spectral reflectance of canopy, and evaporative fraction at grassland and paddy field. *Adv. Space Res.* **2000**, *26*, 1043–1046.

39. Yuan, W.; Liu, S.; Zhou, G.; Zhou, G.; Tieszen, L.L.; Baldocchi, D.; Bernhofer, C.; Gholz, H.; Goldstein, A.H.; Goulden, M.L.; *et al.* Deriving a light use efficiency model from eddy covariance flux data for predicting daily gross primary production across biomes. *Agric. For. Meteorol.* **2007**, *143*, 189–207.

40. Kanniah, K.D.; Beringer, J.; Hutley, L.B.; Tapper, N.J.; Zhu, X. Evaluation of Collections 4 and 5 of the MODIS Gross Primary Productivity product and algorithm improvement at a tropical savanna site in northern Australia. *Remote Sens. Environ.* **2009**, *113*, 1808–1822.

41. Sjöström, M.; Ardö, J.; Arneth, A.; Boulain, N.; Cappelaere, B.; Eklundh, L.; de Grandcourt, A.; Kutsch, W.L.; Merbold, L.; Nouvellon, Y. Exploring the potential of MODIS EVI for modeling gross primary production across African ecosystems. *Remote Sens. Environ.* **2011**, *115*, 1081–1089.

42. Jiang, L.; Islam, S. Estimation of surface evaporation map over southern Great Plains using remote sensing data. *Water Resour. Res.* **2001**, *37*, 329–340.

43. Roerink, G.; Su, Z.; Menenti, M. S-SEBI: A simple remote sensing algorithm to estimate the surface energy balance. *Phys. Chem. Earth Part B* **2000**, *25*, 147–157.

44. Galleguillos, M.; Jacob, F.; Prévot, L.; French, A.; Lagacherie, P. Comparison of two temperature differencing methods to estimate daily evapotranspiration over a Mediterranean vineyard watershed from ASTER data. *Remote Sens. Environ.* **2011**, *115*, 1326–1340.

45. Olivera-Guerra, L.; Mattar, C.; Galleguillos, M. Estimation of real evapotranspiration and its variation in Mediterranean landscapes of central-southern Chile. *Int. J. Appl. Earth Obs. Geoinf.* **2014**, *28*, 160–169.

46. Santos, C.A.C.; Bezerra, B.G.; Silva, B.B.; Rao, T.V.R. Assessment of daily actual evapotranspiration with SEBAL and S-SEBI algorithms in cotton crop. *Rev. Bras. Meteorol.* **2010**, *25*, 383–398.

47. Wang, K.; Dickinson, R. A review of global terrestrial evapotranspiration: Observation, modeling, climatology, and climatic variability. *Rev. Geophys.* **2012**, *50*, 1–54.

48. Ramankutty, N. Croplands in West Africa: A geographically explicit dataset for use in models. *Earth Interact.* **2004**, *8*, 1–22.

49. CRED Emergency Events Database. Available online: http://www.emdat.be/result-country-profile (accessed on 14 April 2014).

50. Brink, A.B.; Eva, H.D. Monitoring 25 years of land cover change dynamics in Africa: A sample based remote sensing approach. *Appl. Geogr.* **2009**, *29*, 501–512.

51. Arino, O.; Gross, D.; Ranera, F.; Leroy, M.; Bicheron, P.; Brockman, C.; Defourny, P.; Vancutsem, C.; Achard, F.; Durieux, L.; *et al.* GlobCover: ESA Service for Global Land Cover from MERIS. In Proceedings of the IEEE International Geoscience and Remote Sensing Symposium, IGARSS 2007, Barcelona, Spain, 23–28 July 2007; pp. 2412–2415.

52. USGS MODIS Data Products Table. Available online: https://lpdaac.usgs.gov/products/modis_products_table (accessed on 14 April 2014).

53. Smets, B.; Eerens, H.; Jacobs, T.; Royer, A. BioPar Dry Matter Productivity (DMP) Product User Manual. Available online: http://web.vgt.vito.be/documents/BioPar/g2-BP-RP-BP053-ProductUserManual-DMPV0-I1.00.pdf (accessed on 14 April 2014).

54. NOAA CPC The NOAA Climate Prediction Center African Rainfall Estimation Algorithm Version 2.0. Available online: http://www.cpc.ncep.noaa.gov/products/fews/RFE2.0_tech.pdf (accessed on 14 April 2014).

55. VITO Low & Medium Resolution EO-Products—Free Data. Available online: www.vito-eodata.be (accessed on 14 April 2014).

56. Justice, C.; Hiernaux, P. Monitoring the grasslands of the Sahel using NOAA AVHRR data: Niger 1983. *Int. J. Remote Sens.* **1986**, *7*, 37–41.

57. Bonifacio, R.; Dugdale, G.; Milford, J. Sahelian rangeland production in relation to rainfall estimates from Meteosat. *Int. J. Remote Sens.* **1993**, *14*, 2695–2711.

58. Mutanga, O.; Skidmore, A. Merging double sampling with remote sensing for a rapid estimation of fuelwood. *Geocarto Int.* **2004**, *19*, doi:10.1080/10106040408542327.

59. Ham, F.; Fillol, E. Pastoral Surveillance System and Feed Inventory in the Sahel. In *Conducting National Feed Assessments*; FAO: Rome, Italy, 2010; Volume 1998, pp. 83–114.

60. Oak Ridge National Laboratory Distributed Active Archive Center FLUXNET Web Page. Available online: http://fluxnet.ornl.gov (accessed on 14 April 2014).

61. Mauder, M.; Liebethal, C.; Göckede, M.; Leps, J.-P.; Beyrich, F.; Foken, T. Processing and quality control of flux data during LITFASS-2003. *Bound. Layer Meteorol.* **2006**, *121*, 67–88.

62. Ramier, D.; Boulain, N.; Cappelaere, B.; Timouk, F.; Rabanit, M.; Lloyd, C.R.; Boubkraoui, S.; Métayer, F.; Descroix, L.; Wawrzyniak, V. Towards an understanding of coupled physical and biological processes in the cultivated Sahel—1. Energy and water. *J. Hydrol.* **2009**, *375*, 204–216.

63. Sturges, H. The choice of a class interval. *J. Am. Stat. Assoc.* **1926**, *21*, 65–66.

64. De Castro Teixeira, A.H.; Bastiaanssen, W.G.M.; Ahmad, M.D.; Moura, M.S.B.; Bos, M.G. Analysis of energy fluxes and vegetation-atmosphere parameters in irrigated and natural ecosystems of semi-arid Brazil. *J. Hydrol.* **2008**, *362*, 110–127.

65. Fensholt, R.; Sandholt, I.; Rasmussen, M.S.; Stisen, S.; Diouf, A. Evaluation of satellite based primary production modelling in the semi-arid Sahel. *Remote Sens. Environ.* **2006**, *105*, 173–188.

66. Seaquist, J. A remote sensing-based primary production model for grassland biomes. *Ecol. Modell.* **2003**, *169*, 131–155.

67. Meroni, M.; Fasbender, D.; Kayitakire, F.; Pini, G.; Rembold, F.; Urbano, F.; Verstraete, M.M. Early detection of biomass production deficit hot-spots in semi-arid environment using FAPAR time series and a probabilistic approach. *Remote Sens. Environ.* **2014**, *142*, 57–68.

68. Loague, K.M.; Green, R.E. Statistical and graphical methods for evaluating solute transport models: Overview and application. *J. Contam. Hydrol.* **1991**, *7*, 51–73.

69. Akaike, H. A new look at the statistical model identification. *IEEE Trans. Autom. Control* **1974**, *19*, 716–723.

70. Bagayoko, F.; Yonkeu, S.; Elbers, J.; van de Giesen, N. Energy partitioning over the West African savanna: Multi-year evaporation and surface conductance measurements in Eastern Burkina Faso. *J. Hydrol.* **2007**, *334*, 545–559.

71. Son, N.T.; Chen, C.F.; Chen, C.R.; Chang, L.Y.; Minh, V.Q. Monitoring agricultural drought in the Lower Mekong Basin using MODIS NDVI and land surface temperature data. *Int. J. Appl. Earth Obs. Geoinf.* **2012**, *18*, 417–427.

72. Coakley, J. Reflectance and Albedo, Surface. Available online: http://curry.eas.gatech.edu/Courses/6140/ency/Chapter9/Ency_Atmos/Reflectance_Albedo_Surface.pdf (accessed on 14 April 2014).

73. ECOWAS-SWAC. The Ecologically Vulnerable Zone of Sahelian Countries. In *Atlas on Regional Integration in West Africa*; ECOWAS—SWAC/OECD: Abuja, Nigeria, 2006; pp. 2–12.

74. Leblanc, M.; Lemoalle, J.; Bader, J.; Tweed, S. Thermal remote sensing of water under flooded vegetation: New observations of inundation patterns for the Small'Lake Chad. *J. Hydrol.* **2011**, *404*, 87–98.

75. Nutini, F.; Boschetti, M.; Brivio, P.; Bocchi, S.; Antoninetti, M. Land-use and land-cover change detection in a semi-arid area of Niger using multi-temporal analysis of Landsat images. *Int. J. Remote Sens.* **2013**, *34*, 37–41.

76. Campbell, B.D.; Stafford Smith, D.M. A synthesis of recent global change research on pasture and rangeland production: Reduced uncertainties and their management implications. *Agric. Ecosyst. Environ.* **2000**, *82*, 39–55.

77. Barbosa, H.A.; Huete, A.R.; Baethgen, W.E. A 20-year study of NDVI variability over the Northeast Region of Brazil. *J. Arid Environ.* **2006**, *67*, 288–307.

Evapotranspiration Variability and Its Association with Vegetation Dynamics in the Nile Basin, 2002–2011

Henok Alemu, Gabriel B. Senay, Armel T. Kaptue and Valeriy Kovalskyy

Abstract: Evapotranspiration (ET) is a vital component in land-atmosphere interactions. In drylands, over 90% of annual rainfall evaporates. The Nile Basin in Africa is about 42% dryland in a region experiencing rapid population growth and development. The relationship of ET with climate, vegetation and land cover in the basin during 2002–2011 is analyzed using thermal-based Simplified Surface Energy Balance Operational (SSEBop) ET, Normalized Difference Vegetation Index (NDVI)-based MODIS Terrestrial (MOD16) ET, MODIS-derived NDVI as a proxy for vegetation productivity and rainfall from Tropical Rainfall Measuring Mission (TRMM). Interannual variability and trends are analyzed using established statistical methods. Analysis based on thermal-based ET revealed that >50% of the study area exhibited negative ET anomalies for 7 years (2009, driest), while >60% exhibited positive ET anomalies for 3 years (2007, wettest). NDVI-based monthly ET correlated strongly ($r > 0.77$) with vegetation than thermal-based ET ($0.52 < r < 0.73$) at $p < 0.001$. Climate-zone averaged thermal-based ET anomalies positively correlated ($r = 0.6$, $p < 0.05$) with rainfall in 4 of the 9 investigated climate zones. Thermal-based and NDVI-based ET estimates revealed minor discrepancies over rainfed croplands (60 mm/yr higher for thermal-based ET), but a significant divergence over wetlands (440 mm/yr higher for thermal-based ET). Only 5% of the study area exhibited statistically significant trends in ET.

Reprinted from *Remote Sens.* Cite as: Alemu, H.; Senay, G.B.; Kaptue, A.T.; Kovalskyy, V. Evapotranspiration Variability and Its Association with Vegetation Dynamics in the Nile Basin, 2002–2011. *Remote Sens.* **2014**, *6*, 5885-5908.

1. Introduction

Drylands are terrestrial ecosystems characterized by the scarcity of water. Rainfall is generally low (<600 mm/yr), and the potential rate of evapotranspiration greatly exceeds rainfall [1]. Drylands occupy about 41% the global surface and provide food, grazing for livestock, energy and forestry products and ecosystem services to about a third of the global population [2,3]. The limitations in water and/or nutrients have made these ecosystems highly sensitive to environmental changes and prone to land degradation, such as desertification [4]. With regards to socioeconomic conditions, dryland populations on average lag significantly behind the rest of the world on human well-being and development indicators [3]. This is particularly true for the Nile Basin of Africa, where a quarter of the continent's population lives in a region where 42% of the basin is dryland and water is scarce [5–7]. Moreover, the population in the region is highly dependent on natural resources for its livelihood, highly vulnerable to food insecurity and exposed to political instability [6–8]. Over the last few years, the Nile Basin has undergone major transformation in land cover/land use change, mainly from expanding urban and agricultural activities, with possible implications for water use and food security in the region [9–13].

At global and continental scales, evapotranspiration (ET) is the second largest component of the terrestrial water budget after precipitation [14]. While this proportion is retained at the basin scale, it is reversed in irrigation schemes and wetlands [15]. Terrestrial ET transfers a large volume of water from soil and vegetation into the atmosphere [16]. About 60% of the global annual land precipitation is lost to ET; while ET from vegetation constitutes about 80% of terrestrial ET [14,17]. In dryland ecosystems where much of the soil is bare, ET can consume as much as 90% or more of the annual precipitation [18]. The high rate of ET (combined with the low rate of rainfall) in dryland ecosystems reduces soil water availability and, subsequently, inhibits the primary productivity of the vegetation [3]. Vegetation productivity is of great economic importance in many regions of dryland ecosystems [2]. The Normalized Difference Vegetation Index (NDVI) is often used as a proxy for vegetation productivity [8,19]. Moreover, human activity, such as agricultural and industrial development, has been a principal factor in the modification of the ecohydrological system [4,20–24].

In order to improve our understanding of ecohydroclimatologic dynamics, several numerical weather prediction systems and land surface models have been developed over the last few decades. The models range in complexity from simple water balance equations to complex physical parameterization of land-atmosphere interactions. In addition, the advent of remote sensing brought the capability to continuously collect time series of spatially-explicit quantitative data on land-atmosphere interactions at regional and global scales at regular time intervals [25]. Model-derived and remote sensing land surface and climatological data have been particularly critical in data-scarce regions of the world. Several remote sensing-based ET methods are currently available that, according to Courault et al. [26], can generally be grouped into: direct methods that use thermal infrared (TIR) directly into simplified semi-empirical models; deterministic methods that use assimilation procedures and combine different remote sensing bands ingested into complex models to estimate ET; the inference (vegetation indices) method that uses remote sensing data to compute the reduction factor, such as the crop coefficient, to compute with reference to evapotranspiration for the estimation of actual evapotranspiration; and the residual methods (of the energy budgets),which use the spatial variability in remote sensing images to calculate the surface energy balance equation and attempt to minimize the use of atmospheric data. Most of the currently operational remote sensing-based ET models such as the Surface Energy Balance Algorithm (SEBAL) [27], Surface Energy Balance System (SEBS) [28], Simplified Surface Energy Balance (SSEB) [29] and Operational Simplified Surface Energy Balance (SSEBop) [30] are in this category. A more detailed discussion on the different ET methods can be found in Calgano et al. [31].

The scarcity of reliable and openly distributed in situ data in the Nile Basin region means that only a few basin-scale studies are available so far; and most of those studies have to rely on model and remote sensing data [32,33]. Previous works on ET in the region include those conducted at field-scales [34,35] or high resolution imagery [16,36,37], or regional/basin-scale [33–50], or the continental/global-scale [17,51,52]. Continental- and regional-scale trend analysis conducted at 0.5° to 1.0° resolution indicated a downward trend in ET over the past few decades in substantial parts of the Nile Basin region [17,51,52]. However, some studies of localized areas, like parts of the Nile Delta [38], that used moderate resolution satellite data suggest an increase in ET over the last few decades. Previous studies in the region focused either on trends in ET dynamics [51] or the relationship between climate and ET [17]. This work attempts to add to this growing scientific literature by analyzing the variability in actual evapotranspiration (referred to as ET in this paper) and its relationship with climate, land cover and vegetation productivity in the Nile Basin using satellite-derived and land surface models during 2002–2011.

Using a hybrid combination of satellite-derived and modeled data, we present basin-wide geographically-distributed ET dynamics and the drivers, a comparison between thermal-based and NDVI-based ET in different climate zones and land cover and the ET-vegetation interaction in the basin during the period of 2002–2011. The data used are long-term records of thermal-based (SSEBop (Simplified Surface Energy Balance Operational), [30]) and NDVI-based ET (MOD16 (MODIS Terrestrial ET Product), [53–55]), satellite-driven rainfall (TRMM (Tropical Rainfall Measuring Mission), [56]) and vegetation (NDVI derived from MODIS Nadir Bidirectional Reflectance Distribution Function (BRDF)-Adjusted Reflectance, NBAR, [57]).

2. Materials and Methods

2.1. Study Area

The Nile Basin (Figure 1a) is located in northeastern Africa and extends from latitude 4°S to 32°N and from longitude 21°30′E to 40°30′E. The basin is home to the River Nile, which drains an area of about 3.3 million km^2 (~10% of the continent's landmass). The basin's land cover is dominated by shrublands and woodlands (37%) and bare soils (30%), while the remaining is irrigated and rainfed agricultural land (11%), grasslands (10%), forest cover (7%), wetlands and lakes (3%) and a fraction of it covered with built-up areas [6]. Subsequently, the spatial distribution of vegetation productivity in the basin is highly variable, as illustrated in Figure 2d. Regions with predominantly high vegetation productivity are the Equatorial Lakes Region (Zones VIII and IX), South Sudan (VI) and the western part of the Ethiopian Highlands (Zone VII). Except for the Nile Delta (Zone I) and the Nile Valley river corridor (Zone II), the rest of the basin in Zones II, III and IV shows very low vegetation productivity.

Rainfall distribution in the basin (Figure 2c) is mainly driven by the seasonal fluctuation of the Inter-Tropical Convergence Zone (ITCZ) and its interaction with topography [58,59]. A pronounced north-south rainfall gradient ranging from almost no rain over Lake Nasser in Egypt (Zone II) to rainfall totaling ~2100 mm/yr in Gore, Ethiopia (Zone VII) lead to the climatic classification of the basin into nine climate zones (Figure 1, [58]). Low mean annual basin rainfall (~1046 mm/yr) in only geographically-limited regions means that the Nile River Basin has one of the lowest discharges compared to other major river basins in the world [5,6].

Evaporative losses in the basin are extremely high because the headwaters of the river primarily originate in the tropics; and because the river stagnates in large lakes (Lake Victoria, Lake Tana, Lake Albert), extensive swamps (the Sudd in South Sudan), artificial impoundments in arid environments (Lake Nasser/Nubia in Egypt/Sudan, Jebel el Aulia in Sudan) and meanders through arid and hyper-arid ecosystems [5,6]. Annual ET estimates (Figure 2a,b) in the basin range from <~20 mm/yr in the hyper-arid and arid regions (Zones II, III and IV) to about >~1400 mm/yr in the equatorial and central basin regions (Zones VI and VIII). In the Nile Delta (Zone 1), where rainfall is minimal and agriculture depends entirely on irrigation, annual ET ranges between ~700–1100 mm/yr. The Nile Basin is drained by two principal tributaries, the Blue Nile (from the Ethiopian Highlands, which overall contribute ~86% of the total inflows to the Main Nile) and the White Nile (from the Equatorial Lakes Region, contributing the rest) that join at the confluence in Khartoum, Sudan, to form the Main Nile River [58–62]. River Atbara drains the northern Ethiopian Highlands and is the last tributary to join the Main Nile (Figure 1a).

Figure 1. The Nile Basin. (**a**) Delineated zones with roman numerals represent rainfall regimes from Camberlin [58]. (**b**) Mean monthly rainfall (mm) from TRMM, 2002–2011 for each climate zone.

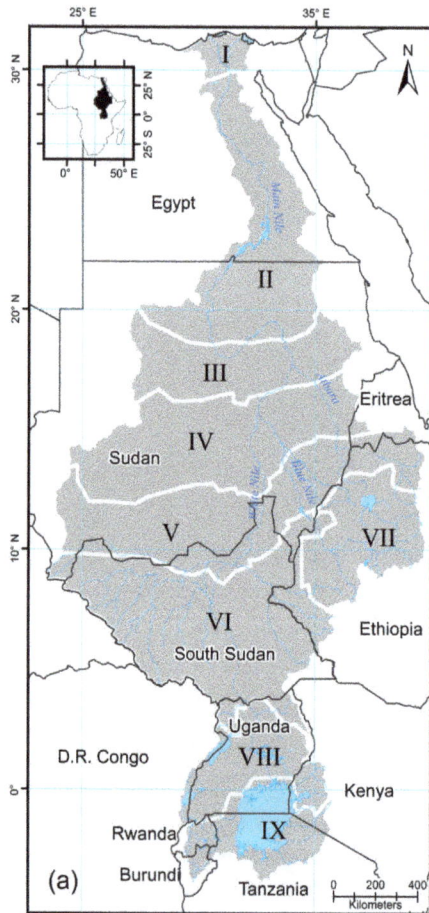

(**a**)

Figure 1. *Cont.*

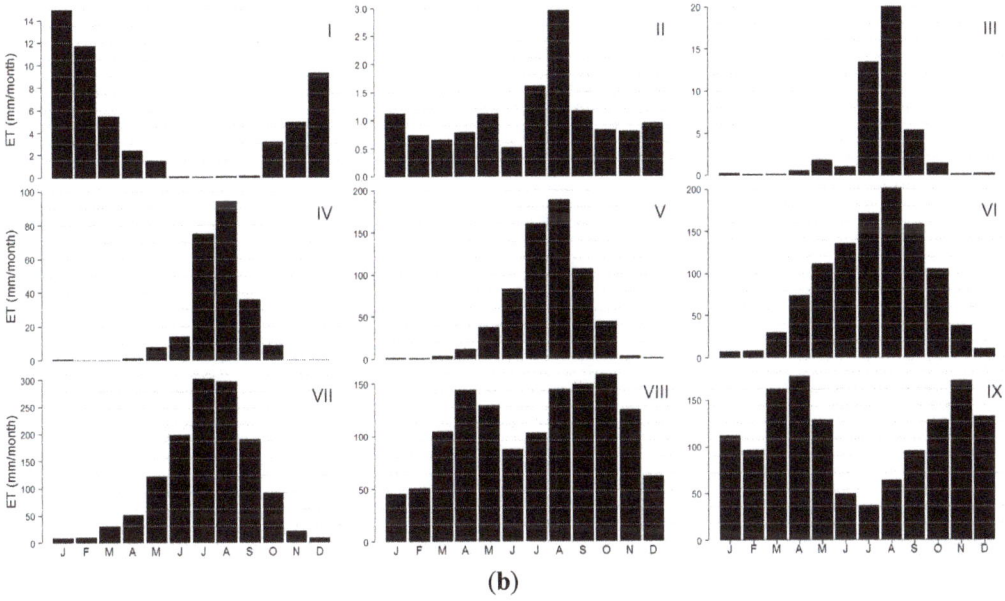

(b)

Figure 2. Long-term mean annual estimates (mm) of (**a**) the Simplified Surface Energy Balance Operational (SSEBop) ET; (**b**) MOD16 ET; (**c**) TRMM rainfall; and (**d**) long-term mean annual NDVI (NDVI > 0); 2002–2011.

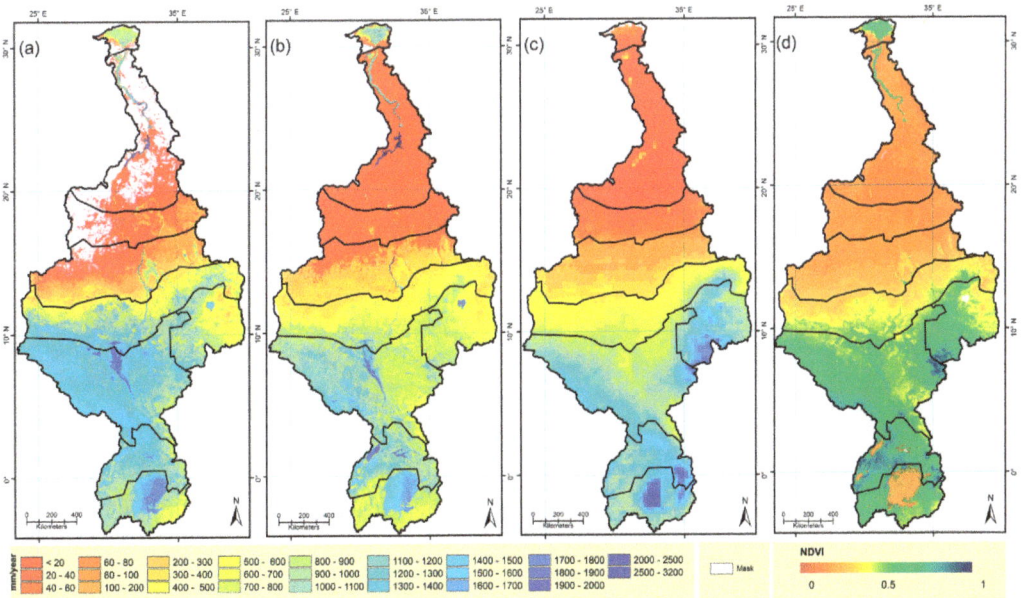

2.2. Data

2.2.1. Thermal-Based Data (SSEBop ET)

Senay *et al.* [29] produced the Simplified Surface Energy Balance (SSEB) model using thermal data for uniform agricultural fields. Later versions integrated additional information on topography, latitude and differences between land surface temperature and air temperature to enhance the model [63,64]. The SSEBop ET algorithm is an operational parameterization of the SSEB model that uses MODIS land surface temperature (LST) and model-derived meteorological parameters to produce a gridded ET product [30]. For a given day and location, the SSEBop approach assumes: (i) a near-constant temperature discontinuity between bare dry surface and atmosphere year to year under clear sky conditions; and (ii) clear sky net radiation as the main driver of surface energy balance [30]. The SSEBop algorithm uses NDVI for a one-time model parameterization to establish the upper and lower boundary conditions for LST; however, the algorithm does not directly include NDVI values in the computation for ET estimation [30]. The method has been tested using 14-year MODIS data from the United States, Africa and Southeast Asia and has been validated comprehensively over the Conterminous U.S. (CONUS) against flux tower observations, water balance ET and MOD16 [65]. The data are available at 1-km resolution 8-day totals from the U.S. Geological Survey (USGS) Earth Resources Observation and Science (EROS) Center.

2.2.2. NDVI-Based Data (MOD16 ET)

The MODIS 1-km spatial resolution Terrestrial ET Product for the Nile Basin [53] is acquired from the Nile Basin Initiative (NBI, [66]). MOD16 data are available at 8-day, monthly and annual time scales. The MOD16 algorithm employs the Penman–Monteith ET model and utilizes MODIS products, including 14 land cover types, Leaf Area Index/Fraction of Photosynthetically Active Radiation (LAI/FPAR), and white sky-albedo for the estimation of ET [54,55]. In addition to the vegetation surface-based algorithm employed in the previous global MOD16 ET algorithm, the improved version uses additional Terra MODIS daytime LST, NDVI and Enhanced Vegetation Index (EVI) data to estimate ET over deserts, urban areas, inland water bodies, such as rivers and lakes, as well as vegetated surfaces, and is produced specifically for the Nile Basin [53].

2.2.3. MODIS NDVI Data

For vegetation data, 1-km spatial resolution, 8-day NDVI composites were derived from the red (0.620–0.670 μm) and near-infrared (0.841–0.876 μm) bands of the MODIS Nadir BRDF-Adjusted Reflectance (NBAR) Product (MCD43B4, version 4) for 46 observations per year for the period from 2002 to 2011. MCD43B4 data are 16-day composites generated using acquisitions from MODIS onboard both Terra and Aqua platforms. The overlapping by 8-days of two successive composites resulted in the availability of one image every 8 days. The NBAR data provide an improved surface reflectance product with reduced cloud and aerosol contamination, with view angle effects removed [57]. The data are freely available via the National Aeronautics and Space Administration (NASA) next generation metadata and service discovery tool, Reverb [67].

2.2.4. TRMM Rainfall Data

In data-scarce regions, such as the Nile Basin, where gauge data are sparse or unevenly distributed, and where weather observation networks are deteriorating, satellite rainfall estimates provide essential, and at times, the only spatiotemporal information data for multiple time periods at a range of spatial scales [68]. In this study, we used the daily TRMM [56] merged high quality infrared precipitation product (3B42, V.7). The TRMM-3B42 algorithm combines geostationary infrared, passive microwaves and also ground-based gauge data [68]. TRMM-3B42 estimates are produced at a 3-hour temporal and a 0.25° spatial resolution; data are acquired from NASA's TRMM site [69].

2.3. Methods

2.3.1. Pre-Processing of Data

Examination of the data (SSEBop, MODIS ET and NDVI) revealed areas where there is a high and low frequency of 8-day time series datasets. Frequency (%) maps of the 8-day time series data from each of the three datasets with values greater than zero are included in Figure 3. For this analysis, only SSEBop ET (ET > 0), MODIS NDVI (NDVI > 0) and MODIS ET (ET > 0) retrievals during 2002–2011 are included. Moreover, while doing inter-comparison and correlations, SSEBop pixels where the corresponding MODIS ET and NDVI values are zero are also excluded.

2.3.2. Standardized Anomalies

To analyze interannual variations, standardized anomalies were computed by subtracting the mean of the annual values from the corresponding individual annual values and dividing by the standard deviation. Dimensionless standard units of the standardized anomalies (SA) facilitate the direct comparison of variations in different geographic locations. Standardized anomalies (SA) were categorized into five classes: (i) SA ≤ −2: severely dry; (ii) −2 ≤ SA < −0.5: moderately dry; (iii) −0.5 ≤ SA ≤ 0.5: normal; (iv) 0.5 < SA ≤ 2: moderately wet; and (v) SA > 2: severely wet. In this paper, we generally define dryness for SA < −0.5 and wetness for SA > 0.5.

2.3.3. Inter-Model Comparisons

Indirect inter-model comparisons of SSEBop and MOD16 estimates were conducted by comparing time series of monthly ET estimates (an average of 30 × 30 pixels) from each ET dataset over selected land cover (using MODIS land cover and Google Earth for visual inspection) with respect to corresponding estimates of vegetation productivity (NDVI). The indirect comparison with respect to rainfall was made using time series of annual ET anomalies averaged per each of the nine climate zones. Pearson's product-moment coefficient of linear correlation (r) is used to measure the relationship between ET from SSEBop and MODIS ET with vegetation productivity.

Figure 3. Frequency (%) of 8-day times series of (**a**) SSEBop (ET > 0); (**b**) MODIS NDVI (NDVI > 0) and (**c**) MOD16 (ET > 0) retrievals during 2002–2011.

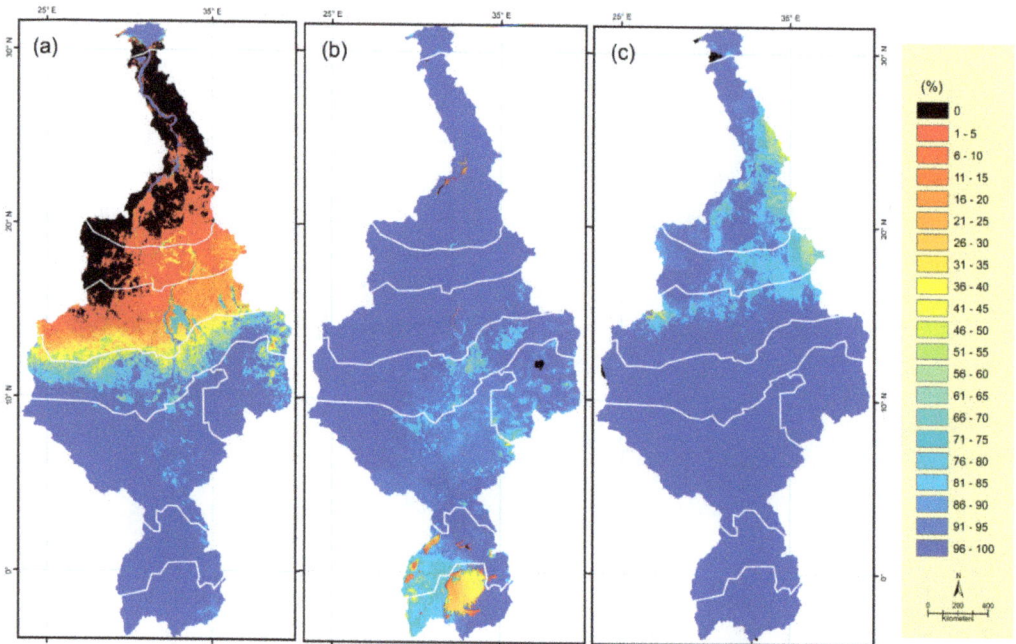

2.3.4. Trend Analysis

In order to detect the presence of temporal trends (consistent, one-directional, long-term changes over time in ET and rainfall), a linear regression trend test was conducted, where time is the independent variable and ET or rainfall is the dependent variable. In order to perform statistical inference regarding the slope (trend through time) of a line fitted using the ordinary least-squares (OLS) method, the following criteria have to be met [70]: (1) the normality of residuals resulting from the linear regression (Shapiro–Wilk test) [71]; (2) homoscedasticity: the variance of the residuals must be constant throughout time (Breusch–Pagan test) [72]; and (3) serial independence: the residuals should be free from autocorrelation (Breusch–Godfrey test) [73] at lags of up to two samples. The linear model in this study was run at an annual time scale for both ET and rainfall for the decadal study period, and statistical significance was chosen at the 95% level.

3. Results

3.1. Basin-Wide ET Dynamics

The results of the interannual anomaly analysis, visually illustrated in Figure 4, conducted using thermal-based ET data (SSEBop) revealed two temporally distinct dry periods (2002–2005 and 2009–2011) and a wet period (2006–2008) that characterized ET dynamics in the Nile Basin. About 30%–50% of the study area (86% of the total basin area) exhibited dryness (negative ET anomalies < −0.5) during the dry periods; while a minimum of 40% exhibited wetness (positive ET anomalies > 0.05) during the three consecutive wet years. Basin-wide, the driest years are 2004, 2005 and 2009 (~40%–50% exhibiting dryness); while 2007 and 2008 constitute the wettest years (~50% exhibiting wetness). The degree of severity of dryness/wetness and proportional areal extent of affected regions basin-wide and for each of the nine climate zones are summarized in Figure 5. Furthermore, the analysis also revealed that the areal extent of high degree severity level dryness (negative ET anomalies < −2) and wetness (positive ET anomalies > 2) affected regions do not exceed 5% of the study area during the study period (Figure 5a).

Figure 4. (a–j) Nile Basin ET dynamics: annual ET (SSEBop) standardized anomalies, 2002–2011. The standardized annual anomalies were computed by subtracting the long-term mean (2002–2011) of the annual values from the corresponding individual annual values (annual mean) and dividing by the standard deviation.

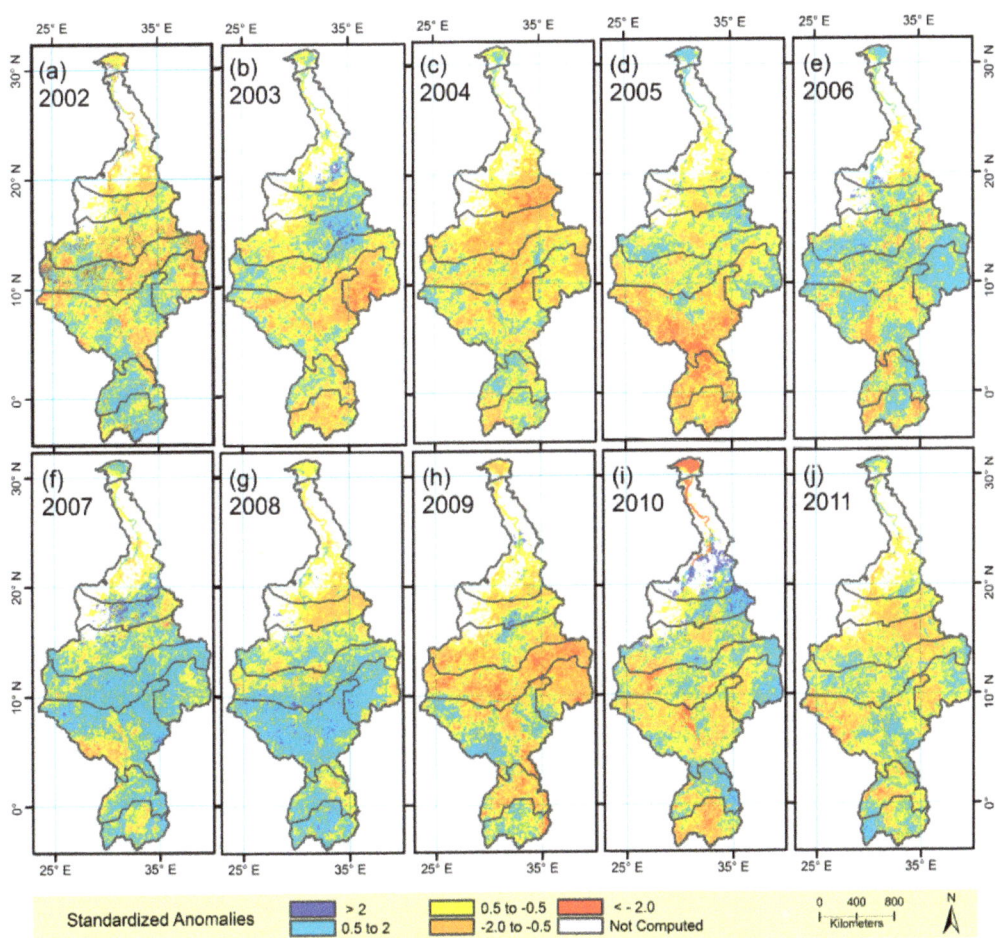

362

Figure 5. (a–k) Areal extent and degree of severity of annual ET anomalies in the Nile Basin during 2002–2011. The bar length represents the proportional area fraction basin-wide (BW) and for each climate zone (1–9, Camberlin [59]). Colors represent the degree of the severity of dryness and wetness of the standardized anomaly (SA): (i) −0.5 < SA ≤ 0.5: normal (in gray); (ii) 0.5 < SA ≤ 2: moderately wet (light green); (iii) SA > 2: severely wet (dark green); (v) −2 < SA ≤ −0.5: moderately dry (orange); (vi) SA ≤ −2: severely dry (red).

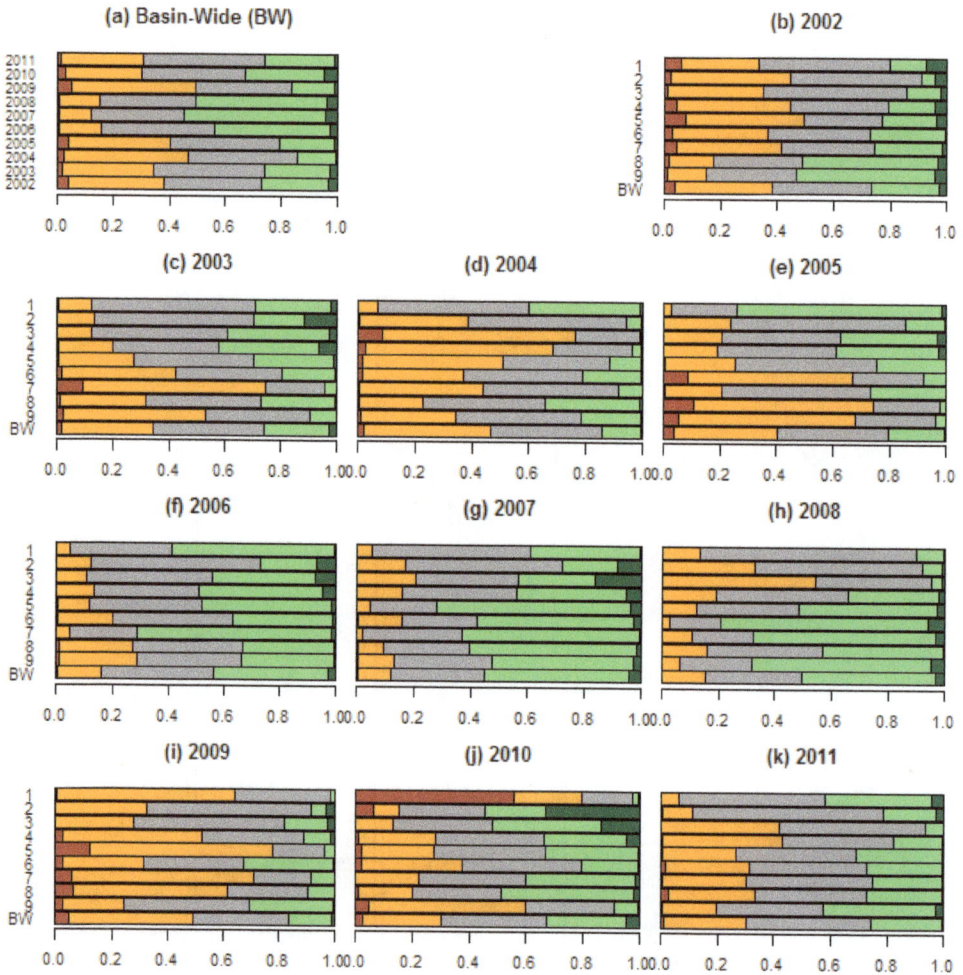

3.2. Drivers of ET

3.2.1. Effects of Climate on ET

Across the basin, the relationships between ET and TRMM rainfall variabilities are influenced by climate, but this relationship also depends on the type of ET data used. Figure 6 shows a time series of standardized annual anomalies for SSEBop, MOD16 and rainfall averaged over each of the nine climate zones. Strong interannual variabilities are observed (especially in SSEBop ET and rainfall) in temporal harmony with the dry and wet periods across the arid, tropical and equatorial regions of the basin. In Zone I, where rainfall is minimal except at northern coastlines, a strong negative anomaly was observed in SSEBop ET and rainfall in 2010, while MOD16 ET appears to show relatively minimum variability throughout the study period (Figure 6a). Nonetheless, the disagreement in the variability between MOD16 ET and rainfall in Zone I is not unexpected, as this zone is an extensively irrigated region with minimal rainfall. In Zone II (Figure 6b), the anomalies from SSEBop, MOD16 and rainfall all peaked in 2010, while showing relatively minimum variability for the rest of the time period. In Zones III and IV, MOD16 and TRMM rainfall show temporal harmony during the wet period, both peaking in 2007; while this peak is absent in SSEBop (Figure 6c,d). However, MOD16 exhibited temporal disharmony with rainfall anomalies for the rest of the year in those zones. In Zones V and VI (Figure 6e,f), with the exception of a few discrepancies, SSEBop and rainfall showed agreement. Overall, the variability in annual MOD16 anomalies in the mainly arid to rainy parts of the basin (Zones V–IX) failed to agree with the variability observed from annual SSEBop ET and TRMM rainfall anomalies. On the other hand, a statistically significant relationship was observed between zonally-averaged SSEBop ET and rainfall variabilities.

Figure 6. (a–i) Interannual ET and rainfall variabilities per climate zones in the Nile Basin: Time series standardized annual anomalies of SSEBop and MOD16 and TRMM rainfall averaged over climate zones I–IX, 2002–2011. The standardized annual anomalies were computed by subtracting the long-term mean (2002–2011) of the annual values from the corresponding individual annual values (annual mean) and dividing by the standard deviation.

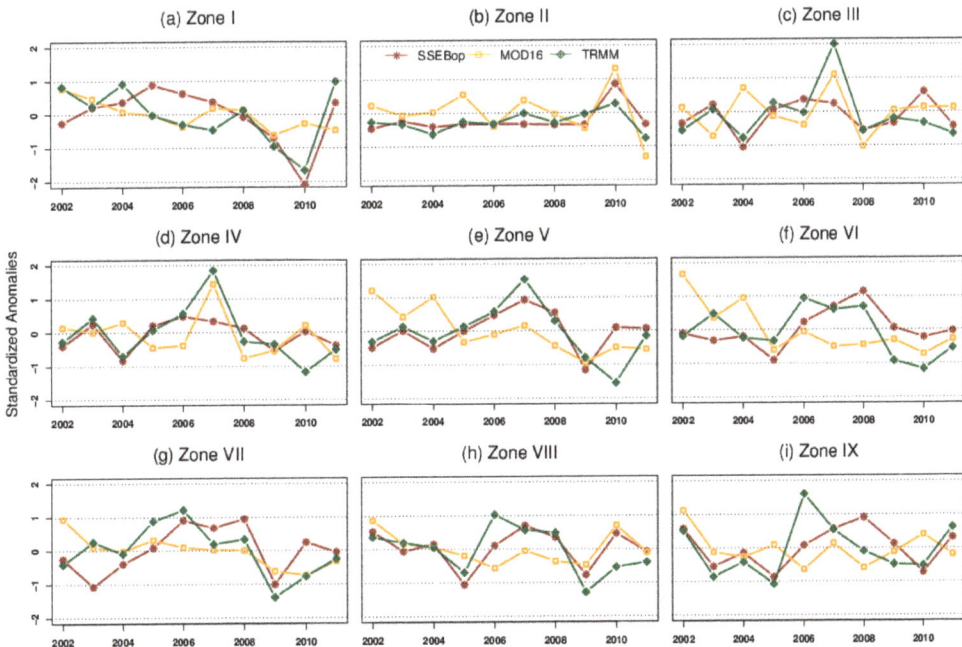

Table 1 shows correlation results between zonally-averaged annual anomalies of SSEBop ET *versus* TRMM rainfall, as well as MOD16 ET *versus* TRMM rainfall. SSEBop ET anomalies showed a weak positive temporal association with corresponding rainfall anomalies in four zones (I, II, V and VIII) with the correlation coefficients ranging between r = 0.61 to 0.67 at a statistical significance of $p < 0.05$. On the other hand, corresponding MOD16 ET anomalies showed statistically significant correlation with rainfall only in Zone II. The fact that zonal anomalies of SSEBop and TRMM exhibited a statistically significant relationship in four of the nine climate zones, while that of MOD16 exhibited in only one, indicates that results of the analysis are affected by data type.

Table 1. Relationship between zonally-averaged standardized annual anomalies of SSEBop *versus* TRMM rainfall and MOD16 *versus* TRMM rainfall for the nine climate zones. Pearson's correlation coefficient (r) and statistical significance (*p*) is presented.

Climate Zones	SSEBop vs TRMM	MOD16 vs TRMM	Climate Zones	SSEBop vs TRMM	MOD16 vs TRMM
	r, p	r, p		r, p	r, p
I	r = 0.65, p = 0.04	r = 0.43, p = 0.21	V	r = 0.51, p = 0.12	r = 0.09, p = 0.80
II	r = 0.63, p = 0.04	r = 0.73, p = 0.01	VII	r = 0.55, p = 0.10	r = 0.47, p = 0.16
III	r = 0.51, p = 0.13	r = 0.43, p = 0.21	VIII	r = 0.67, p = 0.03	r = −0.01, p = 0.97
IV	r = 0.61, p = 0.06	r = 0.53, p = 0.11	IX	r = 0.59, p = 0.07	r = −0.12, p = 0.74
V	r = 0.66, p = 0.04	r = 0.19, p = 0.60			

3.2.2. Effects of Vegetation on ET

The relationship between vegetation productivity dynamics and ET (SSEBop and MOD16) in different climate zones is presented in Figure 7. Both similarities and discrepancies were observed in a manner that the SSEBop and MOD16 monthly values relate with the seasonal variability of vegetation productivity. SSEBop and MOD16 temporally correlate with vegetation productivity in Zones VII and V (Figure 7b,d). However, they reveal discrepancies in Zones I and VI (Figure 7a,c). In Zone I in particular, MOD16 clearly captures the seasonality in vegetation dynamics by capturing the two peak seasons, while SSEBop distinctly captures only a single peak season (Figure 7a). Generally, MOD16 shows better correlation with vegetation productivity in different climate zones with statistical significance of $p < 0.001$.

3.2.3. Effects of Land Cover on ET

Time series of monthly ET estimates from SSEBop and MOD16 from selected land cover/land use types (irrigated and rainfed croplands, wetlands and grasslands) is presented in Figure 7. In irrigated croplands, monthly estimates of SSEBop and MOD16 show significant inconsistencies in seasonality and magnitude (a difference of ~80–90 mm/month) during the peak season (months of July and August), but agree during the rest of the seasons (Figure 5a). However, in rainfed croplands (Figure 7b), they show similarity both in magnitude and seasonality. Estimates from SSEBop and MOD16 also agree in seasonality in wetlands (Figure 5c) and grasslands (Figure 5d), but differ significantly during peak seasons.

Figure 7. Time series of monthly ET derived from the median of 30 × 30 non-zero pixels of SSEBop and MOD16 for selected sites in different land cover types and climate zones across the Nile Basin, 2002–2011. Correlation coefficients (r) between ET and vegetation (*** ($p < 0.001$), ** ($p < 0.01$),* ($p < 0.05$)); and long-term mean annual ET estimates (mm/yr) are presented for (**a**) irrigated croplands in climate zone I, (**b**) rainfed croplands in zone VII, (**c**) wetlands in zone VI and (**d**) grasslands in zone V.

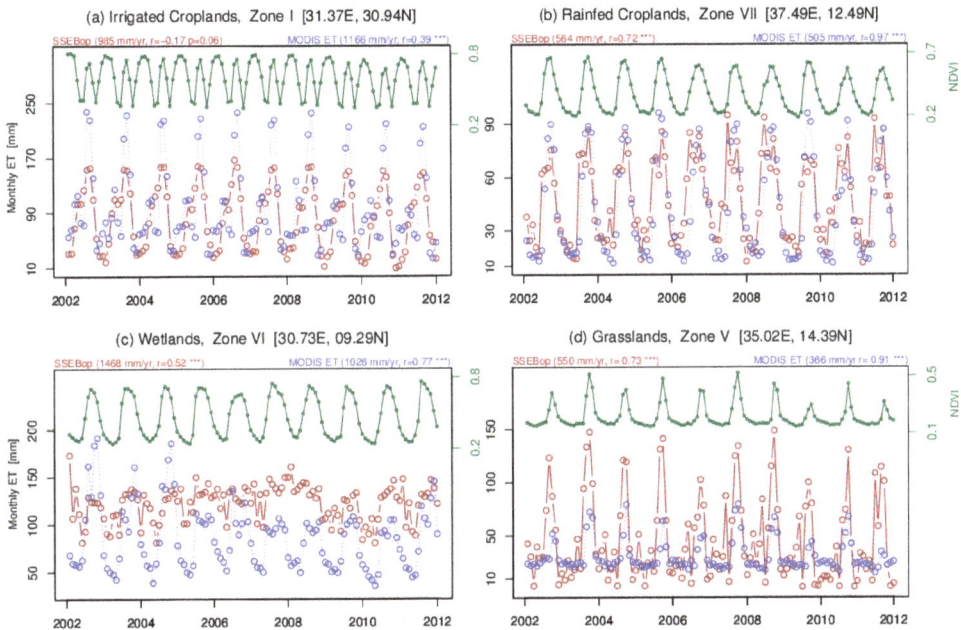

3.3. Trends in ET and Rainfall

The linear regression temporal trend analysis related annual ET and rainfall to time for the decade of 2002–2011 as illustrated in Figure 8. The ET decadal trend map (Figure 8a) revealed a non-significant trend at 95% statistical significance level for nearly the entire basin, except for a few small, mostly scattered, localized areas.

Of the total basin region with valid ET pixels, only 2.3% and 3.4% of the pixels were characterized as significantly downward and upward trends, respectively, at the 95% statistical significance level. The regression test also detected trends in the remaining regions of the study area (~94%, 40% negative, 54% positive), but those are not statistically significant. Nonetheless, interpretation of the results from the linear regression trend test should cautiously take into account the relatively short study period of 10 years.

The regions of central Uganda in the Equatorial Lakes Region in Zone VIII indicated the most conspicuous area characterized by downward trends (Figure 8), while limited and scattered areas in the western Ethiopian Highlands (Zone VII) and eastern Sudan (Zone V) also revealed downward trends. Significant upward trends characterized very limited and localized areas and were scattered across the regions of central Sudan and South Sudan (Zones V and VI). In the northern part of the basin, the Nile Delta (Zone I) and the Nile Valley (Zone II) regions showed largely no-significant trend, except in the eastern/western fringes of the Nile Delta, which showed significant upward trends. Figure 8b illustrates the trend analysis results for rainfall across the basin. The rainfall trend map reveals detected downward rainfall trends in Zone VIII, southern parts of Zone VI, eastern parts of Zone V and scattered areas in western parts of Zone VII. However, no statistically significant upward trends in rainfall were detected in the basin (Figure 8b).

Figure 8. ET and rainfall trends. Linear regression trends for annual (**a**) SSEBop ET and (**b**) TRMM rainfall in the Nile Basin during 2002–2011. Colors represent the trend direction and statistical significance: (i) significantly upward, ($p < 0.05$, dark green); (ii) upward, not significant ($p \geq 0.05$, light green); (iii) significantly downward trend ($p < 0.05$, red); and (v) downward, not significant ($p \geq 0.05$, yellow).

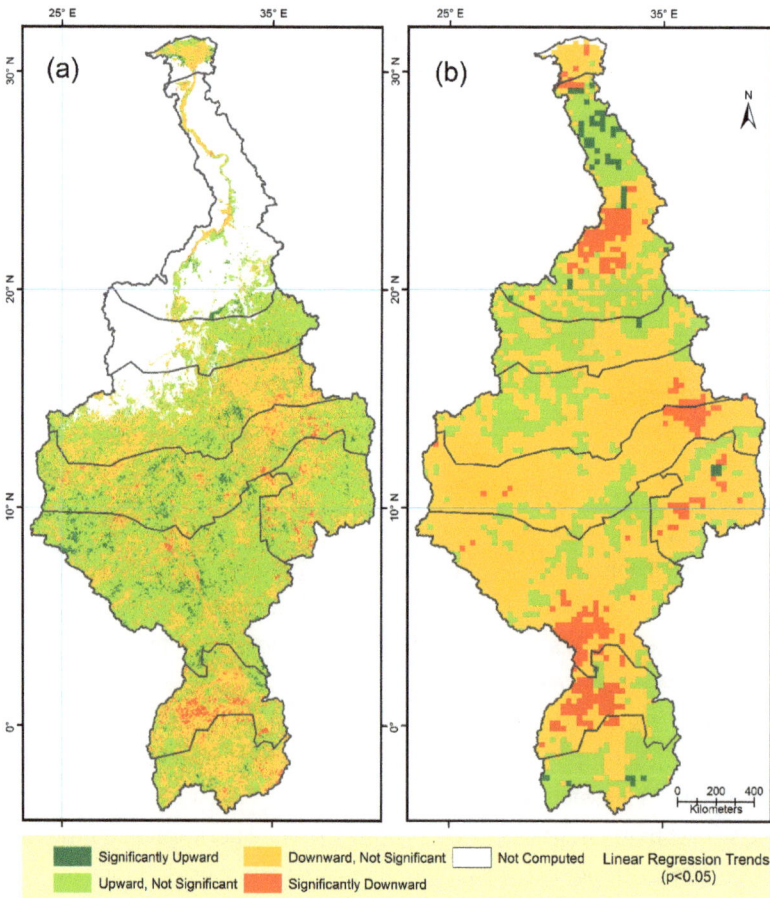

4. Discussion

4.1. ET Dynamics

The interannual SSEBop ET anomaly maps presented in Figure 4 showed basin-wide ET dynamics during the study period. While 2004 and 2009 were found to be relatively the most dryness-dominated years of the decade (2002–2011) basin-wide, 2009 particularly stood out as the driest year when key headwater regions of the basin exhibited considerable dryness (Figures 4c,h and 5a). Viste *et al.* [74] recently noted 2009 as the driest year of the region in nearly three decades where large-scale drought patterns dominated large parts of east Africa. Key headwater regions of the basin that were considerably affected by dryness in 2009 included the Blue Nile basin (Zone VII, ~70% exhibiting dryness), the Tekezze/Sobat basin (eastern parts of Zone V, ~80% exhibiting dryness) and the Equatorial Lakes Region that includes Lake Kyoga-Albert-Aswa basin (Zone VIII, >60% exhibiting dryness) and Lake Victoria basin (Zone IX, >20% exhibiting dryness). Furthermore, the Blue Nile basin region of the Ethiopian Highlands (Zone VII) exhibited dryness in 2002/2004 (>40% exhibiting dryness) and 2003/2009 (>70% exhibiting dryness). As the Blue Nile basin region provides a substantial portion of the Nile water to the basin, substantial droughts in Zone VII could have consequences downstream. The Equatorial Lakes Region exhibited considerable dryness in 2005 (>70% of Zones VIII/IX exhibiting dryness), 2009 (>60% of Zone VIII exhibiting dryness) and 2010 (~60% of Zone IX exhibiting dryness). The extensive dryness (>70% of each zone) in the lower parts of the basin (Zones VI, VIII and IX) in 2005 is probably the consequence of a rainfall deficit in the region [75].

The consequence of extensive dryness basin-wide and, particularly, in the key headwaters of the Nile Basin in 2009 manifested downstream in Zone I in the Nile Delta in 2009 (60% exhibiting dryness) and 2010 (80% exhibiting dryness and 60% severe dryness), as illustrated in Figures 4h,i and 5i,j. Because rainfall in Zone I is minimal (Figure 1b), a plausible explanation for the considerable dryness in 2009 and 2010 in the Nile Delta is a possible economization of irrigation water, as inflow into the Aswan High Dam Reservoir declined in 2009.

4.2. ET Drivers

The relationship of the zonal average of annual anomalies of SSEBop ET and TRMM rainfall per each climate zone (Figure 6) overall suggest that rainfall drives ET variability in the region, albeit there are a few discrepancies, depending on the climate zone. However, this is not consistent with the results observed from MOD16 ET to rainfall relationship. Moreover, because climate Zone I is principally and heavily irrigated agricultural field, where rainfall is minimal, no correlation is expected between rainfall and ET. Nonetheless, SSEBop anomalies showed statistically significant strong temporal harmony with rainfall anomalies in both arid and rainy regions (in four of the nine climate zones with r = 0.6 at $p < 0.05$); while MOD16 anomalies showed no statistically significant correlation with rainfall in eight of the nine climate zones (Table 1). The analysis clearly showed that the temporal relationship between ET and rainfall anomalies is more pronounced with SSEBop ET than MOD16 ET anomalies. This generally positive relationship between SSEBop ET and rainfall anomalies is in general agreement with Jung *et al.* [17], who recently showed a strong relationship between ET and soil moisture anomalies in east Africa.

The correlations between monthly ET *versus* vegetation productivity (Figure 7) depict the degree to which seasonal and interannual variabilities between the two variables are related. The strongest and statistically significant ($p < 0.001$) correlations, on the order of $r = 0.7$ (with SSEBop) and $r = 0.9$ (with MOD16), were observed in climate Zones V and VII; while the weakest correlations were in climate Zone I. The SSEBop monthly ET showed a negative correlation in Zone I and failed to temporally correlate with one of the two peak vegetation productivity seasons. This negative correlation could be the result of one or a combination of multiple reasons including climate (rain in Zone I is minimal), types of primary inputs data and the parameterization used in the model. In climate zones that get a fair amount of rain, however, both the SSEBop and MOD16 temporally correlated well with the seasonal variability in vegetation productivity with little discrepancies and stronger correlations.

Investigation of monthly ET variabilities from SSEBop and MOD16 over selected land cover types is presented in Figure 7. An analysis of monthly ET variability with respect to land cover revealed that the SSEBop and MOD16 estimates differed over irrigated croplands, grasslands and wetlands, but agreed in rainfed croplands. SSEBop showed that wetlands have the highest mean annual ET (1468 mm/yr), followed by irrigated croplands (985 mm/yr), rainfed croplands (564 mm/yr) and grasslands (550 mm/yr). On the other hand, for MOD16, the highest annual ET is from irrigated croplands (1166 mm/yr) followed by wetlands (1026 mm/yr), rainfed croplands (505 mm/yr) and grasslands (366 mm/yr). The largest discrepancy between SSEBop and MOD16 in annual estimates (a difference of ~440 mm/yr) was observed over wetlands (Figure 7). On the other hand, monthly ET variability appeared to be high over cropped fields (very high for MOD16 over irrigated croplands), while small over wetlands for SSEBop and over grasslands for MOD16. This discrepancy in estimates of ET between SSEBop and MOD16 over different land cover types has also been shown by other researchers. Velpuri *et al.* [65] validated the two products in the CONUS and showed that both SSEBop and MOD16 underestimated over croplands. However, we found that over irrigated croplands, only SSEBop underestimated, while both gave similar estimates over rainfed croplands. Further, we found that monthly MOD16 estimates over irrigated croplands in Zone I generally agreed quantitatively with monthly estimates available in the literature [15]. On the other hand, SSEBop has significantly higher estimates than MOD16 over grasslands, especially during the peak seasons (Figure 7d). Velpuri *et al.* indicated that SSEBop provides a better estimate over grasslands than MOD16. Moreover, over deep-rooted vegetation cover, such as forest, Velpuri *et al.* found that SSEBop and MOD16 have an accuracy of $R^2 = 0.72$ and $R^2 = 0.56$, respectively, compared with ground-based observations. The discrepancy in monthly estimates between the two estimates persisted over wetlands (Figure 7c). While MOD16 estimates significantly and irregularly vary seasonally and annually from ~200 mm/month to ~100 mm/month, MOD16 estimates consistently stayed within 80–120 mm/month. A plausible reason for the significant difference between the two estimates could be in the input datasets that the two ET algorithms use: SSEBop uses land surface temperature as its primary input and is independent of the impact that frequent flooding has on vegetation that may affect the ET. On the other hand, MOD16 uses vegetation information in its algorithm, and consequently, the ET results could reflect the changes in vegetation.

4.3. Trends in ET and Rainfall

The detected temporal trend results reveal that the overall annual ET trend during 2002 to 2011 in the Nile Basin can be characterized as no-significant trend, except for a few parts (Figure 8). The most conspicuously affected region is the Equatorial Lakes Region, where statistically significant downward trends were observed in substantial areas of Zone VIII (the region in Uganda surrounded by the tri-lakes: Lakes Victoria, Albert and Kyoga). Western parts of the Ethiopian Highlands and eastern Sudan (Zones V and VII) also showed a few defragmented localized regions of a downward trend. In the northern part of the basin, no significant trends are detected, except around the fringes of the Nile Delta (Zone I). The fringes of the Nile Delta normally have less ET compared to the inland regions of the delta, as recently shown by Simonneaux et al. [41], but show an upward trend. This finding is in agreement with other studies that indicated increasing ET and vegetation productivity in newly reclaimed desert lands in the fringes of the Nile Delta following the expansion of irrigation agriculture in the last few decades [9,39,40,76,77]. On the other hand, the trend analysis on rainfall data showed no significant upward trend, but parts of Uganda (Zone VIII), South Sudan (Zone VI), the western Ethiopian Highlands (Zone VII) and eastern parts of Sudan (Zone V) showed a statistically significant downward trend in rainfall.

4.4. Uncertainties, Errors and Accuracies

Some degree of uncertainty in any model-based or satellite-derived parameter estimates is inevitable. Sources of uncertainties in remotely sensed ET estimates could be attributed to uncertainties in input data that can introduce biases in ET estimates, limitations and biases in the parameterization in the algorithm, cloud cover, errors arising from spatial and temporal scaling approaches, as well as influences from biophysical and geophysical factors, such as land cover and climate. Comparisons with ground-based measurements (which themselves have a reported uncertainty of 10%–15%) indicate that the various remote sensing techniques for estimating ET have uncertainties of 15%–30% [78,79]. A review of about 30 published validations of remotely sensed ET against ground-based flux towers reported an average Root Mean Square Error (RMSE) of just over 50 W/m^2 and relative errors of 15%–30% [65]. In this study, the major sources of uncertainties come from satellite-derived evapotranspiration estimates, rainfall estimates, vegetation indices, the limitations of the linear regression trend method and the short time available for the trend analysis.

A comprehensive review of the accuracies and uncertainties of SSEBop ET and MOD16 ET over different land cover types and climate zones with respect to field-based measurements in the U.S. are provided by Velpuri et al. [65]. The mean basin-scale uncertainty levels in SSEBop ET data are much lower than the reported uncertainty levels (up to 50%) of the mean land ET obtained from remote sensing data, up-scaled tower measurements, land surface models and reanalysis datasets [79,80], illustrating the reliability of monthly SSEBop products for basin-scale ET estimation [65]. Mu et al. [53] provided a performance evaluation of the MOD16 estimates for the Nile Basin, using the basin-scale average of runoff and gridded precipitation data and running the improved algorithm in other regions of the world where there is sufficient availability of the flux tower data.

The relationship between annual anomalies of ET and rainfall in different climate zones was investigated using datasets of varying spatial scales. The ET datasets have a spatial resolution of 1-km, which is much higher than the 25-km resolution of the rainfall data. However, the magnitudes of the uncertainties that could arise from using data of different spatial resolutions diminish, as the anomalies were averaged per each climate zone. Moreover, remote sensing data are subject to systematic errors with consistent bias that overall have little impact on the long-term anomalies. With regards to the linear regression trend analysis, readers should be cautious in interpreting the significance of the linear trend results, since we did not test the normality of the residuals, the homoscedasticity and serial independence of the data before performing the statistical inference regarding the slope, as recommended by de Beurs and Henebry [70]. Nonetheless, the trends generally remain unbiased, even if those assumptions are not met; the limitations with respect to uncertainties in the significance of the estimated parameters, however, remain [70].

5. Conclusions

This paper characterizes variation in actual evapotranspiration (ET) and investigates its relation with vegetation productivity in the Nile Basin for the period of 2002–2011. Hybrids of both satellite-derived and modeled ET datasets, the NDVI-based MOD16 and the thermal-based SSEBop ET datasets were used to comparatively analyze ET variability in relation to vegetation productivity (NDVI), climate (rainfall) and land cover. The analysis of interannual anomalies using thermal-based ET revealed temporally distinct mini-episodes of dry (2002–2005, 2009–2011) and wet (2006–2008) periods that dominated 40%–50% of the study area; with 2007 and 2009 being the wettest and driest years, respectively. An investigation of the relationship between monthly ET variability with vegetation productivity indicated that NDVI-based ET had stronger positive correlations (r = 0.77 to r = 0.97) with vegetation than thermal-based ET (r = 0.52 to r = 0.73) at a statistical significance of $p < 0.001$, particularly in rainfed regions. This finding is not unexpected, as NDVI data are the primary input in the NDVI-based ET data. The analysis of the relationship between annual anomalies of ET and rainfall in different climate zones showed that thermal-based ET anomalies correlated positively (r = 0.6 at $p < 0.05$) with corresponding rainfall anomalies in two of the six investigated rainfed climate zones; whereas NDVI-based ET showed no significant relationship. A comparison of thermal-based and NDVI-based ET estimates over selected land cover types revealed minor disagreements over rainfed croplands (60 mm/yr higher for thermal-based ET), but a significant divergence over wetlands (440 mm/yr higher for thermal-based ET).

The results in this study confirm previous regional-scale drying and greening periods in the region. This paper used rainfall as a proxy for soil moisture, but this may not always be valid, particularly in the wetter parts of the basin; as a result, the use of soil moisture data, such as Soil Moisture and Ocean Salinity (SMOS, Kerr et al. [81]), should be considered. The trend analysis conducted in this study is limited to 10 years based on the availability of the dataset used, which begins at the start of the millennium. However, the trend test could reveal long-term changes in the basin if the analysis could be extended using long-term evapotranspiration data, such as the 1983–2006 global ET dataset produced by Zhang *et al.* [82]. In order to improve the accuracy of remote sensing data, the findings need to be verified with long-term field measurement data.

Acknowledgments

This work was partially supported by: the Global Livestock Collaborative Research Support Program (CRSP): PCE-G-00-98-00036-00 Sub-grant Number 144-29-29; the Applied Science Program of NASA Earth-Sun System Division contract# NNA06CH751; and FEWS-NET. The authors would like to thank Pierre Camberlin (University of Bourgogne, Dijon, France) for the permission to reuse the climate stratification data, and Milly Mbuliro (Nile Basin Initiative (NBI), Entebbe, Uganda) for the provision of the MOD16 data. Finally, we thank the Journal Editor and the anonymous reviewers for their useful and constructive reviews that greatly improved the manuscript.

Author Contributions

All of the authors contributed (in part or in full) to the concept design and development of the research and improvement of the manuscript. Henok Alemu performed the research and prepared the manuscript.

Conflicts of Interest

The authors declare no conflict of interest. Any use of trade, firm or product names is for descriptive purposes only and does not imply endorsement by the U.S. Government.

References

1. Jenerette, G.D.; Barron-Gafford, G.A.; Guswa, A.J.; McDonnell, J.J.; Villegas, J.C. Organization of complexity in water limited ecohydrology. *Ecohydrology* **2012**, *5*, 184–199.
2. Fensholt, R.; Langanke, T.; Rasmussen, K.; Reenberg, A.; Prince, S.D.; Tucker, C.; Scholes, R.J.; Le, Q.B.; Bondeau A.; Eastman, R.; *et al.* Greenness in semi-arid areas across the globe 1981–2007—An Earth observing satellite based analysis of trends and drivers. *Remote Sens. Environ.* **2012**, *121*, 144–158.
3. Safriel, U.; Adeel, Z.; Niemeijer, D.; Puigdefabregas, J.; White, R.; Lal, R.; Winslow, M.; Ziedler, J.; Prince, S.; Archer, E.; *et al.* Dryland Systems. In *Millennium Ecosystem Assessment, 2005. Ecosystems and Human Well-Being: Current State and Trends*; Hassan, R.; Scholes, R.; Ash, N., Eds.; World Ressources Institute: Washington, DC, USA, 2005; pp. 623–662.

4. Newman, B.D.; Wilcox, B.P.; Archer, S.R.; Breshears, D.D.; Dahm, C.N.; Duffy, C.J.; McDowell, N.G.; Phillips, F.M.; Scanlon, B.R.; Vivoni, E.R. Ecohydrology of water-limited environments: A scientific vision. *Water Resour. Res.* **2006**, *42*, W06302.

5. Sutcliffe J.V.; Parks Y.P. *The Hydrology of the Nile*; IAHS Special Publication No.5; International Association of Hydrological Sciences: Wallingford, UK, 1999.

6. Nile Basin Initiative (NBI). State of the River Nile Basin 2012. Available online: www.nbi.org (accessed on 12 October 2013).

7. Food and Agriculture Organization (FAO). Population Prospects in the Nile Basin, 2011. Available online: www.faonile.org (accessed on 15 June 2013).

8. Pricope, N.G.; Husak, G.; Lopez-Carr, D.; Funk, C.; Michaelsen, J. The climate-population nexus in the East African Horn: Emerging degradation trends in rangeland and pastoral livelihood zones. *Glob. Environ. Chang.* **2013**, *23*, 1525–1541.

9. Bakr, N.; Weindorf, D.C.; Bahnassy, M.H.; Marei, S.M.; El-Badawi, M.M. Monitoring land cover changes in a newly reclaimed area of Egypt using multi-temporal Landsat data. *Appl. Geogr.* **2010**, *30*, 592–605.

10. Jägerskog, A.; Cascão, A.; Hårsmar, M.; Kim, K. *Land Acquisitions: How will They Impact Transboundary Waters?* Report Nr 30; Stockholm International Water Institute (SIWI): Stockholm, Sweden, 2012.

11. Cotula, L.; Vermeulen, S.; Leonard, R.; Keeley, J. *Land Grab or Development Opportunity? Agricultural Investment and International Land Deals in Africa*; International Institute for Environment and Development(IIED)/Food and Agriculture Organization of the United Nations (FAO), International Fund for Agricultural Development (IFAD): London, UK/Rome, Italy, 2009. Available online: http://www.ifad.org/pub/land/land_grab.pdf (accessed on 15 July 2013).

12. Whittington, D.; Wu, X.; Sadoff, C. Water resource management in the Nile Basin: The economic value of cooperation. *Water Policy* **2005**, *7*, 227–252.

13. Arsano, Y. *Ethiopia and the Nile, Dilemmas of National and Regional Hydropolitics*; Swiss Federal Institute of Technology Zurich: Zurich, Switzerland, 2007. Available online: http://www.css.ethz.ch/publications/pdfs/Ethiopia-and-the-Nile.pdf (accessed on 11 May 2013).

14. Glenn, E.P.; Huete, A.R.; Nagler, P.L.; Hirschboeck, K.K.; Brown, P. Integrating remote sensing and ground methods to estimate evapotranspiration. *Crit. Rev. Plant. Sci.* **2007**, *26*, 139–168.

15. Droogers, P.; Immerzeel, W.; Perry, C. *Application of Remote Sensing in National Water Plans: Demonstration Cases for Egypt, Saudi-Arabia and Tunisia*; FutureWater Report 80; World Bank: Wageningen, The Netherlands, 2009. Available online: http://www.futurewater.nl/downloads/2008_Droogers_FW80.pdf (accessed on 12 July 2013).

16. Anderson, M.C.; Allen, R.G.; Morse, A.; Kustas, W.P. Use of Landsat thermal imagery in monitoring evapotranspiration and managing water resources. *Remote Sens. Environ.* **2012**, *122*, 50–65.

17. Jung, M.; Reichstein, M.; Ciais, P.; Seneviratne, S.I.; Sheffield, J.; Goulden, M.L.; Bonan, G.; Cescatti, A.; Chen, J.; de Jeu, R.; *et al.* Recent decline in the global land evapotranspiration trend due to limited moisture supply. *Nature* **2010**, *467*, 951–954.

18. Wilcox, B.; Seyfried, M.; Breshears, D.; Stewart, B.; Howell, T. The Water Balance on Rangelands. In *Encyclopedia of Water Science*; Stewart, B.A., Howell, T.A., Eds.; Mercel Dekker: New York, NY, USA, 2003; pp. 791–794.

19. Pettorelli, N.; Vik, J.O.; Mysterud, A.; Gaillard, J.-M.; Tucker, C.J.; Stenseth, N.C. Using the satellite-derived NDVI to assess ecological responses to environmental change. *Trends Ecol. Evolut.* **2005**, *20*, 503–510.

20. Oberg, J.W.; Melesse, A.M. Evapotranspiration dynamics at an ecohydrological restoration site: An energy balance and remote sensing approach. *J. Am. Water Resour. Assoc.* **2006**, *42*, 565–582.

21. Melesse, A.M.; Nangia, V. Estimation of spatially distributed surface energy fluxes using remotely-sensed data for agricultural fields. *Hydrol. Process.* **2005**, *19*, 2653–2670.

22. Melesse, A.M.; Frank, A.; Nangia, V.; Hanson, J. Analysis of energy fluxes and land surface parameters in a grassland ecosystem: A remote sensing perspective. *Int. J. Remote Sens.* **2008**, *29*, 3325–3341.

23. Melesse, A.M.; Abtew, W.; Dessalegne, T. Evaporation estimation of Rift Valley lakes: Comparison of models. *Sensors* **2009**, *9*, 9603–9615.

24. Melesse, A.; Weng, Q.; Thenkabail, P.; Senay, G. Remote sensing sensors and applications in environmental resources mapping and modelling. *Sensors* **2007**, *7*, 3209–3241.

25. Justice, C.O.; Townshend, J.R.G.; Vermote, E.F.; Masuoka, E.; Wolfe, R.E.; Saleous, N.; Roy, D.P.; Morisette, J.T. An overview of MODIS Land data processing and product status, *Remote Sens. Environ.* **2002**, *83*, 3–15.

26. Courault, D.; Seguin, B.; Olioso, A. Review on estimation of evapotranspiration from remote sensing data: From empirical to numerical modeling approaches. *Irrig. Drain. Syst.* **2005**, *19*, 223–249.

27. Bastiaanssen, W.G.M.; Menenti, M.; Feddes, R.A.; Holtslag, A.A.M. A remote sensing Surface Energy Balance Algorithm for Land (SEBAL), Part 1: Formulation. *J. Hydrol.* **1998**, *212–213*, 198–212.

28. Su, Z. The Surface Energy Balance System (SEBS) for estimation of turbulent heat fluxes. *Hydrol. Earth Syst. Sci.* **2002**, *6*, 85–99.

29. Senay, G.; Budde, M.; Verdin, J.; Melesse, A. A coupled remote sensing and simplified surface energy balance approach to estimate actual evapotranspiration from irrigated fields. *Sensors* **2007**, *7*, 979–1000.

30. Senay B.G.; Bohms, S.; Singh, R.K.; Gowda, P.H.; Velpuri, N.M.; Alemu, H.; Verdin, J.P. Operational evapotranspiration mapping using remote sensing and weather datasets: A new parameterization for the SSEB approach. *J. Am. Water Resour. Assoc.* **2013**, *49*, 577–591.

31. Calcagno, G.; Mendicino, G.; Monacelli, G.; Senatore, A.; Versace, P. Distributed Estimation of Actual Evapotranspiration through Remote Sensing Techniques. In *Methods and Tools for Drought Analysis and Management*; Rossi, G., Vega, T., Bonaccorso, B., Eds.; Springer: Amsterdam, The Netherlands, 2007; Volume 62, pp. 125–147.

32. Droogers, P.; Immerzeel, W. *Managing the Real Water Consumer: Evapotranspiration*; FutureWater Report 78; World Bank: Wageningen, The Netherlands, 2008. Available online: http://www.futurewater.nl/downloads/2008_Droogers_FW78.pdf (accessed on 12 July 2013).

33. Yilmaz, M.T.; Anderson, M.C.; Zaitchik, B.; Hain, C.R.; Crow, W.T.; Ozdogan, M.; Chun, J.A.; Evans, J. Comparison of prognostic and diagnostic surface flux modeling approaches over the Nile River basin. *Water Resour. Res.* **2014**, *50*, 386–408.

34. Bashir, M.; Tanakamaru, H.; Tada, A. Remote Sensing-Based Estimates of Evapotranspiration for Managing Scarce Water Resources in the Gezira Scheme, Sudan. In *From Headwaters to the Ocean*; Taniguchi, M., Burnett, W.C., Fukushima, Y., Haigh, M., Umezawa, Y., Eds.; Hydrological Change and Water Management: Kyoto, Japan, 2008; pp. 381–386.

35. Petersen, G.; Fohrer, N. Flooding and drying mechanisms of the seasonal Sudd flood plains along the Bahr el Jebel in southern Sudan. *Hydrol. Sci. J.* **2010**, *55*, 4–16.

36. Ayenew, T. Evapotranspiration estimation using thematic mapper spectral satellite data in the Ethiopian rift and adjacent highlands. *J. Hydrol.* **2003**, *279*, 83–93.

37. Anderson, M.C.; Kustas, W.P.; Norman, J.M.; Hain, C.R.; Mecikalski, J.R.; Schultz, L.; Gonzalez-Dugo, M.P.; Cammalleri, C.; Pimstein, A.; Gao, F. *et al.* Mapping daily evapotranspiration at field to continental scales using geostationary and polar orbiting satellite imagery. *Hydrol. Earth Syst. Sci.* **2011**, *15*, 223–239.

38. Bashir, M.A.; Hata, T.; Tanakamaru, H.; Abdelhadi, A.W.; Tada, A. Satellite-based energy balance model to estimate seasonal evapotranspiration for irrigated sorghum: A case study from the Gezira scheme, Sudan. *Hydrol. Earth Syst. Sci.* **2008**, *12*, 1129–1139.

39. El-Shirbeny, M.A.; Aboelghar, M.A.; Arafat, S.M.; El-Gindy, A.-G.M. Assessment of the mutual impact between climate and vegetation cover using NOAA-AVHRR and Landsat data in Egypt. *Arab J. Geosci.* **2013**, *7*, 1287–1296.

40. Elhag, M.; Psilovikos, A.; Manakos, I.; Perakis, K. Application of the SEBS water balance model in estimating daily evapotranspiration and evaporative fraction from remote sensing data over the Nile delta. *Water Resour. Manag.* **2011**, *25*, 2731–2742.

41. Simonneaux, V.; Abdrabbo, M.A.A.; Saleh, S.M.; Hassanein, M.K.; Abou-Hadid, A.F.; Chehbouni, A. MODIS estimates of annual evapotranspiration of irrigated crops in the Nile Delta based on the FAO method: Application to the Nile river budget. *Proc. SPIE* **2010**, doi:10.1117/12.865066.

42. El Tahir, M.; Wang, W.; Xu, C.; Zhang, Y.; Singh, V. Comparison of methods for estimation of regional actual evapotranspiration in data scarce regions: Blue Nile region, eastern Sudan. *J. Hydrologic Eng.* **2012**, *17*, 578–589.

43. Mohamed, Y.A.; Bastiaanssen, W.G.M.; Savenije, H.H.G. Spatial variability of evaporation and moisture storage in the swamps of the upper Nile studied by remote sensing techniques. *J. Hydrol.* **2004**, *289*, 145–164.

44. Rebelo, L.M.; Senay, G.B.; McCartney, M.P. Flood pulsing in the Sudd wetland: Analysis of seasonal variations in inundation and evaporation in South Sudan. *Earth Interac.* **2012**, *16*, 1–19.

45. Psilovikos, A.; Elhag, M. Forecasting of remotely sensed daily evapotranspiration data over Nile Delta region, Egypt. *Water Resour. Manag.* **2013**, *27*, 4115–4130.

46. De Bruin, H.A.R.; Trigo, I.F.; Jitan, M.A.; Temesgen Enku, N.; van der Tol, C.; Gieske, A.S.M. Reference crop evapotranspiration derived from geo-stationary satellite imagery: A case study for the Fogera flood plain, NW-Ethiopia and the Jordan Valley, Jordan. *Hydrol. Earth Syst. Sci.* **2010**, *14*, 2219–2228.

47. Senay, G.B.; Asante, K.; Artan, G. Water balance dynamics in the Nile Basin. *Hydrol. Process.* **2009**, *23*, 3675–3681.

48. Mohamed, Y.A.; van den Hurk, B.J.J.M.; Savenije, H.H.G.; Bastiaanssen, W.G.M. Hydroclimatology of the Nile: Results from a regional climate model. *Hydrol. Earth Syst. Sci.* **2005**, *9*, 263–278.

49. Sun, Z.; Gebremichael, M.; Ardö, J.; de Bruin, H.A.R. Mapping daily evapotranspiration and dryness index in the East African highlands using MODIS and SEVIRI data. *Hydrol. Earth Syst. Sci.* **2011**, *15*, 163–170.

50. Zhang, Z.; Xu, C.-Y.; Yong, B.; Hu, J.; Sun, Z. Understanding the changing characteristics of droughts in Sudan and the corresponding components of the hydrologic cycle. *J. Hydrometeorol.* **2012**, *13*, 1520–1535.

51. Marshall, M.; Funk, C.; Michaelsen, J. Examining evapotranspiration trends in Africa. *Clim. Dyn.* **2012**, *38*, 1849–1865.

52. Sun, Z.; Gebremichael, M.; Ardo, J.; Nickless, A.; Caquet, B.; Merboldh, L.; Kutschi, W. Estimation of daily evapotranspiration over Africa using MODIS/Terra and SEVIRI/MSG data. *Atmos. Res.* **2012**, *112*, 35–44.

53. Mu, Q.M.; Zhao, M.S.; Running, S.W. *MOD16 1-km² Terrestrial Evapotranspiration (ET) Product for the Nile Basin*; Algorithm Theoretical Basis Document; Numerical Terradynamic Simulation Group, College of Forestry and Conservation, University of Montana: Missoula, MT, USA, 29 May 2013.

54. Mu, Q.M.; Zhao, M.S. Running, S.W. Improvements to a MODIS global terrestrial evapotranspiration algorithm. *Remote Sens. Environ.* **2011**, *115*, 1781–1800.

55. Mu, Q.; Zhao, M.; Running, S.W. *MODIS Global Terrestrial Evapotranspiration (ET) Product (NASA MOD16A2/A3)*; Algorithm Theoretical Basis Document, Collection 5; NASA HQ, Numerical Terradynamic Simulation Group, University of Montana: Missoula, MT,USA, 20 November 2013.

56. Huffman, G.J.; Adler, R.F.; Stocker, E.; Bolvin, D.T.; Nelkin, E.J. Analysis of TRMM 3-Hourly Multi-Satellite Precipitation Estimates Computed in Both Real and Post-Real Time. In Proceedings of 12th Conference on Satellite Meteorology and Oceanography, Long Beach, CA, USA, 9–13 February 2003.

57. Schaaf, C.B.; Gao, F.; Strahler, A.H.; Lucht, W.; Li, X.; Tsang, T.; Strugnell, N.C.; Zhang, X.; Jin, Y.; Muller, J.P.; *et al.* First operational BRDF, albedo nadir reflectance products from MODIS. *Remote Sens. Environ.* **2002**, *83*, 135–148.

58. Camberlin, P. Nile Basin Climates. In *The Nile: Origins, Environments, Limnology and Human Use*; Dumont, H.J., Ed.; Springer: Houten, The Netherlands, 2009; pp. 307–333.

59. Beyene, T.; Lettenmaier, D.P.; Kabat, P. Hydrological impacts of climate change on the Nile River basin: Implications of the 2007 IPCC scenarios. *Clim. Chang.* **2010**, *100*, 433–461.

60. Conway, D. From headwater tributaries to international river: Observing and adapting to climate variability and change in the Nile basin. *Glob. Environ. Chang.* **2005**, *15*, 99–114.

61. Zaitchik, B.F.; Simane, B.; Habib, S.; Anderson, M.C.; Ozdogan, M.; Foltz, J.D. Building climate resilience in the Blue Nile/Abay Highlands: A role for earth system sciences. *Int. J. Environ. Res. Publ. Health* **2012**, *9*, 435–446.

62. Waterbury, J. *The Nile Basin: National Determinants of Collective Action*; Yale University Press: Ann Arbor, MI, USA, 2002.

63. Senay, G.; Budde, M.; Verdin, J. Enhancing the Simplified Surface Energy Balance (SSEB) approach for estimating landscape ET: Validation with the METRIC model. *Agric. Water Manag.* **2011**, *98*, 606–618.

64. Senay, G.; Leake, S.; Nagler, P.; Artan, G.; Dickinson, J.; Cordova, J.; Glenn, E. Estimating basin scale evapotranspiration (ET) by water balance and remote sensing methods. *Hydrol. Process.* **2011**, *25*, 4037–4049.

65. Velpuri, N.M.; Senay, G.B.; Singh, R.K.; Bohms, S.; Verdin, J.P. A comprehensive evaluation of two MODIS evapotranspiration products over the conterminous United States: Using point and gridded FLUXNET and water balance ET. *Remote Sens. Environ.* **2013**, *139*, 35–49.

66. Nile Basin Initiative (NBI). Available online: http://nileis.nilebasin.org/ (accessed on 2 February 2014).

67. Reverb|ECHO-NASA's Next Generation Earth Science Tool. Available online: http://earthdata.nasa.gov/reverb (accessed on 15 May 2013).

68. Dinku, T.; Ceccato, P.; Grover-Kopec, E.; Lemma, M.; Connor, S.J.; Ropelewski, C.F. Validation of satellite rainfall products over East Africa's complex topography. *Int. J. Remote Sens.* **2007**, *28*, 1503–1526.

69. NASA's Tropical Rainfall Measuring Mission (TRMM). Available online: http://trmm.gsfc.nasa.gov/ (accessed on 3 September 2013).

70. De Beurs, K.; Henebry, G. A statistical framework for the analysis of long image time series. *Int. J. Remote Sens.* **2005**, *26*, 1551–1573.

71. Shapiro, S.S.; Wilk, M.B. An analysis of variance test for normality (complete samples). *Biometrika* **1965**, *52*, 591–611.

72. Breusch, T.S.; Pagan, A.R. A simple test for heteroscedasticity and random coefficient variation. *Econometrica* **1979**, *47*, 1287–1294.

73. Breusch, T.S. Testing for autocorrelation in dynamic linear models. *Aust. Econ. Papers* **1978**, *17*, 334–355.

74. Viste, E.; Korecha, D.; Sorteberg, A. Recent drought and precipitation tendencies in Ethiopia. *Theor. Appl. Climatol.* **2013**, *112*, 535–551.

75. Hastenrath, S.; Polzin, D.; Mutai, C. Diagnosing the droughts and floods in equatorial East Africa during boreal autumn 2005–08. *J. Clim.* **2010**, *23*, 813–817.

76. Elhag, M; Psilovikos, A; Makrantonaki, M.S. Land use changes and its impacts on water resources in Nile Delta region using remote sensing techniques. *Environ. Dev. Sustain.* **2013**, *15*, 1189–1204.

77. Abd El-Kawy, O.R.; Rød, J.K.; Ismail, H.A.; Suliman, A.S. Land use and land cover change detection in the western Nile delta of Egypt using remote sensing data. *Appl. Geogr.* **2011**, *31*, 483–494.

78. Kalma, J.D.; McVicar, T.R.; McCabe, M.F. Estimating land surface evaporation: A review of methods using remotely sensed surface temperature data. *Surv. Geophy.* **2008**, *29*, 421–469.

79. Mueller, B.; Seneviratne, S.I.; Jimenez, C.; Corti, T.; Hirschi, M.; Balsamo, G. Evaluation of global observations-based evapotranspiration datasets and IPCC AR4 simulations. *Geophys. Res. Lett.* **2011**, *38*, L06402.

80. Vinukollu, R.K.; Meynadier, R.; Sheffield, J.; Wood, E.F. Multi-model, multisensory estimates of global evapotranspiration: Climatology, uncertainties and trends. *Hydrol. Process.* **2011**, *25*, 3993–4010.

81. Kerr, Y.H.; Waldteufel, P.; Richaume, P.; Wigneron, J.P.; Ferrazzoli, P.; Mahmoodi, A.; Al Bitar, A.; Cabot, F.; Gruhier, C.; Juglea, S.E.; *et al.* The SMOS soil moisture retrieval algorithm. *IEEE Trans. Geosci. Remote Sens.* **2012**, *50*, 1384–1403.

82. Zhang, K.; Kimball, J.S.; Nemani, R.R.; Running S.W. A continuous satellite-derived global record of land surface evapotranspiration from 1983 to 2006, *Water Resour. Res.* **2010**, *46*, W09522.

Chapter 4:
Surface Water Hydrology

Remotely Sensed Monitoring of Small Reservoir Dynamics: A Bayesian Approach

Dirk Eilander, Frank O. Annor, Lorenzo Iannini and Nick van de Giesen

Abstract: Multipurpose small reservoirs are important for livelihoods in rural semi-arid regions. To manage and plan these reservoirs and to assess their hydrological impact at a river basin scale, it is important to monitor their water storage dynamics. This paper introduces a Bayesian approach for monitoring small reservoirs with radar satellite images. The newly developed growing Bayesian classifier has a high degree of automation, can readily be extended with auxiliary information and reduces the confusion error to the land-water boundary pixels. A case study has been performed in the Upper East Region of Ghana, based on Radarsat-2 data from November 2012 until April 2013. Results show that the growing Bayesian classifier can deal with the spatial and temporal variability in synthetic aperture radar (SAR) backscatter intensities from small reservoirs. Due to its ability to incorporate auxiliary information, the algorithm is able to delineate open water from SAR imagery with a low land-water contrast in the case of wind-induced Bragg scattering or limited vegetation on the land surrounding a small reservoir.

Reprinted from *Remote Sens*. Cite as: Eilander, D.; Annor, F.O.; Iannini, L.; van de Giesen, N. Remotely Sensed Monitoring of Small Reservoir Dynamics: A Bayesian Approach. *Remote Sens* **2014**, *6*, 1191–1210.

1. Introduction

To overcome droughts and ensure water availability, the rural population in many semi-arid areas of the world relies on small reservoirs [1]. In this context, small reservoirs are defined as reservoirs with a surface area smaller than 100 hectares. Typically, these reservoirs are embanked streams that capture water in the wet season, to be made available during the dry season. Small reservoirs are used for year-round irrigation, fishery, cattle and domestic purposes.

Currently, the cumulative impact of small reservoirs on water resources at a river basin scale is still debated [2], and the sustainability of small reservoirs under climate change is unknown [3]. In order to plan, manage and improve our understanding of small reservoirs, it is important to monitor the water storage dynamics. Ground-based surveys are both labor intensive and time consuming. Alternatively, small reservoir storage can be measured from space, based on remotely sensed surface area measurements in combination with regional area-volume equations, which can be derived from *in situ* bathymetric measurements [4–6]. For large lakes and reservoirs, water stage measurements from space have recently become available [7,8], which enable the estimation of water storage changes using only remote sensing observations.

Water surface areas can be delineated through optical imagery (e.g., MODIS, SPOTand Landsat), as well as synthetic aperture radar (SAR) imagery (e.g., Envisat, ALOSand Radarsat). A common practice for optical-based water surface delineation is to put a threshold on a vegetation [9] or water index [10,11] for decision making. For large water bodies, MODIS yields good results, but the spatial resolution is too low for small reservoirs. Instead Landsat imagery (30-m spatial resolution) has successfully been applied using various techniques in Ghana [5], Zimbabwe [6], India [12] and Brazil [13]. A strong limitation of optical imagery is its dependence on cloud- and smoke-free day acquisitions, which makes its application for operational monitoring very limited. For the detection and the creation of a base map, optical imagery is, however, very suitable [14].

The application of SAR imagery for small reservoir monitoring has recently been studied based on Envisat ASARimages [4,14]. Smooth open water acts as a specular reflector, reflecting most of the radar signal away from the sensor. Radar backscatter intensities from open water are therefore generally lower than backscatter intensities from the surrounding land, which enables the delineation of open water. The roughness of the water is very variable and influences backscatter intensities over time and space. Difficulties arise when wind-induced Bragg scattering enhances backscatter from the open water or when the contrast between land and water deteriorates, due to the absence of vegetation on the land surrounding a small reservoir at the end of the dry season [14]. Vegetation in the tail-end of small reservoirs has a different signature from open water, which may result in an error in the delineation [4]. Based on an earlier study [14], SAR imagery is found to be suitable for the delineation of small reservoirs in the wet season, but to be affected by wind and a low land-water contrast in the dry season. To date, no weather-independent method that yields good year-round results has been developed for remotely sensed areal measurements of small reservoirs.

In this paper, we propose a new Bayesian algorithm to delineate small reservoirs. The algorithm can deal with a large variability in backscatter intensities from open water, exploits information contained in multi-polarized SAR imagery and readily allows for the input of auxiliary information, e.g., temporal information about the small reservoir area. For this study, Radarsat-2 SAR images of the Upper East Region (UER), Ghana, were acquired. The Radarsat-2 has an improved resolution compared to Envisat ASAR imagery and offers full polarimetric data.

2. Datasets

2.1. Ground Truth

The study area is located in the Upper East Region (UER) of Ghana; see Figure 1. The UER has a semi-arid climate, characterized by a five month, mono-modal wet season and an average rainfall of 1,044 mm/yr over the past 40 years.

Fieldwork was conducted in November 2012. This period is at the start of the dry season, when the water levels in the small reservoirs are still around their upper limits. In total, 29 small reservoirs in the Kasena Nankana West, Kasena Nankana East, Bongo and Bolgatanga districts were visited, of which 26 are covered by all acquired images. All reservoirs were delineated in the field using a Garmin eTrex HCx handheld GPS, with an accuracy of 10 m (95% typical). With a simple walk around the reservoir, waypoints were taken at the land-water boundary in such a way that interpolating between the points yielded a good delineation of the reservoir. The boundary of the reservoir at the tail-end streams was defined as the point of a 10-m stream width. The area of the visited small reservoirs varied from 2.5 ha to 22.6 ha.

Figure 1. The study area in the Upper East Region of Ghana, overlaid with a base map of small reservoirs in the region.

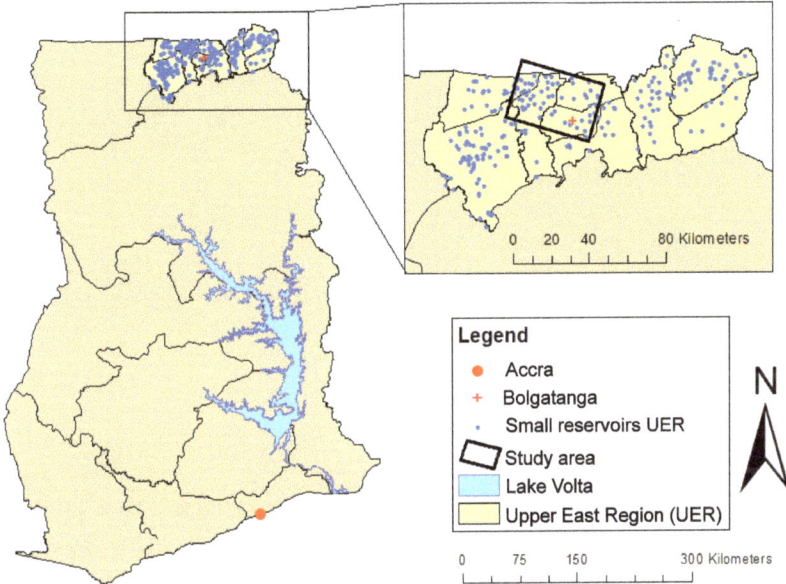

2.2. Precipitation Data

Rainfall series with a 15-min temporal resolution were obtained from a meteorological station in Navrongo (UER, Ghana) for the period from January until April 2013. The rainfall series were converted to daily rainfall and are presented in Section 4.3.

2.3. Radarsat-2 SAR Data

In this study, the small reservoir delineation is based on one or more polarizations from full polarimetric fine resolution Radarsat-2 data. First, two fine quad-pol images covering the study area were acquired at the start of the dry season in November 2012, followed by a time series of Wide fine quad-pol images from January until April 2013. The 2012 acquisitions allow for comparison with the ground truth data, while the time series allows for a temporal analysis and covers different dry season backscatter scenarios. Details about the acquired images are given in Table 1.

Table 1. Radarsat-2 imagery acquired for this study.

Date	Year	Time/Pass	Beam Mode	Incidence Angle (degree)	Pixel Spacing (rg× ax) (m)
18 November	2012	05:44:13/desc	FQ31	48.3–49.4	5.14 × 6.28
21 November	2012	05:56:37/desc	FQ10	29.1–30.9	5.19 × 9.26
15 January	2013	05:52:27/desc	FQ17W	35.7–8.6	5.6 × 7.83
25 January	2013	06:00:44/desc	FQ4W	21.3–24.8	4.6 × 11.94
8 February	2013	05:52:27/desc	FQ17W	35.7–38.6	5.6 × 7.83
18 February	2013	06:00:43/desc	FQ4W	21.3–24.8	4.6 × 11.94
4 March	2013	05:52:27/desc	FQ17W	35.7–38.6	5.6 × 7.83
14 March	2013	06:00:43/desc	FQ4W	21.3–24.8	4.6 × 11.94
28 March	2013	05:52:27/desc	FQ17W	35.7–38.6	5.6 × 7.83
7 April	2013	06:00:44/desc	FQ4W	21.3–24.8	4.6 × 11.94
21 April	2013	05:52:27/desc	FQ17W	35.7–38.6	5.6 × 7.83

3. Methods

3.1. Pre-Processing SAR Imagery

All images of one beam mode were co-registered and resampled to a grid of 5 m × 5 m using the open source Next ESA SAR Toolbox (NEST) software by the European Space Agency. Some additional dedicated effort in MATLAB was required to co-register the stacks from different beam modes. This was done in two steps. Firstly, absolute verification of image geolocation was carried out on the November acquisitions (FQ10 and FQ31), hereby referred to as the reference, which showed consistent agreement with the GPS ground-truth data. Secondly, the residual shifts between the FQ04W and FQ17W datasets and the reference acquisitions were retrieved by incoherent speckle correlation [15] procedures between the reference images and the first image of each dataset. Cross-correlation of intensities performed block-wise throughout the image returned almost uniform patterns of a few pixel shifts, due also to the relatively small elevation dynamics (100 m) of the scene. The shifts were averaged and used to achieve stack co-registration. A simple moving average filter (3 × 3 pixels) was applied on backscatter intensities to remove speckle by increasing the number of looks.

From the literature [16], it was found that the delineation is best performed within a reservoir mask in which the reservoir area and total area of the surrounding land are of a similar size. Here, the masks were created by manually selecting a rectangular area around a small reservoir at full capacity, in such a way that it contained a similar number of water and land pixels.

3.2. Growing Bayesian Classifier

The newly developed growing Bayesian classifier (gBC) [17] is used for the delineation of small reservoirs. The gBC classifies a pixel based on the maximum *a posteriori* probability (MAP), which is calculated from a multivariate Gaussian likelihood function, multiplied by one or more conditional priors. The Gaussian model is justified by the application of the algorithm to logarithmic intensities. In previous work [18], it was outlined that the Gamma distribution, typical of homogeneous areas of fully developed speckle, tends toward a log-normal behavior for an increasing number of looks. The multivariate likelihood function exploits information contained in multi-polarized SAR imagery, while the conditional priors update the likelihood with auxiliary information. The basic gBC makes use of one conditional prior, the so-called growing prior, which includes information about neighboring pixels in the classification. The gBC can readily be extended with auxiliary information, e.g., temporal information, in the form of a conditional prior, according to the principle of naive Bayesian image classification [19,20]. Contrary to traditional maximum likelihood, the gBC does not require *a priori* training data to calculate the classes' signatures. Instead, signatures are developed during the iterative classification procedure based on the growing land and water seeds. The gBC flow scheme is given in Figure 2.

3.2.1. Basic Growing Bayesian Classifier

The gBC is automatically initiated within the reservoir mask. The land seed is initiated at the two outer rows and columns. The water seed at the area with the minimum average backscatter intensity, of a minimum of 3×3 pixels, is derived from a moving average filter. If a delineation at the previous time step is available, the potential water seed area is restricted to the small reservoir area from the previous time step. All other pixels are initially unclassified. The water seed and land seed are then iteratively grown, until both classes converge at the land-water boundary, according to the Bayesian decision rule:

$$\mathbf{x}_i \in \begin{cases} \omega_l; & \Phi_l^*(\mathbf{x}_i) > max(\Phi_w^*(\mathbf{x}_i), \Phi_u^*(\mathbf{x}_i)) \\ \omega_w; & \Phi_w^*(\mathbf{x}_i) > max(\Phi_l^*(\mathbf{x}_i), \Phi_u^*(\mathbf{x}_i)) \end{cases} \tag{1}$$

where ω_k is the class with $k \in l, w, u$ for the land, water and unclassified classes, respectively, and $\Phi_k^*(\mathbf{x}_i)$ the likelihood, $\Phi_k(\mathbf{x}_i)$, based on the pixel intensity vector, $\mathbf{x}_i$, for pixel i multiplied with the growing prior, $P(\omega_k|\nu_j)$, which is proportional to the posterior probability, according to:

$$\Phi_k^* = \Phi_k(\mathbf{x}_i)P(\omega_k|\nu_j) \tag{2}$$

where ν_j is a conditional variable of the growing prior with state j, which is based on the classification of neighboring pixels; see Table 2. Here, the assumption is made that the likelihood function is independent of the state of the conditional variable, ν. Note that a pixel is only classified if its probability of being a member of a class is larger than the probability of remaining unclassified. The likelihood, $\Phi_u(\mathbf{x}_i)$, for a pixel to remain unclassified is defined as the minimum of the land and water likelihood for that pixel. The classification is therefore not governed by the priors alone, but based on the likelihood computed from the pixel intensity vector.

Figure 2. Flow diagram for the growing Bayesian classifier: first, the seeds are initialized (**top right**) for which a SAR reservoir image is required (**top left**); then, the iterative Bayesian classification is performed (**right middle**); finally, a growing filter is applied (**right bottom**); the algorithm can readily be extended with additional information (**left middle**).

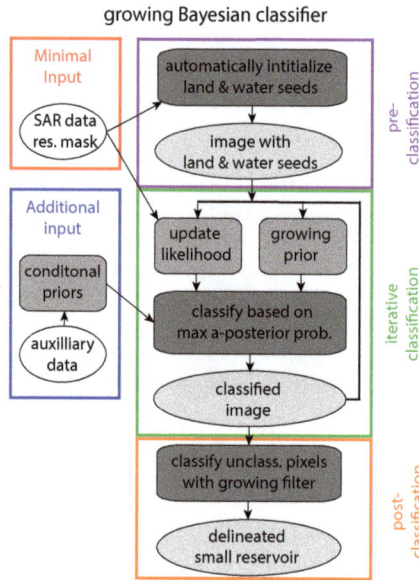

growing Bayesian classifier

Minimal Input

SAR data res. mask

automatically intitialize land & water seeds

image with land & water seeds

pre-classification

Additional input

conditonal priors

auxilliary data

update likelihood

growing prior

classify based on max a-posterior prob.

classified image

iterative classification

classify unclass. pixels with growing filter

delineated small reservoir

post-classification

Table 2. Growing conditional prior probabilities based on the classification of neighboring pixels.

Growing Prior	ν_1	ν_2	ν_3	ν_4
land pixels	>=1	>=1	0	0
water pixels	0	>=1	>=1	0
$P(\omega_{land})$	0.5	0.5	0	0
$P(\omega_{water})$	0	0.5	0.5	0
$P(\omega_{unclassified})$	0.5	0	0.5	1.0

The growing conditional variable, ν (see Table 2), is based on the assumption that, within the reservoir mask, all water pixels are one connected area, *i.e.*, the small reservoir. The state of the variable is defined according to the number of neighboring land and water pixels. The prior only allows for a new water pixel next to an already classified water pixel (the water seed) and for a new land pixel next to an already classified land pixel (the land seed). The growing prior reduces the confusion error (the error from incorrect classification of pixels within the spectral area of overlap between two classes) to the land water boundary, where both classes have an equal prior probability.

After the classification, a growing filter is applied to classify the pixels that remained unclassified after convergence of the land and water seeds. Starting from the seeds, all unclassified pixels connected to the water seed by water pixels are classified as water and all pixels connected to the land seed by land pixels are classified as land. Land and water pixels that are not connected to the seed are then reclassified using the same method.

3.2.2. Extended Growing Bayesian Classifier

The gBC can readily be extended with auxiliary information in the form of conditional priors according to the principle of naive Bayesian classification. In the general case that the basic gBC is extended with P prior probabilities, Equation (2) becomes:

$$\Phi_k^* = \Phi_k((\mathbf{x}_i))P(\omega_k|\nu_j)\prod_{p=1}^{P}P(\omega_k|\beta_j^p) \tag{3}$$

where $P(\omega_k|\beta_j^p)$ is prior probability p with conditional variable β_j^p determined by its state, j.

For this study, an additional conditional variable based on temporal information was developed. The classification from a previous or subsequent time step is used to update the delineation at the current time step. This can be very useful information, especially when the land-water contrast deteriorates. Two temporal conditional variables, τ_i^{t-1} and τ_i^{t+1}, were developed based on the strong seasonal behavior of small reservoirs, *i.e.*, small reservoirs are replenished in the wet season, and the water is released for use in the dry season. The states of the temporal variables are based on the classification of a pixel, i, in the previous and subsequent time step, respectively; see Tables 3 and 4. Equation (3) then becomes:

$$\Phi_k^* = \Phi_k(\mathbf{x})P(\omega_k|\nu_j)P(\omega_k|\tau_i^{t-1})P(\omega_k|\tau_i^{t+1}) \tag{4}$$

Table 3. Temporal conditional prior probabilities based on the classification of a pixel in the previous time step.

Prior τ^{t-1}	τ_1^{t-1}	τ_2^{t-1}	τ_3^{t-1}
Classification in Time Step t-1	Land	Water	Unclassified
$P(\omega_{land})$	0.6	0.25	1/3
$P(\omega_{water})$	0.2	0.5	1/3
$P(\omega_{unclassified})$	0.2	0.25	1/3

The temporal priors increase the posterior probability for a pixel of being a member of the same class as in the previous and/or subsequent time step. If the previous and subsequent classification of a pixel are equal, the pixel is given a relatively high prior probability of being a member of the same class. Similar prior probabilities are given for land and water if the class changes from the previous to the subsequent time step. One exception is formed by pixels classified as water in the subsequent time step during the dry season. Small reservoir areas do not increase, as long as there is no rain during the dry season. Based on this knowledge, a prior probability of being a member of the water class 1 is given to pixels that are classified as water in the subsequent time step. The temporal priors allow for the delineation of images with a very low land-water contrast if the previous and/or subsequent images are correctly classified.

Table 4. Temporal conditional prior probabilities based on the classification of a pixel in the subsequent time step.

Prior τ^{t+1}	τ_1^{t+1}	τ_2^{t+1}	τ_3^{t+1}
Classification in time step t+1	Land	Water	Unclassified
prior τ^{t+1} dry season			
$P(\omega_{land})$	0.5	0	1/3
$P(\omega_{water})$	0.25	1	1/3
$P(\omega_{unclassified})$	0.25	0	1/3
prior τ^{t+1} rainy season/after rain			
$P(\omega_{land})$	0.5	0.25	1/3
$P(\omega_{water})$	0.25	0.5	1/3
$P(\omega_{unclassified})$	0.25	0.25	1/3

In an operational setting, reservoirs at the current time step, t, can be delineated using prior $P(\omega_k|\tau_k^{t-1})$ based on the classification from the previous time step. Then, the delineation at time step t-1 can be updated with temporal priors $P(\omega_k|\tau_k^{t-1})$ and $P(\omega_k|\tau_k^{t+1})$ based on the classified images at time step t and t-2 The classifications at time step t until t-3 could then further be updated with the same procedure. However, this only yields a small improvement and would allow for a classification error to propagate back in time.

4. Results and Discussion

4.1. Polarimetric SAR Remote Sensing of Small Reservoirs for Different Backscatter Scenarios

Four distinct backscatter scenarios for backscatter from small reservoirs are found within the acquired data series of SAR imagery, *i.e.*, Smooth open water, water with vegetation, wind-induced Bragg scatter and backscatter during a rain-event. For every backscatter scenario, a sample of land and water pixels was taken from different reservoirs and its backscatter intensity distribution plotted (Figure 3). The backscatter intensities, as well as the contrast between land and water are different for every scenario. This calls for a flexible classification method (a method without fixed thresholds) and optimal use of the available polarizations.

The optimal combinations of polarizations were evaluated based on the separability between the land and water class. The separability was calculated from the Jeffries–Matusita (JM) distance [21], based on the land and water samples for the different scenarios. A JM distance of two indicates that there is no confusion area (*i.e.*, the spectral area of overlap between two classes) and can thus perfectly be separated based on backscatter intensity alone. A JM distance of zero means that the backscatter distributions for both classes completely overlap. The presented combinations (see Figure 4) are chosen for comparability and based on operational considerations. Depending on the choice of SAR mission, single, dual or full polarimetric images can be acquired. In total, 12 combinations were presented: three single polarizations, eight combinations from dual polarization modes, of which five use polarization intensity ratios, and one full polarimetric combination, in which the average of the cross-polarizations is used to increase radiometric resolution. The presented tests were applied on indicators based on channel intensities or intensity ratios only. Further research should address the use of the coherent polarimetric information for decompositions, such as alpha-entropy [22] and refined polarization synthesis [23–25], to achieve optimal contrast.

Figure 3. Backscatter intensity distributions and scatter plots for the land and water classes from the samples of four distinct small reservoir backscatter scenarios, *i.e.*, smooth open water, water with vegetation, Bragg scattering and backscatter during a rain event.

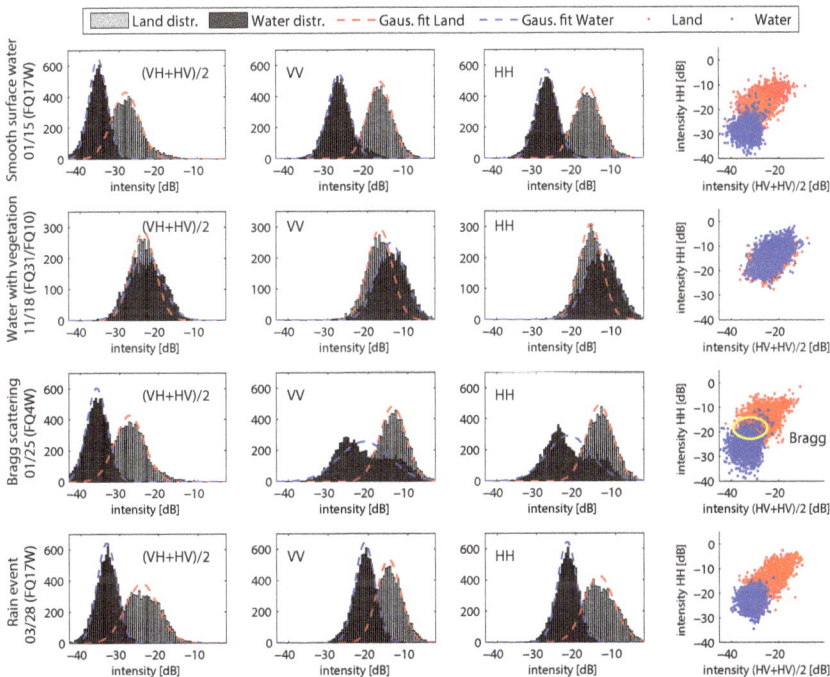

4.1.1. Smooth Open Water

Compared to land, smooth open water shows lower backscatter intensities, as it acts as a specular reflector, reflecting most of the radar signal away from the sensor. The high dielectric constant of water also decreases the penetration depth of the signal, which results in low volume scattering and, thus, predominantly co-polarized reflection [26]. The discrimination of smooth open water from land is therefore a simple task [27]. This is in agreement with the results presented here. The lowest backscatter intensities and smallest confusion area are found in the co-polarized 'HH' polarization, while the cross-polarized 'HV' shows the largest confusion area; see the top row in Figure 3. The JM distances for open water show that a high separability can be obtained from single co-polarized polarizations (1.6 on average in 'HH'), and only a small improvement (up to 1.8 for 'HH, VV, HH/VV') is found from adding more polarizations; see also Figure 4. The contrast is optimal in the case of reeds on the water boundary and land with vegetation, which enhances the delineation accuracy [14].

Figure 4. Jeffries–Matusita (JM) distances for the samples of three distinct backscatter scenarios from small reservoirs, where the error bars show the mean, minimum and maximum JM distances from the different samples.

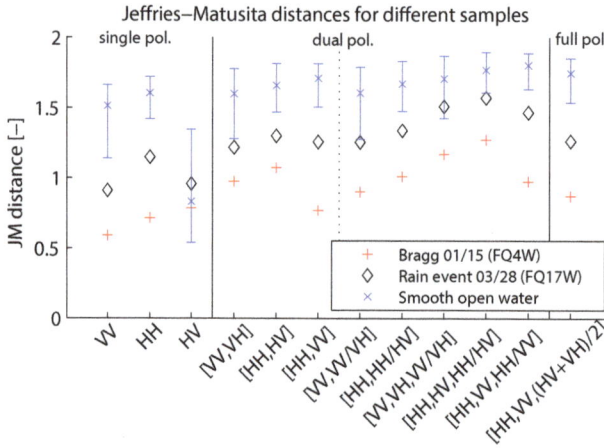

4.1.2. Water With Vegetation

This backscatter scenario refers to vegetation within the small reservoir, mainly at the tail-ends, where grasses and other weeds increase the local surface roughness. This results in higher backscatter intensities; see the second row in Figure 3. Depending on the type of the vegetation, the double bounce scattering can also be enhanced. The mean backscatter intensities from these areas are similar to land in the cross-polarized 'HV' and higher than land in the co-polarized 'HH' and 'VV'. Water with vegetation is therefore difficult to include in the small reservoir delineation [4].

4.1.3. Wind-Induced Bragg Scatter

Bragg scattering occurs when the position of scatterers are aligned parallel with the line of flight with regular spacing. In this case, the radar backscatter is coherently reinforced depending on the incidence angle, wavelength and spacing of the scatterers [27,28]. This type of scattering can be induced by wind waves on the surface of the water, depending on the wind direction and speed. According to [14,29], Bragg scattering from open water is significant with wind velocities over 9–10 km/h and in specific combinations of wind direction and polarization. Bragg scatter is most apparent in co-polarized polarizations and small in cross-polarized polarizations, which is in agreement with the histograms in the third row of Figure 3. Compared to smooth open water, the water distribution shows a long tail with high back scatter intensities and, thus, a larger confusion area between the land and water class. The scatterplot shows the decreased separability between the land and water class in the case of dual polarization, when the water surface shows patches with Bragg scattering. The separability in the single co-polarized 'HH' deteriorates compared to smooth open water (0.72). The separability for the cross-polarized polarization is similar to open water; see Figure 4. The value of a second polarization and polarization combinations becomes visible here, as a clear improvement in separability is found (up to 1.27 for 'HH, HV, HH/HV').

4.1.4. Rain Event

Backscatter intensities from the surface show a significant change during rain events, due to increased surface wetness. For bare soil, where the dominant backscatter mechanism is surface scattering, increased backscatter intensities are expected, due to the increased surface wetness. However, if pools of water start to form, backscatter intensities will decrease, as a larger portion of the signal will be scattered away. The effect of rainfall on vegetation is smaller, as vegetation already contains 'a layer of water', and there are different operating scattering mechanism on which rainfall has different effects. If volume scattering is the dominant mechanism, backscatter intensities can be reduced. Wetness of the top layer increases the portion scattered away from the sensor, and reduced power is available for the volume scattering mechanism. Areas where surface scattering is the dominant mechanism show an increase in backscatter intensities [30]. As surface scattering is the dominant mechanism in the land surrounding small reservoirs, a small increase in backscatter is observed. These effects are most significant during the rain event, when no evaporation has occurred yet. At the water surface, rain droplets can cause an increase in surface roughness, which results in increased backscatter intensities [28]. This is also observed in the sampled reservoirs; see the bottom row in Figure 3. Since no ground truth is available for the March 28 acquisition, the land sample is taken from the land outside the known maximum boundaries of the small reservoir, where vegetation is present. The bare soil surrounding the small reservoirs at this date is not included in the land sample. Larger confusion areas and, thus, lower JM distances for all polarization combinations are measured; see Figure 4. This is caused by a larger increase in backscatter intensities from open water compared to the land with vegetation. The separability between the open water and the surrounding bare land mainly depends on whether pools of water are formed, in which case, the contrast can strongly decrease.

4.2. Comparison with Ground Truth

All small reservoirs that were visited during the fieldwork are delineated from the 'HH, HV' polarization combination from two Radarsat-2 images, which were acquired within three days from the ground truth. Since substantial areas of grass and weed vegetation were found inside the small reservoirs during the fieldwork, the reservoirs were divided into three vegetation content classes: 'low vegetation', 'tail-end vegetation' and 'all boundary vegetation'. This classification was based on photos of the small reservoirs taken during the fieldwork. The delineation (red line) and ground truth (yellow line) of two typical reservoirs from each category are projected on Pauli RGBcolor composite images (Figure 5). Open water (dark areas, low backscatter intensities) are very clear in the images and easily delineated from the surrounding land (blue and green areas, higher single bounce and volume scattering). Patches of Bragg scatter (dark blue areas, increased surface scattering) are also classified as small reservoir, because of the dual polarization combination. The second and third row show the reservoirs with vegetation in the tail-end and at all boundaries, respectively. The small reservoir areas with vegetation (red and bright areas, high backscatter intensities and double bounce scattering) are not classified as small reservoir, because of their very different polarimetric signature. In all the reservoir images, the open water areas are well delineated by the gBC based on visual inspection.

Figure 5. Ground truth (yellow line) and delineation (red line) based on the 'HH, HV' polarization combination overlaid on Pauli RGB-images, with red colors for double bounce, green for volume scatter and blue for single bounce; note that the different color scales are used for the different Pauli components to enhance the image contrast.

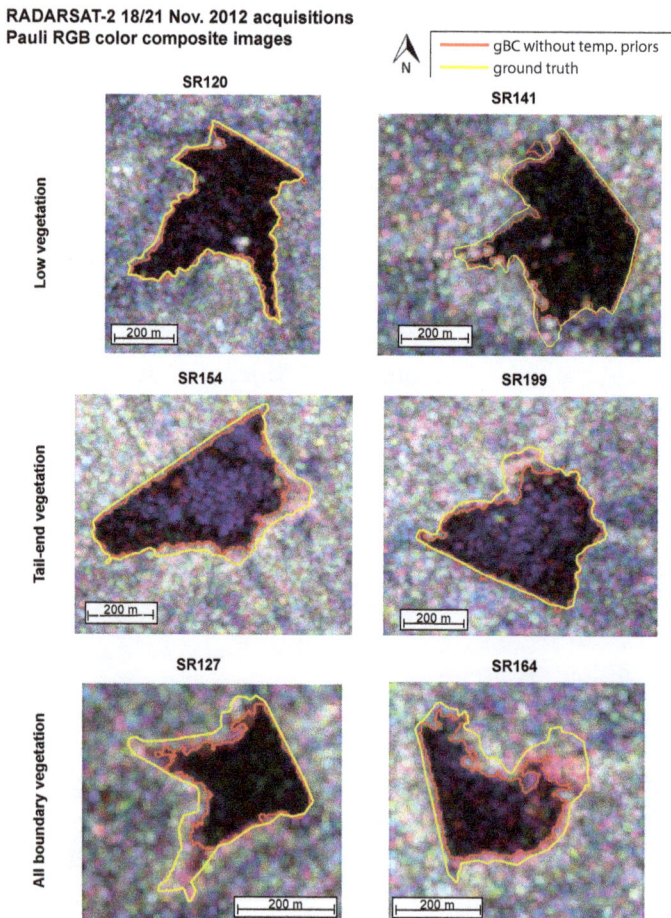

The results are summarized in Figure 6 with the Differential Area Index (DAI), as used by [6], and the JM distance based on the ground truth. The figure shows an overall underestimation of the classified small reservoir area. Compared to the ground truth, an underestimation of 12.8% to 14.8%, depending on the polarization combination, of the small reservoir area is found for small reservoirs with low vegetation. This underestimation is larger for the classes with more vegetation. The error made for delineating open water is expected to be smaller than suggested by these numbers, as the underestimation is due to different classification errors and a bias towards the land in the ground truth. First of all, there is an error due to vegetation in the small reservoirs, even in the reservoirs with 'low vegetation'; see, e.g., SR141 in Figure 5, where some trees within the reservoir cause an incorrect classification at the tail-ends. A smaller error is due to the moving average filter, which reduces noise, but might also cause some boundary pixels to be classified as land instead of water. Furthermore, there is an error in the ground truth from the inaccuracy of the GPS device. The ground truth also shows a bias towards the land, due to the fact that the measurements were all taken while walking around the reservoir as close to the land-water boundary as possible, but on the land side. The JM distances show a clear trend between the increase in vegetation in the small reservoir and a decrease in separability. This is due to the increasing confusion area with an increasing vegetation area in the small reservoir.

Figure 6. Comparison between classified and ground truth areas of all 29 small reservoir from November 2012, based on the Differential Area Index (DAI) and the Jeffries Matusita (JM) distance.

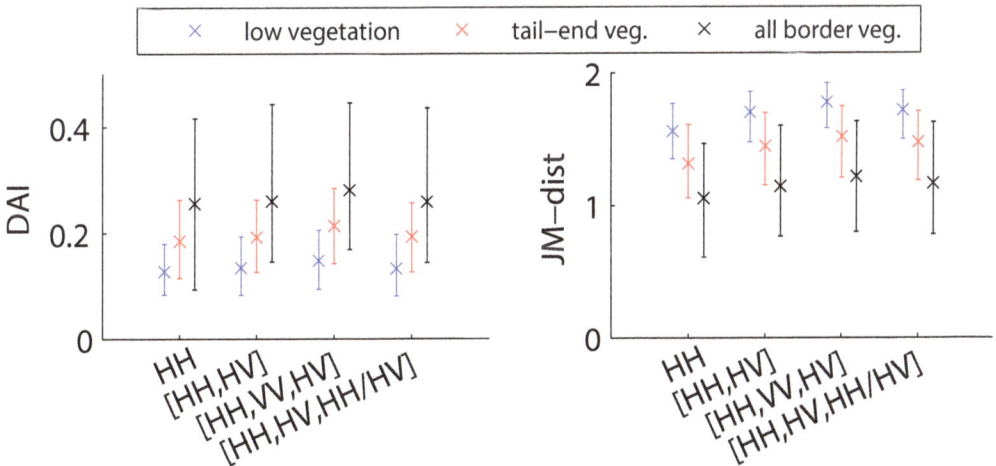

4.3. Image Quality

The quality of the images is determined based on the contrast between the land and water class, which is calculated from the JM distance based on the classified images. This method for determining image quality is similar to the numbers of peaks in the backscatter intensity histogram of an image, as used by [14]. Two quality classes were used, with a threshold for high image quality images set to 1.5. This roughly corresponds to the minimum JM distance found for smooth open water for the 'HH, HV' polarization combination; see Figure 4. Within the time series, 170 reservoir images with high quality and 64 with low quality, of which 18 had substantial Bragg scatter, were found. Only two high quality images with substantial Bragg scatter were found. The images with Bragg scatter have an average JM distance of 1.22, while all other images have an average JM distance of 1.58. The acquisitions with Bragg scatter are 25 January, 14 March and 7 April, of which especially the last two have a significant amount of low quality reservoir images (Figure 7). The acquisitions with Bragg scatter show higher mean backscatter intensities, but also a larger variability in the 'HH' polarization and no significant change in the 'HV' polarization. A large variability in backscatter intensities from open water pixels within the same acquisition is typical for Bragg scattering. Most reservoir images from 28 March, the acquisition during a rain event (see the top graph), are of high quality. The delineations for this date are likely to be overestimated; see the next section. Part of the low backscatter area that is classified as small reservoir is from the bare soil with water pools surrounding the small reservoir. The presented JM distances are, therefore, also likely to overestimate the actual image quality. The rain event has a clear impact on the mean backscatter intensities. In both the 'HH' and 'HV' polarizations, the backscatter intensities are elevated, while the variation is similar. Acquisitions during a rainfall event can easily be detected based on its open water backscatter intensities alone. A substantial number of reservoir images on 21 April have a low quality, although no Bragg scatter is found in this date. Here, the contrast is low, due to reduced vegetation on the land surrounding the small reservoir at the end of the dry season. Visual inspection of the Pauli images from the time series in comparison to images from November indicated a strong decrease in vegetated area inside the small reservoir. Vegetation in the small reservoirs is therefore expected to have a smaller influence on the classification accuracy compared to the start of the dry season.

4.4. Time Series Analysis

To demonstrate the Bayesian approach, two temporal priors were introduced. The priors update the classification at the current time step based on temporal information from the previous and subsequent classifications. Because of the rain event on 19 March (Figure 7), the temporal prior probabilities for prior τ^{t+1} change as well as from the 14 March acquisition; see Table 4.

Figure 8 shows the delineations based on the gBC with (blue and green lines) and without temporal priors (red line) for small reservoirs SR120 and SR154. From the delineations of both reservoirs, it can be seen that the classifications with and without temporal priors are similar for high quality images. Temporal priors improve the classification when the image has a low quality. A relative difference in the classified area between the delineations with and without temporal priors of more than 5% is found for 42 (55%) of the low quality reservoir images and only for 34 (22%) of the high quality reservoir images.

Figure 7. Rainfall time series (**top**), the quality of an acquisition based on the Jeffries Matusita (JM) distance (**middle**) and the mean backscatter intensity of the minimum delineated small reservoir area (**bottom**), where the boxplots show the median, first and second quartile boundaries and the red crosses are outliers; the error bars show the mean and one standard deviation boundaries.

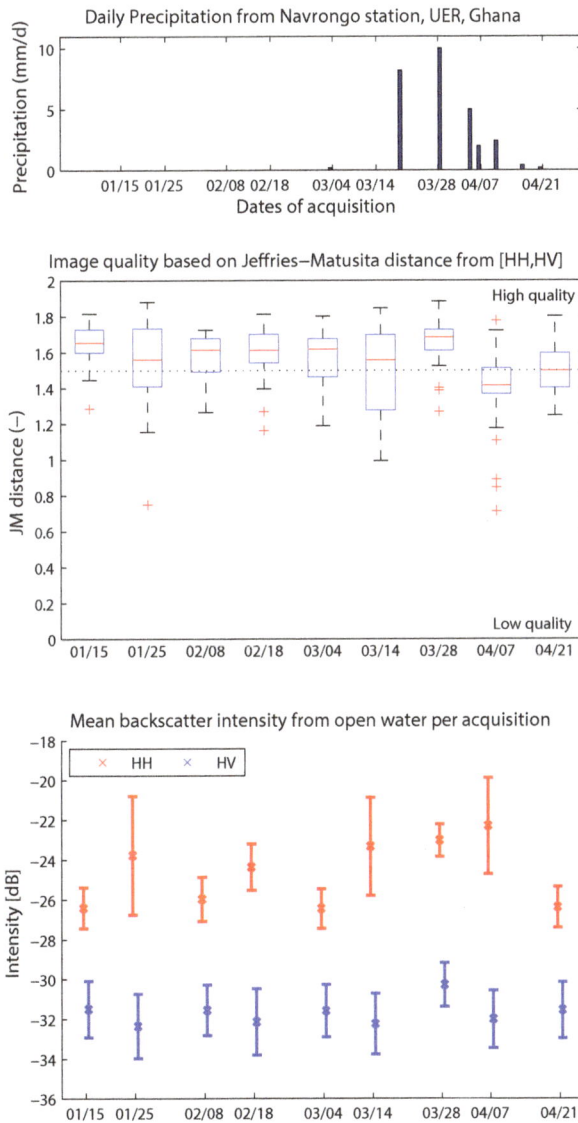

Figure 8. Time series of small reservoir delineation based on the 'HH, HV' polarization combination and the basic gBC (red line), the gBC updated with temporal prior τ^{t-1} (blue line) and the gBC updated with both priors τ^{t-1} and τ^{t+1} (green line) overlaid on HH backscatter intensity images; the bottom graphs show the areal variation in time for the same reservoirs, where the crosses show the filtered time series without the rain-affected March 28 acquisition.

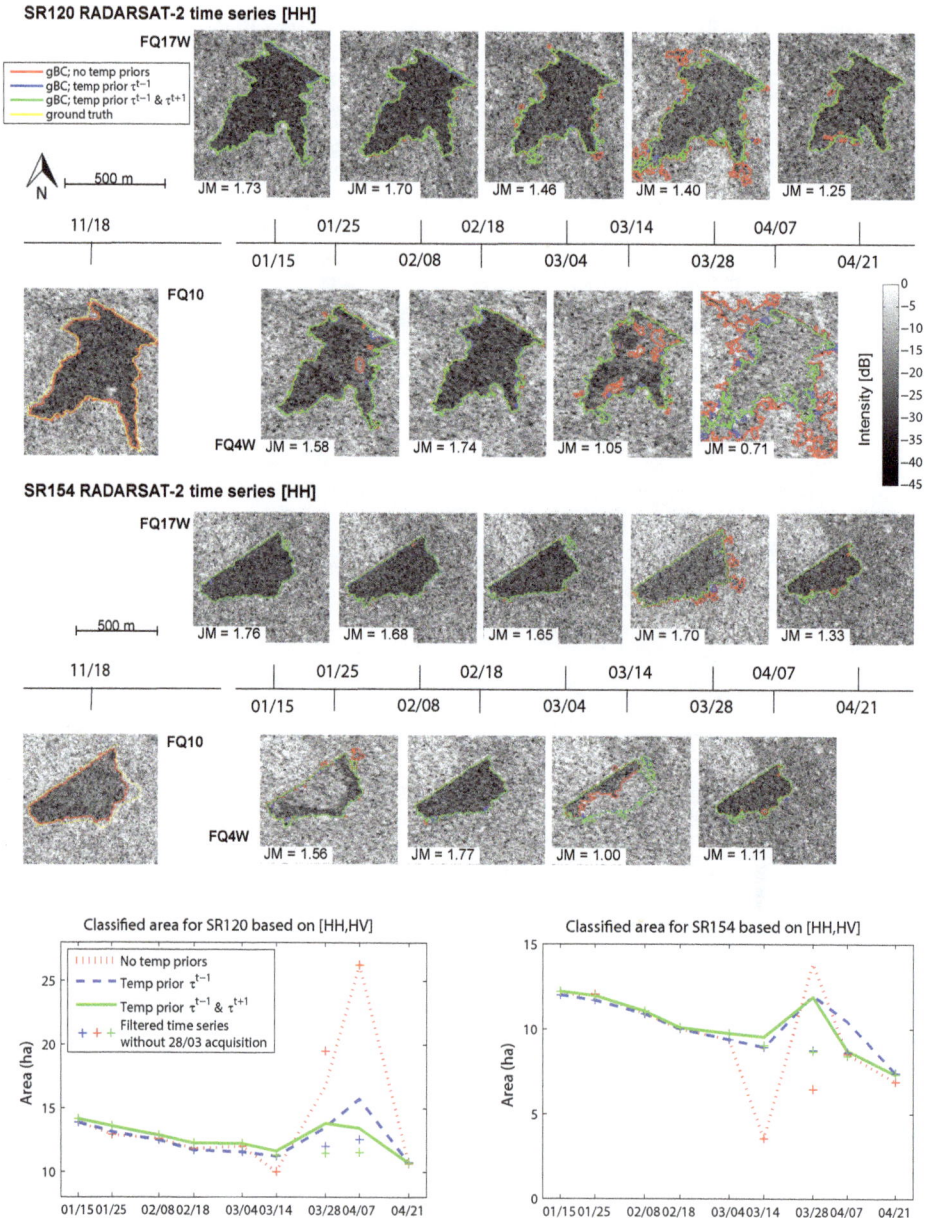

Depending on the location within the small reservoir and the change of the polarimetric signature, some areas with Bragg scatter are classified as small reservoir, even without the temporal priors. This is because of the multi-polarized input data; see, e.g., SR154 on 25 January. The Bragg scatter at 14 March is not captured without the temporal priors in both reservoirs. The graphs in Figure 8 show that most of the water patches with Bragg scatter are delineated based on the classification at the previous time step (prior τ^{t-1}). Additional updating based on the subsequent classification (prior τ^{t+1}) only improves the classification in some cases. An extreme case of Bragg scatter is found in the 7 April image for SR120, where the full reservoir is affected and the contrast with the surrounding land significantly deteriorates. Here, the classification is mostly governed by the temporal priors, as can be seen from the large difference in the delineations with and without temporal priors. The classification at the previous time step is affected by rainfall; see the next paragraph. In the case that the delineation is updated with the classification from this time step, the small reservoir area is still overestimated. When updated with the classification from the subsequent time step, the overestimation is further limited. The discrepancy between the normal and the filtered time series (the 28 March acquisition is filtered out) of small reservoir areas for 7 April shows that the temporal priors are less effective when two low quality images follow each other (Figure 8).

The classified small reservoir areas from images acquired during a rain event are likely to be overestimated. The 28 March acquisition shows increased average backscatter intensities from the full images, but low backscatter intensities from land area within the maximum small reservoir boundary. This is probably because of the formation of water pools on the bare ground surrounding the small reservoirs. The delineation for both reservoirs on this date is not able to separate between the elevated backscatter intensities from the roughened water surface and the decreased backscatter intensities from the bare ground with water pools surrounding the small reservoir. This results in an overestimated small reservoir area. This hypothesis is strengthened by the total rainfall amount from these dates, which amounts to 18 mm (10 mm on 18 March and 8 mm on 19 March). Based on the regional area-volume equations [4] and the classified small reservoir areas, this total rainfall causes an increase of the water level of 94 mm in SR154 and 76 mm in SR120. This increase in the water level is unlikely, given that the first rains after a long dry spell in the region do not create much runoff [31]. The most accurate estimation of the small reservoir area for this date is found through interpolating between the areas from subsequent and previous time steps in a time series, where 28 March is filtered out (Figure 8).

The results for the time series analysis of the gBC are summarized in Figure 9. The cumulative time series, as presented in the top graph in Figure 9, shows a strong decreasing trend of the total small reservoir area in the study area, which is expected during the dry season. The cumulative area on 28 March is a clear outlier, because of rainfall during the acquisition; see the previous paragraph. The filtered time series (the 28 March acquisition is filtered out) is in agreement with the expected trend. The largest influence of the temporal priors is, as expected, found for acquisitions with Bragg scatter (25 January, 14 March and 7 April) and the acquisition during a rain event (28 March). The bottom graph shows the effect of the chosen polarization combination on the delineated small reservoir areas. For acquisitions where most reservoir images are of good quality, the differences are minimal. The single polarization 'HH' tends to underestimate the small reservoir areas for all images with Bragg scatter. Multi-polarized combinations improve the delineation in some cases of Bragg scatter, e.g., the 25 January acquisition. The dual polarization combinations perform similar to the full polarized combination and are thus sufficient for small reservoir delineation. The combination with the backscatter intensity ratio in general results in the largest classified area for small reservoirs. During the 14 March, 28 March and 7 April acquisitions, which have the most low quality reservoir images, the temporal priors are needed to improve the delineations regardless of the polarization combination.

Figure 9. Time series of the cumulative classified area for 26 small reservoirs, based on the growing Bayesian classifier (gBC) with and without temporal priors (**top**) and the gBC without temporal priors for different polarization combinations (**bottom**).

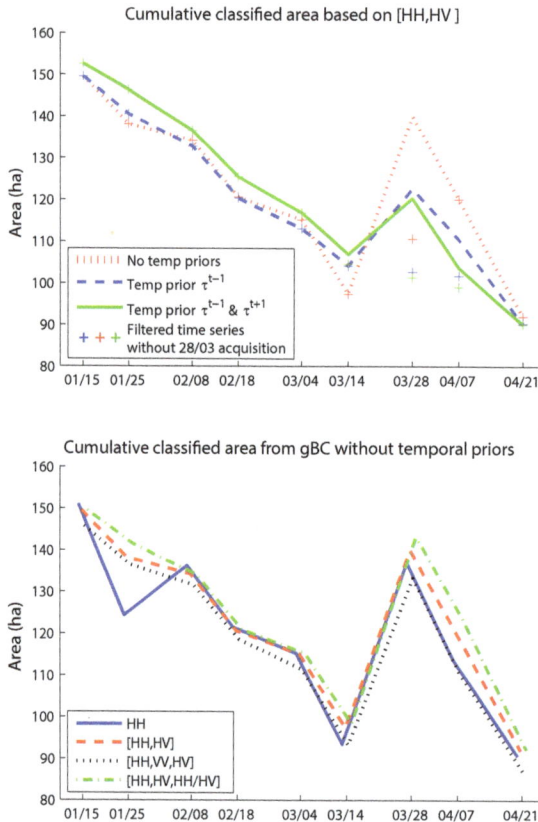

5. Conclusions

A Bayesian approach to monitoring small reservoirs was successfully applied. Despite temporal and spatial variation in backscatter intensities from small reservoirs, the newly developed algorithm is able to delineate open water throughout the dry season. The algorithm has a high accuracy, as the confusion area is restricted to the land-water boundary. Due to auxiliary temporal information, images with a low land-water contrast are resolved, even in the case of wind-induced Bragg scattering. One exception was the images acquired during a rain event, when water pools started forming on the bare ground surrounding the small reservoirs in the dry season and the land-water contrast deteriorates. In such cases, the use of time-series was able to mitigate the segmentation error, but not to completely resolve the land-water ambiguity.

The land-water contrast decreases with increasing roughness of the water surface or decreasing roughness of the surrounding land. The water surface roughness was found to increase due to wind-induced Bragg scattering and during rain events. The roughness of the land surrounding the small reservoir decreases towards the end of the dry season, when the water level in the small reservoirs is low and bare ground surrounds it. Areas with vegetation inside the small reservoir at the start of the dry season have a backscatter signature similar to land and were, therefore, not included in the delineation, causing an underestimation of the actual small reservoir area.

Single co-polarized backscatter intensities are sufficient in the case of high quality images. The dual polarization combinations 'HH, HV' and 'HH, HV, HH/HV' improve the land-water contrast significantly in the case of Bragg scattering and are, therefore, the preferred combinations of backscatter intensities for small reservoir monitoring. This is also relevant in the light of the European Space Agency's (ESA) Sentinel 1 satellite, which does not produce full polarimetric images. Further research should address the exploitation of the coherent polarimetric information.

The overall conclusion of the paper is that, due to a Bayesian approach, the dynamics of small reservoirs can be monitored from SAR data with a high level of automation and without the restriction of cloud-free days. The suggested approach is to create a base map of small reservoirs first, after which small reservoir dynamics can be monitored with SAR data using a Bayesian time series approach. The base map is best created at the onset of the dry season, when the reservoirs are at full capacity and the reservoir masks can be determined.

Acknowledgments

The authors would like to acknowledge the support of the European Space Agency (ESA) ALCANTARAproject AO-1-7102 A11and TIGERproject No. 32, the RADARSAT-2 Science and Operational Applications Research and Development Program ("SOAR Program") No. 5116 of the Canadian Space Agency, the Challenge Program for Water and Food (CPWF-Volta) and the Trans-African Hydro-Meteorological Observatory (TAHMO) for providing the meteorological data for this study.

Author Contributions

All authors contributed extensively to the work presented in this paper. Specific contributions include development of the concept, provision of data and data acquisition capacity (Nick van de Giesen, Frank O. Annor); SAR data pre-processing and signature interpretation (Lorenzo Iannini); fieldwork, development of algorithm, data analysis and preparation manuscript and figures (Dirk Eilander).

400

Conflicts of Interest

The authors declare no conflict of interest.

References

1. Venot, J.P.; Fraiture, C.D.; Nti Acheampong, E. *Revisiting Dominant Notions: A Review of Costs, Performance and Institutions of Small Reservoirs in Sub-Saharan Africa*; IWMI Research Report 144; International Water Management Institute: Colombo, Sri Lanka, 2012; p. 39.
2. Leemhuis, C.; Jung, G.; Kasei, R.; Liebe, J. The Volta Basin Water Allocation System: Assessing the impact of small-scale reservoir development on the water resources of the Volta basin, West Africa. *Adv. Geosci.* **2009**, *21*, 57–62.
3. Krol, M.S.; de Vries, M.J.; van Oel, P.R.; de Araújo, J.C. Sustainability of small reservoirs and large scale water availability under current conditions and climate change. *Water Resour. Manag.* **2011**, *25*, 3017–3026.
4. Annor, F.; van de Giesen, N.; Liebe, J.; van de Zaag, P.; Tilmant, A.; Odai, S. Delineation of small reservoirs using radar imagery in a semi-arid environment: A case study in the Upper East Region of Ghana. *Phys. Chem. Earth, Parts A/B/C* **2009**, *34*, 309–315.
5. Liebe, J.; van de Giesen, N.; Andreini, M. Estimation of small reservoir storage capacities in a semi-arid environment: A case study in the Upper East Region of Ghana. *Phys. Chem. Earth* **2005**, *30*, 448–454.
6. Sawunyama, T.; Senzanje, A.; Mhizha, A. Estimation of small reservoir storage capacities in Limpopo River Basin using geographical information systems (GIS) and remotely sensed surface areas: Case of Mzingwane catchment. *Phys. Chem. Earth* **2006**, *31*, 935–943.
7. Crétaux, J.F.; Birkett, C. Lake studies from satellite radar altimetry. *Comptes Rendus Geosci.* **2006**, *338*, 1098–1112.
8. Gao, H.; Birkett, C.; Lettenmaier, D. Global monitoring of large reservoir storage from satellite remote sensing. *Water Resour. Res.* **2012**, *48*, W09504.
9. Islam, A.; Bala, S.; Haque, M. Flood inundation map of Bangladesh using MODIS time-series images. *J. Flood Risk Manag.* **2010**, *3*, 210–222.
10. McFeeters, S. The use of the Normalized Difference Water Index (NDWI) in the delineation of open water features. *Int. J. Remote Sens.* **1996**, *17*, 1425–1432.
11. Xu, H. Modification of Normalised Difference Water Index (NDWI) to enhance open water features in remotely sensed imagery. *Int. J. Remote Sens.* **2006**, *27*, 3025–3033.
12. Mialhe, F.; Gunnell, Y.; Mering, C. Synoptic assessment of water resource variability in reservoirs by remote sensing: General approach and application to the runoff harvesting systems of south India. *Water Resour. Res.* **2008**, *44*, W05411.
13. Rodrigues, L.; Sano, E.; Steenhuis, T.; Passo, D. Estimation of small reservoir storage capacities with remote sensing in the Brazilian Savannah Region. *Water Resour. Manag.* **2012**, *26*, 1–10.

14. Liebe, J.; van de Giesen, N.; Andreini, M.; Steenhuis, T.; Walter, M. Suitability and limitations of ENVISAT ASAR for monitoring small reservoirs in a semiarid area. *Geosci. Remote Sens.* **2009a**, *47*, 1536–1547.

15. Bamler, R.; Eineder, M. Accuracy of differential shift estimation by correlation and split-bandwidth interferometry for wideband and delta-k SAR systems. *IEEE Geosci. Remote Sens. Lett.* **2005**, *2*, 151–155.

16. Van de Giesen, N. *Characterization of West African Shallow Flood Plains with L-and C-Band Radar*; Owe, M., Brubaker, K., Ritchie, J., Rango A., Eds; IAHS Publication: Wallingford, UK, 2001.

17. Eilander, D.M. Remotely Sensed Small Reservoir Monitoring: A Bayesian Approach. M.sc. Thesis, Delft University of Technology, Delft, The Netherlands, 2013.

18. Hoekman, D.H.; Vissers, M.A. A new polarimetric classification approach evaluated for agricultural crops. *IEEE Trans. Geosci. Remote Sens.* **2003**, *41*, 2881–2889.

19. Besag, J.; York, J.; Mollié, A. Bayesian image restoration, with two applications in spatial statistics. *Ann. Inst. Stat. Math.* **1991**, *43*, 1–20.

20. Strahler, A.H. The use of prior probabilities in maximum likelihood classification of remotely sensed data. *Remote Sens. Environ.* **1980**, *10*, 135–163.

21. Kailath, T. The divergence and Bhattacharyya distance measures in signal selection. *IEEE Trans. Commun. Technol.* **1967**, *15*, 52–60.

22. Cloude, S.; Pottier, E. An entropy based classification scheme for land applications of polarimetric SAR. *IEEE Trans. Geosci. Remote Sens.* **1997**, *35*, 68–78.

23. Yang, J.; Yamaguchi, Y.; Yamada, H.; Sengoku, M.; Lin, S. Optimal problem for contrast enhancement in polarimetric radar remote sensing. *IEICE Trans. Commun.* **1999**, *82*, 174–183.

24. Yang, J.; Yamaguchi, Y.; Boerner, W.M.; Lin, S. Numerical methods for solving the optimal problem of contrast enhancement. *IEEE Trans. Geosci. Remote Sens.* **2000**, *38*, 965–971.

25. Yang, J.; Dong, G.; Peng, Y.; Yamaguchi, Y.; Yamada, H. Generalized optimization of polarimetric contrast enhancement. *IEEE Geosci. Remote Sens. Lett.* **2004**, *1*, 171–174.

26. Henderson, F. Environmental factors and the detection of open surface water areas with X-band radar imagery. *Int. J. Remote Sens.* **1995**, *16*, 2423–2437.

27. Henderson, F.; Lewis, A. *Principles and Applications of Imaging Radar. Manual of Remote Sensing*; John Wiley and Sons: New York, USA, 1998; Volume 2.

28. Valenzuela, G.R. Theories for the interaction of electromagnetic and oceanic waves—A review. *Bound.-Layer Meteorol.* **1978**, *13*, 61–85.

29. Brisco, B.; Short, N.; van der Sanden, J.; Landry, R.; Raymond, D. A semi-automated tool for surface water mapping with RADARSAT-1. *Can. J. Remote Sens.* **2009**, *35*, 336–344.

30. Dubois, P.C.; van Zyl, J.; Engman, T. Measuring soil moisture with imaging radars. *IEEE Trans. Geosci. Remote Sens.* **1995**, *33*, 915–926.

31. Liebe, J.; van de Giesen, N.; Andreini, M.; Walter, M.; Steenhuis, T. Determining watershed response in data poor environments with remotely sensed small reservoirs as runoff gauges. *Water Resour. Res.* **2009**, *45*, W07410.

Sentinel-1 for Monitoring Reservoirs:
A Performance Analysis

Donato Amitrano, Gerardo Di Martino, Antonio Iodice, Francesco Mitidieri,
Maria Nicolina Papa, Daniele Riccio and Giuseppe Ruello

Abstract: In this paper we explore the performances and the opportunities provided by the European satellite Sentinel-1 for water resource management applications in low-income countries. The analysis is supported by a synthetic aperture radar (SAR) simulator, which allowed the quantification of the expected characteristics of Sentinel-1 products in three applications: interferometric digital elevation models (DEMs) generation, land cover mapping and estimation of water volumes retained by small reservoirs. The obtained results quantitatively show that Sentinel-1 data characteristics are fully suitable for most of the application already explored in the recent SAR literature.

Reprinted from *Remote Sens.* Cite as: Amitrano, D.; Di Martino, G.; Iodice, A.; Mitidieri, F.; Papa, M.N.; Riccio, D.; Ruello, G. Sentinel-1 for Monitoring Reservoirs: A Performance Analysis. *Remote Sens.* **2014**, *6*, 10676-10693.

1. Introduction

The United Nations estimated that, by 2025, 1.8 billion people will live in regions characterized by water scarcity and two-thirds of the world's population will be faced with a lack of water [1]. According to recent studies about the growth of global population (up to 8.3 billion in 2030 [2]), in ten years, water scarcity could affect almost 5.5 billion people. This scenario is also exacerbated by desertification (principally caused by unsustainable land management practices) which increases the pressure on water resources worldwide [3]. In fact, desertification, land degradation and drought globally affect 1.5 billion people, 24% of which are in Sub-Saharan Africa [4].

Today, about 66% of Africa is arid or semi-arid and more than one-third of the Sub-Saharan population lives in a water-scarce environment with less than 1000 m^3 available per capita [5]. In these areas, rural population mainly relies on small reservoirs for water harvesting in the rainy season [6]. As an example, in Burkina Faso, it is estimated that 1700 reservoirs are actually employed for agricultural activities, livestock watering and human consumption [7]. They are used to cultivate in counter-season, incrementing the food production and, therefore, the resilience to famine.

Despite their crucial importance, many small reservoirs are neither monitored nor surveyed, due to high costs required by the setup and management of a sensor network. Hence, a complete and up-to-date catalog of location and maximum capacity of the reservoirs is not available, even because they are often built by an initiative of farmers' associations, without general planning.

In this context, remote sensing could be a powerful instrument, allowing a strong reduction of costs and time necessary to obtain relevant information for an effective management of water resources. Thus far, remote-sensing projects have mainly embraced the use of optical data (even thanks to the availability of free LANDSAT imagery), accepting the risk of acquisitions affected by cloud cover, a frequent event in the rainy season.

The use of Synthetic Aperture Radar (SAR) data was rather limited in past applications because of the high costs of data and the complexity of their interpretation by non-expert users. Moreover, the recent SAR literature about water resources monitoring in semi-arid regions is mainly related to sensors (ENVISAT, ERS and more) with low resolution (almost 20 m) imaging capability [8–10].

The most relevant project from the last few years in terms of water resources monitoring is the TIGER initiative, promoted by the European Space Agency (ESA) [11], which aims at assisting African countries to collect, analyze and disseminate water-related geo-information through the exploitation of remotely sensed data. Under the aegis of TIGER, several activities and projects regarding water-related issues have been supported. TIGER thus represented a framework within which European institutions and local African partners shared experiences and expertise stimulating the dissemination and the enforcement of respective knowledge. This led to the achievement of remarkable results as, for example, in the fields of trans-boundary groundwater management [12], land cover mapping [13], water bodies detection [14] and small reservoirs bathymetry [8].

Valuable results have been obtained relating basins' surface areas with retained water volume for northeastern Ghana [9]. These relations are extremely useful since they allow for the extraction of the available water volume through the estimation of a reservoir's surface. This activity can be successfully carried out with remotely sensed data, thereby avoiding expensive bathymetric surveys.

These expressions can be extended to other morphologically similar areas, where bathymetric surveys and/or a suitable DEM are not available. In particular, the almost uniform morphology of the Sahelian region gives the opportunity to use few calibration gauges. This possibility was also supported by the results of the comparison between area–volume relationships extracted from two datasets belonging to different areas of the Sahelian region [15].

The introduction of the new generation of high-resolution sensors (such as COSMO-SkyMed, TerraSAR-X and Sentinel-1) allows for an effective monitoring of small reservoirs. In fact, as discussed in [15], COSMO-SkyMed imagery has been successfully employed for the study of basins with extension of few thousand square meters. The data on water retention at reservoirs can be used also for the implementation and calibration of hydrological models, as suggested in [16]. One of the main limits of COSMO-SkyMed imagery is the significant cost required for acquiring a complete dataset.

The recent launch of Sentinel-1, the new ESA SAR sensor, solves the problems related to the cost of images, since ESA proposes a free data distribution policy. The interpretation of the SAR images is still a limit for the diffusion of this technology, because the image characteristics depend on the geometrical and electromagnetic properties of the observed surfaces. The comprehension of the geometric distortions introduced by the side-looking acquisition mode [17] and of the non-linear electromagnetic scattering phenomena that contribute to the SAR signal formation requires the quantitative knowledge of the interactions between the transmitted electromagnetic field and the physical surfaces of the imaged scene. Therefore, several works in the past literature expressed the necessity of remote sensing processing chains devoted to produce results that could be easily interpreted by the potential end-users [18–20].

In this paper, we present an analysis of the potentiality of Sentinel-1 in water resource monitoring activities. The analysis is supported by a SAR raw signal simulator and is focused on applications concerning water-related problems in semi-arid regions.

The work is organized as follows. In Section 2, we recall the main characteristics and results of the Water Resources Management in Semi-Arid Regions (WARM-SAR) project, which produced significant results for water resource management from high-resolution SAR data. Sections 3–5 deal with an analysis of the new opportunities offered by Sentinel-1: in Section 3, the main characteristics of the mission are recalled; Section 4 is devoted to introducing the SAR simulator used for the quantitative analysis of the Sentinel-1 performances; and, in Section 5, we present the results obtained by processing the simulated Sentinel-1 images. We provide specific results aimed at the production of DEM and related products, land cover maps and monitoring of water volumes. Implications of the use of this new sensor in the hydrological modeling of the study area are discussed throughout the examples. Conclusions are drawn at the end of the work.

2. Water Resources Monitoring in Semi-Arid Regions: The WARM-SAR Project

In this section, we recall the results of the WARM-SAR project, which is devoted to exploring the possible uses of SAR images for water resource monitoring.

WARM-SAR exploited a set of 16 stripmap (3 m resolution) and 7 spotlight (1 m resolution) images with coverage of almost one year and a half, provided at no cost by the Italian Space Agency (ASI) under the aegis of the 2007 COSMO-SkyMed Announcement of Opportunities [21]. The SAR images cover a rectangular area of almost 1600 km² of the Yatenga district in the north of Burkina Faso, a small West African country where nearly 80% of the 14 million inhabitants live in rural areas and the main economic activity consists of subsistence farming and ranching. More than two million people are food insecure and about 34% of the population is subject to chronic malnutrition [22]. The region is characterized by a semi-arid climate, with a rainy season lasting three months.

In the frame of the WARM-SAR project, we developed specific applications for low-income countries characterized by a semi-arid climate. In order to estimate the performance expected by the use of Sentinel-1 data in a similar context, in this section we recall the basic principles of the following applications:

1. Digital Elevation Models (DEMs) generation and related products;
2. Land cover mapping;
3. Monitoring of water volumes retained at reservoirs.

2.1. Digital Elevation Models (DEM) Estimation and Related Products

In [23], a reliable DEM was extracted by means of an interferometric processing of two images acquired at the end of the dry season, when the interferometric coherence is expected to be sufficiently high. The resolution of the DEM (9 m, obtained by the 3-multilook of the 3 m input SAR data) is significantly higher with respect to those previously available: SRTM (resolution 90 m) and ASTER (resolution 30 m). The availability of such a high-resolution DEM allows the estimation of the bathymetry of the small reservoirs that dry up completely at the end of the arid season [15] and the derivation of an analytical relation between reservoirs' surface area and retained volume.

The estimate of the reservoirs bathymetry from remotely sensed imagery provides valuable information about reservoir capacity. In fact, in Sahel, where sedimentation of the reservoirs due to strong soil erosion is extremely fast [24] and quickly changes the topography, there is a lack of updated topographic information. The average sedimentation rates for six reservoirs of the study area were estimated, concluding that, in about 20–30 years, most of the studied reservoirs lost more than 70% of their original capacity [25].

2.2. Land Cover Map Production

The availability of repeated acquisitions offered by new-generation sensors opens new perspectives for remote-sensing applications in semi-arid environments through the definition of new products which fully exploit the particular Sahelian climate [23]. In fact, as discussed in [19], the occurrence of a condition of aridity of the scene at the peak of the dry season makes it possible to establish a reference scenario for the other available images. The procedure for building the Level-1α products is presented in [19] and an example relevant to a small reservoir of the Yatenga district is depicted in Figure 1. The reference image, loaded on the blue band, was acquired at the end of April, i.e., at the peak of the dry season. The test images and the interferometric coherence are assigned to the green and red band, respectively.

This combination of bands makes easy the users' pre-attentive processing, i.e., the unconscious accumulation of information from the environment [26], since it allows for an association between relevant physical features of the scene and the colors restituted by the RGB map, which is consistent with human expectation. With reference to Figure 1, this happens, as an example, for seasonal water, and for seasonal vegetation, which are displayed in blue and green, respectively.

As for the seasonal water, the blue color results from the prevalence of the backscattering from rough soil emerging in the dry season image (blue band) when there is no water in the intake, with respect to the backscattering from flat water surface occurring in wet season images (green band).

As for the seasonal vegetation, the green color results from the prevalence of the backscattering from the vegetation emerging in the wet season images (green band) with respect to the backscattering from rough surfaces occurring in the dry season image (blue band), in absence of vegetation [27].

In both the cases, the coherence (red band) is almost null.

When the electromagnetic response of the reference image and of the test image is comparable, the map exhibits a cyan tonality. Where both the reference and test image are covered by surface water, due to the weak electromagnetic response in these two bands, the composition restitutes the black color (see the dark area in the immediate proximity of the dam in Figure 1).

The main characteristics of the Level-1α products are the congruence with human vision, which allows for an immediate understanding of the more relevant natural cycles of a semi-arid environment (i.e., those of water and of vegetation), and the class detachability. The latter, in particular, makes these products very attractive for supervised classification procedures, even using extremely simple algorithms.

The aforementioned characteristics of Level-1α imagery provide a tool for a quick estimation of the reservoirs' surface areas for several basins, from which it is possible to retrieve the retained water volume, as is explained in Section 5. Thus, these products can be managed by a large variety of scientists and researchers, even non-experts in SAR issues, and could, therefore, potentially increase the use of such data, thereby bringing benefits to water-related research activities.

Figure 1. Level-1α product detail of a small basin near to the Aoérama settlement (test image 31 August 2010).

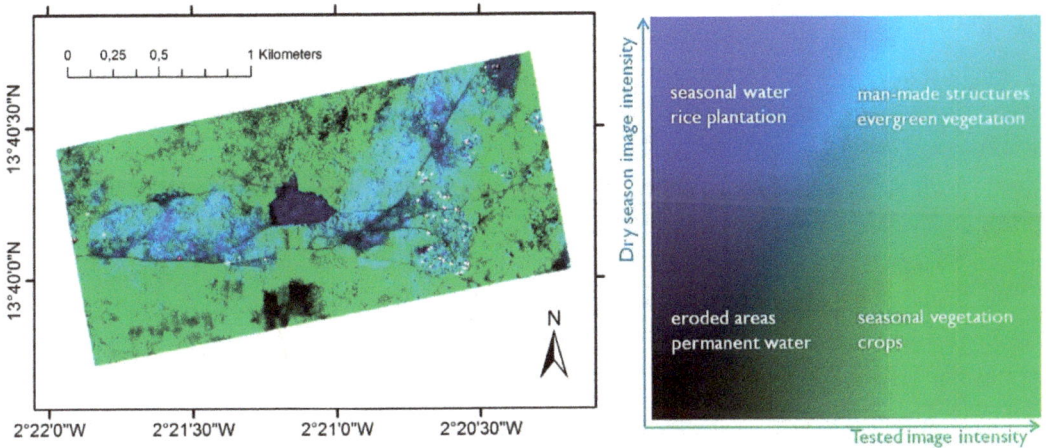

2.3. Monitoring of Water Volumes Retained at Reservoirs

The areas covered by water were extracted by the intensity SAR images despeckled using an optimal weighting multitemporal De Grandi filter [28]. This technique allows amplifying the water—land contrast and producing more accurate information on the extension of water surfaces. A further speckle reduction was obtained by a spatial multilooking, which reduced the images' resolution to 9 m, in accordance with DEM resolution.

The coupled measures of water surface and DEM led to the estimation of water volumes retained by the dam. The retrieved data were used to validate a simple hydrological model [15]. The time evolution of the water storage was computed by the implementation of a balance between the incoming and outgoing water flows.

In order to validate the model, the computed storages are compared with those extracted from SAR data. In spite of the assumptions made and of the uncertainties linked to the estimation of some of the input data, the model was able to catch the overall behavior of the system and, therefore, can be used for the simulation of different scenarios of water abstraction from the reservoirs and, therefore, to optimize the resource management. Another important application of the proposed model is the estimation of the impact of small reservoirs on the downstream flow, which a crucial information in case of water conflicts [8].

3. Sentinel-1: The Mission

Despite the great potentiality offered by WARM-SAR and similar projects, the scaling and the capillary diffusion of the derived products is still constrained by technical and economical limits, especially in low-income countries, where most of the potential beneficiaries of the proposed products can not sustain the cost of commercial data.

The aim of the Sentinel mission is to support the Copernicus program, offering a significant scientific opportunity thanks to: (i) a broad variety of sensing methods, (ii) the compatibility with past ESA mission (in the case of SAR this holds for ERS and ENVISAT) [29], and (iii) the distribution policy, which will guarantee free, full and open access to data. Therefore, the capacity of acquiring, interpreting and processing Sentinels' data will be crucial for the success of projects involving geographical information.

Sentinel-1 is a two-SAR satellite constellation designed to guarantee global coverage with a revisit time of 6 days. The first satellite (Santinel-1A) was launched on 3 April 2014. A 12-meter long radar working in C-band was successfully deployed and is starting to acquire images all over the world. Sentinel-1A is placed in a near-polar, sun-synchronous orbit with a 12-day repeat cycle and 175 orbits per cycle. Both the satellites of the Sentinel-1 constellation share the same orbit plane with a 180° orbital phasing difference.

Sentinel-1 is designed to work with four different possible operative modes: (i) stripmap, (ii) wave, (iii) interferometric wide swath and (iv) extra wide swath. It supports dual polarization acquisition (HH + HV or VV + VH) for all the modes, with the exception of the wave mode (see Table 1). The medium resolution of the stripmap is very close to the resolution of the images used for the development of the products introduced in Section 2. In the following sections, also with the use of a SAR raw signal simulation, we present a study of the expected performances of Sentinel-1 in the design and development of the products presented in the frame of the WARM-SAR project.

Table 1. Sentinel-1 operative modes.

Mode	Swath (km)	Resolution (m × m)	Polarization
Stripmap	80	5 × 5	Dual
Wave	20	5 × 5	Dual
Interferometric WS	250	5 × 20	Dual
Extra WS	400	20 × 40	Single

4. Sentinel-1: SAR simulation

In this section, we introduce the basic concepts of SAR raw signal simulation, which is here used to simulate both Sentinel-1 and COSMO-SkyMed data and to discuss the expected performances of Sentinel-1 in the development of the WARM-SAR applications described in Section 2.

Let x and r be the independent space variables, standing respectively for azimuth and range. By using primed coordinates for the independent variables of the SAR raw signal, $s(x',r')$, this can be expressed as [30]:

$$s(x',r') = \iint \gamma(x,r)g(x'-x,r'-r;r)dxdr, \tag{1}$$

where $\gamma(x,r)$ is the reflectivity pattern of the scene, and $g(x'-x,r'-r;r)$ the unit impulse response of the SAR system [30]. Therefore, in order to obtain $s(x',r')$ we need to evaluate both the reflectivity function and the impulse response of the system.

The evaluation of the reflectivity function requires the use of adequate electromagnetic scattering models, able to provide the solution of interest as a function of the considered sensor and surface parameters. Hence, we need a description of the macroscopic aspects of the surface at the scale of the sensor resolution: this is accomplished by providing as input to the simulator a DEM. The behavior of the DEM is then approximated using a two-scale model [30], *i.e.*, using plane facets, over which a microscopic random roughness is superimposed. Then, the mean square value of the field backscattered from each facet can be evaluated providing an adequate stochastic description of the microscopic roughness. The roughness can be described using parameters resulting from the introduction of different models for the shape of the surface [31]: in the present paper, the roughness is described through a fractal *fractional Brownian motion* (fBm) process [31], *i.e.*, using only two independent parameters, the *Hurst coefficient H* and the *topothesy T* [m]. Finally, to complete the description of the surface, the relative dielectric constant ε and the conductivity σ [S/m] of the observed surface must be provided as input to the simulator. The small perturbation model, with the appropriate fractal power law spectrum, can be then used for the evaluation of the reflectivity function of the surface [31].

Note that the reflectivity function is evaluated in a ground range–azimuth reference system and is necessary to project it in the sensor-centered slant range–azimuth reference system. After this transformation, the obtained reflectivity function is filtered according to the impulse response of the SAR system, providing as output the raw signal, as shown in Equation (1). In order to compute the impulse response, the radar and orbital parameters of the sensor are needed: satellite height h and velocity v, the sensor look angle θ and frequency f_0, the chirp duration τ and bandwidth B, the sampling rate f_s, and the pulse repetition frequency PRF. Finally, after standard focusing, the obtained raw signal provides the final simulated complex SAR image. The block diagram of the algorithm employed to evaluate the raw signal is depicted in Figure 2.

Another important aspect of the simulation procedure is the appropriate inclusion of the speckle effect [32]. Its presence is accounted for thanks to the aforementioned two-scale model. In fact, thanks to this approach, the spatial scales both smaller and larger than the resolution can be treated differently. In particular, the signal macroscopic behavior is described evaluating the mean square value of the field scattered from the plane facets locally approximating the considered surface, assuming knowledge of the microscopic roughness parameters (H and T) and the electromagnetic parameters (ε and σ) of the surface [31]. Conversely, the microscopic behavior, which determines the presence of speckle, is introduced via a statistical model: in this paper, we assume a fully developed speckle [32,33], and the amplitude–signal value obtained for each facet is multiplied by one specific realization of a Rayleigh random variable [30].

In Table 2, the parameters used for the simulation of Sentinel-1 and COSMO-SkyMed images are reported. With these parameters, we obtain a pixel spacing in azimuth–ground range on the final images of 4.66 m × 5 m for Sentinel-1 and 2.08 m × 2.33 m for COSMO-SkyMed.

Figure 2. Block diagram of the SAR raw signal simulation. Radar (RD) and orbital (OD) data, along with the geometric (z) and electromagnetic (ε and σ) parameters of the surface are the required inputs.

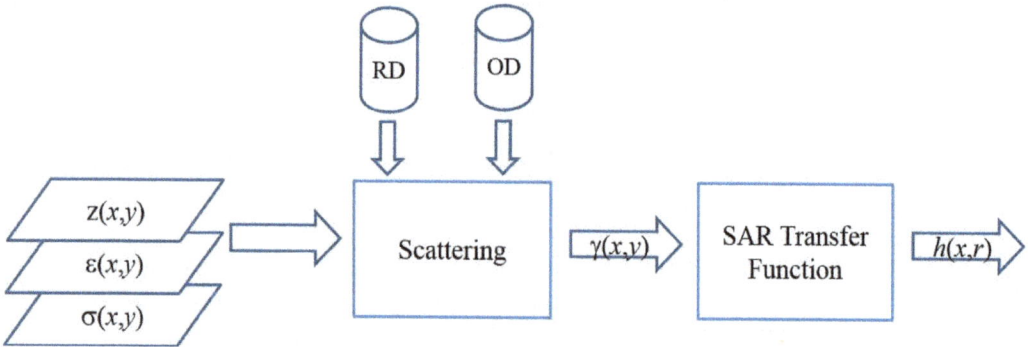

Table 2. Simulation parameters.

Simulation Parameter	Sentinel-1	COSMO-SkyMed
h [km]	693	619.6
v [m/s]	7	7.5
θ [°]	30	33.6
f_0 [GHz]	5.405	9.3
τ [μs]	35	35
B [MHz]	60	65.64
f_s [MHz]	60	116.25
PRF [Hz]	1500	3612.7

5. Sentinel-1: Opportunities

The simulator presented in the previous section allows for a quantitative estimation of the performances of Sentinel-1 mission in water resource management applications. In the following, some applications will be discussed and the quality of the expected products compared with those obtained in the framework of the WARM-SAR project (see Section 2).

5.1. DEM Estimation and Related Products

The use of interferometric techniques for the evaluation of a DEM is in principle replicable with Sentinel-1 data. The difference in spatial resolution in stripmap mode (3 m for COSMO-SkyMed, 5 m for Sentinel-1) will cause a corresponding reduction of the resolution of the DEM from 9 m to 15 m. Such a limitation is not significant for many of the applications proposed in the WARM-SAR project.

As discussed in Section 2, both optical and SAR data can provide observations of the reservoirs' retention area, while, in the absence of bathymetric information, the retained volumes remain unknown. To overcome this limitation, it is possible to use relationships between reservoirs' storage volumes and areas, which in case of homogeneous morphology can be valid for an entire region. For instance, an area–volume relation was developed in literature [34] performing an extensive bathymetric survey in the Upper East Region of Ghana:

$$Volume = 0.00857\ area^{1.4367} \qquad (2)$$

Starting from SAR-derived data, a regression analysis was applied in order to obtain an area–volume relationship for the Yatenga district [15]:

$$Volume = 0.1012\ area^{1.167} \qquad (3)$$

In order to assess the potentiality of Sentinel-1 data for this kind of application, the DEM acquired in the WARM-SAR project was resampled at the Sentinel-1 scale and a new dataset of area–volume for the Laaba reservoir was built. This dataset was compared to the one obtained for the same reservoir using the DEM extracted from COSMO-SkyMed satellite imagery in the WARM-SAR project framework.

The analysis of the DEMs was performed computing the areas of water surfaces at fixed heights, with a step of 0.3 m. The water volumes stored in the reservoir for each step were calculated as a sum of those contained in each pixel of the water surface, which is assimilated to a water column, whose height h_{wc} is given by:

$$h_{wc} = h_c - h \qquad (4)$$

where h_c is the elevation of the equipotential surface traced by the basin, and h is the DEM height corresponding to the considered pixel. The water volume V contained in the whole reservoir is then estimated as the summation of all the volume contributions, given as the product of the pixel area S_i and the water column height h_{wc}:

$$V = \sum_i S_i \times h_{wc} \qquad (5)$$

In Figure 3, the results obtained through the DEM analysis and the application of the area–volume relationship proposed in [15] for the Yatenga region are plotted. The overall trend is similar for both the analyzed DEMs. An apparent anomaly occurs for low water surface areas in the Sentinel-like DEM (for example, three blue dots refer to the same area but to different volumes). This is because the lowest area values correspond to few pixels, which in the Sentinel-1 DEM have a surface of 225 m². The higher resolution of COSMO-SkyMed-derived DEM made possible to compute more reliable values of the retained volume for such areas. This effect depends both on the morphology of the region and on the interpolation that produced differences in height between adjacent pixels greater than 0.3 m.

The above described effect allows the estimation of the range of areas that can be safely retrieved by Sentinel imagery. In principle, reservoirs covered by some thousands of square meters can be effectively monitored. We can also remark that the area–volume relationship proposed in [15] underestimates the reservoir volume when the water surface is in the range between 1000 and 20,000 km².

Figure 3. Laaba's storage volumes as a function of the corresponding surface areas.

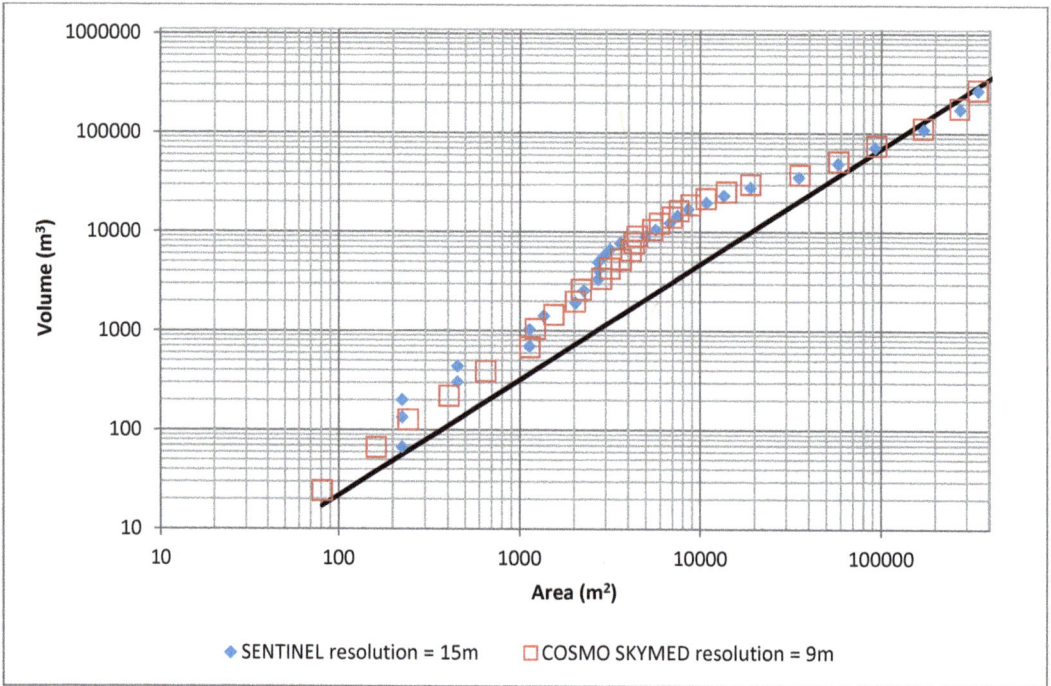

◆ SENTINEL resolution = 15m   ☐ COSMO SKYMED resolution = 9m

5.2. Land Cover Map Production

In this paragraph, we use the simulator presented in Section 4 to quantitatively evaluate how the land cover products developed in the frame of WARM-SAR can be replicated with Sentinel-1 data.

Table 3. Parameters used in the simulations.

Simulation Parameter	Terrain	Water
ε	4	40
σ [S/m]	0.001	1
H	0.8	0.75
s [m^{1-H}]	0.1	0.01

As an example, we simulate the temporal evolution of the Laaba basin's water levels, by using an approach similar to that implemented in [34] for the simulation of flooded scenes. In particular, we modify the initial DEM (acquired when the basin is empty, see Section 2.2) reproducing the progressive filling of the basin with water-height steps of 0.5 m. We set the roughness (Hurst coefficient H and standard deviation at unitary distance s) and electromagnetic parameters (dielectric permittivity ε and conductivity σ) of the terrain and water-filled area to typical values (see Table 3). In Figure 4, we show the DEMs with different water levels that were given as input to the simulator. It is worthwhile to note that with respect to the DEM presented in [23], a local adaptive filter (see [35]) has been applied in order to reduce the noise due to the presence of vegetation on the boundary of the water surface.

Figure 4. Laaba basin, 3D representation of the DEM used for SAR simulation. Water contour height: 341 m (**a**), 340.5 m (**b**), 340 m (**c**), 339.5 m (**d**), 339 m (**e**) and 338.5 m (**f**). A vertical stretching factor has been applied for visualization purposes.

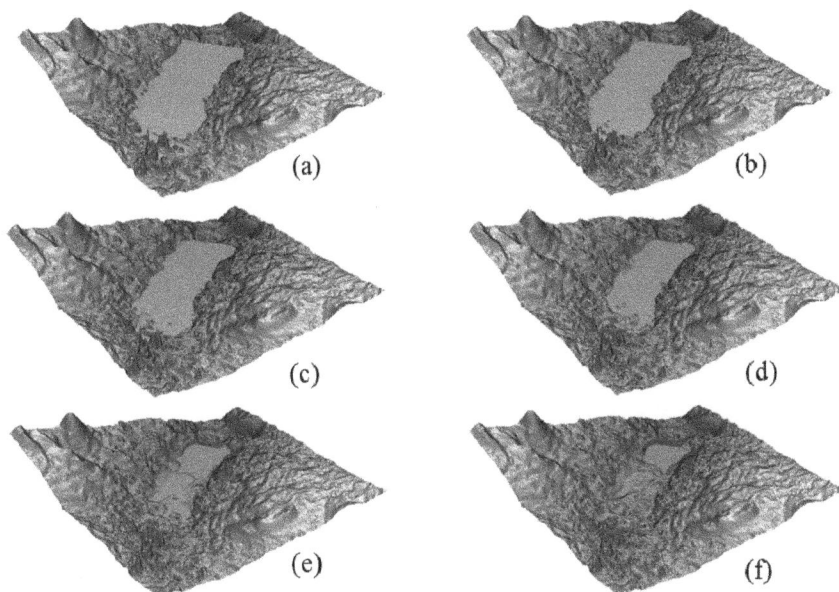

The simulated data have been processed in accordance with the approach presented in [19], where Level-1α products were defined. In Figure 5, we show the simulated Level-1α products for a scenario relevant to the Laaba reservoir, filled with different water levels. A simple classification rule, as explained in Section 2.4, allows the extraction of the water surface from the different acquisitions. It is worthwhile noting that, except for the area covered by water, the Level-1α products exhibit a substantial balance between the test and reference intensity channels, providing an almost homogeneous cyan background. This is due to the fact that the simulator does not take into account the presence of vegetation, and the dielectric constant of terrains outside the reservoir was fixed to a constant value in all the simulated images.

The contrast between water and terrain backscattering is very similar to that of the actual COSMO-SkyMed data. This result is supported by the analysis of the expected backscattering at the X- and C-bands.

In Figure 6, we provide the backscattering coefficient of a typical rough soil, characterized by Hurst coefficient $H = 0.8$ and standard deviation at unitary distance $s = 0.1$ m^{1-H}, as a function of the incidence angles typical of remote-sensing applications (from 15 to 45 degrees). The curves have been evaluated with the small perturbation method (SPM) [31] for the HH polarization. The backscattering behavior is similar at the X- and C-bands. Similar results can be obtained at VV polarization.

Figure 5. Laaba basin, simulated Sentinel-1 Level 1α products. Water contour height: (**a**) 341 m, (**b**) 340.5 m, (**c**) 340 m, (**d**) 339.5 m, (**e**) 339 m and (**f**) 338.5 m. The spatial resolution is 5 × 5 m.

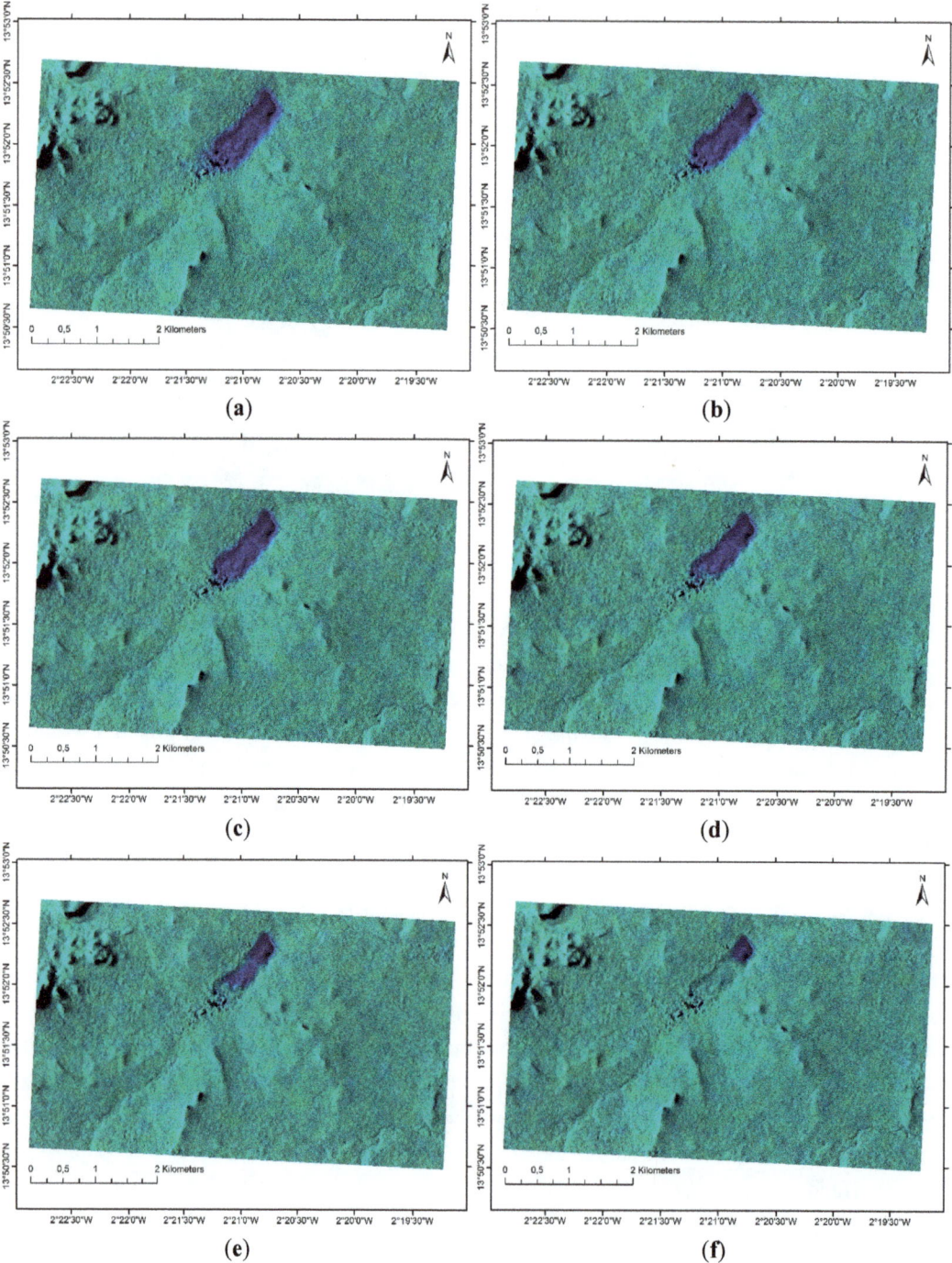

Figure 6. Backscattering coefficient from a rough terrain as a function of the incidence angle, evaluated with the SPM at X- (blue line) and C- (red line) bands.

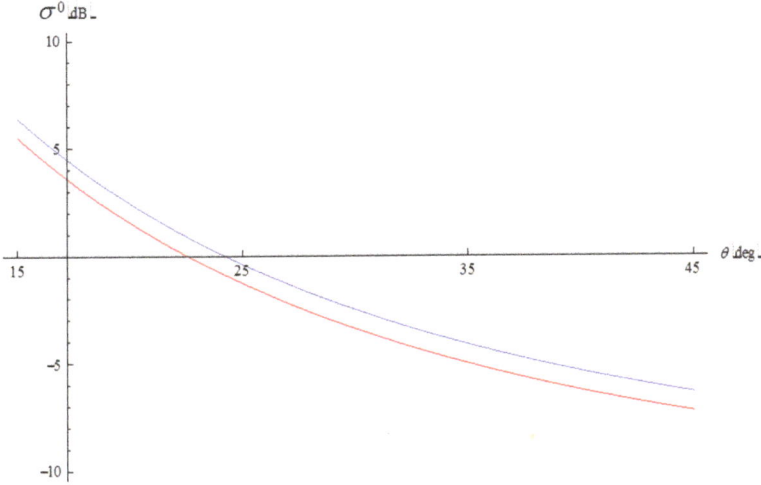

For water surface identification purposes, it is reasonable to analyze how the contrast between the water and the terrain backscattering is influenced by the frequency. In Figure 7, we plot the ratio between the backscattering coefficient of a rough terrain and of a water surface (see the parameters defined in Table 3) as a function of the incidence angle. The result shown in Figure 7 shows that, despite a small reduction, the contrast is still sufficient (almost 17 dB in the considered range of angles) to identify the area covered by water with the techniques presented in the frame of the WARM-SAR project.

Figure 7. Ratio between the backscattering coefficient of a rough terrain and of surface water as a function of the incidence angle evaluated with the SPM at X- (blue line) and C- (red line) bands.

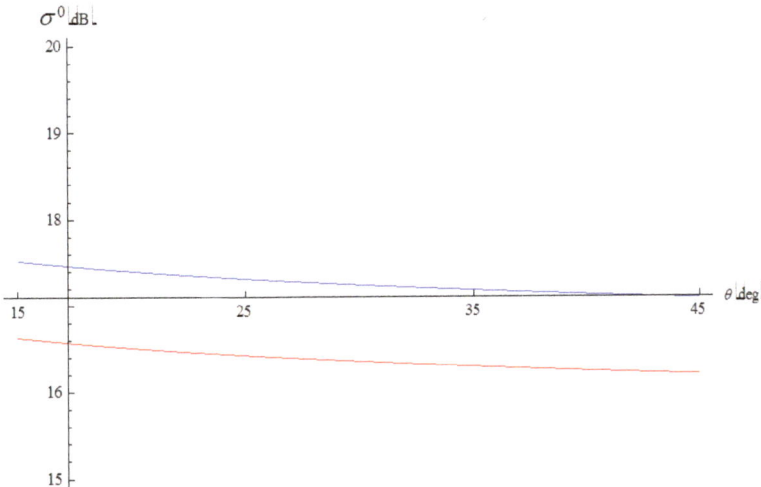

5.3. Monitoring of Water Volumes Retained at Reservoirs

In order to evaluate the potentiality of Sentinel-1 in measuring water volumes in semi-arid regions, we compared a series of simulated COSMO-SkyMed images superimposed to a 9 m spatial resolution DEM and a series of simulated Sentinel-1 images superimposed to a 15 m spatial resolution DEM for different levels of the water contour height. The results of the comparison are shown in Figure 8 for the Laaba reservoir. Such a comparison provides an estimation of the effect of the loss in spatial resolution introduced by Sentinel-1 with respect to the COSMO-SkyMed products used in the WARM-SAR project.

Figure 8. Laaba basin: Surface water as function of the contour height.

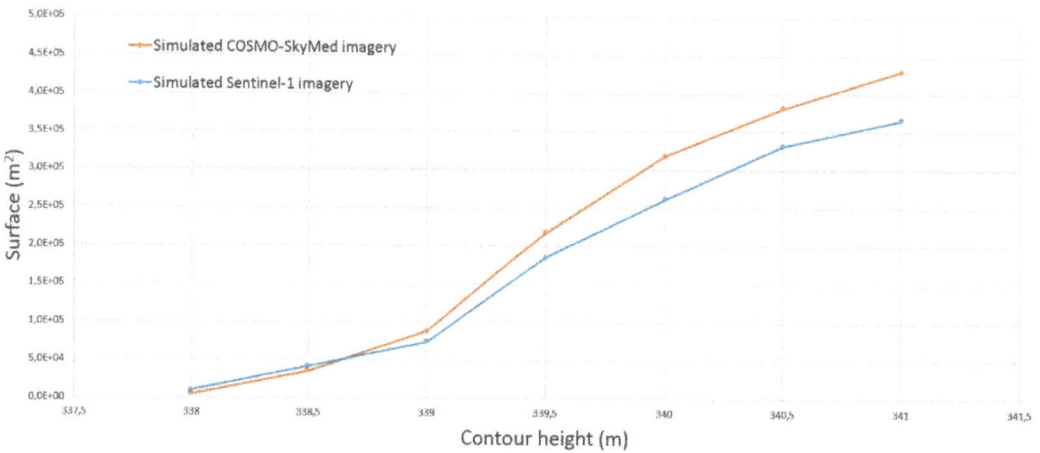

Figure 9. Laaba basin: retained water volume as function of the contour height.

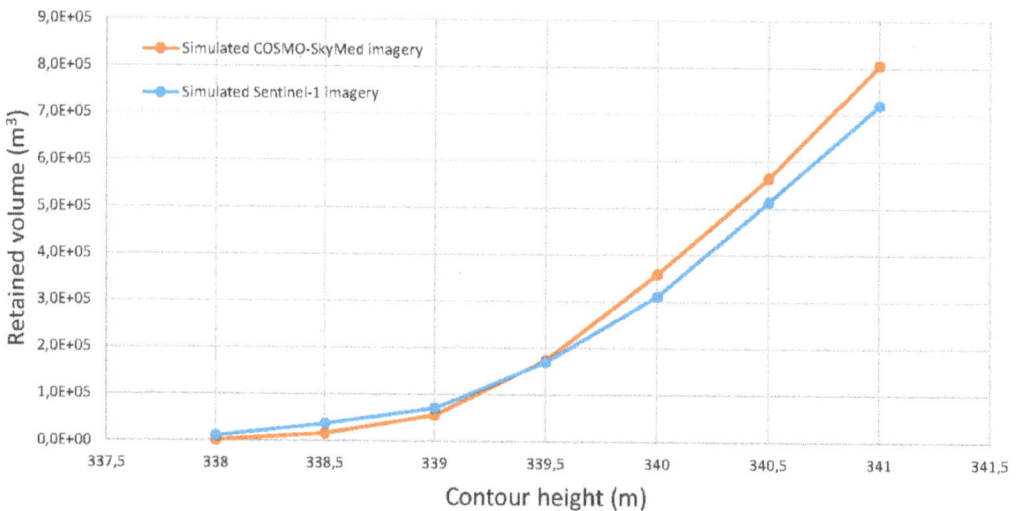

As expected, the discrepancy between Sentinel-1 (blue curve) and COSMO-SkyMed (orange curve) arises especially when the basin tends to fill up. In fact, when the water level increases, its contour is better delineated by the higher COSMO-SkyMed resolution. The error introduced by the loss in spatial resolution is in the order of 15%–20% and it is almost constant for water contour height ∈ [339.5, 341].

The estimated surface is then used for computing the water volume retained into the basin considering each pixel belonging to the water mask extracted from the simulated Level-1α products as a water column, as explained in Section 5.1.

Results of this analysis are shown in Figure 9. The errors in the estimation of the surface areas do not affect significantly the retained volume calculation since they are mainly located at the boundary of the basin, where the contributions to the summation of (4) are small.

6. Conclusions

In this paper we explored the potentiality of the Sentinel-1 mission in providing data and value-added information for water resource management applications. The study has been supported by the use of a SAR raw signal simulator, achieving an innovative framework for the analysis of Sentinel-1 performances.

The results expected by the use of 5-meter resolution Sentinel-1 images in water-related projects were compared to those obtained by the use of 3-meter resolution COSMO-SkyMed imagery in the previously developed WARM-SAR project. The SAR raw signal simulator allowed the estimation of the expected differences of performances due to resolution and frequency. The quantification of the differences in the scattering mechanisms between the X- and C-bands was also presented.

Three classes of applications were investigated: (i) the creation of an interferometric DEM and related products, (ii) the production of land cover maps, and (iii) the monitoring of water volumes retained at reservoirs.

As for the interferometric applications, the ratio between Cosmo-SkyMed and Sentinel-1 resolution is 0.6. Such a factor quantifies also the loss of resolution on the production of the DEM. In addition, we achieved a quantitative estimation of the degradation of performances that would be obtained by replicating most of the products obtained in WARM-SAR with Sentinel-1 data. In particular, we experimented that the Sentinel-1 and COSMO SkyMed derived DEM provide comparable results in the extraction of the relation between reservoir surface area and retained water volume for all the reservoirs whose extension is higher than one thousand square meters.

As for the production of land cover maps, the SAR raw signal simulator led us to simulate, present and interpret Sentinel-1 Level-1α products. Appropriate electromagnetic models were used to demonstrate that the radiometric contrast between water and rough surfaces at the C band is about 2 dB lower than at X band for incidence angles in the range between 15° and 45°. Such a result demonstrates that land cover maps obtained by Sentinel-1 will guarantee class separation comparable to that obtained by Cosmo-SkyMed images.

As for the monitoring of water volumes retained by reservoirs, we experimented that the loss in spatial resolution introduced by Sentinel-1 imagery produces a discrepancy in the order of 15%–20% in the estimation of reservoir surface area with respect to COSMO-SkyMed images. However, this error is reduced to about 12% in the estimation of retained volumes since it is limited at the reservoirs' borders.

In synthesis, this work presented a new framework for simulating Sentinel-1 data and related products. It provides the opportunity to predict and control quality parameters of the Sentinel-1 SAR data and products, whose high informative content, conjugated with the open access policy, represents an extraordinary opportunity for future projects, mainly in low-income countries.

Acknowledgments

The SAR images, at the basis of the study, were provided by the Italian Space Agency (ASI) under the aegis of the 2007 COSMO-SkyMed AO Project "Use of High Resolution SAR Data for Water Resource Management in Semi Arid Regions" and of the HydroCIDOT project.

Author Contributions

All authors contributed extensively to the work presented in this paper. Donato Amitrano, Gerardo Di Martino, Antonio Iodice, Daniele Riccio and Giuseppe Ruello were in charge of modeling, simulating and processing SAR data. Maria Nicolina Papa and Francesco Mitidieri were in charge of implementing the hydrological models.

Conflicts of Interest

The authors declare no conflict of interest.

References

1. UNWater Water Scarcity. Available online: http://www.unwater.org/fileadmin/user_upload/watercooperation2013/doc/Factsheets/water_scarcity.pdf (accessed on 28 October 2013).
2. United Nations Department of Economic and Social Affairs (UNDESA). *Population Division World Population Prospects*; The 2008 Revision, Highlights; UNDESA: New York, NY, USA, 2009.
3. UNESCO. *The United Nations World Water Development Report 4: Managing Water under Uncertainty and Risk*; UNESCO: Paris, France, 2012.
4. International Crops Research Institute for the Semi-Arid Tropics (ICRISAT). *Climate Change and Desertification Put One Billion Poor People at Risk*; ICRISAT: Hyderabad, India, 2008.
5. New Partnership for Africa's Development (NEPAD). *Water in Africa: Management Options to Enhance Survival and Growth*; United Nations Economic Commission for Africa (UNECA): Addis Ababa, Ethiopia, 2006.
6. Eilander, D.; Annor, F.O.; Iannini, L.; van de Giesen, N. Remotely sensed monitoring of small reservoir dynamics : A bayesian approach. *Remote Sens.* **2014**, *6*, 1191–1210.
7. Towards an Atlas of Lakes and Reservoirs in Burkina Faso. Available online: http://www.smallreservoirs.org/full/toolkit/docs/IIa%2002%20Faso%20MAB_ML.pdf (accessed on 4 November 2014).

8. Annor, F.O.; van de Giesen, N.; Liebe, J. Monitoring of small reservoirs storage using envisat ASAR and spot imagery in the upper east region of Ghana. In *Application of Satellite Remote Sensing to Support Water Resources Management in Africa: Results from the TIGER Initiative*; IHP-VII Technical Documents in Hydrology; UNESCO: Paris, France, 2010.

9. Liebe, J.R.; van de Giesen, N.; Andreini, M.S.; Steenhuis, T.S.; Walter, M.T. Suitability and limitations of ENVISAT ASAR for monitoring small reservoirs in a Semiarid Area. *IEEE Trans. Geosci. Remote Sens.* **2009**, *47*, 1536–1547.

10. Alsdorf, D.E.; Rodriguez, E.; Lettenmaier, D.P. Measuring surface water from space. *Rev. Geophys.* **2007**, *45*, 1–24.

11. The TIGER Initiative 2009–2012 Report. Looking for Warter in Africa. Available online: http://www.tiger.esa.int/files/pdf/tiger_report_single_pages_lowres.pdf (accessed on 28 October 2014).

12. Saradeth, S.; Dodo, A.K.; Latrech, D. Earth observation in support of management of internationally shared groundwater resources in Africa: The AQUIFER project. In *Application of Satellite Remote Sensing to Support Water Resources Management in Africa: Results from the TIGER Initiative*; IHP-VII Technical Documents in Hydrology; UNESCO: Paris, France, 2010.

13. Thomas, A.; Ayuk, J. Land use/land cover mapping of the Kuils-Eerste River catchment (Western Cape) through an integrated approach using remote sensing and GIS. In *Application of Satellite Remote Sensing to Support Water Resources Management in Africa: Results from the TIGER Initiative*; IHP-VII Technical Documents in Hydrology; UNESCO: Paris, France, 2010.

14. Arledler, A.; Castracane, P.; Marin, A.; Mica, S.; Pace, G.; Quartulli, M.; Vaglio Laurin, G.; Alfari, I.; Trebossen, H. Detecting water bodies and water related features in the Niger basin area by SAR data: The ESA TIGER WADE project. In *Application of Satellite Remote Sensing to Support Water Resources Management in Africa: Results from the TIGER Initiative*; IHP-VII Technical Documents in Hydrology; UNESCO: Paris, France, 2010.

15. Amitrano, D.; Ciervo, F.; Di Martino, G.; Papa, M.N.; Iodice, A.; Koussoube, Y.; Mitidieri, F.; Riccio, D.; Ruello, G. Modeling watershed response in semi-arid regions with high resolution synthetic aperture radars. *IEEE J. Sel. Top. Appl. Earth Obs.* **2014**, *7*, 2732–2745.

16. Liebe, J.R.; van de Giesen, N.; Andreini, M.; Walter, M.T.; Steenhuis, T.S. Determining watershed response in data poor environments with remotely sensed small reservoirs as runoff gauges. *Water Resour. Res.* **2009**, *45*, doi:10.1029/2008WR007369.

17. Toutin, T. Geometric processing of remote sensing images: Models, algorithms and methods. *Int. J. Remote Sens.* **2004**, *25*, 1893–1924.

18. Gaetano, R.; Amitrano, D.; Masi, G.; Poggi, G.; Verdoliva, A.; Ruello, G.; Scarpa, G. Exploration of Multitemporal COSMO-SkyMed data via tree-structured MRF segmentation. *IEEE J. Sel. Top. Appl. Earth Obs.* **2014**, *7*, 2763–2775.

19. Amitrano, D.; Di Martino, G.; Iodice, A.; Riccio, D.; Ruello, G. A New framework for SAR multitemporal data RGB representation: Rationale and products. *IEEE Trans. Geosci. Remote Sens.* **2015**, *53*, 117–133.

20. Datcu, M.; Seidel, K. Human centered concepts for exploration and understanding of earth observation images. *IEEE Trans. Geosci. Remote Sens.* **2005**, *43*, 52–59.

21. Di Martino, G.; Iodice, A.; Natale, A.; Riccio, D.; Ruello, G.; Zinno, I.; Koussouben, Y.; Papa, M.N.; Ciervo, F. COSMO-SkyMed AO Projects—Use of high resolution SAR data for water resource management in Semi Arid Regions. In Proceedings of the IEEE International Geoscience and Remote Sensing Symposium, Munich, Germany, 22–27 July 2012; pp. 1212–1215.

22. The Food and Nutrition Crisis in the Sahel. Urgent Action to Support the Resilience of Vulnerable Populations. Available online: http://www.fao.org/fileadmin/user_upload/emergencies/docs/DocProg%20FAO%20SAHEL%20EN%20short%20version.pdf (accessed on 28 October 2014).

23. Amitrano, D.; Di Martino, G.; Iodice, A.; Ruello, G.; Ciervo, F.; Papa, M.N.; Koussoube, Y. Effectiveness of high-resolution SAR for water resource management in low-income semi-arid countries. *Int. J. Remote Sens.* **2014**, *35*, 70–88.

24. Grimaldi, S.; Angeluccetti, I.V.; Coviello, V.; Vezza, P. Cost-effectiveness of soil and water conservation measures on the catchment sediment budget—The Laaba watershed case study, Burkina Faso. *Land Degrad. Dev.* **2013**, *2013*, doi:10.1002/ldr.2212.

25. Warren, A.; Batterbury, S.; Osbahr, H. Soil erosion in the West African Sahel: A review and an application of a "local political ecology" approach in South West Niger. *Glob. Environ. Chang.* **2001**, *11*, 79–95.

26. Healey, C.; Booth, K.S.; Enns, J. Visualizing real-time multivariate data using preattentive processing. *ACM Trans. Model. Comput. Simul.* **1995**, *5*, 190–221.

27. Fung, A.K. Scattering from a vegetation layer. *IEEE Trans. Geosci. Elect.* **1979**, *17*, 1–6.

28. De Grandi, G.F.; Leysen, M.; Lee, J.-S.; Schuler, D. Radar reflectivity estimation using multiple SAR scenes of the same target: Technique and applications. In Proceedings of the IEEE International Geoscience and Remote Sensing Symposium, Singapore, 3–8 August 1997; pp. 1047–1050.

29. Malenovský, Z.; Rott, H.; Cihlar, J.; Schaepman, M.E.; García-Santos, G.; Fernandes, R.; Berger, M. Sentinels for science: Potential of Sentinel-1, -2, and -3 missions for scientific observations of ocean, cryosphere, and land. *Remote Sens. Environ.* **2012**, *120*, 91–101.

30. Franceschetti, G.; Migliaccio, M.; Riccio, D.; Schirinzi, G. SARAS: A synthetic aperture radar (SAR) raw signal simulator. *IEEE Trans. Geosci. Remote Sens.* **1992**, *30*, 110–123.

31. Franceschetti, G.; Riccio, D. *Scattering, Natural Surfaces and Fractals*; Academic Press: Burlington, MA, USA, 2007.

32. Goodman, J.W. Some fundamental properties of speckle. *J. Opt. Soc. Am.* **1976**, *66*, doi:10.1364/JOSA.66.001145.

33. Di Martino, G.; Iodice, A.; Riccio, D.; Ruello, G. A physical approach for SAR speckle simulation: First results. *Eur. J. Remote Sens.* **2013**, *46*, 823–836.

34. Liebe, J.R.; van de Giesen, N.; Andreini, M.S. Estimation of small reservoir storage capacities in semi-arid environment: A case study in the Upper East Region of Ghana. *Phys. Chem. Earth.* **2005**, *30*, 448–454.

Evaluating MERIS-Based Aquatic Vegetation Mapping in Lake Victoria

Elijah K. Cheruiyot, Collins Mito, Massimo Menenti, Ben Gorte, Roderik Koenders and Nadia Akdim

Abstract: Delineation of aquatic plants and estimation of its surface extent are crucial to the efficient control of its proliferation, and this information can be derived accurately with fine resolution remote sensing products. However, small swath and low observation frequency associated with them may be prohibitive for application to large water bodies with rapid proliferation and dynamic floating aquatic plants. The information can be derived from products with large swath and high observation frequency, but with coarse resolution; and the quality of so derived information must be eventually assessed using finer resolution data. In this study, we evaluate two methods: Normalized Difference Vegetation Index (NDVI) slicing and maximum likelihood in terms of delineation; and two methods: Gutman and Ignatov's NDVI-based fractional cover retrieval and linear spectral unmixing in terms of area estimation of aquatic plants from 300 m Medium Resolution Imaging Spectrometer (MERIS) data, using as reference results obtained with 30 m Landsat-7 ETM+. Our results show for delineation, that maximum likelihood with an average classification accuracy of 80% is better than NDVI slicing at 75%, both methods showing larger errors over sparse vegetation. In area estimation, we found that Gutman and Ignatov's method and spectral unmixing produce almost the same root mean square (RMS) error of about 0.10, but the former shows larger errors of about 0.15 over sparse vegetation while the latter remains invariant. Where an endmember spectral library is available, we recommend the spectral unmixing approach to estimate extent of vegetation with coarse resolution data, as its performance is relatively invariant to the fragmentation of aquatic vegetation cover.

Reprinted from *Remote Sens.* Cite as: Cheruiyot, E.K.; Mito, C.; Menenti, M.; Gorte, B.; Koenders, R.; Akdim, N. Evaluating MERIS-Based Aquatic Vegetation Mapping in Lake Victoria. *Remote Sens.* **2014**, *6*, 7762-7782.

1. Introduction

Aquatic weed infestation is one of the major environmental challenges globally. The weeds, which mostly comprise water hyacinth and hippo-grass, are associated with many adverse effects. Continuous observation and monitoring of their proliferation is essential for proper lake management and weed control [1]. Remote sensing information has increasingly become essential for water resource management. It is a powerful tool for studying large scale phenomena in aquatic vegetation communities, and is capable of delivering timely information unmatched by any other surveying technique [2]. High resolution data such as those acquired by IKONOS and Korea Multi-Purpose Satellite (KOMPSAT) satellites provide detailed information but are associated with low observation frequency and small swath, and their cost may be prohibitively expensive for large area assessments [3]. A common remote sensing practice is to mosaic spatially adjacent images that are acquired within a short temporal range to produce extensive cover maps. Rapid proliferation and the dynamic nature of floating aquatic vegetation have the implication that mosaicking is not suitable for aquatic environments with freely floating vegetation. Coarse resolution products such as Medium Resolution Imaging Spectrometer (MERIS) and MODIS provide a wider view and higher data frequency at the expense of spatial details. Because of this inherent trade-off, it may be appropriate to use coarse resolution data for continuous frequent observation of aquatic vegetation, and occasionally use the high resolution data to assess the quality of derived information.

Frequent and accurate monitoring of aquatic vegetation is not only essential in providing reliable and timely information to the lake management authorities for sustained water resource management [4], but also in improving the quality of related studies which rely on this information for their analyses. For example a study to evaluate the effect of nutrient influx on vegetation proliferation would require adequately accurate information on the extent and location of aquatic plants. Assessing the accuracy of remotely derived information allows users to ascertain their reliability, and it is a means through which the producers communicate product limitations to users, leading to appropriate use of the information [5]. According to [6], accuracy assessment of remote sensing map products has evolved in four developmental stages. It started with visual assessment of images to determine whether the classification results were good or not. It improved to the stage where an overall non-site-specific percentage accuracy was provided, and further to a site-specific accuracy assessment. Finally, a more detailed analysis of the site-specific accuracy assessments emerged, for example the use of error/confusion matrix and kappa coefficients. Error matrix has become one of the most commonly used method of classification accuracy assessment, with several applications in land use/land cover mapping, for example [7] and [8]. This method requires manual identification of reference sites/pixels from homogeneous surfaces, which are assumed to represent pure feature classes (endmembers) [9]. Error matrix also allows the use of "Pareto Boundary" analysis of the trade-off between commission and omission error, in order to determine the optimal classification performance that can be obtained for a specific low resolution data.

Several algorithms have been developed for the remote retrieval of biophysical characteristics of vegetation. The authors of [1] used an unsupervised clustering technique with thresholds based on a "wetness index", to identify water hyacinth and water hyacinth-free areas in Lake Victoria. The authors of [10] applied Minimum Distance—a simple parametric classification algorithm to identify floating vegetation areas, before applying a spectral linear mixture model—a sub-pixel analysis to discriminate different vegetation species according to the weed spectral behaviour. The most widely used method, however, is the mathematical combination of visible and near-infrared reflectance bands in the form of spectral vegetation indices [11]. Vegetation indices are widely used because of their computational simplicity. Many studies on Lake Victoria have used vegetation indices, for example, [12] used NDVI to investigate the dependency of hyacinth biomass production on nutrients levels. In [13], the authors used a time-series of NDVI to evaluate the link between the occurrence of El Nino events in East Africa and water hyacinth blooms in Winam Gulf section of Lake Victoria. The authors of [14] used NDVI and several of its derivatives to monitor Lake Victoria's water level and drought conditions. More recently [15] used NDVI to map vegetation distribution in the lake, and to develop a floating vegetation index for quantifying its surface extent.

Vegetation cover estimates obtained with remote sensing methods can provide useful decision support information required for the control of aquatic plants proliferation. This information can only be useful if the practitioners have a way of ascertaining its accuracy. Little information is available on the accuracy of aquatic vegetation cover estimates derived at coarse resolution. The accuracy of a method in detection of terrestrial vegetation and aquatic vegetation may be different because of the difference in backgrounds. Water, which forms the background to aquatic vegetation, has a stronger absorption of electromagnetic radiation than soil. Further, the dynamic nature of the floating aquatic vegetation introduces a unique case to the evaluation of detection accuracy of aquatic vegetation. Floating vegetation is carried away by tides and wind, making it difficult to identify sampling sites where manual vegetation cover estimates can be made. Unlike in the case of terrestrial vegetation where the target is stationary, it is technically challenging to collect reference *in situ* measurements. A viable alternative is to use as reference results obtained with finer resolution remote sensing products. In addition, the reference image must have its acquisition time as close as possible, ideally in the order of minutes, to that of the classified image. Obtaining such pair of data is perhaps the biggest challenge in the assessment of floating aquatic vegetation classifications.

Our objective is to evaluate the performance of algorithms commonly used to monitor aquatic plants in extensive water bodies, in terms of their accuracy in detecting aquatic plants from coarse resolution remote sensing data—in this case the 300 m resolution MERIS data. MERIS sensor ended data acquisition in March 2013, but its products are good test data for the anticipated Sentinel 2/3 products. In terms of delineation, we evaluate two methods: NDVI slicing—with special focus on the empirical slicing proposed by [15], and maximum likelihood classifier. In terms of area estimation, we evaluate two methods: NDVI-based fractional cover retrieval model proposed by [16], and linear spectral unmixing (LSU)—which is a form of spectral mixture analysis. We also aim at assessing the suitability of using higher resolution data as reference in assessing the quality of aquatic vegetation cover information obtained with coarse resolution products.

2. Study Area

Our study area is the Winam Gulf section of Lake Victoria. Lake Victoria is a large fresh water body in East Africa. It stretches 412 km from north to south between 0°30′N and 3°12′S and 355 km from west to east between 31°37′E and 34°53′E. The lake, which is the largest of all African lakes, is also the second largest freshwater body in the world by area, with an extensive surface area of 68,800 km^2. Figure 1 shows the geographic location of Lake Victoria.

We focus on the Winam Gulf because this almost enclosed shallow section of the lake is more vulnerable to vegetation invasion perhaps due to high levels of eutrophication. Earlier work of [15] show that vegetation proliferation is preceded by about two months by high levels of water quality indicators such as total suspended matter (TSM) and phytoplankton measured as a Chlorophyll-a (Chl-a) index. The most prevalent of these invasive weeds include the non-native water hyacinth and hippo-grass. The weeds are associated with many adverse effects which include obstruction to fishing, navigation and irrigation, interference with the aquatic biodiversity [17,18], water quality deterioration and a general risk to public health [19]. The lake is an important economic resource to the three riparian countries, Kenya, Tanzania and Uganda, through fishing and transport, as well as providing a livelihood for the local communities.

Figure 1. Geographic location of Lake Victoria in East Africa (**a**) and Africa (**b**), (**c**) Zoomed-in Winam Gulf section of the lake. Source: *Google Maps*.

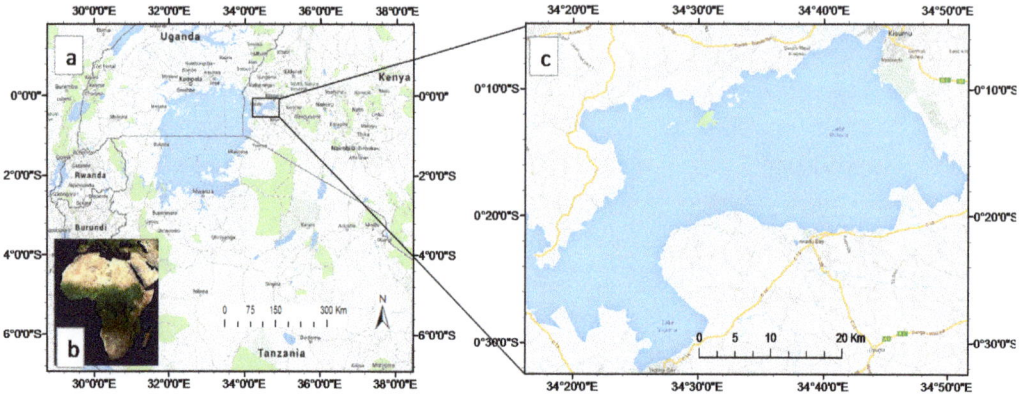

3. Materials and Method

3.1. Data

We use two pairs of coarse and fine resolution images, acquired almost simultaneously, to evaluate the performance of four algorithms in retrieval of aquatic vegetation cover information. MERIS (Medium Resolution Imaging Spectrometer) image in its full resolution mode (MERIS FR) has a spatial resolution of 300 m. We use as reference the results obtained with Landsat-7 ETM+ imagery at 30 m spatial resolution. Due to the dynamic nature of floating aquatic vegetation, the key consideration in selecting data for use in assessing the classification accuracy of aquatic vegetation is the temporal proximity of the image pair. During a field survey, it was estimated to take about an hour for a floating vegetation mat to move across a length approximately equal to the size of a MERIS pixel (300 m). Allowing vegetation displacement to a maximum of 0.25 of the pixel, then the interval between the acquisitions of the image pair should not be longer than 15 min. For the period 2003–2012, the entire lifetime of MERIS sensor, there are just seven scenes of Winam Gulf whose acquisition time coincides with those of ETM+, with acquisition intervals of the image pairs ranging from two to fifteen minutes. We selected two of these image pairs for our analysis, and their specifications are summarised in Table 1. The choice of these image pairs is based primarily on the short acquisition interval, and secondarily on the amount and distribution of aquatic vegetation in the images. Central acquisition time for each image is indicated. Since the image pairs were acquired almost simultaneously, we assume similar conditions of cloud, haze and water surface roughness (due to wind conditions). One of the greatest confounding factors limiting the quantity and accuracy of remotely sensed data from water bodies is sun glint, the specular reflection of directly transmitted sunlight from the upper side of the air-water interface [20]. Sun glint is a function of the state of the water surface (surface roughness), sun position and satellite viewing angle. Sun zeniths of 30°–60° degrees are optimal for minimizing sunglint [21]. Sun zenith angles are respectively 34.6° and 35.4° for MERIS and ETM+. The range of sensor viewing angles across the study area is indicated for each image. Since the radiance received by sensor is inversely proportional to the cosine of the sensor view angle, then sunglint effect on the images is negligible.

Table 1. A summary of satellite data used in the study.

Image Pair	Sensor	Acquisition Time	Spatial Resolution	Spectral Resolution (Visible and NIR)	Sensor Viewing Angles
Pair 1	MERIS	15 December 2010 07:49	300 m	15 bands	7°–11°
	ETM+	15 December 2010 07:48	30 m	5 bands	0°–5°
Pair 2	MERIS	27 July 2011 07:41	300 m	15 bands	6°–10°
	ETM+	27 July 2011 07:48	30 m	5 bands	0°–5°

3.2. Pre-Processing

Before using the satellite data, we convert the sensor radiance values into reflectance values and perform atmospheric corrections as here described. Atmospheric corrections of MERIS data were performed using SMAC Processor 1.5.203 (a Simplified Method for Atmospheric Corrections of satellite measurements) [22], incorporated in the software package BEAM (Basic ERS and Envisat (A) ATSR and MERIS Toolbox). It is a semi-empirical approximation of the radiative transfer in the atmosphere which takes into account the attenuation due to atmospheric absorption and radiance of the scattered skylight. We used FLAASH (Fast Line-of-sight Atmospheric Analysis of Spectral Hypercubes), an atmospheric correction code based on the MODTRAN 4 (MODerate resolution atmospheric TRANsmission) radiative transfer model, to convert Landsat-7 ETM+ sensor radiance to surface reflectance. A certain degree of geolocation errors is inevitable when dealing with multiple data sets. We co-registered the image pairs using an image to image first order polynomial transformation and nearest neighbour resampling, with RMSE = 0.0745 and RMSE = 0.1218 for image pair 1 and 2 respectively. This represents a geolocation error of about 22.35 m for image pair 1, and 36.54 m for image pair 2, which is about the pixel size of the reference image. This may have an impact on the margins of the sample areas, but minimal.

3.3. Sampling

Evaluation of methods was carried out with two hundred (300 m × 300 m) samples. Each MERIS sample pixel corresponds to a square area covered by 10 × 10 ETM+ (30 m × 30 m) pixels. Sample MERIS pixels were selected such that their corresponding location in the ETM+ image would fall right in the middle of the stripes occasioned by the scan line corrector (SLC) failure in Landsat-7, so as to avoid the no-data pixels. Under these restrictive circumstances, a limited number of samples were selected. The samples were selected in Winam Gulf to include both the high and low vegetation density areas, as well as areas along the vegetation-water edges. Figure 2a shows the selected sample pixels for image pair 1; Figure 2b is a close-up (300 m × 300 m) MERIS pixel, while Figure 2c shows the corresponding one hundred (30 m × 30 m) ETM+ pixels. Although we use a shoreline derived from high resolution data to isolate our study area, we avoided samples too close to the shore, to avoid a possible confusion with terrestrial vegetation due to an imperfect shoreline.

Figure 2. (a) Location of the 200 selected sample pixels for the image pair acquired on 15 December 2010, **(b)** A close-up MERIS (300 m) pixel, **(c)** Corresponding 100 Landsat-7 ETM+ (30 m) pixels.

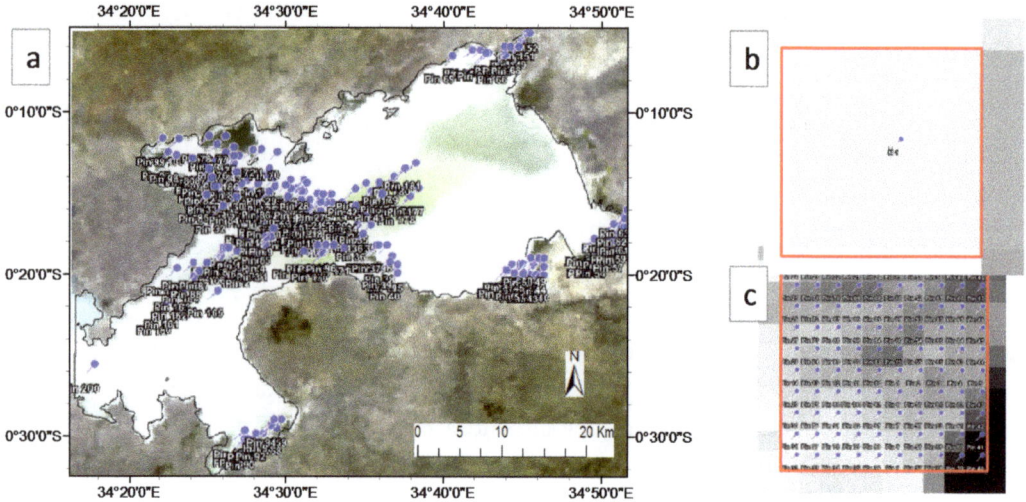

3.4. Experimental Design

Our objective is to determine and monitor the lake area covered by aquatic vegetation: because of the size of Lake Victoria and the temporal variability of the extent of aquatic vegetation, satellite data should provide observations at daily intervals or shorter, with spatial resolution limited to 1 km or worse. We aim at determining the total lake area covered by aquatic vegetation and at delineating it.

In this study we regard MERIS as our primary source of observations and we want to assess the accuracy of both estimated total area and of the delineation of it. We used Landsat ETM+ data as a reference.

In summary we have evaluated different ways to determine the two variables of interest using the methods listed below and described in the following sections:

Delineation: slicing of NDVI, maximum likelihood classifier;
Total area: retrieval of fractional abundance from NDVI and by linear spectral un-mixing;

All four methods have been applied to MERIS data and evaluated against results obtained with ETM+. The list of image data has been given in Table 1, while the design of the evaluation experiment is given in Table 2.

Table 2. Design of the evaluation experiment: classification of MERIS and ETM+ data provides delineation of the area covered by aquatic vegetation; area integral of fractional abundance (f_v) provides estimates of the total area covered by aquatic vegetation. The right column shows the comparison method applied in each case.

MERIS		ETM+		Comparison method
Method	*Retrievals*	*Method*	*Retrievals*	
--	--	Delineation	--	--
Sliced NDVI	3 classes	Sliced NDVI	3 classes	Error matrix
Maximum Likelihood	2 classes	Maximum Likelihood	2 classes	Error matrix
--	--	Total area	--	--
NDVI	f_v	NDVI	f_v	RMSE, Linear regression
Spectral unmixing	f_v	Spectral unmixing	f_v	RMSE, Linear regression

3.5. Empirical Slicing of Vegetation Indices

Though many vegetation indices have been developed [23], in this study we focus on Normalized Difference Vegetation Index (NDVI) [24]. It is a dimensionless quantity which is an indicator of the greenness of vegetation, and is based on the contrast between the maximum reflection in the near infrared (ρ_{nir}) caused by leaf cellular structure and the maximum absorption in the red (ρ_r) due to chlorophyll pigments [25]. It is expressed as a ratio of the difference and the sum of ρ_{nir} and ρ_r:

$$NDVI = \frac{\rho_{nir} - \rho_r}{\rho_{nir} + \rho_r} \tag{1}$$

NDVI is the most commonly used indicator of vegetation parameters in remotely sensed data for global vegetation mapping [25,26]. It has been applied to quantify the vegetation cover in various studies, both in terrestrial environment [27,28] as well as aquatic environment [29]. Empirical slicing of vegetation indices is commonly used to discriminate vegetation from other cover classes. The challenge, however, is the correct identification of suitable thresholds separating various feature classes in the scene. The authors of [13] applied NDVI = 0.1 as a threshold to detect presence of vegetation, while [15] estimated the aquatic vegetation cover in Lake Victoria using a three-level NDVI scale:

$$NDVI = \begin{cases} > 0.4 \ Floating \ vegetation \ (FV) \\ 0.2 - 0.4 \ Sparse \ vegetation \ (SV) \\ < 0.2 \ Open \ water \ surface \ (OW) \end{cases} \tag{2}$$

We give a special reference to the NDVI slicing in Equation (2) to assess accuracy of aquatic vegetation classification with NDVI slicing. While this slicing clearly provides an excellent display of the spatial distribution of vegetation in the lake as seen in [15], in this study we evaluate the impact of limiting NDVI to a few classes. NDVI was computed from Landsat-7 ETM+ using red and near infrared bands 3 and 4 centred at 660 nm and 825 nm respectively; while MERIS NDVI was computed using bands 7 and 13 centred at 664 nm and 865 nm.

3.6. Maximum Likelihood Classification

Maximum likelihood is a conventional classifier, which assigns a pixel to the class to which it most probably belongs according to a Bayesian probability function. Based on statistics (mean; variance/covariance), the probability function is calculated from the inputs for classes established from training sites. In this study, binary images consisting of water and vegetation classes were achieved by applying maximum likelihood classifier to MERIS and ETM+. The training sites for water and vegetation were obtained from the same images. This was implemented in the software package ENVI.

3.7. Estimating Vegetation Fractional Cover (f_v) from NDVI

Vegetation amount is usually parameterized through the fractional area (f_v) of the vegetation occupying each grid cell, which gives its horizontal density [30]. f_v is sometimes estimated from vegetation indices. NDVI does not directly give f_v, and some methods have been developed to derive it from vegetation indices. A commonly-used linear model for deriving f_v from vegetation indices [31] is described by [16] as:

$$f_v = \frac{NDVI - NDVI_o}{NDVI_\infty - NDVI_o} \tag{3}$$

where NDVI$_\infty$ and NDVI$_o$ respectively correspond to $NDVI$ of reference vegetation ($f_v = 1$) and reference soil ($f_v = 0$). We apply a modified version of this model to estimate aquatic vegetation f_v with $NDVI_\infty$ being the highest NDVI value (NDVI of a pure vegetation pixel) and NDVI$_o$ being the lowest NDVI value (NDVI of an open water pixel). Selecting the highest and lowest NDVI values ensures that the derived f_v are non-negative and not greater than one. The constants used for MERIS data are: $NDVI_\infty = 0.96$ and $NDVI_o = -0.58$ while for ETM+ data: $NDVI_\infty = 0.85$ and $NDVI_o = -0.23$.

3.8. Linear Spectral Unmixing (LSU)

Linear spectral unmixing is one of the spectral mixture analysis (SMA) techniques which decompose a mixed pixel into various distinct components. It is most suitable where the spatial resolution of the satellite data is relatively coarse. It has been applied in various studies including analysis of rock and soil types [32], desert vegetation [33], land-cover changes [34], estimation of urban vegetation abundance [35], and delineating potential erosion areas in tropical watersheds [9]. Non-linear mixture models also exist [36], but linear spectral unmixing is by far the most common type of SMA, and is widely used because of its simplicity and interpretability [9]. It is a supervised classification technique which is based on the assumption that the observed reflectance of a pixel (ρ_k) at wavelength (k) is a linear combination of the reflectance ($\rho_{i,k}$) of individual class features represented in that pixel, and the contribution of each depends on its respective abundance (f_i). The basic physical assumption is that there is not a significant amount of photon multiple scattering between the macroscopic materials. For a specified number of endmembers (n), linear spectral unmixing can be expressed as:

$$\rho_k = \sum_{i=1}^{n} f_i \cdot \rho_{i,k} + \varepsilon_k \tag{4}$$

ε_k is the residual error. The unknown value in the expression is the fractional abundance f_i and which the model estimates. We used a fully constrained model which requires that for all i, f_i must sum to unity and is non-negative. The number of spectral bands in an image introduces a limitation to the number of endmembers that can be used for unmixing [37], so that it must always be less than the number of available bands in the multispectral image. The model retrieves the spectral characteristics ρ_i of endmembers from an input endmember spectral library.

In order to build an endmember library for our study region, we first need to identify the appropriate number of endmembers. Spectral characteristics of water in the lake vary spatially according to the concentrations of dissolved or suspended sediments in it, indicating the extent of nutrient enrichment. Clearer and deeper water in the centre of the lake displays low reflectance values, while that near the shores displays generally higher reflectance values. In order to understand the spectral variability of the study area, we performed K-Means clustering [38,39], an unsupervised classification of clouds-free MERIS image whose acquisition date (15 December 2010) coincides with a field survey. K-Means is a statistical clustering method which follows the following procedure for a specified cluster number (k): (1) randomly choose k pixels whose samples define the initial cluster centres; (2) assign each pixel to the nearest cluster centre as defined by the Euclidean distance; (3) recalculate the cluster centres as the arithmetic means of all samples from all pixels in a cluster; and (4) repeat steps 2 and 3 until the convergence criterion is met. The convergence criterion is met when the maximum number of iterations specified by the user is exceeded or when the cluster centres did not change between two iterations. We begin by setting k first to 14 and specifying 30 iterations. A spectral plot of the 14 resultant classes revealed about five significantly unique classes. K-Means was repeated with $k = 5$, resulting in five classes; one confirmed by a field survey as vegetation and four different 'water species' confirmed by their relatively low reflectance values. A scatter plot of red and near infrared spectral bands for selected regions of the resulting five classes (Figure 3) showed that spectral variability among the four water classes was small with respect to the vegetation. Since vegetation was our target class, we reduced the number of endmembers to two by obtaining an average spectrum for the four water classes. Figure 3 shows good separability between vegetation and water classes, with vegetated pixels clustered at the top-left corner of the two dimensional space, due to vegetation's strong absorption of the red and high reflection of the near infrared radiation. Water pixels are clustered at the bottom right corner, due to water's strong absorption of the near infrared radiation.

To ensure consistency, endmember spectral libraries for MERIS and ETM+ shown in Figure 4 were compiled from the same area sampling.

Figure 3. Scatter plots of the reflectance values at red and near infrared spectral bands to showing separability of vegetation and water classes from MERIS (**left**) and ETM+ (**right**).

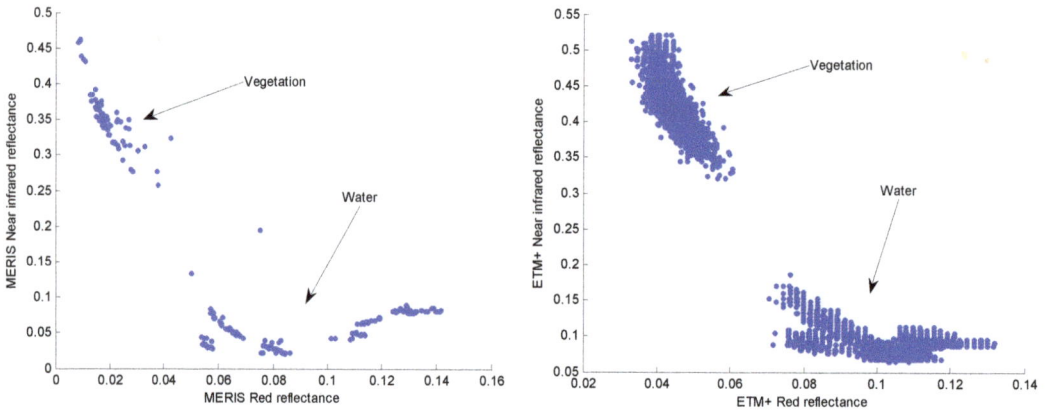

Figure 4. The MERIS endmember spectral library (**left**) and ETM+ endmember spectral library (**right**) used in classification.

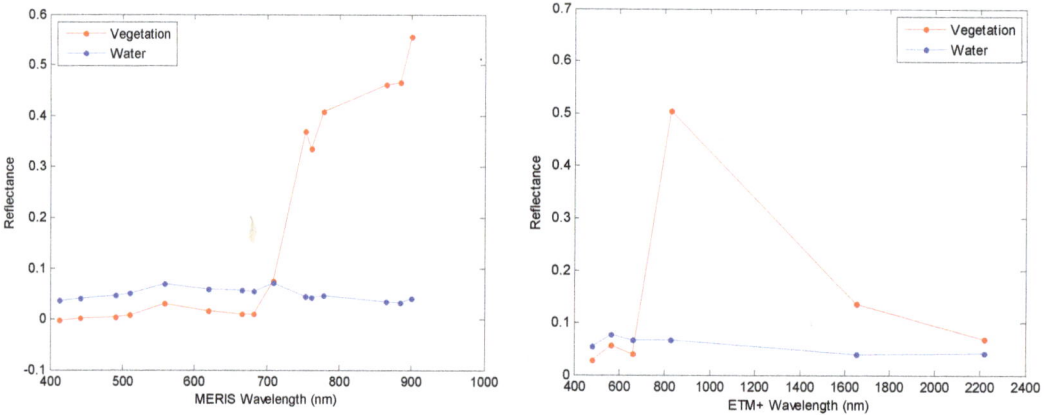

We apply a fully constrained linear spectral unmixing model (Equation (4)), using as input parameters the two endmembers spectral libraries shown in Figure 4, to estimate aquatic (f_v) from MERIS and ETM+ images respectively. This was implemented in the software package BEAM. The model outputs for each endmember a grey scale image, with pixel values indicating the class densities (f_i) in the range of 0–1. We pick the vegetation density (f_v) image for our analysis. We then assess the performance of spectral unmixing in vegetation detection using exactly the same 200 sample pixels that were used for the NDVI accuracy assessment.

4. Results and Discussion

4.1. Area-Averaging of NDVI

Due to non-linearity nature of NDVI, area-averaged NDVI obtained by averaging NDVIs of high resolution pixels is usually different from that obtained with averaged radiances of the high resolution pixels. Figure 5 shows a comparison of the two sets of area-averaged NDVI for 200 samples, each averaged from a hundred high resolution pixels. Even though the two sets of average NDVIs seem to compare well with $R^2 = 0.98$ and RMSE = 0.04, variations of mean NDVI computed from the two methods can be as high as 0.17 (see the green dots in Figure 5). This difference is quite significant especially for a study where averaged NDVI is an important variable used to derive another parameter. In this study, the NDVI of a square area represented by 100 ETM+ pixels corresponding to one MERIS pixel was obtained by first averaging the reflectance of 100 ETM+ pixels in the red and near infrared, before using them to compute NDVI.

Figure 5. A comparison between two ways of averaging NDVI. Blue dots show a scatter plot of NDVI computed from average high resolution radiances *versus* NDVI computed as an average of high resolution NDVIs. The green dots show the difference between the two sets of NDVI.

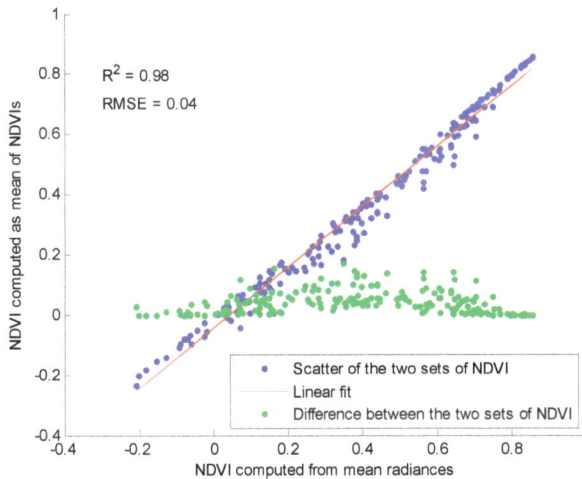

4.2. Sliced Normalized Difference Vegetation Index (NDVI)

Figure 6 shows NDVI images of the Winam Gulf section of Lake Victoria as derived from MERIS and ETM+ data. Figure 6a is MERIS NDVI while Figure 6c is ETM+ NDVI. The large green area in the centre of the lake is a floating mat of aquatic vegetation. The image shows the three classes described in Equation (2); open water (OW), sparse vegetation (SV) and floating vegetation (FV). The stripes in ETM+ image are a consequence of the SLC failure in the Landsat-7 satellite, resulting in no-data pixels. As expected, it is clear from these images that the higher resolution ETM+ (30 m) displays more vegetation cover details than MERIS (300 m), as seen from the close-up portions of the images.

Figure 6. NDVI sliced to 3 levels: open water (OW), sparse vegetation (SV) and floating vegetation (FV), showing the distribution of vegetation as derived from (**a,b**) MERIS and (**c,d**) ETM+.

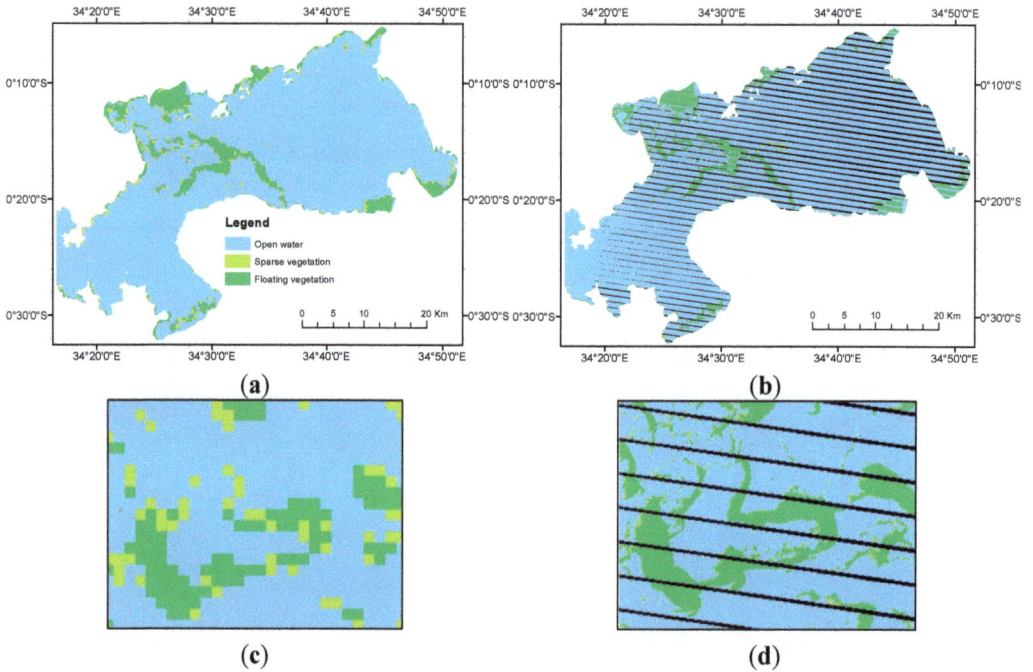

(a)

(b)

(c)

(d)

Error matrices in Table 3 show the quantitative assessment of the classification accuracy of NDVI slicing. The rows show classification of the 200 MERIS sample pixels to the three classes of Equation (2); FV, SV and OW. The columns show classification of the reference pixels (ETM+) to the same classes according to Equation (2).

Table 3. Error matrices for image pair 1 (**left**) and pair 2 (**right**) showing classification performance of NDVI sliced into floating vegetation (FV), sparse vegetation (SV) and open water (OW).

		Reference (ETM+)				Commission Error
		FV	SV	OW	Row Total	
Classified (MERIS)	FV	85	12	1	98	13%
	SV	16	17	6	39	56%
	OW	3	9	51	63	19%
	Column Total	104	38	58	200	
Omission Error		18%	55%	12%		

		Reference (ETM+)				Commission Error
		FV	SV	OW	Row Total	
Classified (MERIS)	FV	89	9	3	101	12%
	SV	10	17	16	43	60%
	OW	1	15	40	56	29%
	Column Total	100	41	59	200	
Omission Error		11%	59%	32%		

The diagonal matrix of the error matrix of the first image pair, Table 3(left), shows that 153 of the 200 sample pixels are correctly classified, which gives an overall classification accuracy of 76.5%. The error matrix of the second image pair, Table 3(right), shows an overall classification accuracy of 73%. In both cases, the commission and omission errors in the classification of SV are higher than those of FV and OW, indicating that major misclassifications occurred along the water-vegetation boundaries. These results seem to confirm the findings of [31], where the reported classification accuracies for NDVI of pixels along the edges of distinct endmembers were generally lower than those of non-edges. This is a common weakness to all "hard" classifiers which output discrete feature classes, and only gets worse with lower resolutions. Since NDVI slicing is heavily biased along the water-vegetation boundaries, it may have a significant impact on detection of aquatic vegetation from low resolution data. Slicing NDVI to a few levels also limits its sensitivity to vegetation density variations.

It is however known that some errors are due to the low resolution of the data rather than the weakness of the classification algorithm, a concept known as "low resolution bias" explained by [40] using Pareto Boundary. Using the Pareto Boundary, it is possible to determine the optimal classification accuracies that can be obtained by an algorithm, beyond which it is impossible to reduce the commission errors without increasing the omission errors, and vice versa. If all the pure pixels are correctly classified, then the irreducible errors are assumed to be due to the low resolution of the data. We refer readers to article [40] for a detailed description of the low resolution bias.

Figure 7 shows Pareto Boundaries for the optimal classification of floating vegetation, sparse vegetation and open water. This was obtained by setting a series of thresholds for the number of high resolution pixels of a specific class that are required to assign a low resolution pixel to that class, and computing the commission and omission errors incurred at each threshold level. Though Pareto Boundary applies only to dichotomic classifications, these results were obtained by first considering FV and collapsing SV and OW into the background, and repeating the procedure for SV and OW; resulting in three Pareto Boundaries, one for each class. Positions of the commission and omission errors obtained from the error matrix are shown in the commission error—omission error space; indicating how close the classification results are to the optimal levels.

In both image pairs, the positions of the commission and omission errors of SV are clearly farther away from their Pareto Boundaries, further confirming observations made from Table 3, that slicing of NDVI results in major misclassifications especially along the water-vegetation boundary. The radiance from this multi-class boundary received by a low resolution sensor is a combination of the spectral responses of the representative classes, and a "hard" classification of such a pixel results in high commission and omission errors.

4.3. Maximum Likelihood Classification

Error matrices in Table 4 show accuracy assessment of vegetation delineation obtained with binary Maximum Likelihood classifier for image pair 1 (left) and image pair 2 (right).

The rows show classification of MERIS sample pixels to two classes; FV and OW. The columns show classification of regions corresponding to MERIS sample pixels to two classes; FV class for regions half or more of ETM+ pixels are classified as vegetation, and OW class where less than half of ETM+ pixels are classified as vegetation. These results show an overall classification accuracy of 81.5% and 78.5% for pair 1 and pair 2 respectively. Pareto boundary analyses of the trade-off between commission and omission errors in these classifications are presented in Figure 8. In both cases, the positions of the omission and commission errors in the classification of vegetation and water are close to their respective Pareto boundaries, indicating a good performance of the classifier.

Figure 7. An analysis of the trade-off between commission and omission errors in the classification of floating vegetation (FV), sparse vegetation (SV) and open water (OW) by NDVI slicing for image pair 1 (**left**) and image pair 2 (**right**). Pareto Boundaries show the optimal classifications that can be achieved with low resolution MERIS data using reference obtained with higher resolution ETM+. Positions of the omission and commission errors in the classification of FV, SV and OW are shown in the error space.

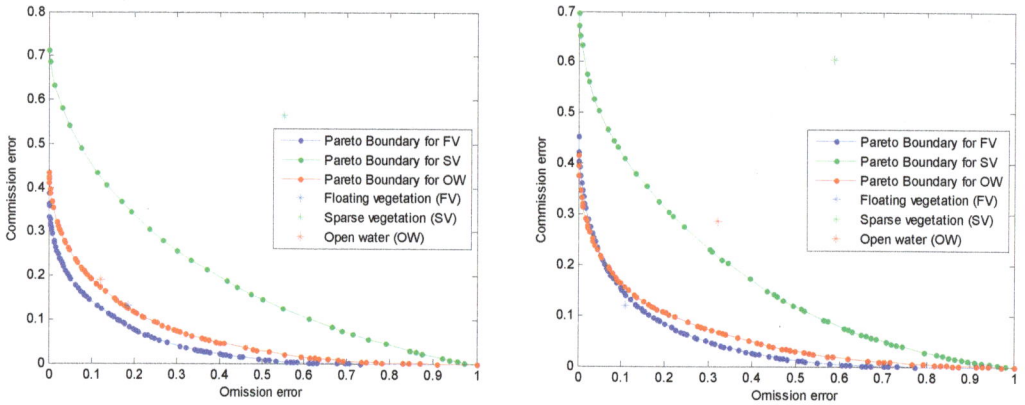

Table 4. Error matrices for image pair 1 (**left**) and pair 2 (**right**) showing performances of Maximum Likelihood in the classification of floating vegetation (FV) and open water (OW).

		Reference (ETM+)			Commission Error			Reference (ETM+)			Commission Error
		FV	OW	Row Total				FV	OW	Row Total	
Classified (MERIS)	FV	99	34	133	26%	Classified (MERIS)	FV	96	38	134	28%
	OW	3	64	67	4%		OW	5	61	66	8%
	Column Total	102	98	200			Column Total	101	99	200	
Omission Error		3%	35%			Omission Error		5%	38%		

4.4. Estimating Vegetation Fractional Cover (fᵥ) from NDVI

Figure 9 shows a comparison between f_v derived from MERIS NDVI using the NDVI-based f_v retrieval model (Equation (3)) with reference f_v derived from ETM+ NDVI with the same method.

The results show an RMSE of 0.11 and 0.09 for image pair 1 and 2 respectively, indicating the level of errors incurred in estimating f_v from low resolution MERIS data using NDVI-based f_v retrieval model.

Figure 8. An analysis of the trade-off between commission and omission errors in the classification of floating vegetation (FV) and water (OW) by maximum likelihood. Pareto Boundary shows the optimal classification that can be achieved with coarse resolution MERIS data with reference obtained from ETM+. Positions of the omission and commission errors in the classification of FV and OW is shown in the error space for image pair 1 (**left**) and image pair 2 (**right**).

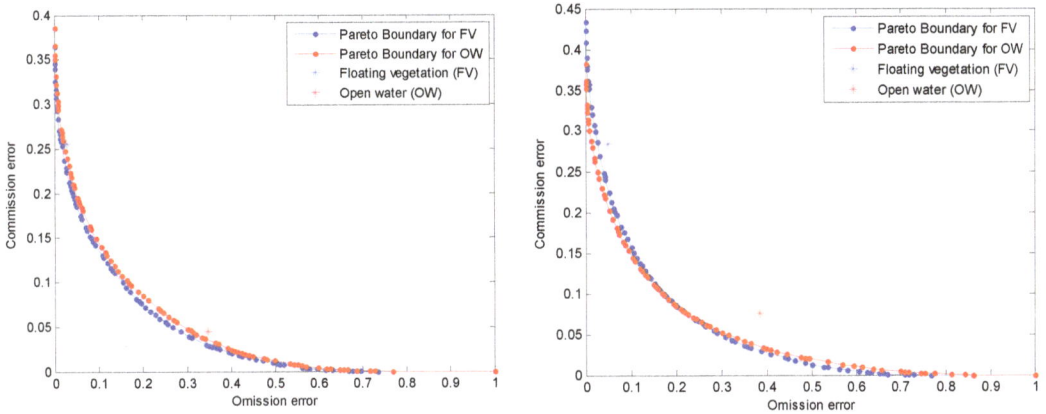

Figure 9. Correlation between f_v derived from MERIS NDVI with reference f_v derived from ETM+ NDVI, for image pair 1 (**left**), and image pair 2 (**right**).

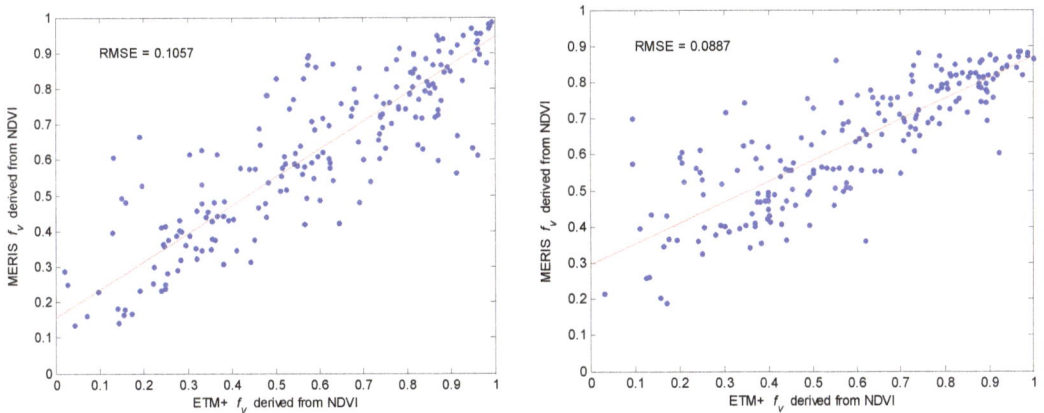

4.5. Spectral Unmixing

Figure 10 shows f_v results of spectral unmixing of MERIS and ETM+ images. The scale indicates f_v increasing from blue (open water surface) to green (fully dense vegetation cover).

436

Figure 10. The spectral unmixing classification results of (**a,b**) MERIS and (**c,d**) ETM+. The images show increasing f_v from blue (open water surface) to green (fully vegetated).

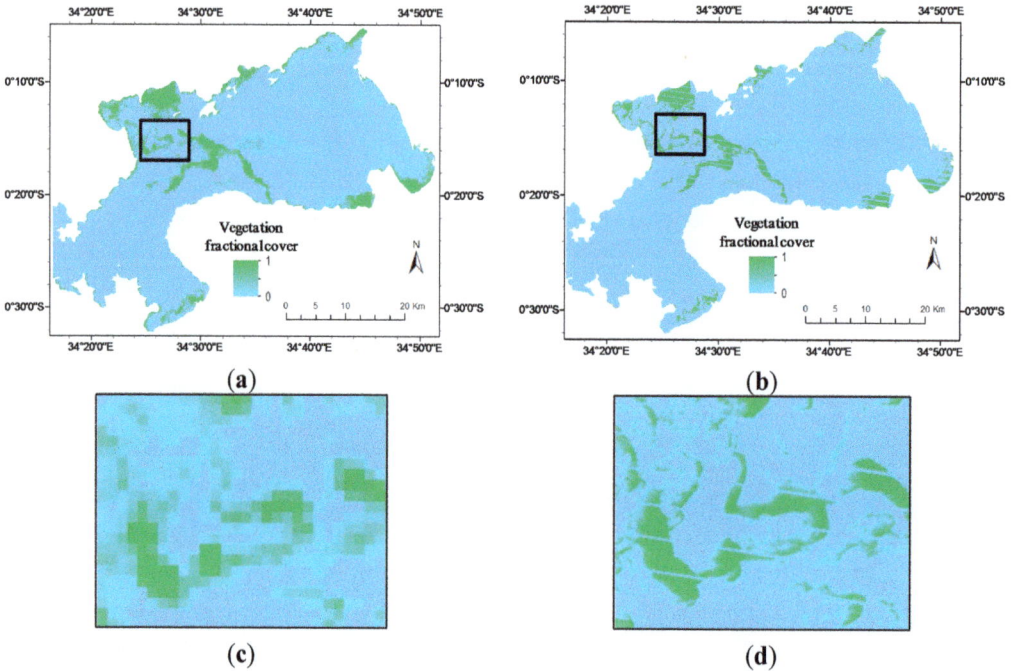

(a)

(b)

(c)

(d)

As seen in the close-up of the two images, both the low and high resolution data display f_v even in the sparsely populated areas. This is the advantage of spectral unmixing and similar methods which output f_v, so that no vegetation information is lost however small the proportion of the pixel it covers. Of course the accuracy of the model in estimating these densities decreases with reduced resolution, but this problem is not unique to spectral unmixing.

A comparison of f_v derived with spectral unmixing from MERIS imagery with those derived from the reference ETM+ data (Figure 11), shows an RMSE of 0.10 for both image pair 1 and 2, indicating the level of accuracy conceded for deriving f_v at lower resolution.

The methods showed varying results when tested with two sets of samples; dense vegetation (case 1), and sparse vegetation (case 2). Sliced NDVI showed classification accuracies of 96% and 52% for case 1 and 2 respectively, indicating better performance at high vegetation densities. This heavy bias along the water-vegetation boundaries may have a significant impact on detection of sparse aquatic vegetation from coarse resolution data. Maximum likelihood classifier showed accuracies of 98% and 72% for case 1 and 2 respectively, also indicating better performance at high vegetation densities. The method of deriving f_v from NDVI also showed better performance at higher vegetation densities with RMSE of 0.04 and 0.15 for case 1 and 2 respectively, because this method was designed as a dense vegetation model. Spectral unmixing showed minimal variation in the two vegetation density scales, with RMSE of 0.10 and 0.09 for case 1 and 2 respectively. These results show that detection accuracy of vegetation may vary with the density scale of vegetation cover, but with moderate effect on the results of maximum likelihood and minimal effect on the results of spectral unmixing. Spectral unmixing decomposes the pixel into various class features according to their relative abundances, and no vegetation cover information is discarded even if it constitutes a small proportion of the pixel. Spectral unmixing appears to be the most suitable method of estimating the extent of vegetation cover with coarse resolution products, but if an endmember spectral library is unavailable—due to technicalities involved in compiling it—the NDVI-based approach may be appropriate alternative, but only over dense vegetation.

Figure 11. A scatter plot showing a correlation between f_v obtained from MERIS pixels and the corresponding mean f_v obtained from ETM+ pixels for image pair 1 (**left**) and image pair 2 (**right**).

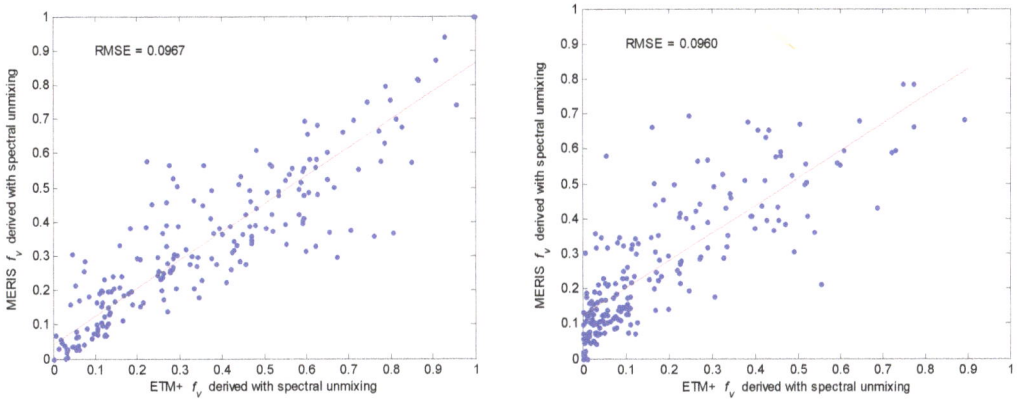

Remote sensing has a huge potential of providing crucial decision support information required for the control of aquatic plants proliferation. Changes in the status of aquatic plants in inland waters sometimes occur rapidly, and thus require regular and frequent monitoring. Due to the dynamic nature of floating aquatic plants, the technique of mosaicking small pieces of high resolution remote sensing products is not feasible. For large water bodies, vegetation cover information is best derived with remote sensing products with sufficiently large swath and high acquisition frequency. Most of these products are associated with coarse spatial resolution. For each of the methods evaluated in this study, the accuracy of obtaining vegetation cover information at coarse resolution has been analysed. It is worth noting that some of the vegetation detection errors discussed may be due to geolocation errors as a result of misregistration of the image pairs, and some due to displacement of floating plants. Considering the dynamic nature of floating plants, the acquisition interval between the image pairs is crucial in assessing the accuracy of aquatic vegetation information derived at coarse resolution. The appropriate interval can be determined by considering the spatial resolution of the image pairs as well as the rate of vegetation displacement; the latter can be estimated by considering wind speed.

5. Conclusions

Timely and frequent observations of large water bodies can provide the information needed to reposition the resources at hand for interventions (e.g., mechanical removal) to mitigate the impact of aquatic vegetation. While it is desirable to use finer resolution products to accurately detect small changes in the proliferation of aquatic plants, coarse resolution products remain best suited for the management of the plants in extensive water bodies. Understanding which remote sensing techniques work best with these coarse resolution products is thus necessary. In this study we analyzed the accuracy of vegetation cover information derived from coarse resolution MERIS product in terms of delineating aquatic plants, as well as estimating its surface extent, in both cases using as reference the results obtained with Landsat-7 ETM+ acquired almost simultaneously with MERIS.

In terms of delineation of aquatic plants, we evaluated two methods: NDVI slicing and Maximum Likelihood classifier. NDVI slicing produced an average classification accuracy of 75%, but showed a lower performance of 52% over sparse vegetation, with Pareto Boundary analysis showing largest commission and omission errors in these regions. Maximum likelihood classifier showed an average classification accuracy of 80% and a lower performance of 72% over sparse vegetation. In general, maximum likelihood classifier showed better performance than NDVI slicing, and the fragmentation of vegetation cover showed lesser effect on the performance of maximum likelihood than NDVI slicing.

In terms of total area estimation, we evaluated two methods: NDVI-based vegetation fractional cover retrieval method suggested by Gutman and Ignatov [16], and linear spectral unmixing. NDVI-based approach showed an average root mean square (RMS) error of 0.097, but larger errors of 0.15 over sparse vegetation. Linear spectral unmixing showed an average RMS error of 0.096, with similar performance over dense and sparse vegetation. The two methods seem to have similar performance over dense vegetation, but while the performance of NDVI-based approach significantly drops at sparse vegetation that of spectral unmixing remains invariant with the scale of vegetation density.

In summary, among the methods evaluated in this study, we recommend maximum likelihood for the delineation of aquatic plants and spectral unmixing for estimation of its surface extent, as the methods produce more accurate results and their performances are less sensitive to the fragmentation of aquatic vegetation cover.

Acknowledgments

This research was carried out under the framework of ESA's (European Space Agency) ALCANTARA Initiative, and was facilitated by Delft University of Technology, The Netherlands, and the University of Nairobi, Kenya. The authors acknowledge Isaiah Cheruiyot for assisting with field campaign.

Author Contributions

All authors contributed extensively to the work presented in this paper. Collins Mito, Massimo Menenti, Ben Gorte and Elijah Cheruiyot contributed to the concept design and research development. Field data was acquired by Elijah Cheruiyot, Collins Mito and Ben Gorte. Satellite data was processed by Elijah Cheruiyot and Nadia Akdim. Analysis was carried out by all authors with significant contribution from Elijah Cheruiyot and Roderik Koenders. Elijah Cheruiyot prepared the manuscript. All authors read and approved the manuscript.

Conflicts of Interest

The authors declare no conflict of interest.

References

1. Albright, T.P.; Moorhouse, T.G.; McNabb, T.J. The rise and fall of water hyacinth in Lake Victoria and the Kagera River Basin, 1989–2001. *J. Aquat. Plant Manag.* **2004**, *42*, 73–84.
2. Silva, T.S.F.; Costa, M.P.F.; Melack, J.M.; Novo, E.M.L.M. Remote sensing of aquatic vegetation: Theory and applications. *Environ. Monit. Assess.* **2008**, *140*, 131–145.
3. Sawaya, K.E.; Olmanson, L.G.; Heinert, N.J.; Brezonik, P.L.; Bauer, M.E. Extending satellite remote sensing to local scales: Land and water resource monitoring using high-resolution imagery. *Remote Sens. Environ.* **2003**, *88*, 144–156.
4. Govender, M.; Chetty, K.; Bulcock, H. A review of hyperspectral remote sensing and its application in vegetation and water resource studies. *Water SA* **2007**, *33*, 145–152.
5. Latifovic, R.; Olthof, I. Accuracy assessment using sub-pixel fractional error matrices of global land cover products derived from satellite data. *Remote Sens. Environ.* **2004**, *90*, 153–165.
6. Congalton, R.G. A review of assessing the accuracy of classifications of remotely sensed data. *Remote Sens. Environ.* **1991**, *37*, 35–46.
7. Kitada, K.; Fukuyama, K. Land-use and land-cover mapping using a gradable classification method. *Remote Sens.* **2012**, *4*, 1544–1558.
8. Kindu, M.; Schneider, T.; Teketay, D.; Knoke, T. Land use/land cover change analysis using object-based classification approach in munessa-shashemene landscape of the ethiopian highlands. *Remote Sens.* **2013**, *5*, 2411–2435.
9. De Asis, A.M.; Omasa, K.; Oki, K.; Shimizu, Y. Accuracy and applicability of linear spectral unmixing in delineating potential erosion areas in tropical watersheds. *Int. J. Remote Sens.* **2008**, *29*, 4151–4171.
10. Cavalli, R.M.; Laneve, G.; Fusilli, L.; Pignatti, S.; Santini, F. Remote sensing water observation for supporting Lake Victoria weed management. *J. Environ. Manag.* **2009**, *90*, 2199–2211.
11. Viña, A.; Gitelson, A.A.; Nguy-Robertson, L.A.; Peng, Y. Comparison of different vegetation indices for the remote assessment of green leaf area index of crops. *Remote Sens. Environ.* **2011**, *115*, 3468–3478.

12. Ouma, G.; Omeny, P.A.; Kabubi, J. Remote sensing application on eutrophication monitoring in Kavirondo Gulf of Lake Victoria Kenya. *J. Afr. Meteorol. Soc.* **2003**, *6*, 11–18.

13. Kiage, L.M.; Obuoyo, J. The potential link between el nino and water hyacinth blooms in winam gulf of Lake Victoria, East Africa: Evidence from satellite imagery. *Water Resour. Manag.* **2011**, *25*, 3931–3945.

14. Omute, P.; Corner, R.; Awange, J.L. The use of NDVI and its derivatives for monitoring Lake Victoria's water level and drought conditions. *Water Resour. Manag.* **2012**, *26*, 1591–1613.

15. Fusilli, L.; Collins, M.O.; Laneve, G.; Palombo, A.; Pignatti, S.; Santini, F. Assessment of the abnormal growth of floating macrophytes in Winam Gulf (Kenya) by using MODIS imagery time series. *Int. J. Appl. Earth Obs. Geoinf.* **2013**, *20*, 33–41.

16. Gutman, G.; Ignatov, A. The derivation of the green vegetation fraction from NOAA/AVHRR data for use in numerical weather prediction models. *Int. J. Remote Sens.* **1998**, *19*, 1533–1543.

17. Aloo, P.; Ojwang, W.; Omondi, R.; Njiru, J.M.; Oyugi, D. A review of the impacts of invasive aquatic weeds on the bio-diversity of some tropical water bodies with special reference to Lake Victoria (Kenya). *Biodivers. J.* **2013**, *4*, 471–482.

18. Plummer, M.L. Impact of invasive water hyacinth (eichhornia crassipes) on snail hosts of schistosomiasis in Lake Victoria, East Africa. *EcoHealth* **2005**, *2*, 81–86.

19. Mailu, A.M.; Ochiel, G.R.S.; Gitonga, W.; Njoka, S.W. Water hyacinth: An environmental disaster in the Winam Gulf of Lake Victoria and its control. In Proceedings of the First IOBC Global Working Group Meeting for the Biological and Integrated Control of Water Hyacinth, Harare, Zimbabwe, 16–19 November 1998; Hill, M.P., Julien, M.H., Center, T.D., Eds.; pp. 101–105.

20. Kay, S.; Hedley, J.D.; Lavender, S. Sun glint correction of high and low spatial resolution images of aquatic scenes: A review of methods for visible and near-infrared wavelengths. *Remote Sens.* **2009**, *1*, 697–730.

21. Dekker, A.G.; Byrne, G.T.; Brando, V.E.; Anstee, J.M. *Hyperspectral Mapping of Intertidal Rock Platform Vegetation as a Tool for Adaptive Management*; CSIRO Land and Water: Acton, Australia, 2003.

22. Rahman, H.; Dedieu, G. SMAC: A simplified method for the atmospheric correction of satellite measurements in the solar spectrum. *Int. J. Remote Sens.* **1994**, *15*, 123–143.

23. Jackson, R.D.; Huete, A.R. Interpreting vegetation indices. *Prev. Vet. Med.* **1991**, *11*, 185–200.

24. Rouse, J.W.; Haas, R.H.; Schell, J.A. *Monitoring the Vernal Advancement and Retrogradation (Greenwave Effect) of Natural Vegetation*; Texas A&M University: College Station, TX, USA, 1974.

25. Haboudane, D. Hyperspectral vegetation indices and novel algorithms for predicting green LAI of crop canopies: Modeling and validation in the context of precision agriculture. *Remote Sens. Environ.* **2004**, *90*, 337–352.

26. Elmore, A.J.; Mustard, J.F.; Manning, S.J.; Lobell, D.B. Quantifying vegetation change in semiarid environments: Precision and accuracy of spectral mixture analysis and the normalized difference vegetation index. *Remote Sens. Environ.* **2000**, *73*, 87–102.

27. Lunetta, R.S.; Knight, J.F.; Ediriwickrema, J.; Lyon, J.G.; Worthy, L.D. Land-cover change detection using multi-temporal MODIS NDVI data. *Remote Sens. Environ.* **2006**, *105*, 142–154.

28. Zhang, J.; Liu, Z.; Sun, X. Changing landscape in the three gorges reservoir area of Yangtze River from 1977 to 2005: Land use/land cover, vegetation cover changes estimated using multi-source satellite data. *Int. J. Appl. Earth Obs. Geoinf.* **2009**, *11*, 403–412.

29. Ma, R.; Duan, H.; Gu, X.; Zhang, S. Detecting aquatic vegetation changes in Taihu Lake, China using multi-temporal Satellite Imagery. *Sensors* **2008**, *8*, 3988–4005.

30. Jiang, Z.; Huete, A.R.; Chen, J.; Chen, Y.; Li, J.; Yan, G.; Zhang, X. Analysis of NDVI and scaled difference vegetation index retrievals of vegetation fraction. *Remote Sens. Environ.* **2006**, *101*, 366–378.

31. Johnson, B.; Tateishi, R.; Kobayashi, T. Remote Sensing of fractional green vegetation cover using spatially-interpolated endmembers. *Remote Sens.* **2012**, *4*, 2619–2634.

32. Adams, J.B.; Smith, M.O.; Johnson, P.E. Spectral mixture modeling: A new analysis of rock and soil types at the Viking Lander 1 site. *J. Geophys. Res. Solid Earth* **1986**, *91*, 8098–8112.

33. Smith, M.O.; Ustin, S.L.; Adams, J.B.; Gillespie, A.R. Vegetation in deserts: I. A regional measure of abundance from multispectral images. *Remote Sens. Environ.* **1990**, *31*, 1–26.

34. Adams, J.B.; Sabol, D.E.; Kapos, V.; Filho, R.A.; Roberts, D.A.; Smith, M.O.; Gillespie, A.R. Classification of multispectral images based on fractions of endmembers: Application to land-cover change in the Brazilian Amazon. *Remote Sens. Environ.* **1995**, *52*, 137–154.

35. Small, C. Estimation of urban vegetation abundance by spectral mixture analysis. *Int. J. Remote Sens.* **2001**, *22*, 1305–1334.

36. Liu, W.; Wu, E.Y. Comparison of non-linear mixture models: Sub-pixel classification. *Remote Sens. Environ.* **2005**, *94*, 145–154.

37. Theseira, M.A.; Thomas, G.; Sannier, C.A.D. An evaluation of spectral mixture modelling applied to a semi-arid environment. *Int. J. Remote Sens.* **2002**, *23*, 687–700.

38. Hartigan, A. *Clustering Algorithms*; Wiley: Hoboken, NJ, USA, 1975.

39. Hartigan, J.A.; Wong, M.A. Algorithm AS 136: A k-means clustering algorithm. *Appl. Stat.* **1979**, *28*, 100–108.

40. Boschetti, L.; Flasse, S.P.; Brivio, P.A. Analysis of the conflict between omission and commission in low spatial resolution dichotomic thematic products: The Pareto Boundary. *Remote Sens. Environ.* **2004**, *91*, 280–292.

Water Level Fluctuations in the Congo Basin Derived from ENVISAT Satellite Altimetry

Mélanie Becker, Joecila Santos da Silva, Stéphane Calmant, Vivien Robinet, Laurent Linguet and Frédérique Seyler

Abstract: In the Congo Basin, the elevated vulnerability of food security and the water supply implies that sustainable development strategies must incorporate the effects of climate change on hydrological regimes. However, the lack of observational hydro-climatic data over the past decades strongly limits the number of studies investigating the effects of climate change in the Congo Basin. We present the largest altimetry-based dataset of water levels ever constituted over the entire Congo Basin. This dataset of water levels illuminates the hydrological regimes of various tributaries of the Congo River. A total of 140 water level time series are extracted using ENVISAT altimetry over the period of 2003 to 2009. To improve the understanding of the physical phenomena dominating the region, we perform a K-means cluster analysis of the altimeter-derived river level height variations to identify groups of hydrologically similar catchments. This analysis reveals nine distinct hydrological regions. The proposed regionalization scheme is validated and therefore considered reliable for estimating monthly water level variations in the Congo Basin. This result confirms the potential of satellite altimetry in monitoring spatio-temporal water level variations as a promising and unprecedented means for improved representation of the hydrologic characteristics in large ungauged river basins.

Reprinted from *Remote Sens.* Cite as: Becker, M.; da Silva, J.S.; Calmant, S.; Robinet, V.; Linguet, L.; Seyler, F. Water Level Fluctuations in the Congo Basin Derived from ENVISAT Satellite Altimetry. *Remote Sens.* **2014**, *6*, 9340-9358.

1. Introduction

Despite the global importance of the Congo Basin, which is the second largest river basin in the world, only a small number of studies to date have focused on the potential impact of climate change on the hydro-climatic variability over the Congo Basin using *in situ* data and/or hydrological models. The limited understanding of climate dynamics in the Congo Basin is in part due to the lack of the *in situ* monitoring of climate variables in that area. Climate and hydrological station networks are sparse and poorly maintained; the small number of networks that were implemented during the colonial period has shrunk considerably [1–3]. The Congo Basin has experienced a turbulent history since pre-colonial times [4,5]. The resultant political instability, social unrest, and poor infrastructure may partly explain the lack of scientific attention [6]. Another great obstacle is the substantial difficulty of performing fieldwork in the Congo swamps. This large gap in understanding hydro-climate processes in this region increases the uncertainties in the evaluation of risks associated with decision making for major water resource development plans [7]. Conversely, recent improvements in remote sensing technology provide more observations than ever before that can advance hydrological studies [8,9], particularly in tropical regions. Given the vast size of the Congo Basin, remote sensing observations provide the only viable approach to understanding the spatial and temporal variability of the basin's hydro-climatic patterns. Several studies have therefore begun to address this topic by using remote sensing observations with a particular focus on hydrology. The following paragraph summarizes the results obtained by previous investigations.

Rosenqvist and Birkett [10] showed that temporal changes in river water levels in the Congo Basin can be derived from radar imagery. Eltahir *et al.* [11] inferred an anti-correlation in runoff anomalies between the Amazon Basin and the Congo Basin using two *in situ* time series of river flow from records at Manaus and Kinshasa, respectively, coupled with satellite-derived estimates of rainfall from the Tropical Rainfall Measuring Mission (TRMM). These authors argued for a climatic "see-saw oscillation" from one side of the Atlantic to the other. Crowley *et al.* [12] estimated terrestrial water storage within the Congo Basin from 2002 to 2006 from Gravity Recovery and Climate Experiment (GRACE) data. This estimate showed significant seasonal and long-term trends, with a total loss of approximately 280 km^3 of water over the study period. Jung *et al.* [13] evaluated the potential of space-borne radar for monitoring the large, subcontinental-scale river basins of the Amazon and Congo Rivers. The authors documented temporal changes in water surface elevations over time to reveal strikingly different flood behaviors in the Amazon and the largely undocumented Congo systems. The Congo system displayed less connectivity between the main and floodplain channels than did the Amazon system and exhibited more subtle changes during rising and falling limbs of the seasonal hydrograph. Lee *et al.* [14] used remote sensing measurements (*i.e.*, GRACE, satellite radar altimetry, GPCP, JERS-1, SRTM, and MODIS) to estimate the amount of water entering and exiting Congo wetlands and to determine the source of that water. O'Loughlin *et al.* [15] produced the first detailed hydraulic characterization of the middle reach of the Congo River utilizing mostly remotely sensed datasets (Landsat imagery, ICESat).

Our paper contributes to this body of work by providing an investigation of the ENVISAT altimetry data to analyze contemporary river dynamics in the Congo Basin over the period 2003–2009, for which *in situ* level measurements are insufficient or non-existent. The paper is organized as follows. Section 2 describes the main characteristics of the study area. Section 3 presents the different datasets and the methods used in the study. Section 4 presents the resulting classification of the river water level signatures. Section 5 validates the regionalization and discusses the seasonal dynamics of river water levels. Finally, Section 6 discusses several issues concerning the applicability of the altimeter-based techniques for the Congo Basin.

2. Study Area: Congo Basin

2.1. Location

The Congo River Basin is a transboundary basin located in western equatorial Africa that extends over 3.7 million km^2 (Figures 1 and 2). This shallow depression along the equator in the heart of Africa, named "Cuvette Central Caongolaise" [16], is bordered by higher areas (Figure 1): the Chaillu Mountains (900 m) and the Batéké Plateau (600–800 m) lie to the west and southwest.

Figure 1. Elevation map based on the HydroSHEDS (Hydrological data and maps based on SHuttle Elevation Derivatives at multiple Scales).

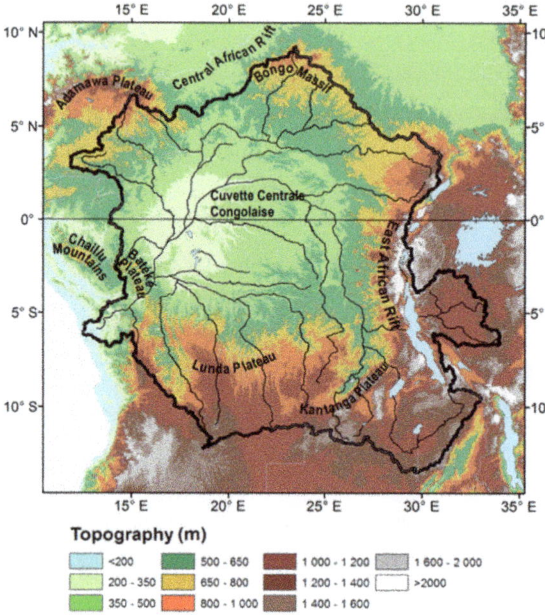

Figure 2. The principal tributaries and lakes in the Congo Basin.

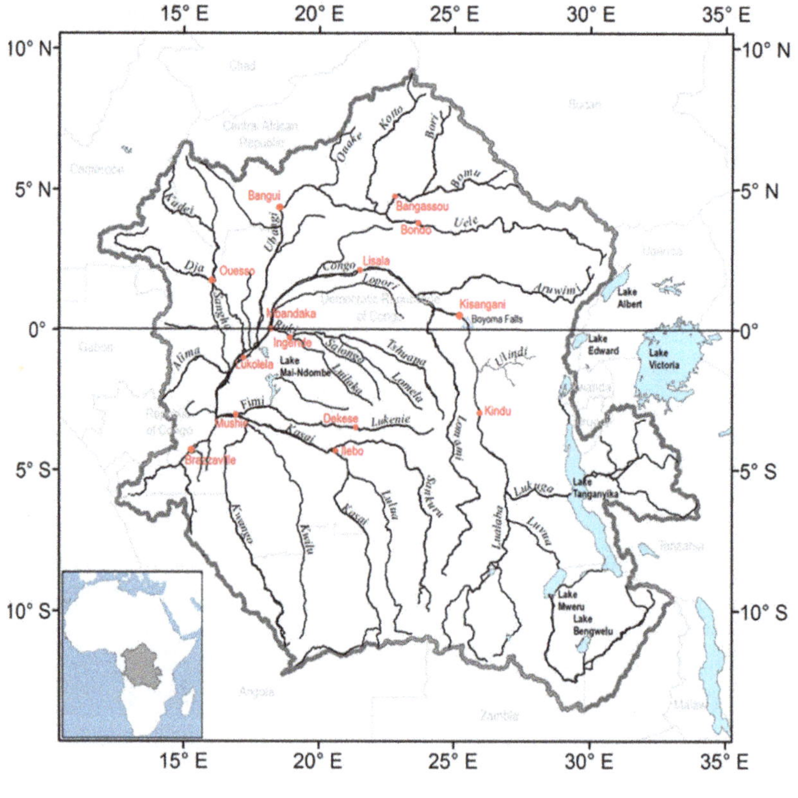

North of the basin are the Adamawa Plateau (1500 m) and the flanks of the Central African Rift (600–700 m), the boundary between the Congo and the Chad Basins, and the Bongo Massif (1300 m and higher). To the east, the most important relief is that of the volcanic foothills of the East African Rift, which reach altitudes of 2000–3000 m. The Katanga and Lunda Plateaus (1000–1500 m) bound the southern part of this vast watershed.

2.2. Hydrological System

The Congo Basin features a complex hydrological system composed of the Congo River, its many tributaries, and extensive swamps. The sources of the Congo River are on the highlands of the East African Rift in Lake Tanganyika, which feeds the Lukuga and Lualaba Rivers; these become the Congo River at Kisangani, below Boyoma Falls (Figure 2). The other two principal tributaries of the Congo are the Kasai River from the south and the Ubangi River from the north.

The Ubangi River is formed by the confluence of the Uele and Bomu rivers. Other main tributaries of the Ubangi River are the Bori River, the Kotto River and the Ouake River. The major tributaries of the Kasai River are the Kwango River and the Lulua River. They join to form the Kasai River from the south and drain a large part of the southern and southwestern Democratic Republic of Congo and northern Angola. The Fimi/Lukenie system runs parallel to and just north of the main Kasai River. Water draining from Lake Mai-Ndombe empties south through the Fimi River into the Kasai River. The Sangha River is a second-order tributary of the Congo River.

2.3. Climate

The Congo River receives year-round rainfall from the migration of the Inter-Tropical Convergence Zone (ITCZ). The northern part of the basin experiences a minor rainy season from September to November and a major one from the first half of March to early May; in the south, the minor rainy season lasts from February to May, and the major rainy season occurs between September and December. The source regions receive an average annual rainfall of 1200 mm. The middle and the downstream parts of the watershed receive 1800–2500 mm of rainfall per year and experience almost no dry season [16–18].

3. Primary Datasets and Methods

3.1. Altimetry and Virtual Station Data

The European Space Agency launched the ENVIronmental SATellite (ENVISAT) in March 2002 as part of its Earth Observation Program. This mission concluded in April 2012. ENVISAT carried ten instruments [19], including a nadir radar altimeter (RA-2 or Advanced Radar Altimeter). The ground track of the nominal ENVISAT orbit over the Congo Basin is shown in Figure 3. At all points where a satellite track intersects a water body, or "virtual stations", we extracted a water level time series, which allowed us to measure the successive water levels at each pass of the satellite over the large rivers channels, smaller tributaries and wetlands within sub-basins of the Congo Basin. The raw ENVISAT data are freely distributed by the Center for

Topographic studies of the Ocean and Hydrosphere (CTOH, [20]) in the standardized format of along-track Geophysical Data Records (GDRs). These data include four estimates of the distance between the satellite antenna and the ground, or the range. These four ranges are obtained by processing the radar echo with a dedicated algorithm called a retracker. Although none of the four retrackers had been tuned for echoes from river surfaces, Frappart *et al.* [21] and Santos da Silva *et al.* [22] showed that the ice-1 algorithm [23] performed well over rivers. Therefore, in this study, we used the ice-1 ranges when processing the raw ENVISAT data to compute water level time series at each virtual station. Our corrections are of two types: propagation corrections and geophysical corrections. The geophysical corrections are designed to remove instantaneous crustal movements. We applied corrections for solid earth and polar tides. Propagation corrections are designed to correct for the propagation of the electromagnetic wave as the radar travels through an ionized medium, the ionosphere, and a dense medium, the troposphere. We applied the corrections derived from global models as provided in the GDRs, in particular the tropospheric corrections derived from the European Centre for Medium-Range Weather Forecasts (ECMWF) meteorological model. We obtained the water stage time series between 2003 and 2009 (complete years) at 140 virtual stations (Figure 3) using the Virtual ALtimetry Station Tool (VALS) [24] for the ENVISAT tracks crossing the Congo Basin.

Figure 3. Location of ENVISAT tracks over the Congo Basin. Points represent the 140 virtual stations, with the 99 selected stations shown in black.

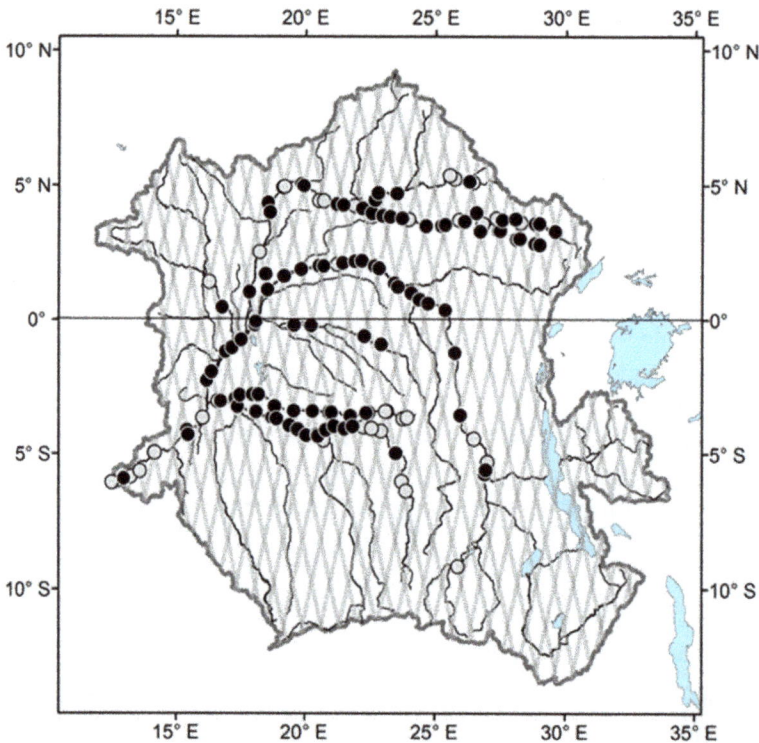

Details regarding the procedure used to process the data using VALS can be found in Santos da Silva *et al.* [22]. In this study, the water level data are referenced to the EGM2008 geoid model [25]. The water level time series at every virtual station passed a quality control test for gaps and/or shifts in the data. All time series with gaps greater than 3 consecutive months were deleted. All time series with a visually detectable spurious strong shift were also deleted. For the remaining time series, outliers were identified using Rosner's test [26] and removed. When small gaps (≤3 consecutive months) were observed, we reintroduced missing data by linearly interpolating the time series. Only 99 time series from the initial dataset of the 140 river water level series (RWL) satisfied these requirements (Figure 3). In this study, we use water level data rather than river discharge data. Because direct measurements of discharge in river channels can be time-consuming and costly, flow is often estimated indirectly by the rating curve method [27]. According to this technique, measurements of a river stage are converted to river discharge by a function (rating curve), which is preliminarily estimated by using a set of stage and flow measurements. Hence, uncertainties in measurements and the rating curve method increase the final uncertainty. Using the river level data is thus more straightforward in this region because we lack the data needed to calculate the rating curve at each virtual station. Moreover, this study uses water levels rather than discharge because, unlike *in situ* data, altimetry-derived stages are related to a common geoidal reference; the classification process used and described hereafter allows us to separate and analyze the section morphology effect while also not increasing the corresponding uncertainties by estimating discharge from stages.

3.2. Brazzaville Gauging Station

The time series of monthly water levels at Brazzaville (Figure 2. 15.3°E and 4.3°S) over the period 2003–2009 is selected from the Environmental Research Observatory HYBAM [28] Station: 50800000 Rio Congo at Congo Beach Brazzaville, covering the period from 1990 to the latest available year). This time series is used to evaluate our classification method. The same quality control process applied to the virtual stations was applied to this time series.

3.3. Lake Water Level

The monthly water level time series of Lake Mweru (Figure 2, 29.8°E and 8.7°S) and Lake Tanganyika (Figure 2, 29.5°E and 6.5°S) are available through Hydroweb [29]. Hydroweb is developed by LEGOS (Laboratoire d'Etudes en Géophysique Océanographie Spatiales) in France and provides water level time series of large rivers, approximately 150 lakes and reservoirs, and wetlands around the world using the merged data from the Topex/Poseidon, Jason-1, Jason-2, ENVISAT, European remote sensing satellite (ERS) and Geosat Follow-On (GFO) satellite missions. The processing procedures of Hydroweb are described in Crétaux *et al.* [30]. The Hydroweb lake water levels are monthly values obtained by merging measurements from different tracks of different altimeter satellites overflying the same lake in the same month [30]. These time series will be used to evaluate our classification method.

3.4. K-Means Clustering

The K-means is a common algorithm for classifying objects into K clusters, with K being a positive integer number. The classification is performed by minimizing the sum of squares of distances between data and the corresponding cluster centroid. Thus, the sample is assigned to a cluster based on minimizing, in its simplest form, the Euclidean distance between the vector of its variables and the means of the variables within a cluster. The K-means algorithm proceeds by updating the mean and grouping the data again. This procedure continues until all samples no longer change clusters. Given a dataset, a desired number of clusters K, and a set of K initial starting points, the K-means clustering algorithm finds the desired number of distinct clusters and their centroids. The K-means algorithm is described in more detail by Hartigan [31] and Hartigan and Wong [32]. In hydrology, the K-means algorithm and its variants have been used primarily in the regionalization of watersheds [33–37]. In this study, K-means analysis is performed for predefined cluster numbers varying from 5 to 15, where 15 is the maximum number of groups that maintains sufficient sample sizes in each group. To choose the initial cluster centroid positions, we select K uniform points at random from the range of the normalized parameters. The chosen parameter vectors are elevation data based on the HydroSHEDS DEM data at 30 arc-second resolution [38], river water level anomaly (RWLA) amplitude, dates of low and high stages and interannual correlation structure (lag-1), representing the dynamic component of the process. For example, if the autocorrelation in a time series at lag-1 is high (>0.6) the values are highly correlated with the value in the previous month. We run 10,000 replicates from randomly chosen starting parameter vectors; all runs converge to the same solution. The optimal number of clusters to retain is determined with the aid of the Davies–Bouldin index, a cluster validity measure that is a function of the ratio of the sum of within-cluster dispersion to between-cluster separation [39]. We calculate the separation measure for numbers of clusters ranging from 5 to 15. We filter our results according to certain specific criteria, such as a homogeneous distribution of observations within each cluster and no single-member clusters.

4. Results of the RWLA K-Means Clustering

The first step of the proposed approach is to cluster the 99 time series of altimeter-derived RWLA to identify groups with similar characteristics, defined by a conservative set of morphometric and hydrologic parameters. This study is developed for the RWLA is hereafter defined as the difference between the water level value and the temporal mean of the time series. Finally, according to these requirements presented in Section 3.4, the RWLA dataset is divided into 9 clusters exhibiting similar features. The optimal cluster locations are shown in Figure 4, and a topology map of RWLA signature vectors is shown in Figure 5. The topology highlights the variation of the RWLA features along the different classes, characterizing the behavior of the input variables and their interrelations. The RWLA time series composing each cluster are shown in Figure 6.

Figure 4. Optimal locations of the 9 clusters over the Congo Basin. Each circle represents the location of a virtual station and is color-coded to indicate its affiliation to a particular cluster. Light-gray circles with black crosses inside represent the "Outliers" cluster. These data are not used in the K-means clustering.

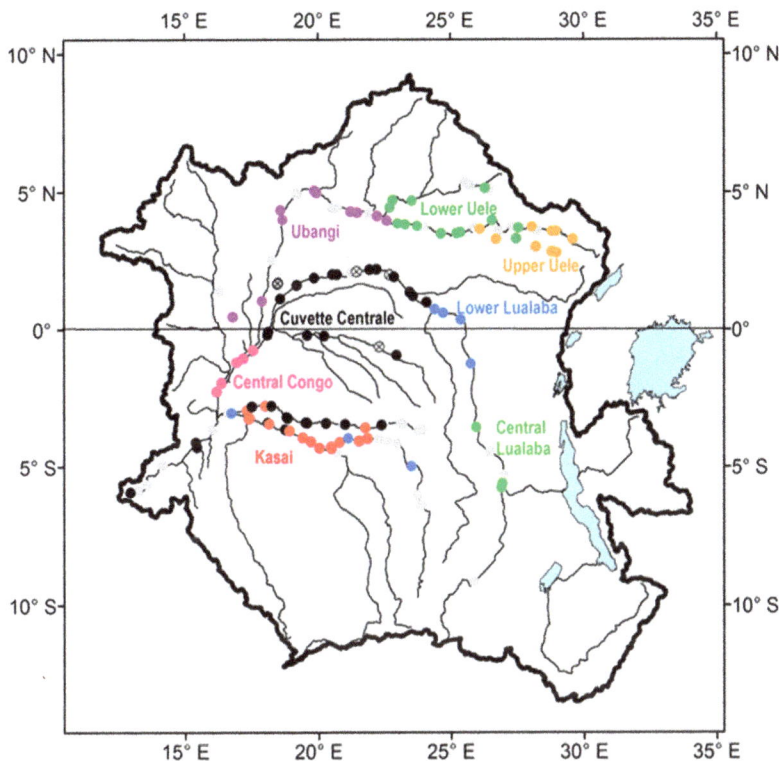

Figure 5. Optimal cluster topology of RWLA signature vectors.

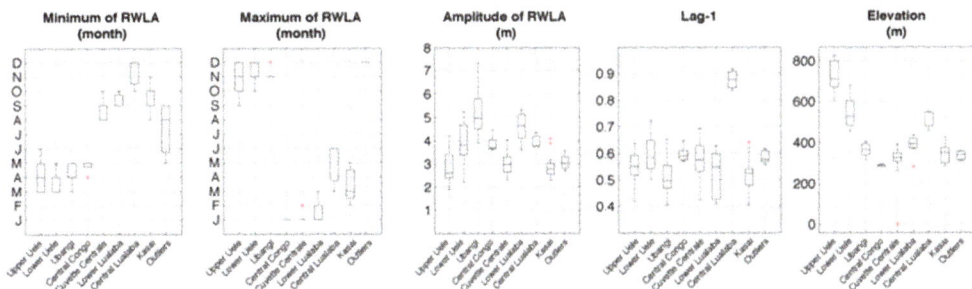

450

Figure 6. **Left panel**: The nine groups found by using K-means time series clustering in the 99 RWLA time series (black lines). The number of RWLA time series in each cluster is presented in parentheses. **Right panel**: The bold line represents the mean of the RWLA for each cluster, and the envelope (gray) shows the 5% and 95% quartile of the mean.

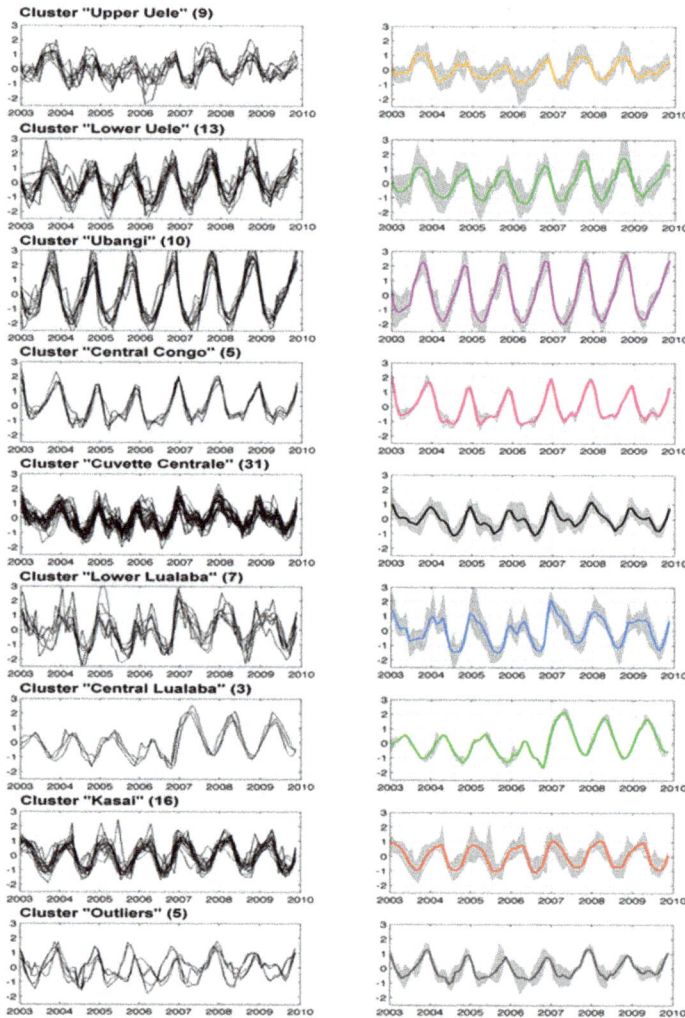

Cluster "Upper Uele", in the extreme northeast of the Congo Basin, contains 9 RWLA time series. Cluster "Lower Uele" contains the downstream part of the Uele River and its confluence with the Bomu River; the cluster includes 13 RWLA time series. Cluster "Ubangi" is composed of 10 RWLA time series. Cluster "Central Congo" contains 5 time series located along the Congo River after its confluence with the Ubangi River and prior to its confluence with the Kasai River. Cluster "Central Lualaba" includes 3 RWLA time series. Cluster "Lower Lualaba" is formed by 7 RWLA time series, 4 of which are located in the eastern part of the basin along the Lualaba River between the confluences with the Ulindi River and the Lomami River. The remaining 3 times series of this cluster are located along the Kasai River. For this cluster, we observe a large dispersion of the lag-1 coefficient, most likely because the RWLA time series are not located on the same rivers and therefore have different temporal correlation structures. Cluster "Cuvette Centrale" contains 31 RWLA time series. This cluster includes RWLA time series located in three regions: on the main stream of the Congo River between the confluences of the Lomami River and the Ubangi River, along the Ruki and Tshuapa Rivers, and along the Fimi and Lukenie Rivers. The 16 RWLA time series that compose Cluster "Kasai" are located in the meridional part of the Congo Basin. Cluster "Outliers" contains 4 RWLA time series spread throughout the basin. From the comparison shown in Figure 6, we can conclude that there is no good consistency between the time series in this cluster and we therefore removed it from consideration in the remainder of the study.

5. Validation of the RWLA Regionalization

Mahé [40] defined four great climatic zones over the Congo Basin: the North (Ubangi River Basin), where the influence of the North African continental air mass is prominent; the South (Kasai River Basin), which is influenced by South African air masses; the eastern and south-eastern parts of the basin (Lualaba River Upper Basin), which are influenced by the humid Indian Ocean air masses; and the Center-West, where the climate is controlled by the Atlantic Ocean. The seasonal partition of rainfall is bimodal along the equator and becomes unimodal farther north and south. We should therefore typically observe two-peak hydrographs (bimodal) for rivers near the equator and a gradual transformation into one-peak hydrographs (unimodal) farther north and south of the equator. We use this hypothesis to validate our RWLA regionalization. Figure 7 shows the hydrographs of monthly mean RWLA from 2003 to 2009 for each cluster. For each of the clusters, we verify that their time series show the same seasonal dynamics.

5.1. The North-Ubangi River Basin

Cluster "Upper Uele": We observe a unimodal distribution of the RWLA in all years except 2009. This cluster is characterized by a high water level from September to November and a low water level from February to March. The transition period, May to June, is very short. The years 2004, 2005, 2006 and 2009 were particularly dry (average RWLA < 0.5 m), whereas in 2003, the RWLA was greater than 1 m for 4 months (August through November). The seasonal variability contributed between 1 to 1.8 m over this period.

Cluster "Lower Uele": The distribution of RWLA is unimodal, with low water levels from March to April and high water levels from September to October. These dynamics are comparable to the monthly maxima and minima recorded by the historical river gauge at Bondo in 1956, located on the Uele River (Rosenqvist and Birkett [10], in Table 2). The seasonal amplitude is approximately 3 m over our study period.

Figure 7. Hydrograph of the RWLA mean for each cluster.

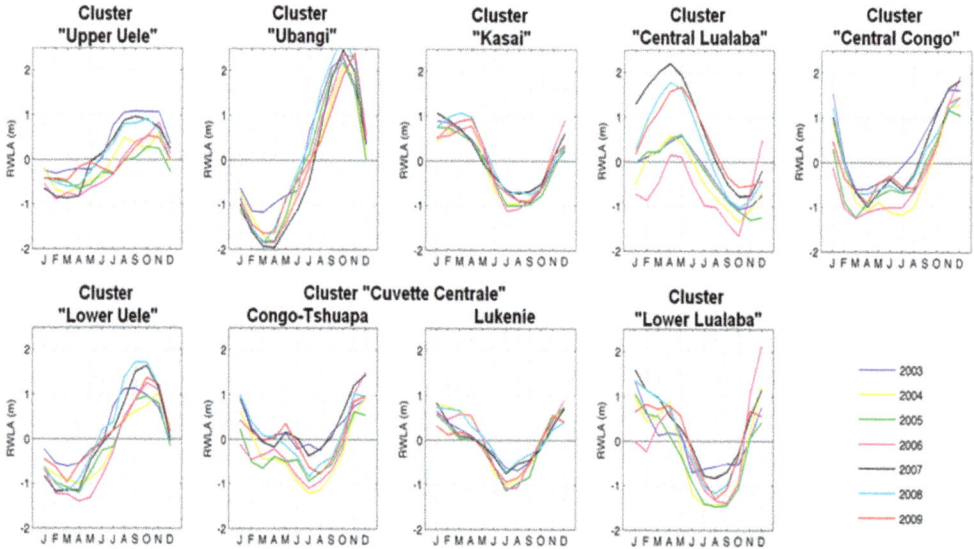

Cluster "Ubangi": The RWLA time series of this cluster are relatively homogeneous from 2003 to 2009. The distribution is unimodal. The dry season occurs from December through March (4 months), followed by rising water in May and a high water level in October. From November to January (3 months), we observe a rapid decrease in water level. These dynamics are similar to the monthly maxima and minima recorded by the historical river gauge at Bangui over 1890–1955, located on the Ubangi River ([10], in Table 2). This cluster has the largest seasonal variability in terms of RWLA, approximately 4.1 m over our study period.

5.2. The Southeast–Central and Upper Basins of the Lualaba River

Cluster "Central Lualaba": The RWLA in this cluster shows marked variability from one year to another. The distribution of the RWLA is unimodal, with high water levels from April to May and low water levels in October. These dynamics are consistent with the changes in water level recorded by the historical river gauge at Kindu over 1912–1955, located on the Lualaba River (Table 2 in Rosenqvist and Birkett 2002 [10]). We can observe 2 different periods in RWLA: (1) the seasonal variability from 2003 to 2006 shows well-marked minima and maxima but slight amplitude variations; (2) the seasonal variability over 2007–2009 shows well-marked minima and maxima and very large amplitude variations. In 2007, the extreme anomaly (2.2 m) was, on average, 4 times greater than the minimum values during the first period (~0.5 m). We observe evidence of completely different behavior of the RWLA in the year 2006. The temporal structure of low and high water levels is retained, but all the values are either near 0 m or negative, except in December. During the 3 months from October to December, we observe an increase in water level by more than 2.2 m. This extreme RWLA variability in 2006 and 2007 can be explained by hydro-climatic changes occurring in the East African Rift region. This cluster is located downstream on the Lualaba River (elevation ~500 m), which is fed by Lake Tanganyika and Lake

Mweru. The extreme water levels are most likely related to the 2005 severe drought reported in Equatorial East Africa [41] and to the positive strong Indian Ocean Dipole (IOD) in 2006. Similar behavior has been noted for other continental water cycle parameters. For example, Becker *et al.* [42] confirmed that precipitation and terrestrial water storage in the East African great lakes region have a common mode of variability, with a minimum in late 2005 and a sharp rise in 2006–2007. The authors showed that this event was due to forcing by the 2006 IOD on East African rainfall. As expected, we observe asymmetry in RWLA seasonal variability between the northern and the southern regions due to their locations on both sides of the equator. A comparison between the RWLA mean time series from the "Central Lualaba" cluster and the water level (WL) time series of Lake Mweru and Lake Tanganyika, computed from the T/P, Jason-1, Jason-2, ENVISAT, ERS and GFO satellite missions and provided by Hydroweb is presented in Figure 8a. The Luvua River exits Lake Mweru and flows northwest, and the Lukuga River drains Lake Tanganyika. These 2 rivers join the cluster on the Lualaba River.

Figure 8. (a) Comparison of Cluster 4 RWLA time series with Lake Tanganyika and Lake Mweru Lake water level time series obtained from Hydroweb over 2003–2009. The time series are normalized to place them on the same scale; (b) Comparison of Cluster 7.a RWLA time series with RWLA at Brazzaville gauging station obtained from ORE-HYBAM over 2003–2009.

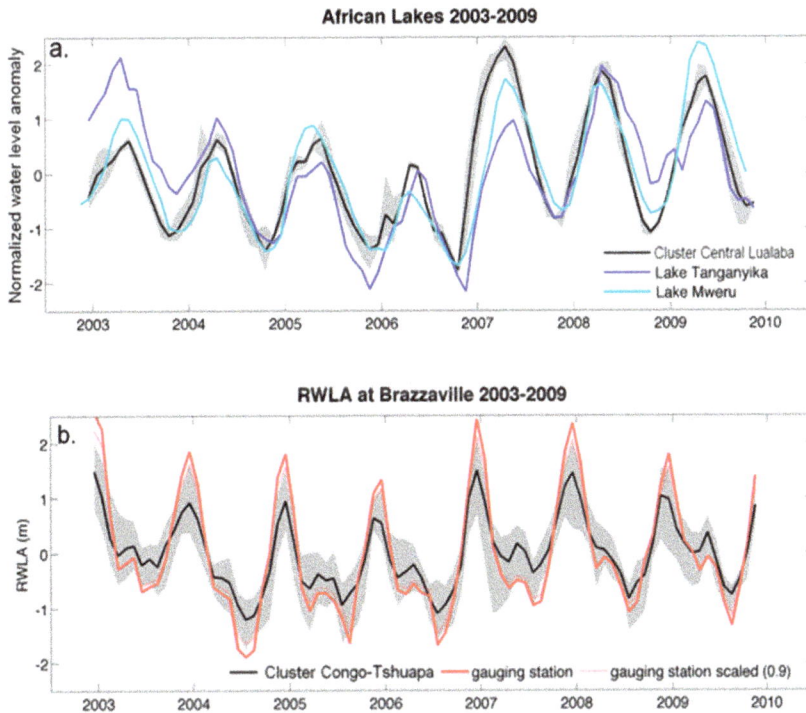

The RWLA seasonal variability from the cluster agrees well with the WL seasonal variability of the two lakes. We observe a lagged correlation coefficient of 0.9 between the cluster RWLA and

Lake Mweru WL with a delay of 1–2 months for the Lake Mweru WL. Further work concerning the hydrology of this region is necessary to explain the 1–2 month delay observed between the cluster and Lake Mweru. The correlation coefficient is 0.7 (p-value < 0.001) between the RWLA cluster and the Lake Tanganyika WL; no significant delay is detected between these two curves. A slight trend is observed in the Lake Tanganyika WL before 2007, but it is not observed in the RWLA of the cluster and Lake Mweru. Such consistency between the RWLA and WL time series enables us to validate this cluster.

5.3. The South-Kasai River Basin

Cluster "Kasai": We observe a unimodal RWLA distribution of this cluster time series over the studied period. The RWLA minima occur from August to September, and the maxima occur from January to April. RWLA maxima occur in January (2003, 2005 and 2007) or April (2004, 2006, 2008 and 2009), usually in alternating years. The seasonal variability is between 1.8 and 2 m over the period. These dynamics are comparable with the monthly maxima and minima distribution recorded by the historical river gauge at Mushie in 1932–1955 and at Ilebo in 1924–1955, both located on the Kasai River ([10], in Table 2).

5.4. The Center-West–Congo River Basin

Cluster "Cuvette Centrale": This cluster holds the largest number of RWLA time series (31) and has the largest latitudinal variability (from 2.5°N to 6°S). To avoid over-parameterization, we do not include prior information regarding the latitude coordinate or the RWLA bimodal/unimodal seasonal signature in the K-means clustering method. It is thus prudent to check the homogeneity of the RWLA seasonal variability within this cluster. As might be expected, we clearly observe 2 sub-clusters: (1) 20 RWLA time series located on the main stream of the Congo River and the Tshuapa River (hereafter named Cluster "Congo-Tshuapa"); (2) 11 RWLA time series located on the Lukenie River (hereafter named Cluster "Lukenie"). The RWLA time series of the cluster "Congo-Tshuapa" has a bimodal distribution. The water levels begin to rise in August and September due to rainfall intensification in the southern hemisphere. The high-water period is reached in December and lasts a relatively short time. The secondary low-water period occurs in March, during the dry season that prevails in the northern hemisphere tributaries. The primary low-water period in July and August corresponds to the dry season that prevails in the southern hemisphere [43]. These results are validated by the dynamics of the historical river gauge records at Mbandaka over 1913–1955 and at Lisala over 1914–1955, both located on the Congo River, and at Ingende over 1933–1955, located on the Ruki River ([10], in Table 2). We notice a contrast between the very dry years 2004, 2005 and 2006, when low water levels lasted 8, 9 and 10 months, respectively, and the years 2003, 2007, 2008 and 2009, when the low-water periods were almost non-existent. The average seasonal variability over our study period is on average 1.8 m, except in 2006 when it was 2.6 m, twice the amplitude observed in 2003.

Figure 8b shows a comparison between the RWLA time series from the cluster "Congo-Tshuapa" and the RWLA time series recorded at the Brazzaville gauging station over the period 2003–2009.

These two RWLA time series are remarkably synchronized and have a correlation coefficient of 0.96 (p-value < 0.001). However, the RWLA time series recorded at Brazzaville shows an amplitude 10% greater than the 95% confidence upper and lower bounds. Although not shown in the figure, the Brazzaville gauge also has a significant correlation coefficient (p-value < 0.001) with respect to the RWLA of Cluster "Central Congo" (0.9), Cluster "Lower Lualaba" (0.75), Cluster "Lukenie" (0.75) and Cluster "Kasai" (0.5).

The RWLA time series of Cluster "Lukenie" has unimodal dynamics and is relatively homogeneous from 2003 to 2007. The dates of extreme water levels coincide with those of the Cluster "Congo-Tshuapa": minimum in July and maximum in December and January. The seasonal variability over our study period averages 1.8 m. These results are similar to the historical water level time series records at Dekese over 1932–1955, located on the Lukenie River ([10], in Table 2).

Cluster "Central Congo": The RWLA stations are located on the Congo River downstream of its confluence with the Ubangi River and before its confluence with the Kasai River. This cluster is located in a hydrographically complex region and is influenced by three major rivers: the Ubangi, the Upper Congo and the Sangha [44]. The bimodal water level dynamics are very similar to those of Cluster "Congo-Tshuapa": high water levels from November to December and a second high-water period from May to June. However, in Cluster "Central Congo", low water levels occur in March and another more extreme low occurs in July, which is nearly the opposite of the Cluster "Congo-Tshuapa" dynamics. This finding can be explained by the strong influence of the Ubangi River (Cluster "Ubangi"), which is positive in the wet season (September) and negative in the dry season (March). Moreover, the RWLA seasonal variability is consistent with the dynamics at the Ouesso historical river gauge on the Sangha River (from the Global Runoff Data Center, [45]) and with the monthly maxima and minima recorded by the historical river gauge at Lukolela over 1909–1955, located on the Congo River ([10], in Table 2). The low-water period is very long, lasting 7 to 8 months. The seasonal variability is, on average, 2.4 m over our study period, except in 2006, when it was 3.2 m.

Cluster "Lower Lualaba": We apply the same methodology as that applied to Cluster "Cuvette Centrale" to the 7 time series that make up this cluster, but we do not observe any significant difference in seasonal variability between the RWLA time series from the Kasai River and the RWLA time series located on the lower Lualaba River. The RWLA hydrograph for Cluster 8 has a unimodal distribution, except for the years 2004 and 2006, for which a second high-water period occurs in April. The maximum occurs in December-January and the minimum in August. These results are validated by the historical water level time series recorded at Kisangani from 1907–1955, located on the Upper Congo River ([10], in Table 2). We note that in this region, the RWLA seasonal variability is, on average, 2.5 m over our study period, except in 2006, when it was 3.5 m.

We investigate the regionalization that these clusters suggest. Clusters "Upper Uele", "Lower Uele" and "Ubangi" spatially match the northern region as described by Mahé [36]. The increasing amplitude from upstream to downstream is coherent with the gathering of water along the river system. Cluster "Central Lualaba" represents the eastern and southeastern parts of the basin and is very distinct from the other clusters. The southern and central western regions are not very well represented by the clusters considered in the present study. Cluster "Lower Lualaba" appears intermediate between Cluster "Central Lualaba" and all other clusters. Clusters "Central Congo" and "Congo-Tshuapa" are bimodal, similar to each other, and separated from Clusters "Kasai" and "Lukenie". Our data suggest that regionalization in the central part of the catchment follows more of an east-west gradient than a north-south one.

6. Further Research

The dataset used in the present study is currently being expanded. In terms of spatial extent, many other virtual stations are currently being computed from ENVISAT to sample more rivers, such as the Kwango and Kwilu Rivers in the southwestern part of the basin, the Dja River in the northwestern part of the basin and in the east, and the Lukuga and Luvua Rivers, which drain the Tanganika and Mweru Lakes, respectively, into the Congo River. In terms of temporal extent, the 7-year ENVISAT time series will soon be extended with data from new satellites: Jason-2, launched in June 2008, and SARAL, launched in February 2013. Combining Jason-2 and SARAL observations for land water monitoring will take advantage of the 10-day temporal resolution of Jason-2 and the high geographical coverage of SARAL, which flies along the same orbit as ENVISAT. An example of a long series over the Congo River that can be obtained by combining ENVISAT, Jason-2 and SARAL data is shown in Figure 9.

Figure 9. An extended time series of water levels obtained by combining successive ENVISAT track (blue), Jason-2 track (red) and SARAL track (black) measurements over the Congo River. The Jason-2 and SARAL measurement series are adjusted for biases relative to the ENVISAT series. This virtual station is located at [1°08′S; 18°33′30″E].

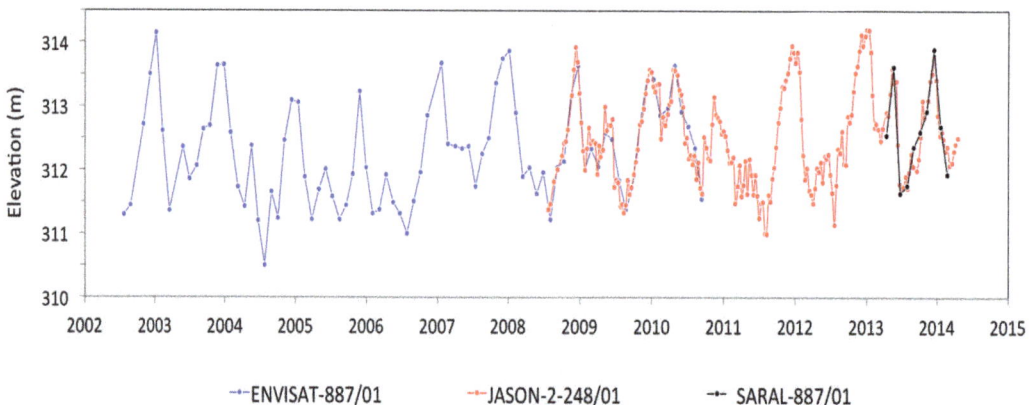

7. Conclusion

This study was conducted using stage rather than discharge measurements, which makes it unusual within the field of hydrology. The utility of altimeter-derived information is illustrated by finding the spatial and temporal signatures of climate variability in water level variations within the Congo Basin. Studies of this type have been traditionally based on historical *in situ* gauging station records, when and where available. However, climate and hydrological networks are sparse within the Congo Basin. Using satellite altimetry, we constructed a very large number of virtual stations across the Congo Basin to obtain information on the regional variability of surface water level anomalies in places where no *in situ* data are available over the period 2003–2009. This study shows that water levels can be measured throughout the basin, even in remote places, including the upstream, narrow parts of rivers. The study yielded interesting insights into the regionalization and characterization of the hydrological regime of the Congo Basin. Our analyses show an east-west gradient that has not previously been identified. The central western region is limited to a small region near the Congo swamp and represents the only bimodal hydrological regime of the basin. The Kasai region is similar to the central eastern region and is a progressive transition zone with the southeastern region. In conclusion, we have been validated the proposed regionalization scheme. Therefore considered reliable for estimating monthly water level variations in the Congo Basin. This result confirms the potential of satellite altimetry in monitoring spatio-temporal water level variations as a promising and unprecedented means for improved representation of the hydrologic characteristics in large ungauged river basins.

Acknowledgments

We thank the CTOH for providing the ENVISAT GDR data. M. Becker is supported by the Centre National d'Etudes Spatiales (CNES) and a Fonds social européen (FSE) fellowship. This study was supported by the AforA project within the CNES/TOSCA fund. We thank the three anonymous reviewers for their careful reading of our manuscript and their many insightful comments and suggestions.

Authors' Contributions

Mélanie Becker conducted the data analysis and wrote the majority of the paper. Joecila Santos da Silva was responsible for the processing of the ENVISAT observations. Stéphane Calmant supervised the research and contributed to manuscript organization and writing. Vivien Robinet contributed to the design of the regionalization method. Laurent Linguet and Frédérique Seyler helped with discussions and manuscript revisions.

Conflicts of Interest

The authors declare no conflict of interest.

References

1. Devroey, E. Observations Hydrographiques du Bassin Congolais (1932–1947). Available online: http://www.worldcat.org/title/observations-hydrographiques-du-bassin-congolais-1932-1947/oclc/8209249 (accessed on 16 May 2014).
2. Charlier, J. *Études Hydrographiques Dans Le Bassin Du Lualaba, Congo Belge, 1952–1954*; Comité Hydrographique du Bassin Congolais: Bruxelles, Belgium, 1955.

3. Snel, M.J. *Contribution à l'étude Hydrogéologique du Congo Belge. Service Géologique, Bulletin No. 7*; Democratic Republic of the Congo: Kinshasa, Democratic Republic of the Congo, 1957; Volume 2.

4. Hochschild, A. *King Leopold's Ghost: A Story of Greed, Terror and Heroism in Colonial Africa*; Houghton Mifflin: Boston, MA, USA, 1998.

5. Ndaywel è Nziem, I.; Obenga, T.; Salmon, P. *Histoire du Zaïre: De L'héritage ancien à L'âge Contemporain*; Duculot: Louvain-la-Neuve, Belgique, 1997.

6. Campbell, D. The Congo River basin. In *The World largest Wetlands: Ecology and Conservation*; Cambridge University Press: Cambridge, UK, 2005; pp. 149–165.

7. Shem, O.W.; Dickinson, R.E. How the Congo basin deforestation and the equatorial monsoonal circulation influences the regional hydrological cycle. In Proceedings of the 86th Annual American Meteorological Society Meeting, Tuesday, 31 January 2006, Atlanta, GA, USA, 2006.

8. Alsdorf, D.E.; Rodrfgue, E.; Lettenmaier, D. Measuring Surface Water from Space. Available online: http://bprc.osu.edu/hydro/publications/2007a_Alsdorf.pdf (accessed on 16 May 2014).

9. Tang, Q.; Gao, H.; Lu, H.; Lettenmaier, D.P. Remote sensing: Hydrology. *Prog. Phys. Geogr.* **2009**, *33*, 490–509.

10. Rosenqvist, A.A.; Birkett, C.M. Evaluation of JERS-1 SAR mosaics for hydrological applications in the Congo River Basin. *Int. J. Remote Sens.* **2002**, *23*, 1283–1302.

11. Eltahir, E.A.; Loux, B.; Yamana, T.K.; Bomblies, A. A see-saw oscillation between the Amazon and Congo basins. *Geophys. Res. Lett.* **2004**, *31*, doi:10.1029/2004GL021160.

12. Crowley, J.W.; Mitrovica, J.X.; Bailey, R.C.; Tamisiea, M.E.; Davis, J.L. Land water storage within the Congo Basin inferred from GRACE satellite gravity data. *Geophys. Res. Lett.* **2006**, *33*, doi:10.1029/2006GL027070.

13. Jung, H.C.; Hamski, J.; Durand, M.; Alsdorf, D.; Hossain, F.; Lee, H.; Hossain, A.K.M.; Hasan, K.; Khan, A.S.; Hoque, A.K.M. Characterization of complex fluvial systems using remote sensing of spatial and temporal water level variations in the Amazon, Congo, and Brahmaputra Rivers. *Earth Surf. Process. Landf.* **2010**, *35*, 294–304.

14. Lee, H.; Beighley, R.E.; Alsdorf, D.; Jung, H.C.; Shum, C.K.; Duan, J.; Guo, J.; Yamazaki, D.; Andreadis, K. Characterization of terrestrial water dynamics in the Congo Basin using GRACE and satellite radar altimetry. *Remote Sens. Environ.* **2011**, *115*, 3530–3538.

15. O'Loughlin, F.; Trigg, M.A.; Schumann, G.-P.; Bates, P.D. Hydraulic characterization of the middle reach of the Congo River. *Water Resour. Res.* **2013**, *49*, 5059–5070.

16. Bernard, E., Le Climat Écologique: De la Cuvette Centrale Congolaise. Available online: http://www.persee.fr/web/revues/home/prescript/article/geo_0003-4010_1948_num_57_305_12165 (accessed on 16 May 2014).

17. Bultot, F. Atlas Climatique du Bassin Congolais. Available online: http://lib.ugent.be/catalog/rug01:001906688?i=0&q=000000869163 (accessed on 16 May 2014).

18. Mahe, G.; L'hote, Y.; Olivry, J.C.; Wotling, G. Trends and discontinuities in regional rainfall of West and Central Africa: 1951–1989. *Hydrol. Sci. J.* **2001**, *46*, 211–226.

19. Wehr, T.; Attema, E. Geophysical validation of ENVISAT data products. *Adv. Space Res.* **2001**, *28*, 83–91.

20. Center for Topographic studies of the Ocean and Hydrosphere (CTOH). Available online: http://ctoh.legos.obs-mip.fr/ (accessed on 16 September 2014).

21. Frappart, F.; Calmant, S. Cauhopvalidation of ENVISAT data products and Central Africa: 1951s in the Congo Basin using GRACE and satellite radar altimetry. *Remote Sens. Environ.* **2006**, *100*, 252–264.

22. Santos da Silva, J.; Calmant, S.; Seyler, F.; Rotunno Filho, O.C.; Cochonneau, G.; Mansur, W.J. Water levels in the Amazon basin derived from the ERS 2 and ENVISAT radar altimetry missions. *Remote Sens. Environ.* **2010**, *114*, 2160–2181.

23. Bamber, J.L. Ice sheet altimeter processing scheme. *Int. J. Remote Sens.* **1994**, *15*, 925–938.

24. VALS Tool Virtual ALtimetry Station. VALS Version 0.5.7, 2009. Available online: http://www.ore-hybam.org/index.php/eng/Software/VALS (accessed on 16 May 2014).

25. Pavlis, N.K.; Holmes, S.A.; Kenyon, S.C.; Factor, J.K. The development and evaluation of the Earth Gravitational Model 2008 (EGM2008). *J. Geophys. Res. Solid Earth* **2012**, *117*, doi:10.1029/2011JB008916.

26. Rosner, B. On the detection of many outliers. *Technometrics* **1975**, *17*, 221–227.

27. Clarke, R.T. Uncertainty in the estimation of mean annual flood due to rating-curve indefinition. *J. Hydrol.* **1999**, *222*, 185–190.

28. ORE HYBAM—The Environmental Research Observatory on the Rivers of the Amazon Basin. Available online: http://www.ore-hybam.org/ (accessed on 18 September 2014).

29. Hydroweb—Hydrology from Space. Available online: http://www.legos.obs-mip.fr/fr/soa/hydrologie/hydroweb/Page_2.html (accessed on 18 September 2014).

30. Crétaux, J.-F.; Jelinski, W.; Calmant, S.; Kouraev, A.; Vuglinski, V.; Bergé-Nguyen, M.; Gennero, M.-C.; Nino, F.; Abarca Del Rio, R.; Cazenave, A. SOLS: A lake database to monitor in the Near Real Time water level and storage variations from remote sensing data. *Adv. Space Res.* **2011**, *47*, 1497–1507.

31. Hartigan, J.A. *Clustering Algorithms*; Wiley: New York, NY, USA, 1975.

32. Hartigan, J.A.; Wong, M.A. Algorithm AS 136: A k-means clustering algorithm. *J. R. Stat. Soc. Ser. C Appl. Stat.* **1979**, *28*, 100–108.

33. Bhaskar, N.R.; O'Connor, C.A. Comparison of method of residuals and cluster analysis for flood regionalization. *J. Water Resour. Plan. Manag.* **1989**, *115*, 793–808.

34. Burn, D.H.; Goel, N.K. The formation of groups for regional flood frequency analysis. *Hydrol. Sci. J.* **2000**, *45*, 97–112.

35. Rao, A.R.; Srinivas, V.V. Regionalization of watersheds by hybrid-cluster analysis. *J. Hydrol.* **2006**, *318*, 37–56.

36. Isik, S.; Singh, V.P. Hydrologic regionalization of watersheds in Turkey. *J. Hydrol. Eng.* **2008**, *13*, 824–834.

37. Toth, E. Catchment classification based on characterisation of streamflow and precipitation time series. *Hydrol. Earth Syst. Sci.* **2013**, *17*, doi:10.5194/hess-17-1149-2013.

38. Hydrological Data and Maps Based on SHuttle Elevation Derivatives at Multiple Scales USGS HydroSHEDS. Available online: http://hydrosheds.cr.usgs.gov/index.php (accessed on 18 September 2014).

39. Davies, D.L.; Bouldin, D.W. A cluster separation measure. *IEEE Trans. Pattern Anal. Mach. Intell.* **1979**, *PAMI-1*, 224–227.

40. Mahé, G. Modulation annuelle et fluctuations interannuelles des précipitations sur le bassin versant du Congo. In *Grands bassins fluviaux périatlantiques*; ORSTOM: Paris, France, 1995; pp. 13–26.

41. Hastenrath, S.; Polzin, D.; Mutai, C. Diagnosing the 2005 drought in equatorial East Africa. *J. Clim.* **2007**, *20*, doi:10.1175/2009JCLI3094.1.

42. Becker, M.; Llovel, W.; Cazenave, A.; Güntner, A.; Crétaux, J.-F. Recent hydrological behavior of the East African great lakes region inferred from GRACE, satellite altimetry and rainfall observations. *Comptes Rendus Geosci.* **2010**, *342*, 223–233.

43. Vennetier, P. Géographie du Congo-Brazzaville. Available online: http://horizon.documentation.ird.fr/exl-doc/pleins_textes/divers11-11/01471.pdf (accessed on 16 May 2014).

44. Laraque, A.; Orange, D.; Maziezoula, B.; Olivry, J.-C. Origine des variations de debits du Congo à Brazzaville durant le XXème siècle. In *Water Resources Variability in Africa during the 20th Century*; Servat, E., Hughes, D., Fritsch, J.-M., Hulme, M., Eds.; AISH: Wallingford, UK, 1998; pp. 171–179.

45. GRDC—Global Runoff Data Center, Zaire—Ouesso Station. Available online: http://www.grdc.sr.unh.edu/html/Polygons/P1448100.html (accessed on 18 September 2014).

Estimation of Reservoir Discharges from Lake Nasser and Roseires Reservoir in the Nile Basin Using Satellite Altimetry and Imagery Data

Eric Muala, Yasir A. Mohamed, Zheng Duan and Pieter van der Zaag

Abstract: This paper presents the feasibility of estimating discharges from Roseires Reservoir (Sudan) for the period from 2002 to 2010 and Aswan High Dam/Lake Nasser (Egypt) for the periods 1999–2002 and 2005–2009 using satellite altimetry and imagery with limited *in situ* data. Discharges were computed using the water balance of the reservoirs. Rainfall and evaporation data were obtained from public domain data sources. *In situ* measurements of inflow and outflow (for validation) were obtained, as well. The other water balance components, such as the water level and surface area, for derivation of the change of storage volume were derived from satellite measurements. Water levels were obtained from Hydroweb for Roseires Reservoir and Hydroweb and Global Reservoir and Lake Monitor (GRLM) for Lake Nasser. Water surface areas were derived from Landsat TM/ETM+ images using the Normalized Difference Water Index (NDWI). The water volume variations were estimated by integrating the area-level relationship of each reservoir. For Roseires Reservoir, the water levels from Hydroweb agreed well with *in situ* water levels (RMSE = 0.92 m; R^2 = 0.96). Good agreement with *in situ* measurements were also obtained for estimated water volume (RMSE = 23%; R^2 = 0.94) and computed discharge (RMSE = 18%; R^2 = 0.98). The accuracy of the computed discharge was considered acceptable for typical reservoir operation applications. For Lake Nasser, the altimetry water levels also agreed well with *in situ* levels, both for Hydroweb (RMSE = 0.72 m; R^2 = 0.81) and GRLM (RMSE = 0.62 m; R^2 = 0.96) data. Similar agreements were also observed for the estimated water volumes (RMSE = 10%–15%). However, the estimated discharge from satellite data agreed poorly with observed discharge, Hydroweb (RMSE = 70%; R^2 = 0.09) and GRLM (RMSE = 139%; R^2 = 0.36). The error could be attributed to the high sensitivity of discharge to errors in storage volume because of the immense reservoir compared to inflow/outflow series. It may also be related to unaccounted spills into the Toshka Depression, overestimation of water inflow and errors in open water evaporation. Therefore, altimetry water levels and satellite imagery data can be used as a source of information for monitoring the operation of Roseires Reservoir with a fairly low uncertainty, while the errors of Lake Nasser are too large to allow for the monitoring of its operation.

Reprinted from *Remote Sens.* Cite as: Muala, E.; Mohamed, Y.A.; Duan, Z.; van der Zaag, P. Estimation of Reservoir Discharges from Lake Nasser and Roseires Reservoir in the Nile Basin Using Satellite Altimetry and Imagery Data. *Remote Sens.* **2014**, *6*, 7522-7545.

1. Introduction

Hydrological data are key information for water resources management. However, such data are frequently not readily available, particularly in transboundary river basins, either because of not being measured or limited accessibility to the data by the riparian states. Information of river flow, reservoir storage and water use in a given riparian country is obviously of high importance for the whole basin. Such data is often not fully shared, particularly in water-scarce basins, e.g., the Nile, Indus, Tigris and Euphrates river basins [1]. The decline of hydrological networks in the world, particularly in developing countries, adds to the challenges of having accurate and representative hydrological data in river basins [2].

The Nile Basin covers an area of 3.3 million km², is 6500 km long and is shared by 11 countries (Figure 1a). From south to north, the Nile traverses through varying climates, including the equatorial lakes, savannah, Sahara and, ultimately, the Mediterranean climate at its outlet. The basin is experiencing increasing water demands by the growing population, creating strong competition over the (fixed) water resource. Large dams have been constructed in the lower part of the basin (Egypt and Sudan), and new dams are planned or under construction in many locations, e.g., Bujagali Dam in Uganda, the Grand Ethiopia Renaissance Dam (GERD) in Ethiopia, and the Setit Dam in Sudan. Large dams change the water regime and availability not only locally, but at the basin scale, which then necessitates transboundary water management for optimal utilization of the resources. In fact, this has been the trigger for the formation of the Nile Basin Initiative (NBI), started in 1999, to support sustainable development and the equitable utilization of, and benefits from, the Nile water resources [3].

As at 2013, the riparian countries had not yet reached consensus on a data sharing protocol. The Comprehensive Framework Agreement (CFA) was signed by six out of the eleven Nile countries. The CFA intends to provide a legal and institutional framework for basin water resources management, including data sharing protocols [4]. The countries share only few hydrological data among themselves [5].

Satellite remote sensing is emerging as a potential technique to support hydrological monitoring and, hence, inform water resources management in river basins [1,6]. Satellite altimetry, a remote sensing technique, has been successfully used to derive water level data in lakes, reservoirs, rivers, floodplains and wetlands, providing data for more than 15 years [7].

The water level data derived from satellite altimetry have been combined with *in situ* measurements to estimate water storage in lakes and reservoirs, with successful applications in different parts of the world. The water volume variation of Lake Dongting in China was estimated using a relation developed between water level from satellite altimetry and *in situ* water storage [8]. Medina *et al.* [9] also estimated the volume variation of Lake Izabal in Guatemala from relations utilizing *in situ* measurements, satellite altimetry and imagery data. Duan and Bastiaanssen [10] proposed a method using only satellite altimetry and imagery data to estimate the volume variations of Lake Tana (Ethiopia) and Lake Mead (USA). This latter method can be used in the absence of *in situ* data. This method has been used in this study to estimate volume variation for Roseires Reservoir and Lake Nasser.

Figure 1. Locations of (**a**) Nile Basin; (**b**) Lake Nasser, Egypt; (**c**) Roseires Reservoir, Sudan and the gauging stations.

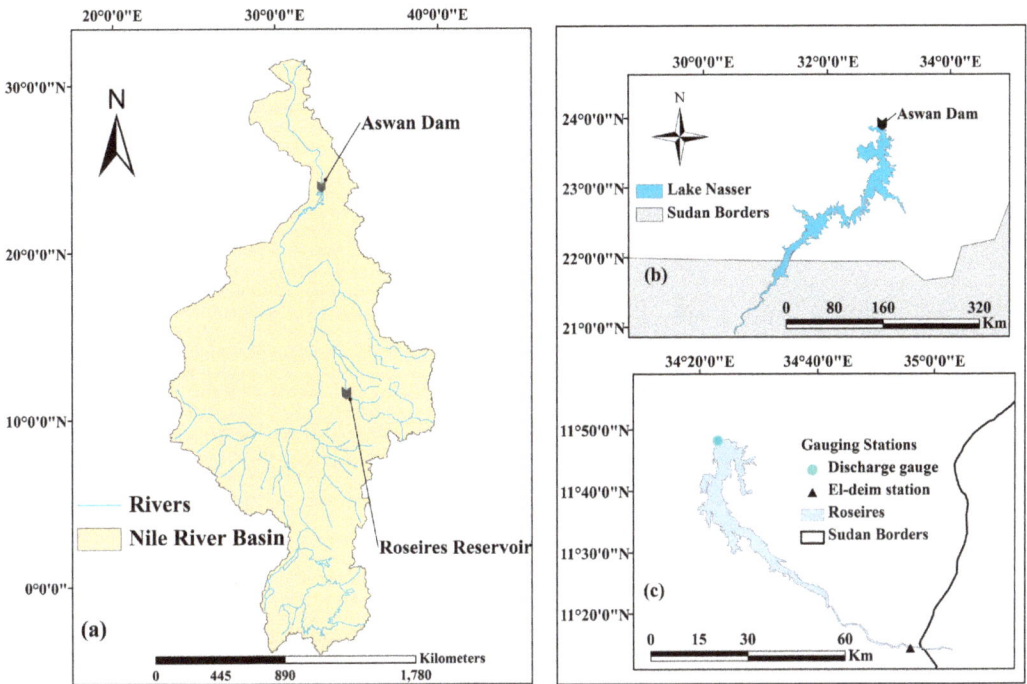

Altimetry data have also been used to estimate river discharges, e.g., for the Ob River [11], Chari/Ouham confluence near Lake Chad Basin [12], Amazon [13] and Ganga-Brahmaputra rivers [14]. In these studies, the river discharges were derived from rating curves developed from altimetry water levels and *in situ* discharge measurements. In the absence of *in situ* measurements, Leon *et al.* [15] utilized altimeter water level in a flow routing model (Muskingum–Cunge) to estimate the discharge of the upper Negro River in the Amazon basin (Brazil). The literature shows that discharge of rivers can be derived from rating curves. However, for lakes, the water level-discharge relationship does not significantly exist or, even if it exists, it may not be available for sharing. Swenson and Wahr [16] estimated the various components of Lake Victoria in East Africa using water balance. The lake's level/storage, evaporation and precipitation were derived from satellite (e.g., altimetry, TRMM and GRACE) and the water inflow from models. The study presented here also uses the water balance, but uses satellite imagery and altimetry, public data on rainfall and evaporation and limited *in situ* data to estimate the discharge of a lake/reservoir. Therefore, this research seeks to add to the methods of estimating lake/reservoir discharges.

In summary, the literature shows many applications of using satellite altimetry and imagery data in combination with *in situ* measurements to derive hydrological information for water resources management. The degree of success is case specific, depending on the frequency and quality of satellite data in relation to the size and shape of the given water body.

This study aims at testing satellite data (altimetry and imagery) combined with limited *in situ* measurements for the operation of two large reservoirs in the Nile Basin: Roseires Reservoir and

Aswan High Dam/Lake Nasser (Figure 1). The key question was: How accurate can satellite altimetry and imagery data estimate changes in storage volume and, combined with *in situ* measurements of reservoir inflow, reservoir discharges?

The paper is organized as follows: Section 2 gives a brief description of the two study sites. Section 3 presents all of the data used and their processing for the two reservoirs (altimetry, imagery and *in situ* measurements). The methods used for water balance and data validation are discussed in Section 4, and Section 5 presents the results and discussion. Finally, key conclusions and lessons learned are reported in Section 6.

2. Description of the Study Areas

Roseires Reservoir (Sudan) and Lake Nasser (Egypt) serve crucial functions to the population of both countries (Figure 1). They have been selected for this study because: (i) altimetry data have not been used yet to derive reservoir discharge; (ii) they are located in a transboundary river basin (Nile), where data exchange among riparian countries is limited, even though Sudan and Egypt have an agreement on data sharing (1959 agreement); and (iii) altimetry and satellite imagery data, as well as *in situ* data on inflows and outflows (for validation) were available for both sites.

2.1. Roseires Reservoir

Roseires Reservoir is located on the Blue Nile at Damazin, 550 km southeast of Sudan's capital, Khartoum, and 110 km from the Ethiopia-Sudan border (Figure 1c). The reservoir is located at the southeastern part of Sudan, whose topography is made up of steppes and low mountains. The altitude ranges from about 350 m at Khartoum to 490 m above mean sea level (AMSL) at the Ethiopian-Sudan border. The reservoir was constructed in 1966 for irrigation (Gezira Scheme) and hydropower generation (280 MW). The physical characteristics of Roseires Reservoir are presented in Table 1. The reservoir capacity has decreased by 40% due to sedimentation [17]. However, in January 2013, works were completed that heightened the dam wall by 10 m, increasing the reservoir's storage capacity to 7.4 billion m^3 [18].

Table 1. Characteristics of the two reservoirs studied: Roseires Reservoir and Lake Nasser.

Characteristics	Roseires Reservoir (Sudan)	Lake Nasser (Egypt)
Max length (km)	80	500
Max width (km)	9	12
Maximum depth (m)	68	110
Mean depth (m)	50	70
Reservoir area (km^2)	290	6000
Water volume (km^3)	3	162
Average annual inflow (km^3/y)	49	70
Average residence time (y)	0.06	2.30
Major water uses	Irrigation, hydropower	Irrigation, hydropower

The temperature within the environs of the reservoir ranges from 27 to 46 °C. Rainfall normally occurs between June and October, with an annual average of 0.7 m/y (Damazin station). The annual average water inflow at El-deim is 49 km^3/y.

The operation of Roseires Reservoir distinguishes four stages. During the first stage (rising flood) from July to August, the reservoir level is drawn to a minimum level of 470 m (AMSL). In the second stage, between 1 and 26 September, the reservoir is filled, depending on the water inflow from El-deim. During the third state, the full retention level is maintained, while the fourth stage marks the start of the emptying of the reservoir [17]. Now that the dam wall has been heightened, filling is expected to start earlier each year.

2.2. Lake Nasser

Lake Nasser is one of the greatest man-made lakes in the world, formed after the creation of the Aswan High Dam (AHD) in 1971 on the Nile River (Figure 1b). The lake is located in a desert region. The Arabian/Eastern Desert is located east of the Nile, while the low-lying sand dunes and depressions are found in the Western Desert. The dam was built to provide hydropower (2100 MW) and a steady water supply for irrigation in Egypt (55.5 km^3/y). The lake has a length of about 500 km, 330 km in Egypt and 170 km in Sudan. Table 1 gives further characteristics of Lake Nasser.

The lake is vital to Egypt, as it stores and regulates Nile water, being the main source of freshwater for about 85% of its population. The lake is located in a very hot, dry climate with an annual evaporation ranging from 2.1 to 2.6 m/y [19]. The annual rainfall over the lake is negligible [20].

The operation rules of Nasser reservoir aim at ensuring adequate water supply and the safety of the Aswan High Dam. At the beginning of the water year (1 August), the water level is kept at 175 m AMSL to fulfil high and low flow requirement. When the water level upstream reaches an elevation between 178 m and 183 m, excess water is directed to the Toshka Depression and, if necessary, by means of the emergency spillways on the western bank of the Nile [21]. The maximum retention at 180 m AMSL is obtained in November and, subsequently, the reservoir levels decrease from January to July as water is released.

3. Input Datasets

3.1. In Situ *Data for Roseires Reservoir*

In situ daily water levels (h), water inflows (Q_{in}), outflows (Q_{out}) and volumes (V) for Roseires Reservoir for the period 2002–2010 were obtained from the Ministry of Water Resources, Sudan. The *in situ* water level is referenced to Alexandria Datum (*i.e.*, above mean sea level). The level is taken from graduations on the dam wall. The inflow data is measured at El-deim station close to the Sudan-Ethiopia border. The (observed) outflow from the reservoir is monitored from a short distance downstream of the dam. The bathymetry survey of 2005 with reservoir levels and water volumes were used to derive the volume-level relationship, as given by Equation (1). Note that Equations (1) and (2) below (Section 5.2) have very sensitive decimal places and must be used as such. The bathymetric table shows that the minimum level is 467.00 m, with a corresponding volume of 13.70 m³ million, while the maximum level was 481.00 m, with a volume of 1934.73 m³ million. The data nicely spread between the minimum and maximum level every 1.00 m. The equation gave an excellent fit ($R^2 = 0.9995$).

$$V = 0.16565h^3 - 223.30002h^2 + 100,166.89815h - 14,949,451.43946 \tag{1}$$

where:

V = Storage volume in Mm³,

h = observed reservoir level in m AMSL.

3.2. In Situ Data for Lake Nasser

Ten-day mean measurements for *in situ* water levels (h), water inflows (Q_{in}) and outflows (Q_{out}) for the periods 1999–2002 and 2005–2009 and daily *in situ* water levels and volumes for the period 2007–2009 were obtained from the Ministry of Water Resources and Irrigation (MoWR), Egypt. *In situ* data were not received for the period 2003–2004. The *in situ* data are with respect to Alexandria Datum (*i.e.*, AMSL). To allow for daily interpolation, the 10-day mean measurements were assumed to occur on Days 5, 15 and 25 of each month, and linear interpolation was performed for the intermediate days. This interpolation may introduce some uncertainty in the validation results, but this is expected to be small. The water inflow is recorded at Dongola station in Sudan and is computed by a rating curve equation.

The daily *in situ* water levels and volumes for the period 2007–2009 were used to derive the volume-level relationship for the lake. The relation was further used to convert *in situ* water levels to *in situ* volumes for the whole period of 1992–2002 and 2005–2009. The converted *in situ* volumes were used to validate the volumes from satellite measurements. The derived relation is given in Equation (2). This is based on 731 data points, with a minimum level of 173.30 m (volume: 112,600 million m³) and maximum level of 180.11 m (volume: 150,193 million m³). The data gave an excellent fit ($R^2 = 1.0$)

$$V = 5.56806h^3 - 2,858.00945h^2 + 493,925.51557h + 28,630,490.83329 \tag{2}$$

3.3. Altimetry Water Level Datasets

Altimetry water levels from the Hydroweb and Global Reservoir and Lake Monitoring (GRLM) databases were used in this study. These databases were chosen because of the temporal resolution, level of processing and data availability for the two study areas. Readers are referred to Duan and Bastiaanssen [10] for a discussion of all four satellite altimetry water level databases for lakes and reservoirs.

Hydroweb is prepared by LEGOS/GOHS (Laboratoire d'Études en Géophysique et Océanographie Spatiale/Equipe Geodesie, Oceanograhie et Hydrologie Spatiale) in Toulouse, France. The altimetry data were derived from Topex/Poseidon, ERS-1 and 2, Envisat, Jason-1 and GFO satellites. The data are average monthly water level time series [22]. The reference of the water level is the GRACE Gravity Model 02 (GGM02) geoid. The procedure for water level processing in Hydroweb is described in detail by Cretaux *et al.* [23].

The Global Reservoir and Lake Monitor data (GRLM) are prepared by the United States Department of Agriculture's Foreign Agricultural Service (USDA-FAS) in collaboration with NASA and the University of Maryland. The database uses data from Topex/Poseidon (T/P), Jason-1, Jason-2 and Envisat, and the data are at time interval of 10-days [24]. The reference of water levels from GRLM is with respect to the mean 9-year T/P water level. Therefore, they are expressed in relative water levels. The procedure for water level processing in GRLM can be found in [25]. The water levels of Lake Nasser are available in both databases, while water levels of Roseires can be found only in Hydroweb.

3.4. Landsat TM/ETM+ Imagery Data

Landsat TM/ETM+ imagery data were used to extract the water surface areas of the two reservoirs: Roseires and Lake Nasser. TM/ETM+ imagery data were chosen because of their long-term data availability (since 1984), free access and high spatial resolution (30 m). One scene of a Landsat image can entirely cover Roseires Reservoir, while three scenes are needed to completely cover Lake Nasser. The acquisition dates of images for the two reservoirs are given in Section 5. The data were downloaded freely [26].

Two problems were encountered when using Landsat images: cloud cover and domain, and were worked around as discussed in this section. Landsat images with dates coinciding with that of altimetry water level measurements were chosen to extract the reservoir's water extent. The selection of coinciding dates was not always possible, because of high cloud cover and different revisit periods of the Landsat and altimeter satellites. In such cases, the closest dates were selected since the climate variations in some month(s) would not change much. In Hydroweb, water levels in a close or the same month and year with images were chosen, while in the GRLM database, specific days were chosen. Some of these dates had a big difference (*i.e.*, 10 to 60 days) with respect to the date of altimetry water levels . Undoubtedly, this introduces an error in the estimation of the reservoir area. In the acquisition of the images of Roseires, it was observed that the boundary of Roseires Reservoir overlapped with the edge of some image scenes, reducing the availability of one complete scene. This situation also limited the acquisition of close or coinciding dates to derive the area-level relationship. The merging of three Landsat images to cover Lake Nasser with different acquisition dates also introduces errors in the delineation of the lake area.

The Landsat-7 satellite has had a problem with its Scan Line Corrector (SLC) since 31 May 2003, resulting in SLC-off ETM+. This failure has led to about a 22% data loss due to the increased scan gap [27]. Therefore, gap filling was done for images after May 2003, using Local Linear Histogram Matching (LLHM). The LLHM uses a Landsat TM or ETM+ SLC-on image to fill the SLC-off image [28]. The images chosen for the gap filling were cloud-free images and had comparable seasonal conditions. The Landsat-7 image gap filling was done for both Roseires Reservoir and Lake Nasser, since some of the acquisition dates were after May 2003.

3.5. Rainfall and Evaporation

Long-term mean values of monthly rainfall data for Roseires Reservoir and Lake Nasser were obtained from the International Water Management Institute (IWMI) On-line Climate Summary Service Model [29]. Generally, the IWMI On-line Service Model data is based on data obtained from weather stations around the world for the period 1961–1990. Using mean values of rainfall and evaporation is expected to generate negligible error in the water balance of the two reservoirs. Rainfall volume on Roseires Reservoir is very small, because the reservoir area is at a minimum during the rainy season, while on Lake Nasser, the rainfall rate is negligible. Open water evaporations from the two reservoirs were calculated using the Penman [30] formula. All parameters in the Penman formula, *i.e.*, temperature, relative humidity, wind speed and relative sunshine duration, were obtained from long-term mean monthly values of the IWMI online climate summary service model. This is expected to cause a small error in evaporation volume, which is at least much smaller than the uncertainty of the reservoir area.

4. Methods

4.1. Altimetry Water Level Measurements

As given in Section 3.2, altimetry water levels for the two reservoirs were acquired from two databases (Hydroweb and GRLM). The mean difference (constant shift) between the *in situ* and altimetry water levels was computed and then simply added to altimetry water levels [10,31]. This ensures the attainment of a common datum, which allows the comparison of the two data series. For validation, the commonly used indicators, *i.e.*, the coefficient of determination (R^2) and the root mean square error (RMSE), were computed.

Alternatively, Global Navigation Satellite System (GNSS) data provide a datum shift for altimetry data for certain specific locations. However, there were no network stations of GNSS for our studies of the two reservoirs, and therefore, these were not used [32].

4.2. Delineation of Reservoir Surface Area

The extents (surface areas) of the two reservoirs were delineated from Landsat satellite images using the Normalized Difference Water Index (*NDWI*) [33] as given by Equation (3):

$$NDWI = \frac{(GREEN - NIR)}{(GREEN + NIR)} \tag{3}$$

where *GREEN* and *NIR* are the green and near-infrared red bands, respectively. The water features have positive values due to their higher reflectance of the green band compared to the *NIR* band, while vegetation and soil features have zero or negative values because of their higher reflectance of the *NIR* band compared to the green band. The Modified Noramlized Difference Water Index (MNDWI), which replaces the *NIR* band in Equation (3) with the mid-infrared red (MIR) band, has been reported to perform better than *NDWI* [34]. However, we found that the MIR band (Band 5) for both study areas had poor qualities; the boundaries of the image scenes shift when the MNDWI is applied. Therefore, *NDWI* was used to delineate the reservoir areas in this study. Band 2 of

Landsat TM/ETM+ (green), and Band 4 (*NIR*) were used in Equation (3). With the aid of visual inspection and *NDWI* ranging from 0.01 to 1, water bodies of both study areas were extracted. The range was based on the general observation of detecting water bodies within a trial and error range from 0 to a positive value. The reservoir surface area was then calculated as the sum of the areas of the pixels identified as water bodies.

4.3. Storage Volume Estimation

A surface area-water level relation (based on satellite measurements) was integrated to obtain volume-level relations for each reservoir. First, the lowest water level (h_{min}) of the altimetry time series data was identified. The lowest water level was then subtracted from all water levels (h) obtained from each satellite altimetry data (*i.e.*, $h - h_{min}$). The ($h - h_{min}$) can also be known as the water depth *d* above h_{min} [10]. It is assumed that the storage volume is zero at h_{min}, *i.e.*, when $d = 0$, $V = 0$, but $A \neq 0$, where *A* is surface area of the reservoir/lake. However, in reality, there is a storage volume in the reservoir at *d*, at least equal or larger than the dead storage. This is to allow water volume computations independent of the dead storage. For comparison, the *in situ* volumes have also been converted to volumes above the lowest water level (h_{min}). The conversion was done by subtracting the *in situ* volume for the same date that the lowest water level occurred in the satellite altimetry products.

Therefore, time series of two variables (surface area and water level) were prepared. The surface area (*A*) of a reservoir/lake delineated from the TM/ETM+ images for a given date is associated with altimetry water level measurements (h), converted to ($h - h_{min}$) of the same (or closest) date. A second-polynomial function ($A = f(d) = ad^2 + bd + c$) was obtained by correlating the surface area (*A*) in Mm² and water depth (*d*) in m, a, b, c being constants determined by regression analysis. The *A-d* relation was then integrated to obtain the volume-level relation ($V = f(d) = ad^3/3 + bd^2/2 + cd + e$); where *V* is the water volume above h_{min}. The *A-d* relation was integrated with the condition that the water volume (*V*) is equal to zero when water depth (*d*) is zero. The constants a, b, and c are the same values as in the *A-d* relation, and e is solved as zero (0) given the condition $V = 0$ when $d = 0$.

4.4. Water Balance of Reservoirs

The discharge from a reservoir/lake has been computed from the water balance equation of Equation (4), assuming negligible groundwater interactions:

$$Q_{in} + A\,(P - E) - dS/dt = Q_{out} \tag{4}$$

where:

Q_{in} = inflow in Mm³/day
Q_{out} = reservoir discharge in Mm³/day
P and *E* = precipitation and open water evaporation in m/day, respectively
dS/dt = change in storage volume with time in Mm³/day
A = the reservoir's water surface area in Mm² (=km²).

The inflow, Q_{in}, was obtained from *in situ* measurements. The reservoir discharge, Q_{out}, was computed based on altimetry water level measurements, Area-level and volume-level relations (*i.e.*, dS/dt) were derived from Landsat and altimetry data. Q_{out} is computed on decadal time steps (10-day) for GRLM and monthly for Hydroweb. The computed discharges Q_{out} of Equation (4) were then validated against observed discharges. The objective here is to assess the accuracy of the discharge computed from a reservoir if it is based on satellite data of water levels and storage volume.

5. Results and Discussion

5.1. Results for Roseires Reservoir

5.1.1. Altimetry Water Level Measurements

The time-series of monthly water levels from Hydroweb and *in situ* measurements for Roseires Reservoir is shown in Figure 2. The Hydroweb water levels have been shifted vertically to the datum of the *in situ* measurements by adding a constant shift [31]. The shift of −1.54 m is the mean difference between the two data series for the period of nine years from 2002 to 2010 (Table 2). Figure 2 shows that water levels from Hydroweb agreed well with the *in situ* water level measurements ($R^2 = 0.96$), in particular for high reservoir levels. However, water levels from Hydroweb overestimate reservoir levels during the flood season (when water levels are kept low), as was the case in 2004 and 2005, but not during the 2006 flood. Note that usually, the reservoir is at the maximum level by the end/beginning of the year and at minimum level during the flood season, June, July and August. Because of the very high flow during the flood season of 2006, which exceeded the gate capacity of Roseires Dam, the reservoir level rose above minimum levels, the so-called compulsory storage [35]. The under estimation at high water levels and over estimation at low levels could be attributed to the adjustment from the constant shift.

Table 2. Statistics of altimetry-derived water levels for Roseires Reservoir and Lake Nasser.

Study Areas	Dataset	Period	No. *	Interval	R^2	RMSE (m)	Shift Constant (m)	Mean Shifted Water Level (m)
Roseires	Hydroweb	2002–2010	63	monthly	0.96	0.92	−1.54	479.75
Lake Nasser	Hydroweb	1999–2002 2005–2009	89	monthly	0.81	0.72	−0.12	176.89
	GRLM	1999–2002 2005–2009	215	10-day	0.94	0.62	179.43	177.12

* Refers to the number of data points used.

471

Figure 2. Time series of altimetry water levels from Hydroweb (blue), ranging from 473 m to 485 m, compared to *in situ* measurements (red), ranging from 470 m to 484 m, for Roseires Reservoir during the period 2002–2010. The water levels from Hydroweb have been shifted vertically by −1.54 m to correct for the datum. The error bars in blue represent the standard deviation of altimetry water levels from the Hydroweb database.

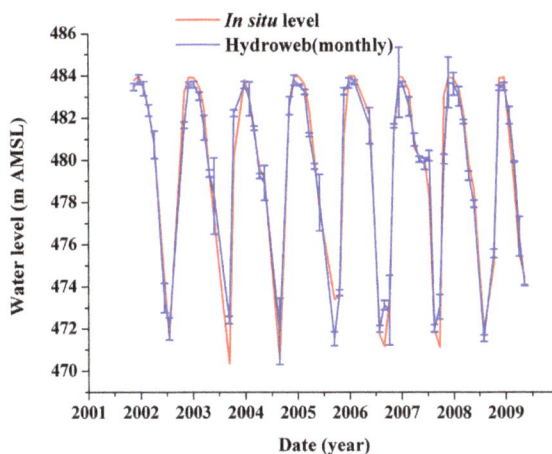

The RMSE of Hydroweb water levels against *in situ* water levels for Roseires was 0.92 m, which is about 7% of the seasonal variation (~14 m), and R^2 is 0.96, showing good agreement with *in situ* measurements. The literature reports a wide range of RMSE for different lakes and reservoirs worldwide. In general, the RMSE is small for large lakes, e.g., 3 to 7 cm for Lake Victoria, East Africa [31], but increases to several decimeters for smaller lakes, e.g., 26 cm for Lake Woods and 105 cm for Lake Powel [23]. The accuracy of altimetry water levels has been attributed to the size of the target water body, the surrounding topography and the roughness of the surface [23]. With larger rivers, RMSE ranges from 10 to 20 cm, e.g., Amazon River [36], and increases over narrower rivers and/or in the presence of vegetation [6]. The satellite laser altimetry, Ice, Cloud and Elevation Satellite (ICESat) derived water levels for Roseires Reservoir reveal an RMSE of 17 cm [37]. The improved accuracy of water levels from ICESat as compared to Hydroweb could be due to the smaller footprint of satellite laser altimeter (ICESat) than satellite radar altimeter (Hydroweb).

5.1.2. Reservoir Area and Volume

Nine pairs of coincident water levels from Hydroweb and Landsat TM/ETM+ imagery data were selected to determine surface area-level and further volume-level relations for Roseires (Table 3). Based on these nine pairs, the area-level relationship for Roseires was derived by regression analysis, as shown in Figure 3. The scatter could be attributed to two reasons: (i) different acquisition dates of altimetry data and satellite images, in particular during transition periods (filling or emptying of the reservoir); and (ii) the inherent uncertainties of both altimetry

472

measurements (discussed in Section 5.1.1) and the area delineation from the *NDWI* images. However, a representative *A-d* relation could be derived ($R^2 = 0.87$). The volume-level relation was obtained by the integration of Equation (5). The resulting function, as shown in Equation (6), was used to convert water levels from Hydroweb to calculated volumes for Roseires Reservoir.

$$A = 0.34d^2 - 5.04d + 147.30 \tag{5}$$

$$V = 0.11d^3 - 2.52d^2 + 147.30d \tag{6}$$

where:
A = area of reservoir derived from Landsat imagery (km²).
d = water depth (m).
V = volume in Mm³.

Table 3. Water depth with corresponding imagery data and delineated surface area for Roseires Reservoir.

No.	Altimetry Water Levels			Landsat TM/ETM+ Images		
	Date	Original (m)	d (m) [a]	Date	Sensor	Area (km²)
1	July 2009	473.09	0.02	14 September 2009	ETM+	148.10
2	August 2008	474.59	1.52	27 September 2008	ETM+	162.00
3	March 2010	477.93	4.87	9 March 2010	ETM+	167.90
4	April 2009	479.48	6.41	23 April 2009	ETM+	175.70
5	March 2009	480.80	7.74	22 March 2009	ETM+	211.90
6	February 2010	481.48	8.41	12 January 2010	TM	239.00
7	March 2008	482.52	9.45	4 April 2008	ETM+	217.30
8	January 2010	483.68	10.61	4 January 2010	ETM+	252.20
9	November 2007	485.24	12.17	28 November 2007	ETM+	246.60

"Original" is the original value obtained from Hydroweb. [a] The water depth referred to above is lowest altimetry water level, h_{min}.

Figure 3. The area (*A*) and water depth (*d*) relationship of Roseires Reservoir from paired altimetry water levels and the area derived from Landsat images.

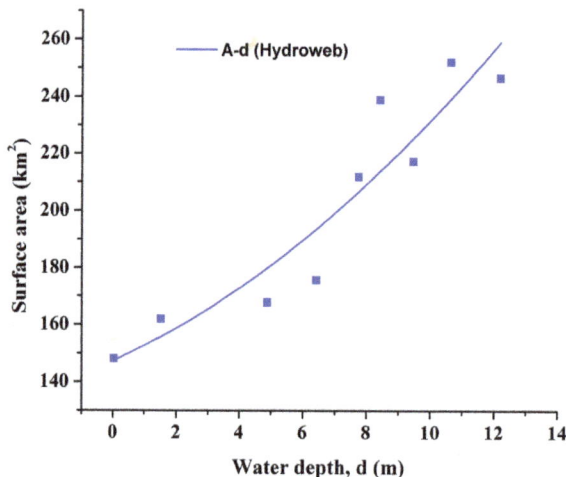

The monthly series of calculated from volume satellite data and calculated volume from *in situ* data for Roseires Reservoir from 2002 to 2010 are shown in Figure 4. The calculated volumes from satellite data were obtained by applying Equation (6) to altimetry water level data. The calculated volumes from satellite data were relative to the reference volume, which corresponds to the lowest water level (h_{min}) from Hydroweb. The calculated volumes from *in situ* data were derived by subtracting the *in situ* water volume that corresponds to the lowest altimetry water level from all *in situ* water volumes obtained from Equation (1). As can be seen from Figure 4, the operation pattern could be reproduced by satellite-derived data. The RMSE was 355 Mm^3 (*i.e.*, 23% of the mean volume of 1529 Mm^3), while R^2 was 0.94. The calculated volume based on satellite data consistently underestimated water volumes when the reservoir was at a maximum and overestimated at minimum levels. At low water levels, all of the calculated volumes (*in situ*) had negative values, except for August 2006, when the lowest water level occurred. August 2006, was used as the reference water level for the *in situ* water volume, since the date corresponds to the lowest water level (h_{min}) from Hydroweb. However, for other dates, even lower *in situ* water levels were measured. This reference water level value is thus greater than the other low water levels, hence the existence of negative volumes at low levels. These negative volumes do not affect our discharge estimation, since this is calculated from water volume differences between two time periods rather than the absolute values of water volumes for a specific time (see Section 5.1.3 on discharge estimation).

Figure 4. Monthly series of calculated volume (*in situ* data and satellite data) above the reference level for Roseires Reservoir between 2002 and 2010. The *in situ* volume (red) was derived from the *V-h* relationship obtained from a bathymetry survey of 2005, (Equation (1)), but relative to the reference volume. The satellite volume (blue) is derived from the *V-d* relationship using satellite measurements, *i.e.*, Landsat imagery and altimetry water level (Equation (6)).

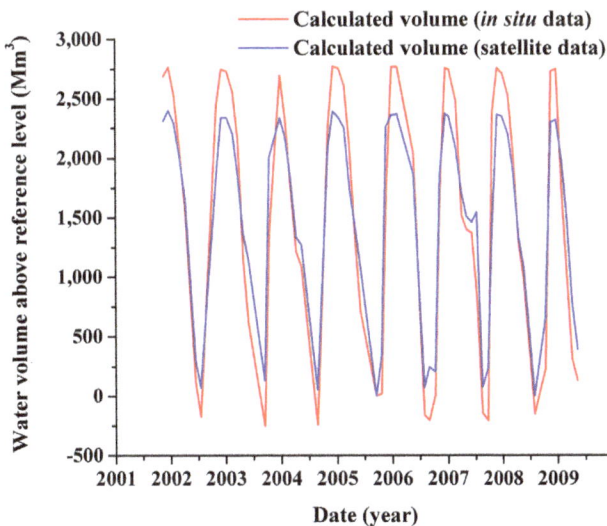

474

5.1.3. Reservoir Discharge

For validation, observed discharges from Roseires Reservoir based on a rating curve were compared with those computed by the water balance (wb) equation (Equation (4)). First, Q_{out} was computed using all *in situ* measurements (Q_{in}, P, E, A, dS/dt), labelled as Q (wb *in situ* data) in Figure 5. Here, dS/dt is based on *in situ* water level measurements, and the volume-level- relation is derived from bathymetric survey. Subsequently, Q_{out} was computed using satellite (sat.) data of A and dS/dt and *in situ* data of Q_{in}, P, E, labelled as Q (wb sat. data-Hydroweb). Here, dS/dt is based on altimetry water level measurements and the volume-level relation from satellite data with respect to the time step of the altimetry database (10-day or monthly). As shown, the three differently computed outflows are in excellent agreement with each other, e.g., Q (wb sat. data-Hydroweb) correlates well with Q (observed), $R^2 = 0.98$. The RMSE between the two datasets is 671 Mm^3/month, corresponding to 18% of the observed mean discharge. This can be considered an acceptable accuracy, given that the uncertainty of discharge measurement is between 5% and 10%, while errors up to 15% to 20% were reported for large rivers [14]. Discharge measurement errors of 4%–17% were reported for the Amazon River [13], and 17% for the Ob' river [11].

Figure 5. Roseires Reservoir discharges between 2002 and 2010. Red is Q (observed) based on rating curve; green is Q (wb *in situ* data) based on *in situ* data; blue is Q (wb sat. data-Hydroweb) based on satellite data.

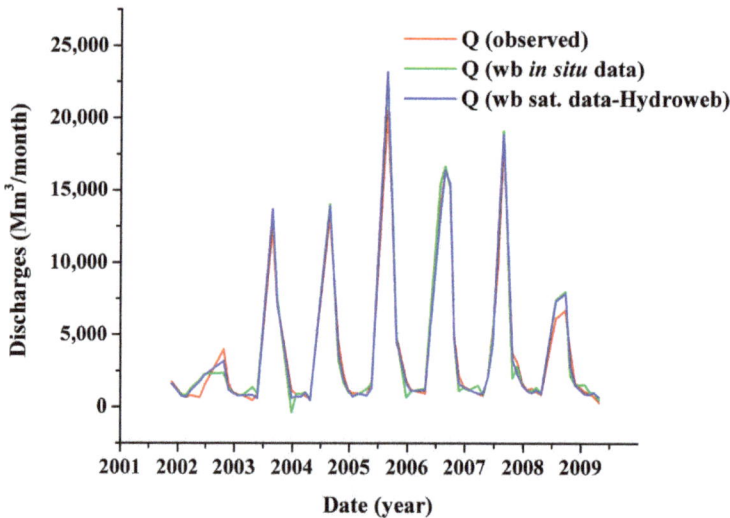

To understand the errors introduced by satellite *versus* other sources of errors, the two observed discharges were compared, Q (observed) from the rating curve and Q (wb *in situ* data). The mean difference between the two discharges resulted in 16%. This shows that the error of the discharge using satellite data relative to the discharge (all *in situ* data) can be in the order of 2%, assuming linear error propagation.

5.2. Results of Lake Nasser

5.2.1. Altimetry Water Level Measurements

The water level data for Lake Nasser from Hydroweb (monthly) and GRLM (10-day) covered the periods 1999 to 2002 and 2005 to 2009. To attain a common datum for comparison, the water levels of Hydroweb and GRLM were shifted vertically with −0.12 m and 179.43 m, respectively (Table 2). The shifted values varied largely in magnitude, because the Hydroweb products were absolute water levels, while the GRLM products were water level variations referenced to the mean nine-year T/P water level. Figure 6 shows the time-series of *in situ* measurements, Hydroweb and the GRLM. It compares the *in situ* levels and shifted altimetry water levels, and the gap is a result of the lack of *in situ* data from 2003 to 2005. In the absence of *in situ* data for this period, Section 5.2.2 shows the possibility of estimating water volume from Hydroweb water levels to cover this gap.

Figure 6. Time series of altimetry water levels from Hydroweb (blue) and GRLM (green) compared to *in situ* measurements (red) for Lake Nasser during the periods 1999–2002 and 2005–2009. The error bars in blue and green represent the standard deviation of water levels for Hydroweb and GRLM, respectively.

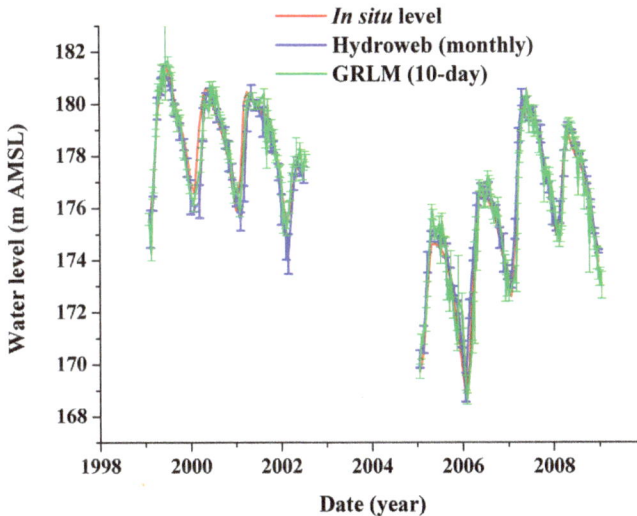

In Figure 6, both altimetry water levels are in good agreement in phase and amplitude, especially at high reservoir levels. It could be observed that the Lake Nasser levels showed a clear decline between 2002 and 2007. This could be attributed to the drought in East Africa affecting the major sources of the Nile River, *i.e.*, the White Nile from Lake Victoria and the Blue Nile from the Ethiopian Plateau [16]. The huge size of Lake Nasser reservoir (two-times annual flows) allows a long memory of response to inflow variability. Inflow was declining until 2005, but higher flows entered the lake from 2006 onward.

The RMSE of monthly data from Hydroweb against *in situ* water levels was 0.72 m, which is about 6% of the annual fluctuation (~13 m), while R^2 is 0.81 (Table 2). Similarly, the RMSE of altimetry data from GRLM against *in situ* water levels was 1.22 m, and R^2 is 0.82. In general, the raw GRLM data has outliers, which were identified by comparison with *in situ* data. The mean square errors (MSE) between the two datasets exceeding 3 m were considered major outliers and were removed. Consequently, 20 outliers were taken out of the 235 data points of GRLM. The RMSE after the removal was 0.62 m, which is about 5% of the annual fluctuation, and the R^2 was 0.94 (Table 2). The lower accuracy for Lake Nasser compared to Roseires could be attributed to the narrow and long shape of the former, which allows only small areas to be covered in the footprint of satellite altimetry, leading to the altimetry product being contaminated by land [25]. Land contamination for inland water is normally removed through customized processing of altimetry data, which is beyond the scope of this current study. Details on the removal of land contaminations have been discussed by developers of GRLM [25] and Hydroweb [23]. Furthermore, the difference in the places where the *in situ* gauge and altimetry satellites measure water levels would add some uncertainty. The single point-based *in situ* gauge station cannot reflect the spatial variation of water levels. The satellite altimetry measures along the track, which has a certain distance from the *in situ* gauge station.

5.2.2. Reservoir Area and Volume

Table 4 presents the selection of Landsat images matched with altimetry water level dates and the derived surface areas for Lake Nasser. Six monthly water levels from Hydroweb and areas derived from Landsat Images were paired to estimate the *A-d* relationship. Similarly, eight GRLM water levels and areas were paired to determine the *A-d*. Due to the narrow and long shape of Lake Nasser, three scenes were merged to obtain one complete scene for the lake. Two of the image scenes have similar dates, and each scene has its specific water level and extent. The acquisition dates for two of the scenes, Path 175 Row 44 and Path 175 Row 45, are the same; hence only one column is maintained as for both in Table 4. To reduce the error of water extent (area) arising from scene merging, the maximum allowed reservoir water level difference between the selected images scenes was assumed to be 0.6 m. Landsat images spanning from 1999 to 2002 were mostly used for area estimation, as they were free of image gaps, *i.e.*, easier to process compared to the SLC gap-filling procedure. Few Landsat SLC-off images were used, since the gap filling of three SLC-off image scenes using SLC-on images for Lake Nasser was time consuming.

The Equations (7), (8), (9) and (10) represent the *A-d* and *V-d* relationships derived from Hydroweb and GRLM. Equation (7) and (9) were integrated to obtain Equation (8) and (10) respectively. The lowest water levels (as the reference for water volume estimation) for Hydroweb and GRLM were different; therefore, their derived equations are database specific. Equations (8) and (10) were used to convert the water levels to water volumes.

For Hydroweb, the *A-d* relation is given by Equation (7), represented by the blue curve in Figure 7. The R^2 was 0.96.

$$A = 2.49d^2 + 141.52d + 3824.11 \tag{7}$$

$$V = 0.83d^3 + 70.76d^2 + 3824.11d \tag{8}$$

For GRLM, the A-d relation with R^2 of 0.99 is given by Equation (9), represented by the red curve in Figure 7.

$$A = 5.43d^2 + 99.06d + 3836.13 \tag{9}$$

$$V = 1.81d^3 + 49.53d^2 + 3836.13d \tag{10}$$

Table 4. Water depth and the corresponding estimated surface areas for Lake Nasser.

Water Level Sources	No.	Altimetry Water Level			Landsat TM/ETM+ Images		
		Date	Original (m)	d (m) [a]	Path 174 Row 44	Path 175 Row 45	Area (km²)
Hydroweb	1	July 2006	169.24	0.00	9 July 2006	16 July 2006	3846.08
	2	April 2006	172.94	3.70	20 April 2006	27 April 2006	4214.58
	3	July 2002	174.88	5.64	12 June 2002	21 July 2002	4881.45
	4	June 2002	176.07	6.84	12 June 2002	11 June 2002	5012.57
	5	August 1999	176.31	7.07	7 August 1999	6 August 1999	4919.08
	6	December 2002	177.40	8.16	December 2002	28 December 2002	5102.32
	7	March 2002	178.71	9.47	8 March 2002	15 March 2002	5476.42
	8	September 2000	179.35	10.11	10 September 2000	9 September 2000	5274.76
	9	November 1999	181.41	12.17	3 November 1999	10 November 1999	5998.30
GRLM	1	12 July 2006	−10.76	0.00	9 July 2006	16 July 2006	3846.08
	2	4 May 2006	−6.96	3.80	20 April 2006	27 April 2006	4214.58
	3	26 July 2002	−4.17	6.59	12 June 2002	21 July 2002	4881.45
	4	10 August 1999	−2.85	7.91	7 August 1999	6 August 1999	4919.08
	5	16 June 2002	−2.69	8.07	12 June 2002	11 June 2002	5012.57
	6	16 April 2001	−1.52	9.24	8 May 2001	13 April 2001	5177.57
	7	10 September 2000	−0.94	9.82	10 September 2000	9 September 2000	5274.76
	8	19 March 2002	−0.21	10.55	24 March 2002	15 March 2002	5476.42
	9	5 February 2001	0.11	10.87	5 March 2001	8 February 2001	5552.88
	10	17 November 1999	1.92	12.68	3 November 1999	10 November 1999	5998.30

"Original" is the original value obtained from the satellite altimetry products. [a] The water depth refers to above the lowest water level (h_{min}).

Figure 8 shows time-series of calculated volume from *in situ* data (Equation (2)) and calculated volumes from Hydroweb data (Equation (8)) and GRLM data (Equation (10)) for Lake Nasser during the periods 1999 to 2002 and 2005 to 2007. The calculated volumes from *in situ* data were relative to the reference volume, which corresponds to the lowest water level (h_{min}) from Hydroweb and GRLM. The operation pattern of Lake Nasser has been agreeably reproduced, although Hydroweb and GRLM underestimated the volume at minimum and maximum reservoir levels. The RMSE from Hydroweb data was 5720 Mm³ (*i.e.*, 15% of the mean volume of 38,847 Mm³), while R^2 was 0.93. The RMSE based on GRLM data was 3858 Mm³ (*i.e.*, 10% of the mean volume of 39,377 Mm³), while R^2 was 0.94. GRLM is slightly better than Hydroweb in terms of smaller relative RMSE for Lake Nasser.

Figure 7. The area (*A*)-water depth (*d*) relationships for Lake Nasser were derived from altimetry water levels and surface area (Landsat) data of Table 4 for Hydroweb (blue) and GRLM (red).

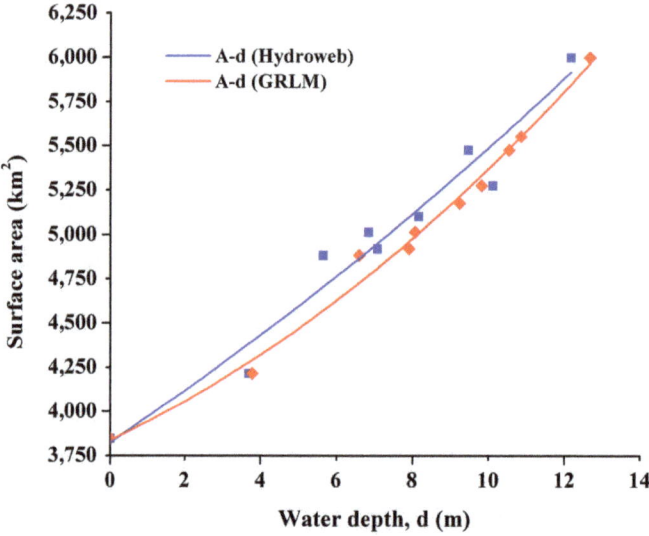

Figure 8. Calculated volume for Lake Nasser above the reference level from *in situ* data and GRLM during the period 1999–2000 and from Hydroweb during the period 1999–2007. Calculated volumes for the gap in the period (2003–2005) show the use of the Hydroweb water level in the absence of *in situ* water levels. The time steps for the volumes (Mm^3) in GRLM and Hydroweb are 10-day (decadal) and monthly, respectively.

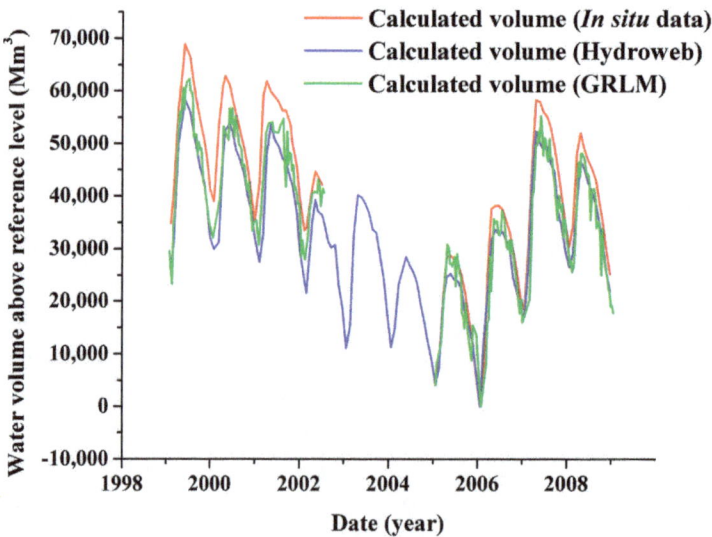

5.2.3. Reservoir Discharge

The discharges of Lake Nasser have been computed by the water balance equation. The comparison of the three discharges is given in Figure 9. Q (observed) is the observed discharge based on the rating curve, Q(wb sat. data-GRLM) is based on satellite data, while Q (wb *in situ* data) is based on *in situ* data.

There was good agreement between Q (wb sat. data-GRLM) and Q (wb *in situ* data), but large differences between Q(observed) and either Q (wb sat. data-GRLM) or Q (wb *in situ* data). The estimated discharges were overestimated at both low and high water levels. The 10-day discharge Q (wb sat. data-GRLM) poorly agreed with the Q (observed), giving an RMSE of 139% and an R^2 of 0.36 (Table 5). This error far exceeds the acceptable error in discharge measurements in large rivers.

Figure 9. Time series of observed and computed discharges for Lake Nasser during the periods 1999–2002 and 2005–2007. Red is Q (observed) based on observed discharge; green is Q (wb *in situ* data); and blue is Q (wb sat. data-GRLM).

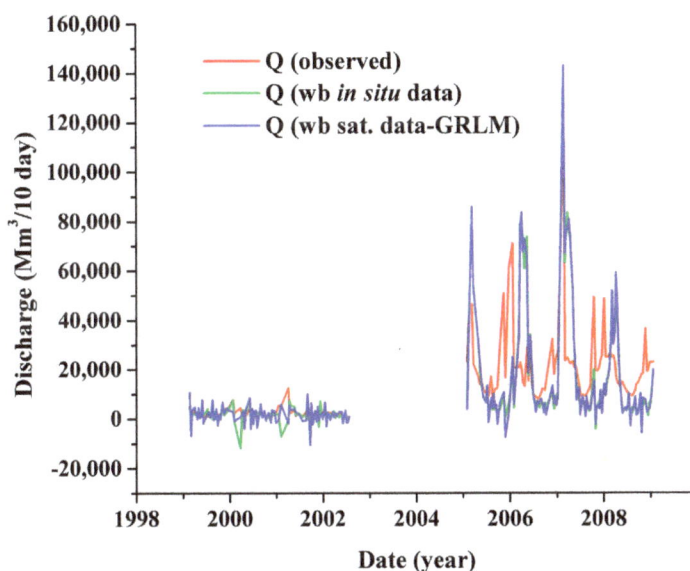

The three sets of discharge data were compared to understand error sources. A comparison of Q (wb *in situ* data) and Q (observed) also yielded a relatively large error (RMSE = 137%) and poor agreement (R^2 = 0.36). Comparing Q (wb sat. data-GRLM) and Q (wb *in situ* data) resulted in a much smaller, but still unacceptably high, error (RMSE = 30%), with a very high R^2 of 0.97. This may imply that satellite measurements might not be the major source of error. Discharges derived using Hydroweb (monthly) data in the water balance computations gave a relatively low error compared to GRLM (Table 5).

Table 5. Statistics of discharges using satellite data and *in situ* data in a water balance.

Study Areas	Data	Period	No. *	Observed_mean [a]	Estimated_mean [a]	R^2	RMSE [a]	%RMSE [b]
Roseires	Hydroweb	2002–2010	63	3769	3833	0.99	652	17.79%
	In situ data	"	101	3792	3869	0.98	611	16.11%
Lake Nasser	Hydroweb	1999–2002	89	5467	4235	0.09	3826	69.98%
		2005–2009						
	In situ data	"	89	5467	4127	0.26	3566	65.21%
	GRLM	1999–2002	215	11,947	11,409	0.36	16,704	138.92%
		2005–2009						
	In situ data	"	215	11,947	11,358	0.37	16,416	137.41%

* Refers to the number of datasets used. [a] Mm³/month for Hydroweb and *in situ* data and Mm³/10-day for GRLM and *in situ* data. Observed_mean is the mean of the observed discharges, and Estimated_mean is the mean of the estimated discharges. [b] Percentage of the RMSE relative to the mean observed data, *i.e.*, (RMSE/the mean observed discharge) × 100%.

The difference between Q (observed) and Q (wb) might be due to several reasons. Firstly, it could be due to unaccounted spills from the reservoir into the Toshka Depression and to other desert streams. It is known that when the reservoir level exceeds 178 m, water may flow into Toshka. Bastawesy *et al.* [38] confirm that there were water releases into Toshka within the period 1998–2002. This indicates that Toshka receives excess water in certain times of the year when water level exceeds 178 m. Secondly, the inflow Q_{in} might have been overestimated in this study, as it was measured at about 250 km upstream of the lake and more than 750 km from the outflow gates of the dam. Inflow in this reach may not be accurately captured. Finally, the large size of Lake Nasser compared to its inflow (*i.e.*, its large residence time) increases the uncertainty in computed reservoir releases, compared with that of a reservoir with a smaller residence time (such as Roseires Reservoir); *i.e.*, the same error in water level estimation introduces, in the case of a reservoir with a large residence time, a large absolute error in the computed volume of the water stored, which will translate into a large absolute error in computed outflow, which may result in a very large relative error.

6. Conclusions

As data on reservoirs are often scarce or not in the public domain, this study evaluated the feasibility and accuracy of using satellite altimetry and imagery data to estimate stored water volumes and, combined with limited *in situ* data, discharges from Roseires Reservoir (Sudan) and Lake Nasser (Egypt) in the Nile Basin. These estimated discharges would not only provide information on water releases to downstream users, but also give insight into the reservoir operation strategies.

The water volume of the reservoirs was derived from the integration of a lake-specific area-level relationship. The surface area was estimated from Landsat images using *NDWI*, while the water level was from satellite altimetry. The discharge was computed using the water balance of the reservoir. The other components of the water balance (evaporation and rainfall data) were obtained from the IWMI online database. The *in situ* water levels, *V-d* relations, water inflows and outflows were obtained from the responsible ministries of Sudan and Egypt. The obtained *in situ* water levels, water volumes and discharges were used for the validation of satellite-derived results.

For Roseires Reservoir, monthly water levels from Hydroweb showed a good agreement with *in situ* water levels (RMSE = 0.92 m and R^2 = 0.96). The RMSE of the calculated volume and discharge Q (wb, sat. data) were 23% and 18%, respectively. The discharge is within the acceptable error of 15%–20% for single discharge measurements in large rivers. The outcome shows the potential to use satellite information for reservoir operation, which could be very useful for the contexts of no *in situ* data. The results showed that satellite-derived data can be used as a fairly reliable source of information for water resources management at a river basin scale.

For Lake Nasser, water levels from both GRLM (10-day) and Hydroweb (monthly) agreed well with *in situ* water levels (*i.e.*, RMSE = 0.62 m, R^2 = 0.96 and RMSE = 0.72 m, R^2 = 0.81, respectively). The RMSE of the calculated volume from GRLM and Hydroweb were 10% and 15%, respectively. However, the error of the estimated discharge based on the water balance was quite high. The RMSE of the estimated discharge from GRLM and Hydroweb were 139% and 70%, respectively. Similarly, the same order of magnitude of error was obtained when the discharge was calculated with the water balance equation using *in situ* measurements. Sources of errors could include unaccounted outflows (e.g., reservoir spills to the Toshka Depression and water flows to desert streams) and overestimation of the water flowing into Lake Nasser. These errors would affect the water balance and influence the computed reservoir discharge.

This research contributes to the derivation of water discharges/releases in reservoir operations where there are limited or no *in situ* data. Specifically, the stored water volumes of Roseires Reservoir and Lake Nasser were successfully estimated, deriving their respective area-level and volume-level relationships using satellite data. Furthermore, by combining satellite-derived information on storage changes with *in situ* inflow data, the water releases from the two reservoirs could be adequately estimated for Roseires Reservoir, but with major limitations in the case of Lake Nasser. In the latter case, some major error sources may in fact not be related to satellite-derived information. It is also concluded that in general, error propagation in the estimation of reservoir releases for reservoir systems with shorter residence times will be smaller than in systems with longer residence times.

These findings are valuable for water resources management, particularly in a transboundary basin, such as the Nile Basin, where data sharing is still limited.

Acknowledgments

We wish to thank the Ministry of Water Resources and Irrigation, Egypt, and the Ministry of Water Resources, Sudan, for providing data on *in situ* measurements for this research. Eric Muala is grateful to the Joint Japan World Bank Graduate Scholarship Program (JJ/WBGSP) for its financial support of this research at UNESCO-IHE Institute for Water Education, Delft, The Netherlands. Zheng Duan is also grateful for financial support from China Scholarship Council (CSC). We also acknowledge financial support from the DGIS-UNESCO|IHE Programmatic Cooperation (DUPC) program of UNESCO-IHE.

Author Contributions

Eric Muala collected, analysed data and processed most data, including the satellite imagery. All *in-situ* measurements were facilitated, obtained and analysed by Yasir Mohamed. Zheng Duan analysed some data, gave guidance for volume methodology and satellite data processing. Eric Muala was responsible for the research and the write up of the manuscript with contributions from all co-authors. Yasir Mohamed, Zheng Duan and Pieter van der Zaag critically revised the manuscript. All authors read and approved the final manuscript.

Conflicts of Interest

The authors declare no conflicts of interest.

References

1. Alsdorf, D.E.; Rodriguez, E.; Lettenmaier, D.P. Measuring surface water from space. *Rev. Geophys.* **2007**, *45*, 1–24.
2. Report of the GCOS/GTOS/HWRP Expert Meeting on Hydrological Data for Global Studies. Available online: http://www.fao.org/gtos/doc/pub32.pdf (accessed on 8 August 2014).
3. Arsano, Y.; Tamrat, I. Ethiopia and the Eastern nile Basin. *Aquat. Sci.* **2005**, *67*, 15–27.
4. Gerlak, A.; Lautze, J.; Giordano, M. Water resources data and information exchange in transboundary water treaties. *Int. Environ. Agreem.: Polit. Law Econ.* **2011**, *11*, 179–199.
5. Water Resources Management, Nile Basin Initiative. Available online: http://nilebasin.org/newsite/ (accessed on 8 August 2014).
6. Calmant, S.; Seyler, F.; Cretaux, J.F. Monitoring continental surface waters by satellite altimetry. *Surv. Geophys.* **2008**, *29*, 247–269.
7. Cretaux, J.F.; Birkett, C. Lake studies from satellite radar altimetry. *Comp. Rend. Geosci.* **2006**, *338*, 1098–1112.
8. Zhang, J.Q.; Xu, K.Q.; Yang, Y.H.; Qi, L.H.; Hayashi, S.; Watanabe, M. Measuring water storage fluctuations in Lake Dongting, China, by Topex/Poseidon satellite altimetry. *Environ. Monit. Assess.* **2006**, *115*, 23–37.
9. Medina, C.; Gomez-Enri, J.; Alonso, J.J.; Villares, P. Water volume variations in Lake Izabal (Guatemala) from *in situ* measurements and ENVISAT Radar Altimeter (RA-2) and Advanced Synthetic Aperture Radar (ASAR) data products. *J. Hydrol.* **2010**, *382*, 34–48.
10. Duan, Z.; Bastiaanssen, W.G.M. Estimating water volume variations in lakes and reservoirs from four operational satellite altimetry databases and satellite imagery data. *Remote Sens. Environ.* **2013**, *134*, 403–416.
11. Kouraev, A.V.; Zakharova, E.A.; Samain, O.; Mognard, N.M.; Cazenave, A. Ob' river discharge from TOPEX/Poseidon satellite altimetry (1992–2002). *Remote Sens. Environ.* **2004**, *93*, 238–245.
12. Coe, M.T.; Birkett, C.M. Calculation of river discharge and prediction of lake height from satellite radar altimetry: Example for the Lake Chad basin. *Water Resour. Res.* **2004**, *40*, 1–11.

13. Zakharova, E.A.; Kouraev, A.V.; Cazenave, A.; Seyler, F. Amazon river discharge estimated from TOPEX/Poseidon altimetry. *Comp. Rend. Geosci.* **2006**, *338*, 188–196.

14. Papa, F.; Durand, F.; Rossow, W.B.; Rahman, A.; Bala, S.K. Satellite altimeter-derived monthly discharge of the Ganga-Brahmaputra River and its seasonal to interannual variations from 1993 to 2008. *J. Geophys. Res.: Oceans* **2010**, *115*, 1–19.

15. Leon, J.G.; Calmant, S.; Seyler, F.; Bonnet, M.P.; Cauhope, M.; Frappart, F.; Filizola, N.; Fraizy, P. Rating curves and estimation of average water depth at the upper Negro River based on satellite altimeter data and modeled discharges. *J. Hydrol.* **2006**, *328*, 481–496.

16. Swenson, S.; Wahr, J. Monitoring the water balance of Lake Victoria, East Africa, from space. *J. Hydrol.* **2009**, *370*, 163–176.

17. Bashar, K.E.; Ahmed, E. Sediment accummulation in roseires reservoir. *Nile Basin Water Sci. Eng. J.* **2010**, *3*, 46–55.

18. Dams Implementation Unit (DIU). Available online: http://www.roseiresdam.gov.sd/en/index.php (accessed on 8 August 2014).

19. Ebaid, H.M.I.; Ismail, S.S. Lake nasser evaporation reduction study. *J. Adv. Res.* **2010**, *1*, 315–322.

20. Elsawwaf, M.; Willems, P.; Pagano, A.; Berlamont, J. Evaporation estimates from Nasser Lake, Egypt, based on three floating station data and Bowen ratio energy budget. *Theor. Appl.Climatol.* **2010**, *100*, 439–465.

21. Mobasher, A.M.A. Adaptive Reservoir Operation Strategies under Changing Boundary Conditions—The Case of Aswan high Dam Reservoir. Ph.D. Thesis, TU Darmstadt, Darmstadt, Germany, 2010.

22. Hydrology from Space. Lakes, Rivers and Wetlands Water Levels from Satellite Altimetry. Available online: http://www.legos.obs-mip.fr/soa/hydrologie/hydroweb/ (accessed on 8 January 2014).

23. Cretaux, J.F.; Jelinski, W.; Calmant, S.; Kouraev, A.; Vuglinski, V.; Berge-Nguyen, M.; Gennero, M.C.; Nino, F.; del Rio, R.A.; Cazenave, A.; *et al.* SOLS: A lake database to monitor in the Near Real Time water level and storage variations from remote sensing data. *Adv. Space Res.* **2011**, *47*, 1497–1507.

24. USDA/FAS/OGA and NASA Global Agriculture Monitoring (GLAM) Project. Lake and Reservoir Surface Height Variations from the USDA's Global Reservoir and Lake (GRLM). Available online: http://www.pecad.fas.usda.gov/cropexplorer/global_reservoir/ (accessed on 8 January 2014)

25. Birkett, C.; Reynolds, C.; Beckley, B.; Doorn, B. From research to operations: The USDA global reservoir and lake monitor coastal altimetry. In *Coastal Altimetry*; Vignudelli, S., Kostianoy, A.G., Cipollini, P., Benveniste, J., Eds.; Springer: Berlin/Heidelberg, Germany, 2011; pp. 19–50.

26. United States Geological Survey (USGS)/Earth Resources Observation Center (EROS). Available online: http://glovis.usgs.gov/ (accessed on 25 July 2014)

27. Maxwell, S. Filling Landsat ETM+ SLC-off gaps using a segmentation model approach. *Photogramm. Eng. Remote Sens.* **2004**, *70*, 1109–1112.

28. Scaramuzza, P.; Micijevic, E.; Chander, G. Slc gap-filled products: Phase one methodology. Available online: http://landsat.usgs.gov/documents/SLC_Gap_Fill_Methodology.pdf (accessed on 8 August 2014).

29. International Water Management Institute (IWMI). On-Line Climate Summary Service Model.. Available online: http://wcatlas.iwmi.org/Default.asp (accessed on 25 July 2014).

30. Penman, H.L. Natural evaporation from open water, bare soil, and grass. *Proc. R. Soc. Lond. A* **1948**, *193*, 120–145.

31. Birkett, C.M.; Beckley, B. Investigating the performance of the Jason-2/OSTM radar altimeter over lakes and reservoirs. *Mar. Geod.* **2010**, *33*, 204–238.

32. Global Navigation Satellite System (GNSS). Available online : http://igscb.jpl.nasa.gov/network/complete.html (accessed on 6 June 2014)

33. McFeeters, S. The use of the Normalized Difference Water Index (NDWI) in the delineation of open water features. *Int.J. Remote Sens.* **1996**, *17*, 1425–1432

34. Xu, H. Modification of normalised difference water index (NDWI) to enhance open water features in remotely sensed imagery. *Int. J. Remote Sens.* **2006**, *27*, 3025–3033.

35. Muala, E. The Use of Satellite Altimetry for Water Resources Management: Case Study of the Nile Basin. Master's Thesis, UNESCO-IHE, Delft, The Netherlands, 2012.

36. Birkett, C.M.; Mertes, L.A.K.; Dunne, T.; Costa, M.H.; Jasinski, M.J. Surface water dynamics in the Amazon Basin: Application of satellite radar altimetry. *J. Geophys. Res.: Atmos.* **2002**, *107*, 8059–8080.

37. Duan, Z.; Bastiaanssen, W.G.M.; Muala, E. Icesat-derived water level variations of roseires reservoir (Sudan) in the Nile basin. In Proceedings of the 2013 IEEE International Geoscience and Remote Sensing Symposium (IGARSS), Melbourne, VIC, Australia, 21–26 July 2013.

38. Bastawesy, M.A.; Khalaf, F.I.; Arafat, S.M. The use of remote sensing and GIS for the estimation of water loss from Tushka lakes, southwestern desert, Egypt. *J. Afr. Earth Sci.* **2008**, *52*, 73–80.

Inundations in the Inner Niger Delta: Monitoring and Analysis Using MODIS and Global Precipitation Datasets

Muriel Bergé-Nguyen and Jean-François Crétaux

Abstract: A method of wetland mapping and flood survey based on satellite optical imagery from the Moderate Resolution Imaging Spectroradiometer (MODIS) Terra instrument was used over the Inner Niger Delta (IND) from 2000–2013. It has allowed us to describe the phenomenon of inundations in the delta and to decompose the flooded areas in the IND into open water and mixture of water and dry land, and that aquatic vegetation is separated from bare soil and "dry" vegetation. An Empirical Orthogonal Function (EOF) analysis of the MODIS data and precipitation rates from a global gridded data set is carried out. Connections between flood sequence and precipitation patterns from the upstream part of the Niger and Bani river watersheds up to the IND are studied. We have shown that inter-annual variability of flood dominates over the IND and we have estimated that the surface extent of open water varies by a factor of four between dry and wet years. We finally observed an increase in vegetation over the 14 years of study and a slight decrease of open water.

Reprinted from *Remote Sens.* Cite as: Bergé-Nguyen, M.; Crétaux, J.-F. Inundations in the Inner Niger Delta: Monitoring and Analysis Using MODIS and Global Precipitation Datasets. *Remote Sens.* **2015**, *7*, 2127-2151.

1. Introduction

Satellite imagery is a classic tool used to monitor water extent over large areas of the Earth's surface for long periods of time [1]. The current number and variety of space-borne instruments available to observe continental water content is very wide. The choice to use one sensor rather than another is generally driven by the type of targets and objectives of the study.

A flood occurs when a large and usually dry area is covered temporarily by a certain amount of water. It can also happen when a large amount of water flows into a channel that is not large enough to carry it; the excess water consequently spills over into the surrounding areas. In a flood, the lateral water spread dominates the vertical rising, until equilibrium is reached where underground recharge and evaporation compensate the surface water inflow in the floodplain. Floods may be caused by overspill from river channels, high local precipitation or both phenomena acting together. In the first scenario, a significant time gap may occur between the start of the upstream rainfall and the time of maximum inundation in the floodplain.

For example, Zwarts *et al.* [2] have shown that local rainfall is very limited and that water extent over the IND is driven by surface river flow variability in the upstream area of the delta.

The main purpose of this study is to provide a space-based tool for monitoring floods over the IND and to interpret their pluri-annual variability using global precipitation data.

In Section 2, we give an overview of flooding in the IND. In Section 3, we describe remote sensing and precipitation datasets and their processing. In Section 4, we present results for the phenomenology inter-annual water changes, aquatic vegetation and vegetation extent in IND and general rainfall patterns over upstream areas of the Niger and Bani rivers. In Section 5, we analyze the link between rainfall and inundation over the IND. Perspectives and conclusions are drawn in the last section.

2. Floods over the IND

The IND is a vast floodplain of 73,000 km^2 located in the arid (north part of the delta: 15,000 km^2) and semi-arid (south part of the delta: 58,000 km^2) Sahelian zone (Figure 1). It is composed of large numbers of swamps, river channels, permanent lakes and non-permanent flooded areas. The IND is seasonally inundated (from September to November) due to rainfall over the Niger and Bani Rivers' upstream areas (From June to September). This produces (when cumulated) an average annual discharge at the delta's entries of 1490 m^3 s^{-1} (period 1955–1996), which represents 47 km^3.yr^{-1} [3]. Inundations in the southern part are characterized by a regular and rapid flow of water while in the northern part the water stays longer [3]. The evaporation rate over the IND varies both spatially and temporally: it is higher in the northern part (700 mm/year on average) than in the southern part (140 mm/year on average), and depending on the year can vary from 400 to 1300 mm/year [3]. Inundations over the IND are marked by very high inter-annual and long-term variability, which is linked to the amount of rainfall in the upstream rivers' watersheds. Li *et al.* [4] highlighted large inter-decadal variations of rainfall over the last 50 years, with the wettest decade (1950–1960) and the driest (1970–1980) being due to climate variability over the Atlantic Ocean. Based on a hydrological water balance model, Mahé *et al.* [3] calculated that the flooded extent over the IND varied from dry to wet years by a factor of 5; from 40,000 km^2 in 1955 to 9000 km^2 in 1984.

Since the IND is a wetland located in an arid region, it serves as a crucial source for the economic activity in the region. Several million inhabitants are strongly dependent on water resources for agriculture, fishing, and pastoralism. Consequently, inter-annual inundation variability plays a major role in water management over the IND and severely impacts land use. Thus, the economic and ecological health of the region can be considered vulnerable to and dependent on these inter-annual inundation cycles. For example, Liersh *et al.* [5] have shown that a mix of upstream reservoir management and climate warming will provoke a decrease of inflow to the delta in the coming decades.

In another recent study, it has been simulated from Global Climate Models (GCMs) that rainfall over the Bani River catchment area will likely decrease until the end of the 21st century by about 15% to 17%. Consequently, there will be a large reduction in river discharge into the IND [6]. Moreover, inundations in the IND influence local rainfall through vertical energy flux and water vapor from the land to the atmosphere [7]. They observed a resurgence of daytime cloud cover during inundations. This was in response to evaporation over the IND, which in turn increased the likeliness of storms in the mid-afternoon, mainly on the western border of the IND.

As a consequence of rainfall variability over West Africa, land surface occupation (water, vegetation, aquatic vegetation, and bare soil) is also changing from year to year resulting in very complex patterns that are influenced by the topography of river channels in the delta, presence of vegetation, and the total amount of water filling the IND. During periods of flood, the IND land is also subject to noticeable vegetation growth.

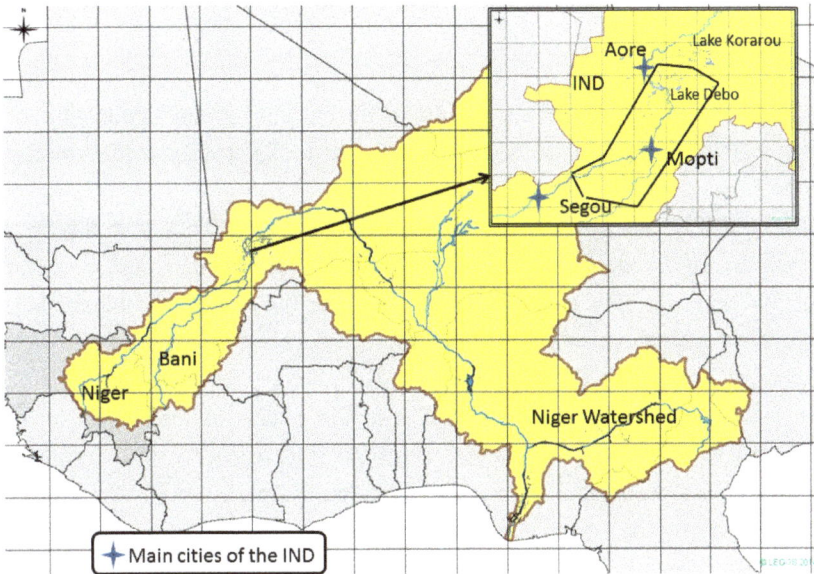

Figure 1. Map of the Niger River watershed (yellow) and Inner Niger Delta (IND) (zoom). Black lines represent country borders.

In this context, reliable spatio-temporal information about the extent of water and vegetation over the IND is useful in understanding the links with climate variability. In the past, *in situ* gauges measuring river discharge were installed principally at the entries and mouth of the delta [3]. Unfortunately, our ability to measure and forecast the total fresh water input in the IND each year is quite impossible because of: (1) economic and infrastructural problems generally affecting non-industrialized nations; (2) water flow physics across vast lowlands that are not permanently inundated; and (3) the fact that measuring water extent variations over such a large area is practically impossible from *in situ* measurements.

Space-based methods can be an answer to these difficulties in studying water extent over wetlands and floodplains.

Moreover, over the IND the seasonal inundations act as a source of significant evaporative losses, which have to be taken into account for regional water cycle modeling. For example, Dadson *et al.* [8] showed that flooded areas over the IND are doubling total losses from land surface over the region, that seasonal and inter-annual inundations control the evaporation rate, and also that it improves predictions of land-atmosphere energy and water vapor fluxes. Finally, knowledge of flooding dynamics in large floodplains within a river basin may also improve global climate simulations [9] and constrain hydrodynamical model parameters [1].

3. Data Sets and Methodologies

3.1. Datasets

3.1.1. Preamble

To characterize the process of inundations and their dynamics across floodplains, it is essential to measure some key variables related to water extent over the IND and regional patterns of rainfall upstream of the delta.

We will show the results obtained for water extent over the IND, and for the spatial and temporal variability of precipitation determined from global gridded data over the upstream watershed of the Niger and the Bani rivers and over the floodplains. As mentioned above, we expect to correlate the inter-annual variability of floods over the IND to surface discharges from upstream rivers (that are directly linked to rainfalls [2].

Water extent and precipitation rates over the upstream rivers and the IND have been mapped and we have produced a time series of the variables from 2000 to 2013. For water mapping, we use the MODIS instrument (Section 3.2) and for precipitation mapping, we use a global gridded dataset: TRMM 3B43 version 7 products (Section 3.3).

3.1.2. MODIS

The MODIS instrument is a multispectral imaging system installed onboard the Terra (launched in December 1999) and Aqua (launched in May 2002) satellites. MODIS provides Earth surface reflectance on thirty-six narrow bands of frequency; seven among them are from the visible to Mid Infrared part of the spectrum with a spatial resolution of 250 (two first bands) and 500 m (all bands):

- Band 1: 620–670 nm.
- Band 2: 841–876 nm.
- Band 3: 459–479 nm.
- Band 4: 545–565 nm.
- Band 5: 1230–1250 nm.
- Band 6: 1628–1652 nm.
- Band 7: 2107–2155 nm.

A limitation of optical remote sensing data for flood monitoring is cloud cover. This is accentuated in tropical countries. In such situations, MODIS images cannot measure the ground surface radiance and consequently cannot provide continuous information on the water extent during a flood. However, the IND region is located at the boundary between the Sahara Desert and western tropical Africa and is not affected by cloud cover, except during the summer months of June to August. Moreover major floods over the IND occur at the end or after the rainy season, which reduces this problem.

The orbits of Terra and Aqua were designed to allow satellites to cross the equator at local time 10:30 AM and 1:30 PM respectively. Considering that cloud formation linked to flooding over the IND is highest during the afternoons [7], we chose to collect and analyze the mosaic of MODIS/Terra instrument on an 8-day time basis. The 8-day composite is a compilation in one image of the best signal observed for each pixel over the following 8-day period, which reduces errors due to cloud, aerosols and viewing angle. It limits the risk of having long periods without valid images. For a detailed description of MODIS algorithms and products, see [10]. Flood mapping is thus enabled thanks to this compositing approach that provides enough days of clear sky even during the rainy period. We observe less than 5% of noisy images due to cloud presence during this period. Moreover, the Terra Instrument allows calculation of the water extent over a longer period.

3.1.3. Rainfall Data

Many different datasets of precipitation at a global scale exist and we have chosen to use the most recent version (7) of the Tropical Rainfall Measuring Mission (TRMM) 3B43 dataset. TRMM is the National Aeronautics and Space Administration (NASA) mission launched in November 1997 and has on-board precipitation radar developed by the Japanese Aerospace Exploration Agency (JAXA). It was designed to measure the intensity and distribution of the rain, and map storm structures. Here we use data from the 3B43 algorithm because it merges satellite and *in situ* data, TRMM 3 hourly data, and the monthly *in situ* data of thousands of rain gauges around the world compiled by the Global Precipitation Climatology Centre (GPCC).

The spatial resolution of these products is $0.25° \times 0.25°$. It was produced over a calendar month-time resolution. The data was averaged over the IND (156 pixels from the TRMM datasets) on a monthly basis, and limited to the period 2000–2013 to correspond in time with MODIS data used in this study. An average climatology of precipitation was calculated every month (Each map represents the mean precipitation of the given month based on precipitation from the whole period data set). Monthly maps of anomalies are also produced: anomalies are calculated by the difference between monthly and average precipitation for each month over the whole period. The maps allow detection of geographical patterns for precipitation over the region of interest (ROI) and its inter-annual variability. It is then used to determine if the gradient of precipitation between the Niger and Bani rivers can explain the flood dynamics measured by MODIS. EOFs over the IND and upstream Niger and Bani rivers were also calculated to be compared to a similar analysis of MODIS images. It allows the separation of spatio-temporal modes of variability of rainfall (Section 3.2).

To summarize, we have produced:

- Maps of monthly average climatology, and monthly and yearly anomalies.
- Time series for total precipitation over both rivers on the upstream part of the IND including the IND.
- EOF spatio-temporal modes for 14 years of data over the ROI.

A comparison of the results obtained using MODIS and precipitation data sets was also performed.

3.2. Methodology to Detect Water over the IND with MODIS and Validation

A method of land surface classification for hydrology has been developed using a combination of three of the seven MODIS bands and is acquired from MOD09A1 (MODIS/Terra Surface Reflectance 8-Day L3 Global 500 m SIN Grid V005), Version 5 product [11]. This method was applied to many different regions over the Earth and published in several works: the Aral Sea [12], The Andean Altiplano ([13,14]), Lake Tchad [11] , The IND ([15,16]), and the Ganga basin [17]. Other methods using MODIS images were published for similar studies [18–23].

Several other remote sensing instruments had already been used in the literature for flood mapping ([1,24–31]). Some of them have much higher spatial resolution (Landsat imagery, ASAR on Envisat, Radarsat instrument) but none has a combination of the following advantages for flood monitoring: a continuity of homogeneous data over a long period of time, a global coverage of the Earth over a short recurrent period (daily/weekly), and a free service.

Shallow depths and a high suspended sediment concentration, such as those observed along the IND considerably increase the amount of solar energy reflected by a body of water ([32–34]) have shown that strong water absorption at wavelength >1 µm in MODIS bands (5–7) does not allow illumination of sediments in the water or at the shallow bottom of a water column. Consequently, in order to avoid the problem of suspended particles, the most appropriate band to detect open water during a flood event is number 5 in Mid Infra-Red (MIR: 1230–1250 nm). A threshold value for reflectance in the IR band of MODIS should be attributed to discriminate water pixel from "no" water pixel.

In this paper, a simple combination of a threshold technique was performed on the MODIS Band 5 and Normalized Difference Vegetation Index (NDVI) to delineate the shallow, sediment laden, open waters of the IND flood plain. It was also used to discriminate between the mixture of water and dry land, aquatic vegetation and vegetation on dry land. It has been assumed that a small value of surface reflectance in band 5 allows characterizing open water. For MOD09GHK normalized products used in this study, the value of 1200 (a reflectance of 0.12 scale by a factor of 10,000 for distribution as integer) has been chosen as a cut-off value under which a pixel is supposed to be fully inundated.

When surface reflectance in band 5 increases to the threshold value of 2700, a test is made on NDVI to discriminate the pixel covered by a mixture of water, dry land and aquatic vegetation. NDVI is a robust index for monitoring temporal changes of vegetation photosynthetic activity ([35,36]). In the arid environment of our study area, a high level of vegetation photosynthetic activity can only be sustained by the presence of surface water or groundwater discharge. A threshold technique is used to select high NDVI values and to detect areas of high photosynthetic activity from aquatic vegetation and hydrophilic plants. NDVI ranges from negative values (generally considered as open water [37]) to >0.4 for dense vegetation.

To detect the presence of vegetation on dry land, NDVI is tested. For high reflectance in band 5 (superior to 2700): if the NDVI index is superior to 0.4, the pixel is considered to be covered with vegetation and if it is below 0.4 then the pixel is classified as dry land [11].

To validate this approach and to define the threshold values of NDVI and reflectance in band 5, a ground calibration was performed over the Diamantina floodplain in Central Australia in 2006. It was determined through aerial photography, laser and radar altimetry, and field measurements of different surface types taken from GPS measurements. The transition zones between different types of surfaces were precisely delineated and compared to the reflectance on the MODIS data which allowed fixing the threshold values given in Table 1 that then have been validated on other regions.

Table 1. Threshold value used for qualification of ground type is applied to monitor a flood event on the flood plain. Units are reflectance scaled by 10,000 as delivered in the MOD09GHK product (validation of this method is given in Section 3.2.3).

Index	Open Water	Mix Water/Dry Land	Aquatic Vegetation	Vegetation	Dry Land
Band 5	<1200	>1200 & <2700	>1200 & <2700	>2700	>2700
NDVI	No Test	<0.4	>0.4	>0.4	<0.4

MODIS images are given in HDF format. They are georeferenced and processed over the ROI centered on the IND area. The 8-day mosaic images were acquired over the period February 2000–December 2013 and processed with the algorithm described above. Three main products were obtained from this data processing:

- A synthetic map of ROI with classification obtained from MODIS images every 8 days.
- A map representing the average flood duration (in days) over the whole ROI, annual maps of flood duration anomalies with respect to average duration, and the Empirical Orthogonal Functions (EOFs, see Section 4.3) providing the spatio-temporal modes for each of the classes over the whole ROI.
- Evolution in time of the surface area of each class over the whole ROI.

The surface time series for each class was correlated with the precipitation rate over the Niger and Bani river basins upstream of the IND.

Validation was performed over the Aral Sea in Central Asia using previous results obtained in [38,39]. Variations in height, surface and volume of the Aral Sea were calculated for 1992 to 2005, with a combination of altimetry data from Topex/Poseidon and Jason satellites, and a digitized bathymetry map of the lake's basin. We extended this calculation until 2007 for this study. Using a selection of 55 MODIS images over the period from 2000 to 2007, we calculated the surface variations of Small and Large Aral and compared with the results obtained by altimetry and bathymetry. The results are shown in Figure 2. The agreement between both methods of surface variation calculations is marked with a correlation coefficient of 0.996 and RMS differences of detected water extents less than 2%.

The method was also validated over Lake Poopo in Bolivia. MODIS products were compared to Landsat imagery and resulted in a correlation of 0.97 [14]. They were used to study links between Lake Poopo and Lake Titicaca [13].

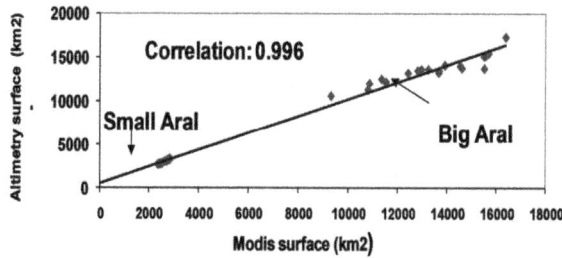

Figure 2. Scatter plot of classification of open water extent from Moderate Resolution Imaging Spectroradiometer and from a combination of satellite altimetry and digitized bathymetry of Aral Sea.

We compared our classification of the IND with results published in [40] for water and vegetation classifications. This was done by visual analysis of maps of inundated surfaces reproduced in [40] for 3 different dates and they showed good agreement qualitatively, but we did not assess this quantitatively. Images processed with our method were also used to validate the flood propagation model along the Niger River in [16]. Using our MODIS classification, Pedinotti *et al.* [16] observed improved comparison of their model outputs with downstream *in situ* discharges.

Furthermore, the MODIS data analysis done in this study allows downscaling low-resolution data as proposed in [17]. They used the MODIS classification presented here for water over the IND in synergy with the GIEMS database [17]. It provided water extent for the whole Earth on a monthly basis from 1993 to 2007 at a spatial resolution of 25 km.

Comparison of rainfall datasets to MODIS data products over the IND have been performed using the EOFs. The EOF analysis decomposes the spatio-temporal data in orthogonal modes of decreasing variance, expressed by spatial patterns and associated variations in time (also called principal component analysis (PCA)). In this method, we calculate the eigenvalues and eigenvectors of the covariance matrix of the data. We first calculated the 14-year average of MODIS data for each class and then subtracted it from the yearly average from 2000 to 2013. This allowed emphasis on inter-annual variability of the different modes. The theory of this method is fully described in [41,42] and the algorithm used in [43].

4. Results and Interpretation

4.1. Land Surface Classification over the IND

Processing MODIS images has allowed the classification of Earth's surface using the method described in Section 3.2.2 and also the monitoring of time variations for different surface types, from open water to vegetation on dry land. The maps are used principally as indicators to monitor floods over the IND. They will allow the examination of the flood process over the IND as a first step (Section 4.1.1) and second, to detect specific geographical patterns of very wet years compared to dry years using the multi-year analysis of MODIS data (Section 4.1.2). Phenomenology and inter-annual variability of vegetation and aquatic vegetation dynamics is also examined.

4.1.1. Mean Annual Flooding over the IND

For the period of our study and using the eight-day mosaic images, we classified the land over the ROI (Section 3.2). This is shown in Figure 3a,b and represents one full year (2003) using one selected image per month with the classification deduced from MODIS data. This figure illustrates the sequence of inundation over the IND. The first month selected in the set of images is June, just at the beginning of the rainy season, and the last is May of the following year.

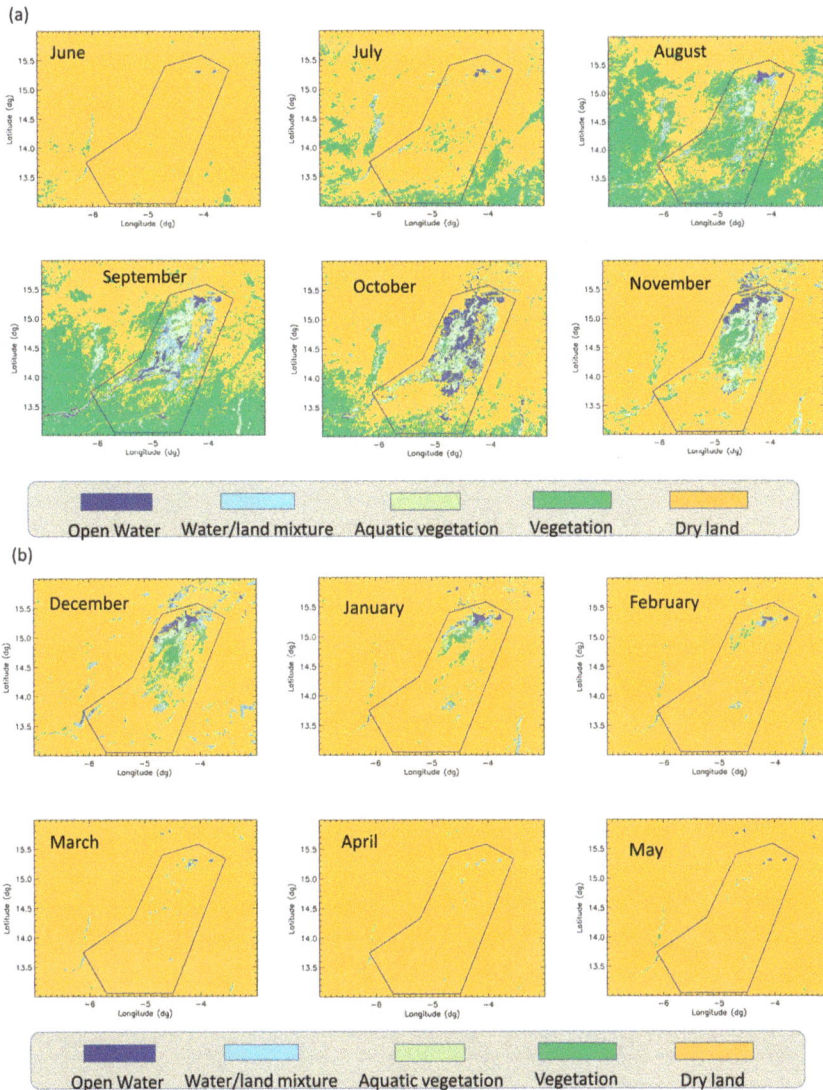

Figure 3. Moderate Resolution Imaging Spectroradiometer classification of the Inner Niger Delta (IND) with one image per month in 2003–2004 being selected for illustration of the inundation phenomenon over the Inner Niger Delta (IND) (**a**) from June to November 2003, (**b**) from December 2003 to May 2004.

In June, permanent lakes located in northern IND areas start to grow, with July marking the apparition of vegetation in the southern IND and on the delta itself. Vegetation grows regularly and the whole region is covered by the end of August. At the same time, free water is still limited to small areas like permanent lakes, the main parts of rivers, and some areas of the IND southwest of Lake Debo that are also covered with aquatic vegetation (Figure 3a,b). The end of August or beginning of September is marked by flow increase over the IND, which continues for around six weeks and reaches its peak by mid-October. Meanwhile, surrounding regions of the IND lose their vegetation cover. From November to January, water over the IND evaporates and remaining vegetation located over the delta simply dries out. We observe that surfaces are first covered with aquatic vegetation, and only then followed by vegetation. In January, the IND is almost completely dried up except for some very small areas near Lake Debo where it remains wet until June.

This inundation sequence is also illustrated in Figure 4. It represents (over the 14 years of this study) an average surface extent of different classes detected by MODIS over the ROI. In order to be comparable, these average surfaces were normalized. This figure shows that the peak of vegetation occurs at the beginning of September, whereas the peak for open water, mixture of water and dry land and aquatic vegetation appears one and a half months later.

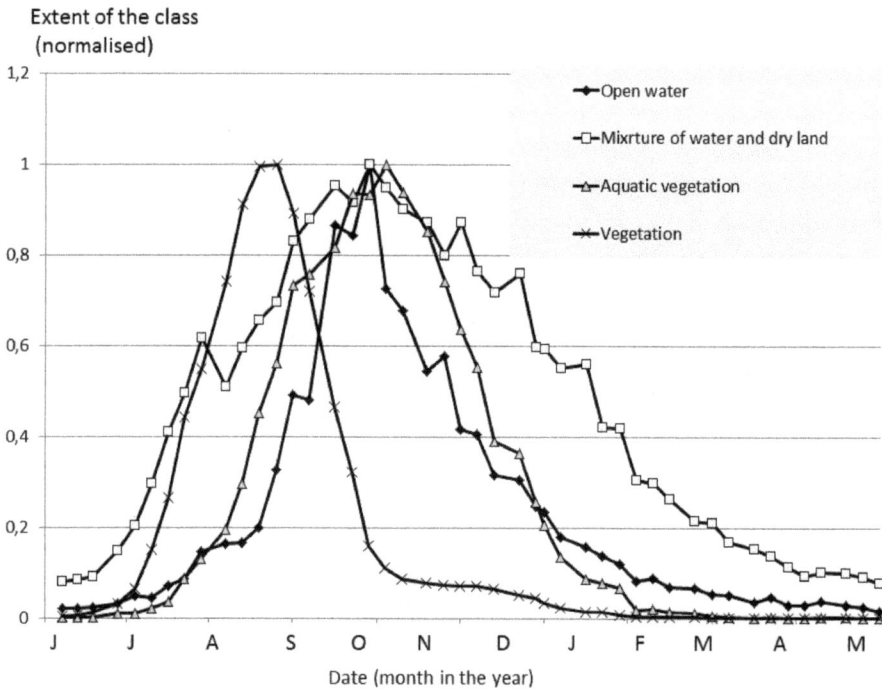

Figure 4. Normalized total extent of each of the classes over the Inner Niger Delta (IND) during a hydrological year (starting in June, finishing in May). All years have been averaged over each time portion of eight days.

Aquatic vegetation and open water are also in phase, which signifies that when the main flood occurs a small part of the IND is already covered by water or by aquatic vegetation (probably due to local rainfall). The peak of open water is also significantly narrower than that of aquatic vegetation. Mixture of water and dry land most likely indicates that:

- The inundated surface is shallow.
- The vertical topography slope of the IND is small.
- Downstream water flow is probably slower due to the presence of vegetation.

From January to May, the IND and surrounding regions dry out. Surface water evaporates completely except for some permanent small lakes such as Lake Debo in the northern part of the IND. Vegetation fully disappears from the region, most notably in the Delta.

The average duration of time over the period 2001–2013 of each class is also given in Figure 5a–d (2000 is excluded because the first images were collected in mid February 2000).

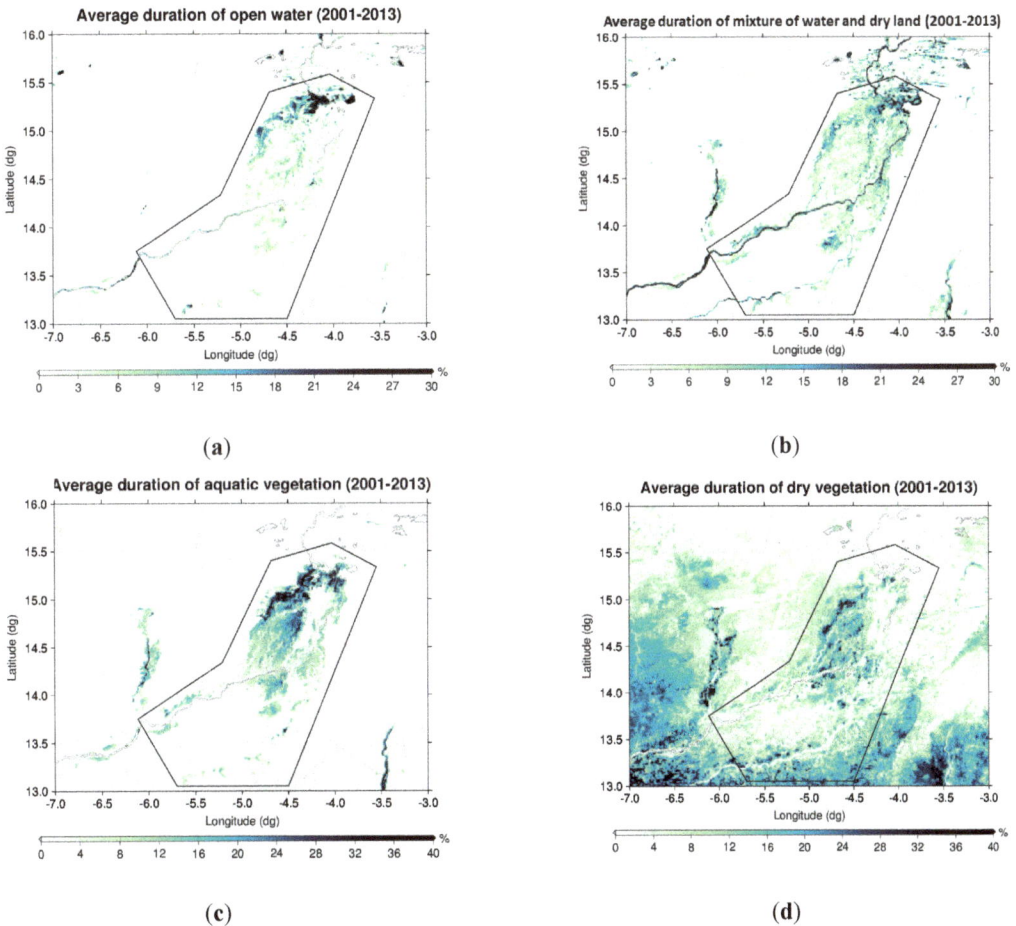

(a)

(b)

(c)

(d)

Figure 5. Average duration (in percentage per year) of (**a**) open water; (**b**) mixture of water and dry land; (**c**) aquatic vegetation; and (**d**) vegetation calculated from 2001 to 2013 (2000 has not been taken into account because the first MODIS data was collected in February of this year).

Figure 5a shows that flow duration in the southern part of the delta is rather short (less than a month) and very heterogeneously distributed. The "NW belt" of long duration (more than two months) is highlighted in dark blue. Due to the images' spatial resolution both the Niger and Bani rivers do not appear as permanently covered with water. These can be better viewed in Figure 5b, which shows the duration of mixed water (pixel covered both with water and dry land). This is confirmation that the southern part of the delta is a zone of rapid water flow and that the northern part behaves as a pool where water stays longer [3]. A much larger surface is covered with aquatic vegetation and mixture of water and dry land for periods exceeding one month (Figure 5b,c). Over the IND aquatic vegetation duration is also noticeably longer on the left bank of the Niger River than on its right bank where the Bani River is located. Along the NW belt mentioned above, aquatic vegetation duration often exceeds three months (Figure 5c). Dry vegetation (Figure 5d) is present almost everywhere over the IND for long period of time (more than three months) and up to three to four months on the delta's west bank.

4.1.2. Inter-Annual Variability of Floods over the IND

Satellite images may be used to classify the land surface in time. In the case of the IND it will allow monitoring the inter-annual dynamics of different classes, and their correlation with climatic condition changes over IND and Niger and Bani river watersheds.

High variability of inundated surfaces from one year to another is well illustrated in Figure 6, which shows a two year period (2001: high inundation year and 2002: low inundation year) when MODIS images were taken at the date of maximum extent of flood. The inundation patterns observed over these two 'extreme' years show a very high spatial variability of inter-annual flooding. This is also illustrated in Figure 7: between the wettest year (2001) and the driest one (2011) where the maximum extent of open water class varies by a factor of four. This is less pronounced for the three other classes with a factor of 2 to 2.5 calculated between wet and dry years. Another visualization of inter-annual variability of floods is given in Figure 8a,b. They show the anomaly of duration in each class for the wet (2001) and dry (2002) years with respect to the average year. They illustrate that not only surface extent varies significantly from the wet to dry years, but also the total duration of open water, mixture of water and dry land, aquatic vegetation, and vegetation. This is particularly marked for aquatic vegetation and vegetation. Indeed, for open water duration anomalies (Figure 8a), we observe up to 60 days of duration differences, principally concentrated on the NW belt. For aquatic vegetation and vegetation, we see that on almost the entire delta the differences are about 60 days between 2001 and 2002 (Figure 8b). It is also noteworthy that we observe the exact opposite behavior with the inter-annual variability of the maximum surface extent for each of the classes. The differences between wet and dry years are more pronounced for the open water class than for the three others. This indicates that for open water the time of residency is less sensitive to the inundation magnitude.

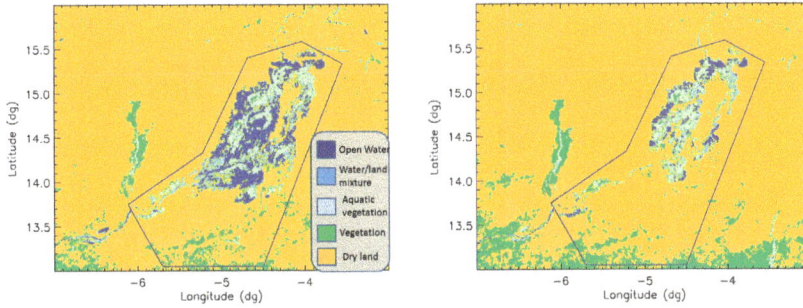

Figure 6. Map of Moderate Resolution Imaging Spectroradiometer classification for the day of maximum inundation for 2001 (high inundation) and 2002 (low inundation).

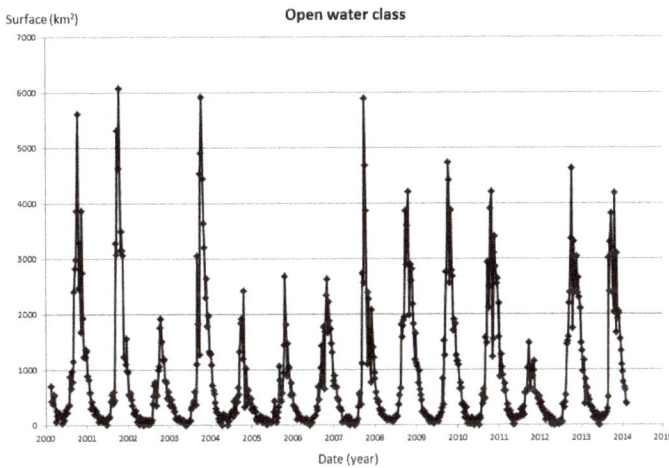

Figure 7. Time series of surface extent of open water class calculated from MODIS data.

Finally, another specific geographical pattern of inter-annual flooding appears in the year 2007. 2007 is one of the four high flood years (the three others being 2000, 2001 and 2003 (Figure 7)). However, the principle origin of the 2007 high flood is the Bani River to the south, which is not the case for the other years. Indeed the anomaly of water duration along the Niger River particularly at the entrance of the delta is very close to zero in 2007 (Figure 9). This can be contrasted with 2008 (another high flood year) in which the Bani River does not seem to supply excess water to the delta, while the Niger 'entrance' to the delta remains flooded much longer (about 30 days) than in an average year. This may also explain why the NW belt presents a deficit in the number of flooded days in 2007, given that 2007 is among the most flooded years. This reveals that flooding in the IND is a combination of high water in both the Bani and Niger rivers, but with changes in their roles from one year to another one. This is a factor of complexity for understanding flood processes in the IND.

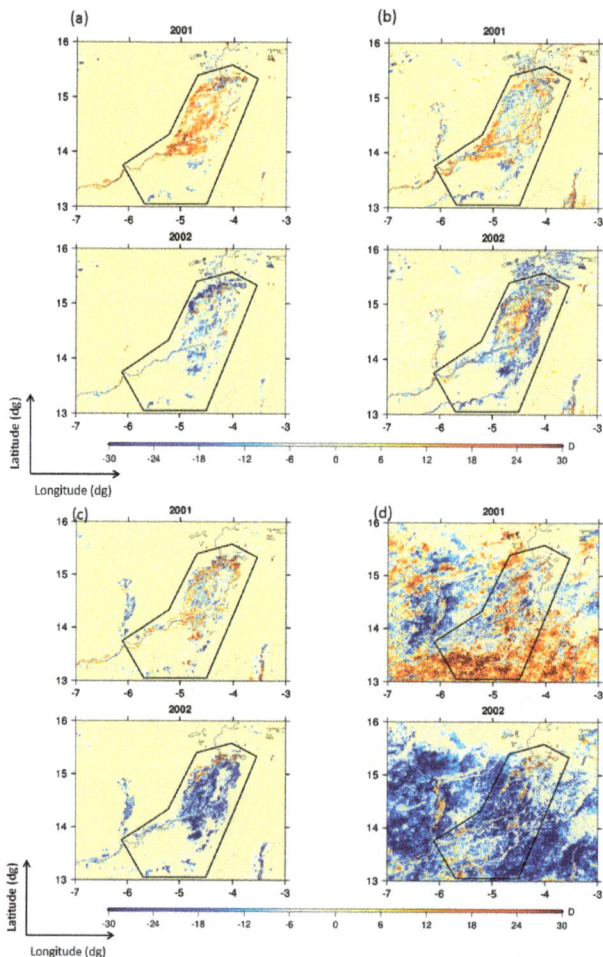

Figure 8. Yearly anomalies of duration (in day) of presence of (**a**) open water and (**b**) mixture of water and dry land; (**c**) aquatic vegetation; and (**d**) vegetation for 2001 and 2002.

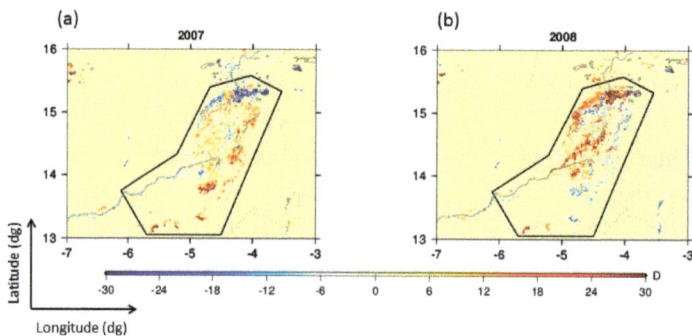

Figure 9. Yearly anomalies of duration (in day) of presence of open water for (**a**) 2007 and (**b**) 2008.

4.2. Rainfall

4.2.1. Seasonal Rainfall over West Africa

Precipitation over this region is the main contributor to inundations over the IND [2]. The exact correlation between the amount of precipitation and its geographical patterns with spatio-temporal variability of the water extent over the IND is the main issue explored in this study.

Figure 10 shows mean annual rainfall over West Africa in the vicinity of the IND from TRMM 3B43 data. This map highlights the latitudinal pattern of annual precipitation with a strong gradient from south to north of the region. In the south, where the two rivers take their source, the annual rainfall is about 1200–1500 mm, while the north, where the IND is located, is a semi-arid region with less than 300 mm/yr of rainfall.

Using the precipitation data sets, we then calculated a so-called-seasonal "climatology": a monthly map of average rainfall deduced from the full period of observation (Figure 11). Firstly, the period from November to April marks the dry season with zero precipitation over both rivers. Over the IND, the dry season continues until June with very low rainfall: less than a few dozen mm in May. In May, more significant rainfall occurs over the source of the Niger and Bani rivers. But the start of the wet season in June is when strong precipitation over the Atlantic coast of West Africa moves northeastward. Meanwhile, it starts to rain over the IND itself, with increasing precipitation in July and August. In August, precipitation is at its maximum over the entire region. It then starts to decrease from the IND to the southwest until November when it is totally dry everywhere (Figure 11).

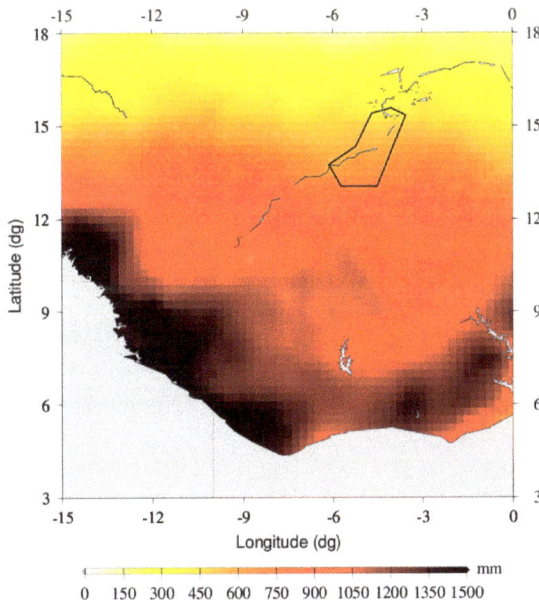

Figure 10. Map of mean annual precipitation over the Inner Niger Delta (IND) and upstream basins of Niger and Bani rivers, taken from Tropical Rainfall Measuring Mission (TRMM) 3B43 data from 2000 to 2013.

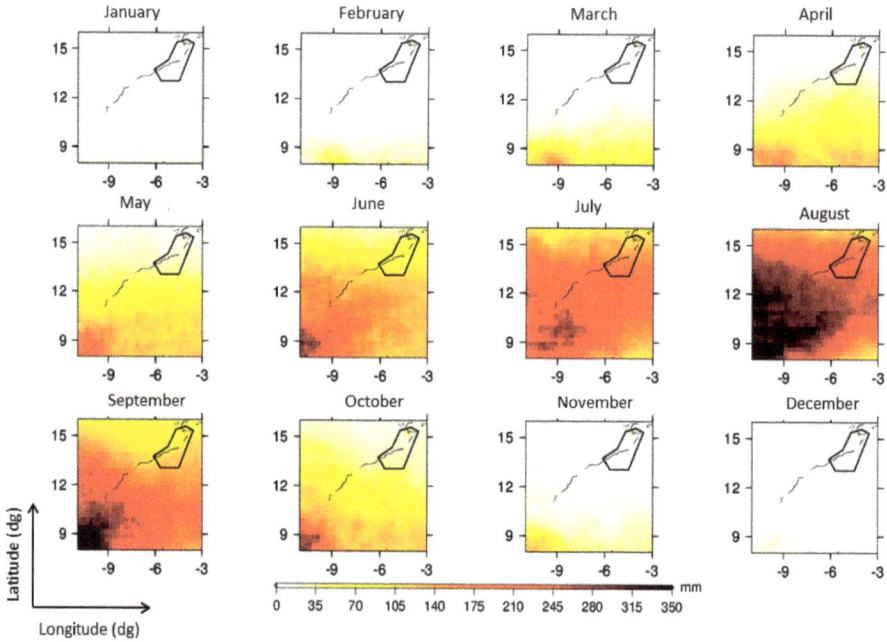

Figure 11. Maps on monthly average precipitation from Tropical Rainfall Measuring Mission (TRMM) 3B43.

4.2.2. Rainfall Inter-Annual Variability

As seen above, the inundation over the IND has strong inter-annual variability. Precipitation over this region also presents significant inter-annual fluctuations from a dry year with around 800 mm/yr to 1100 mm/yr on average for a wet year. From the precipitation data, four years appear as wet years: 2003, 2008, 2010 and 2012 (Figure 12a). Figure 12b shows for each year, from 2001 to 2012, the variations in space of yearly precipitation anomalies (difference of yearly precipitation to global average) from dry to wet years. First, we can see a high spatial heterogeneity of rainfall anomaly distribution from year to year. The year 2003 is very interesting as it can be seen that precipitation data present positive anomalies, but with high spatial heterogeneity. In contrast, years 2010 and 2012 have a very strong positive anomaly near the IND and the upstream part of the Niger and Bani rivers. The year 2008 presents a dipole of a positive anomaly in the west (where the Niger River is located) and a negative one in the east (where the Bani River is located). The year 2002 is a very dry year for the whole area of study.

(a)

(b)

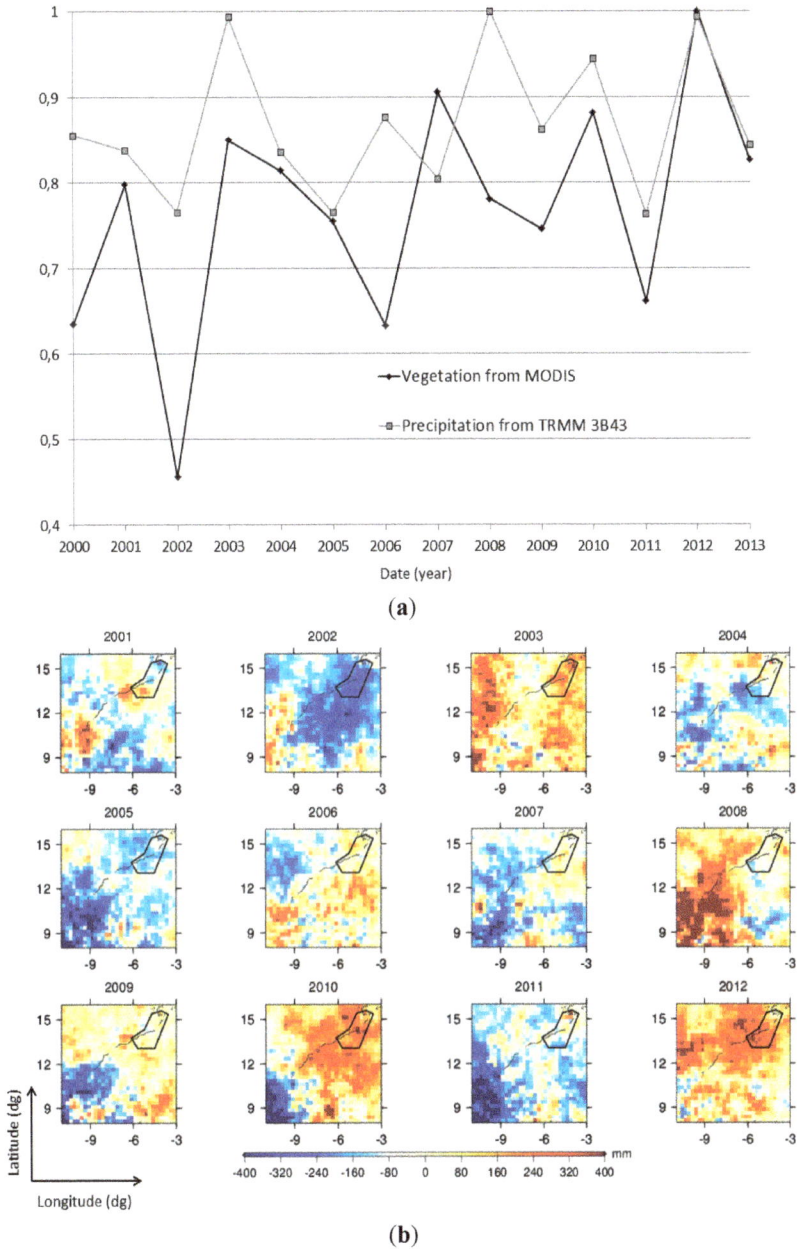

Figure 12. (**a**) Annual normalized precipitation over the Region Of Interest (ROI) from Tropical Rainfall Measuring Mission 3B43 compared to annual normalized coverage area of vegetation over the Inner Niger Delta (IND) measured by Moderate Resolution Imaging Spectroradiometer; (**b**) annual rainfall anomalies with respect to the global mean.

4.3. Links between Rainfall and Floods over the IND

It is already well known that inundations over the IND are linked to precipitation upstream of the delta. If we analyze the annual precipitation anomaly patterns and compare it to flood patterns over the IND, some additional and more detailed conclusions can be drawn.

From MODIS data we also calculated the total surface using the three classes that include water: open water, mixture of water and dry land, and aquatic vegetation. The time series is given in Figure 13 where the total inundated surface calculated every eight days is compared to monthly rainfall. The data was normalized in order to make a comparison. The first conclusion to be drawn from this figure concerns the time shift between maximum rainfall over the Niger and Bani river basins and the maximum flood over the IND. We found one and a half months with a standard deviation of eight days. It tends to confirm that inundations are due to rainfall in the upstream river areas located a few hundred kilometers away from the IND. If we calculate the correlation between annual rainfall and maximum inundation over the IND, we obtain 0.65. No decadal trend was observed from this analysis showing that for inundations (total of the three "water" classes), the inter-annual variability dominates the long-term trend.

In Section 4.1.2 it can be seen that in 2007 the Bani River was possibly the main contributor to the inundation (Figure 9). This is well confirmed by analysis of the precipitation spatio-temporal variability. In Figure 10b, it is clear that positive anomalies of rainfall in 2007 were localized over the Bani River, while the Niger River globally presents a negative anomaly. Therefore, there is an explanation of why an inundation higher than normal was observed in 2007, although this year was not considered as one of the wettest years of the study period. From Figure 13 we also see that 2002, 2004, 2005 and 2011 are the four driest years in terms of inundation. Annual rainfall anomalies (Figure 12a,b) also show that these years were characterized by a large rainfall deficit over the Niger and Bani rivers upstream of the delta.

If the precipitation data is now compared with vegetation cover over the IND, we see an increase in vegetation of 21% over the 14-year study while only a slight 6% total precipitation increase is apparent (Figure 12a). Monthly rainfall is also compared with the eight-day vegetation class from MODIS (Figure 13). It shows that maximum rainfall corresponds exactly to maximum vegetation cover over the IND. But the apparition of vegetation is shifted each year by more than three months after the first rainfall in March and April. This is easily explained by Figure 11 which shows that if rainfall appears earlier in the year over the upstream river basins, it will start only in June or July over the IND.

To complete the comparison between rainfall and inundation over the IND, we performed an EOF analysis over the different datasets.

Figure 14a–d shows the first mode of spatio-temporal variations for the water, mixture of water and dry land, aquatic vegetation and vegetation. The EOF was applied from 2001 to 2013, excluding year 2000 as the first processed MODIS images date back to February 2000. For each of the four classes, the first mode represents 30%–40% of signal variance. Its temporal variation is well correlated with the time series of surface extent for each of these classes. For example, for open water it is directly recognized from mode 1 that the years 2002, 2004, 2005, 2006 and 2011 are considered as globally dry, as was already observed in Figure 13. In order to recompose the total signal we must add the factor of the spatial and temporal modes of the whole decomposition. Here positive values of temporal mode correspond to dry year as the spatial mode is negative over the entire ROI. It can also be seen that for the three first classes, temporal variations are very similar with only slight differences, which enhances the assumption that rainfall inter-annual variability is the main driver for these classes. For vegetation (Figure 14d), a global increase over the study period is observed (with the main signal in the west outside of the delta), which is also confirmed by direct calculation of annual vegetation cover (Figure 12a). It is noteworthy that spatial mode 1 of open water classes clearly exhibits the presence of the alleged NW belt (Section 4.1 Figure 5a). There is a globally distributed increase in aquatic vegetation over the IND, and that for all classes, the dry year of 2011 is well explained by the first mode of variations.

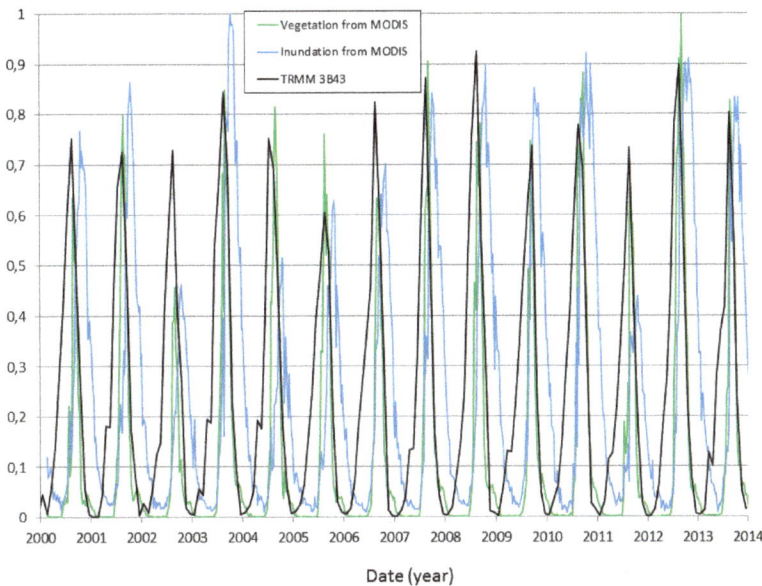

Figure 13. Time series of monthly precipitation over the Region Of Interest (ROI) compared to inundated surfaces over the Inner Niger Delta (IND) including open water, mixture of water and dry land and aquatic vegetation, and to vegetation total extent. All data has been normalized for the comparison.

The same computation was made with the rainfall data. EOFs for the first three modes (explaining approximately 65% of the signal variance) are given in Figure 15a–c. Mode 1 is exclusively representative of rainfall over the upstream part of the Niger River (south west of the watershed: red zone in Figure 15a), while mode 2 exhibits a dipole between the delta area and upstream of the Niger River (Figure 15b). Mode 3 (still explaining 7% of the variance for total rainfall) presents specific spatial patterns with mainly two small regions participating in the signal: the far upstream of the Niger River and the west part of the delta (blue zones in Figure 15c). A large signal on mode 3 is also present in the south (red zones in Figure 15c) but is located outside the watershed of the two rivers.

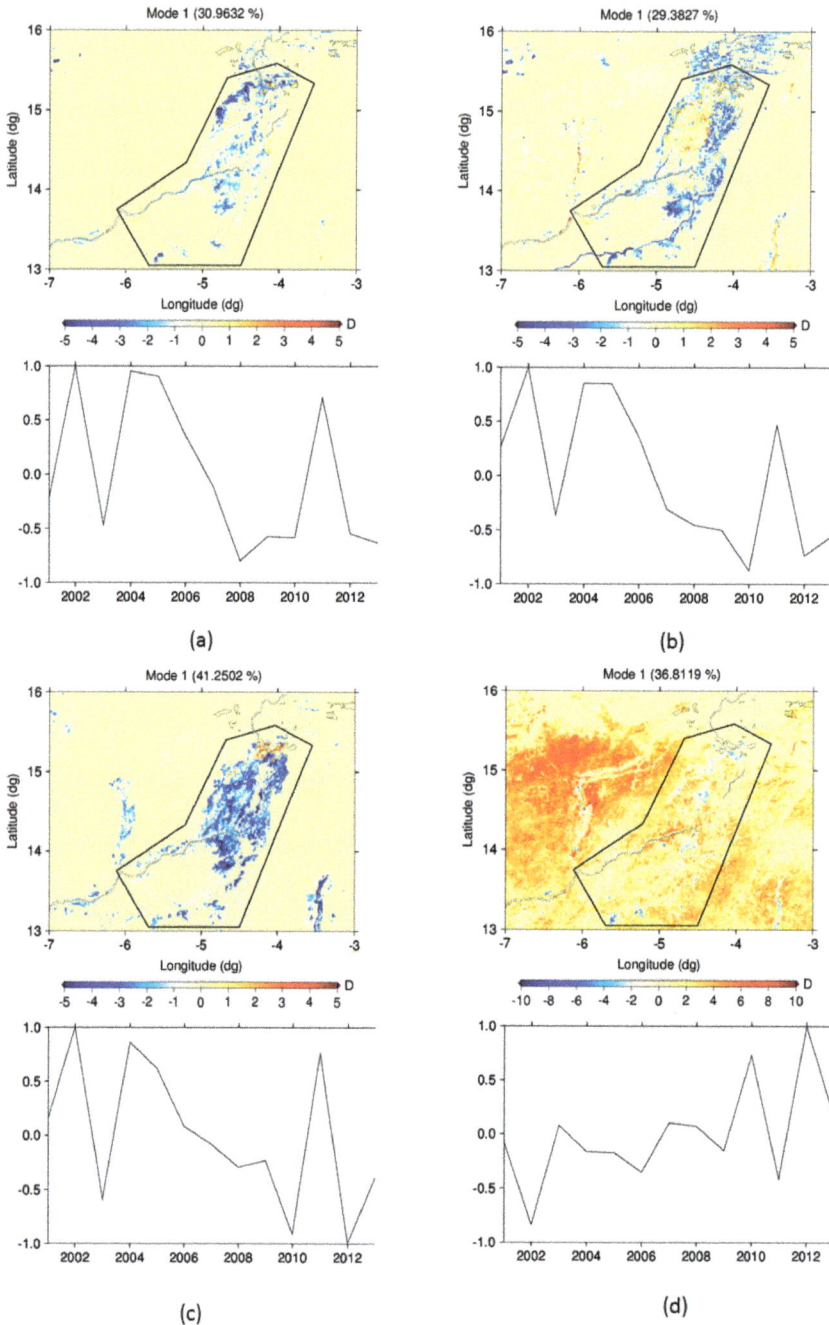

Figure 14. Moderate Resolution Imaging Spectroradiometer Empirical Orthogonal Functions (EOF) mode 1 for (**a**) open water; (**b**) mixture of water and dry land; (**c**) aquatic vegetation and (**d**) vegetation over the Region Of Interest (ROI) (black line's polygon). Units of Empirical Orthogonal Functions time series are normalized.

Mode 1 explains some of the inter-annual inundation variability observed over the IND. We see that the two years of high inundation (2003 and 2008) are quite fully explained by this first mode (Figure 15a). The high peak on mode 2 observed in 2010 with spatial signal closer to the IND (Figure 15b) explains why this was a highly inundated year. In 2012, the conjunction of high precipitation in the entire basin (mode 1 and 2 spatial distribution complete each other for this year) explains the inundation. In 2001, mode 3 seems responsible for the observed excess of open water over the IND. For this year again, it seems that precipitation excess is localized far upstream of the Niger and close to the delta in the west (map of spatial mode 3: Figure 15c). Mode 3 also contributes slightly to the 2008 inundation, in addition to mode 1. The dry years of 2002, 2004, 2005 and 2011 are well explained by mode 1, with a deficit of water over the upstream Niger (Figure 15a). In 2002, the drought is amplified by mode 2, which indicates a general deficit of rainfall (as also seen in Figure 12b). For 2004, amplification comes from the western part of the delta (Figure 15c).

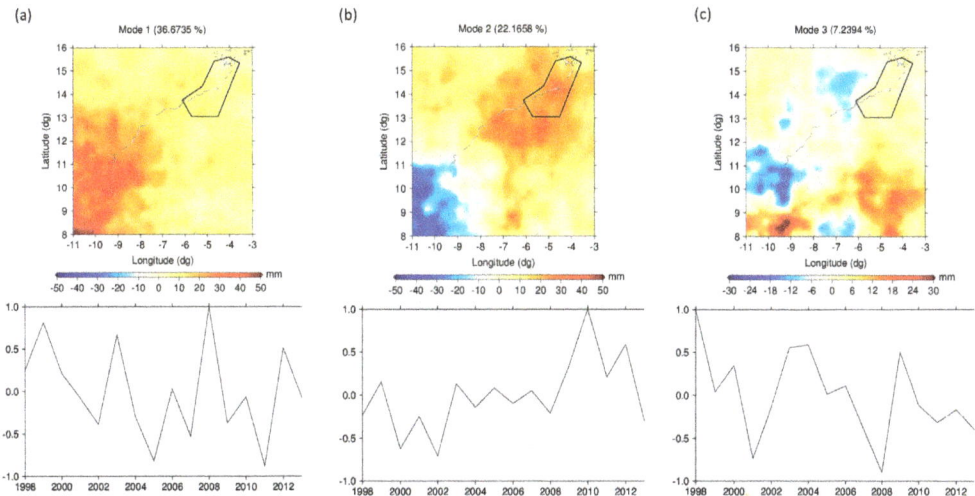

Figure 15. Three first Empirical Orthogonal Functions (EOF) modes (**a–c**) of precipitation over the Region of Interest, calculated with Tropical Rainfall Measuring Mission (TRMM) 3B43 data.

Figure 15a,b also highlights that mode 1 is dominated by inter-annual variability with a slight additional decreasing trend over the study period while mode 2 shows increasing rainfall near the IND. This may explain why inundation over the IND is marked by a trend and why the direct increase in precipitation near the delta may be a cause of vegetation growth in this area. If temporal variability of mode 2 of the rainfall and mode 1 of the vegetation cover are compared over the IND, the correlation is 0.73. Strong vegetation growth was also observed near the western delta (Figure 14d), which is also in good agreement with observations by [7]. It could be a combination of modes 2 and 3, which exhibit rainfall anomalies over this region.

Figure 16 shows the third mode of the open water class over the IND representing 11% of signal variance, and we can observe a dipole between the part of the delta fed by the Niger River and the part of the delta fed by the Bani River. This indicates a slight decrease of several days per year in inundation in the west part of the IND and an increase in the eastern part. The decrease in precipitation over the Niger River observed by mode 1 of rainfall (Figure 15a without any signal over the Bani River) and combined with an increase in precipitation over the Bani (which can be observed from mode 2 of rainfall (Figure 15b)) may explain this dichotomy between western and eastern parts of the delta. The spatial repartition of the signal from mode 2 also includes a part of the Niger River but in the region near the delta. This region has lower rainfall than over the upstream area and therefore does not compensate the general decrease, as observed by mode 1 in the upstream part of Niger River.

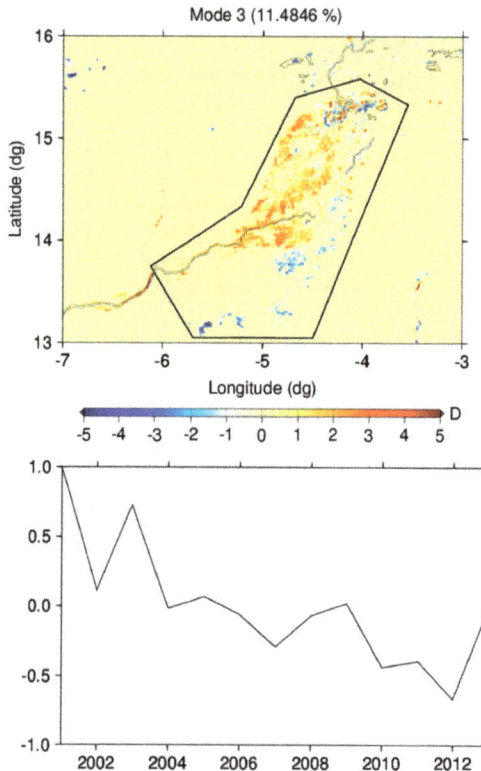

Figure 16. Empirical Orthogonal Functions (EOF) modes of Moderate Resolution Imaging Spectroradiometer open water class over the Region Of Interest (ROI).

The EOF analysis therefore allows a more detailed description of the geographical patterns of inundations over the delta and the dichotomy of both rivers contributing to the delta.

5. Conclusions

In this study, we applied a methodology based on MODIS imagery analysis to detect water and vegetation over the Inner Niger Delta floodplains. Understanding inundation sequences and processes for regions such as the Inner Niger Delta are crucial for economic purposes as land use is affected. Precise yearly flood monitoring is crucial for pastoral agriculture, particularly in the frame of inter-annual climate fluctuations.

We have shown that MODIS images are suitable to achieve global coverage and continuous monitoring of floodplain inundations like that of the Inner Niger Delta and it has allowed us to describe the phenomenon of inundations in the delta in terms of water and vegetation extents. We have demonstrated that inter-annual variability of flood patterns dominates the IND. Another interesting result of this study, deduced from Empirical Orthogonal Functions analysis of MODIS and TRMM 3B43 data, is the characterization of the respective roles of the Niger and Bani rivers in the flooding process over the IND. We have determined the link between spatial patterns of water (including open water, mixture of water and dry land, and aquatic vegetation) and vegetation with rainfall on the upstream part of the two rivers and over the IND. We observed a factor of four on the total open water extent between dry and wet years and estimated the time residency of different types of surface over the Inner Niger Delta presenting high inter-annual anomalies. Moreover, a general increase in vegetation over the study period (2000–2013) and a slight decrease of open water has been revealed. For inundated areas the inter-annual variability is predominant.

In addition to *in situ* observations and hydrological modeling (global/regional climate and hydrodynamical models), space observations may significantly help improve our understanding of hydrological processes in floodplains and their interaction with climate variability. Assimilation of remote sensing data in a model of an ungauged basin is a recurrent issue, especially in a complex geographical system like the Inner Niger Delta, which is a mix of small rivers, channels, swamps and lakes [44].

Acknowledgments

The authors wish to thank the Centre National d'Etudes Spatiales (CNES) for financing our work through the TOSCA project. The MODIS data was downloaded from the National Aeronautics and Space Administration (NASA) and is courtesy of the online Earth data web site named: Earth Observing System Data and Information System EOSDIS, URL: https://earthdata.nasa.gov/. The TRMM dataset was downloaded from the http://mirador.gsfc.nasa.gov/collections/TRMM_3B43__006.shtml web site of Goddard Space Flight Center, USA.

Author Contributions

Muriel Bergé-Nguyen was responsible of the development of the software allowing processing and interpreting the different data sets. She has prepared all figures in the manuscript, and is in charge of data processing of TRMM and MODIS datasets. Jean-François Crétaux had the original idea for the study, led the data analysis in coordination with Muriel Bergé-Nguyen, and drafted the manuscript. All authors read and approved the final manuscript.

Conflicts of Interest

The authors declare no conflict of interest.

References

1. Alsdorf, D.E.; Rodriguez, E.; Lettenmaier, D.P. Measuring surface water from Space. *Rev. Geophys.* **2007**, *45*, doi:10.1029/2006RG000197.
2. Zwarts, L.; Cisse, N.; Diallo, N. Hydrology of the upper Niger. In *The Niger, a Lifeline. Effective Water Management in the Upper Niger Basin*; Zwarts, L., Van Beukering, P., Kone, B., Wymenga, E., Eds.; Inst. for Inland Water Management and Waste Water Treatment: Lelystad, The Netherlands, 2005; pp. 15–41.
3. Mahe, G.; Bamba, F.; Soumaguel, A.; Orange, D.; Olivry, J.-C. Water losses in the inner delta of the River Niger: Water balance and flooded area. *Hydrol. Process.* **2009**, *23*, 3157–3160.
4. Li, K.Y.; Coe, M.T.; Ramankutty, N. Investigation of hydrological variability in West Africa using land surface models. *J. Clim.* **2005**, *18*, 3173–3188.
5. Liersch, S.; Cools, J.; Kone, B.; Koch, H.; Diallo, M.; Reinhardt, J.; Fournet, S.; Aich, V.; Hattermann, F.F. Vulnerability of rice production in the Inner Niger Delta to water resources management under climate variability and change. *Environ. Sci. Policy* **2013**, *34*, 18–33.
6. Ruelland, D.; Collet, L.; Ardoin-Bardin, S.; Roucou, P. How could hydro-climatic conditions evolve in the long term in West Africa? The case study of the Bani River Catchment. Hydro-climatology: Variability and change. In Proceedings of the Symposium J-H02 Held during IUGG2011, Melbourne, Australia, 28 June–7 July 2011; pp. 195–201.
7. Taylor, C.M. Feedbacks on precipitation from an African wetland. *Geophys. Res. Lett.* **2010**, *37*, doi:10.1029/2009GL041652.
8. Dadson, S.J.; Ashpole, I.; Harris, P.; Davies, H.N.; Clark, D.B.; Blyth, E.; Taylor, C.M. Wetland inundation dynamics in a model of land surface climate: Evaluation in the Niger inland delta region. *J. Geophys. Res.* **2010**, doi:10.1029/2010JD014474.
9. Coe, M.T. Simulating continental surface waters: An application to Holocene northern Africa. *J. Clim.* **1997**, *10*, 1680–1689.
10. Justice, C.O.; Townshend, J.R.G.; Vermote, E.F.; Masuoka, E.; Wolfe, R.E.; Saleous, N.; Roy, D.P.; Morisette, J.T. An overview of MODIS Land data processing and product status. *Remote Sens. Environ.* **2002**, *83*, 3–15.
11. Crétaux, J.-F.; Bergé-Nguyen, M.; Leblanc, M.; Abarca-Del-Rio, R.; Delclaux, F.; Mognard, N.; Lion, C.; Pandey, R.K.; Tweed, S.; Calmant, S.; *et al.* Flood mapping inferred from remote sensing data. *Int. Water Technol. J.* **2011** *1*, 48–62.
12. Cretaux, J.-F.; Letolle, R.; Calmant, S. Investigations on Aral Sea regressions from Mirabilite deposits and remote sensing. *Aquat. Geochem.* **2009**, *15*, 277–291.
13. Abarca-del-Rio, R.; Cretaux, J.-F.; Berge-Nguyen, M.; Maisongrande, P. Does the Titicaca Lake still control the Poopo lake water levels? An investigation using satellite altimetry, and MODIS data (2000–2009). *Remote Sens. Lett.* **2012**, *3*, 707–714.
14. Arsen, A.; Crétaux, J.-F.; Berge-Nguyen, M.; Abarca-Del-Rio, R. Remote sensing derived bathymetry of Lake Poopó. *Remote Sens.* **2014**, *6*, 407–420.

15. Pedinotti, V.; Boone, A.; Decharme, B.; Cretaux, J.-F.; Mognard, N.; Panthou, G.; Papa, F.; Tanimoun, B.A. Evaluation of the ISBA-TRIP continental hydrologic system over the Niger Basin using *in situ* and satellite derived datasets. *Hydrol. Earth Syst. Sci.* **2012**, *16*, 1745–1773.

16. Aires, F.; Papa, F.; Prigent, C.; Crétaux, J.-F.; Bergé-Nguyen, M. Characterization and space/time downscaling 1 of the inundation extent over the Inner Niger Delta using GIEMS and MODIS data in preparation to the SWOT mission. *J. Hydrometeorol.* **2014**, *15*, 171–192.

17. Pandey, R.K.; Cretaux, J.-F.; Berge-Nguyen, M.; Tiwari, V.M.; Drolon, V.; Calmant, S. Water level estimation by remote sensing for 2008 flood of Kosi. *Int. J. Remote Sens.* **2014**, *35*, 424–440.

18. Yésou, H.; Huber, C.; Lai, X.; Averty, S.; Li, J.; Daillet, S.; Bergé-Nguyen, M.; Chen, X.; Huang, S.; James, B.; *et al*. Nine years of water resources monitoring over the middle reaches of the Yangtze River, with ENVISAT, MODIS, Beijing-1 time series, altimetric data and field measurements. *Lakes Reserv. Res. Manag.* **2011**, *16*, 231–247.

19. Li, S.; Sun, D.; Goldberg, M.; Stefanidis, A. Derivation of 30-m resolution water maps from TERRA/MODIS and SRTM. *Remote Sens. Environ.* **2013**, *134*, 417–430.

20. Brakenridge, G.R.; Anderson, E. MODIS-based flood detection, mapping and measurement: The potential for operational hydrological applications. *Earth Environ. Sci.* **2006**, *72*, 1–12.

21. Sun, D.L.; Yu, Y.Y.; Goldberg, M.D. Deriving water fraction and flood maps from MODIS images using a decision tree approach. *IEEE J. Selected Topics Appl. Earth Obs. Remote Sens.* **2011**, *4*, 814–825.

22. Sun, D.L.; Yu, Y.Y.; Zhang, R.; Li, S.; Goldberg, M.D. Towards operational automatic flood detection using EOS/MODIS data. *Photogramm. Eng. Remote Sens.* **2012**, *78*, 637–646.

23. Sakamoto, T.; Nguyen, N.V.; Kotera, A.; Ohno, H.; Ishitsuka, N.; Yokozawa, M. Detecting temporal changes in the extent of annual flooding within the Cambodia and the Vietnamese Mekong Delta from MODIS time series imagery. *Remote Sens. Environ.* **2007**, *109*, 295–313.

24. Smith, L.C. Satellite remote sensing of river inundation area, stage, and discharge, a review. *Hydrol. Process.* **1997**, *11*, 1427–1439.

25. Henry, J.-B.; Chastanet, P.; Fellah, K.; Desnos, Y.-L. Envisat multi-polarised ASAR data for flood mapping. *Int. J. Remote Sens.* **2006**, *27*, 1921–1929.

26. Frappart, F.; Do Minh, K.; L'Hermitte, J.; Ramilllien, G.; Cazenave, A.; le Toan, T.; Mognard-Campbell, N. Water volume change in the lower Mekong Basin from satellite altimetry and imagery data. *Geophys. J. Int.* **2006**, *167*, 570–584.

27. Töyrä, J.; Pietroniro, A.; Martz, L.W. Multisensor hydrologic assessment of a freshwater wetland. *Remote Sens. Environ.* **2001**, *75*, 162–173.

28. Töyrä, J.; Pietroniro, A. Towards operational monitoring of a northern wetland using geomatics-based techniques. *Remote Sens. Environ.* **2005**, *97*, 174–191.

29. Matgen, P.; Schumann, G.; Henry, J-B.; Hoffman, L.; Pfister, L. Integration of SAR-derived river inundation areas, high-precision topographic data and a river flow model toward near real-time flood management. *Int. J. Appl. Earth Obs. Geoinf.* **2007**, *9*, 247–263.

30. Martinez, J.-M.; le Toan, T. Mapping of flood dynamics and spatial distribution of vegetation in the amazon floodplain using multitemporal SAR data. *Remote Sens. Environ.* **2007**, *108*, 209–223.

31. Bartsch, A.; Pathe, C.; Scipal, K.; Wagner, W. Detection of permanent open water surfaces in central Siberia with Envisat ASAR wide swath data with special emphasis on the estimation of methane fluxes from tundra wetlands. *Hydrol. Res.* **2008**, *39*, 89–100.

32. Engman, E.T.; Gurney, R.J. *Remote Sensing in Hydrology*; Chapman and Hall: London, UK, 1991; p. 225.

33. Bukata, R.P. Satellite Monitoring of Inland and Coastal Water Quality: Retrospection, Introspection, Future Directions; Taylor and Francis: New York, NY, USA, 2005; p. 246.

34. Li, R.-R.; Kaufman, Y.J.; Gao, B.-C.; Davis, C.O. Remote sensing of suspended sediments and shallow coastal waters. *IEEE Trans. Geosci. Remote Sens.* **2003**, *41*, 559–566.

35. Lyon, J.G.; Yuan, D.; Lunetta, R.S.; Elvidge, C.D. A change detection experiment using vegetation indices. *Photogramm. Eng. Remote Sens.* **1998**, *64*, 143–150.

36. Lunetta, R.S.; Knight, J.F.; Ediriwickrema, J.; Lyon, J.G.; Worthy, L.D. Land-cover change detection using multi-temporal MODIS NDVI data. *Remote Sens. Environ.* **2006**, *105*, 142–154.

37. Mohamed, Y.A.; Bastiaanssen, W.G.M.; Savenije, H.H.G. Spatial variability of evaporation and moisture storage in the swamps of the upper Nile studied by remote sensing techniques. *J. Hydrol.* **2004**, *289*, 145–164.

38. Aladin, N.V.; Crétaux, J.-F.; Plotnikov, I.S.; Kouraev, A.V.; Smurov, A.O.; Cazenave, A.; Egorov, A.N.; Papa, F. Modern hydro-biological state of the Small Aral Sea. *Environmetric* **2005**, *16*, 375–392.

39. Crétaux, J.-F.; Kouraev, A.V.; Papa, F.; Bergé Nguyen, M.; Cazenave, A.; Aladin, N.V.; Plotnikov, I.S. Water balance of the Big Aral sea from satellite remote sensing and *in situ* observations. *J. Great Lakes Res.* **2005**, *31*, 520–534.

40. Seiler, R.; Csaplovis, E.; Vollmer, E. Monitoring land cover changes of the Niger inland delta ecosystem (Mali) by means of envisat-Meris data. In *African Biodivers*; Springer US: New York, NY, USA, 2005; pp. 395–404.

41. Preisendorfer, R.W. *Principal Component Analysis in Meteorology and Oceanography*; Elsevier Science: New York, NY, USA, 1988; p. 425.

42. Von Storch, A.; Zwiers, W. Empirical orthogonal function. In *Statistical Analysis in Climate Research*, 3nd ed.; Cambridge University Press: Cambridge, UK, 2003; pp. 293–312.

43. Toumazou, V.; Crétaux, J.-F. Using a Lanczos eigen solver in the computation of Empirical Orthogonal Functions. *Mon. Weather Rev.* **2001**, *129*, 1243–1250.

44. Bates, P.D.; Horritt, M.S.; Smith, C.N.; Mason, D. Integrating remote sensing observations of flood hydrology and hydraulic modelling. *Hydrol. Process.* **1997**, *11*, 1777–1795.

Seven Years of Advanced Synthetic Aperture Radar (ASAR) Global Monitoring (GM) of Surface Soil Moisture over Africa

Alena Dostálová, Marcela Doubková, Daniel Sabel, Bernhard Bauer-Marschallinger and Wolfgang Wagner

Abstract: A surface soil moisture (SSM) product at a 1-km spatial resolution derived from the Envisat Advanced Synthetic Aperture Radar (ASAR) Global Monitoring (GM) mode data was evaluated over the entire African continent using coarse spatial resolution SSM acquisitions from the Advanced Microwave Scanning Radiometer for Earth Observing System (AMSR-E) and the Noah land surface model from the Global Land Data Assimilation System (GLDAS-NOAH). The evaluation was performed in terms of relative soil moisture values (%), as well as anomalies from the seasonal cycle. Considering the high radiometric noise of the ASAR GM data, the SSM product exhibits a good ability (Pearson correlation coefficient (R) = ~0.6 for relative soil moisture values and root mean square difference (RMSD) = 11% when averaged to 5-km resolution) to monitor temporal soil moisture variability in regions with low to medium density vegetation and yearly rainfall >250 mm. The findings agree with previous evaluation studies performed over Australia and further strengthen the understanding of the quality of the ASAR GM SSM product and its potential for data assimilation. Problems identified in the ASAR GM algorithm over arid regions were explained by azimuthal effects. Diverse backscatter behavior over different soil types was identified. The insights gained about the quality of the data were used to establish a reliable masking of the existing ASAR GM SSM product and the identification of areas where further research is needed for the future Sentinel-1-derived SSM products.

Reprinted from *Remote Sens.* Cite as: Dostálová, A.; Doubková, M.; Sabel, D.; Bauer-Marschallinger, B.; Wagner, W. Seven Years of Advanced Synthetic Aperture Radar (ASAR) Global Monitoring (GM) of Surface Soil Moisture over Africa. *Remote Sens.* **2014**, *6*, 7683-7707.

1. Introduction

The ability of coarse resolution (~25–50 km) microwave remote sensing products from both passive and active satellites to capture the variability of soil moisture was demonstrated by numerous studies (e.g., [1–4]). Their benefits in many research fields, such as numerical weather forecasting [5,6], runoff modeling [7,8], agricultural drought monitoring [9], land data assimilation [10] or studies of land atmospheric feedbacks [11], have been demonstrated. Consequently, these products have become commonly accepted in the past few years.

Surface soil moisture (SSM) products with improved spatial resolution are expected to broaden the number of applications and allow the usage of the SSM data in regional higher spatial resolution models. Motivated by the latter, the use of Synthetic Aperture Radar (SAR) ScanSAR data to monitor SSM was suggested by Wagner *et al.* [12,13].

The Advanced Synthetic Aperture Radar (ASAR) sensor onboard the Envisat satellite was an active microwave system operating at a central frequency of 5.331 GHz (C-band). It offers multiple acquisition modes employing both the conventional stripmap SAR, as well as the ScanSAR technique. The ScanSAR Global Monitoring (GM) mode provided global measurements with a trade-off between spatial (1 km) and temporal resolution (four to seven days, dependent also on the sensor acquisition plan) and, therefore, allows the monitoring of dynamic processes, such as soil moisture, on regional to global scales [13]. The ASAR GM SSM has been derived over Oklahoma and Australia, and the evaluation studies over these regions proved the ability of the product to resolve the spatial details in the soil moisture patterns that were not observable with the coarse resolution scatterometers or radiometers. Nonetheless, spatial averaging to between 3 and 10 km was recommended to reduce the high noise of the ASAR measurements caused by the relatively low radiometric accuracy (~1.2 dB) [14,15] of the GM mode measurements [16–18].

For small-scale applications, also Wide Swath (WS) mode, Image Mode (IM) or Alternating Polarization (AP) mode are used [19–21]. These modes offer even higher spatial resolution (30 m for AP and IM, 150 m for WS) and radiometric accuracy (~0.6 dB in the case of WS [22]) with regional spatial coverage and irregular temporal sampling. Gruber et al. [14] and Baup et al. [19] showed that the WS mode offers better performance in terms of radiometric resolution, radiometric stability and speckle reduction than the GM mode. This is, however, at the cost of lower temporal resolution and reduced spatial coverage of WS when compared to the GM mode.

At the time of writing of this publication, the Sentinel-1 SAR sensor is in the commissioning phase. The Sentinel-1 sensor is an active microwave system operating at a central frequency (5.405 GHz) that is very close to that of ASAR (5.331 GHz). The transfer of the SSM retrieval algorithm to Sentinel-1 has therefore been foreseen and has been discussed in a number of publications [14,15,23,24]. Given the significantly improved radiometric resolution of the Sentinel-1 (0.05–0.07 dB) combined with a regular temporal coverage, soil moisture products derived from Sentinel-1 are expected to be of considerably better quality when compared to the ASAR SSM products [14,15].

Within the European Space Agency's (ESA) Tiger Innovator projects Soil Moisture for Hydrometeorologic Applications (SHARE) and TIGER-NET a 1-km surface soil moisture product was developed and processed over the African continent based on the complete archive of the ASAR GM mode data (December 2004 to April 2012). The production and evaluation of the ASAR GM product over the entire African continent is scientifically valuable given the variability of the climatological, biogeographical, pedological and lithological characteristics over the continent, which is expected to reveal new challenges and opportunities for improvements of the Vienna University of Technology (TU Wien) algorithm [25]. For instance, prior studies using a scatterometer demonstrated some unexpected backscatter behavior and negative correlations between the SSM estimates from active and passive sensors over very dry areas [1]. Similar problems can be expected to occur in the SAR SSM products. However, the higher spatial resolution of SAR data may improve the understanding of the regionalization of such phenomena and link it to other parameters, such as soil types, lithology, vegetation or combinations thereof.

The evaluation step is performed using SSM data from the Advanced Microwave Scanning Radiometer for Earth Observing System (AMSR-E), as well as from the Noah land surface model from the Global Land Data Assimilation System (GLDAS-NOAH). According to the suggestion in [16–18], the 1-km dataset was aggregated to 5-km spatial resolution prior to the evaluation. Due to the scale difference between the aggregated ASAR GM SSM product (5 km) and reference SSM datasets (0.25°), the evaluation cannot take advantage of the high spatial resolution of the ASAR observations. The uncertainties in the reference data together with the spatial sampling error will be included in the bivariate error measures [26]. However, the evaluation against *in situ* and medium resolution datasets was impossible at the continental scale, due to a lack of such data over the African continent. To assess the role of the improved radiometric resolution, the evaluation includes a comparison between the performance of GM and WS SSM product over the Zambezi catchment.

2. Datasets

2.1. ASAR Surface Soil Moisture

The ASAR SSM dataset was retrieved using a TU Wien change detection algorithm [25] and represents the relative surface soil moisture in the upper soil layer (<3 cm) at 1-km spatial resolution. The algorithm was originally developed for data from European Remote-Sensing Satellite (ERS) and Advanced Scatterometer (ASCAT) scatterometers [25] and subsequently adopted for ASAR GM [17] and WS data, respectively. In the case of high-resolution WS data, averaging was performed within the georeferencing step. The spatial resolution of the georeferenced WS dataset is 1 km, and the radiometric resolution is enhanced from ~0.6 dB to ~0.2 dB [14]. The characteristics of the GM and WS mode data are summarized in Table 1.

The change detection algorithm assumes a linear relationship between changes in volumetric soil moisture content and changes of backscatter expressed in decibels. The degree of saturation in the soil pores is estimated by relating each backscatter value to backscatter reference maps representing wet and dry soil conditions. Wet and dry conditions refer to a completely dry soil and saturation of the soil, respectively. For a sufficiently dense multi-year time series of backscatter measurements, the assumption is that measurements for both dry and wet soil conditions are captured, allowing maps of the dry and wet references to be derived from the data. However, over arid and semi-arid areas, a so-called wet correction [27] must be applied to the dataset, as the probability of acquisitions for wet conditions is very low. An empirical correction of biases in the wet reference is applied when the wet reference is below −6 dB and the sensitivity (the difference between wet and dry reference) is less than 10 dB. The wet reference is then increased to a value of maximum −6 dB, under the condition that the sensitivity is not made greater than 10 dB. The unit of the resulting product represents the degree of saturation (%).

Table 1. The characteristics of the Advanced Synthetic Aperture Radar Global Monitoring mode and Wide Swath mode data.

	Global Monitoring Mode	Wide Swath Mode
Central frequency	5.331 GHz (C-Band)	5.331 GHz (C-Band)
Spatial resolution	1000 m	150 m
Radiometric resolution	~1.2 dB	~0.6 dB
Temporal resolution	Irregular, typically 4 to 7 days	Irregular, dependent on the sensor acquisition plan
Spatial coverage	Global	Regional, dependent on the sensor acquisition plan
Polarization used	H/H	H/H
Orbit direction used	Ascending and descending	Ascending and descending

In total, more than 18,000 ASAR GM over the whole continent and 1100 ASAR WS images over Zambezi catchment were processed. For the evaluation of WS data, the Zambezi catchment was chosen due to the high coverage of WS acquisitions in the area. Erroneous datasets (exceptionally high or low backscatter values, strong striping within an image and shifted images) were removed. The resulting ASAR GM SSM data coverage is shown at Figure 1. Limited coverage in some areas is due to conflicting data acquisitions in other modes.

Figure 1. The number of Advanced Synthetic Aperture Radar Global Monitoring mode Surface Soil Moisture measurements between December 2004 and April 2012.

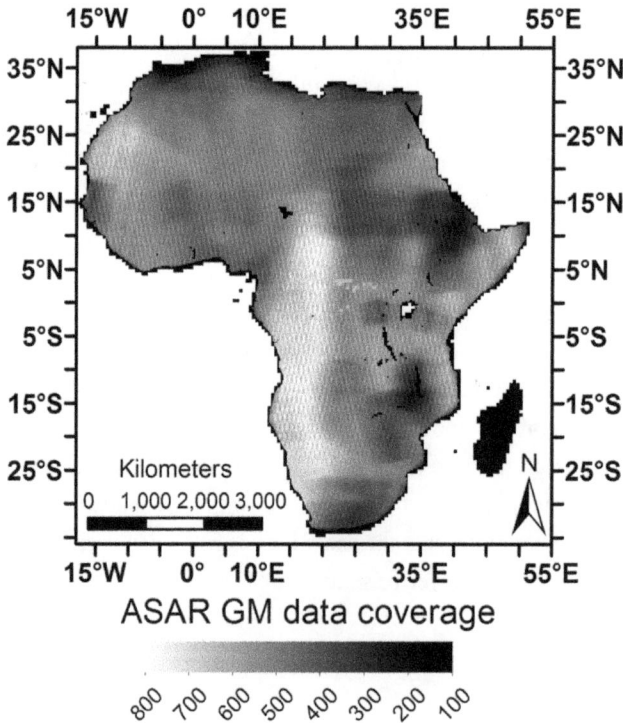

2.2. Reference Datasets

2.2.1. AMSR-E VUA SSM

The AMSR-E SSM is derived from the C-band brightness temperature using Version 3 of the Land Parameter Retrieval Model [28] by Vrije Universiteit Amsterdam (VUA). It represents the volumetric soil moisture (m^3/m^3) in the near-surface soil layer (<3 cm) at the original spatial resolution of 56 km (resampled to a grid with a sampling distance of 0.25°). Only soil moisture retrievals based on descending (night-time) orbit data were used, as these are expected to be more accurate than day-time acquisitions due to the reduced difference between the surface and canopy temperature at night [2].

2.2.3. GLDAS-NOAH SSM

The GLDAS-NOAH model contains land surface parameters simulated from the Noah model in the Global Land Data Assimilation System [29]. The SSM dataset represents the modelled soil moisture information in the upper soil layer (approximately 0–10 cm) at a spatial resolution of 0.25°.

2.3. Ancillary Datasets

The land cover information is retrieved from the U.S. Geological Survey Global Land Cover Characteristics (USGS GLCC) Land Use/Land Cover System (data available from the U.S. Geological Survey) [30]. For the soil type information, the Harmonized World Soil Database (HWSD) was used (data available from the International Institute for Applied Systems Analysis) [31]. Mean yearly precipitation was computed from the Tropical Rainfall Measuring Mission (TRMM) monthly rainfall data (TRMM 3B42 V7 product available from NASA Goddard Earth Sciences Data and Information Services Center) [32] and the mean Normalized Difference Vegetation Index (NDVI) from Moderate-Resolution Imaging Spectroradiometer (MODIS) NDVI measurements between 2005 and 2011 (the MOD13Q1 product available from NASA Land Processes Distributed Active Archive Center, USGS/Earth Resources Observation and Science Center) [33]. The ancillary datasets are shown in Figure 2.

3. Methods

A variety of statistical metrics exists for quantifying the agreement between datasets. Each metric is robust with respect to some attributes and relatively insensitive or incomplete with respect to others. For example, if there is no variation in the real soil moisture content, there may not be any linear correlation between soil moisture datasets, even though the datasets may be accurate in absolute terms. On the other hand, the retrievals can be biased in their mean and dynamic range, but still well reproduce the temporal variability [34] and can be useful in data assimilation if the biases are corrected and the errors are small relative to the model prediction errors. In this study, two common bivariate error measures were used: the Pearson correlation coefficient (R) as a

516

measure of linear dependency and the root mean square difference (RMSD) as a measure of the closeness of the ASAR SSM dataset to the reference dataset.

Figure 2. Ancillary datasets used for the evaluation: (**a**) mean yearly Normalized Difference Vegetation Index (NDVI) value from Moderate-Resolution Imaging Spectroradiometer (MODIS) NDVI measurements; (**b**) mean yearly precipitation from Tropical Rainfall Measuring Mission (TRMM) monthly rainfall data; (**c**) Land Use/Land Cover System from the U.S. Geological Survey Global Land Cover Characteristics; (**d**) Harmonized World Soil Database soil type classification.

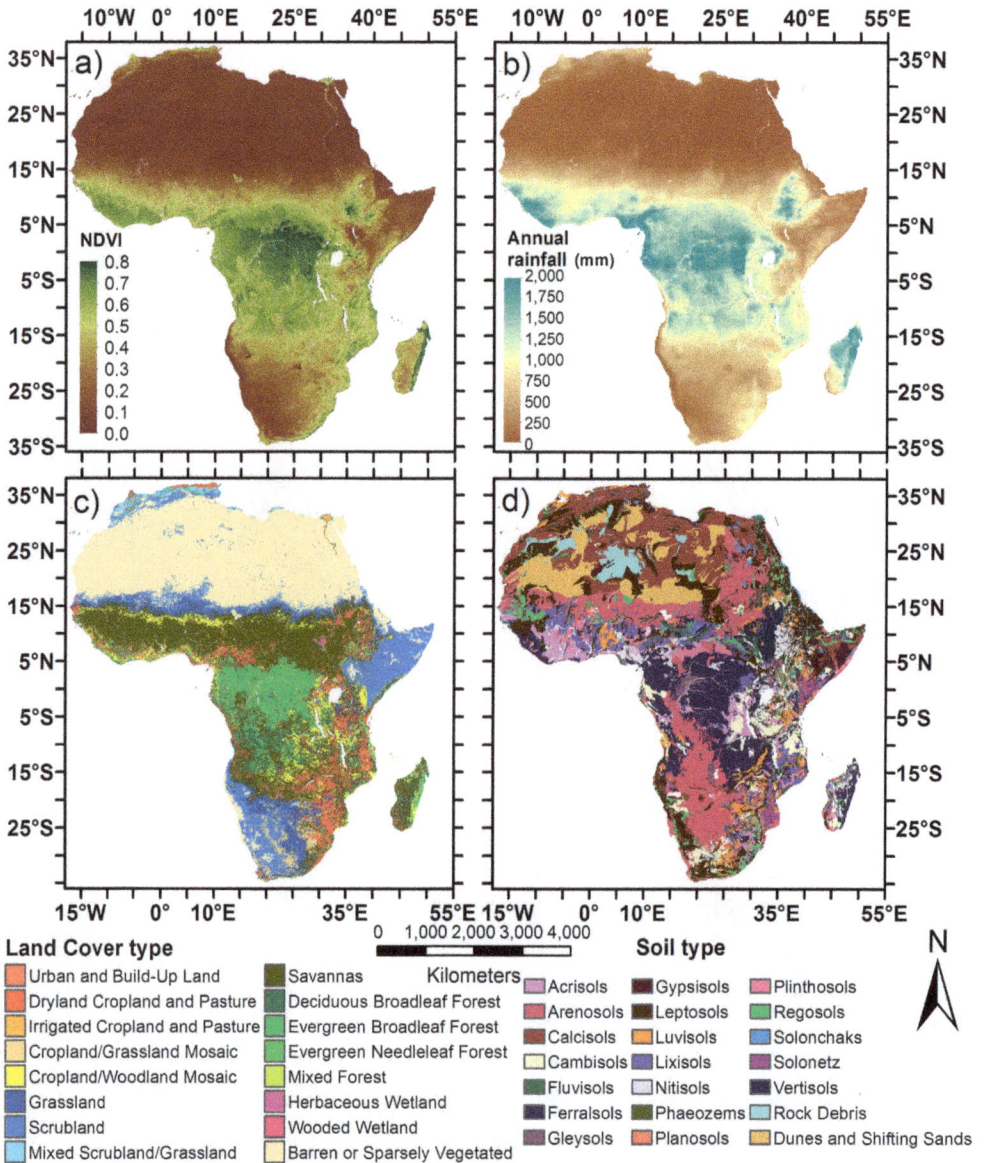

In case of a strong seasonality of soil moisture, R results will be dominated by the seasonal variation and will not reflect the ability of the product to detect single events [5]. To avoid this limitation, R is computed for both the original soil moisture values and the SSM anomalies (SSM_{anom}). The anomalies are computed following Brocca et al. [2] with a modification of a longer time window:

$$SSM_{anom}(t) = \frac{SSM(t) - \overline{SSM(t - 28:t + 28)}}{\sigma[SSM(t)(t - 28:t + 28)]} \tag{1}$$

where $SSM(t)$ corresponds to the surface soil moisture value obtained from the satellite measurement or modeled data at time t and the overbar and σ stand for the temporal mean and standard deviation operators, respectively, for a time window of 8 weeks. All available measurements within the time window were used to compute the temporal mean and standard deviation for each product. A threshold of at least 10 acquisitions within the time window and 50 available data pairs was set. The resulting SSM anomaly is dimensionless.

To reduce the radiometric noise and to provide a better signal-to-noise ratio, the ASAR SSM dataset was spatially averaged to 5-km resolution following the recommendations of preceding studies [17,18]. The R and RMSD were computed between the ASAR 5-km pixel and the nearest acquisition of the AMSR-E and GLDAS-NOAH, respectively. The measures were computed for the entire continent with the exception of land cover classes where the soil moisture retrieval is not possible. These classes were selected using the USGS GLCC Land Use/Land Cover System and include urban areas, water bodies and densely vegetated areas (represented by the class of evergreen broadleaf forest). Temporal matching of the datasets was performed separately for ASAR SSM, GLDAS-NOAH and AMSR-E SSM, respectively. A maximum difference of 12 h between the satellite acquisitions was allowed in the case of ASAR and AMSR-E data. To remove bias and to overcome the problem of different units (%, m^3/m^3, kg/m^3), the linear regression transformation of the reference datasets to the ASAR SSM was applied. The resulting RMSD highlights the random errors between the datasets.

The evaluation metrics were assessed for different land cover classes and soil types. In the case of 1-km resolution ancillary data, the prevailing class within the 5-km resolution evaluation pixel was selected. Based on our results, a new mask was proposed that distinguishes between areas where the TU Wien algorithm is well suited for the soil moisture retrieval from ASAR observations from those where the algorithm fails. Finally, possible causes for algorithm failure were proposed.

In the final section, the possible improvements of the evaluation results are assessed when the ASAR GM algorithm is transferred to data with improved radiometric resolution (i.e., Sentinel-1). The ASAR WS data aggregated to 1-km spatial resolution was used for the evaluation and the results were compared both to the 1-km and aggregated 5-km GM product. Due to the lower temporal resolution of the WS data, the SSM anomalies were not computed, and only GLDAS-NOAH SSM was used as a reference. The above specified spatial and temporal matching of the data, as well as the linear regression transformation applies also for 1-km resolution data.

4. Results and Discussion

4.1. Correlation Results Analysis

The correlation results (Figure 3) indicate a good ability of the ASAR GM to depict SSM variability over areas with mean annual rainfall greater than 300 mm and mean NDVI above 0.2 (Figure 2a,b) The mean R over the entire continent equals 0.35 and 0.34 for SSM values for AMSR-E and GLDAS-NOAH, respectively. As expected, the corresponding mean R for the SSM anomalies is lower (0.23 and 0.2, respectively) due to the lower variability of the SSM anomalies time series. It should be reiterated that urban areas, water bodies and vegetated areas are not included in the latter results. Arid and semi-arid regions (precipitation $<$ 300 mm/year and NDVI $<$ 0.2) are dominated by correlation values below 0.3. In some areas, negative correlation values down to -0.7 are found. Such negative correlations have been previously observed at the C-band between scatterometer acquisitions and modeled SSM [35]. The assumption is that the inverse behavior may be attributed to enhanced backscatter due to the volume scattering over very dry soils [1]. The highest positive correlations of original SSM values (>0.6) were found over areas with sufficient rainfall (500 to 1500 mm/year) and moderate vegetation cover (mean NDVI of 0.3 to 0.6). Correlation values of the SSM anomalies are generally lower (~0.4) in these areas. The results suggest that the ASAR GM SSM product can capture the seasonal cycle of soil moisture well, whereas its ability to represent single precipitation events is lower. The possible reasons may be the low ASAR GM radiometric accuracy. The correlation values sharply increase towards middle NDVI (~0.4) and precipitation values (~800 mm for GLDAS and ~600 mm for AMSR-E) and then stagnate to decrease towards denser vegetation and higher annual rainfall. The dependency of R on precipitation amount and vegetation density is shown in Figure 4. The figure also depicts the lower R values at higher vegetation density when computed with AMSR-E as a reference (Figure 4d). Dorigo *et al.* [36] made similar observations and attributed the behavior to the lower quality of the AMSR-E product over vegetated areas.

The R values close to zero are found over desert areas and also over the irrigated cropland and pasture classes (Figure 5). A significant portion of the latter class is composed of the regularly flooded vegetated areas in the Nile Delta. Similarly, a weak correlation ($R = \sim0.3$) is found over wetlands. In both cases, the change detection algorithm is hampered by the backscatter decrease, due to the regular flooding. The scrubland class according to the USGS GLCC represents a wide variety of regions. In particular, it spreads over areas with average yearly rainfall between 100 and 500 mm/year. This causes the range of R values to be between 0.1 and 0.6.

The R values for anomalies are less stratified by the land cover class when compared to the absolute values. The possible reason is that the impact of the ASAR GM noise on the R values is higher than the impact due to the vegetation attenuation. Furthermore, the influence of the strong seasonal cycle causing the large variability of SSM values in some land cover classes (*i.e.*, savannas) is reduced.

Figure 3. (**a,b**) The correlation coefficient between the Advanced Synthetic Aperture Radar (ASAR) Global Monitoring (GM) (original Surface Soil Moisture (SSM) values) and Noah model from the Global Land Data Assimilation System (GLDAS-NOAH) SSM and Advanced Microwave Scanning Radiometer for Earth Observing System (AMSR-E) SSM, respectively; (**c,d**) the correlation coefficient between ASAR GM (SSM anomalies) and GLDAS-NOAH SSM and AMSR-E SSM, respectively. The grey color represents the masked areas (rain forests and urban areas) or areas with insufficient data coverage (below 50 data pairs).

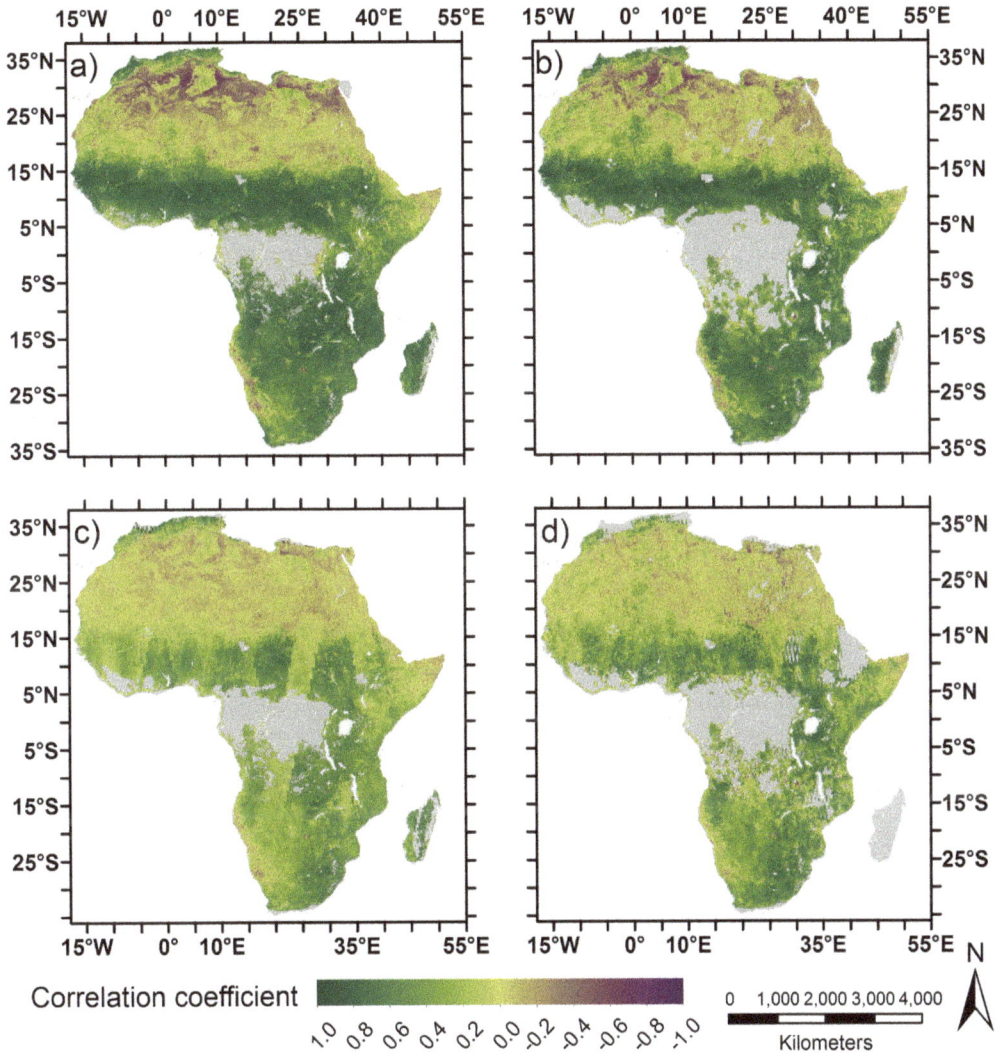

Figure 4. (a,b) The correlation coefficient (R) as a function of average precipitation for the Advanced Synthetic Aperture Radar (ASAR) Global Monitoring (GM) Surface Soil Moisture (SSM) *vs.* the Noah model from the Global Land Data Assimilation System (GLDAS-NOAH) and Advanced Microwave Scanning Radiometer for Earth Observing System (AMSR-E) SSM, respectively; **(c,d)** R as a function of average Normalized Difference Vegetation Index for ASAR GM SSM *vs.* GLDAS-NOAH and AMSR-E SSM, respectively. The solid line represents the median value; dashed lines represent the 25th and the 75th quartile.

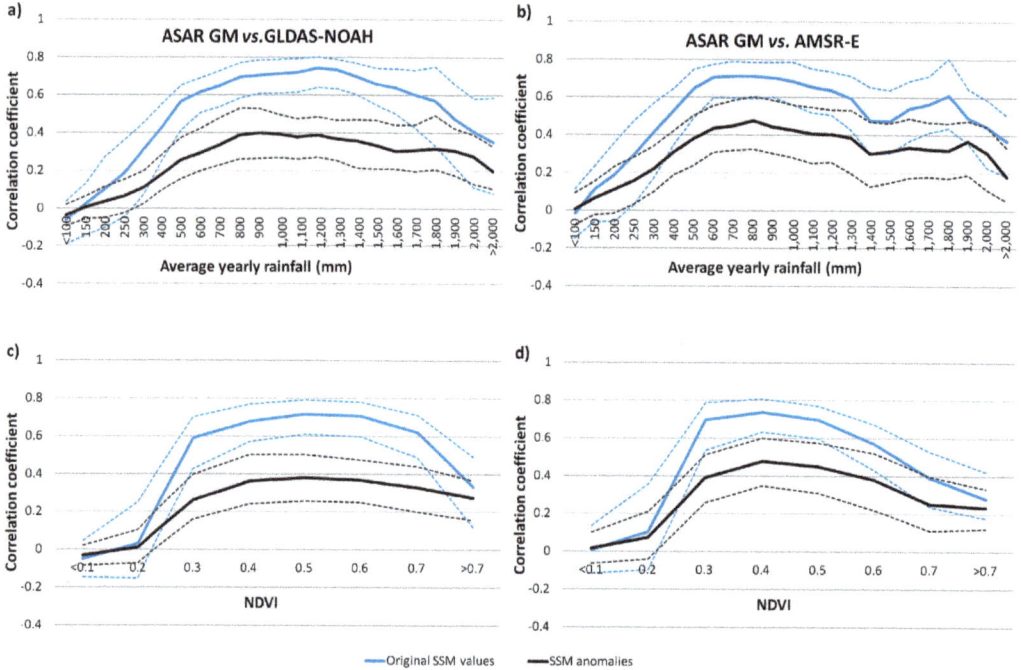

As for the soil types, high correlation values are found over the tropical and sub-tropical soils connected to pronounced dry and wet periods over these regions (*i.e.*, Plinthosols, Lixisols or Vertisols) (Figure 6). Strong seasonality with repeated wetting and drying of these soil types is well captured in ASAR GM SSM data, resulting in median correlation values around 0.7. The lowest R values are found over the soils connected with permanently dry environments, such as Calcisols and Gypsisols, or over Solonchaks, characterized by soluble salt accumulation.

Strikingly low and even negative correlation values can be found in arid regions that appear to be related to soil type composition (Figure 2b). To quantify this relationship, the R values were computed for different soil types over Barren or sparsely vegetated land cover classes with annual rainfall between 100 and 250 mm (Figure 7a). Generally, the correlation values are close to zero, but the distribution over various soil types differs. The lowest correlation values can be found over Calcisols, Leptosols and Solonchaks with a median value of about −0.1, whereas Cambisols and Arenosols show a median correlation of 0.2. Similarly, differences in the correlation results can be observed also in other land cover classes and precipitation ranges. A strong dependency on soil type is observable; for instance, in the case of the land cover class, scrubland, combined with an annual rainfall between 400 and 500 mm, and the median correlation varies between 0.17 in the case of Leptosols and 0.55 in the case of Arenosols (Figure 7b).

Figure 5. The box-plot representation of the correlation results stratified by the Land use/Land Cover system from U.S. Geological Survey Global Land Cover Characteristics. The boxes show the median, 25th and 75th percentiles; the lines represent minimum and maximum values after outlier removal (first and 99th percentile). The amount of 5-km Advanced Synthetic Aperture Radar (ASAR) pixels used for the evaluation for each class is shown in brackets behind the class name. (**Left**) The original Surface Soil Moisture (SSM) values; (**Right**) SSM anomalies. (**Top**) ASAR Global Monitoring (GM) mode *vs.* Advanced Microwave Scanning Radiometer for Earth Observing System (AMSR-E); (**Bottom**) ASAR GM *vs.* the Noah model from the Global Land Data Assimilation System (GLDAS-NOAH).

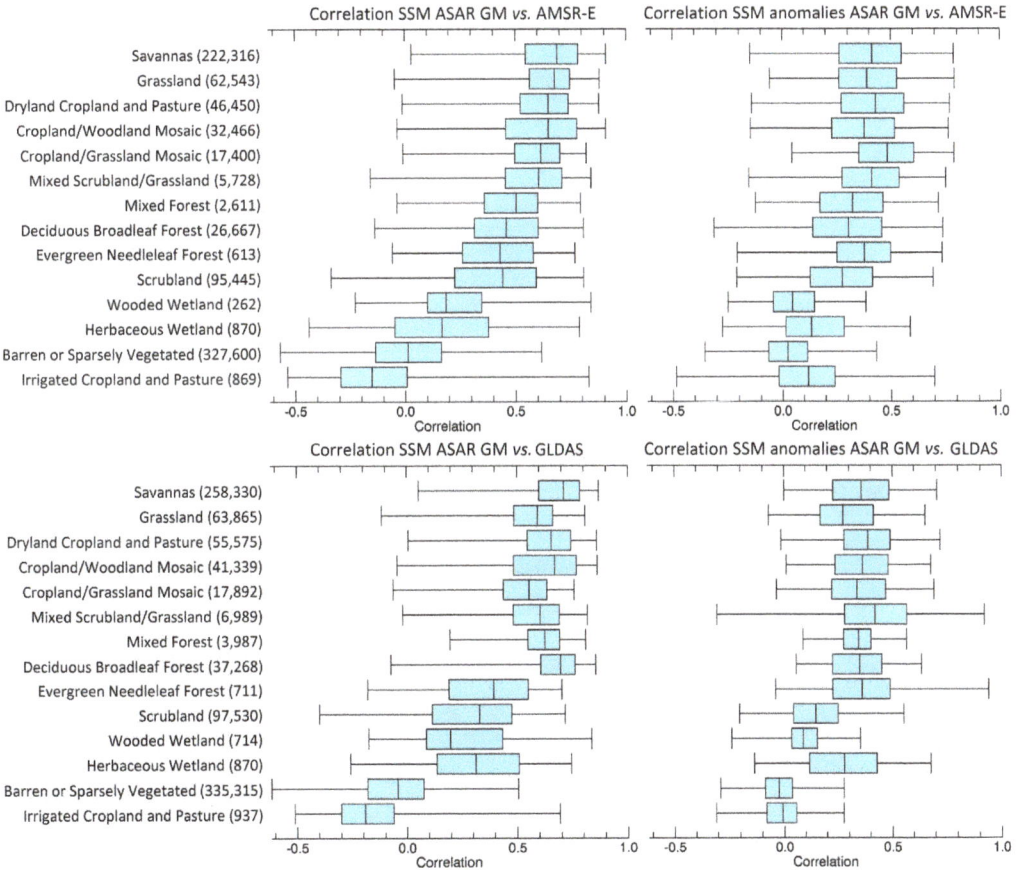

Figure 6. The box-plot representation of the correlation results stratified by the Harmonized World Soil Database soil types. The boxes show the median, 25th and 75th percentiles; the lines represent minimum and maximum values after outlier removal (first and 99th percentile). The amount of 5-km Advanced Synthetic Aperture Radar (ASAR) pixels used for the evaluation for each class is shown in brackets behind the class name. (**Left**) Original Surface Soil Moisture (SSM) values; (**Right**) SSM anomalies. (**Top**) ASAR Global Monitoring (GM) mode *vs.* Advanced Microwave Scanning Radiometer for Earth Observing System (AMSR-E); (**Bottom**) ASAR GM *vs.* the Noah model from the Global Land Data Assimilation System (GLDAS-NOAH).

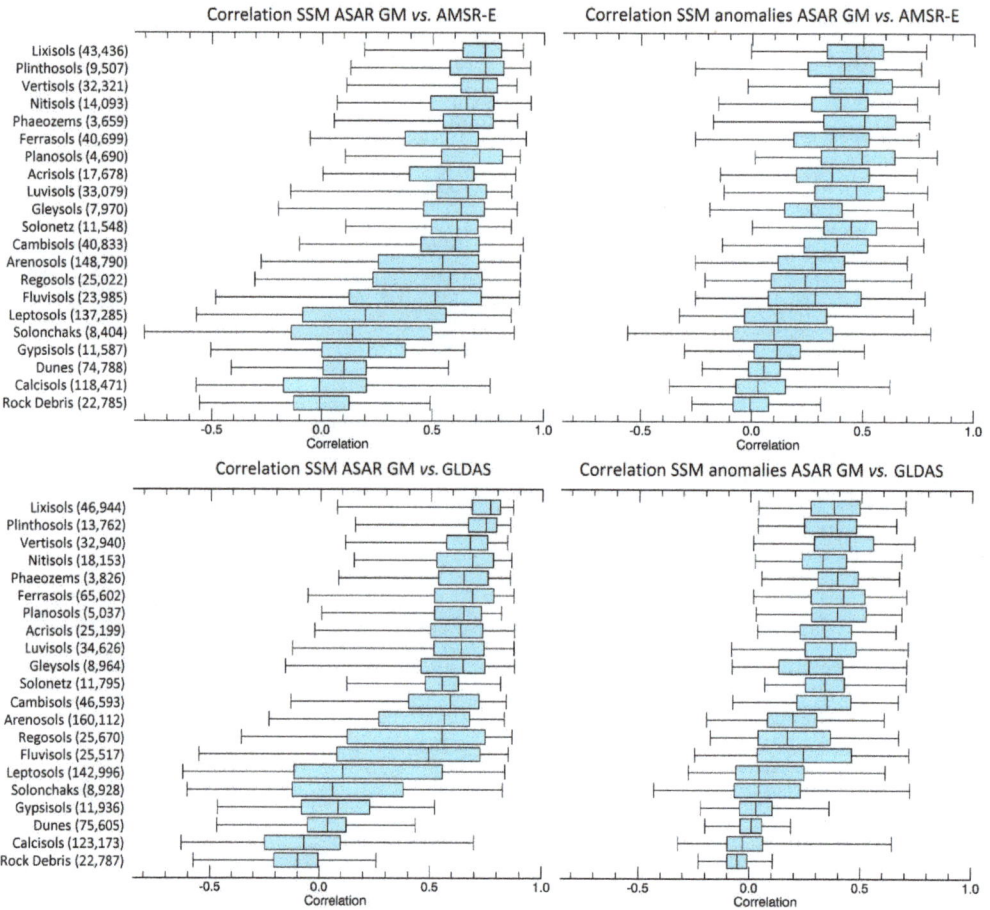

Figure 7. The box-plot representation of the correlation results for 5-km Advanced Synthetic Aperture Radar (ASAR) Global Monitoring (GM) and the Noah model from the Global Land Data Assimilation System (GLDAS-NOAH) Surface Soil Moisture (SSM) stratified by the Harmonized World Soil Database soil types for specific land cover class and precipitation categories. The boxes show the median, 25th and 75th percentiles; the lines represent minimum and maximum values after outlier removal (first and 99th percentile). The amount of 5-km ASAR pixels used for the evaluation for each class is shown in brackets behind the class name. (**Left**) barren or sparsely vegetated land cover classes with annual rainfall of 100 to 250 mm; (**Right**) scrubland land cover class with annual rainfall of 400 to 500 mm.

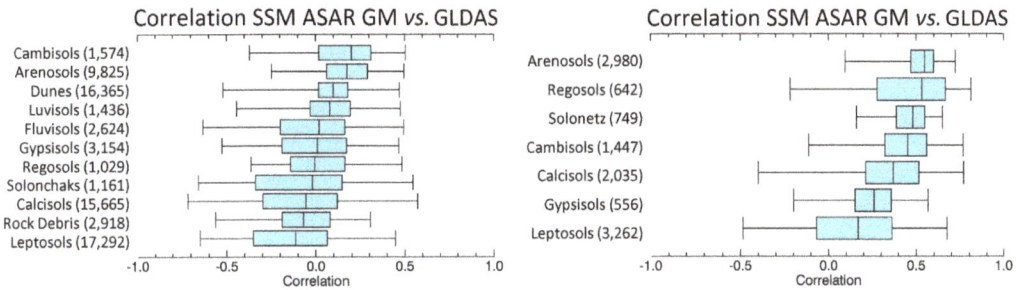

The Calcisol top-layer soil is traditionally crumb or granular. Although it has good water holding properties, slaking and crust formation may hinder the infiltration of rain water and cause surface run-off. The Leptosols soil group is widely spread with different physical and hydrological properties, but generally, it is defined as very shallow (<25 cm) soils over hard rock or extremely gravely and/or stony deeper soils. The Arenosols group consists of sandy soils. It is usually deep and has less than 35% of rock fragments within 100 cm of the soil surface, enabling good sensitivity to surface soil moisture. Cambisols are typically medium-textured and have a high porosity and a good water holding capacity [37]. Clearly, the soil structure and hydrological properties influence the behavior of backscatter over arid areas. This topic requires further research together with detailed and precise soil information.

4.2. RMSD Results Analysis

The overall patterns of the RMSD maps (Figure 8) reflect the large-scale precipitation forcing and the vegetation and geomorphological structures at medium (~5 km) scales. The distribution of RMSD values can be divided into areas with high RMSDs over regions with higher annual rainfall (>250 mm) and those with relatively low RMSD values over dry regions (<250 mm) (Figure 2a). This was expected, as the magnitude of the RMSD is also dependent on the local variability of soil moisture [38]. The RMSD maps correspond quite well over sparse vegetation with lower values of about 1.5% for AMSR-E and differ over vegetated areas (NDVI > 0.5) (Figure 9). Similarly, a decrease in correlation values between ASAR GM and AMSR-E was observed for NDVI > 0.5 (see Figure 4d). In the case of GLDAS-NOAH, the RMSD values remain relatively stable for

NDVI values between 0.3 and 0.7. This discrepancy can be explained by the higher error of the AMSR-E when compared to active microwave sensor acquisitions over vegetated areas [36]. The mean RMSD is, however, identical (11%) for both maps and corresponds also to the mean RMSD at 5 km over Australia, reported in [26].

As expected, the R values remain low over desert areas due to the lack of soil moisture variability. On the other hand, the RMSD maps show the large variability of values in these regions (Figure 6). While in some desert areas, the RMSD remains relatively low (<8%), as expected, given the low soil moisture variations, the RMSD values can reach up to 20% to 35% elsewhere in the desert. Such extremely high values were not found over other continents [17,26] and, therefore, deserve more attention.

Figure 8. (a) The root mean square difference (RMSD) between Advanced Synthetic Aperture Radar (ASAR) Global Monitoring (GM) and the Noah model from the Global Land Data Assimilation System (GLDAS-NOAH) Surface Soil Moisture (SSM); and **(b)** the RMSD between ASAR GM and Advanced Microwave Scanning Radiometer for Earth Observing System (AMSR-E) SSM. The grey color represents the masked areas (rain forests and urban areas) or areas with insufficient data coverage (below 50 data pairs).

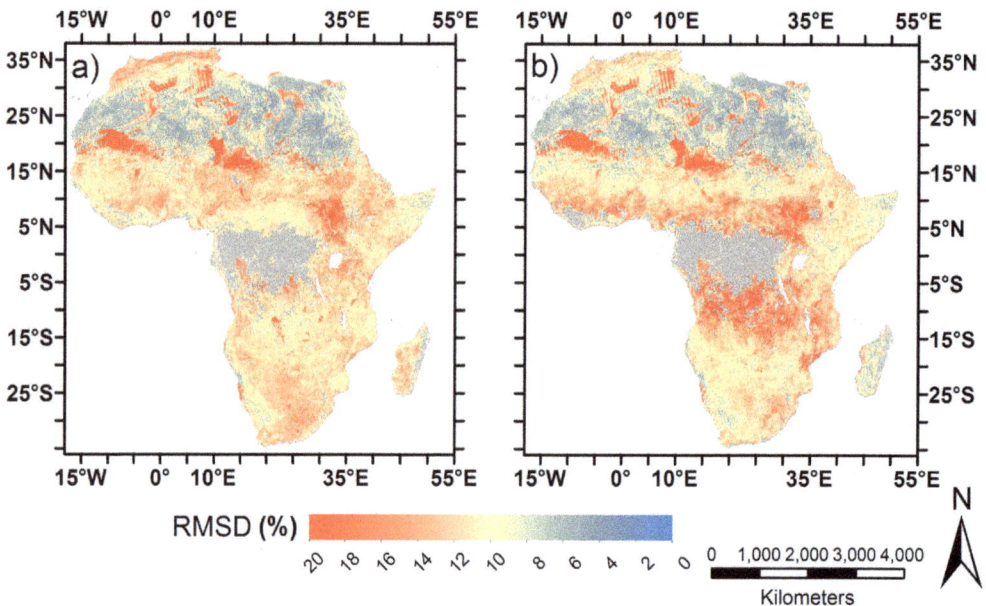

Figure 9. The root mean square difference (RMSD) as a function of average Normalized Difference Vegetation Index for: (**a**) Advanced Synthetic Aperture Radar (ASAR) Global Monitoring (GM) *vs.* the Noah model from the Global Land Data Assimilation System (GLDAS-NOAH) Surface Soil Moisture (SSM); and (**b**) ASAR GM *vs.* Advanced Microwave Scanning Radiometer for Earth Observing System (AMSR-E) SSM. The solid line represents the median value; dashed lines represent the 25th and 75th quartiles.

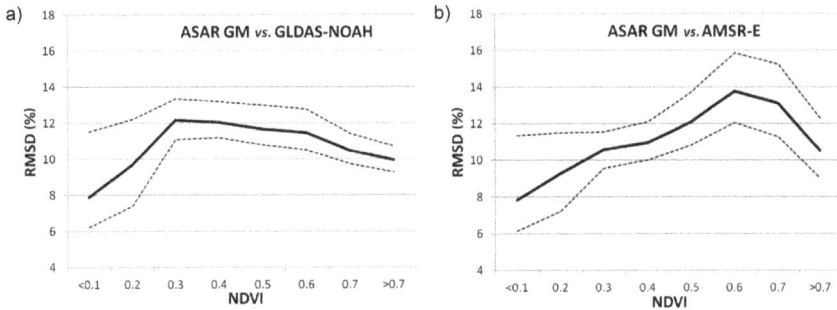

To investigate the origin of the large RMSD variations over deserts, analyses of the backscatter dependence on the local incidence angle were performed. Generally, the change detection algorithm assumes a linear dependency of the backscatter on the local incidence angle and accounts for this by normalizing the backscatter to the local incidence angle of 30° using a regression line [17]. However, this assumption is hampered over desert areas with an RMSD over 20%. The observed limitations can be separated into two groups. Figure 10a represents a location with RMSD values over 20%. In this area, the backscatter from the descending and ascending orbits of the ASAR GM suffer a bias that devaluates the data normalization fit and adds an additional non-random error to the normalized data. The strong bias can be explained by the azimuthal effects that occur due to the spatial orientation of topographic features within the sensor footprint. A similar behavior is observable also over mountainous areas and has been demonstrated in the case of scatterometer acquisitions [39,40]. Usage of only ascending or descending orbit could overcome the problem; this would, however, further reduce the temporal resolution of the product. Next, the exceptionally high RMSD (23%) in Figure 10b is due to high backscatter occurring at an incidence angle of about 30°. This effect forms characteristic striping on the RMSD maps (*i.e.*, around 30°N and 10°E) and can be explained by resonant Bragg scattering. The surface ripples on sand dunes cause constructive interference of the coherent radar signal at certain incidence angles (dependent on the slope of the sand dune) [39]. Stripes of strongly enhanced backscatter values are clearly visible in SAR images over these areas (see Figure 10d). The locations of the illustrative points are shown at Figure 10c.

Figure 11 shows the box-plot representations of RMSD for the USGS Land Use/Land Cover and HWSD soil type classes. The wide inter-quartile range of the non-soil classes, dunes and shifting sands (9% to 19%), indicates that these regions are connected with the above described geometrical distortions in the desert areas. Exceptionally high values can also be found over land cover class herbaceous wetland, with a median of 19%. This class is comprised of the Okavango Delta region

526

and the Ez Zeraf Game Reserve. Both areas are seasonally flooded. The resulting double-bounce effect from water surface and vegetation hampers the soil moisture retrieval and causes large RMSD values.

Figure 10. The relationship between the measured backscatter and the local incidence angle illustrating problems in desert environments: (**a**) dependency of the backscatter value on the azimuth angle and, therefore, on orbit direction (ascending or descending); (**b**) Bragg scattering from sandy dunes at around a 30-degree incidence angle; (**c**) locations of the plotted Advanced Synthetic Aperture Radar (ASAR) Global Monitoring (GM) pixels; (**d**) resonant Bragg scattering effect on the ASAR GM measurements.

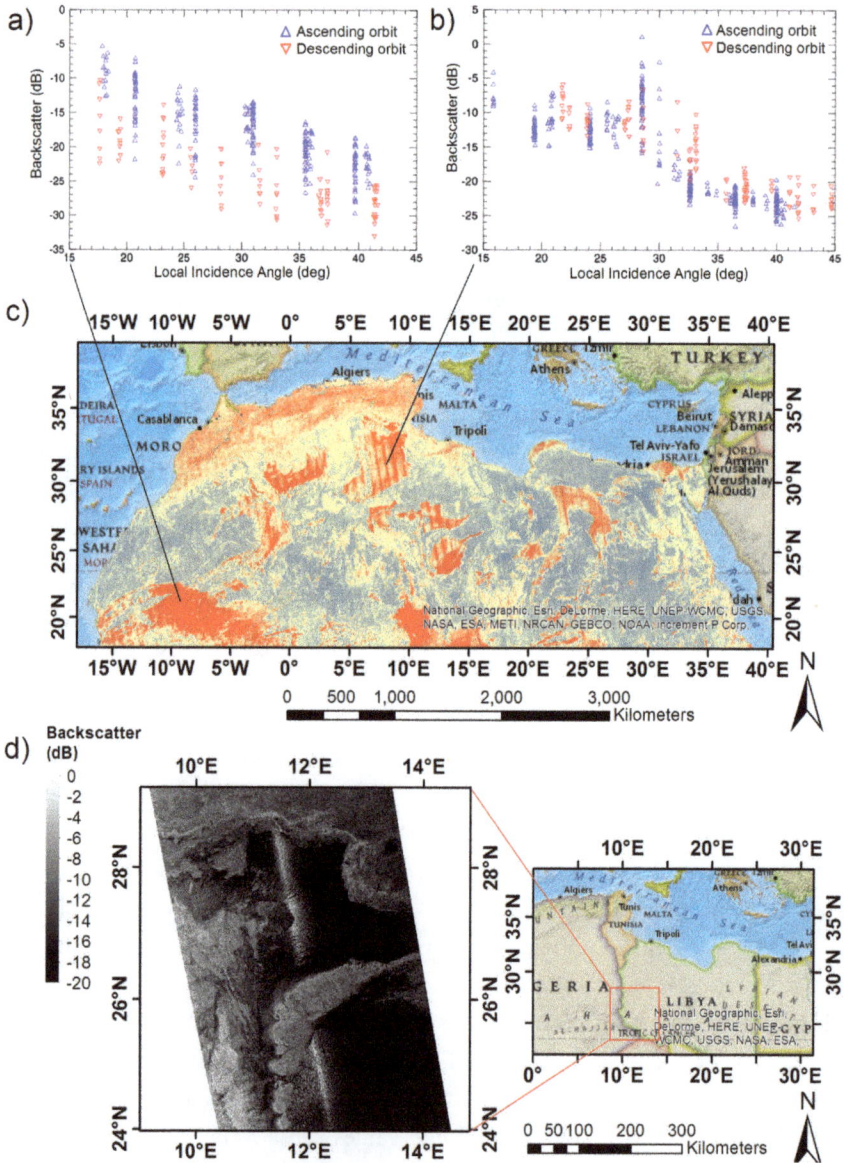

Figure 11. The box-plot representation of the root mean square difference (RMSD) results stratified by the Land Use/Land Cover system from U.S. Geological Survey Global Land Cover Characteristics (**Top**) and Harmonized World Soil Database soil types (**Bottom**). The boxes show the median, 25th and 75th percentiles; the lines represent minimum and maximum values after outlier removal (first and 99th percentile). The amount of 5-km Advanced Synthetic Aperture Radar (ASAR) pixels used for the evaluation for each class is shown in brackets behind the class name. (**Left**) ASAR Global Monitoring (GM) mode *vs.* Advanced Microwave Scanning Radiometer for Earth Observing System (AMSR-E) Surface Soil Moisture (SSM); (**Right**) ASAR GM *vs.* the Noah model from the Global Land Data Assimilation System (GLDAS-NOAH) SSM.

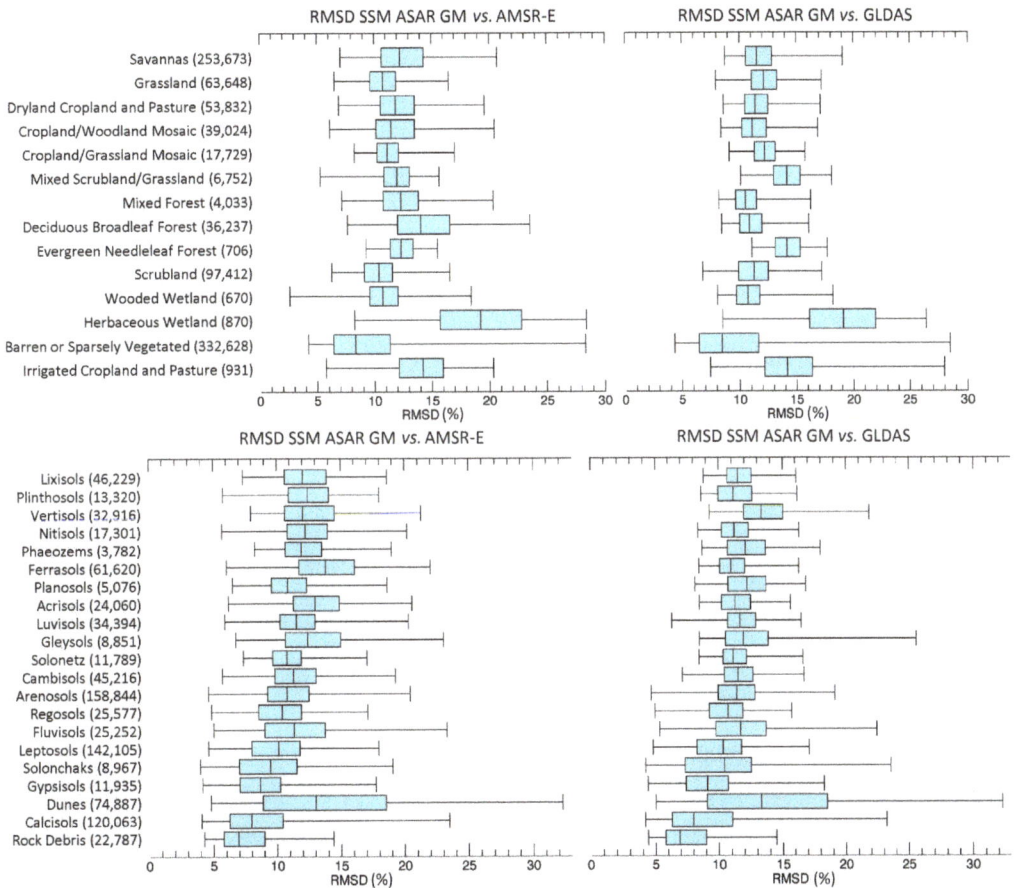

The RMSD results are influenced by the bias correction method applied. Recently, a study by Yilmaz *et al.* [41] suggested that using the triple collocation-based rescaling method results in an optimal solution, whereas regression techniques offer only approximations of this optimal solution. Hence, the investigation on the differences in RMSD maps using the triple collocation-based matching technique could provide additional insights. Even further, triple collocation error assessments removes reference uncertainty and could thus refine the RMSD results.

Apparently, the discussed factors (precipitation, NDVI, land cover class and soil type) have influence on the skill of the retrieval algorithm to represent soil moisture. However, these factors are also inter-correlated. Additional research is needed to assess the influence of the individual factors. This requires detailed combined analysis, such as principal component analysis.

4.3. Mask for the ASAR GM SSM Dataset

Motivated by the results of this study, a mask for the ASAR GM SSM product was created to distinguish the problematic areas. Across the continent, areas covered by surface water, rain forest and urban areas were masked according to the USGS GLCC Land Use/Land Cover System. Areas with a correlation below −0.2 between ASAR GM and GLDAS-NOAH SSM were masked, as well. Additionally, for areas with sufficient rainfall (south of 15°N), masking was based on the ASAR GM scaling layer. The scaling layer quantifies the temporal correlation between the backscatter intensities on the local (1 km) and the regional (25 km) scales [42]. The scaling layer masking is based on the concept of the temporal stability of soil moisture fields [22]. The assumption is that in the case of a low correlation ($R^2 < 0.3$) between the local and regional backscatter intensities, the land cover and soil structure/texture characteristics influence the final ASAR GM product stronger than the temporal variation in soil moisture. However, in arid environments where the temporal dynamics of the soil moisture is strongly limited and, in some areas, backscatter intensities are strongly dependent on the azimuth angle and, thus, on the orbital direction of the satellite, the masking with the help of the scaling layer is not suitable.

In case of scatterometer measurements, the estimated standard deviation (ESD) parameter was used to quantify the effect of azimuthal dependence. This parameter is described in detail in [27]. The areas of high (>0.4 dB) ESD correspond to the high RMSD values between ASAR GM SSM and the reference datasets in the arid regions. Therefore, the mask based on the ERS scatterometer measurements was created to mask the areas with geometrical distortions in arid areas (north of 15°N). An example of the resulting masked surface soil moisture maps are shown in Figure 12.

4.4. Comparison with ASAR WS SSM

Due to the radiometric resolution of about 1.2 dB, the noise in the ASAR GM SSM product is relatively high. The averaging of the product to approximately 3 to 10 km reduces the noise [16–18]; the advantage of high (1 km) resolution is, however, lost. Data with a higher radiometric resolution can provide comparable results to the aggregated 5-km product, also at 1-km spatial resolution. This was demonstrated over Zambezi catchment by comparing 1-km aggregated ASAR WS data with 1-km and 5-km aggregated ASAR GM data. Figure 13 shows box-plot representations of R values between ASAR products and GLDAS-NOAH. Overall, at 89% of the points, the correlation between ASAR and GLDAS-NOAH SSM is significantly improved for ASAR WS when compared to 1-km ASAR GM. The significance level was set to 0.05 using the z-test and Fishers R to z transformation [43]. The average R improvement equals 0.22. Clearly, the change detection algorithm fails to deliver reliable soil moisture retrievals over the herbaceous wetland and barren or sparsely vegetated land cover classes, both in the case of GM, as well as WS mode. For other land cover classes, a significant improvement of the correlation of approximately 0.2 can be observed when using WS mode or 5-km aggregated GM mode measurements instead of the 1-km GM SSM dataset.

Figure 12. The 1-km Advanced Synthetic Aperture Radar (ASAR) Global Monitoring (GM) surface soil moisture monthly composites for (**a**) January and (**b**) October 2011. The grey color represents the masked areas.

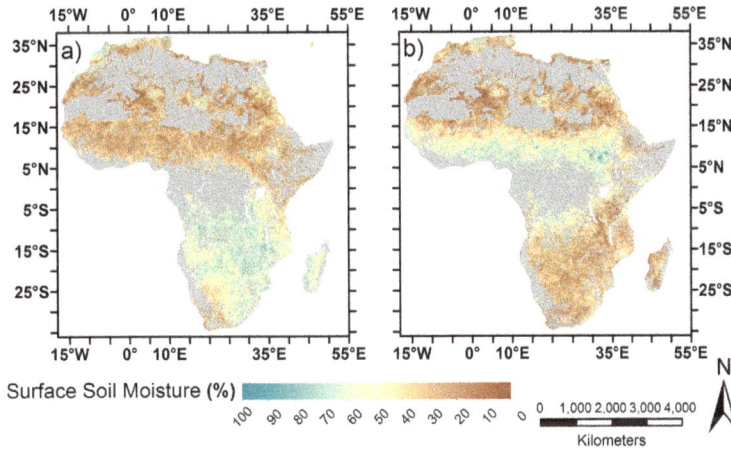

These results indicate that the quality of the soil moisture estimates derived with the TU Wien method can be significantly improved over some landscapes with the use of data with higher radiometric resolution. This is encouraging considering that the Sentinel-1 sensor should provide a three-fold improvement in radiometric resolution compared to ASAR WS [44].

4.5. Limitations of the Evaluation Methodology

The limitations of our study should be reiterated, as they reveal the potential areas for further research. Due to the unavailability of another high or medium resolution SSM dataset over the entire African continent, coarse resolution reference datasets were used for the evaluation. The spatial sampling error together with the uncertainties in the reference data will be included in the bivariate error measures [26]. Furthermore, the use of linear matching using the minimal least squares distance is expected to impact the final RMSD estimates. To overcome the later, an investigation of more complex matching techniques, such as the triple-collocation-based matching technique, is recommended [41]. Even further, triple-collocation error assessments remove reference uncertainty and are therefore expected to refine the RMSD results.

An important limitation of the ASAR SSM product is the relatively low and irregular temporal resolution (typically four to seven days in the case of ASAR GM, but dependent also on the sensor acquisition plan). Especially in the case of SSM anomalies, the spatial differences in temporal resolution are visible in the correlation results. The areas with a lower number of measurements within the time-window correspond to the areas of lower correlation results (*i.e.*, the stripe around

530

10°N and 25°E in Figure 3c). The computation of SSM anomalies as the difference between the absolute soil moisture value and the seasonal cycle of SSM averaged over several years could reduce this effect.

Figure 13. The box-plot representation of the Pearson correlation (*R*) results stratified by the Land Use/Land Cover system from U.S. Geological Survey Global Land Cover Characteristics over Zambezi catchment in southern Africa. The boxes show the median, 25th and 75th percentiles; the lines represent minimum and maximum values after outlier removal (first and 99th percentile). The amount of 5-km Advanced Synthetic Aperture Radar (ASAR) pixels used for the evaluation for each class is shown in brackets behind the class name. The *R* values were computed for the Noah model from the Global Land Data Assimilation System (GLDAS-NOAH). (**a**) One-kilometer resolution ASAR Global Monitoring (GM) Surface Soil Moisture (SSM); (**b**) 5-km aggregated ASAR GM SSM; and (**c**) 1-km aggregated ASAR Wide Swath (WS) SSM.

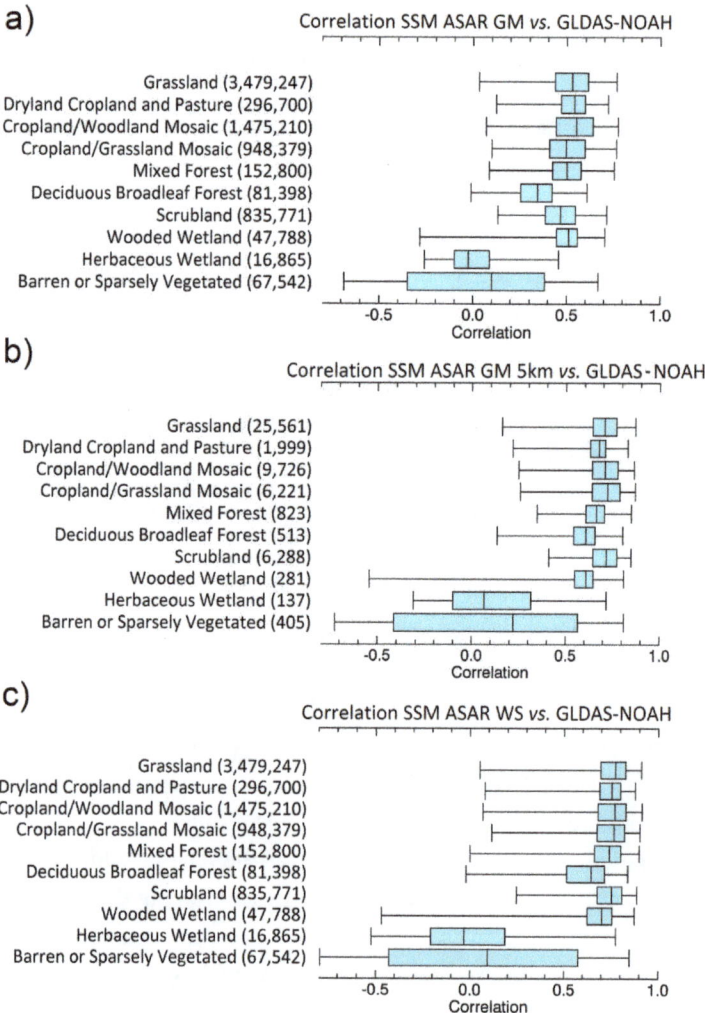

Finally, the influence of various factors (precipitation, NDVI, land cover class and soil type) on the TU Wien algorithm's ability to retrieve surface soil moisture estimates was presented. These factors are however also inter-correlated, and the influence of the individual factors cannot be assessed without detailed combined analysis, such as principal component analysis.

5. Conclusions

The high resolution soil moisture product has the potential to contribute to a number of applications, such as hydrological or runoff modeling. However, the understanding of the quality and limitations of the product is a vital precondition for its usage. This work presents the continental-wide evaluation of the Advanced Synthetic Aperture Radar (ASAR) Global Monitoring (GM) mode Surface Soil Moisture (SSM) product developed at TU Vienna using the change detection algorithm over African continent. The study is unique, as it presents the first long-term and large-scale evaluation of the soil moisture dataset derived from the SAR data over Africa. The results were stratified by the precipitation amount, vegetation cover, land cover classes and soil types and provide insights into the product performance over various environments. Based on the evaluation results, a new mask for the African continent was introduced, covering the areas where the algorithm does not provide reliable SSM estimates.

A comparison with coarse resolution SSM datasets from Advanced Microwave Scanning Radiometer for Earth Observing System (AMSR-E) and the Noah land surface model from the Global Land Data Assimilation system (GLDAS-NOAH) proved the ability of the ASAR GM SSM product to demonstrate the temporal variability of the soil moisture over areas with sufficient rainfall (>250 mm/year) and low to medium density vegetation (a mean Normalized Difference Vegetation Index of 0.2 to 0.6). Correlations over 0.6 were found, $i.e.$, in savannas or croplands, whereas arid regions or wetlands showed low or negative correlations, down to −0.7. Furthermore, differences in performance over various soil types were presented, revealing lower correlations over some soil types ($i.e.$, Leptosols, Calcisols) within the same land cover class and precipitation thresholds.

Three distinct problems in the ASAR GM SSM algorithm were detected during the evaluation process, all of which were located in the arid regions: (i) an inverse relationship between ASAR GM backscatter and soil moisture, causing negative correlation values; (ii) biases between the backscatter from the descending and ascending orbits; and (iii) a distinct bias in the backscatter around a 30° local incidence angle. While the first phenomenon could be explained by the extreme behavior of backscatter over very dry soils, the other two problems could be explained by azimuthal anisotropy effects and Bragg scattering. Further investigation of these problems is expected to bring improvements to soil moisture products based on ASAR GM, scatterometer, as well as future Sentinel-1 data.

At the time of the writing of this publication, the Sentinel-1 sensor is in the commissioning phase. The transfer of the change detection algorithm to Sentinel-1 is therefore foreseen. Given the significantly improved radiometric resolution of Sentinel-1, soil moisture products derived from Sentinel-1 are expected to be of considerably better quality when compared to the ASAR GM SSM products. The impact of enhanced radiometric resolution on the 1-km SSM product was evaluated in this work over Zambezi catchment in Southern Africa using 1-km ASAR Wide Swath (WS)

SSM. Significantly higher correlations (improvements of ~0.2) were obtained over most landscapes using WS data instead of GM data. Further research is required to quantify the robustness and possible areas of improvements of the TU Wien method applied to low-noise SAR data, such as those that will become available through the Sentinel-1 mission.

Acknowledgments

This study has been carried out in the framework of the ESA Data User Element (DUE) Tiger Innovator Projects SHARE (European Space Research Institute (ESRIN)/Contract No. 19420/05/I-EC) and TIGER-NET (ESRIN/Contract No. 4000105732/12/I-NB).

Author Contributions

All authors contributed extensively to the work presented in this paper. Specific contributions included the development of the ASAR GM SSM algorithm (Wolfgang Wagner, Daniel Sabel, Marcela Doubková), SAR data processing and quality checking (Daniel Sabel, Marcela Doubková, Alena Dostálová, Bernhard Bauer-Marschallinger), data evaluation over Africa (Alena Dostálová) and preparation of the manuscript and figures (Alena Dostálová, Marcela Doubková). All co-authors contributed to the editing of the manuscript and to the discussion and interpretation of the results.

Conflicts of Interest

The authors declare no conflict of interest.

References

1. Gruhier, C.; de Rosnay, P.; Hasenauer, S.; Holmes, T.; de Jeu, R.; Kerr, Y.; Mougin, E.; Njoku, E.; Timouk, F.; Wagner, W.; *et al.* Soil moisture active and passive microwave products: Intercomparison and evaluation over a Sahelian site. *Hydrol. Earth Syst. Sci.* **2010**, *14*, 141–156.
2. Brocca, L.; Hasenauer, S.; Lacava, T.; Melone, F.; Moramarco, T.; Wagner, W.; Dorigo, W.; Matgen, P.; Martínez-Fernández, J.; Llorens, P.; *et al.* Soil moisture estimation through ASCAT and AMSR-E sensors: An intercomparison and validation study across Europe. *Remote Sens. Environ.* **2011**, *115*, 3390–3408.
3. Draper, C.S.; Walker, J.P; Steinle, P.J; de Jeu, R.A.M; Holmes, T.R.H. An evaluation of AMSR-E derived soil moisture over Australia. *Remote Sens. Environ.* **2009**, *113*, 703–710.
4. Jackson, T.J.; Cosh, M.H.; Bindlish, R.; Starks, P.J.; Bosch, D.D.; Seyfried, M.; Goodrich, D.C.; Moran, M.S.; Du, J. Validation of advanced microwave scanning radiometer soil moisture products. *IEEE Trans. Geosci. Remote Sens.* **2010**, *48*, 4256–4272.
5. Scipal, K.; Drusch, M.; Wagner, W. Assimilation of a ERS scatterometer derived soil moisture index in the ECMWF numerical weather prediction system. *Adv. Water Resour.* **2008**, *31*, 1101–1112.

6. Mahfouf, J.-F. Assimilation of satellite-derived soil moisture from ASCAT in a limited-area NWP model. *Q. J. R. Meteorol. Soc.* **2010**, *136*, 784–798.

7. Crow, W.T.; Ryu, D. A new data assimilation approach for improving runoff predictions using remotely sensed soil moisture retrievals. *Hydrol. Earth Syst. Sci.* **2002**, *13*, 1–16.

8. Brocca, L.; Melone, F.; Moramarco, T.; Wagner, W.; Naeimi, V.; Bartalis, Z.; Hasenauer, S. Improving runoff prediction through the assimilation of the ASCAT soil moisture product. *Hydrol. Earth Syst. Sci. Discuss.* **2010**, *7*, 4113–4144.

9. Bolten, J.D.; Crow, W.T.; Zhan, X.; Jackson, T.J.; Reynolds, C.A. Evaluating the utility of remotely sensed soil moisture retrievals for operational agricultural drought monitoring. *IEEE J. Sel. Top. Appl. Earth Obs. Remote Sens.* **2010**, *3*, 57–66.

10. Draper, C.S.; Reichle, R.H.; de Lannoy, G.J.M.; Liu, Q. Assimilation of passive and active microwave soil moisture retrievals. *Geophys. Res. Lett.* **2012**, *34*, doi:10.1029/2011GL050655.

11. Taylor, C.M.; de Jeu, R.A.M.; Guichard, F.; Harris, P.P.; Dorigo, W.A. Afternoon rain more likery over drier soils. *Nature* **2002**, *489*, 282–286.

12. Wagner, W.; Pathe, C.; Sabel, D.; Bartsch, A.; Kuenzer, C.; Scipal, K. Experimental 1 km soil moisture products from ENVISAT ASAR for Southern Africa. In Proceedings of the ENVISAT Symposium, Montreux, Switzerland, 23–27 July 2007.

13. Wagner, W.; Scipal, K.; Bartsch, A.; Pathe, C. ENVISAT's capabilities for global monitoring of the hydrosphere. In Proceedings of 2005 IEEE International Geoscience and Remote Sensing Symposium (IGARSS), Seoul, Korea, 25–26 June 2005; pp. 5678–5680.

14. Gruber, A.; Wagner, W.; Hegyiova, A.; Greifeneder, F.; Schlaffer, S. Potential of Sentinel-1 for high-resolution soil moisture monitoring. In Proceedings of 2013 IEEE International Geoscience and Remote Sensing Symposium (IGARSS), Melbourne, Australia, 21–26 July 2013; pp. 4030–4033.

15. Hornacek, M.; Wagner, W.; Sabel, D.; Truong, H.-L.; Snoeij, P.; Hahmann, T.; Diedrich, E.; Doubková, M. Potential for high resolution systematic global surface soil moisture retrieval *via* change detection using Sentinel-1. *IEEE J. Sel. Top. Appl. Earth Obs. Remote Sens.* **2012**, *5*, 1303–1311.

16. Sabel, D.; Bartalis, Z.; Bartsch, A.; Doubková, M.; Hasenauer, S.; Naeimi, V.; Pathe, C.; Wagner, W. Synergistic use of scatterometer and scanSAR data for extraction of surface soil moisture information in Australia. In Proceedings of the 2008 EUMETSAT Meteorological Satellite Conference, Darmstadt, Germany, 8–12 September 2008.

17. Pathe, C.; Wagner, W.; Sabel, D.; Doubková, M.; Basara, J. Using ENVISAT ASAR global mode data for surface soil moisture retrieval over Oklahoma, USA. *IEEE Trans. Geosci. Remote Sens.* **2009**, *47*, 468–480.

18. Doubková, M.; van Dijk, A.I.J.M.; Sabel, D.; Wagner, W.; Blöschl, G. Evaluation of the predicted error of the soil moisture retrieval from C-band SAR by comparison against modelled soil moisture estimates over Australia. *Remote Sens. Environ.* **2012**, *120*, 188–196.

19. Baup, F.; Mougin, E.; Hiernaux, P.; Lopes, A.; de Rosnay, P.; Chenerie, I. Radar signatures of Sahelian surfaces in Mali using ENVISAT-ASAR data. *IEEE Trans. Geosci. Remote Sens.* **2007**, *45*, 2354–2363.

20. Loew, A.; Ludwig, R.; Mauser, W. Derivation of surface soil moisture from ENVISAT ASAR wide swath and image mode data in agricultural areas. *IEEE Trans. Geosci. Remote Sens.* **2006**, *44*, 889–899.

21. Rahman, M.M.; Moran, M.S.; Thoma, D.P.; Bryant, R.; Holifield, C.D.; Jackson, T.; Orr, B.J.; Tischler, M. Mapping surface roughness and soil moisture using multi-angle radar imagery without ancillary data. *Remote Sens. Environ.* **2008**, *112*, 391–402.

22. Wagner, W.; Pathe, C.; Doubková, M.; Sabel, D.; Bartsch, A.; Hasenauer, S.; Blöschl, G.; Scipal, K.; Martínez-Fernández, J.; Löw, A. Temporal stability of soil moisture and radar backscatter observed by the Advanced Synthetic Aperture Radar (ASAR). *Sensors* **2008**, *8*, 1174–1197.

23. Balenzano, A.; Mattia, F.; Satalino, G.; Pauwels, V.; Snoeij, P. SMOSAR algorithm for soil moisture retrieval using Sentinel-1 data. In Proceedings of 2012 IEEE International Geoscience and Remote Sensing Symposium (IGARSS), Munich, Germany, 22–27 July 2012; pp. 1200–1203.

24. Wagner, W.; Sabel, D.; Doubková, M.; Hornacek, M.; Schlaffer, S.; Bartsch, A. Prospects of Sentinel-1 for land applications. In Proceedings of 2012 IEEE International Geoscience and Remote Sensing Symposium (IGARSS), Munich, Germany, 22–27 July 2012; pp. 1741–1744.

25. Wagner, W.; Lemoine, G.; Borgeaud, M.; Rott, H. A study of vegetation cover effects on ERS scatterometer data. *IEEE Trans. Geosci. Remote Sens.* **1999**, *37*, 938–948.

26. Doubková, M.; Dostalova, A.; van Dijk, A.; Blöschl, G.; Wagner, W.; Diego, F. How do extrinsic errors and noise impact the bivariate error measures of the radar derived ASAR 1 km surface soil moisture? *J. Sel. Top. Appl. Earth Obs. Remote Sens.* **2014**. (In print).

27. Naeimi, V.; Scipal, K.; Bartalis, Z.; Hasenauer, S.; Wagner, W. An improved soil moisture retrieval algorithm for ERS and METOP scatterometer observations. *IEEE Trans. Geosci. Remote Sens.* **2009**, *47*, 1999–2013

28. Owe, M.; de Jeu, R.A.M.; Holmes, T. Multi-sensor historical climatology of satellite-derived global land surface. *Hydrol. Earth Syst. Sci.* **2008**. *15*, 425–436.

29. Rodell, M.; Houser, P.R.; Jambor, U.; Gottschalck, J.; Mitchell, K.; Meng, C.-J.; Arsenault, K.; Cosgrove, B.; Radakovich, J.; Bosilovich, M.; *et al.* The global land data assimilation system. *Bull. Am. Meteorol. Soc.* **2004**, *85*, 381–394.

30. Loveland, T.R.; Reed, B.C.; Brown, J.F.; Ohlen, D.O.; Zhu, J, Yang, L.; Merchant, J.W. Development of a global land cover characteristics database and IGBP DISCover from 1-km AVHRR data. *Int. J. Remote Sens.* **2000**, *21*, 1303–1330.

31. FAO; IIASA; ISRIC; ISSCAS; JRC. *Harmonized World Soil Database (Version 1.2)*; FAO: Rome, Italy, 2012.

32. Huffman, G.J.; Bolvin, D.T.; Nelkin, J.; Wolff, D.B.; Adler, R.F.; Gu, G.; Hong, Y.; Bowman, K.P.; Stocker, E.F. The TRMM Multisatellite Precipitation Analysis (TMPA): Quasi-global, multiyear, combined-sensor precipitation estimates at fine scales. *J. Hydrometeorol.* **2007**, *8*, 38–55.

33. NASA Land Processes Distributed Active Archive Center (LP DAAC). *MODIS L3 Vegetation Indices 16-Day L3 Global 250 m (MOD13Q1)*; USGS/Earth Resources Observation and Science (EROS) Center: Sioux Falls, SD, USA, 2006.

34. Entekhabi, D.; Reichle, H.R.; Koster, D.R.; Crow, T.W. Performance metrics for soil moisture retrievals and application requirements. *J. Hydrometeorol.* **2010**, *11*, 832–840.

35. Wagner, W.; Hahn, S.; Gruber, A.; Dorigo, W. Identification of soil moisture retrieval errors: Learning from the comparison of SMOS and ASCAT. In Proceedings of 2012 IEEE International Geoscience and Remote Sensing Symposium (IGARSS), Munich, Germany, 22–27 July 2012; pp. 3795–3798.

36. Dorigo, W.A.; Scipal, K.; Parinussa, R.M.; Liu, Y.Y.; Wagner, W.; de Jeu, R.A.M.; Naeimi, V. Error characterisation of global active and passive microwave soil moisture datasets. *Hydrol. Earth Syst. Sci.* **2010**, *14*, 2605–2616.

37. Driessen, P.; Deckers, J.; Spaargaren, O.; Nachtergaele, F. *Lecture Notes on the Major Soils of the World*; Number 94 in World Soil Resources Reports; Food and Agriculture Organization (FAO): Rome, Italy, 2001.

38. Draper, C.; Reichle, R.; de Jeu, R. Naeimi, V.; Parinussa, R.; Wagner, W. Estimating root mean square errors in remotely sensed soil moisture over continental scale domains. *Remote Sens. Environ.* **2013**, *137*, 288–298.

39. Bartalis, Z.; Scipal, K.; Wagner, W. Azimuthal anisotropy of scatterometer measurements over land. *IEEE Trans. Geosci. Remote Sens.* **2006**, *44*, 2083–2092.

40. Stephen, H.; Long, D.G. Microwave backscatter modeling of erg surfaces in the Sahara desert. *IEEE Trans. Geosci. Remote Sens.* **2005**, *43*, 238–247.

41. Yilmaz, M.T.; Crow, W.T. The optimality of potential rescaling approaches in land data assimilation. *J. Hydrometeorol.* **2012**, *14*, 650–660.

42. Sabel, D.; Pathe, C.; Wagner, W.; Hasenauer, S.; Bartsch, A.; Künzer, C.; Scipal, K. Using ENVISAT Scansar Data for Characterising Scaling Properties of Scatterometer Derived Soil Moisture Information over Southern Africa. Available online: http://earth.esa.int/workshops/envisatsymposium/proceedings/posters/4P9/457362sa.pdf (accessed on 14 May 2014).

43. Fisher, R.A. On the probable error of a coefficient of correlation deduced from a small sample. *Metron* **1921**, *1*, 3–32.

44. Snoeij, P.; Attema, E.; Davidson, M.; Duesmann, B.; Floury, N.; Levrini, G.; Rommen, B.; Rosich, B. The Sentinel-1 radar mission: Status and performance. *IEEE Aerosp. Electron. Syst. Mag.* **2010**, *25*, 32–39.